Biofabrication for Orthopedics

Biofabrication for Orthopedics

Methods, Techniques, and Applications

Edited by
Wenguo Cui
Xin Zhao
Shen Liu

Volume 1

WILEY-VCH

Editors

Prof. Wenguo Cui
Shanghai Jiao Tong University School of Medicine
Shanghai Institute of Traumatology and Orthopaedics, Ruijin Hospital
197 Ruijin 2nd Road
200025 Shanghai
China

Prof. Xin Zhao
Hong Kong Polytechnic University
Department of Biomedical Engineering
11 Yok Choi Road, Hung Hom
999077 Hong Kong
Hong Kong

Prof. Shen Liu
Shanghai Jiao Tong University
Affiliated Sixth People's Hospital
600 Yishan Road
Department of Orthopaedics
200233 Shanghai
China

Cover Image: © SciePro/Adobe Stock Photos

Library of Congress Card No.: applied for

British Library Cataloguing-in-Publication Data
A catalogue record for this book is available from the British Library.

Bibliographic information published by the Deutsche Nationalbibliothek
The Deutsche Nationalbibliothek lists this publication in the Deutsche Nationalbibliografie; detailed bibliographic data are available on the Internet at <http://dnb.d-nb.de>.

Print ISBN: 978-3-527-35169-5
ePDF ISBN: 978-3-527-83136-4
ePub ISBN: 978-3-527-83135-7
oBook ISBN: 978-3-527-83137-1

Typesetting Straive, Chennai, India
Printing and Binding CPI Group (UK) Ltd, Croydon, CR0 4YY

C112261_030922

Contents

Volume 1

Foreword from Prof. Changsheng Liu *xi*
Foreword from Prof. Yingze Zhang *xiii*
Foreword from Prof. Lianfu Deng *xv*
Foreword from Prof. Cato T. Laurencin *xvii*
Preface *xix*

Part I Biofabrication Techniques *1*

1 Current Progress and Technological Challenges in Translational 3D Bioprinting *3*
Greeshma Ratheesh, Natividad G. Cerezo, Weidong Gao, Jayanti Mendhi, Prashant Sonar, and Yin Xiao
1.1 Introduction *3*
1.2 Challenges in the 3D Bioprinting Process *5*
1.2.1 Manufacturing Challenges *5*
1.2.1.1 Choice of Bioink *5*
1.2.1.2 Cell Selection and Optimization *7*
1.2.1.3 Printing Resolution and Mechanical Stability *8*
1.2.1.4 Design and Post-processing *9*
1.2.2 *In Vitro* Challenges *11*
1.2.2.1 Cell Viability *11*
1.2.2.2 Cell Differentiation *12*
1.2.3 Clinical Challenges *13*
1.2.3.1 Remodeling and Maturation *13*
1.2.3.2 Immune Response *13*
1.2.3.3 Vascularization *14*
1.2.3.4 Innervation *16*
1.3 Conclusion and Future Perspectives *18*
Acknowledgments *19*
References *19*

2 Bioceramics for Promoting Bone Regeneration *25*
Meng Zhang, Hongshi Ma, and Chengtie Wu
2.1 Introduction *25*
2.2 Types of Bioceramics *26*
2.2.1 Bioactive Ceramics *26*
2.2.2 Biodegradable Ceramics *26*
2.2.2.1 Bioactive Glass *26*
2.2.2.2 β-Tricalcium Phosphate *30*
2.2.2.3 Calcium Phosphate Cements *32*
2.2.2.4 Silicate Bioceramics *32*
2.3 Mechanical Properties *36*
2.4 Biological Properties *38*
2.4.1 Protein Adsorption *39*
2.4.2 Immunomodulation *40*
2.4.3 Vascularization *42*
2.5 Summary *44*
Acknowledgments *44*
References *45*

3 3D Printing and Bioprinting Strategies Applied Toward Orthopedics *55*
Zongliang Wang, Prajwal Agrawal, and Yu Shrike Zhang
3.1 Introduction *55*
3.2 Biomaterial Inks *58*
3.2.1 Hydrogel Biomaterial Inks *59*
3.2.1.1 Collagens *59*
3.2.1.2 Gelatin *60*
3.2.1.3 Alginate *62*
3.2.1.4 Hyaluronic Acid *63*
3.2.2 Thermoplastic Polymeric Inks *64*
3.2.2.1 PCL *65*
3.2.2.2 PLA *66*
3.2.2.3 Polyetheretherketone (PEEK) *66*
3.2.3 Calcium Phosphate (CaP) Ceramic Inks *67*
3.2.4 Supramolecular and Stimuli-Responsive Inks *68*
3.3 3D Printing and Bioprinting Techniques *71*
3.3.1 Extrusion-Based (Bio)printing *71*
3.3.2 Inkjet-Based (Bio)printing *77*
3.3.3 Light-Based (Bio)printing *81*
3.4 Current Challenges and Future Directions *85*
Acknowledgments *86*
References *86*

4 Stem Cells and Their Application in Orthopedics *93*
Xiao Li, Xiaobo Zhu, Jing Sun, Patrick S.H. Yung, Rocky S. Tuan, and Yangzi Jiang
4.1 Introduction *93*
4.2 Mesenchymal Stem Cells (MSCs) *94*
4.2.1 Tissue Sources of MSCs *94*
4.2.2 Isolation, Identification, and Expansion of MSCs *96*
4.2.3 Current Cell Markers for MSC Identification *99*
4.2.3.1 Differentiation of MSCs both *In Vitro* and *In Vivo* *102*
4.3 MSC-Derived Extracellular Vesicles (MSC-EVs) and Exosomes *108*
4.4 Clinical Application of Stem Cells in Orthopedics *109*
4.4.1 Bone Defects, Non-union, and Osteogenesis Imperfecta *109*
4.4.1.1 Bone Non-union and Defects *109*
4.4.1.2 Osteogenesis Imperfecta (OI) *110*
4.4.2 Osteochondral Defect and Osteoarthritis (OA) *110*
4.4.3 Tendon and Ligament Injury *115*
4.4.3.1 Rotator Cuff Tear *115*
4.4.3.2 Anterior Cruciate Ligament (ACL) Reconstruction *117*
4.4.4 Spine *117*
4.4.4.1 Spinal Fusion *117*
4.4.4.2 Spinal Cord Injuries (SCI) *117*
4.5 Considerations of Stem Cells and Derivations for Clinical Usage *118*
4.6 Conclusion *122*
Acknowledgments *122*
Abbreviations *122*
References *124*

5 Electrospinning Techniques *133*
Yuanfei Wang, Ziyi Zhou, Xiaopei Zhang, Na Liu, Zhendong Feng, Xiumei Mo, Wenguo Cui, and Tong Wu
5.1 Introduction *133*
5.2 Different Types of Electrospinning Techniques *134*
5.2.1 Traditional Electrospinning *134*
5.2.2 Coaxial Electrospinning *137*
5.2.3 Emulsion Electrospinning *138*
5.2.4 Conjugated Electrospinning *138*
5.2.5 Dynamic Liquid Electrospinning *139*
5.2.6 Multi-fluid Electrospinning *140*
5.2.7 Electrospun Fibers Integrated with Post-treatments *141*
5.2.7.1 Coating Physical Components on the Fibers *141*
5.2.7.2 Deposition of Electrosprayed Microparticles on the Fibers *141*
5.2.7.3 Welding of Nanofibers *143*

5.2.7.4 Remolding Electrospun Nanofibers to 3D Scaffolds *143*
5.3 Typical Applications of Electrospun Fibers in Orthopedics *144*
5.3.1 Bone Tissue Repair *144*
5.3.2 Vascularization *147*
5.3.3 Cartilage Repair *148*
5.3.4 Tendon and Ligament Repair *149*
5.3.5 Repair of Tendon-to-Bone Interface *153*
5.4 Conclusion and Future Outlook *155*
References *155*

6 Joint Lubrication and Wear *159*
Yi Wang, Keming Hu, Weiwei Zhao, and Hongyu Zhang
6.1 Introduction *159*
6.2 Natural Joint Structure *160*
6.2.1 Role of Articular Cartilage *160*
6.2.2 Regional Classification of Articular Cartilage *161*
6.3 Joint Lubrication Mechanism *162*
6.3.1 Tribological Theory *162*
6.3.2 Hydration Lubrication *163*
6.4 Joint Lubrication Behavior *164*
6.4.1 Component *164*
6.4.2 Lubrication Performance *164*
6.5 Artificial Biolubricants *168*
6.6 Artificial Joint Prosthesis *174*
6.6.1 Metal Material *175*
6.6.2 Polymer Material *175*
6.6.3 Ceramic Material *176*
Acknowledgments *177*
References *177*

7 Microfluidic Biotechnology for "Bone-on-a-Chip" *181*
Chaenyung Cha
7.1 Introduction *181*
7.2 Basic Principles and Properties of Microfluidics *185*
7.2.1 Fabrication of Microfluidic Devices via Soft Lithography *185*
7.2.2 Fluid Controlling Components of Microfluidic Devices *187*
7.2.2.1 Microvalves *188*
7.2.2.2 Micropumps *189*
7.2.2.3 Micromixers *190*
7.2.2.4 Concentration Gradient *190*
7.2.3 Cell/Tissue Culture in Microfluidic Devices *192*
7.3 Microfluidic "Organ-on-a-Chip" Technology *193*
7.3.1 Cell/Tissue Culture in Microfluidic Devices *193*
7.3.2 Biological Features of Organ-on-a-Chip Devices *194*
7.3.2.1 Tissue Barrier Functions *194*
7.3.2.2 External Stimulation *195*

7.3.2.3 Microvascular Network *195*
7.3.3 Recent Advances in "Bone-on-a-Chip" Technology *197*
7.3.4 Mineralization and Osteogenesis *198*
7.3.5 Osteochondral Tissue *201*
7.3.6 Bone Marrow *203*
7.3.7 Cancer *204*
7.4 Conclusion and Future Perspectives *206*
Acknowledgments *207*
References *207*

8 Bioactive Glasses in Orthopedics *211*
Amel Houaoui and Jonathan Massera
8.1 First Bioactive Glass *211*
8.2 Bioactive Glass Versatility *212*
8.2.1 Osteogenesis *212*
8.2.2 Angiogenesis *213*
8.2.3 Antimicrobial Activity *214*
8.3 Alternative Bioactive Glasses *215*
8.4 Bioactive Glasses in Composites and Hybrid Materials *216*
8.4.1 Materials Used as Organic Phase in Composites or Hybrid Devices for Bone Repair *217*
8.4.1.1 Natural Polymers *217*
8.4.1.2 Synthetic Polymers *219*
8.4.2 Composite Materials *220*
8.4.2.1 Bioactive Glasses Used in Composites *221*
8.4.2.2 Bioactive Glass Influence on Cell Behavior *In Vitro* *222*
8.4.2.3 Composites' Fabrication Methods *224*
8.4.3 Hybrid Materials *226*
8.4.4 Composites in Clinics *229*
8.5 Conclusion *230*
Acknowledgments *231*
References *231*

Volume 2

Foreword from Prof. Changsheng Liu *xiii*
Foreword from Prof. Yingze Zhang *xv*
Foreword from Prof. Lianfu Deng *xvii*
Foreword from Prof. Cato T. Laurencin *xix*
Preface *xxi*

Part II Biomedical Applications in Orthopedics *243*

9 3D Printing for Orthopedics *245*
Lan Li and Qing Jiang

10 **Bone Implants (Bone Regeneration and Bone Cancer Treatments)** *265*
Hangxiang Sun, Xingzhi Zhou, Yongxing Zhang, Lei Zhang, Xiaohua Yu, Zhaoming Ye, and Cato T. Laurencin

11 **Bionic Fixation: Design, Biomechanics, and Clinical Application** *323*
Wei Chen and Qi Zhang

12 **Cartilage Injury and Repair** *341*
Yusheng Li, Ning Wang, Di Liu, and Zijun Cai

13 **Biofabrication for Cartilage Regeneration** *367*
Sien Lin, Boguang Yang, and Gang Li

14 **Intervertebral Disc Repair** *401*
Yong Cao, Tianding Wu, Chunyue Duan, JinSong Li, Yuxin Jing, Jiaqi Xu, Daudi R. Manini, and Jianzhong Hu

15 **New Prospects in Skin Tissue Engineering and Fabrication** *423*
Lei Wang, Xiaonan Liu, and Ke Shen

16 **Techniques for Biofabrication of Vascular and Vascularized Tissue** *461*
Zhe Zhang, Qiang Yang, and Chunming Wang

17 **Vascularization** *481*
Xiao Lu, Jianxi Lu, Fangchun Jin, and Yajie Lu

18 **Muscle Repair** *513*
Rong Bao, Wei Wang, Shen Liu, and Cunyi Fan

19 **Rotator Cuff Repair** *535*
Chongyang Wang and Yaohua He

20 **Tendon Repair** *559*
Xiao Wang

21 **Nerve Regeneration** *581*
Zhiwen Yan, Wei Chen, Yun Qian, and Cunyi Fan

Index *607*

Foreword from Prof. Changsheng Liu

I would like to thank Profs Wenguo Cui, Xin Zhao, and Shen Liu for inviting me to write this foreword. I am honored to take part in the knowledge exchange of our scientific culture, as well as to promote training of our students and professionals.

With the rapid development of biomaterials and biofabrication technologies, a variety of biomaterials have been made for the repair and regeneration of bone, muscle, tendon, and other orthopedic tissues and organs. This book focuses on the current state of biofabrication technologies and materials in orthopedics biomedical applications. The various highlights of this book include a comprehensive analysis of different biofabrication methods for constructing relevant biomaterials in the orthopedics field, with an emphasis on the recent developments of technologies involving 3D printing, bioceramic, electrospinning, microfluidics, bioactive glass, etc. A detailed look into these biofabrication techniques using different biomaterials is also included in terms of functional performance, advantages, and disadvantages for repairing and regeneration in orthopedics, of which the choices of biomaterials are especially critical since their applications must be based on a thorough knowledge of the anatomy and physiology of the bone, muscle, and adjacent tissue. With the content being organized clearly, practically, rigorously, and most importantly, up to date with current knowledge, this review will therefore provide the research community a reference source for current approaches to biomaterials and scaffold preparation in orthopedics, as well as to reinforce our readers' understanding of materials' biomedical applications in orthopedics.

In essence, this book is clear and well written in a way that provides highly useful and relevant content to support trained orthopedic doctors and biomaterial researchers. This helps give the readers the basis of the preparation of a variety of biomaterials and thus enhance the understanding of their application in orthopedics. Through this book, the authors have faith that the readers and researchers can generate next-generation biomaterials for orthopedic applications.

Changsheng Liu
East China University of Science
and Technology, P.R. China

Foreword from Prof. Yingze Zhang

In the past decades, traditional orthopedic metal implants such as plates and rivets have benefited many patients with orthopedic diseases. However, problems such as mechanical failure and poor efficacy always exist, which need to be solved through innovation. Nowadays, interdisciplinary convergence and cross-border integration of several technologies are becoming the norm and will continue to give birth to the forefront of new disciplines. Thanks to the development of tissue engineering and regenerative medicine, a large number of new materials and technologies with great application prospects in orthopedic medicine have emerged, providing new opportunities for the diagnosis and treatment of orthopedic diseases.

This book systematically demonstrates the widely popular and eye-catching biomaterials and biofabrication technologies, explains in detail their advantages and disadvantages in orthopedics, and lists their applications in the repair of bone, muscle, ligament, tendon, and other orthopedic organs.

Each chapter of this book emphasizes the impact of advanced materials and biofabrication technologies in the field of orthopedics, including how biofabrication affects all substances used in bone implants (from 3D printing to electrospinning and microfluidic technology). One of the highlights of this book is apparently to cover the most innovative biomaterial manufacturing technologies and emphasizes the safety of biofabrication that we have not fully understood and need to develop solutions. It also provides concise reasons on why we should consider using these materials to complete the regeneration and repair of orthopedic systems, discussing in-depth beyond the typical trial-and-error thinking of traditional orthopedics.

The editors and authors of this book hope to enrich readers and researchers in the field of orthopedics, orthopedic biomaterials, and their clinical applications, and in my opinion, this book is certainly a must-have for all those who want to create new solutions to the most stubborn problems in orthopedics, so that we can improve the quality of life for patients of implants with novel biofabrication technologies. We believe that new orthopedic biofabrication technologies for bone tissue regeneration and a better understanding of the interface between biomaterials and adjacent tissues will be an important part of orthopedic biomaterials and implants in the future for better orthopedic surgery.

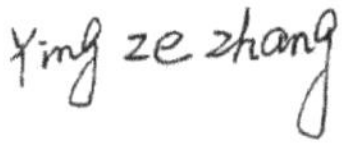

Yingze Zhang
Department of Orthopedic Surgery
Third Hospital of Hebei
Medical University Director
NHC Key Laboratory
of Intelligent Orthopaedic
Equipment, Shijiazhuang
P.R. China

Foreword from Prof. Lianfu Deng

With the rapid development of medical science and engineering technology, the field of orthopedics has also made great progress. In recent years, biomaterial implants have been widely used in the replacement treatment or adjunctive therapy of injured bones to replace, support localization, or repair human bones, joints, and soft tissues. The clinical demand for orthopedic implanted medical devices has been maintaining a rapid growth rate. However, due to the differences in disease or injury types, the demand for personalized design and fabrication of biomaterials and biomaterials with specific functions is also increasing. This also greatly promoted the development of orthopedic materials through biofabrication technology.

The continuous innovation and development of biomaterial fabrication technologies (3D (three-dimensional) bioprinting, electrospinning, and microfluidic) have improved the utilization of biomaterials, including the fabrication of complex 3D structures, the improvement of biomaterial interface and cell interaction, and the development of controlled drug-release systems. The development and biofabrication of advanced biomaterials in orthopedics field covers the interdisciplinary study of materialogy, tissue engineering and regenerative medicine, cell and molecular biology, and clinical medicine. The biofabrication techniques bridge basic research and clinical treatment, help to accelerate the orthopedic basic scientific research to the engineering application of the industrialization process, and promote the development of targeted treatment strategies for different orthopedics diseases or injuries. It is meaningful to advance the development progress of translational medicine.

It is timely to compile this book on advanced fabrication technology of biomaterials and its application in orthopedics. This book summarizes the latest scientific research progress and development trend of orthopedics worldwide, aiming to track the frontier of the discipline, introduce advanced theories and technologies, and promote the development of orthopedics and orthopedics technology. This book will help doctors, researchers, and students of medical specialty to have a good command of the design, fabrication, and application of biomaterials as well as the regulatory mechanism of bone regeneration process and better understand the development direction of biological fabrication and orthopedic technology.

Lianfu Deng
Shanghai Key Laboratory
for Prevention and Treatment
of Bone and Joint Diseases
Shanghai Institute of
Traumatology and Orthopaedics
Ruijin Hospital, Shanghai Jiao Tong
University School of Medicine
P.R. China

Foreword from Prof. Cato T. Laurencin

It is always a great privilege to be invited to write a foreword for a book. Moreover, when the book enlarges our scientific knowledge and viewpoints while promoting the training of our students and professionals, it is even more welcome. I am therefore more than happy to provide my insights regarding such books.

The convergence of the life, physical, and engineering sciences has led to the creation of a new multidisciplinary field, which I have termed as "regenerative engineering." Orthopedic surgery is a prominent field that stands to gain from the insights of regenerative engineering. In particular, biofabrication techniques developed in this field applying the principles of regenerative engineering have the potential of bringing revolutionary treatments to clinical practice. This book summarizes important aspects of biofabrication techniques for orthopedic surgery and offers valuable information of state-of-the-art technological advancements that can ultimately become a practical toolbox for physicians, scientists, and engineers. While traditional orthopedic surgery books cover conventional strategies, this book focuses on newer biological strategies such as bioprinting, stem-cell therapy, and organ-on-a-chip technology. The chapters encourage readers to examine and embrace novel technologies to address clinical challenges. The content is well organized while covering a wide range of related fields. The knowledge and information provided by all the contributors are up to date, covering comprehensively technological aspects as well as clinical applications involved in orthopedic surgery practice.

The editors of this book, Profs Wenguo Cui, Xin Zhao, and Shen Liu, have assembled an excellent team and have created an important contribution to our field. The contributors to this book are respected scholars and professionals that come from a broad background including orthopedic surgery, biomedical engineering, stem-cell biology, and physics. This allows the book to have insights regarding the current status and future trends of this field.

Scientific books must evolve to capture important aspects of the corresponding fields. This book is published at the right time when many exciting new biofabrication technologies have arisen, many of them inspired by the field of regenerative engineering. I congratulate again the editors of the book. This well-organized contribution to the world's literature will be relevant to scientists, engineers, and clinicians everywhere.

Cato T. Laurencin, MD, PhD
Member, National Academy of Science/
National Academy of Engineering
Albert and Wilda Van Dusen Distinguished
Professor of Orthopaedic Surgery
Professor of Chemical and
Biomolecular Engineering
Professor of Materials Science
and Engineering
Professor of Biomedical Engineering

Director, The Raymond and Beverly Sackler
Center for Biomedical, Biological, Physical
and Engineering Sciences

Chief Executive Officer
The Connecticut Convergence Institute
for Translation in Regenerative Engineering
The University of Connecticut

Preface

Wenguo Cui

Xin Zhao

Shen Liu

Bone, as one of the most essential organs in the body, not only supports the stability of body along with protection of other organs, but also provides hematopoietic function and stores minerals such as calcium and phosphorus. Unfortunately, when some damages/diseases, such as fracture, osteoarthritis, and osteoporosis, appear in the bones, these functions will be disrupted, leading to the systemic responses as seen clinically. Researchers are therefore proposing to develop techniques to remodel the damaged/diseased bone functions. Last decades, biofabrication techniques, including three-dimension (3D) bioprinting, 3D printing, electrospinning, microfluidics, and stem-cell therapy, have been widely applied in tissue regeneration. Some products of these biofabrication techniques were even commercialized, such as Absorb GT1 and XinSorb. *Biofabrication for Orthopedics – Special Topic* in Wiley mainly focuses on the development of biofabrication and applications in orthopedics, comprising 20 chapters in total, as detailed below.

Areas Covered in the Special Topic

The book starts with the two popular biofabrication techniques in the recent decades, namely, translational 3D bioprinting and 3D printing, which are comprehensively reviewed by Xiao and colleagues in the choice of bioink, cell types, the remodeling and maturation of bioprinting scaffolds, and by Zhang and colleagues in the choice of 3D printing polymers. Bioceramics, as a class of inorganic and non-metallic materials, is introduced by Wu and colleagues in the preparation of various bioceramics, along with the potential applications for the hard tissue repair. Additionally, Massera and colleagues provide an overview about the synthesis and application of different bioactive glasses. In addition to the bioactive scaffolds, the mesenchymal-stem-cells-derived extracellular vesicles and exosomes can also be regarded as biofabrications, which are thoroughly discussed by Jiang and colleagues. With the use of advanced biomaterials, two common techniques, electrospinning and microfluidics, are also reported by Wu and colleagues, and Cha and colleagues, respectively.

The application of 3D printing in orthopedics, including, but not limited to, bone engineering and cartilage engineering, is provided by Jiang and colleagues. Implantation and fixation are important processes in orthopedics. Laurencin and colleagues review implants that provide structural support and regulate stem-cell behaviors, while Chen and colleagues review the fixation of distal tibiofibular syndesmosis and intertrochanteric fracture of the femur. Besides bones, cartilages are also discussed starting with pathology cartilage injuries and clinical magnetic resonance imaging assessments that are introduced by Wang, followed by the review by Li and colleagues detailing scaffolds and cell-derived products for the cartilage regeneration. Additionally, intervertebral disc, as another special organ, is also reviewed by Cao and colleagues in terms of its current understanding and its repair using biofabrication-based treatments. Moreover, Zhang and colleagues also provide a comprehensive overview on joint lubrication.

In the process of bone regeneration, angiogenesis plays a critical role in accelerating regeneration. In this regard, Lu and colleagues thoroughly describe how biofabrication guided angiogenesis in controlling the pore diameter and interconnection diameter. Furthermore, Wang and colleagues review in detail the applications of different biofabrication techniques in vascularization such as bioprinting, electrospinning, or microfluidics. Since bone injuries usually are accompanied by skin, muscle, nerve, and tendon injuries, Wang and colleagues introduce various types of cells involved in the skin regeneration, and how the cell-derived products regulate the repairing. Fan and colleagues additionally review the functions of muscles, the influences of volumetric muscle loss (VML), and how it can be repaired via different biofabrication techniques. Nerve regeneration including peripheral nerve regeneration, spinal cord regeneration, and the clinical applications of nerve conduit are reviewed by Qian and colleagues. As a tendon complex, He and colleagues introduce the rotator cuff under clinical use and show how biofabrications regulate rotator cuff repairing.

In summary, this book provides a comprehensive review of novel biofabrication techniques including 3D printing, electrospinning, microfluidics, exosomes, and more for applications in bone and bone-related regeneration. We believe this book can serve as the definitive reference of the latest technologies that could inspire medical practitioners and translational researchers to improve the quality of life of patients, whether it is to transform clinical procedures in present or to propose novel procedures of improved biomaterials in the near future.

27 October 2021

Wenguo Cui
Department of Orthopaedics
Shanghai Key Laboratory
for Prevention and Treatment
of Bone and Joint Diseases
Shanghai Institute of Traumatology
and Orthopaedics, Ruijin Hospital
Shanghai Jiao Tong University School
of Medicine, 197 Ruijin 2nd Road
Shanghai 200025
P.R. China

Xin Zhao
Department of Biomedical Engineering
the Hong Kong Polytechnic University
Hung Hom, Hong Kong
P.R. China

Shen Liu
Department of Orthopaedics
Shanghai Jiao Tong University
Affiliated Sixth People's Hospital
600 Yishan Road, Shanghai 200233
P.R. China

Part I

Biofabrication Techniques

1

Current Progress and Technological Challenges in Translational 3D Bioprinting

Greeshma Ratheesh[1], *Natividad G. Cerezo*[2], *Weidong Gao*[1], *Jayanti Mendhi*[1], *Prashant Sonar*[3,4], *and Yin Xiao*[1,5]

[1]*Queensland University of Technology, Institute of Health and Biomedical Innovation, Centre of Biomedical Technology, 60 Musk Avenue, Kelvin Grove, Brisbane, QLD 4059, Australia*
[2]*The University of Queensland (UQ), School of Dentistry, 288 Herston Road, Herston, Brisbane, QLD 4006, Australia*
[3]*Queensland University of Technology, School of Chemistry and Physics, 2 George Street, Brisbane, Brisbane, QLD 4000, Australia*
[4]*Queensland University of Technology, Centre for Materials Science, 2 George Street, Brisbane, Brisbane, QLD 4059, Australia*
[5]*Queensland University of Technology, Australia-China Centre for Tissue Engineering and Regenerative Medicine (ACCTERM), 60 Musk Avenue, Kelvin Grove, Brisbane, QLD 4059, Australia*

1.1 Introduction

In the field of tissue engineering and regenerative medicine, bioprinting has emerged as an advanced tool that enables the creation of 3D functional tissue models with tailored biological and mechanical properties. It is defined as the application of scientific postulates to the design, construction, modification, and growth of living tissues using suitable biomaterials, cells, and growth factors, alone or in combination [1–3]. Usually, it is referred to as the use of computer-aided technique to pattern and assemble living cells and biomaterials in a two-dimensional or three-dimensional orientation by printing successive layers of materials. This process allows the production of bioengineered tissue structures that could serve in tissue engineering and regenerative medicine, pharmacokinetics, and cell biology studies [4]. The approach helps to eliminate problems such as donor site scarcity, immune rejection, and pathogen transfer. Cells such as osteoblasts, chondrocytes, and mesenchymal stem cells obtained from hard and soft tissues proliferate on printed scaffolds that are biodegradable and are eventually resorbed as the tissue construct grows [4]. Scaffold properties, such as porosity, interconnectivity, roughness, etc., play an essential role in the cell proliferation and differentiation [5]. It has been demonstrated that the interconnected pores with size between 0.100 and 1 mm, can facilitate the tissue regeneration, angiogenesis, nutrient transfer, oxygen supply, and withdrawal of metabolic wastes [6]. Other parameters such as scaffold topography, including the composition and the microstructure, aid in cellular adhesion, spreading, proliferation, and cytoskeletal function [7].

Biofabrication for Orthopedics: Methods, Techniques, and Applications, First Edition.
Edited by Wenguo Cui, Xin Zhao, and Shen Liu.

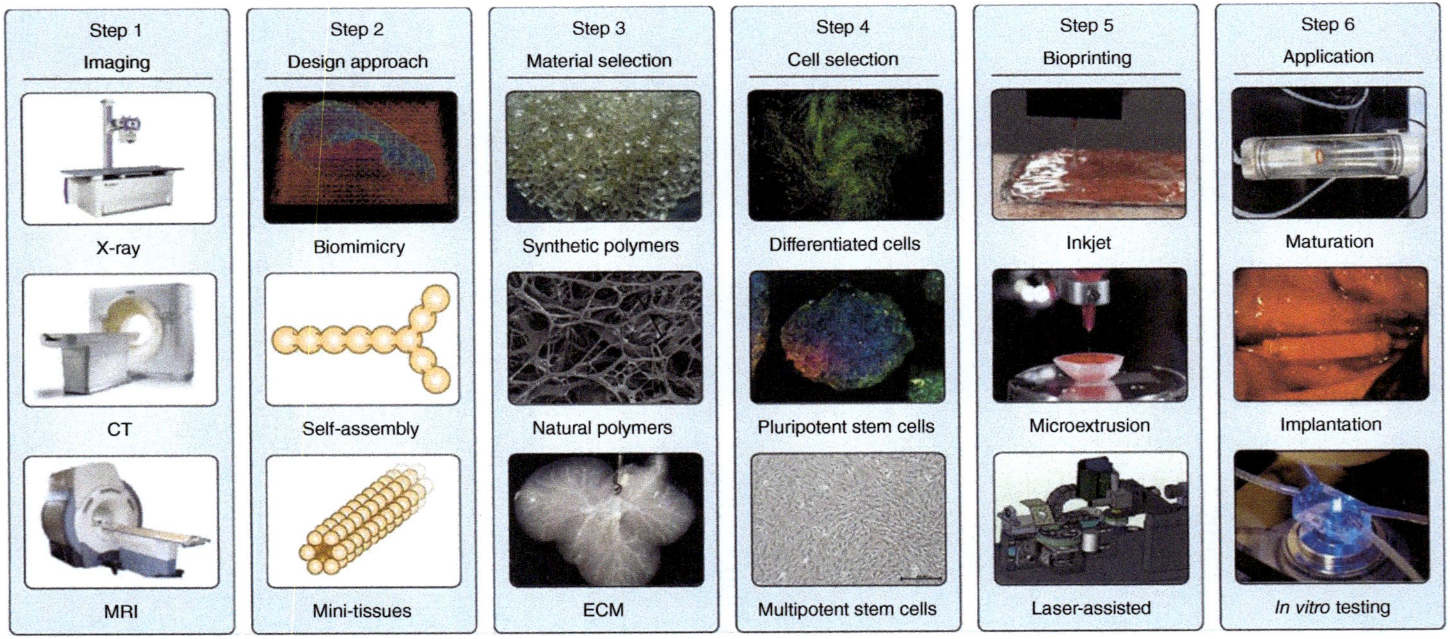

Figure 1.1 Steps in bioprinting process. (1) Imaging of the damaged tissue; (2) design approach using biomimicry, self-assembly, and mini-tissue building blocks; (3) choice of material; (4) choice of cell; (5) choice of desired printing technique; and (6) maturation and application of the printed tissue. Source: Xing et al. [24]. Reprinted with permission of MDPI/CC BY 4.0.

One of the advanced methods in scaffold fabrication is the bioprinting process. Unlike other scaffold fabrication methods, the material used in bioprinting is living cells that are incorporated into soft printable gel precursor solution, commonly referred to as bioink [8–10]. Bioink is defined as the material that has the ability to mimic the extracellular matrix (ECM) environment in order to enhance the proliferation and differentiation of cells [10–22]. The bioink could be natural, synthetic, or a combination of both, which guarantees cell viability and printability. The rheological properties of an ideal bioink should maintain the scaffold structure; hence, the viscosity of the bio-ink and/or its capacity for rapid crosslinking is central to the success of the printing [23]. The process of bioprinting using these bioinks helps to encapsulate and deliver one or several cell types in a precise manner and enables the manufacturing of complex structures that could integrate into living tissues.

The process of bioprinting as illustrated in Figure 1.1 involves a series of steps, namely, imaging of the defective site, designing using a computer-aided design, followed by the selection of an appropriate material and cells, and finally toward the printing. The final success of the construct is determined by the post-processing, which involves regular monitoring with an optimum supply of nutrient and oxygen in the printed construct [25]. The complexity and versatility of this technique guarantee a wide range of applications such as regeneration of tissues like bone [26], cartilage [27], liver [28], and skin [29]. The fabrication methods have been improved by protein coating [30], cell deposition, and drug delivery [31, 32], and the fabricated tissues also assist in studying tumor growth [33], vascular network formation [34, 35], and understanding of stem cell differentiation [36, 37]. This chapter focuses on the current progress and technological challenges in the bioprinting process toward tissue regeneration and present the knowledge gaps in adopting this technology.

1.2 Challenges in the 3D Bioprinting Process

The successful fabrication of functional tissue or organ depends on the advancement in three types of technology, namely, the manufacturing technology, cell technology, and technology for *in vivo* integration [38]. Although the field of 3D bioprinting has demonstrated the remarkable potential for future development in the field of tissue and organ printing, there are several challenges that must be overcome to reach the stage of translation to a biological platform. Some of the critical challenges in the field of 3D bioprinting have been discussed in the following sections.

1.2.1 Manufacturing Challenges

1.2.1.1 Choice of Bioink

Hydrogels are ideal materials for the 3D bioprinting process because of the tunable physicochemical properties and the ability to mimic ECM. The ability of hydrogels to hold high water content that allows cell migration, nutrient/oxygen transfer, and waste removal makes it an important candidate for 3D bioprinting. There are a variety of hydrogels used for the bioprinting process such as alginate, collagen I, methacrylated gelatin, and poly(ethylene glycol) (PEG) [39–43]. Hydrogels are either from a natural source or artificially synthesized or a combination of different materials. Compared to the natural hydrogel, the synthetic materials that possess the ability to modify the chemical structure show more advantages [44, 45]. However, there are certain drawbacks in the use of such hydrogels, such as the mechanical strength, printing resolution, and maintaining the shape of the printed construct [46–48]. The mechanical stability of the scaffold printed using such bioink still remains a critical challenge. In recent times, bioactive or composite materials are developed with the insight of improving the mechanical stability or tissue interaction [49]. The 3D printable bioink can be extended by creating certain chemical modifications on the biomaterials [50–52]. For instance, the multi-armed oligomers/polymers or branched polymers can be acrylated or methacrylated to produce macromeres or prepolymers that can be photocrosslinked. This kind of polymer that is photocrosslinked has high gel content with high degree of crosslinking [51]. On the other hand, in order to achieve high mechanical strength and biological activity, scaffolds are fabricated by combining a biodegradable polymer or other biomaterials which provide the required mechanical stability and a functional bioink loaded with cells for imparting an increased biological activity on the resulting construct [53–55]. The incorporation of such substitutes not only improves the mechanical property but also has an influence on the improved porosity. For example, in a study demonstrated by Dandoy et al. 2011, a silica shell was introduced around alginate-based hydrogel beads by *in situ* silicic acid sol–gel transition by polycations. The resulting composite hydrogel possesses improved mechanical strength and chemical stability. They also revealed regulated porosity to govern mass exchange between the external environment and the interior hydrogel in order to achieve controlled paracrine for cell therapy [56].

Yet another important characteristics that bioink should possess is the ability to improve the biomimetic property that regulates the cell behavior [57, 58]. The development of an optimal biomaterial that possesses a suitable microenvironment for cell response with optimal mechanical properties and printability remains a critical challenge in regenerative engineering. An ideal bioink should possess the ability to improve the biomimetic properties and thereby regulate the cell behavior [57, 58]. The biological properties of bioink can be enhanced by the use of certain growth factors or cytokines that regulate cell proliferation and differentiation by providing suitable biochemical and physiochemical factors for tissue regeneration. The material printability and cytocompatibility are currently the most important criteria of bioprinting process. For example, Kristin et al. 2015 developed a novel bioink using spider silk protein. The group suggests that such kind of bioink can be used without any crosslinking agent or thickeners for increasing the mechanical stability of the scaffold, and the use of cell-binding motif to the silk protein for enhancing the tunability of cell–material interactions [59].

1.2.1.2 Cell Selection and Optimization

Most of the tissues or organs are made up of multiple types of cells that serve different functions such as structural support, role in vascularization, or provide a suitable niche for the differentiation of stem cells. The choice of cells is highly dependent on the targeted tissue that is to be regenerated. The careful selection of cells should fulfill their physiological state *in vivo* and maintain the function under optimum conditions [10, 60]. The type of cells selected should proliferate in a controlled fashion; in other words, less proliferation can lead to the loss of viability, whereas high proliferation leads to hyperplasia or possible apoptosis. Furthermore, the number of cells incorporated in a specific construct is also a crucial parameter that varies with the type of tissue. For example, the estimated number of cells per gram of liver tissue is 1.3×10^8 cells. Hence, it is necessary to replicate this density during the bioprinting process [61]. Studies have reported that the minimum therapeutic threshold in order to serve the solid organ function is 1–10 billion cells [62, 63] (Figure 1.2). However, the cell content depends on the type of the bioprinting process. Extrusion-based bioprinting can fabricate constructs with high cell density (approximately 8000 cells in 300 μm diameter), and on the other hand, inkjet bioprinting/droplet-based bioprinting can print with a lower cell density of $<10^6$ cells/ml [61].

The increase in cell density has both advantages and disadvantages. The higher cell density during the printing process will aid in cell–cell interaction and promote tissue regeneration. However, this can affect the printability due to the change in the viscosity of bioink, which in turn affects the cell distribution [64]. Moreover, cell density also plays a critical role in cell proliferation and cell–cell interactions. In a printed construct, the bioink normally mimics the role of ECM environment for cell support. Studies have proved that an increase in cell density elevates the remodeling process of ECM [65, 66]. It is always advisable to use a high cell density for a

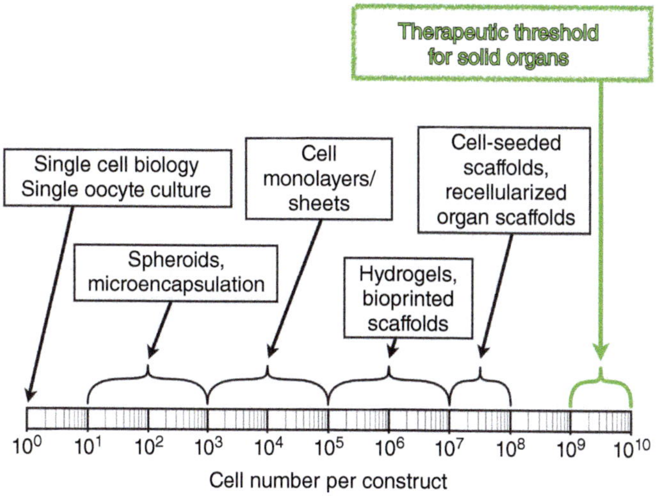

Figure 1.2 Scale representing the number of cells required per construct to form a solid tissue construct. Source: Miller [63] / Plos One / CC BY-4.0.

better cell proliferation; however, cell density should be carefully distributed in the fabricated tissues considering the fact that each organ has different cell densities in different areas of tissue structure; for instance, in pancreas, high density of beta cells is surrounded by other types of cells (Alpha, Delta, Epsilon, and F cells) at a lower cell density, indicating the various cell density in functional tissue structures [67]. All these parameters suggest that more research is still needed for a better understanding in the distribution of cells post-bioprinting and maintaining the optimum cell density to that of the native organ.

1.2.1.3 Printing Resolution and Mechanical Stability

In addition to biocompatibility and biodegradability, an ideal bioink should possess certain physicochemical properties that improve the resolution and mechanical stability of the printed construct [68]. The printed construct must have the optimum mechanical stability and porous environment for cell migration, penetration, proliferation, and vascularity. The rigidity of the native tissue ranges from 0.2 to 15 000 kPa, depending on the kind of the tissue. Similarly, the shear stiffness value also has a broad range from 100 Pa to 20 kPa. A wide range of parameters such as the viscosity, crosslinking strength, temperature, and printing speed govern the mechanical properties of the bioprinted construct. The rheological and mechanical properties of bioink varies with the type of the printing process; for instance, extrusion-based bioprinting uses a highly viscous material ($3–6 \times 10^8$ mPa s); on the other hand, droplet-based bioprinting uses a less viscous material (3.5–12 mPa s) [69]. This in turn influences the mechanical stability of the printed construct.

The printability and mechanical stability of the printed tissue or organ are highly dependent on the type of biomaterials used for printing. Paxton et al. 2017 suggested that assessment of printability of bioink is a two-step process, firstly to check the ability of bioink to form a fiber for 3D construct fabrication and secondly to understand the rheological characteristics [70]. The rheological requirement for extrusion-based printing is influenced by the concentration of materials used for printing. In general, a low-viscous material is preferred so as to avoid the shear stress that can in turn influence the cell viability [71]. However, high viscosity or solidification is necessary to retain the shape and structure of the printed scaffold. The printing resolution of bioprinting can be improved by the thermogelation process on materials such as gelatin [71, 72]. Bioink such as PEG is less viscous that makes it difficult to maintain their shape after printing. In one of the recent studies, Ribeiro et al. 2017 developed an evaluation technique for shape fidelity by testing filament collapse and filament fusion on poloxamer 407 and PEG blend. The group demonstrated a simple theoretical model to relate the collapse of filament with the yield stress of the bioink [73], in which the deflection angle in each gap distance was calculated using the formula

$$\theta = \sin^{-1}\left(\frac{\rho g L}{\sigma_{\text{yield}}}\right)$$

where ρ is the material density, g is the gravitational acceleration, L is the distance from the edge of the pillar to the midpoint of the suspended filament, and σ_{yield} is the yield stress of the hydrogel. Although the theoretical model overestimates the defection angle, the slope of regression shows the same trend as the experimental ones. This is because the model considers only the gravitational force and the yield stress of the hydrogel.

Another important parameter that influences the mechanical property and shape fidelity of the construct is the time and temperature (for temperature-sensitive bioink). For instance, gelatin-based bioink suffers poor shape fidelity until the strategy of two-step crosslinking is introduced. The two-step crosslinking shows a better shape fidelity, in which the first step is the physical crosslinking (lowering the temperature), and the second is a chemical crosslinking [74]. Silke et al. 2014 demonstrated a two-step crosslinking using gelatin and alginate. The gelatin was heated to 40 °C and dispensed onto a cooling plate maintained at 10 °C, which leads to immediate solidification of the gel. This gel was further crosslinked with $CaCl_2$ solution to retain the shape of the construct [74]. GelMA (methacrylated gelatin) is another popular material that can be used for two-step crosslinking, in which the second step is a photocrosslinking. The mechanical stability of such construct relies on the number of crosslinking moieties and the UV exposure time. Cristina et al. 2016 demonstrated that the elastic modulus of the construct increased with UV exposure time [75]. Furthermore, the photocrosslinking property is also reported to enhance the shape fidelity of the printed construct with a suitable supportive polymer [74]. The shape fidelity and printing resolution can be also improved by blending two different materials. For example, studies have shown that the incorporation of supportive polymer such as hyaluronic acid methacrylamide (HAMA) in GelMA improved the shape fidelity and mechanical stability [76, 77]. Although there are lots of research studies in developing new printable biomaterials, there is still a lot of room for the development of a suitable material that improves cell growth and possesses the same mechanical stability as that of native tissue.

1.2.1.4 Design and Post-processing

As stated by Jordan S. Miller, the "Anatomical complexity still remains unsolved" [63]. For the successful translation of the printed construct, the organ or tissue should exactly match with the patient. Advancement in the quality of design requires certain fundamental changes such as cell culturing and seeding [34, 78]. The scaling up of the vascular and neural network in an implant is highly challenging. For example, the biomimicry is crucial for the architectural design of the fabricated tissues such as the brain, which is made up of a macroscopic structure consisting both solid phase and fluid phase [79]. Each step in the process of designing a scaffold should be done with at most care and accuracy. The hurdle in the design of an organ or tissue can be overcome by the advancement in the technology. One of the important parameters that govern the design of the custom shape scaffold is the anatomical 3D imaging. Eduarda et al. 2019 shows a schematic overview of the various pre-clinical and clinical imaging technique (Figure 1.3) [80]. Mahmoud et al. 2015 introduced automated scaffold design (ASD) that helps to design a 3D scaffold that eliminated a geometrical mismatch. A method called k-means clustering algorithm was used to distinguish between different tissues and hence decode the defective portion [81].

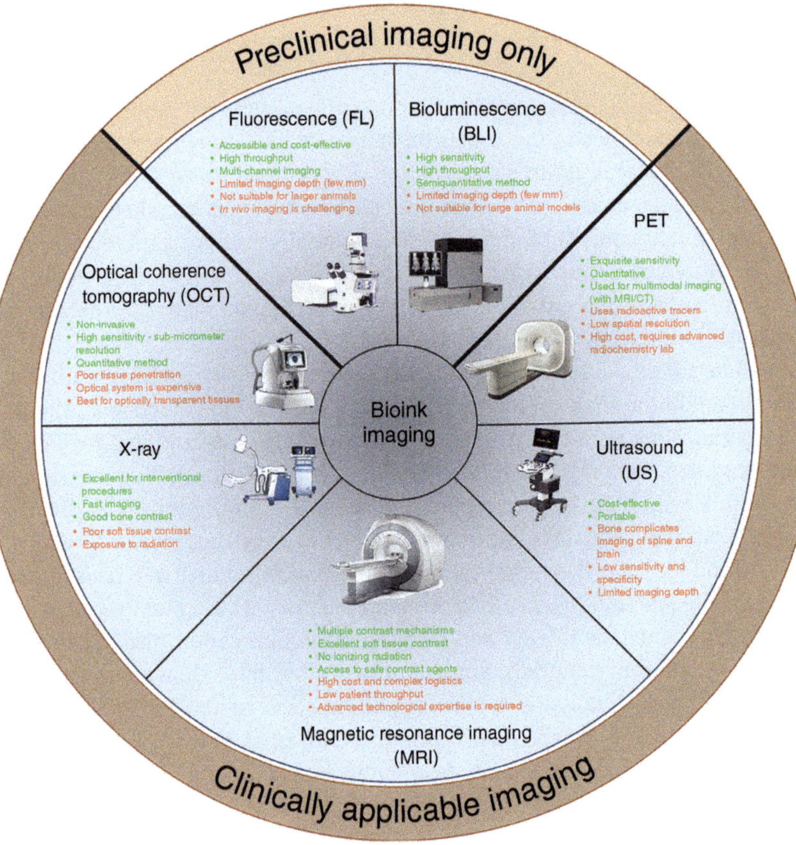

Figure 1.3 Overview of various clinical and pre-clinical imaging techniques. PET, positron emission tomography. Source: Oliveira et al. [80] Reprinted with permission of Elsevier.

Post-processing is an important step in the bioprinting process as most of the bioprinted constructs are fragile soon after the printing process. In order to maintain the structural and mechanical stability, these printed constructs are chemically or physically crosslinked. One of the critical challenges in post-processing is the duration taken for the crosslinking. Any delay in the crosslinking would lead to the loss of structural integrity [82]. Physical crosslinking of the hydrogel is achieved by a non-covalent interaction such as van der Waals forces of interaction, hydrogen bonding, hydrophobic attraction, electrostatic forces, and hence such kind of interaction produces a hydrogel with low mechanical properties that has the ability to disintegrate faster because of the non-covalent molecular interactions [83, 84]. On the other hand, chemical crosslinking produces a hydrogel with high mechanical properties because of the strong covalent bonding [85]. However, such hydrogel is non-biocompatible because of the organic solvents and the polymerization initiators used in the process. In some instance, a dual-stage crosslinking approach that involves both ionic and UV crosslinking are used on low-viscous alginate and gelatin methacryloyl (GelMA) that improve the printability and enhance the cell viability [75].

1.2.2 *In Vitro* Challenges

1.2.2.1 Cell Viability

The consideration of cells is the starting point to the tissue engineering process. Apart from the quantity and quality of cells, an important parameter that has to be taken into consideration is the origin (autologous, allogeneic, or xenogeneic) and its off-the-shelf availability on an emergency situation [86]. The main goal of bioprinting or organ printing process is the ability of the cell-seeded construct to finally mimic the normal tissue. Therefore, the critical issue is the specific or controlled delivery of necessary biological cues in order to achieve a healing environment. The cell viability in the tissue engineering construct depends on various parameters such as the type of hydrogel, printing technique, printing speed, nozzle diameter, and temperature. Besides these different parameters, cell viability is highly dependent on the mechanical environment or in other terms the physical force the cells are exposed to. It is well known that the different bioink behaves differently to the shear force exerted on them. This change in bioink can be particularly observed near the nozzle walls. The viscosity of bioink, pressure, and the nozzle diameter has a direct influence on the shear force exerted on the cell incorporated in the bioink. Amit Panwar and Lay Poh Tan, 2016 has schematically illustrated the influence of various parameters on the cell viability [77] (Figure 1.4). The increase in the nozzle diameter would help to lower the shear stress exerted on the bioink. However, it could affect the printing resolution. It has been reported that high shear stress exerted on the cell by the viscous bioink during the printing process would affect the cell viability and would also cause an alteration in the proliferation pattern and functionality and differentiation in those cells that survived during the printing process [87]. Studies have also shown that the concentration of the hydrogel influences the cell survival. An increase in the gel concentration hinders cell migration and diffusion because of the entangled network of the polymer chain, which in turn affects cell viability [77, 88]. Yet another important parameter that determines the viability of cells is the ability to withstand harsh crosslinking steps after printing that is usually carried out through chemical agents or UV light [10]. Over the past few decades, a lot of studies has to be performed aiming to improve the viability and differential potential of the cells incorporated within the bioprinted construct. Pepper et al. 2010 proposed different methods to process inkjet bioprinted constructs in order to promote cell hydration, viability, and pattern retention. The group proposed the ideal use of collagen as a printing substrate and the incorporation of the cells in Hank's balanced salt solution for the bioprinting process [89]. However, we are still far behind in the development of an ideal method to improve the cell viability in a bioprinted construct. All this explains the need for more research in the development of printable biomaterials that possess mechanical stability and provide a suitable environment for cell proliferation.

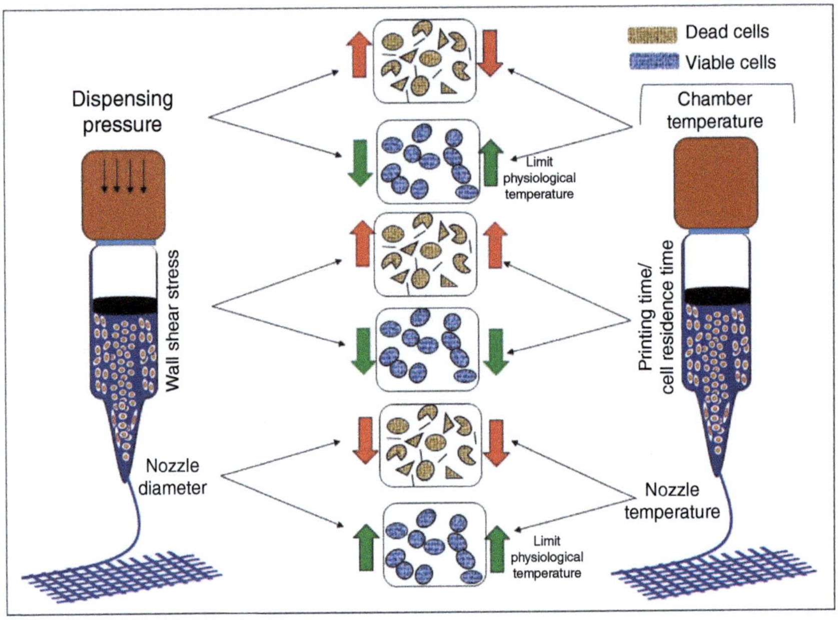

Figure 1.4 Influence of various printing parameters in cell viability. Source: Panwar and Tan [77] / MDPI / CC BY-4.0.

1.2.2.2 Cell Differentiation

The type of cells selected for the bioprinting depends on the target tissue that is being regenerated. In most of the cases, stem cells are used pertaining to their ability to differentiate into specific lineages. The advancement in the stem cell technology is one important achievement that governs the bioprinting process. Stem cells are found in various tissues and they self-renew to produce more stem cells and can be differentiated into a specialized cell type to form a tissue or an organ [38]. The bioprinting is a very sensitive process, as there are high chances of tissue rejection even though a patient-specific stem cell has differentiated into organ-specific cells [86].

The ability of the construct to mimic the target tissue including the extracellular microenvironment has been well understood; however, the 3D biomaterials used for the fabrication of a scaffold can also influence the function of cells [90]. Material properties such as mechanical stiffness [91, 92], nanoscale topography [93–95], and the chemical functionality [96, 97] play a vital role in determining the human mesenchymal stem cell (hMSC) differentiation. Studies have demonstrated that the ions released during the degradation of the biomaterials can influence the phenotype of stem cells [98–100]. The material degradation can also lead to variation in stem cell shape [101] and traction force [102], which in turn leads to the changes in the cell differentiation. Pluripotent stem cells that are incorporated in degradable biomaterials such as agarose, gelatin, or poly(lactide-*co*-glycolide) also affect cell differentiation [103]. Thus, in order to achieve the full advantage of stem cell technology, one must fully understand the differentiation pattern for tissue-specific cells, which requires knowledge about the signaling pathway that leads to a specific type of differentiated tissue cells.

1.2.3 Clinical Challenges

1.2.3.1 Remodeling and Maturation

Maturation of the fabricated construct is an essential step for clinical translation of printed construct. The fabricated construct is usually transferred to a suitable bioreactor to ensure cell growth, proliferation, vascularization, remodeling, and maturation. The process usually takes several months of culture in a bioreactor to achieve increased level of adhesion molecules, ECM, and collagen formation [104]. The maturation of complex-vascularized constructs is dependent on a perfusion culture system to ensure appropriate physical stimulation, nutrient and biochemical factor supply, and the elimination of metabolic waste. Nevertheless, perfusion-based maturation assessment is complex and includes sophisticated equipment. The maturation of printed construct is also dependent on the type of the bioprinting process and the type of bioink used for the fabrication. For instance, scaffold-free construct maturation is different from that of a hydrogel-printed construct.

During organogenesis, different cell aggregate-based bioinks undergo a complex sequence of events [105]. Cells bind to each other when enclosed in a mold to reduce free energy and promote cell-to-cell interactions through cadherin-mediated adhesion [106]. Cells tend to deposit their own ECM over time, which in turn facilitates cell adhesion and creates contractile forces that form intact neo-tissues. ECM components such as elastin and collagen are deposited by the cells that allow increased tissue cohesion and mechanical properties. To reduce the energy configuration during fusion, cell aggregates assume a more rounded geometry, followed by ECM deposition and maturation [107]. In comparison, cells incorporated in hydrogel are exposed to a particular environment and show a distinct sequence of events during organogenesis. In the first stage, the cells attach to the scaffold, followed by proliferation and ECM production. Additionally, proteinases such as matrix metalloproteinase are expressed by the incorporated cells, which help in the degradation of the bioink. This matrix degradation during cell proliferation tends to influence on the morphological and physiological changes in the printed construct [107]. The maturation of a bioprinted construct to a fully functional organ or tissue is still a big challenge in the field of bioprinting and is said to be governed by a wide range of influencing factors such as the cell type, culture environment, or the printing process. The field therefore requires intensive research to understand cell growth and tissue formation.

1.2.3.2 Immune Response

The 3D bioprinting process involves the use of autologous primary or stem cells. The use of these kinds of cells may avoid some negative immune response after implantation; however, the biomaterials used for the incorporation of these cells acquire host proteins, thereby modulating the immune response. There are a series of mechanisms such as delayed-type hypersensitivity, innate immunity, and antibody-mediated and immunoregulatory response that are involved in the rejection of an implant. Macrophages trigger the adhesion of giant cells on the surface, which in turn release mediators for degradation [108]. Studies have shown that many factors are responsible for immunological response on implants. For instance, a recent study by Sridharan et al. demonstrated the influence of material stiffness on the migration mode of macrophages. The group studied the response of macrophages on varying stiffness of polyacrylamide gel coated with collagen and revealed that stiff gel (323 kPa) induces pro-inflammatory response while medium (88 kPa) or soft (11 kPa) gel induces anti-inflammatory and high phagocytic phenotype. They also demonstrated that the stiffness of gel determines the behavior of macrophages, which is independent of biochemical cues and also plays a significant role in the migration of macrophages, wherein the medium or soft gel exhibits a podosome-independent fast amoeboid movement and the stiff gel shows a podosome-dependent slow movement [109]. Similarly, in another study, Abebayehu et al. 2019 demonstrated the role of scaffold architecture in inflammatory responses mediated by the mast cells [110].

1.2.3.3 Vascularization

Vascularization is the most important parameter in any engineered construct. The printed construct must be able to integrate with the host ECM and thereafter aid the vascular ingrowth within the construct. Developing a functional vasculature in a fabricated construct is a critical parameter for tissue survival. Studies have revealed that the maximum distance for nutrient diffusion without vasculature for cell survival is 100–200 μm [10]. Heterogeneity is the current problem in vasculature in the fabricated construct. However, this has been addressed in recent years by printing with different cell type in the same construct [111, 112]. For the translation of the bioprinted construct, immediate perfusion is necessary. It is reported that angiogenesis and the formation of intervening capillaries in a gel construct takes more than 14 days [61, 113]. Various methods have been employed for the fabrication of a vascularized structure. One of the commonly used methods to fabricate a vascularized construct is the combination of direct and indirect bioprinting process, in which fugitive ink is used to print channels. For example, Kang et al. 2016 fabricated a micro-channeled construct for cartilage tissue regeneration [88]. In a recent study, Tijore et al. 2018 constructed a gelatin hydrogel microchannel structure that facilitated vascular smooth muscle cell (vSMC) attachment and elongated along the direction of the microchannels and thereby retaining the contractile phenotypic characteristics [114]. Figure 1.5 shows various successfully bioprinted constructs for vascularization.

Yet another method to fabricate vascular channels is the direct bioprinting process. Dolati et al. 2014 fabricated a larger vascular conduit of 600–700 mm using a coaxial nozzle [117]. One of the critical challenges in printing vascular channels is the architectural complexity. A successful method to answer this complexity is the use of high precision laser-based bioprinting, which prints bifunctional microvasculature [118, 119]. However, the current strategies for fabricating vascular channels fail to mimic the anatomy of a native vessel [120]. Larger blood vessels

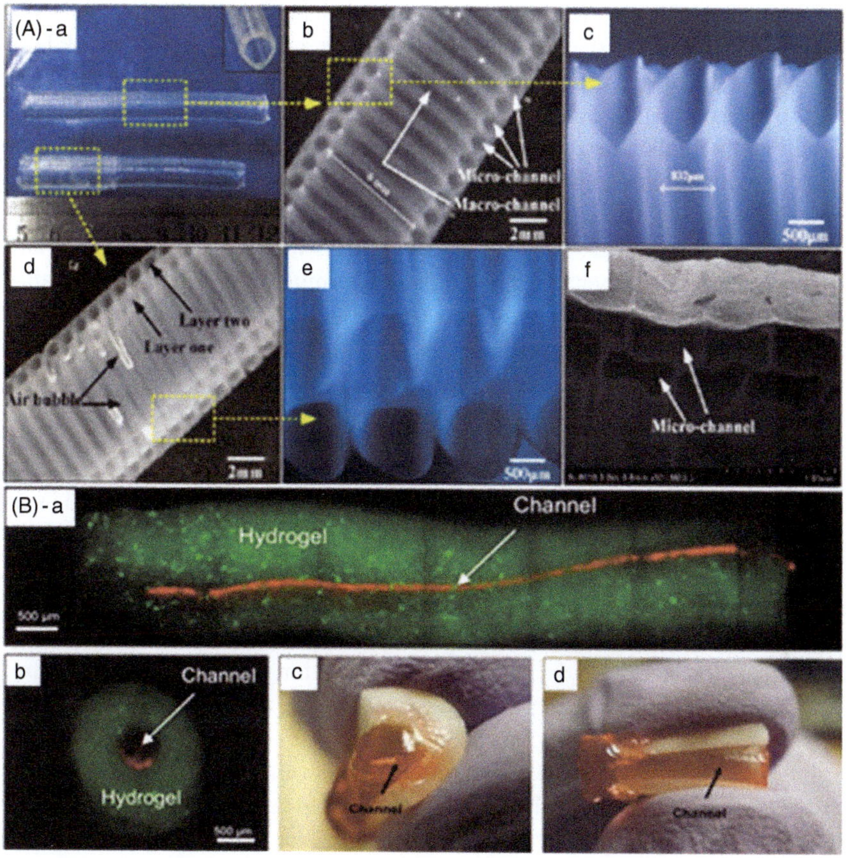

Figure 1.5 Bioprinting of a vascular construct: (A-a) structural feature of alginate vessel, (b and c) longitudinal cross section of the single-layer structure under different magnifications, (d and e) longitudinal section of the double-layer structure, and (f) scanning electron microscopic image of longitudinal section, (B-a and b) longitudinal and cross-sectional views of GelMA hydrogel loaded with green fluorescent microparticles showing the fabricated microchannel being perfused with a solution of red fluorescent microparticle (c and d) and photographic image of GelMA hydrogels from longitudinal and occlusal view. Source: (a) Gao et al. [115] / with permission from American Chemical Society, Athirasala et al. [116] / with permission from Springer Nature.

are made up of three layers, intima, tunica media, and tunica adventita, where capillaries are made up of an endothelial intimal layer [51]. Furthermore, bioprinting the blood capillary is a complex pertaining to its size. Even a high-precision laser-based bioprinter produces droplets of 20 mm size. Although the method can develop a high-resolution print, it is a time-consuming process that would eventually affect cell viability [121]. One of the methods to overcome this limitation is the capillary sprouting via angiogenesis, for which endothelial cells were incorporated within the printed construct [82, 116]. Various growth factors such as vascular endothelial growth factor (VEGF), transforming growth factor (TGF), platelet-derived growth factors (PDGFs), and fibroblast growth factor (FGF) are incorporated in the construct so as to promote angiogenesis. This under an *in vivo* condition results in the angiogenic sprouting at a rate of 1 mm per week [51]. Yet another critical challenge in the bioprinting of vascular tissue is the ability of the construct to support and withstand the mechanical and chemical stimulus exerted by the fluidic flow. This was better demonstrated by a study conducted by Gao et al. 2017, where they fabricated a vascular structure with multilevel fluidic channels that could integrate on devices and stimulate the microenvironment of blood vessels. The printed structure showed an ultimate strength of 0.184 MPa and sustainability of both mechanical and chemical stimulation [115]. However, after five days of incubation, the ultimate strength of the construct decreased from 0.184 to 0.105 MPa. All these finding explains the difficulties in fabricating and the scope for further research in developing a vascularized tissue construct.

1.2.3.4 Innervation

Innervation or nerve distribution is an essential parameter in tissues such as skeletal muscles, esophagus, stomach, and bladder. The lack of innervation in these kinds of tissue leads to a non-functional and atrophy tissue. Studies have reported that the nervous system plays a major role in regeneration, sensation, adaptation, and immune interaction [122]. Lee et al. 2009 were one of the early groups that demonstrated the bioprinting process for the fabrication of neural tissue. The group used a collagen-based precursor that was subjected to nebulize crosslinking [123] by sodium bicarbonate [124]. Studies have also been conducted to understand the nerve tissue regeneration by bioprinting Schwann cells. These cells are non-neuronal cells that serve as a support and protective element in the homeostasis of the peripheral nervous system. In a recent study, Liqun et al. 2019 fabricated scaffolds using Schwann cells and a low-viscous hydrogel such as alginate, hyaluronic acid, and fibrin. The group demonstrated that the scaffold promoted the elongation of the Schwann cells and showed aligned dorsal root ganglia outgrowth [125]. Figure 1.6 shows various successfully bioprinted construct for innervation. Fabrication of cellularized conduits is also one of the methods in nerve regeneration through bioprinting. Hu et al. 2016 fabricated a bioconduit that promoted reinnervation of a 1 cm gap [128]. Although there are many advances in the conduit technology, the reinnervation for axon elongation is restricted to a gap of less than 3–4 cm [61, 107].

Similar to blood vessels, nerves also have a number of layers with a specific function. For the successful integration of a bioprinted nerve, two important factors such as the electrical and biochemical stimulus should be addressed. Recently, lot of research has been performed, which aims the improvement of cell attachment to the scaffold fabricated using conductive polymer [129–131]. Although there are a lot of studies that focus on electrical stimulation for tissue engineering, the mechanism by which it can be replicated using bioprinting process is still not clear. Moreover, the scope of clinical translation of such a construct fabricated using conductive polymer is unpredictable. In spite of all these developments, more research is still needed to improve the innervation for printing complex organs.

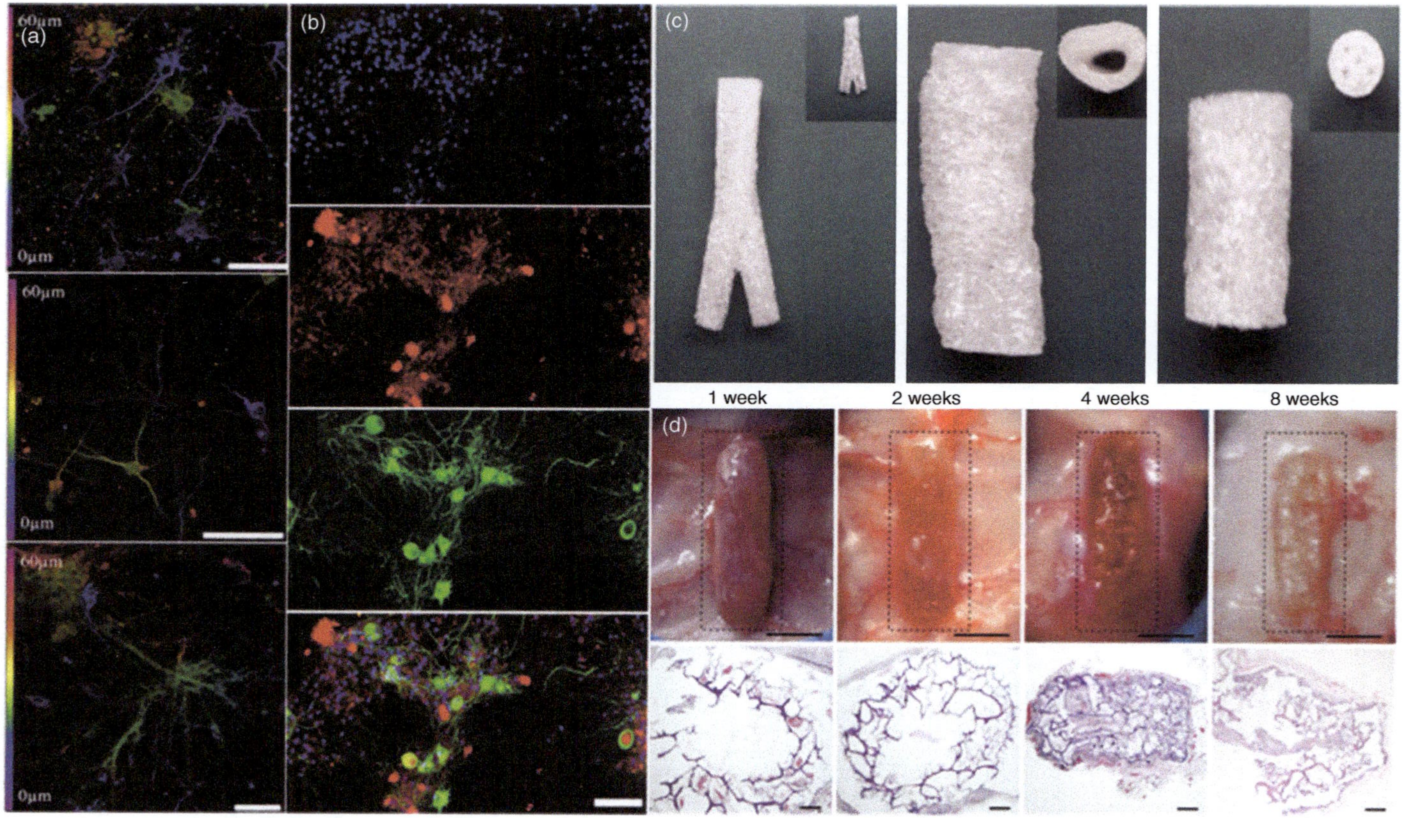

Figure 1.6 Bioprinted constructs for innervation: (a) Confocal microscopic image of the 3D cultural neuronal model after five days of culture, (b) immunocytochemical staining of Schwann cell-laden scaffold for dorsal root ganglia neurite outgrowth, (c) cellularized conduits for peripheral nerve regeneration, and (d) degradation and histological evaluation of nerve guidance conduits implanted on the dorsal site of rats. Source: (a) Lozano et al. [126] / with permission from Elsevier, (b) Ning et al. [125] / with permission from Royal Society of Chemistry, and (c) Hu et al. [127] / with permission from Scientific Reports / CC BY 4.0.

1.3 Conclusion and Future Perspectives

The process of 3D bioprinting has demonstrated the remarkable potential for future development in the field of organ printing. However, there are several existing challenges and shortcomings in this technology that has to be overcome to reach the stage of translation to a biological platform. The future development in the field of bioprinting will involve the use of a wide range of bioink that improves cell survival rate, cellular interaction, and tissue forming capacity with increased mechanical properties and functionality. The personalized treatment strategy by bioprinting can be revolutionized by the use of patient-specific bioink incorporated with the patient-derived stem cells. In a recent study, Filardo et al. 2019 demonstrated the use of primary mesenchymal stem cells with collagen-based-bioink as a treatment strategy of knee meniscal lesion [132]. Similarly, studies have shown the use of patient-specific growth factor. Faramarzi et al. 2018 studied the use of platelet-rich plasma as a patient-specific source of autologous growth factor for the fabrication of bioprinted scaffold. The printed construct acts as a platform for the control release of growth factors for vascularization and stem cell migration [133]. In a recent study, patient-specific bone particle bioprinting demonstrated the use of autogenously viable bone particles in combination with GelMA. The study demonstrated the feasibility of using the autogenous bone particles, which may have the potential for chairside clinical translation [134, 135].

The rapid growth of bioprinting technology and the promise in tissue engineering and regenerative medicine can visualize the possibility of fabricating volumetric composite tissues in the near future. The success in bioprinting will emerge from a joint effort from various fields such as materials science, engineering, biology, and clinical sciences. The fast-paced evolution of multiple head bioprinters and high-resolution printing will help to develop the complex heterocellular tissues with multi-scale innervation and vascularization. The recent development in the integrated tissue and organ printer (ITOP) with a pressure controller and multiple dispensing heads will overcome the current challenges in size and mechanical stability of the tissue construct [88]. However, there is an urgent need in the development of a new technique to fabricate vasculature channels for volumetric construct in order to maintain cell viability [34, 136]. One of the latest developments in the field of bioprinting is the use of 4D technology, which allows high-resolution shape fidelity and might serve as a promising tool for the development of functional organs and tissues [137]. The development in the new printable bioink using advanced biomaterials and self-assembling peptide will help to improve the structural stability and mechanical integrity of the printed construct [138, 139]. Furthermore, the development of a new bioreactor would enhance the post-printing maturation that provides necessary chemical stimuli that aids in the tissue growth and remodeling.

Bioprinted tissues such as skin and cartilage require minimal vascularization and are expected to be the first 3D-printed transplantable tissues, whereas tissue that has a complex structure (e.g. heart, liver, and pancreas) requires a lot of extensive research for a successful transplantation. Several advances have already been reported in the field of bioprinting such as skin tissue that serves similar characteristics of a human skin [140], small liver, heart, and pancreas. Although we are still in the early stage of clinical translation of bioprinted tissue, the technology already serves as a guide for non-biological surgical implants [107]. However, we are far behind from creating a replicate of native organ. There is a lack of proper understanding in the different key factors of the bioprinting; for this purpose, this paper gives an insight into the critical challenges in the bioprinting process that has to be given at most priority for the printing of a functional tissue.

Acknowledgments

The author gratefully acknowledges Queensland University of Technology (QUT) for the PhD research scholarship.

References

1 Langer, R. and Vacanti, J. (1993). *Science* 260 (5110): 920–926.
2 Sittinger, M., Bujia, J., Rotter, N. et al. (1996). *Biomaterials* 17 (3): 237.
3 Rose, F.R. and Oreffo, R.O. (2002). *Biochem. Biophys. Res. Commun.* 292 (1): 1.
4 Matai, I., Kaur, G., Seyedsalehi, A. et al. (2020). *Biomaterials* 226: 119536.
5 Guillotin, B., Souquet, A., Catros, S. et al. (2010). *Biomaterials* 31 (28): 7250.
6 Ashammakhi, N., Ahadian, S., Xu, C. et al. (2019). *Mater. Today Bio* 1: 100008.
7 Guillemot, F., Souquet, A., Catros, S. et al. (2010). *Acta Biomater.* 6 (7): 2494.
8 Kačarević, Ž.P., Rider, P.M., Alkildani, S. et al. (2018). *Materials* 11 (11): 2199.
9 Hospodiuk, M., Dey, M., Sosnoski, D., and Ozbolat, I.T. (2017). *Biotechnol. Adv.* 35 (2): 217.
10 Murphy, S.V. and Atala, A. (2014). *Nat. Biotechnol.* 32 (8): 773.
11 Hollister, S.J. (2005). *Nat. Mater.* 4 (7): 518.
12 Engel, E., Michiardi, A., Navarro, M. et al. (2008). *Trends Biotechnol.* 26 (1): 39.
13 Sears, N.A., Seshadri, D.R., Dhavalikar, P.S., and Cosgriff-Hernandez, E. (2016). *Tissue Eng. Part B* 22 (4): 298.
14 Hammer, J., Han, L.H., Tong, X., and Yang, F. (2014). *Tissue Eng. Part C* 20 (2): 169.
15 Shanjani, Y., Pan, C., Elomaa, L., and Yang, Y. (2015). *Biofabrication* 7 (4): 045008.
16 Hinton, T.J., Jallerat, Q., Palchesko, R.N. et al. (2015). *Sci. Adv.* 1 (9): e1500758.
17 Foty, R.A., Pfleger, C.M., Forgacs, G., and Steinberg, M.S. (1996). *Development* 122 (5): 1611.
18 Jakab, K., Norotte, C., Damon, B. et al. (2008). *Tissue Eng. Part A* 14 (3): 413.
19 Yu, Y., Moncal, K.K., Li, J. et al. (2016). *Sci. Rep.* 6: 28714.
20 Jose, R.R., Brown, J.E., Polido, K.E. et al. (2015). *ACS Biomater. Sci. Eng.* 1 (9): 780.
21 Thomas, D., Jessop, Z.M., and Whitaker, I.S. (2017). *3D Bioprinting for Reconstructive Surgery: Techniques and Applications*. Woodhead Publishing.

22 Faramarzi, N., Yazdi, I.K., Nabavinia, M. et al. (2018). *Adv. Healthc. Mater.* 7 (11): 1701347.

23 Peak, C.W., Singh, K.A., Adlouni, M.a. et al. (2019). *Adv. Healthc. Mater.* 8: 1801553.

24 Xing, F., Xiang, Z., Rommens, P.M., and Ritz, U. (2020). *Dent. Mater.* 13 (10): 2278.

25 Navarro, J., Calderon, G.A., Miller, J.S., and Fisher, J.P. (2019). Bioinks for three-dimensional printing in regenerative medicine. In: *Principles of Regenerative Medicine* (ed. A. Atala, R. Lanza, A.G. Mikos and R. Nerem), 805. Elsevier.

26 Loozen, L.D., Wegman, F., Öner, F.C. et al. (2013). *J. Mater. Chem. B* 1 (48): 6619.

27 Cui, X., Breitenkamp, K., Finn, M.G. et al. (2012). *Tissue Eng. Part A* 18 (11, 12): 1304.

28 Chang, R., Emami, K., Wu, H., and Sun, W. (2010). *Biofabrication* 2 (4): 045004.

29 Michael, S., Sorg, H., Peck, C.T. et al. (2013). *PLoS One* 8 (3): e57741.

30 Derby, B. (2008). *J. Mater. Chem.* 18 (47): 5717.

31 Snyder, J.E., Hamid, Q., Wang, C. et al. (2011). *Biofabrication* 3 (3): 034112.

32 Rodriguez-Devora, J.I., Zhang, B., Reyna, D. et al. (2012). *Biofabrication* 4 (3): 035001.

33 Zhao, Y., Yao, R., Ouyang, L. et al. (2014). *Biofabrication* 6 (3): 035001.

34 Kolesky, D.B., Truby, R.L., Gladman, A. et al. (2014). *Adv. Mater.* 26 (19): 3124.

35 Bertassoni, L.E., Cecconi, M., Manoharan, V. et al. (2014). *Lab Chip* 14 (13): 2202.

36 Phillippi, J.A., Miller, E., Weiss, L. et al. (2008). *Stem Cells* 26 (1): 127.

37 Wang, Z., Abdulla, R., Parker, B. et al. (2015). *Biofabrication* 7 (4): 045009.

38 Ozbolat, I.T. and Yu, Y. (2013). *IEEE Trans. Biomed. Eng.* 60 (3): 691.

39 Blandino, A., Macias, M., and Cantero, D. (1999). *J. Biosci. Bioeng.* 88 (6): 686.

40 Yang, X., Lu, Z., Wu, H. et al. (2018). *Mater. Sci. Eng., C* 83: 195.

41 Nichol, J.W., Koshy, S.T., Bae, H. et al. (2010). *Biomaterials* 31 (21): 5536.

42 Skardal, A., Mack, D., Kapetanovic, E. et al. (2012). *Stem Cells Transl. Med.* 1 (11): 792.

43 Gao, G., Yonezawa, T., Hubbell, K. et al. (2015). *Biotechnol. J.* 10 (10): 1568.

44 Tan, H., Li, H., Rubin, J.P., and Marra, K.G. (2011). *J. Tissue Eng. Regener. Med.* 5 (10): 790.

45 Mehrban, N., Teoh, G.Z., and Birchall, M.A. (2016). *Int. J. Bioprint.* 2 (1): 6–19.

46 Levato, R., Visser, J., Planell, J.A. et al. (2014). *Biofabrication* 6 (3): 035020.

47 Skardal, A., Devarasetty, M., Kang, H.-W. et al. (2015). *Acta Biomater.* 25: 24.

48 Ahmed, E.M. (2015). *J. Adv. Res.* 6 (2): 105, 105–121.

49 Do, A.V., Khorsand, B., Geary, S.M., and Salem, A.K. (2015). *Adv. Healthc. Mater.* 4 (12): 1742.

50 Melchels, F.P., Domingos, M.A., Klein, T.J. et al. (2012). *Prog. Polym. Sci.* 37 (8): 1079.

51 Cui, H., Nowicki, M., Fisher, J.P., and Zhang, L.G. (2017). *Adv. Healthc. Mater.* 6 (1).

52 Roppolo, I., Chiappone, A., Angelini, A. et al. (2017). *Mater. Horiz.* 4 (3): 396.

53 Campos, D.F.D., Philip, M.A., Gürzing, S. et al. (2019). *3D Print. Addit. Manuf.* 6 (2): 63.

54 Bhattacharyya, A., Janarthanan, G., and Noh, I. (2021). *Addit. Manuf.* 37: 101639.

55 Wang, X., Tolba, E., Schröder, H.C. et al. (2014). *PLOS One* 9 (11): e112497.

56 Dandoy, P., Meunier, C.F., Michiels, C., and Su, B.-L.J.P.O. (2011). *PLOS One* 6 (6): e20983.

57 Place, E.S., Evans, N.D., and Stevens, M.M. (2009). *Nat. Mater.* 8 (6): 457.

58 Jungst, T., Smolan, W., Schacht, K. et al. (2015). *Chem. Rev.* 116 (3): 1496.

59 Schacht, K., Jüngst, T., Schweinlin, M. et al. (2015). *Angew. Chem. Int. Ed.* 54 (9): 2816.

60 Ambesi-Impiombato, F., Parks, L., and Coon, H. (1980). *Proc. Natl. Acad. Sci. U.S.A.* 77 (6): 3455.

61 Leberfinger, A.N., Dinda, S., Wu, Y. et al. (2019). *Acta Biomater.* 95: 32.

62 Zhang, Y.S., Yue, K., Aleman, J. et al. (2017). *Ann. Biomed. Eng.* 45 (1): 148.

63 Miller, J.S. (2014). *PLoS Biol.* 12 (6): e1001882.

64 Hölzl, K., Lin, S., Tytgat, L. et al. (2016). *Biofabrication* 8 (3): 032002.

65 Mauck, R., Wang, C.-B., Oswald, E. et al. (2003). *Osteoarthritis Cartilage* 11 (12): 879.

66 Chang, S.C., Rowley, J.A., Tobias, G. et al. (2001). *J. Biomed. Mater. Res. Part A* 55 (4): 503.

67 Hospodiuk, M., Dey, M., Ayan, B. et al. (2018). *Biofabrication* 10 (3): 035003.

68 Garreta, E., Oria, R., Tarantino, C. et al. (2017). *Mater. Today Bio* 20 (4): 166.

69 Mandrycky, C., Wang, Z., Kim, K., and Kim, D.-H. (2016). *Biotechnol. Adv.* 34 (4): 422.

70 Paxton, N., Smolan, W., Böck, T. et al. (2017). *Biofabrication* 9 (4): 044107.

71 Chimene, D., Lennox, K.K., Kaunas, R.R., and Gaharwar, A.K. (2016). *Ann. Biomed. Eng.* 44 (6): 2090.

72 Bertassoni, L.E., Cardoso, J.C., Manoharan, V. et al. (2014). *Biofabrication* 6 (2): 024105.

73 Ribeiro, A., Blokzijl, M.M., Levato, R. et al. (2017). *Biofabrication* 10 (1): 014102.

74 Wüst, S., Godla, M.E., Müller, R., and Hofmann, S. (2014). *Acta Biomater.* 10 (2): 630.

75 Colosi, C., Shin, S.R., Manoharan, V. et al. (2016). *Adv. Mater.* 28 (4): 677.

76 Skardal, A., Zhang, J., McCoard, L. et al. (2010). *Tissue Eng. Part A* 16 (8): 2675.

77 Panwar, A. and Tan, L.P. (2016). *Molecules* 21 (6): 685.

78 Shao, Y., Jia, H.Y., Cao, T.Y., and Liu, D.S. (2017). *Acc. Chem. Res.* 50 (4): 659.

79 Mahumane, G.D., Kumar, P., du Toit, L.C. et al. (2018). *Biomater. Sci.* 6 (11): 2812.

80 Oliveira, E.P., Malysz-Cymborska, I., Golubczyk, D. et al. (2019). *Acta Biomater.* 95: 60–72.

81 Mahmoud, S., Eldeib, A., and Samy, S. (2015). *Australas. Phys. Eng. Sci. Med.* 38 (2): 223.

82 Lee, J.M. and Yeong, W.Y. (2016). *Adv. Healthc. Mater.* 5 (22): 2856.

83 Hennink, W. and Van Nostrum, C.F. (2012). *Adv. Drug Delivery Rev.* 64: 223.

84 Bae, K.H., Wang, L.-S., and Kurisawa, M. (2013). *J. Mater. Chem. B* 1 (40): 5371.

85 Wang, H. and Heilshorn, S.C. (2015). *Adv. Mater.* 27 (25): 3717.

86 Lanza, R., Langer, R., and Vacanti, J.P. (2011). *Principles of Tissue Engineering*. Academic Press.

87 Blaeser, A., Duarte Campos, D.F., Puster, U. et al. (2016). *Adv. Healthc. Mater.* 5 (3): 326.

88 Kang, H.-W., Lee, S.J., Ko, I.K. et al. (2016). *Nat. Biotechnol.* 34: 312.

89 Pepper, M.E., Cass, C.A., Mattimore, J.P. et al. (2010). *IEEE* 259. https://doi.org/10.1109/IEMBS.2010.5627467.

90 Murphy, W.L., McDevitt, T.C., and Engler, A.J. (2014). *Nat. Mater.* 13 (6): 547.

91 Engler, A.J., Sen, S., Sweeney, H.L., and Discher, D.E. (2006). *Cell* 126 (4): 677.

92 Lee, J., Abdeen, A.A., Zhang, D., and Kilian, K.A. (2013). *Biomaterials* 34 (33): 8140.

93 Dalby, M.J., Gadegaard, N., Tare, R. et al. (2007). *Nat. Mater.* 6 (12): 997.

94 McMurray, R.J., Gadegaard, N., Tsimbouri, P.M. et al. (2011). *Nat. Mater.* 10 (8): 637.

95 Yim, E.K., Darling, E.M., Kulangara, K. et al. (2010). *Biomaterials* 31 (6): 1299.

96 Benoit, D.S., Schwartz, M.P., Durney, A.R., and Anseth, K.S. (2008). *Nat. Mater.* 7 (10): 816.

97 Saha, K., Mei, Y., Reisterer, C.M. et al. (2011). *Proc. Natl. Acad. Sci. U.S.A.* 108 (46): 18714.

98 Barradas, A.M., Fernandes, H.A., Groen, N. et al. (2012). *Biomaterials* 33 (11): 3205.

99 Zhang, W., Wang, G., Liu, Y. et al. (2013). *Biomaterials* 34 (13): 3184.

100 Yang, F., Yang, D., Tu, J. et al. (2011). *Stem Cells* 29 (6): 981.

101 Yang, P.J., Levenston, M.E., and Temenoff, J.S. (2012). *Tissue Eng. Part A* 18 (21, 22): 2365.

102 Khetan, S., Guvendiren, M., Legant, W.R. et al. (2013). *Nat. Mater.* 12 (5): 458.

103 Bratt-Leal, A.M., Carpenedo, R.L., Ungrin, M.D. et al. (2011). *Biomaterials* 32 (1): 48.

104 Hajdu, Z., Mironov, V., Mehesz, A.N. et al. (2010). *J. Tissue Eng. Regener. Med.* 4 (8): 659.

105 Ozbolat, I.T. and Hospodiuk, M. (2016). *Biomaterials* 76: 321.

106 Khojasteh, A., Motamedian, S.R., Rad, M.R. et al. (2015). *World J. Stem Cells* 7 (10): 1215.

107 Ravnic, D.J., Leberfinger, A.N., Koduru, S.V. et al. (2017). *Ann. Surg.* 266 (1): 48.

108 Boratynska, M. and Patrzalek, D. (2016). *Arch. Immunol. Ther. Exp.* 64 (Suppl 1): 37.

109 Sridharan, R., Cavanagh, B., Cameron, A.R. et al. (2019). *Acta Biomater.* 89: 47.

110 Abebayehu, D., Spence, A.J., McClure, M.J. et al. (2019). *J. Biomed. Mater. Res. Part A* 107 (4): 884.

111 Duan, B., Hockaday, L.A., Kang, K.H., and Butcher, J.T. (2013). *J. Biomed. Mater. Res. Part A* 101 (5): 1255.

112 Hockaday, L., Kang, K., Colangelo, N. et al. (2012). *Biofabrication* 4 (3): 035005.

113 Dew, L., English, W.R., Ortega, I. et al. (2016). *Cells Tissues Organs* 202 (5, 6): 319.

114 Tijore, A., Behr, J.M., Irvine, S.A. et al. (2018). *Biomed. Microdevices* 20 (2): 32.

115 Gao, Q., Liu, Z., Lin, Z. et al. (2017). *ACS Biomater. Sci. Eng.* 3 (3): 399.

116 Athirasala, A., Lins, F., Tahayeri, A. et al. (2017). *Sci. Rep.* 7 (1): 3323.

117 Dolati, F., Yu, Y., Zhang, Y. et al. (2014). *Nanotechnology* 25 (14): 145101.

118 Hribar, K.C., Meggs, K., Liu, J. et al. (2015). *Sci. Rep.* 5: 17203.

119 Xiong, R., Zhang, Z., Chai, W. et al. (2015). *Biofabrication* 7 (4): 045011.

120 Ozbolat, I.T. (2016). *3D Bioprinting: Fundamentals, Principles and Applications.* Academic Press.

121 Bishop, E.S., Mostafa, S., Pakvasa, M. et al. (2017). *Genes Dis.* 4 (4): 185.

122 Keast, J.R. (2013). *Organogenesis* 9 (3): 168.

123 Lee, W., Debasitis, J.C., Lee, V.K. et al. (2009). *Biomaterials* 30 (8): 1587.

124 Lee, Y.-B., Polio, S., Lee, W. et al. (2010). *Exp. Neurol.* 223 (2): 645.

125 Ning, L., Zhu, N., Mohabatpour, F. et al. (2019). *J. Mater. Chem. B* 7 (29): 4538.

126 Lozano, R., Stevens, L., Thompson, B.C. et al. (2015). *Biomaterials* 67: 264.

127 Hu, Y., Wu, Y., Gou, Z. et al. (2016). *Sci. Rep.* 6: 32184.

128 Hu, Y., Wu, Y., Gou, Z. et al. (2016). *Sci. Rep.* 6: 32184.

129 Su, D., Zhou, J., Ahmed, K.S. et al. (2019). *Int. J. Biol. Macromol.* 129: 895.

130 Dai, W., Shi, J., Liu, S. et al. (2019). *Chin. J. Reparative Reconstr. Surg.* 33 (3): 356.

131 Xu, Y., Huang, Z., Pu, X. et al. (2019). *Cell Prolif.* 52: 3.

132 Filardo, G., Petretta, M., Cavallo, C. et al. (2019). *Bone Joint Res.* 8 (2): 101.

133 Faramarzi, N., Yazdi, I.K., Nabavinia, M. et al. (2018). *Adv. Healthc. Mater.* 7 (11): e1701347.

134 Ratheesh, G., Vaquette, C., and Xiao, Y. (2020). *Adv. Healthc. Mater.* 9: 23.

135 Duarte Campos, D.F., Blaeser, A., Buellesbach, K. et al. (2016). *Adv. Healthc. Mater.* 5 (11): 1336.

136 Datta, P., Ayan, B., and Ozbolat, I.T. (2017). *Acta Biomater.* 51: 1.

137 Li, Y.-C., Zhang, Y.S., Akpek, A. et al. (2016). *Biofabrication* 9 (1): 012001.

138 Raphael, B., Khalil, T., Workman, V.L. et al. (2017). *Mater. Lett.* 190: 103.

139 Huang, C.-T., Kumar Shrestha, L., Ariga, K., and Hsu, S.-h. (2017). *J. Mater. Chem. B* 5 (44): 8854.

140 Pourchet, L.J., Thepot, A., Albouy, M. et al. (2017). *Adv. Healthc. Mater.* 6 (4): 1601101.

2

Bioceramics for Promoting Bone Regeneration

Meng Zhang, Hongshi Ma, and Chengtie Wu

Shanghai Institute of Ceramics, Chinese Academy of Sciences, State Key Laboratory of High Performance Ceramics and Superfine Microstructure, Shanghai 200050, P.R. China. University of Chinese Academy of Sciences, No. 19(A) Yuquan Road, Shijingshan District, Beijing, P.R. China

2.1 Introduction

Bioceramics represent a class of inorganic, non-metallic materials applied for the repair and regeneration of the damaged tissue, especially for hard tissue repair. After decades of development in clinical application, bioceramics have been classified as three basic types: bioinert ceramics, bioactive ceramics, and bioresorbable ceramics, among which the chemical composition, mechanical property, and biological behaviors changed a lot [1, 2]. Bioinert ceramics, such as alumina and zirconia, have been extensively applied in artificial joints and dental implants because of their low wear, high strength, corrosion resistance, and no reaction with living tissues. Bioactive ceramics, such as hydroxyapatite (HA), serve as bone fillers, bone scaffolds, and coatings on bone implants [3, 4]. These bioactive ceramics react with the surrounding environment, thus bonding directly with bone. Bioresorbable ceramics, including bioactive glass (BG), tricalcium phosphate (TCP), calcium phosphate cements (CPCs), and some silicates, were good candidates for bone tissue engineering scaffolds or bone cements as they could be absorbed after degradation and gradually substituted by the new bone [5–7].

As the demand for bioceramics has changed from mere mechanical support to multifunctions, many efforts have been made to introduce essential trace elements or combine bioceramics with other ceramics, polymers, and metals for better osteogenesis-related functions [8, 9]. In addition to compositional variation, structural orchestration of bioceramics, such as nanobioceramics and porous bioceramic scaffolds, has attracted full attention in bone tissue engineering [10, 11]. It is of great significance to tailor bioceramics via nanotechnology as the large specific surface area promoted the ionic release, biomineralization, protein adsorption, and further cellular processes such as adhesion, proliferation, and differentiation of mesenchymal stem cells. In terms of further applications in bone regeneration, bioceramic scaffolds play a vital role in meeting the comprehensive requirements of structural and mechanical properties, as well as biological properties [12]. Bioceramic

Biofabrication for Orthopedics: Methods, Techniques, and Applications, First Edition.
Edited by Wenguo Cui, Xin Zhao, and Shen Liu.

scaffolds with tailored macrostructure provide a desirable three-dimensional environment for nutrient transportation, cell penetration, and distribution, while specific micro/nanostructures regulate the cell fate via mechanical stimulation or cell surface protein targeting.

This chapter intends to review on the latest research of bioceramics for promoting bone regeneration from the perspective of chemical composition, mechanical properties, and biological properties. Emphasis was put on the compositional and structural influences on biological functions, especially protein adsorption, immunomodulation, vascularization, and osteogenesis.

2.2 Types of Bioceramics

2.2.1 Bioactive Ceramics

Similar to the main inorganic component of natural bone, HA is a biocompatible and bioactive ceramic extensively used as the bone graft, the coating on implants, and the drug delivery vehicle. As an implantable material, HA directly bonds to the bone tissue, with carbonated apatite generated on the interface, which ensures the firm fixation. Thanks to the osteoconductivity, HA supports tissue ingrowth without any inflammatory responses happening. For example, Yang et al. reported that micro-patterned HA nanoparticles were prepared as periosteum-mimic membranes to induce vascularized bone regeneration [13]. The mineralized HA micropatterns promoted the formation of blood vessels and extracellular matrix via coprecipitating the minerals and proteins. Micro-patterned HA participated in cell recruitment, regulation of cell differentiation, and vascularized bone regeneration through mimicking the ossification process. Although HA presents desirable bioactivity, brittleness and low fracture toughness restrain the application of HA in load-bearing implants. Feng et al. synthesized a HA-based multilayered material combined with graphene oxide and chitosan to optimize the mechanical properties (Figure 2.1) [14]. The concentric circular structure and the inorganic–organic compound endowed the HA-based scaffolds with similar mechanical strength with cortical bone, with 95 MPa of compressive strength, 161 MPa of flexural strength, and 1.1 MJ/m^3 of toughness. In addition, the hierarchical structure of HA-based scaffolds induced blood vessels ingrowth to promote bone regeneration. As a coating material, HA possesses considerable bonding strength to the implant surface, remarkable thermodynamic stability in the body fluid, and excellent bone bonding ability during the long-term service, which enhanced the bone integration and corrosion resistance of metallic implants.

2.2.2 Biodegradable Ceramics

2.2.2.1 Bioactive Glass

Unlike other bioceramics, bioactive glass is, to a large extent, variable in composition because of the amorphous phase. In general, silicate glasses, borate glasses, and

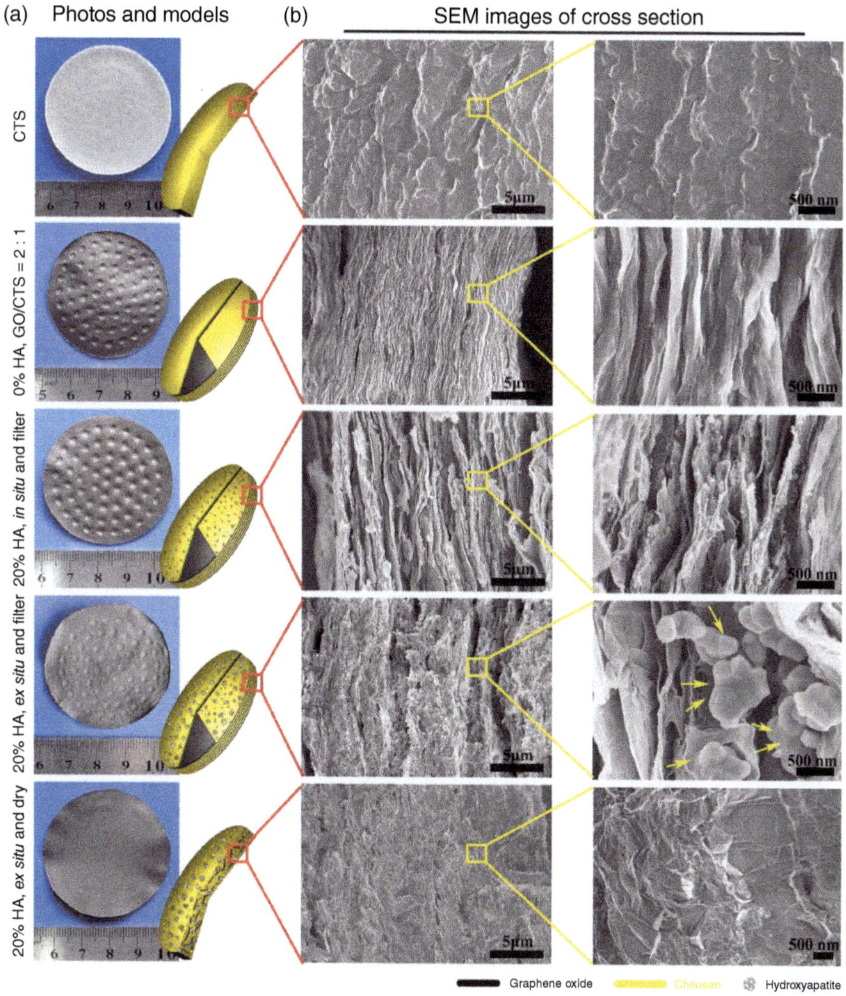

Figure 2.1 Optical photographs (a) and scanning electron microscope (SEM) images (b) of the HA-based multilayered films. GO, Graphene oxide. Source: Feng et al. [14] / with permission of Elsevier.

phosphate glasses constitute the main types of bioactive glasses, whereas coatings, mesoporous bioactive glasses (MBGs), fibers, and scaffolds are the predominant states of applications. Typically, 45S5 bioactive glass, as a silicate bioactive glass, has attracted much attention in tissue engineering since firstly discovered by Prof. Larry Hench in 1969. It can not only bond to bone tissue via HA formation but also release Ca, Si, and P ions in the body fluids. Once BGs contact with the body fluids, the bilayer of hydrated silica and hydroxy carbonate apatite forms on the surface, followed by the adsorption of proteins and growth factors. However, the degradation of BG causes instability of implants in the clinical application [15]. The design of ion doping [16] and composites [17] has been brought into consideration to optimize the degradation rate and osteointegration performance.

BGs can be shaped into various sizes for different purposes. MBGs have been developed as a porous nanostructured material on the basis of their pore size (2–50 nm). Because of the interconnected mesoporous channel, MBGs show a larger specific surface area and a higher degradation rate compared with BGs. Moreover, MBGs show higher apatite formation ability than BGs. For example, $CaO–P_2O_5–SiO_2$ MBGs showed improved bioactivity, degradation rate, and drug delivery ability both *in vitro* and *in vivo* compared with BGs [18]. Other strategies such as ion doping are applied for improving the osteogenic ability and multifunctionality of MBGs (Table 2.1). Ion-doped MBG, such as

Table 2.1 Ionic substitution of bioceramics.

Bioceramics	Doped elements	Doping content (mol%)	Doping effects	References
MBG	Cu	5	Stimulated the expression of HIF-1α and VEGF in BMSCs	[19]
	Sr	2.5, 5, and 10	Enhanced the expression of osteogenesis-related genes and ALP activity of periodontal ligament cells	[20]
	Co	2	Secreted more VEGF protein and induced hypoxic cascade via stabilizing HIF-1α expression	[21]
	Fe	5	Promoted the OCN and ALP expressions of BMSCs as well as the mitochondrial activity	[22]
β-TCP	Zn	1.4	Exhibited a decreased dissolution rate and more stable cell–biomaterial interface	[23]
	Cu	Up to 15	Reduced the unit cell volume, improved the antimicrobial activity, and induced the expression of VEGF via activating HIF-1 to promote the angiogenesis	[24–26]
	Mg	1 and 6	Resisted the acid-etching process and the osteoclastic resorption	[27]
	Li	5	Enhanced the osteogenic differentiation of HPDLCs and hBMSCs via Wnt signaling pathways	[28]
	Sr	Up to 80	Linearly enlarged the lattice parameters, facilitated the tube formation of endothelial cells, stimulates the protein secretion of VEGF and bFGF, and downregulates the expression of TNF-α and IL-8	[12, 29, 30]
	Mn	0.02, 0.76, and 2	Inhibited the bacterial growth, accelerated the bone mineralization, and stimulated the osteogenic differentiation adipogenic differentiation, and chondrogenic differentiation	[31, 32]

Source: Chengtie Wu et al.

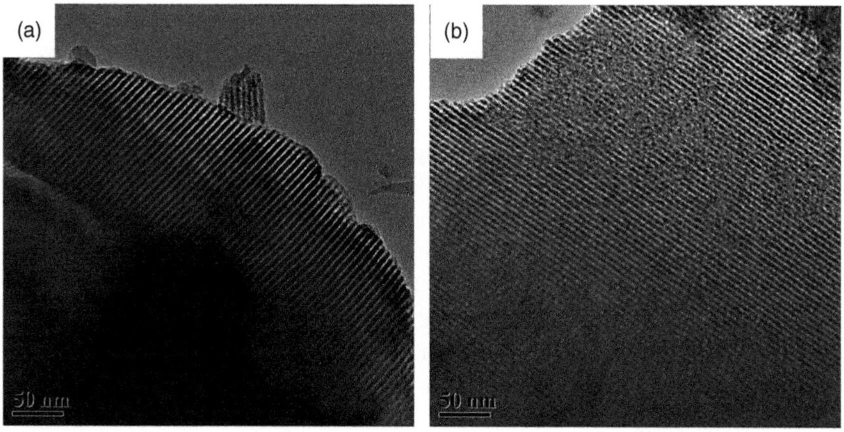

Figure 2.2 Transmission electron microscope (TEM) analysis for 1Cu-MBG (a) and 5Cu-MBG (b) scaffolds. Source: Wu et al. [19] / with permission of Elsevier.

copper-containing mesoporous bioactive glass (Cu-MBG), could stimulate the expression of hypoxia-inducible factor (HIF-1α), as well as the expression of vascular endothelial growth factor (VEGF) in bone marrow stromal cells (BMSCs), promoting hypoxia-like response, which resulted in enhanced angiogenesis and osteogenesis (Figure 2.2) [19]. It has been demonstrated that Cu is a cofactor of lysyl oxidase, which is related to the regulation of collaged crosslinking. Cu deficiency can promote osteoporosis via reduction of collagen crosslinking, and Cu overload can inhibit collagen synthesis [33]. In addition, strontium-containing mesoporous bioactive glass (Sr-MBG) showed the decreased order of mesopores and controlled release of Sr^{2+} ions. The incorporation of Sr enhanced the expression of osteogenesis-related genes and alkaline phosphatase (ALP) activity of periodontal ligament cells [20]. Besides, low-dose cobalt (Co), less than 5%, incorporated into MBGs showed higher secretion of VEGF protein, as well as osteogenesis-related genes with no significant cytotoxicity in BMSCs. The Co^{2+}-doped MBG was also able to induce hypoxic cascade via stabilizing HIF-1α expression [21]. It was probably because that Co^{2+} as a hypoxia-mimicking agent can cause the differentiation of acute promyelocytic leukemia cells and activate HIF-1α, which stimulates the transcription of VEGF to increase oxygen availability [34]. To obtain multifunctional MBG, iron-containing mesoporous bioactive glass (Fe-MBG) was developed to achieve magnetic property for hyperthermic application in repairing bone tumor-induced large bone defects [22]. The incorporation of Fe promoted the osteogenic gene expressions of BMSCs, as well as the mitochondrial activity. Compared with pure MBG, the incorporation of Fe in Fe-MBG leads to the pH value closer to 7.4, which might be beneficial to the mitochondrial activity and gene expression. These studies suggested that the biological function of ion-doped bioactive glass was mainly depended on the concentration and variety of elements. Relatively low concentration of elements showed better performance on the stimulation of bone regeneration while high concentration of metal ions could lead to acute pH rise and subsequent cell apoptosis. It is of great importance to design the ionic composition and control the ionic release in service to enhance the osteogenic and angiogenic ability of bioactive glass.

2.2.2.2 β-Tricalcium Phosphate

TCP, a main component of calcium phosphate along with HA, has been widely applied in bone cements and bone engineering scaffolds. TCP is divided into α-phase (α-TCP) and β-phase (β-TCP), and the latter attracts more attention in bone regeneration because of the better degradability. As the Ca/P ratio of β-TCP and HA is 1.5 and 1.67, respectively, β-TCP exhibits a higher resorption rate and ion-releasing rate. This effect improves the aggregation of calcium and phosphate ions on site and promotes the mineralization on the surface of β-TCP. Calcium ions participate in bone matrix formation by means of calcification and regulate the life cycle of osteoblasts through PI3K/protein kinase B (Akt) pathways, as well as the bone resorption function of osteoclasts. Phosphate ions promote the osteoblast differentiation through insulin-like growth factors-1 (IGF-1) and ERK1/2 (extracellular signal-related kinases) pathways and suppress the osteoclast differentiation via regulating the receptor activator of nuclear factor-κB (RANK)-ligand/osteoprotegerin (OPG) ratio.

Ionic substitution of β-TCP has been implemented to modify the sintering ability, mechanical property [9], resorption property, and biological properties such as osteogenesis and angiogenesis. Zinc (Zn) substitution for Ca in β-TCP results in the unit cell diminution with the doping concentration no more than 20 mol%, which stabilize the structure of β-TCP [35]. Zn-doped β-TCP exhibits a decreased dissolution rate and a more stable cell–biomaterial interface as compared to β-TCP, which provides the substrates for cell adhesion and stimulates osteoblast activity [23]. Incorporation of Cu under the amount of 15 mol% into β-TCP could be achieved, which causes the reduction of unit cell volume [24]. Cu substitution helps for improving the antimicrobial activity and angiogenic activity [25]. The released Cu^{2+} ions induce the expression of VEGF via activating HIF-1 to promote the angiogenesis [26]. Magnesium (Mg) substitution helps to increase 260 °C of the transition temperature from β-TCP to α-TCP, which contributes to the densification of sintered materials [36]. The doped Mg (1 and 6 mol%) into the lattice improves the stabilization of β-TCP, resisting the acid-etching process and the osteoclastic resorption (Figure 2.3) [27]. Lithium (Li) substitution of β-TCP leads to the liner reduction of *a*-axis parameter and slight reduction of *c*-axis parameter with the doping amount no more than 9 mol% [37]. Li contributes to downregulate the fracture risks via improving the bone mineral density and bone quality [28]. 5 mol% of Li substitution of β-TCP enhances the osteogenic differentiation of human periodontal ligament cells (HPDLCs) and human bone marrow stromal cells (hBMSCs) via Wnt signaling pathways compared with pure β-TCP [38]. Strontium substitution shows a great doping amount at up to 80 mol% content in β-TCP, whereas Mg and Zn can replace Ca at no more than 20 mol% content. High doping amount of Sr linearly enlarges the lattice parameters, resulting in unremarkable unit cell reorganization [35]. Sr consists in the high bone metabolic turnover area and acts on targets similar to that of Ca through the activation of calcium sensing

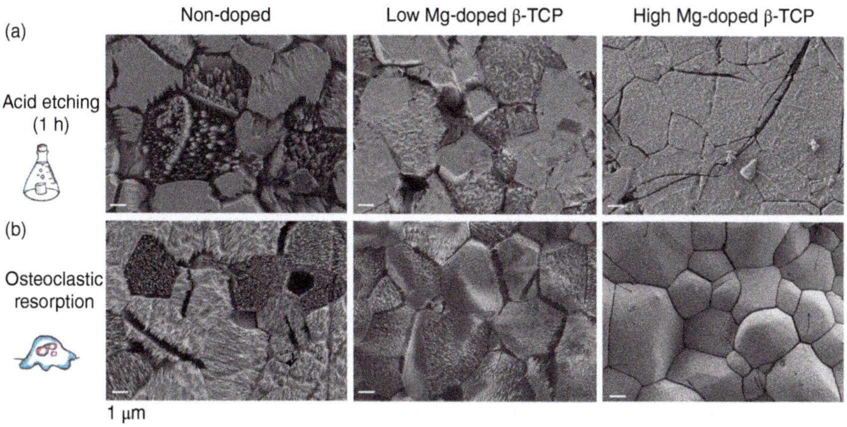

Figure 2.3 SEM images of the non-doped and Mg-doped β-TCP surfaces: (a) after 1 hour immersion in the acidic solution and (b) after 24 hours of contact with osteoclast cells. Source: Gallo et al. [27] / with permission of Elsevier.

receptor (CaSR) for further regulation of bone metabolism [39]. The side effect of the lowered Ca ion concentrations in the apatite precipitation process can be offset by Sr ions because 0.4–0.9 mM of Sr ions are able to facilitate the differentiation of BMSCs [40]. Incorporation of Sr in calcium phosphate also makes contribution to stimulate osteogenic differentiation, regulate the immune response, and promote angiogenesis at the early stage [41]. Sr-doped calcium polyphosphate (SCPP) facilitates the tube formation of endothelial cells and stimulates the protein secretion of VEGF and basic fibroblast grow factor (bFGF) for cell infiltration and angiogenesis [29]. Moreover, incorporation of Sr downregulates the expression of inflammatory cytokines (tumor necrosis factor [TNF-α]) and chemotactic (interleukin [IL-8]) to suppress the inflammatory cell recruitment [30]. Manganese (Mn)-doped β-TCP was investigated to deliver Mn ions for improving biological performance [42]. It was reported that up to 10 mol% of Mn content in β-TCP caused the reduction of unit cell parameters, as well as the morphological transformation [31]. As the doping concentration of Mn increased, the lattice parameter of a-axis and c-axis declined. A small amount (0.76 mol%) of incorporation led to the acceleration of bone mineralization. In addition, rather low level of doping concentration seems to perform better in biological properties. In another research, Mn-doped β-TCPs, $Mn_xCa_{3-x}(PO_4)_2$ ($x = 0, 0.001, 0.01$, and 0.1), were investigated [32]. β-TCP doped with the highest amount of Mn^{2+} showed significant inhibition of bacterial growth, whereas β-TCP incorporated with the lowest concentration of Mn^{2+} had the best performance on stimulating the osteogenic differentiation, adipogenic differentiation, and chondrogenic differentiation of adipose-derived mesenchymal stem cells (ADMSCs).

β-TCP can also be combined with HA to form biphasic calcium phosphate (BCP), which is a homogeneous mixture utilizing their respective characters. Optimization of degradation rate of BCP could be achieved through the manipulation of HA/TCP ratio. It was reported that BCP containing 30% HA and 70% TCP showed the best performance in bone morphogenic protein 2 (BMP-2) expression and *in vivo* osteoinduction among different compositions [43]. In another report, BCP containing 60% HA and 40% TCP showed long-term stability and exhibited steady and homogeneous bone growth via retarding the resorption process [44]. However, there is wide agreement that BCP shows higher bone regeneration rate and quality than HA and TCP in spite of the HA/TCP ratio.

2.2.2.3 Calcium Phosphate Cements

CPCs have been widely applied in orthopedics since 1980s. CPC is generally divided into apatite (HA and calcium-deficient hydroxyapatite, CDHA) CPC and brush (dicalcium phosphate dehydrate, DCPD) CPC, both of which are generated from mixing CPC powders and aqueous solution. The dissolution of calcium orthophosphates and precipitation of calcium phosphate crystals (HA, CDHA, or DCPD) result in the final set of CPC. The indurated CPCs need to initially perform connection and stabilization at the implant site, and the manipulation of CPCs should be for clinical convenience.

2.2.2.4 Silicate Bioceramics

Silicate bioceramics have attracted great attention in orthopedic applications because of their remarkable osteogenic and angiogenic abilities. Silicate bioceramics are generally composed of calcium silicates, calcium magnesium silicates, calcium zinc silicates, and calcium zirconium silicates. Calcium silicate, the most commonly used silicate bioceramic, is divided by monocalcium silicate ($CaSiO_3$), dicalcium silicate (Ca_2SiO_4), and tricalcium silicate (Ca_3SiO_5).

Monocalcium silicate has two phases, pseudowollastonite (α-$CaSiO_3$) and wollastonite (β-$CaSiO_3$). Both phases are bioactive and bioresorbable, whereas β-$CaSiO_3$ shows higher solubility and bioactivity than α-$CaSiO_3$ [45]. β-$CaSiO_3$ releases Ca and silicate ions in biological solution, influencing the cell metabolism, proliferation, and differentiation. The investigation of toxicity of β-$CaSiO_3$ extract showed that there were no significant cytotoxic effects on osteoblastic cells [46]. In addition, cell metabolism was increased for cells cultured in the extracts with the extraction time of 168 hours. The effect of ionic products of β-$CaSiO_3$ on cell proliferation was in a dose-dependent manner. A high concentration of silicate ions (above 120 ppm) released from β-$CaSiO_3$ showed initial cytotoxicity on human mesenchymal stem cells (hMSCs), while low concentration of silicate ions (about 80 ppm) were proved to promote the cell proliferation [47]. In a study of β-$CaSiO_3$ substrates, osteogenic differentiation of hMSCs was observed [48]. The soluble factors from the substrates played a more important role than surface toughness in the osteogenic differentiation process. Individual soluble factors showed a positive effect on the cell proliferation and differentiation in an optimized concentration range while multiple soluble factors showed complex, synergetic, or antagonistic effect on cellular behaviors. Apart from osteogenic ability, $CaSiO_3$ possesses angiogenic ability because of the release of silicate ions. It was reported that $CaSiO_3$ extracts with the Si ion concentration of 0.7–1.8 μg/ml stimulated the proliferation and proangiogenic factors expression of human umbilical vein endothelial cells (HUVECs), as well as nitric oxide (NO) secretion [49]. $CaSiO_3$ extracts stimulated the overexpression of VEGF and bFGF for initiating the proliferation and migration of HUVECs, leading to an enhanced angiogenic effect compared with the supplement of single recombinant angiogenic factor. *In vivo* studies on $CaSiO_3$ have been conducted. In an experiment of implantation of α-$CaSiO_3$ cylinders in rat tibias, the implants directly contacted with the new bone, which was entirely mineralized [50]. During the implantation of 12 weeks, the interface between the implants and bone tissue kept biologically and chemically active. Additionally, the osteogenic capacity of $CaSiO_3$ is more remarkable than that of β-TCP both *in vitro* and *in vivo*. It was reported that α-$CaSiO_3$ and β-$CaSiO_3$ performed better in attachment and proliferation of osteoblastic-like cells compared with β-TCP [51]. The ALP activity of cells incubated on $CaSiO_3$ enhanced as well because of the release of Si and Ca ions. In a calvarial defect model, porous β-$CaSiO_3$ ceramics degraded faster than β-TCP after 16 weeks of implantation [52]. More new bone formation was observed in porous β-$CaSiO_3$ ceramics with an apatite layer formed on the interface between materials and bone tissue compared with β-TCP.

Dicalcium silicate and tricalcium silicate are generally used for Portland cement and coatings as they hydrate and solidify in aqueous solutions. The hydration rate of Ca_2SiO_4 is much slower than that of Ca_3SiO_5. β-Ca_2SiO_4 cements, of which the liquid-to-powder (L/P) ratio was 1.0–1.2 ml/g, had an initial setting time of 60–100 minutes and a final setting time of 220–280 minutes when the nozzle diameter was 2.0 mm [53]. By contrast, Ca_3SiO_5 cements with an L/P ratio of 1.0–1.2 ml/g had an initial setting time of 150–225 minutes and a final setting time of 240–390 minutes with a nozzle diameter of 2.0 mm [54]. Both Ca_2SiO_4 and Ca_3SiO_5 cements proved to be bioactive and biodegradable, with ionic products stimulating the proliferation of fibroblast. In addition, the maximal heat liberation rate of Ca_3SiO_5 cements is 6.3 mW/g, whereas that of CPC is 8.4 mW/g, suggesting that Ca_3SiO_5 cements are able to reduce the risks of thermal necrosis and local inflammation. In terms of the application as a coating material, Ca_2SiO_4 exhibits remarkable bioactivity because of the activated Si 3-ring sites with negative charges when incubated in simulated body fluid (SBF) solution [55]. Moreover, Ca_2SiO_4 coatings release Ca and Si ions to promote the proliferation and differentiation of osteoblast-like cells via regulating the ratio between OPG and receptor activator of nuclear factor-κB ligand (RANKL).

Calcium magnesium silicates, including diopside ($CaMgSi_2O_6$), monticellite ($CaMgSiO_4$), akermanite ($Ca_2MgSi_2O_7$), merwinite ($Ca_3MgSi_2O_8$), and bredigite ($Ca_7MgSi_4O_{16}$), are potential candidates as bone substitute materials. As the energy of Mg—O bond is higher than that of Ca—O bond, calcium magnesium silicates possess enhanced crystal structure stability, mechanical strength, and biodegradability compared with calcium silicates [56]. The bending strength is improved from merwinite, akermanite to monticellite with the increase of Mg content [57]. The degradation rate of calcium magnesium silicates is regulated by the content of Mg. It was reported that the degradation rate and the apatite formation capacity decreased from diopside, akermanite to bredigite [58]. Different from Mg ions, Si ions accounted for the stimulation of osteoblast proliferation. The extracts from diopside, akermanite, and bredigite significantly stimulate the proliferation, NO synthesis, and angiogenic gene expression of human aortic endothelial cells (HAECs) compared with β-TCP, suggesting that Si ions play an important role in inducing angiogenesis [59].

Diopside, as a typical calcium magnesium silicate, possesses excellent mechanical properties, with a bending strength of 300 MPa and a fracture toughness of 3.5 MPa $m^{1/2}$ [60]. The compressive strength of diopside scaffolds, prepared via polymer sponge template method, is 0.63–1.36 MPa, with the porosities of 75–80% [61]. The mechanical strength of diopside scaffolds is higher than that of bioactive glass and $CaSiO_3$ scaffolds comparable to that of human sponge bone. High Mg content of diopside promotes the apatite formation capacity and dissolution rate while low Si content reduces the osteogenic and angiogenic potential of diopside [59]. *In vivo* bone and periodontal regeneration ability of diopside was verified. Diopside scaffolds promoted the expression of type I collagen (Col I), runt-related transcription factor 2 (Runx 2), transforming growth factor beta 1 (TGF β1), and bone morphogenic protein 2 (BMP-2) of periodontal ligament cells (PDLCs) compared with β-TCP scaffolds [62]. The *in vivo* study showed that more new bone and cementum formation was observed in diopside scaffolds than that of β-TCP scaffolds. To improve the mechanical and biological properties of diopside, various methods such as materials compositing, ionic substitution, and structuralization are implemented. The design of composite materials is applied for the mechanical reinforcement of diopside. It was reported that 2 wt% of multiwalled carbon nanotubes (MWCNTs) could enhance the compressive strength and fracture toughness of diopside scaffolds by 106% and 21%, respectively [63]. The MWCNT bridging and crack deflection contributed to the energy absorption, which improved the strength and toughness. Ionic substitution is an effective method to improve the biological properties. The incorporation of Sr promoted the apatite formation capacity of diopside because of the reduction in the release of Mg^{2+} [64]. Fluorine (F) substitution showed a similar enhancement on the apatite-forming ability with Sr substitution [65]. Porous diopside microspheres were investigated for drug delivery. The drug-loading/releasing ability of dexamethasone (DEX) from diopside microspheres could be regulated by controlling the porosity and size as well as the surface modification with poly(lactide-*co*-glycolide) [66].

Monticellite has a comparable bending strength and enhanced fracture toughness compared with HA. The bending strength and fracture toughness of monticellite are 159.7 MPa and 1.63 MPa $m^{1/2}$, respectively [67]. The Young modulus of monticellite is 51 GPa, which is slightly higher than Young's modulus of cortical bone (7–30 GPa). In addition, the coefficient of thermal expansion of monticellite is 10.76×10^6 °C, which is close to that of Ti-6Al-4V, about 10.03×10^6 °C, indicating that monticellite is a good candidate for the bioactive coating material. Monticellite enhances the proliferation and ALP activity of osteoblast-like G292 cells compared with HA, attributing to the existence of Mg and Si ions [68]. The increase of Mg^{2+} concentration enhances the pH value of natural medium, killing bacteria via destroying the cytoplasmic membrane and denaturing proteins, which is accountable for the antibacterial activity of monticellite [69].

Akermanite has been widely investigated as an attractive candidate for bone tissue engineering. Akermanite shows excellent apatite formation ability after soaking in SBF for 10 days [70]. After soaking in SBF for 20 days, bone-like apatite with a thickness of 110 μm was formed on wollastonite, while the thickness of the apatite formed on akermanite reaches 130 μm [71]. The mechanism of apatite formation on akermanite is similar to that of wollastonite, involving the dissolution of Ca and P ions, formation of hydrated silica, phosphate nucleation, formation of amorphous calcium, and the final formation of HA. In addition, Akermanite shows superiority over β-TCP both *in vitro* and *in vivo*. The extracts of akermanite enhanced the ALP activity and the expression of osteopontin (OPN), osteocalcin (OCN), and bone sialoprotein (BSP) of hBMSCs compared with the extracts of β-TCP [72]. Similar results were also obtained in the study of akermanite and β-TCP disks, whereas akermanite disks could promote osteoblastic differentiation of hBMSCs even without osteogenic reagents [73]. The adhesion, proliferation, and osteogenic differentiation of human adipose-derived stem cells (hASCs) were also enhanced on akermanite disks compared with β-TCP disks [74]. As the effect of akermanite extracts on cell behaviors is in a dose-dependent manner, different concentrations of the original extract were investigated. It was reported that 1/4 extract of akermanite with 2.36 mM of Ca ions, 11.1 mM of Mg ions, and 1.03 mM of Si ions was the optimized concentration to promote the osteogenic differentiation of hASCs via the ERK pathway [75]. Moreover, porous akermanite implants show higher bone volume (BV)/total volume (TV) value than that of β-TCP in rabbit femoral defects, with more newly formed bone penetrated into the akermanite implants, indicating that akermanite has the advantage over β-TCP in terms of bone regeneration. The subchronic systemic toxicity of akermanite has been investigated *in vivo*. The extracts of akermanite were injected into Wistar rats to simulate the ionic products of akermanite implants [76]. No significant difference was observed in the body weight, cytological and biochemical examination of blood, and organ-to-body weight ratios. Additionally, the extracts of akermanite did not induce the pathological changes to heart, liver, and kidneys, indicating that the ionic products from akermanite did not result in subchronic systemic toxicity.

Merwinite possesses considerable mechanical strength and apatite formation ability. The bending strength and fracture toughness of merwinite is 151 MPa and 1.72 MPa $m^{1/2}$, respectively [77]. Young's modulus of merwinite is 31 GPa, which is close to that of the cortical bone. The coefficient of thermal expansion (CTE) of merwinite is 9.87×10^6 °C, which is comparable to that of Ti-6Al-4V alloy. After soaking in SBF for seven days, HA was formed on the surface of merwinite, indicating the remarkable bioactivity of merwinite [78]. Moreover, the ionic products of merwinite significantly stimulated the proliferation of osteoblasts. The study of merwinite granules repairing the rat femoral defects indicated that merwinite was more effective in promoting new bone formation compared with HA [79]. The soluble factors of merwinite had positive chemotactic actions on osteogenic stromal cells, leading to the activation of osteogenic cells and the formation of osteoblasts.

Bredigite has advantages on the mechanical strength and degradation rate over bioactive glass. The bending strength and fracture toughness of bredigite are 156 MPa and 1.57 MPa $m^{1/2}$, respectively [80]. The Young modulus of bredigite is 43 GPa, which is between monticellite and merwinite. After soaking in SBF for 10 days, a layer of nanocrystalline HA was formed on the surface of bredigite. Reducing the particle size is an effective method to enhance the apatite formation ability. Bredigite with a particle size of 38–48 nm induced bone-like apatite formation after soaking for three days. In addition, bredigite nanopowder showed higher dissolution rate than microsize powder, indicating the higher surface activity of nanopowder. The ionic products from bredigite enhanced the proliferation and cementogenic differentiation of PDLCs, involving the mineralization module formation and the expression of ALP, OPN, OCN, BSP, cementum attachment protein (CAP), and cementum protein 1 (CEMP1), via the activation of Wnt/β-catenin signaling pathway [81]. Bredigite extracts promoted the expression of pluripotency-related genes and the multilineage differentiation of human dental pulp cells (hDPCs) in comparison to β-TCP extracts, leading to the stimulation on hDPCs pluripotency comparable to bFGF [82].

Silicates play important roles in mineralization, osteogenesis, and angiogenesis, whose mechanisms have not been fully understood. The effect of Si on the apatite formation promotes the integration of silicates and bone tissue, which is of great significance in bone tissue engineering. The high mechanical strength and desirable osteoinductivity make silicates as promising bone substitutes for large bone defects.

2.3 Mechanical Properties

The mechanical properties of bioceramics, such as compressive strength, bending strength, fracture toughness, and fatigue resistance, are of great significance in the load-bearing applications. The compressive strength and bending strength of bioceramics are quite satisfactory in comparison with that of cortical bone and cancellous bone, while the facture toughness and fatigue resistance of bioceramics are not competitive because of the inherent brittleness. Various approaches, such as adding sintering additives, compounding with other biocompatible materials, and structural design, have been explored to modulate the mechanical properties of bioceramics.

Adding sintering additives has been generally applied to reduce the sintering temperature and promote the densification. Sintering additives need to have low melting points to lower the sintering temperature and possess considerable biocompatibility. For example, B_2O_3 is proved to be effective in the control of densification of bioceramics via the formation of solid solution and effective in the promotion of wound healing. 1 wt% of B_2O_3-incorporated β-$CaSiO_3$ bioceramics sintered at 1000 °C enhanced the flexural strength from 30.2 to 168.4 MPa. The fracture toughness was enhanced from 0.5 to 2.1 MPa $m^{1/2}$, accompanied with the formation of the liquid phase and a dense microstructure [83]. In another research, B_2O_3-doped BG promoted the mechanical properties of bioactive glass ceramics (BGCs). The compressive and flexural strength of 10 mol% B_2O_3-doped BG-reinforced 45S5 BGC sintered at 850 °C showed four times than those of 45S5 BGC [84]. 5 wt% of B_2O_3/SiO_2 (with a mass ratio of 92.5 : 7.5)-incorporated β-$CaSiO_3$ bioceramics were sintered at 1000 °C and showed a flexural strength of 228.14 MPa, which was 3.4 times higher than that of pure β-$CaSiO_3$ and met the strength requirement of cortical bone (135–193 MPa), indicating the great potential in the repair of load-bearing bone [85]. ZnO is another additive for bioceramics with desirable antibacterial activity and considerable osteogenic ability. Calcium phosphate silicate (CPS) incorporated with 1 wt% ZnO showed a bending strength of 80.8 MPa, while the bending strength of pure CPS ceramic was 40.9 MPa. The improved bending strength might be the result of the formation of liquid phase and the $Ca_2ZnSi_2O_7$ nanograins on the grain boundaries [86]. CuO is selected as an additive because of the positive effect of Cu on the formation of collagen and angiogenesis. CPS incorporated with 3 wt% of CuO sintered at 1200 °C exhibited a bending strength of 63.4 MPa, while pure CPS ceramic showed only 29.3 MPa. The bending strength of CuO-incorporated CPS was comparable to that of CPS processed by hot-pressing technology [87].

Nature bone is composed of inorganic minerals and organic matrix. The combination of bioceramics and organics is effective in the enhancement of mechanical strength and fracture toughness, avoiding the inherent brittleness of bioceramics. Agarose is a biocompatible and biodegradable gelling agent with fast room temperature polymerization, which is a suitable candidate as a binder for bioceramics. Agarose–BCP composite scaffolds showed a rubbery consistency with a strain of 60%, leading to a compression strength of 10–50 MPa. However, the strain of pure BCP scaffolds was no more than 2%, indicating that agarose conferred excellent ductility to the agarose–BCP system [88]. Poly(vinyl alcohol) (PVA) is a synthetic polymer with suitable biocompatibility and tunable hydrophilicity. The elastic modulus of PVA-BG hybrids was 6 MPa when the degree of hydrolysis was 98.5%, which was suitable for the repair of cancellous bone with a demand of modulus from 2 to 12 MPa [89]. Polylactic acid (PLA) is an artificial polymer with high modulus, high strength, biodegradability, and biocompatibility. Poly(L-lactic acid) (PLLA)–HA composite scaffolds exhibited significant enhancement of mechanical properties. The compressive strength of HA scaffolds increased from 0.34 to 3.03 MPa after immersing in PLLA solution with a concentration of 1 g/10 ml [90]. Coating poly-DL-lactic acid (PDLLA) on bioactive glass scaffolds did not influence the pore interconnectivity but increased the work of fracture of scaffolds [91]. 5 wt% PDLLA-coated bioactive glass scaffolds showed 40% improvement in compressive strength, with a porosity of 80% [92]. Infiltration of PLA into β-TCP scaffolds resulted in more than sixfold increase in compressive strength, which was comparable to the cortical bone at the same density [93]. The polymer not only acts as a healing agent at the surface of scaffolds, but also provides high elastic modulus to sustain the load. Collagen is the main organic component of natural bone with superior biocompatibility and biodegradability. Direct mixing method and co-precipitation method are generally applied to fabricate collagen–bioceramic composites. Incorporating nanosized hydroxyapatite (nHA) into collagen scaffolds increased the compressive modulus by 18 times with 500 wt% nHA addition [94]. The effect of β-TCP on the mechanical properties of collagen-β-TCP composite was in a dose-dependent manner. Higher dose of β-TCP nanoparticles led to higher mechanical stiffness of collagen-β-TCP composite in a certain range, and the range of 5–10 wt% was the optimal range of β-TCP content [95]. The mechanical properties of collagen-β-TCP composite blocks exhibited higher yield strength than β-TCP after 12 weeks of implantation in a rabbit posterior spinal fusion model, which was comparable to autologous bone [96].

Design of the architecture of bioceramics at different scales is essential for the application of bone regeneration. Macrostructures, such as geometry, porosity, and pore size, have direct effects on the mechanical properties. Bioceramic scaffolds with different pore geometries possess different mechanical behaviors. In a study of Sr-doped $Ca_2ZnSi_2O_7$ scaffold, four kinds of pore geometries (rectangular, hexagonal, curved, and zigzag) were designed. Scaffolds with a hexagonal pattern exhibited the highest flexural strength (30 MPa) compared with other patterns because of the maximum contact area between different layers at intersections [97]. In another study, nagelschmidtite (NAGEL, $Ca_7Si_2P_2O_{16}$) scaffolds with three macropore geometries (square, triangular, and parallelogram) were fabricated by 3D printing. NAGEL scaffolds with a square pore geometry showed higher modulus and compressive strength than those with triangular and parallelogram pore geometries [98]. Porosity is a main feature of bioceramic scaffolds with a significant effect on mechanical properties. It was investigated that there was a linear decrease of the average compressive strength of HA scaffolds from 25.3 to 3.86 MPa when the average porosity increased from 43.44% to 73.79%. However, a nonlinear decrease of compressive strength was observed at 1.27 MPa when the porosity increased to 85.77%. Similarly, there was a linear decrease of compressive strength in HA-other ceramics composite scaffolds from 16.13 to 2.31 MPa when the porosity increased from 54.62% to 72.09% and a nonlinear decrease of compressive strength to 0.92 MPa when the porosity increased to 89.92%. Nevertheless, only a linear relationship was observed in TCP scaffolds. The compressive strength decreased linearly from 9.09 to 0.65 MPa, while the porosity increased from 50% to 90%. Pore size is another factor with an impact on the mechanical properties. It was reported that the compressive strength was nearly linear with macropore size for a given porosity and a smaller pore size resulted in a higher compressive strength [99]. Moreover, there is a less significant change in compressive strength of scaffolds with pore size from 200 to 400 μm compared with scaffolds with the other pore sizes. It is believed that porosity has a more significant effect on mechanical properties than the average pore size [100].

2.4 Biological Properties

Bioceramics are superior to other biomaterials in the aspects of participation in the metabolic processes and the integration of bone tissue with implants. Since the implantation of bioceramics, at least 12 reaction stages will be processed.

Firstly, fast ionic exchange occurs on the surface of bioceramics accompanied with dissolution of bioceramics and the formation of carbonate hydroxyapatite (HCA). Subsequently, the adsorption of growth factors proceeds and the macrophages are recruited. After that, the stem cells attach and proliferate on the surface of bioceramics, and osteogenic differentiation is processed along with invasion of capillaries. Protein adsorption, immunomodulation, and vascularization are the main biological properties that play key roles in bone regeneration because abundant physiochemical and biological processes occur during the whole process of the implantation of bioceramics.

2.4.1 Protein Adsorption

Protein adsorption is the first process after the implantation of bioceramics, which directly work on the subsequent cascade that occur between bioceramics and body fluids. The proteins adsorbed on the surface of bioceramics provide a biological layer for cell attachment and further implant-tissue integration. Protein adsorption is generally influenced by chemical composition, surface topography, surface charge, and hydrophobicity.

Variation of chemical composition leads to the change of surface charge, hydrophobicity, and dissolution rate. For example, increasing the CaO content in BGs results in the reduction of specific surface area and surface charge density and the increase of the average pore size [101]. Incorporation of Ag into BGs helps to increase the adsorption of bovine serum albumin (BSA) through the binding of silver by BSA [102]. In addition, BGCs with 20 wt% iron oxide show desirable adsorption of BSA because of the increase of non-bridging oxygen and the enhancement of surface activity [103].

Surface topography, including particle size, pore size, porosity, and specific surface area dominate protein adsorption via modulating the surface area. Generally, smaller particle size, reasonable pore size to proteins, higher porosity, larger specific surface area, and greater roughness contribute to protein adsorption for the increased active sites. For instance, HA powder with a particle size of 100 nm showed higher protein adsorption than HA powder with a particle size of 1 μm [104]. Reasonable pore size is required since pore sizes larger than 15 nm are effective to adsorb proteins with high molecular weight, which is able to induce bone growth [105]. The penetration of fibrinogen is promoted by the pore size larger than 6 nm and inhibited by the pore size smaller than 2 nm [106]. Micropores play an important role in the rearrangement of proteins after adsorption on bioceramics, which is the hold-back effect of porous structure [107]. Higher porosity results in higher surface area as protein adsorbed on the porous bioceramic surface is multilayer while that on the dense bioceramic surface is monolayer.

Surface charge has effects on protein adsorption through the electrostatic interactions between the bioceramic surface and proteins. Bioceramics such as BGs, HA, and TCP generally exhibit negative charge. Oppositely charged proteins such as lysozyme (LSZ) had higher affinity for bioceramics than negatively charged proteins such as BSA because the isoelectric points (IEPs) of LSZ are approximately

11 while that of BSA closes to 4.5 in phosphate-buffered saline (PBS) solution with a pH of 7.4 [108]. However, there are both positive charge sites such as Ca^{2+} sites and negative charge sites such as PO_4^{3-} and OH^- sites in bioceramics, which can be modulated to enhance the adsorption of positively charged or negatively charged proteins [109].

Hydrophobicity generally benefits protein adsorption compared with hydrophilic surface. For calcium phosphate bioceramics, BCP has higher protein adsorption ability than HA because of the higher hydrophobicity. BGs with lower hydrophilicity tend to adsorb more proteins because the fewer water molecules between proteins and bioceramic surfaces need to be expelled [105].

2.4.2 Immunomodulation

Inflammatory responses occur when bioceramics are implanted because of the exogenous properties of bioceramics. After implantation, proteins from blood adsorb to bioceramic surfaces and lead to thrombus formation and complement activation [110]. Subsequently, monocytes adhere to the bioceramic surface and differentiate to macrophages, releasing cytokines and chemokines to recruit tissue repair cells such as fibroblasts and mesenchymal stem cells (MSCs) to the inflammation sites [111]. The foreign body reaction dominates the fate of implants and further bone regeneration, which is of great significance to investigate and modulate via the surface modification, and regulation of ionic products of bioceramics.

Calcium phosphate bioceramics are effective to modulate inflammation via influencing the behaviors of immune cells. For example, β-TCP could induce the maturation/activation of phenotypes and functions of dendritic cells (DCs) and macrophages by enhancing the expression of costimulatory surface molecules in DCs and macrophages, indicating that β-TCP shows mild immunomodulatory compared with the authentic toll-like receptor lipopolysaccharide (LPS) [112]. TCP bioceramics with submicron surface topography induced the polarization of macrophages to tissue repair macrophages via PI3K/Akt pathways while TCP bioceramics with micron surface topography induced the polarization of macrophages to the pro-inflammatory phenotype, demonstrating the importance of surface topography in immunomodulation and further bone formation (Figure 2.4) [113]. In addition, β-TCP extracts with a concentration of 25 mg/ml were proved to induce the polarization of macrophages to the M2 phenotype and further enhanced osteogenic differentiation of MSCs via the Wnt signaling pathway [114]. Besides β-TCP, BCP was found to support the attachment and proliferation of macrophages and promoted the expressions of inflammatory-related cytokines and growth factors, which further facilitated the osteogenic gene expressions of MSCs [115]. In addition, BCP regulated the polarization of macrophages, enhancing the fraction of $CD206^+$ M2 macrophages and promoting the expression of M2 phenotypic markers (Figure 2.5) [116]. Similarly, macrophages incubated on CDHA bioceramics showed a declined expression of pro-inflammatory cytokines (IL-6, TNF-α, and inducible nitric oxide synthase [iNOS]), resulting in enhanced osteogenic differentiation of osteoblasts and more collagen type I production [117].

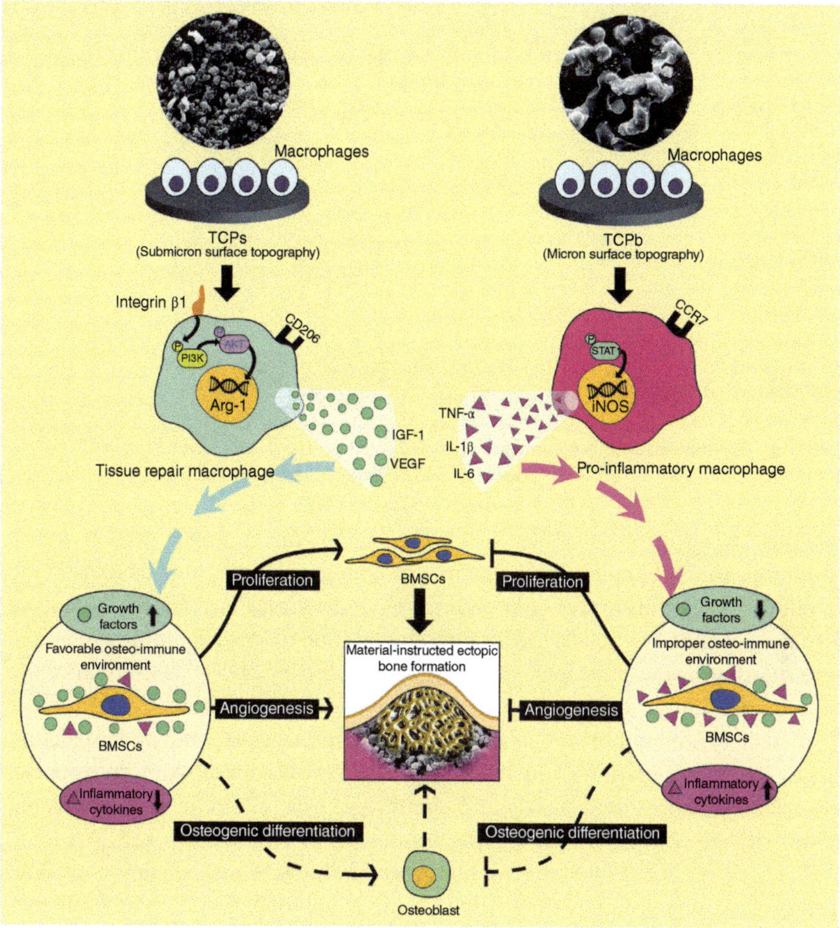

Figure 2.4 The surface topography of TCP bioceramics induced macrophage polarization and further bone formation. Source: Li et al. [113] / with permission of Royal Society of Chemistry.

Silicate bioceramics showed immunomodulatory capacity on macrophage polarization. Calcium silicate extract stimulated the expression of M2 surface marker, CD206, and anti-inflammatory cytokines, IL-10 and TGF-β, in macrophages [118]. Moreover, the calcium silicate-pretreated macrophage-conditioned medium stimulated the osteogenic differentiation of BMSCs via activating the oncostatin M (OSM) and relevant pathway. NAGEL and akermanite bioceramics inhibited inflammatory cytokine secretion of macrophages compared with TCP bioceramics. The Ca, Mg, and Si ions released from NAGEL and akermanite inhibited the macrophage inflammatory responses via suppressing the mitogen-activated protein kinase (MAPK) and nuclear factor-B (NF-B) signaling pathway and facilitating the caspase-dependent apoptosis of macrophages [119]. Lithium calcium silicate ($Li_2Ca_4Si_4O_{13}$, LCS) induced macrophages to anti-inflammatory M2 phenotype via

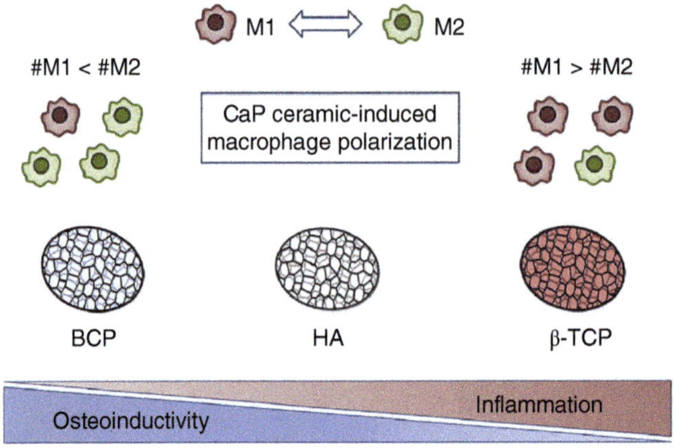

Figure 2.5 Calcium phosphate ceramics with different phase compositions regulate macrophage polarization. Source: Chen et al. [116] / with permission of Elsevier.

reducing the expression of TNF-α, IL-6, and IL-1β and facilitating IL-10 expression. In addition, the LCS-pretreated macrophage-conditioned medium facilitated the proliferation and maturation of chondrocytes, indicating that LCS promoted the maturation of chondrocyte via the immunomodulation of M2 macrophage polarization [120]. Doping bioactive ions into bioceramics is an effective way to enhance the immunomodulatory effects. Strontium-incorporated calcium silicate could promote M2 polarization and inhibit M1 polarization of macrophages as compared to calcium silicate. The incorporation of Sr could suppress synovial inflammatory response and further enhance osteogenesis and chondrogenesis [121]. Boron-incorporated calcium silicate ($Ca_{11}Si_4B_2O_{22}$) coating inhibited M1 polarization and promoted macrophages to M2 phenotype by suppressing the toll-like receptor signaling pathway. The incorporation of B helped inhibit macrophage differentiation into osteoclasts and stimulate osteogenic differentiation of BMSCs via the BMP-2 signaling pathway [122]. These liens of evidence show that the osteoinductive potential of bioceramics is closely related to their immunomodulatory ability and it is a novel strategy to take good advantage of the immunomodulatory capacity of bioceramics in the application of bone repair.

2.4.3 Vascularization

Vascularization is crucial for the supply of nutrients and oxygen for individual cells in bone regeneration. Insufficient vascularization causes nutrient deficiencies and hypoxia in bone tissue, resulting in the invalidation of bone substitutes [123]. Several strategies of designing bioceramics have been proposed for promoting vascularization.

Structural design of bioceramic scaffolds has a profound effect on the vascularization. Conventional scaffold fabrication methods, such as gas foaming, freeze drying, and particle leaching, have been widely used to regulate the pore size and porosity of bioceramic scaffolds. However, compared with conventional methods,

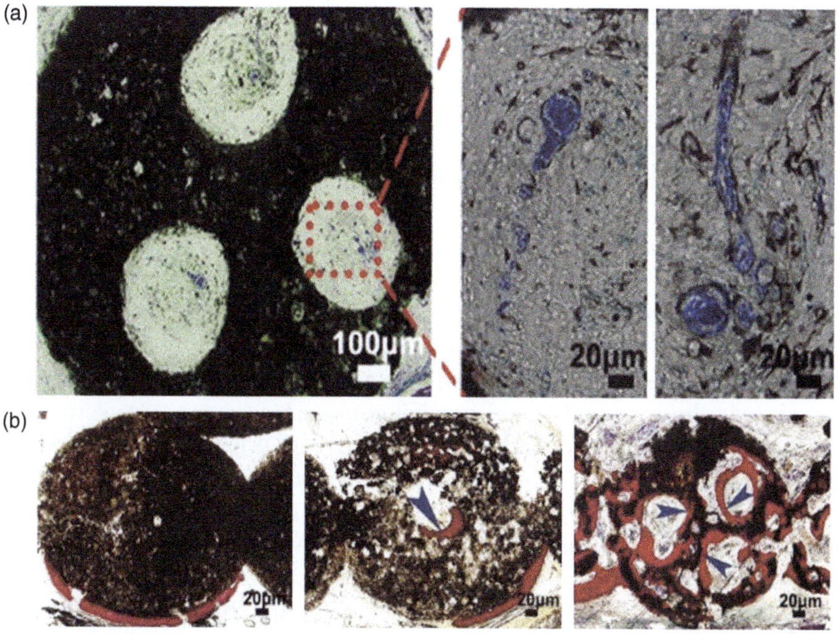

Figure 2.6 *In vivo* angiogenesis (a) and new bone formation (b) of the lotus root-like biomimetic scaffolds. The blue arrows indicate the newly formed blood vessels. Source: Feng et al. [125] / with permission of John Wiley & Sons.

3D printing technology can offer better control of interconnectivity of scaffolds by means of the parameterized pore channels, ensuring the ingrowth of blood vessels. 3D-printed hollow-pipe-packed bredigite scaffolds enhanced early vascularization in rabbit radius segmental defects. The hollow pipes with an internal diameter of 500 μm promoted the fast infiltration of host blood vessels compared with the solid struts [124]. Similarly, 3D-printed lotus root-like akermanite scaffolds enhanced the angiogenic process in the repair of rabbit calvarial defects. The lotus root-like channels with an internal diameter from 400 to 600 μm induced the invasion of blood vessels, which was beneficial to further bone regeneration (Figure 2.6) [125].

Incorporation of essential trace elements, such as Si, Sr, Cu, and Co, has been extensively used to promote the angiogenic ability of bioceramics. A suitable Si ion concentration from 0.6 to 2 μg/ml stimulated the proliferation, angiogenic gene (kinase insert domain receptor [KDR], basic Fibroblast Growth Factor Receptor 1 [bFGFR1], and TGF-βR3) expression, and NO synthesis [126]. Si ions from the extracts of calcium silicate with the concentration from 0.7 to 1.82 μg/ml stimulated the HUVEC proliferation, promoted the expression of VEGF and bFGF and their receptors, activated the expression of endothelial nitric oxide synthase (eNOS), and enhanced the expression of NO from HUVECs [49]. Sr ions were found to possess more significant stimulatory effects on the expression of VEGF and angiopoietin-1 (Ang-1) than Si ions in the extracts of Sr-substituted silicate bioceramics ($Sr_2MgSi_2O_7$) [127]. Moreover, the combination of Si and Sr ions possessed synergistic effects on angiogenesis. Cu ions have been reported to induce angiogenesis. Cu ions doped in MBG with the concentration from 14 to 56.6 μg/ml could induce HIF-1α and VEGF secretion from hBMSCs. Cu-doped calcium silicate with 0.3 μg/ml Cu ions obviously stimulated angiogenic growth factor expression [128]. In addition, Cu ions in Cu-doped calcium silicate extracts dominated the HIF-1α stabilization effects, inducing the hypoxia microenvironment. Co ions have been known as the chemical inducer of HIF-1α, stimulating the hypoxia-like response. The extracts of Co-containing β-tricalcium phosphate (Co-TCP) ceramics stimulated the expression of VEGF in hBMSCs and promoted the network formation of HUVECs compared with the extracts of pure TCP [129]. The effective concentration range of Co^{2+} ions for enhancing VEGF expression is 5–10 μM. In another study, the extracts of cobalt-containing silicate bioceramics, Co-akermanite ($Ca_2CoSi_2O_7$), significantly promoted the VEGF and eNOS expression in HUVECs, enhancing the formation of the capillary-like structure of HUVECs in a HDF (human dermal fibroblast)-HUVEC coculture model [130].

Vascularization is a key process in large bone repair. Although trace element-doped bioceramics can induce vascularization, the design of scaffolds in multiscales determines the vascular ingrowth. Taking all these factors into consideration increases the chances of sufficient vascularization, especially for the formation of vessels with several millimeters.

2.5 Summary

Bioceramics for bone regeneration have been investigated for many decades, and nowadays, many bioceramic products are available to patients. As bioinert ceramics revealed the limitations in bioactivities, bioactive and biodegradable ceramics were developed to induce bone regeneration. However, bioceramics suffer from intrinsic brittleness in the load-bearing applications, and various efforts have been made to regulate their composition and structure. Continuous mechanical support of bioceramics and bioceramic–bone interactions is indispensable in the whole process of bone repair. Nature-inspired approaches and 3D printing technologies make a breakthrough in the architecture design and fabrication of bioceramics. Moreover, ionic substitution and materials compositing result in the enhancement of the mechanical properties and biological properties of bioceramics. Further studies should be devoted on developing the multifunctional and smart bioceramics with the integration of mechanical properties, protein adsorption, immunomodulation, vascularization, and osteogenesis.

Acknowledgments

This work was supported by the National Key Research and Development Program of China (2018YFC1105201), National Natural Science Foundation of China (81771989), Innovation Cross Team of Chinese Academy of Sciences (JCTD-2018-13), and Science and Technology Commission of Shanghai Municipality (20442420300).

References

1 Dubok, V.A. (2000). Bioceramics – yesterday, today, tomorrow. *Powder Metall. Met. Ceram.* 39 (7, 8): 381–394.

2 Vallet-Regi, M. and Ruiz-Hernandez, E. (2011). Bioceramics: from bone regeneration to cancer nanomedicine. *Adv. Mater.* 23 (44): 5177–5218.

3 Ghiasi, B., Sefidbakht, Y., Mozaffari-Jovin, S. et al. (2020). Hydroxyapatite as a biomaterial – a gift that keeps on giving. *Drug Dev. Ind. Pharm.* 46 (7): 1035–1062.

4 Gomes, D.S., Santos, A.M.C., Neves, G.A., and Menezes, R.R. (2019). A brief review on hydroxyapatite production and use in biomedicine. *Cerâmica* 65 (374): 282–302.

5 Mehrabi, T., Mesgar, A.S., and Mohammadi, Z. (2020). Bioactive glasses: a promising therapeutic ion release strategy for enhancing wound healing. *ACS Biomater. Sci. Eng.* 6 (10): 5399–5430.

6 Bohner, M., Santoni, B.L.G., and Dobelin, N. (2020). β-tricalcium phosphate for bone substitution: synthesis and properties. *Acta Biomater.* 113: 23–41.

7 Srinath, P., Abdul Azeem, P., and Venugopal Reddy, K. (2020). Review on calcium silicate-based bioceramics in bone tissue engineering. *Int. J. Appl. Ceram. Technol.* 17 (5): 2450–2464.

8 Yoshida, K., Kondo, N., Kita, H. et al. (2005). Effect of substitutional monovalent and divalent metal ions on mechanical properties of β-tricalcium phosphate. *J. Am. Ceram. Soc.* 88 (8): 2315–2318.

9 Obadia, L., Deniard, P., Alonso, B. et al. (2006). Effect of sodium doping in β-tricalcium phosphate on its structure and properties. *Chem. Mater.* 18 (6): 1425–1433.

10 Xu, M., Li, H., Zhai, D. et al. (2015). Hierarchically porous nagelschmidtite bioceramic-silk scaffolds for bone tissue engineering. *J. Mater. Chem. B* 3 (18): 3799–3809.

11 Kumar, P., Saini, M., Dehiya, B.S. et al. (2020). Comprehensive survey on nanobiomaterials for bone tissue engineering applications. *Nanomaterials* 10 (10): 2019.

12 Ma, H., Feng, C., Chang, J., and Wu, C. (2018). 3D-printed bioceramic scaffolds: from bone tissue engineering to tumor therapy. *Acta Biomater.* 79: 37–59.

13 Yang, G., Liu, H., Cui, Y. et al. (2020). Bioinspired membrane provides periosteum-mimetic microenvironment for accelerating vascularized bone regeneration. *Biomaterials* 268: 120561.

14 Feng, C., Xue, J., Yu, X. et al. (2021). Co-inspired hydroxyapatite-based scaffolds for vascularized bone regeneration. *Acta Biomater.* 119: 419–431.

15 Baino, F. and Verné, E. (2017). Glass-based coatings on biomedical implants: a state-of-the-art review. *Biomed. Glasses* 3 (1): 1–17.
16 Rau, J.V., Curcio, M., Raucci, M.G. et al. (2019). Cu-releasing bioactive glass coatings and their in vitro properties. *ACS Appl. Mater. Interfaces* 11 (6): 5812–5820.
17 Sergi, R., Bellucci, D., and Cannillo, V. (2020). A comprehensive review of bioactive glass coatings: state of the art, challenges and future perspectives. *Coatings* 10 (8): 757.
18 Wu, C., Zhang, Y., Zhou, Y. et al. (2011). A comparative study of mesoporous glass/silk and non-mesoporous glass/silk scaffolds: physiochemistry and in vivo osteogenesis. *Acta Biomater.* 7 (5): 2229–2236.
19 Wu, C., Zhou, Y., Xu, M. et al. (2013). Copper-containing mesoporous bioactive glass scaffolds with multifunctional properties of angiogenesis capacity, osteostimulation and antibacterial activity. *Biomaterials* 34 (2): 422–433.
20 Wu, C., Zhou, Y., Lin, C. et al. (2012). Strontium-containing mesoporous bioactive glass scaffolds with improved osteogenic/cementogenic differentiation of periodontal ligament cells for periodontal tissue engineering. *Acta Biomater.* 8 (10): 3805–3815.
21 Wu, C., Zhou, Y., Fan, W. et al. (2012). Hypoxia-mimicking mesoporous bioactive glass scaffolds with controllable cobalt ion release for bone tissue engineering. *Biomaterials* 33 (7): 2076–2085.
22 Wu, C., Fan, W., Zhu, Y. et al. (2011). Multifunctional magnetic mesoporous bioactive glass scaffolds with a hierarchical pore structure. *Acta Biomater.* 7 (10): 3563–3572.
23 Xue, W., Dahlquist, K., Banerjee, A. et al. (2008). Synthesis and characterization of tricalcium phosphate with Zn and Mg based dopants. *J. Mater. Sci.-Mater. Med.* 19 (7): 2669–2677.
24 Spaeth, K., Goetz-Neunhoeffer, F., and Hurle, K. (2020). Cu^{2+} doped β-tricalcium phosphate: solid solution limit and crystallographic characterization by rietveld refinement. *J. Solid State Chem.* 285: 121225.
25 Jacobs, A., Renaudin, G., Forestier, C. et al. (2020). Biological properties of copper-doped biomaterials for orthopedic applications: a review of antibacterial, angiogenic and osteogenic aspects. *Acta Biomater.* 117: 21–39.
26 Kornblatt, A.P., Nicoletti, V.G., and Travaglia, A. (2016). The neglected role of copper ions in wound healing. *J. Inorg. Biochem.* 161: 1–8.
27 Gallo, M., Le Gars Santoni, B., Douillard, T. et al. (2019). Effect of grain orientation and magnesium doping on β-tricalcium phosphate resorption behavior. *Acta Biomater.* 89: 391–402.
28 Zamani, A., Omrani, G.R., and Nasab, M.M. (2009). Lithium's effect on bone mineral density. *Bone* 44 (2): 331–334.
29 Saberianpour, S., Heidarzadeh, M., Geranmayeh, M.H. et al. (2018). Tissue engineering strategies for the induction of angiogenesis using biomaterials. *J. Biol. Eng.* 12: 36.
30 Buache, E., Velard, F., Bauden, E. et al. (2012). Effect of strontium-substituted biphasic calcium phosphate on inflammatory mediators production by human monocytes. *Acta Biomater.* 8 (8): 3113–3119.

31 Torres, P.M., Vieira, S.I., Cerqueira, A.R. et al. (2014). Effects of Mn-doping on the structure and biological properties of β-tricalcium phosphate. *J. Inorg. Biochem.* 136: 57–66.

32 Rau, J.V., Fadeeva, I.V., Fomin, A.S. et al. (2019). Sic parvis magna: manganese-substituted tricalcium phosphate and its biophysical properties. *ACS Biomater. Sci. Eng.* 5 (12): 6632–6644.

33 Beattie, J.H. and Avenell, A. (1992). Trace element nutrition and bone metabolism. *Nutr. Res. Rev.* 5 (1): 167–188.

34 Ren, H.Y., Cao, Y., Zhao, Q.J. et al. (2006). Proliferation and differentiation of bone marrow stromal cells under hypoxic conditions. *Biochem. Biophys. Res. Commun.* 347 (1): 12–21.

35 Bigi, A., Foresti, E., Gandolfi, M. et al. (1997). Isomorphous substitutions in β-tricalcium phosphate: the different effects of zinc and strontium. *J. Inorg. Biochem.* 66 (4): 259–265.

36 Famery, R., Richard, N., and Boch, P. (1994). Preparation of α-tricalcium and β-tricalcium phosphate ceramics, with and without magnesium addition. *Ceram. Int.* 20 (5): 327–336.

37 Yoshida, K., Hyuga, H., Kondo, N. et al. (2006). Substitution model of monovalent (Li, Na, and K), divalent (Mg), and trivalent (Al) metal ions for β-tricalcium phosphate. *J. Am. Ceram. Soc.* 89 (2): 688–690.

38 Han, P., Xu, M., Chang, J. et al. (2014). Lithium release from β-tricalcium phosphate inducing cementogenic and osteogenic differentiation of both hPDLCs and hBMSCs. *Biomater. Sci.* 2 (9): 1230–1243.

39 Saidak, Z. and Marie, P.J. (2012). Strontium signaling: molecular mechanisms and therapeutic implications in osteoporosis. *Pharmacol. Ther.* 136 (2): 216–226.

40 Kruppke, B., Heinemann, C., Wagner, A.S. et al. (2019). Strontium ions promote in vitro human bone marrow stromal cell proliferation and differentiation in calcium-lacking media. *Dev. Growth Differ.* 61 (2): 166–175.

41 Wan, B., Wang, R., Sun, Y. et al. (2020). Building osteogenic microenvironments with strontium-substituted calcium phosphate ceramics. *Front. Bioeng. Biotechnol.* 8: 591467.

42 Cheng, G., Deng, C., Wu, C. et al. (2019). Effects of Mn-doping on the structural evolution of β-tricalcium phosphate by Rietveld refinement and Raman spectroscopy. *Mater. Lett.* 235: 236–238.

43 Kim, S.E. and Park, K. (2020). Recent advances of biphasic calcium phosphate bioceramics for bone tissue regeneration. In: *Biomimicked Biomaterials: Advances in Tissue Engineering and Regenerative Medicine* (ed. H.J. Chun, R.L. Reis, A. Motta and G. Khang), 177–188. Singapore: Springer-Verlag Singapore Pte Ltd.

44 Puttini, I.O., Poli, P.P., Maiorana, C. et al. (2019). Evaluation of osteoconduction of biphasic calcium phosphate ceramic in the calvaria of rats: microscopic and histometric analysis. *J. Funct. Biomater.* 10 (1): 7.

45 Bunpetch, V., Zhang, X., Li, T. et al. (2019). Silicate-based bioceramic scaffolds for dual-lineage regeneration of osteochondral defect. *Biomaterials* 192: 323–333.

46 Dufrane, D., Delloye, C., McKay, I.J. et al. (2003). Indirect cytotoxicity evaluation of pseudowollastonite. *J. Mater. Sci. - Mater. Med.* 14 (1): 33–38.

47 Zhang, N., Molenda, J.A., Mankoci, S. et al. (2013). Crystal structures of $CaSiO_3$ polymorphs control growth and osteogenic differentiation of human mesenchymal stem cells on bioceramic surfaces. *Biomater. Sci.* 1 (10): 1101–1110.

48 Zhang, N., Molenda, J.A., Fournelle, J.H. et al. (2010). Effects of pseudowollastonite ($CaSiO_3$) bioceramic on in vitro activity of human mesenchymal stem cells. *Biomaterials* 31 (30): 7653–7665.

49 Li, H. and Chang, J. (2013). Stimulation of proangiogenesis by calcium silicate bioactive ceramic. *Acta Biomater.* 9 (2): 5379–5389.

50 de Aza, P.N., Luklinska, Z.B., Martinez, A. et al. (2000). Morphological and structural study of pseudowollastonite implants in bone. *J. Microsc.* 197: 60–67.

51 Ni, S., Chang, J., Chou, L., and Zhai, W. (2007). Comparison of osteoblast-like cell responses to calcium silicate and tricalcium phosphate ceramics in vitro. *J. Biomed. Mater. Res. Part B* 80 (1): 174–183.

52 Xu, S., Lin, K., Wang, Z. et al. (2008). Reconstruction of calvarial defect of rabbits using porous calcium silicate bioactive ceramics. *Biomaterials* 29 (17): 2588–2596.

53 Gou, Z., Chang, J., Zhai, W., and Wang, J. (2005). Study on the self-setting property and the in vitro bioactivity of β-Ca_2SiO_4. *J. Biomed. Mater. Res. Part B* 73 (2): 244–251.

54 Zhao, W., Wang, J., Zhai, W. et al. (2005). The self-setting properties and in vitro bioactivity of tricalcium silicate. *Biomaterials* 26 (31): 6113–6121.

55 Cao, H. and Liu, X. (2013). Plasma-sprayed ceramic coatings for osseointegration. *Int. J. Appl. Ceram. Technol.* 10 (1): 1–10.

56 Vallet-Regi, M., Salinas, A.J., Roman, J., and Gil, M. (1999). Effect of magnesium content on the in vitro bioactivity of $CaO–MgO–SiO_2–P_2O_5$ sol–gel glasses. *J. Mater. Chem.* 9 (2): 515–518.

57 Chen, X., Ou, J., Wei, Y. et al. (2010). Effect of MgO contents on the mechanical properties and biological performances of bioceramics in the $MgO-CaO–SiO_2$ system. *J. Mater. Sci. - Mater. Med.* 21 (5): 1463–1471.

58 Wu, C. and Chang, J. (2007). Degradation, bioactivity, and cytocompatibility of diopside, akermanite, and bredigite ceramics. *J. Biomed. Mater. Res. Part B* 83 (1): 153–160.

59 Zhai, W., Lu, H., Wu, C. et al. (2013). Stimulatory effects of the ionic products from Ca–Mg–Si bioceramics on both osteogenesis and angiogenesis in vitro. *Acta Biomater.* 9 (8): 8004–8014.

60 Nonami, T. and Tsutsumi, S. (1999). Study of diopside ceramics for biomaterials. *J. Mater. Sci. - Mater. Med.* 10 (8): 475–479.

61 Wu, C., Ramaswamy, Y., and Zreiqat, H. (2010). Porous diopside ($CaMgSi_2O_6$) scaffold: a promising bioactive material for bone tissue engineering. *Acta Biomater.* 6 (6): 2237–2245.

62 Zhang, Y., Li, S., and Wu, C. (2014). The in vitro and in vivo cementogenesis of $CaMgSi_2O_6$ bioceramic scaffolds. *J. Biomed. Mater. Res. Part A* 102 (1): 105–116.

63 Shuai, C., Liu, T., Gao, C. et al. (2014). Mechanical reinforcement of diopside bone scaffolds with carbon nanotubes. *Int. J. Mol. Sci.* 15 (10): 19319–19329.

64 Shahrouzifar, M.R. and Salahinejad, E. (2019). Strontium doping into diopside tissue engineering scaffolds. *Ceram. Int.* 45 (8): 10176–10181.

65 Shahrouzifar, M.R., Salahinejad, E., and Sharifi, E. (2019). Co-incorporation of strontium and fluorine into diopside scaffolds: bioactivity, biodegradation and cytocompatibility evaluations. *Mater. Sci. Eng., C* 103: 109752.

66 Wu, C. and Zreiqat, H. (2010). Porous bioactive diopside ($CaMgSi_2O_6$) ceramic microspheres for drug delivery. *Acta Biomater.* 6 (3): 820–829.

67 Chen, X., Ou, J., Kang, Y. et al. (2008). Synthesis and characteristics of monticellite bioactive ceramic. *J. Mater. Sci. - Mater. Med.* 19 (3): 1257–1263.

68 Kalantari, E. and Naghib, S.M. (2019). A comparative study on biological properties of novel nanostructured monticellite-based composites with hydroxyapatite bioceramic. *Mater. Sci. Eng., C* 98: 1087–1096.

69 Kalantari, E., Naghib, S.M., Iravani, N.J. et al. (2018). Nanostructured monticellite for tissue engineering applications – part II: molecular and biological characteristics. *Ceram. Int.* 44 (12): 14704–14711.

70 Wu, C. and Chang, J. (2004). Synthesis and apatite-formation ability of akermanite. *Mater. Lett.* 58 (19): 2415–2417.

71 Wu, C., Chang, J., Ni, S., and Wang, J. (2006). In vitro bioactivity of akermanite ceramics. *J. Biomed. Mater. Res. Part A* 76 (1): 73–80.

72 Huang, Y., Jin, X., Zhang, X. et al. (2009). In vitro and in vivo evaluation of akermanite bioceramics for bone regeneration. *Biomaterials* 30 (28): 5041–5048.

73 Sun, H., Wu, C., Dai, K. et al. (2006). Proliferation and osteoblastic differentiation of human bone marrow-derived stromal cells on akermanite-bioactive ceramics. *Biomaterials* 27 (33): 5651–5657.

74 Liu, Q., Cen, L., Yin, S. et al. (2008). A comparative study of proliferation and osteogenic differentiation of adipose-derived stem cells on akermanite and β-TCP ceramics. *Biomaterials* 29 (36): 4792–4799.

75 Gu, H., Guo, F., Zhou, X. et al. (2011). The stimulation of osteogenic differentiation of human adipose-derived stem cells by ionic products from akermanite dissolution via activation of the ERK pathway. *Biomaterials* 32 (29): 7023–7033.

76 Ma, N., Ma, B., Zhou, Y. et al. (2019). In vivo evaluation of the subchronic systemic toxicity of akermanite bioceramic for bone regeneration following ISO standard methods. *RSC Adv.* 9 (31): 17530–17536.

77 Ou, J., Kang, Y.Q., Huang, Z.B. et al. (2008). Preparation and in vitro bioactivity of novel merwinite ceramic. *Biomed. Mater.* 3 (1): 8.

78 Hafezi-Ardakani, M., Moztarzadeh, F., Rabiee, M. et al. (2010). Sol-gel synthesis and apatite-formation ability of nanostructure merwinite ($Ca_3MgSi_2O_8$) as a novel bioceramic. *J. Ceram. Process. Res.* 11 (6): 765–768.

79 Hafezi, M., Reza Talebi, A., Mohsen Miresmaeili, S. et al. (2013). Histological analysis of bone repair in rat femur via nanostructured merwinite granules. *Ceram. Int.* 39 (4): 4575–4580.

80 Wu, C., Chang, J., Wang, J. et al. (2005). Preparation and characteristics of a calcium magnesium silicate (bredigite) bioactive ceramic. *Biomaterials* 26 (16): 2925–2931.

81 Zhou, Y., Wu, C., Zhang, X. et al. (2013). The ionic products from bredigite bioceramics induced cementogenic differentiation of periodontal ligament cells via activation of the Wnt/β-catenin signalling pathway. *J. Mater. Chem. B* 1 (27): 3380–3389.

82 Chen, L., Liu, L., Wu, C. et al. (2017). The extracts of bredigite bioceramics enhanced the pluripotency of human dental pulp cells. *J. Biomed. Mater. Res. Part A* 105 (12): 3465–3474.

83 Pan, Y., Yin, J., Zuo, K. et al. (2016). The sintering behavior and mechanical properties of $CaSiO_3$ bioceramics with B_2O_3 addition. *Ceram. Int.* 42 (7): 9222–9226.

84 Yang, X., Zhang, L., Chen, X. et al. (2012). Incorporation of B_2O_3 in $CaO–SiO_2–P_2O_5$ bioactive glass system for improving strength of low-temperature co-fired porous glass ceramics. *J. Non-Cryst. Solids* 358 (9): 1171–1179.

85 Mei, L., Yin, J., Xia, Y. et al. (2020). Preparation of high-strength β-$CaSiO_3$ bioceramic with B_2O_3 and SiO_2 sintering additives. *Ceram. Int.* 46 (16): 25970–25978.

86 Deng, F.Y., Wang, F., Liu, Z.W. et al. (2018). Enhanced mechanical property of $Ca_5(PO_4)_2SiO_4$ bioceramic by a biocompatible sintering aid of zinc oxide. *Ceram. Int.* 44 (15): 18352–18362.

87 Xu, S., Wu, Q., Guo, Y. et al. (2021). Copper containing silicocarnotite bioceramic with improved mechanical strength and antibacterial activity. *Mater. Sci. Eng., C* 118: 111493.

88 Puertolas, J.A., Vadillo, J.L., Sanchez-Salcedo, S. et al. (2011). Compression behaviour of biphasic calcium phosphate and biphasic calcium phosphate-agarose scaffolds for bone regeneration. *Acta Biomater.* 7 (2): 841–847.

89 Mansur, H.S. and Costa, H.S. (2008). Nanostructured poly(vinyl alcohol)/bioactive glass and poly(vinyl alcohol)/chitosan/bioactive glass hybrid scaffolds for biomedical applications. *Chem. Eng. J.* 137 (1): 72–83.

90 Tian, T., Jiang, D.L., Zhang, J.X., and Lin, Q.L. (2008). Fabrication of bioactive composite by developing PLLA onto the framework of sintered HA scaffold. *Mater. Sci. Eng., C* 28 (1): 51–56.

91 Bretcanu, O., Boccaccini, A.R., and Salih, V. (2012). Poly-DL-lactic acid coated bioglass® scaffolds: toughening effects and osteosarcoma cell proliferation. *J. Mater. Sci.* 47 (15): 5661–5672.

92 Tajbakhsh, S. and Hajiali, F. (2017). A comprehensive study on the fabrication and properties of biocomposites of poly(lactic acid)/ceramics for bone tissue engineering. *Mater. Sci. Eng., C* 70: 897–912.

93 Martinez-Vazquez, F.J., Perera, F.H., Miranda, P. et al. (2010). Improving the compressive strength of bioceramic robocast scaffolds by polymer infiltration. *Acta Biomater.* 6 (11): 4361–4368.

94 Lickorish, D., Ramshaw, J.A.M., Werkmeister, J.A. et al. (2004). Collagen-hydroxyapatite composite prepared by biomimetic process. *J. Biomed. Mater. Res. Part A* 68A (1): 19–27.

95 Zheng, H., Bai, Y., Shih, M.-S. et al. (2014). Effect of a β-TCP collagen composite bone substitute on healing of drilled bone voids in the distal femoral condyle of rabbits. *J. Biomed. Mater. Res. Part B* 102 (2): 376–383.

96 Lu, Y.J., Li, M.H., Long, Z.Y. et al. (2019). Collagen/β-TCP composite as a bone-graft substitute for posterior spinal fusion in rabbit model: a comparison study. *Biomed. Mater.* 14 (4): 11.

97 Roohani-Esfahani, S.I., Newman, P., and Zreiqat, H. (2016). Design and fabrication of 3D printed scaffolds with a mechanical strength comparable to cortical bone to repair large bone defects. *Sci. Rep.* 6: 8.

98 Xu, M.C., Zhai, D., Chang, J., and Wu, C.T. (2014). In vitro assessment of three-dimensionally plotted nagelschmidtite bioceramic scaffolds with varied macropore morphologies. *Acta Biomater.* 10 (1): 463–476.

99 Liu, D.M. (1997). Influence of porosity and pore size on the compressive strength of porous hydroxyapatite ceramic. *Ceram. Int.* 23 (2): 135–139.

100 Jodati, H., Yılmaz, B., and Evis, Z. (2020). A review of bioceramic porous scaffolds for hard tissue applications: effects of structural features. *Ceram. Int.* 46 (10): 15725–15739.

101 Li, R., Clark, A.E., and Hench, L.L. (1991). An investigation of bioactive glass powders by sol-gel processing. *J. Appl. Biomater.* 2 (4): 231–239.

102 Kittler, S., Greulich, C., Gebauer, J.S. et al. (2010). The influence of proteins on the dispersability and cell-biological activity of silver nanoparticles. *J. Mater. Chem.* 20 (3): 512–518.

103 Sharma, K., Bhattacharya, S., Murali, C. et al. (2012). Interactions of ferrimagnetic glass/glass-ceramics with bovine serum albumin. *Appl. Surf. Sci.* 258 (7): 2356–2361.

104 Rouahi, M., Champion, E., Gallet, O. et al. (2006). Physico-chemical characteristics and protein adsorption potential of hydroxyapatite particles: influence on in vitro biocompatibility of ceramics after sintering. *Colloids Surf., B* 47 (1): 10–19.

105 Zheng, K., Kapp, M., and Boccaccini, A.R. (2019). Protein interactions with bioactive glass surfaces: a review. *Appl. Mater. Today* 15: 350–371.

106 Lin, S., Van den Bergh, W., Baker, S., and Jones, J.R. (2011). Protein interactions with nanoporous sol–gel derived bioactive glasses. *Acta Biomater.* 7 (10): 3606–3615.

107 Zhu, X.D., Fan, H.S., Xiao, Y.M. et al. (2009). Effect of surface structure on protein adsorption to biphasic calcium-phosphate ceramics in vitro and in vivo. *Acta Biomater.* 5 (4): 1311–1318.

108 Zhu, X.D., Fan, H.S., Zhao, C.Y. et al. (2007). Competitive adsorption of bovine serum albumin and lysozyme on characterized calcium phosphates by polyacrylamide gel electrophoresis method. *J. Mater. Sci. - Mater. Med.* 18 (11): 2243–2249.

109 Ohta, K., Monma, H., and Takahashi, S. (2001). Adsorption characteristics of proteins on calcium phosphates using liquid chromatography. *J. Biomed. Mater. Res. Part A* 55 (3): 409–414.

110 Franz, S., Rammelt, S., Scharnweber, D., and Simon, J.C. (2011). Immune responses to implants – a review of the implications for the design of immunomodulatory biomaterials. *Biomaterials* 32 (28): 6692–6709.

111 Anderson, J.M., Rodriguez, A., and Chang, D.T. (2008). Foreign body reaction to biomaterials. *Semin. Immunol.* 20 (2): 86–100.

112 Tai, S., Cheng, J.Y., Ishii, H. et al. (2016). Effects of β-tricalcium phosphate particles on primary cultured murine dendritic cells and macrophages. *Int. Immunopharmacol.* 40: 419–427.

113 Li, M.Z., Guo, X.D., Qi, W.T. et al. (2020). Macrophage polarization plays roles in bone formation instructed by calcium phosphate ceramics. *J. Mater. Chem. B* 8 (9): 1863–1877.

114 Zheng, M., Weng, M., Zhang, X. et al. (2021). β-tricalcium phosphate promotes osteogenic differentiation of bone marrow-derived mesenchymal stem cells through macrophages. *Biomed. Mater.* 16: 025005.

115 Wang, J., Chen, X.N., Yang, X. et al. (2020). Positive role of calcium phosphate ceramics regulated inflammation in the osteogenic differentiation of mesenchymal stem cells. *J. Biomed. Mater. Res. Part A* 108 (6): 1305–1320.

116 Chen, X.N., Wang, M.L., Chen, F.Y. et al. (2020). Correlations between macrophage polarization and osteoinduction of porous calcium phosphate ceramics. *Acta Biomater.* 103: 318–332.

117 Sadowska, J.M., Wei, F., Guo, J. et al. (2019). The effect of biomimetic calcium deficient hydroxyapatite and sintered β-tricalcium phosphate on osteoimmune reaction and osteogenesis. *Acta Biomater.* 96: 605–618.

118 Zhou, P., Xia, D., Ni, Z. et al. (2021). Calcium silicate bioactive ceramics induce osteogenesis through oncostatin M. *Bioact. Mater.* 6 (3): 810–822.

119 Huang, Y., Wu, C., Zhang, X. et al. (2018). Regulation of immune response by bioactive ions released from silicate bioceramics for bone regeneration. *Acta Biomater.* 66: 81–92.

120 Zhai, D., Chen, L., Chen, Y. et al. (2020). Lithium silicate-based bioceramics promoting chondrocyte maturation by immunomodulating M2 macrophage polarization. *Biomater. Sci.* 8 (16): 4521–4534.

121 Wang, C.Y., Chen, B., Wang, W. et al. (2019). Strontium released bi-lineage scaffolds with immunomodulatory properties induce a pro-regenerative environment for osteochondral regeneration. *Mater. Sci. Eng., C* 103: 12.

122 Lu, X., Li, K., Xie, Y.T. et al. (2019). Improved osteogenesis of boron incorporated calcium silicate coatings via immunomodulatory effects. *J. Biomed. Mater. Res. Part A* 107 (1): 12–24.

123 Rouwkema, J., Rivron, N.C., and van Blitterswijk, C.A. (2008). Vascularization in tissue engineering. *Trends Biotechnol.* 26 (8): 434–441.

124 Zhang, W., Feng, C., Yang, G. et al. (2017). 3D-printed scaffolds with synergistic effect of hollow-pipe structure and bioactive ions for vascularized bone regeneration. *Biomaterials* 135: 85–95.

125 Feng, C., Zhang, W., Deng, C. et al. (2017). 3D printing of lotus root-like biomimetic materials for cell delivery and tissue regeneration. *Adv. Sci.* 4: 1700401.

126 Zhai, W., Lu, H., Chen, L. et al. (2012). Silicate bioceramics induce angiogenesis during bone regeneration. *Acta Biomater.* 8 (1): 341–349.

127 Mao, L., Xia, L., Chang, J. et al. (2017). The synergistic effects of Sr and Si bioactive ions on osteogenesis, osteoclastogenesis and angiogenesis for osteoporotic bone regeneration. *Acta Biomater.* 61: 217–232.

128 Kong, N., Lin, K.L., Li, H.Y., and Chang, J. (2014). Synergy effects of copper and silicon ions on stimulation of vascularization by copper-doped calcium silicate. *J. Mater. Chem. B* 2 (8): 1100–1110.

129 Zhang, M., Wu, C., Li, H. et al. (2012). Preparation, characterization and in vitro angiogenic capacity of cobalt substituted β-tricalcium phosphate ceramics. *J. Mater. Chem.* 22 (40): 21686.

130 Tian, T., Han, Y., Ma, B. et al. (2015). Novel Co-akermanite ($Ca_2CoSi_2O_7$) bioceramics with the activity to stimulate osteogenesis and angiogenesis. *J. Mater. Chem. B* 3 (33): 6773–6782.

3

3D Printing and Bioprinting Strategies Applied Toward Orthopedics

Zongliang Wang, Prajwal Agrawal, and Yu Shrike Zhang

Harvard Medical School, Division of Engineering in Medicine, Department of Medicine, Brigham and Women's Hospital, Cambridge, MA 02139, USA

3.1 Introduction

The emergence and rapid development of additive manufacturing (AM) have provided a prospective option to produce personalized grafts for applications in tissue regeneration and precision medicine by utilizing three-dimensional (3D) printing and bioprinting techniques. The advancements in these fields open new doors for bone tissue regeneration to address the growing demand for customized bone implants for bone defects or bone loss caused by trauma, infection, surgical resection of cancer, or degenerative bone diseases [1]. Also, these techniques help to overcome the native tissue-mimicking structural limitation of pre-formed scaffolds by precisely integrating them into the surrounding defective sites having hierarchical internal structures, mechanical properties, and biological cellular compositions [2].

3D bioprinting has also enabled direct cell encapsulation (possibly with other bioactive components) into various types of bioinks utilized to generate patient-specific bone implants, featuring controllable spatial patterns and structures. In correspondence, an additional cell-seeding process becomes unnecessary, minimizing the possible inhomogeneous cell distributions within the constructs [3]. Moreover, the porosity and pore size of the internal porous microstructures can be precisely controlled by adjusting the (bio)printing parameters [4]. Besides, the mechanical properties, vascularization, and osteogenesis can also be regulated by incorporating nanomaterials or by combining them with nanotechnologies or microfluidic devices [5]. All of these advantages help in making 3D printing and bioprinting techniques superior to conventional autografts or allografts, which possess disadvantages such as limited supply, adverse immune reactions, and risks of disease transmission and other complications [6].

Apart from the usage of biomaterials and the advantages of (bio)printing process, (bio)printing has another main advantage of personalizable design procedures, which act as a critical step for a successful (bio)printing. The personalized

Biofabrication for Orthopedics: Methods, Techniques, and Applications, First Edition.
Edited by Wenguo Cui, Xin Zhao, and Shen Liu.

Table 3.1 Summary of 3D printing and bioprinting techniques utilized in orthopedics.

Methods		Materials	Cells	Gelation	Comments	References
Extrusion based	FDM	PLA, gold nanoparticles, GelMA	hADSCs	UV	High mechanical strength and osteogenic differentiation	[9a]
	Melt extrusion	PCL, alginate/PEO	C2C12 cells	$CaCl_2$	Potential myogenic tissue regeneration	[10]
	3D plotting	CPC, alg/mc	hBMSCs	$CaCl_2$	Very minor decrease in cell viability	[11b]
	Room temperature extrusion using binder materials	BG/CSn and PVA	Monkey BMSCs	—	Nutrition and metabolic waste transport and promotion of bone tissue regeneration	[12]
	MEW	PCL, α-TCP, nHA, GelMA, and poloxamer	AC chondroprogenitor cells	UV	Structural stability during handling and surgical implantation in osteochondral defects	[13]
	Room temperature liquid ink extrusion	HA, PCL, PLGA	hBMSCs	Evaporation	Vascularization, quick integration with surrounding tissues, rapid ossification, and new bone growth	[14]
	Melt extrusion and electrospun	PCL and PLGA	hBMSCs	—	Mimicking the structural environment with differently scaled biomaterials	[15a]
	IOP	PU, PCL, hyaluronic acid, gelatin, and fibrinogen	C2C12 and 3T3 cells	Thrombin in $CaCl_2$ solution	Precise shape fidelity, elastic PU-C2C12 muscle side, and stiff PCL-3T3 tendon side	[5d]
	Microfluidic extrusion-based bioprinting	Alginate, GelMA, CS-AEMA, HAMA, and TCP	hBMSC, hAC	UV, $CaCl_2$	High resolution, shape fidelity, and cell viability	[15b]

Inkjet-based	Electrified inkjet	PCL	MC3T3-E1 cells	—	Improved cell attachment, activity, and differentiation by nanoscale topography	[16a]
	Piezoelectric drop-on-demand inkjet	Fibrin and BMP-2	MDSCs	—	Spatially controlled multilineage differentiation of stem cells	[16b]
Vat polymer-ization based	DLP	AKT and BG	hBMSCs and HUVECs	Laser	Accelerated blood vessels ingrowth and new bone formation *in vivo*	[6a]
	SLA	ECM, GelMA, and exosome	Chondrocytes	Visible light	Sustainable release of exosomes and regeneration of cartilage	[17]

Abbreviations: 3D, three-dimensional; alg/mc, alginate/methylcellulose; α-TCP, alpha-tricalcium phosphate; AKT, $Ca_2MgSi_2O_7$; BG, bioactive glass; BMSCs, bone mesenchymal stem cells; BMP-2, bone morphogenetic protein 2; C2C12 cell, mouse myoblast; $CaCl_2$, calcium chloride; CPC, calcium phosphate cement; CS-AEMA, 2-aminoethyl methacrylated chondroitin sulfate; CSn, chitosan nanoparticles; DLP, digital light processing; ECM, extracellular matrix; FDM, fused-deposition modeling, GelMA, gelatin methacryloyl; HA, hydroxyapatite; hAC, human articular chondrocyte; hADSC, human adipose-derived stem cell; HAMA, methacrylated hyaluronic acid; hBMSC, human bone mesenchymal stem cell; HUVEC, human umbilical vein endothelial cell; IOP, integrated organ printing; MC3T3-E1 cell, mouse embryonic osteoblast precursor cell; MDSC, mouse primary muscle-derived stem cell; MEW, melt electrowriting; nHA, nanohydroxyapatite; 3T3 cell, mouse embryonic fibroblast; PCL, poly(ε-caprolactone); PEO, poly(ethylene oxide); PLA, polylactic acid; PLGA, poly(lactic-*co*-glycolic acid); PU, polyurethane; PVA, polyvinyl alcohol; SLA, stereolithography apparatus; and UV, ultraviolet.

design procedure enables the achievement of customized 3D implants using computer-aided design (CAD) software, such as Solidworks, SketchUp, 3ds Max, Rhinoceros 5.0, Mach3, and Image ware 13.1, among others [7]. These designs (usually in the Standard Triangle Language [STL] file format) are given as an input to slicing software, which as the name suggests, "slices" the STL files into different formats such as geometric code (GCode) (*e.g.* extrusion-based, inkjet-based, stereolithography [SLA] (bio)printing) or layers (*e.g.* digital light processing [DLP] (bio)printing) that the (bio)printer firmware can read. As in most cases, the firmware for a (bio)printer understands GCode; for its generation, slicing software such as Slic3r, Repetier Host, OctoPrint, Simplify3D, and others are utilized [7f]. For DLP and related (bio)printing procedures, slicing software such as Moneyprint and others can be adopted to generate the sequence of projection files and define the set of instructions [7c, 8]. These customizable design procedures play a vital role in the advancement of 3D printing and bioprinting toward translational applications in orthopedics.

In this chapter, first, an overview of 3D printing and bioprinting techniques and their potential applications in orthopedics is briefly given (Table 3.1). Later, a summary on the formulations of inks and bioinks, printing and bioprinting platforms, as well as innovations that are currently available and progresses are laid out. Subsequently, the challenges that remain to be addressed are identified. Finally, highlights related to future directions of 3D printing and bioprinting techniques for clinical bone regeneration are stated.

3.2 Biomaterial Inks

The biomaterials utilized in 3D printing and bioprinting techniques are usually termed "inks" or "bioinks" because of their outstanding contributions in specifically facilitating precise 3D structures and high fidelities of the fabricated bone grafts [18]. The biomaterial inks should provide both the structural support and bioactive functions to ensure the activities of post-seeded or encapsulated cells eliciting minimal inflammatory responses of the host tissues [18c]. The increasing attention in the 3D (bio)printing fields has paved the way for the rapid development and modification of biomaterials that can be used as inks or bioinks [19]. The (bio)inks currently range from soft hydrogels such as collagen [20], gelatin [21], alginate [22], and hyaluronic acid (HA) [23] to hard thermoplastic polymers such as poly(ε-caprolactone) (PCL) [14] and polylactic acid (PLA) [24], as well as inorganic calcium phosphate cement (CPC) [25]. Because of the rheological characteristics of soft hydrogel-based biomaterials, they are suitable for 3D bioprinting and hence enable encapsulating cells or bioactive substances in the bioinks for regulating desired functions such as differentiation [26]. By contrast, thermoplastic polymer-based biomaterials are usually utilized as inks in 3D printing because of their high melting temperatures. Cells are subsequently seeded after printing rather than directly formulating within the inks to generate intrinsic structural supports within the constructs [27]. In addition, the solutions of thermoplastic polymers in organic solvents can be used as liquid inks for extrusion-based printing and helps to produce constructs such as hyperelastic bones [28]. Also, stimuli-responsive materials or hydrogels possessing shape memory properties can trigger a dynamic response to the changes in temperature, potential of hydrogen (pH), light, electric, acoustic, or magnetic field and therefore hold new possibilities to print four-dimensional (4D) constructs along with self-constructing structures, soft robotics, and medical devices because of their vast customizability and also because of the absence of additional mechanical elements [29]. A brief introduction of some commonly used biomaterial inks in 3D printing and bioprinting processes for orthopedic applications is summarized in this section.

3.2.1 Hydrogel Biomaterial Inks

Hydrogels provide a 3D hydrophilic and nutrient-rich environment similar to the native extracellular matrix (ECM) for cells [30]. This is largely because of their superior water retention capacity and their ability to encapsulate bioactive molecules. Also, the natural polymeric hydrogels are mostly bioabsorbable and have an excellent ability to integrate with the surrounding tissues along with controlled degradation rates and porosities, which is achieved via appropriate chemical modifications and cross-linking controls [31].

3.2.1.1 Collagens

Collagens, especially type I (Col-I), are the most abundant structural proteins in the human body and are also an important component of the bone tissue [32]. Research has shown that incorporating Col-I into mineralized bone cement for 3D printing could enhance their biomechanical properties, promote osteoblast adhesion, and induce mineral deposition [20, 33]. To mimic the natural bone in structure and composition, an ink was developed by dispersing deproteinized bovine bone and nano-hydroxyapatite (nHA) into Col-I for 3D bone printing [20]. The manufactured porous architecture facilitated the proliferation and osteogenic differentiation of human bone mesenchymal stem cells (hBMSCs), suggesting that it would be a prospective choice in clinical applications. To improve the angiogenesis and osteogenesis ability, Ma et al. induced the *in situ* formation of mineralized Col-I (MC) that consisted of hydroxyapatite (HAp) nanocrystals and Col-I fibers in the 3D-printed scaffolds of titanium alloy (Ti6Al4V, PT) [33]. The Young's modulus of obtained MC/PT scaffolds was similar to that of the natural bone, and their biocompatibility and osteogenic differentiation were optimal with a mass ratio of 1 : 1 between HA nanocrystals and Col-I fibers in MC. The MC/PT scaffolds could improve the vascularization of new bone tissue and also enhance their integration.

In another study by Diogo et al., a stable bioink possessing mineralized blue shark collagen and alginate was developed in order to 3D-bioprint cell-encapsulated hydrogels for bone regeneration [34]. The addition of calcium ions (Ca^{2+}) enabled *in situ* mineralization of collagen and ionically cross-linked with alginate, and it avoided the further addition of harmful cross-linking agents. The presence of shark collagen favored the survival of loaded cells during bioprinting. This work presents an inspired marine by-product valorization for 3D bioprinting of *in situ* mineralized collagen to engineer customized mineralized tissues.

3.2.1.2 Gelatin

Gelatin is prepared from collagen hydrolysis, warm water solubilization, and physical gelation below the melting temperature (about physiological 37 °C). This results in the wide use of gelatin in combination with other natural polymers [13, 21, 22]. To effectively form stable hydrogel scaffolds with proper mechanical supports, Khademhosseini and coworkers synthesized gelatin methacryloyl (GelMA) with a controllable degree of methacryloyl modification and a tunable range of mechanical and swelling properties, which could be achieved on demand by ultraviolet (UV) cross-linking [35]. Since then, GelMA has been widely utilized in bone and cartilage (bio)printing to mimic different compositions and structures of natural tissues in combination with nanomaterials or microfibers [36]. For instance, injectable osteoinductive composite hydrogels composed of GelMA and nanosilicate (nSi) were developed by Gaharwar and coworkers for bone tissue engineering [37]. The nSi could strongly interact with GelMA because of the surface charge and unique two-dimensional (2D) ultrathin shape. Shear-thinning properties were achieved through a dynamic self-assembly process. A highly cross-linked network was formed with the self-assembly of GelMA and nSi when exposed to UV irradiation (Figure 3.1a). The compressive modulus of the composite had a fourfold increment because of the incorporation of nSi with a larger pore size than that of the pristine GelMA hydrogel. The nanocomposite hydrogels were able to promote osteogenesis without any addition of osteoinductive growth factors. Also, when nSi was added to the prepolymer of GelMA, it resulted in a shear-thinning characteristic that allowed it to be 3D printed as a mechanically stiff designed complex structure after covalently cross-linking using UV irradiation (Figure 3.1b).

Subsequently, to improve the printability and structural stability of the inks, Chimene *et al.* developed a GelMA-based ink by combining ionic–covalent entanglement with nanoscale reinforcement (abbreviated as NICE) approaches for 3D bone printing [38]. This NICE ink formulation consisted of GelMA, kappa-carrageenan (κCA), and nSi. GelMA provided an environment similar to ECM containing cell adhesion motifs. However, the low-viscosity properties of GelMA resulted in a poorly printed construct. As a polysaccharide derived from seaweed, the κCA network was quite brittle after ionical cross-linking by potassium ions. Also, this κCA network was different from the elastic GelMA network because of the covalent cross-linking. An ionic–covalent entanglement network could be formed by combining GelMA with κCA (Figure 3.1c). The ductile network of GelMA could disperse stress within the damaged area during the deformation process, while the brittle network of κCA tended to be disrupted [38a]. Therefore, the mechanical strength and toughness of NICE hydrogels

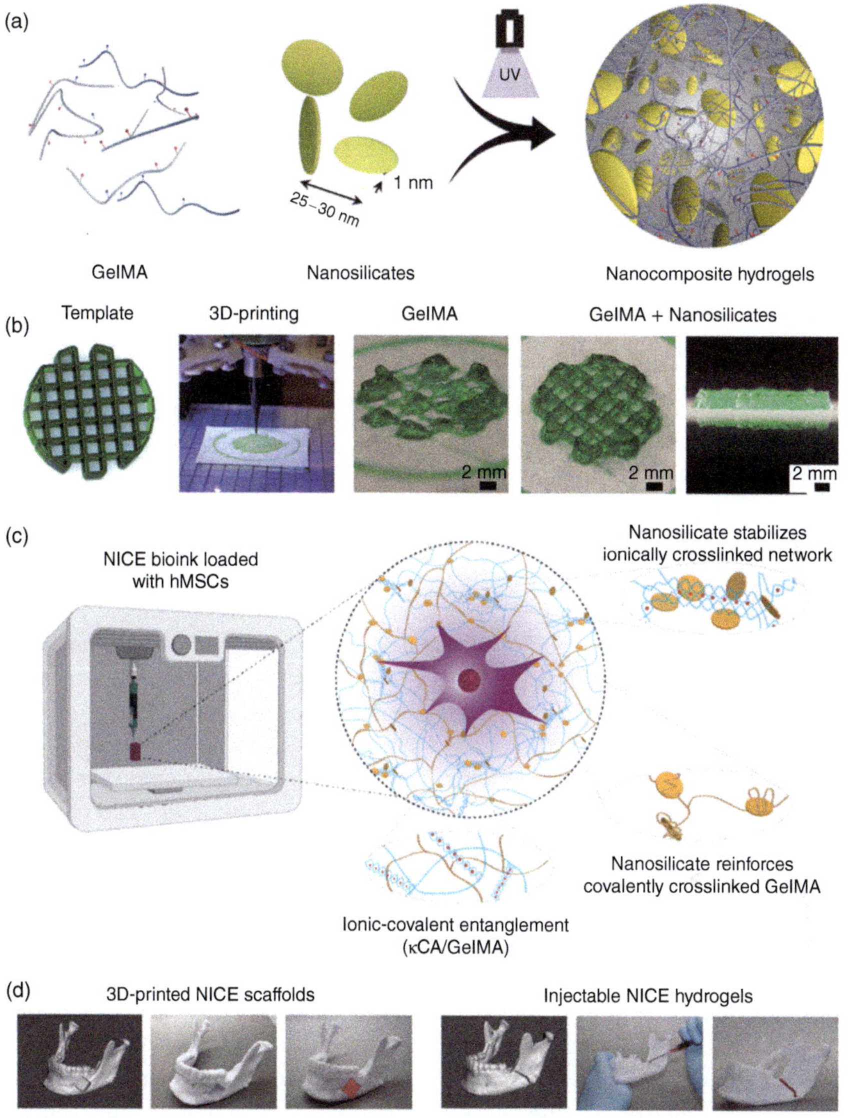

Figure 3.1 (a) Illustration of fabricating nSi-incorporated GelMA hydrogels by UV radiation. (b) Shear-thinning characteristic of nSi-incorporated GelMA hydrogels resulting in 3D printed complex structures. (c) NICE ink consisted of GelMA, κCA, and nSi. (d) The NICE ink could be 3D printed into custom scaffolds or injected into the defective sites. 3D, three-dimensional; GelMA, gelatin methacryloyl; κCA, kappa-carrageenan; NICE, nanoengineered-ionic-covalent-entanglement; nSi, nanosilicate; and UV, ultraviolet. Source: (a) Xavier *et al.* [37] / with permission of American Chemical Society and (b) Chimene et al. [38b] / with permission of American Chemical Society.

were greatly increased by combining dispersion of stress and dissipation of energy in the entanglement process [39]. Also, κCA could help in improving the viscosity of hydrogels and preventing the flow of the cooled bioinks. Finally, patient-specific, implantable 3D scaffolds were fabricated utilizing the NICE ink for craniomaxillofacial bone repair (Figure 3.1d), indicating that the NICE ink would provide a promising candidate in 3D printing of bone grafts for precision medicine.

In Diloksumpan *et al.*'s work, articular cartilage chondroprogenitor cells were laden in extrusion-bioprinted soft GelMA hydrogel mimicking the cartilage region of the bone-to-cartilage interface. This interface was combined with melt electrowriting (MEW) PCL microfiber meshes as a bone-biomimetic ceramic bioink consisting of α-tricalcium phosphate (α-TCP), nHA, and a biodegradable cross-linkable poloxamer (C-PCaP) [40]. Ceramic-integrated MEW meshes were able to be protruded into the osteochondral plug with >6.5-fold of adhesion strength. It could enable a stable connection among bone and cartilage regions by the biphasic hydrogel–bioceramic rather than pure GelMA. This 3D hybrid (bio)printing technique with multitype materials provided a meaningful platform to engineer biomimetic constructs for regenerating hard-to-soft interfaces.

3.2.1.3 Alginate

Hydrogels provide an opportunity to combine different materials for hard and soft tissues by distributing cells and bioactive factors in a specific pattern [41]. Alginate-based hydrogel scaffolds were evaluated for bone, cartilage, and muscle regeneration either with or without cell encapsulation [42]. To overcome the limitation caused due to the low viscosity of alginate for 3D extrusion-based (bio)printing while possessing a good shape fidelity, Schütz *et al.* incorporated methylcellulose (mc) into alginate, and precise plotting of volumetric constructs was obtained by using the alginate/mc (alg/mc) composite hydrogels encapsulated with hBMSCs [43]. Alginate dialdehyde (ADA) hydrogels were widely utilized to 3D-(bio)print bone [44] and cartilage [45] and in other (bio)printing applications [46]. For instance, the plotting of cross-linked ADA-gelatin (ADA-GEL) multicomponent hydrogel enabled migration, proliferation, and specific protein expression of osteoblastic and osteoclastic progenitor cells [44]. Additionally, gelatin containing cell-adherent arginine–glycine–aspartate (RGD) peptides was also utilized to improve 3D printability and to provide cell adhesion motifs within ADA hydrogels [10]. In another study, Distler et al. prepared ADA-GEL hydrogel using an extrusion-based 3D bioprinting approach for mouse myoblast (C2C12) growth [22]. They discovered that the bioprinting process generated shear forces that could enable orienting the microstructure of hydrogel, and the subsequent cell growth would be further oriented by cross-linking of the hydrogel. The study provided promising potential for engineering the muscle tissues by utilizing a developed hydrogel system inducing myogenesis.

A cell-laden alg/mc bioink was combined with pasty CPC to engineer the osteochondral tissue grafts using 3D bioplotting [47]. The shape fidelity of

3D-bioplotted alg/mc and CPC was achieved by using various needle sizes. In this study, Ahlfeld et al. prevented the formation of microcrack in CPC by establishing a post-plotting regime while conserving the viability of cells. The surface charge, microroughness, and stiffness of CPC caused the migration of hBMSCs from alg/mc. The results implied that CPC scaffolds could be produced with cell-laden hydrogels in clinically regenerating bone and its interfaces.

3.2.1.4 Hyaluronic Acid

Hyaluronic acid is the main critical polymeric glycosaminoglycan (GAG) component in articular cartilage ECM that helps in maintaining chondrocyte functions and also in the secretion of matrix components. Because of these added benefits, it was paid close attention to in 3D (bio)print of bone and cartilage [48]. In one study, to overcome the lack of physical properties and gelation abilities required for 3D extrusion-based bioprinting [49], Antich *et al.* designed chondrocyte-loaded bioink by combining 1 wt% of hyaluronic acid with 2 wt% of alginate for 3D bioprinting of implants to regenerate the articular cartilage [23]. To improve the physical behavior of the cartilage constructs, the hyaluronic acid-based bioink and PLA were further co-bioprinted, and the hybrid construct was able to facilitate the growth of chondrocytes. Cell functions of chondrocytes, deposition of a specific matrix, and formation of new tissues were improved by the hyaluronic acid-based bioink.

A collagen and hyaluronic acid hydrogel bioink containing human turbinate-derived mesenchymal stem cells (MSCs) were prepared to 3D-bioprint a multilayered 3D osteochondral construct with a predetermined PCL framework providing mechanical stability [50]. The composite hydrogel fulfilled easily bioprinting and stacking of multilayered construct, avoiding the addition of other harmful reagents or treatments for post-crosslinking. Another hyaluronic acid-based osteoblast-laden hydrogel bioink was developed along with poly(ethylene glycol)-diacrylate (PEGDA)/laponite XLG nanoclay ($[Mg_{5.34}Li_{0.66}Si_8O_{20}(OH)_4]Na_{0.66}$) for dual-channel 3D bone bioprinting (Figure 3.2) [51]. The bioink A that consisted of poly(ethylene glycol) (PEG)-clay nanocomposite hydrogel was extruded from one channel, facilitating oxygen and nutrient delivery, whereas the bioink B, which was encapsulated with rat osteoblasts (ROBs) in 20 wt% of hyaluronic acid solution, was extruded from the other channel. The ROBs were uniformly dispersed in the hyaluronic acid component and were protected for high cell viability when crosslinked under UV irradiation. The proliferation and differentiation of ROBs in the bioprinted composite scaffolds were better than those of scaffolds without hyaluronic acid.

Also, the methacrylated hyaluronic acid (HAMA) is a synthetic hybrid polymer generated by photochemically crosslinking hyaluronic acid and methacrylate under a UV light source showing altered physical properties than that of hyaluronic acid [52]. The osteocyte-laden hybrid bioinks with different proportions of HAMA and GelMA have been evaluated for 3D bioprinting of cartilage and bone tissues [53]. For instance, a formulation of HAMA or GelMA bioinks was used for stereolithography bioprinting that also supported the formation of cartilage ECM and chondrocyte

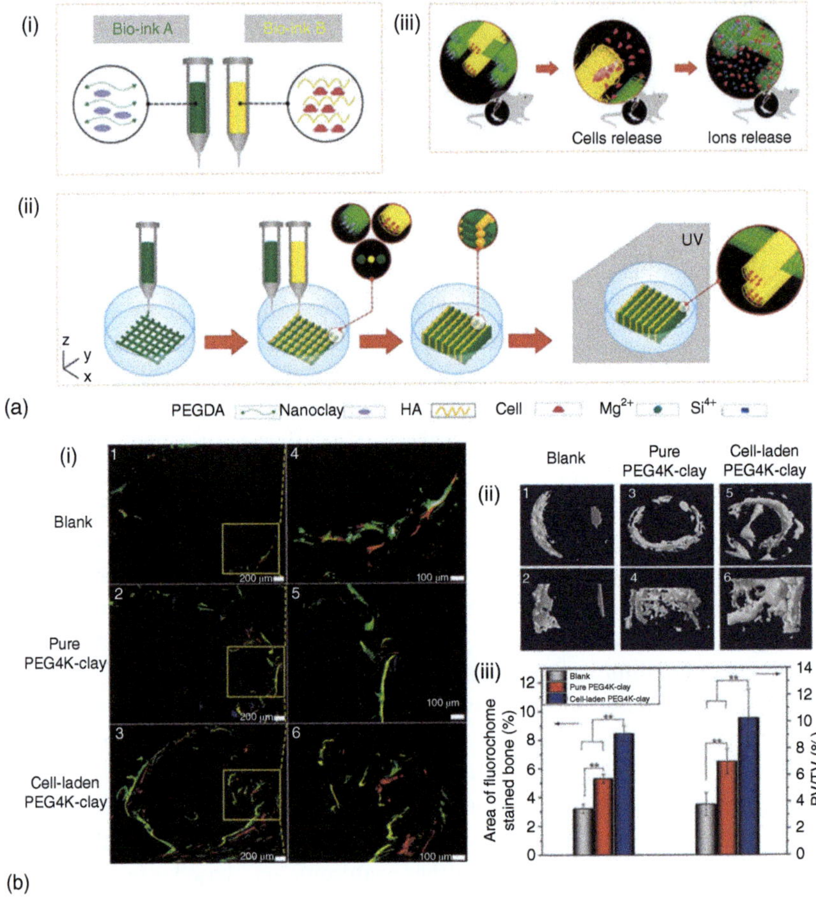

Figure 3.2 (a) Illustration of dual-channel bioink system for 3D bioprinting and *in vivo* implantation. (i) Bioinks A and B, (ii) 3D bioprinting of ROB-laden constructs, and (iii) *in vivo* implantation. (b) Fluorescence labeling of 3D-bioprinted scaffolds containing ROBs, (i) tetracycline hydrochloride (yellow), alizarin red S (red), and calcein (green) labeling. (ii) 3D micro-CT analysis of implants and bone formation. (iii) Area percentages of fluorochrome-stained bones and BV/TV ratios (**p <0.01, blank as a control). 3D, three-dimensional; BV/TV, bone volume/total volume; HA, hyaluronic acid; micro-CT, micro-computed tomography; PEG4K, poly(ethylene glycol) Mw-4000; and ROB, rat osteoblast. Source: Zhai *et al.* [51] / with permission of John Wiley & Sons.

phenotype recovery [54]. Besides, Yang *et al.* have also developed a 3D-bioprinted osteocyte-laden hydrogel composed of Col-I, GelMA, and HAMA for biomimetic mineralization [53]. A high level of density and viability of the osteocyte cell line could be achieved within this bioprinted composite construct.

3.2.2 Thermoplastic Polymeric Inks

The printable thermoplastic polymers that possess high stress and melts upon extrusion through a nozzle have gained increasing attention for manufacturing sophisticated 3D constructs in bone tissue engineering [55]. After being extruded, the polymer rapidly solidifies and holds its shape as the filament cools, and at the same time, a sharp increase in viscosity is observed. Later in this section, some thermoplastic polymeric materials are summarized for the applications in 3D bone printing.

3.2.2.1 PCL

PCL is unarguably one of the most popular polymeric biomaterials in 3D printing of the bone because of its good biocompatibility, suitable rheology, and viscoelasticity, as well as proper degradability [56]. Even though 3D-printed PCL scaffolds facilitate certain physical supportability and porosity, the lack of biological activities makes them not suitable for promoting cellular functions [57]. To improve bone regeneration outcomes, Li et al. fabricated a 3D-printed PCL scaffold and then immobilized liposomes loaded with aspirin (Asp@Lipo) together with bone-forming peptide-1 (BFP-1) on the surfaces of the scaffolds [14]. *In vitro* experiments and *in vivo* cranial defect repair showed that a 3 : 7 ratio of Asp@Lipo/BFP-1 exhibited the best osteogenic efficiency of hBMSCs. Furthermore, the mechanism exploration using RNA-sequencing (RNA-Seq) analysis implied that these composite scaffolds improved bone regeneration via the signaling pathway of phosphatidylinositol 3 kinase/protein kinase B (PI3K/AKT).

In another study by Yeo et al., multilayered PCL struts were fabricated for developing a hierarchical scaffold by combining alginate bioink loaded with cells and PCL micro/nanofibers to investigate whether the myoblast proliferation and myotube formation were affected by the topology and bioprinting ability [58]. The authors first prepared a PCL hierarchical structure with 3D-melt-plotted PCL struts that provided physical support and then electrospun aligned PCL fibers on the printed struts. Subsequently, the C2C12 myoblast-embedded alginate bioink was bioprinted onto the hierarchical PCL structure by using polyethylene oxide (PEO) as a sacrificing component in the bioink in order to generate channels efficiently releasing the laden cells. Apart from the good growth of myoblasts on the cell-laden scaffolds, fiber alignment strongly influenced the morphology of the proliferated cells as well as the formation of myotubes. This work provided a new platform to combine scaffold design with a novel cell-bioprinting strategy for myogenic tissue regeneration. Besides, as the United States Food and Drug Administration (FDA)-approved organic and inert polymer for certain devices, PCL was utilized to develop a new biohybrid material by combining with a morphogenetically active ink containing calcium polyphosphate microparticles (Ca-polyP-MPs) for 3D printing of soft (cartilage, tendon, or ligament)-to-hard (bone) tissue transition regions [59]. The granular hybrid PCL/Ca-polyP-MP was quickly heated to 100 °C for fabricating 3D-printed scaffolds, and it was found that 400 μm of thickness as well as roughly 0.5-mm^2 size of open holes were beneficial for cell migration. It was observed that 1.60 ± 0.1 GPa of Young's modulus and 153 ± 28 MPa of Martens hardness for the 3D-printed morphogenetic scaffolds were comparable with those of trabecular and cortical bones. The growth of human osteosarcoma cells (SaOS-2) was promoted by the PCL-based composite scaffolds, indicating the extreme suitability of PCL/Ca-polyP-MP for fabricating printable bone scaffolds.

3.2.2.2 PLA

As approved by the FDA for some biomedical applications, PLA is another widely used synthetic polymer for 3D printing of bone grafts owing to its biocompatibility and physicochemical characteristics similar to PCL [60]. Fairag *et al.* designed and manufactured 3D-printed PLA scaffolds by utilizing a desktop 3D printer [24]. Among the pore sizes of 500, 750, and 1000 μm, the 750-μm scaffolds showed the best cytocompatibility and bioactivity for primary human osteoblasts as well as strong osteogenic differentiation capacity for hBMSCs. The results indicated that the prepared PLA scaffolds could be an on-site customized product for push-fitting in bone defect for hBMSC infiltration, adhesion, proliferation, and formation of new bone.

Velioglua et al. fabricated a 3D-printed PLA bone substitute with proper pore sizes, porosity, and appropriate mechanical properties that were similar to the native trabecular bone [61]. The authors found that the increase of pore dimensions from 0.5 to 1.0 and 1.25 mm decreased the compressive moduli and contact angles of the 3D-printed scaffolds. The highest proliferation ratio of hBMSCs was observed on the pore surface of the scaffold with a 1.25 mm of pore size. However, the analysis of compressive mechanical property demonstrated that the 3D-printed PLA scaffolds exhibited lower modulus values compared with that of the natural bovine bone. The ingrowth of host cells into the printed scaffold pores and the production of ECM components would further be benefitted from the physical behavior of the new bone structure. Therefore, the 3D-printed PLA scaffold is a potentially outstanding candidate substitute in bone repair. Also, a bioink of alginate hydrogel reinforced by PLA nanofibers containing human adipose-derived stem cells (hADSCs) was developed for engineering musculoskeletal soft tissue grafts using pneumatic 3D extrusion-based bioprinting [62]. These innovative patient-specific constructs could facilitate specific needs of deficient meniscuses by manipulating and controlling the alignment and distribution of PLA nanofibers in the 3D-bioprinted scaffolds and also promote the formation of anisotropically oriented ECM mimicking the natural fibrous musculoskeletal soft tissues.

3.2.2.3 Polyetheretherketone (PEEK)

Polyetheretherketone (PEEK) belongs to the polyaryletherketone family with excellent thermostability, mechanical properties, and biocompatibility [63]. The possession of thermoplastic characteristics and a comparable modulus to the human cortical bone make the PEEK materials suitable for applications in 3D printing of bone implants for personalized medicine. Tseng *et al.* developed a screw-based extrusion system, which achieved controllable layering of melted PEEK onto a heating substrate [64]. The nozzle temperature ranged from 370 to 390 °C, while the plate temperature was maintained at 280 °C, which was considered ideal for maintaining the extrusion rate at 1.5 g/min. The 3D-printed porous intervertebral cages utilizing this system were the potential candidates for orthopedic utilities.

Additionally, the flexural strength of 3D-printed PEEK grafts was improved by the incorporation of short carbon fibers (SCFs) into PEEK composites [65]. In the 3D printing process, the SCFs achieved the uniform nucleation of PEEK, with 17.7% higher flexural strength for vertically printed SCF/PEEK composites than that of the horizontally printed ones. To overcome the inertness of pure PEEK, and also for the integration with the host bone (for efficient regeneration), HA was used to produce 3D-printed PEEK/HA composites [66]. The HA particles were uniformly distributed in the composites. The increase in HA content decreased the strength of the PEEK/HA composites while increased their modulus. In comparison with pure PEEK, the 30 wt% of HA in PEEK resulted in 48.2% lower tensile strength and 68.6% higher tensile modulus. For clinical applications, Chen et al. reported that the 3D-printed PEEK prosthesis was effective for reconstructing chronic clavicle osteomyelitis and healing bone defects [67]. In this case, the 3D-printed PEEK prosthesis was positioned and fixed into the desired site of clavicle resection. During two years of continuous follow-up, the patient reported satisfied pain relief, as well as significant improvement of appearance and shoulder function.

To sum up, the thermoplastic material characteristics and mechanical properties of these polymeric inks make them suitable for applications in 3D (bio)printing of bone implants. However, when compared with hydrogel biomaterials, their inability to encapsulate cells hinders their potential applications in building tissue-like constructs via 3D bioprinting techniques [68]. In contrast, hydrogels contain a large amount of water in their 3D networks that support cell viability and growth. Yet, from another perspective, the thermoplastic polymers are usually mechanically strong and can easily be further strengthened, modified, or doped with other bioactive substances to achieve mechanical properties that are much more physiological than hydrogel biomaterials [69].

3.2.3 Calcium Phosphate (CaP) Ceramic Inks

In addition to the organic matrix of Col-I, calcium phosphate (CaP) mineral is the main inorganic component of native bone with an exquisite structural organization that can be imitated by utilizing a 3D (bio)printing technique [25]. The extruded CaP cement or slurries that are passed through a nozzle without heating can be solidified under mild conditions [25]. For example, the biocompatibility, osteoconductivity, and self-setting capacity by hydrolysis of α-TCP make it suitable for developing customized printable bone cement [70]. In one previously mentioned study, to overcome the high solubility and fast degradation of α-TCP, nHA was incorporated into α-TCP and biodegradable C-PCaP in order to form a printable bioceramic setting at the physiological temperature that mimicked the mineral phase of bone [40]. Moreover, in this study, a mechanically reinforced bone-to-cartilage interface was produced by combining extrusion-based printing of the bioceramic ink shaped as a subchondral bone substitute anchoring with spatially organized MEW PCL polymeric microfibers, followed by embedding in the GelMA hydrogel loaded with cells to represent the cartilage component.

On the other side, β-tricalcium phosphate (β-TCP) is another extensively utilized CaP ceramic in 3D bone printing benefitting from its biocompatibility, osteoconductivity, and possession of similar chemical composition with native bone apatite [71]. However, the *in vivo* bone formation and degradation rate of β-TCP is so slow that its clinical application has been restricted [72]. To this end, the rapid degradation of calcium silicate (CS) makes it possible to not only control the degradation rate of β-TCP/CS composite but to also stimulate osteogenic differentiation of osteoblasts [73]. For instance, Dong *et al.* fabricated a 3D-printed porous β-TCP/CS hybrid scaffold possessing controllable inside and outside structures [74]. The shrinkage, *in vitro* mineralization ability, and degradation rate of the printed β-TCP/CS scaffolds were improved by the introduction of CS within the composite. Furthermore, the cell attachment, spreading, and osteogenic differentiation of mouse bone mesenchymal stem cells (mBMSCs) of β-TCP/CS hybrid scaffolds were promoted when compared with the pristine β-TCP scaffolds.

The formulations of pasty CPC can also be (bio)printed as different designed shapes through extrusion at low temperature in which the bioactive growth factors can also be integrated without affecting their biological activities, while cell functions may be influenced when encapsulated in a solid and stiff CPC matrix. Therefore, Ahlfeld et al. combined 3D-plotted CPC with bioprinted hBMSC-alg/mc bioink to generate a biphasic scaffold as an osteochondral tissue graft [47]. This study prevented the formation of microcrack inside the CPC strands conserving cell viability and also made it possible to produce CPC scaffolds with controlled seeding and survival of cells for engineering bone and its interfaces.

In addition to the above-mentioned CaP ceramics, HA is also a highly bioactive ceramic and is widely used as one component of the bioceramic ink formulations for 3D bone printing [20, 28, 75]. Furthermore, biphasic calcium phosphates (BCPs) containing HA and TCP also exhibit a promising potential capacity for bone remodeling [76]. In this study, a novel powder system consisting of HA and TCP was developed for 3D printing of complex-shaped BCP scaffolds exhibiting >96.5% of dimensional accuracy and >300 μm of macropore size. The improved shape fidelity and tunable phase composition of the BCP scaffolds provided a candidate for bioactive bone implants that could be utilized for the reparation of non-load-bearing defects.

3.2.4 Supramolecular and Stimuli-Responsive Inks

Even with the advantages of resembling native ECM and tunable performance, the intrinsic poor mechanical properties of previously mentioned hydrogels still restrain their further application in load-bearing bone regeneration [77]. Meanwhile, supramolecular polymer hydrogel has emerged as a high-strength ink in printing load-bearing scaffolds because of its outstanding characteristics of the extrudable and 3D-printable monomer solution. This was due to the fact that the quick gelation of the extruded solution could self-support the layer-by-layer construct, form tight bounding of the multilayer stacks that helped to prevent delamination, and provide enough stability to the printed hydrogel-based scaffolds in the physiological aqueous condition and aid in avoiding further swelling for bear-loading support [78]. For example, Dai *et al.* have developed poly(*N*-acryloyl glycinamide) (PNAGA) hydrogel, which was a supramolecular polymer with dual pendant amide motifs, and the hydrogen bond-assisted crosslinking resulted in improved physical properties and stable swelling behavior [78c]. The same authors further incorporated nanoclay (Laponite XLG) in the PNAGA hydrogel to 3D print a high-strength nanocomposite scaffold (Figure 3.3) [79]. The viscosity of this hybrid ink could be finely tuned by the addition of nanoclay in order to prevent the collapse of printed multilayered scaffolds. After polymerizing under UV light irradiation, the PNAGA-clay composite hydrogel scaffolds achieved high strengths and stable swelling, which was due to the interactions of a dual amide hydrogen bond with the crosslinked nanoclay in order to provide reliable loading support. The Young's modulus, tensile strength, and break stain of the PNAGA 30% clay hydrogel increased up to 0.18 MPa, 1.17 MPa, and 1300%, respectively. The compressive strengths of PNAGA-clay composite hydrogels were up to 3.50 MPa in comparison with that of pristine PNAGA 30% at 2.66 MPa. Magnesium ions (Mg^{2+}) and silicon ions (Si^{4+}) could be sustainably released from the PNAGA-clay scaffolds, and the osteogenic differentiation of ROB cells could be correspondently promoted.

The new bone formation could efficiently be elicited by implanting the PNAGA-clay scaffolds into rat tibia defects. Findings from this study presented new strategies to develop and strengthen bioactive inks for 3D printing of load-bearing scaffolds and devices.

The 4D (bio)printing techniques that are generated by integrating time as the fourth dimension within 3D (bio)printing techniques are facilitating the fabrications of complex, functional dynamic biological architectures by using stimuli-responsive biomaterial inks in the next generation of bone tissue engineering [29a, b, 80]. The stimuli-responsive biomaterials used as (bio)inks in the 4D (bio)printing technique mainly include physical, chemical, and biological responsive materials [9, 80]. A series of stimuli-responsive shape recovery materials have been widely investigated as suitable scaffolds and hydrogel (bio)inks for bone regeneration [12]. For instance, to treat the irregular shape of bone critical-size defect by minimal invasive surgery, Wang et al. incorporated osteogenic peptide and black phosphorus (BP) nanosheets into a β-TCP/poly(lactic acid-*co*-trimethylene carbonate) photothermal-responsive shape memory scaffold by using 4D printing [81]. The shape reconfiguration of the customized scaffolds was achieved by rapidly increasing the temperature to 45 °C from on-demand near-infrared irradiation that enabled irregular bone defects to be easily and precisely fitted. After implantation, the temperature of the scaffolds rapidly decreased to 37 °C, displaying mechanical properties similar to those of the human cancellous bone. The scaffolds showed a potential platform for treating irregular-shaped bone defects.

The transforming shape and function in 4D (bio)printing would provide promising potential in designing and controlling the (bio)printed constructs with desired

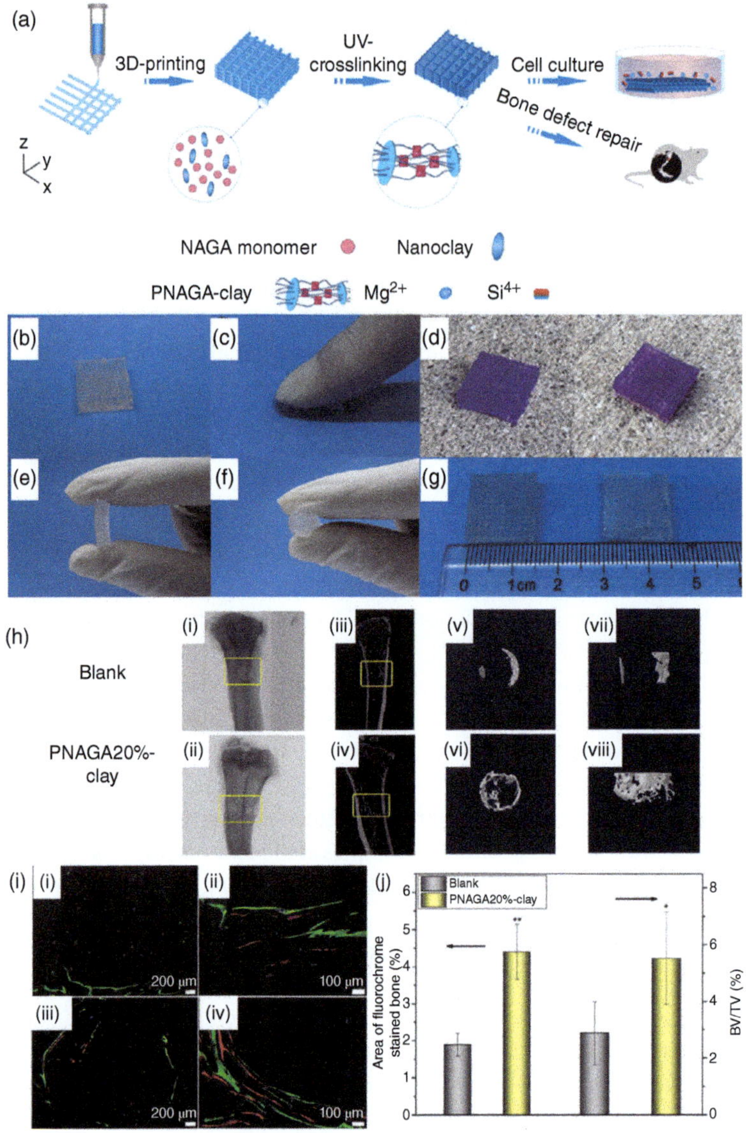

Figure 3.3 (a) 3D printing of PNAGA20%-clay scaffolds and resistance to compression by (b and c) finger, (d) car wheel, and (e and f) hand folding. (g) The scaffolds were stable when immersing in water for three months (left, before; right, after immersing). (h) (i and ii) Micro-CT, (iii and iv) micro-CT reconstruction images, and (v–viii) reconstructed 3D models of implants and new bone formation. (i) Fluorescence labeling of (i and ii) blank and (iii and iv) PNAGA20%-clay. Colors: tetracycline hydrochloride (yellow), alizarin red S (red), and calcein labeling (green). (j) Area percentages of fluorochrome-stained bones and new BV/TV ratios ($^{*}p < 0.05$, $^{**}p < 0.01$, blank as a control). 3D, three-dimensional; BV/TV, bone volume/total volume; Mg^{2+}, magnesium ion; micro-CT, micro-computed tomography; NAGA, *N*-acryloyl glycinamide; PNAGA, poly(*N*-acryloyl glycinamide); Si^{4+}, silicon ion; and UV, ultraviolet. Source: Zhai *et al.* [79] / with permission of American Chemical Society.

specific unique characteristics fulfilling the requirements in bone tissue engineering and clinical orthopedic applications. Still, more detailed studies on biomaterials and fabrication techniques for 4D (bio)printing are needed.

3.3 3D Printing and Bioprinting Techniques

Extrusion-based, inkjet-based, and vat polymerization-based (bio)printing are the three commonly used 3D (bio)printing techniques for orthopedic applications [6a, 15]. Extrusion-based 3D (bio)printing technique enables the melted polymeric materials, polymer solutions, ceramic slurries, or hydrogel (bio)inks to be extruded via nozzles that are driven by pneumatic or mechanical pressures [28, 55, 82]. Bio(ink) droplets are deposited by the thermally or electrically driven nozzles in the inkjet-based 3D (bio)printing technique [16]. In vat polymerization-based (bio)printing, a light source such as laser, UV, or other are used to crosslink, melt, or sinter the biomaterial inks in the vat for generating the designed constructs [7b, 26, 83]. To create bone tissues or related tissue interfaces, some other techniques are usually combined for producing complex or biomimetic spatial structures, such as electrospinning, integrated organ printing (IOP) systems, and microfluidics, among others [7a, 84]. The 3D (bio)printing techniques used for orthopedic applications will be briefly summarized in this part.

3.3.1 Extrusion-Based (Bio)printing

Extrusion is one of the most utilized (bio)printing techniques that compromises a robotically moving printhead and/or stages usually in the x–y–z direction [17]. In the extrusion-based process, melted polymeric materials, ceramic pastes, or slurry materials that are dissolved in organic solvents or cell-laden hydrogel bioinks can be continually flowed from the extrusion nozzle(s) and are deposited in a layer-by-layer manner to form a designed spatial 3D structure [85]. Despite the advantages, it currently possesses several limitations including but not limited to relatively low resolutions in the range of over 100 μm [6a, 49], and cell deformation induced by shear stresses that can be overcome by regulating parameters such as bioink behaviors, extrusion pressure, and loaded cell density [49].

With the rapid development of 3D (bio)printing techniques, extrusion-based systems have enabled both the material melting and non-melting process [85, 86]. The melting process mainly includes powder melt extrusion [76, 86b], fused deposition modeling (FDM) [87], and fiber deposition [58, 75a], among others. The biomaterial inks can also be extruded by heating rather than melting in the non-melting process, such as 3D bioplotting [20, 88], direct ink writing (DIW) [40, 89], solid or solvent-based extrusion free-forming (SEF) [28, 90], and others.

In one study, biodegradable thermoplastic PLA microstructure was 3D printed using the FDM technique, and it was further infused by the UV-induced chemically crosslinked GelMA hydrogels laden with hADSCs and incorporated with RGD cyclic peptide-conjugated gold nanoparticles (RGNPs) for reinforcing mechanical properties and enhancing cellular behaviors as well as bone remodeling [91]. The compressive moduli for PLA, GelMA/PLA, and GelMA/GNPs/PLA scaffolds were 401.0 ± 29.2, 378.9 ± 27.7, and 402.0 ± 26.9 MPa, and they were similar to the human native mandibular bones (431.0 ± 217.0 MPa). The addition of RGNPs in the composite constructs enhanced cellular behavior, as indicated by better cell attachment, migration, and growth. The findings from this study demonstrated that the ADSC-encapsulated composite hydrogels could regulate the function of stem cells and regeneration of bone defects by providing 3D biomimetic structures with mechanical support and osteogenesis potential.

Yeo et al. 3D printed microscale PCL struts by extruding melt PCL from a heating cartridge by using a 400 μm inner diameter nozzle, which was driven by 240 ± 10 kPa of pneumatic pressure and moved at a speed of 5 mm/s [58]. Then, PCL dissolved in a 4 : 1 ratio of methylene chloride and dimethylformamide was electrospun by ejecting the solution through 0.20 ml/h of flow rate and 12 kV of voltage. Finally, alginate/PEO loaded with C2C12 cells was 3D bioprinted on PCL struts to form a hierarchical structure (Figure 3.4). The generated micro/nanotopological cues and efficient cell release of the scaffolds facilitated potential myogenic tissue regeneration.

3D (bio)plotting is widely used to develop bone or cartilage constructs with pasty CPC or hydrogel formulations using low-temperature extrusion-based AM techniques [92]. For instance, CPC and a cell-laden alg/mc bioink were 3D bioplotted using a 410-μm inner diameter of the needle to produce a 250-μm layer thickness of osteochondral tissue grafts (Figure 3.5) [47]. The formation of microcrack in the CPC strands could be prevented by optimizing the post-plotting regime while maintaining the cell vitality. The biphasic osteochondral scaffolds enabled a very minor cell viability decrease only when directly contacted at the CPC and alg/mc interface. It demonstrated that the CPC ceramic would be an alternative ink to the thermoplastic polymers in 3D (bio)plotting for bone regeneration. Moreover, a fiber deposition technology was utilized at room temperature to 3D print bioactive glass/chitosan nanoparticle (BG/CSn) composite porous scaffolds of 350 nm diameter of BG microspheres as the printing material and bound with polyvinyl alcohol (PVA) [93]. Also, NEL-like type I molecular-1 DNA and monkey bone mesenchymal stem cells (BMSCs) were co-loaded to generate a bone implant for the bone defect reparation in rhesus monkeys. The homogeneously dispersed BG microspheres and the formation of a porous structure with connected pores contributed to nutrition and metabolic waste transportation and thus promoted the bone tissue regeneration as observed by radiography (X-ray) and microcomputed tomography (micro-CT) analyses of newly formed bone mass, density, hardness, and structure. In another hydrogel-based 3D plotting work, Zhai et al. printed a nanocomposite ink comprising NAGA monomer and nanoclay using a three-axis precisely positioned system (Figure 3.3) [79]. The scaffolds were fabricated using 25–35 kPa of syringe pump pressure, 15 mm/s of moving speed, 250 μm of nozzle size, approximately 1 mm of strand distance, and 10 layers were deposited. The

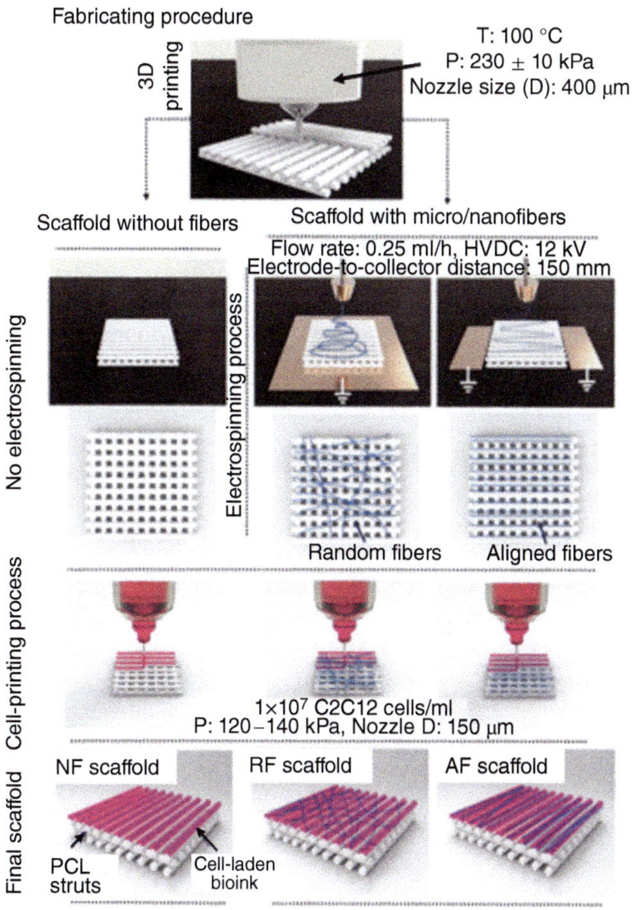

Figure 3.4 The fabrication process of melt-plotted PCL struts and inserting fibers through an electrospinning process, as well as cell printing using an alginate/PEO bioink loaded with mouse myoblasts. 3D, three-dimensional; AF, aligned fibers; C2C12, mouse myoblasts; D, diameter; HVDC, high-voltage direct current; NF, nanofibers; P, pressure; PCL, poly(ε-caprolactone); PEO, polyethylene oxide; RF, random fibers; and T, temperature. Source: Yeo *et al.* [58] / with permission of Institute of Physics (IOP) Publishing Group.

shear-thinned pre-gel solution in the printed cartridges was easily extruded on the printing substrate, and its viscosity was quickly recovered. Afterward, the pre-scaffold was crosslinked by polymerization under UV light illumination for 30 minutes at 365 nm.

MEW of spatially organized PCL microfibers was integrated with extrusion-based printing of a ceramic ink for multimaterial 3D printing of a bone-to-cartilage interface [40]. The printable bioceramic ink setting at physiological temperature as bone support was composed of α-TCP, nHA, and crosslinkable poloxamer. This enabled directly printing within the MEW-fabricated PCL meshes, preserving their microarchitecture (Figure 3.6). GelMA hydrogels encapsulating with articular

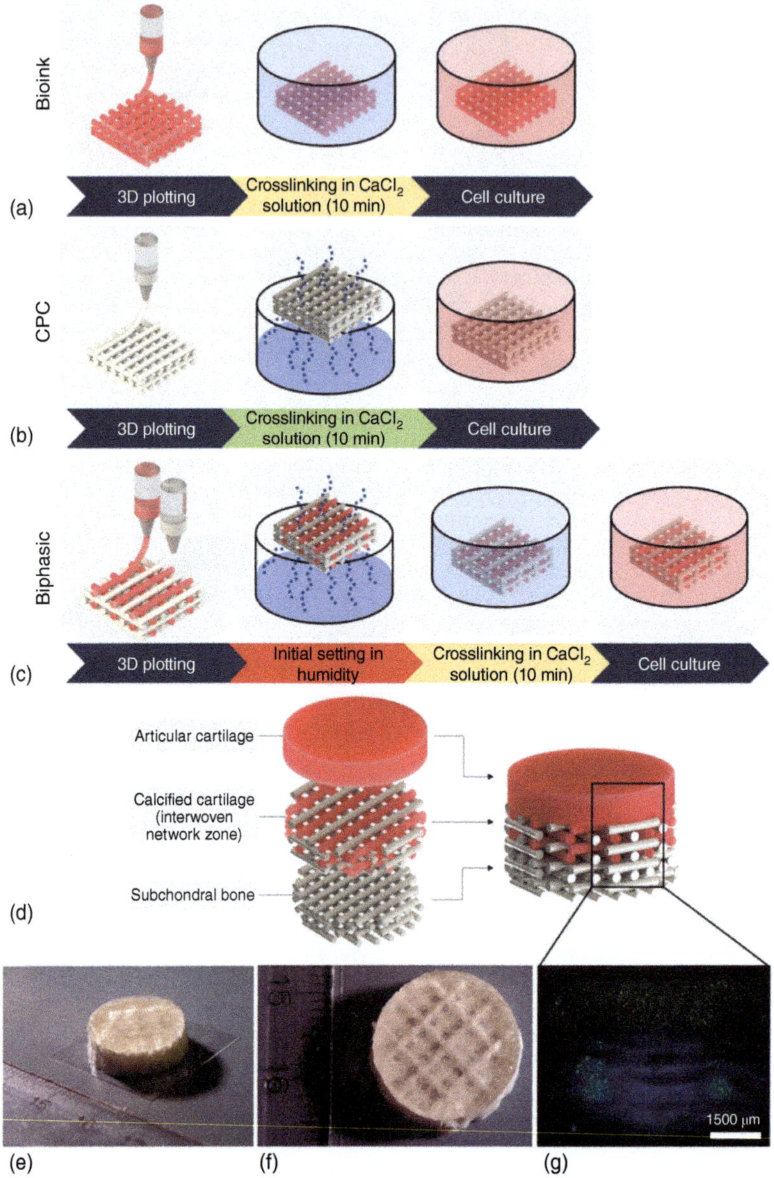

Figure 3.5 (a–c) Fabrication of different scaffolds. (a) 3D plotting of alg/mc scaffolds, crosslinking by immersing in $CaCl_2$, and cell culture. (b) 3D plotting of CPC scaffolds, incubation in humidity, and cell culture. (c) Fabrication of cell-laden alg/mc and CPC biphasic scaffolds, incubation in humidity, crosslinking in $CaCl_2$, and cell culture. (d–g) Osteochondral defect repair model. (d) Illustration of 3D-plotted osteochondral scaffolds consisted of three zones. (e, f) Side and top views of the 3D-plotted scaffolds. (g) Cross section of a fluorescently labeled osteochondral scaffold (bioink, green; CPC, blue). 3D, three-dimensional; alg/mc, alginate/methylcellulose; $CaCl_2$, calcium chloride; and CPC, calcium phosphate cement. Source: Ahlfeld *et al.* [47] / with permission of IOP publishing.

cartilage chondroprogenitor cells could be embedded within the cartilage plug of ceramic-integrated MEW meshes achieving >6.5-fold of adhesion strength for the hydrogel-to-ceramic and hence enabling stable structure for implanting in osteochondral defects. In addition to that, the DIW technique was also evaluated for direct paste-writing based on the layer-wise extrusion of moldable liquid or paste inks using high-consistency enzymatically fibrillated cellulose nanofibers [89].

A hyperelastic "bone" containing 90 wt% of HA and 10 wt% of PCL or poly(lactic-*co*-glycolic acid) (PLGA) was 3D printed at a high speed of 275 cm^3/h by extruding liquid inks at room temperature using the SEF technique (Figure 3.7) [28]. The composite ink of HA, PCL, and PLGA was stirred to evaporate the solvent and achieve the desired viscosity of 30–35 Pa s for 3D printing. The hyperelastic bone showed ~32–67% of strain to failure, ~4–11 MPa of elastic modulus, 50% of porosity, and facilitated the viable, growth, and osteogenic differentiation of hBMSCs. Furthermore, a negative immune response was elicited by the hyperelastic bone together with desirable vascularization, quick integration with the surrounding tissues, and ossification, as well as new bone formation in the absence of additional bioactive factors.

Additionally, some other techniques including electrospinning, the IOP system, and microfluidics have been combined to produce complex or biomimetic musculoskeletal transition structures or interfaces [94]. In one of these studies, Criscenti et al. fabricated triphasic scaffolds by the combination of a 3D-printed PCL structure with PLGA electrospun nanofibers. It mimicked the bone–ligament interface along with a linear gradient of material and also mimicked the mechanical and structural properties to regenerate the bone–ligament interfaces (Figure 3.8) [95]. Observation from scanning electron microscopy (SEM) demonstrated that 208 ± 7.1 μm of fiber diameter, 734 ± 15 μm of fiber spacing, 202 ± 9 μm of layer thickness, and $77.1 \pm 1.7\%$ of porosity for the PCL region were achieved (Figure 3.8b(i)). The fiber diameter of the PLGA region was 0.588 ± 0.02 μm with a porosity of $95.9 \pm 1.6\%$ (Figure 3.8b(ii,iii)). Interconnected porous structures were obtained possessing a porosity of $80.2 \pm 2.7\%$ for the interface region without layer delamination. The biological response of hBMSCs was elicited by the gradient of physical and mechanical properties. The results indicated that integrating electrospinning with AM would be a potential platform to fabricate scaffolds for regenerating bone-to-ligament tissue interfaces by structurally mimicking the native tissues.

The IOP technology provides another option to produce engineered tissue grafts by structurally and functionally mimicking the natural musculoskeletal tissues [7a]. In this study, Merceron *et al.* used the IOP system to deposit four different component inks for fabricating a construct of muscle tendon unit (MTU). As shown in Figure 3.9, the elastic muscle side was developed by co-bioprinting of thermoplastic polyurethane (PU) with hydrogel bioink encapsulated with C2C12 cells. On the other side, the stiff tendon was prepared by co-bioprinting of PCL with hydrogel bioink laden with mouse embryonic fibroblasts (NIH/3T3). Precise shape fidelity was found along with the elastic PU-C2C12 muscle side exhibiting 0.39 ± 0.05 MPa

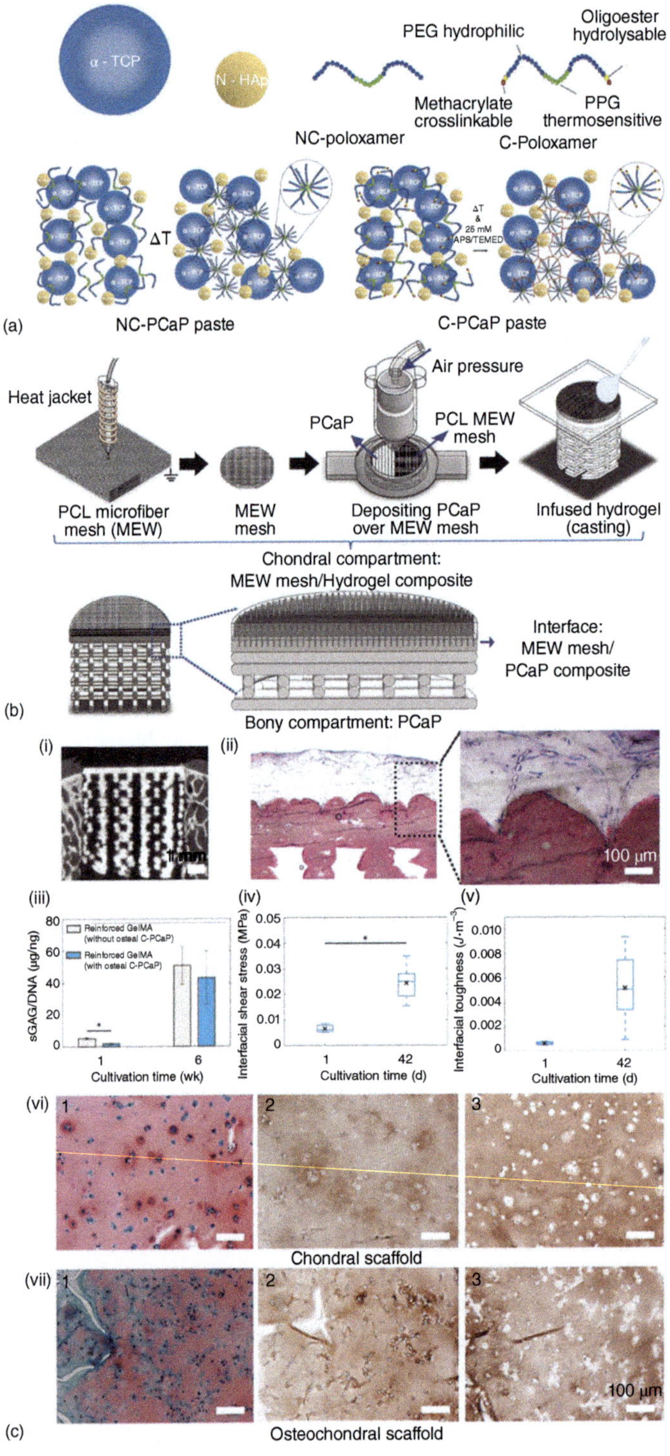

α-TCP
N-HAp
PEG hydrophilic
Oligoester hydrolysable
Methacrylate crosslinkable
PPG thermosensitive
NC-poloxamer
C-Poloxamer
ΔT
(a)
NC-PCaP paste
C-PCaP paste
Heat jacket
Air pressure
PCaP
PCL MEW mesh
PCL microfiber mesh (MEW)
MEW mesh
Depositing PCaP over MEW mesh
Infused hydrogel (casting)
Chondral compartment: MEW mesh/Hydrogel composite
Interface: MEW mesh/ PCaP composite
(b)
Bony compartment: PCaP
(i)
(ii)
100 µm
(iii)
(iv)
(v)
Reinforced GelMA (without osteal C-PCaP)
Reinforced GelMA (with osteal C-PCaP)
sGAG/DNA (µg/ng)
Cultivation time (wk)
Interfacial shear stress (MPa)
Cultivation time (d)
Interfacial toughness (J·m⁻³)
Cultivation time (d)
(vi)
Chondral scaffold
(vii)
100 µm
(c)
Osteochondral scaffold

Figure 3.6 (a) Material composition of PCaP pastes. (b) Fabricating process of the osteochondral construct through combining different 3D printing techniques. (c) Cartilage deposition in the osteochondral construct. (i) Micro-CT image of the PCaP scaffold; (ii) basic fuchsin and methylene blue staining of the C-PCaP scaffold; (iii) quantification of GAG per DNA content; (iv) the interfacial adhesion strength; (v) interfacial toughness; (vi-1 and vii-1) Safranin-O staining, (vi-2 and vii-2) Col-II immunostaining, and (vi-3 and vii-3) Col-I immunostaining, (vi) without or (vii) with C-PCaP, cultivated for 42 days. $^{*}p < 0.05$. 3D, three-dimensional; α-TCP, alpha-tricalcium phosphate microparticles; Col-I, collagen type I; Col-II, collagen type II; C-PCaP, crosslinkable printable calcium phosphate; DNA, deoxyribonucleic acid; GAG, glycosaminoglycan; GelMA, gelatin methacryloyl; MEW, melt electrowriting; micro-CT, micro-computed tomography; NC-PCaP, non-crosslinkable printable calcium phosphate; N-HAp, nano-hydroxyapatite; PCaP, printable calcium phosphate; PCL, poly(ε-caprolactone); PEG, poly(ethylene glycol); and PPG, poly(propylene glycol). Source: Diloksumpan *et al.* [40] / Institute of Physics (IOP) Publishing Group / CC BY 4.0.

of elastic modulus, the stiff PCL-3T3 tendon side showing 46.67 ± 2.67 MPa of Young's modulus, and 1.03 ± 0.14 MPa of Young's modulus for a region of the interface. The embedded cells exhibited good viability (>80%) at one and seven days with extremely aligned muscle and tendon morphological characteristics, as well as upregulated MTU-associated gene expressions. It demonstrated that the IOP system was versatile to produce MTU tissue grafts.

In another converging study, to mimic the cartilage ECM in organization and composition, Idaszek *et al.* combined an extrusion-based bioprinter with microfluidic (Figure 3.10) [96]. In this system, multiple inks of 4 wt% alginate, 6 wt% GelMA, 4 wt% 2-aminoethyl methacrylated chondroitin sulfate (CS-AEMA) for hyaline cartilage ink and 4 wt% alginate, 6 wt% GelMA, 4 wt% CS-AEMA, 0.5 wt% HAMA, and 0.5 wt% TCP for calcified cartilage ink could be extruded individually or simultaneously. Then, they were mixed and deposited using a coaxial nozzle enabling layered deposition or generation of biological, mechanical, and chemical cue gradients. The layered hyaline cartilage and calcified cartilage constructs were 3D bioprinted with the cell-encapsulated hydrogel bioinks, exhibiting high cell viability and shape fidelity. The hypertrophic marker expressions were upregulated in the calcified cartilage rather than in the construct with heterogeneous gradients. It was concluded that high printing resolution and cell viability could be obtained through combining microfluidics with 3D bioprinting, and it also presented as a promising future for fabricating heterogeneous osteochondral interfaces.

3.3.2 Inkjet-Based (Bio)printing

The inkjet-based 3D (bio)printing technique precisely disposes individual droplets of (bio)inks with quite small volumes from the nozzle(s) laden with or without cells. They are generally driven by using thermal or piezoelectric energy to form predefined structures by post-solidification. However, it is not used as much as extrusion-based (bio)printing to generate bone tissue engineering scaffolds even if it is also cost-effective and fulfills the quality of higher resolution for approximately

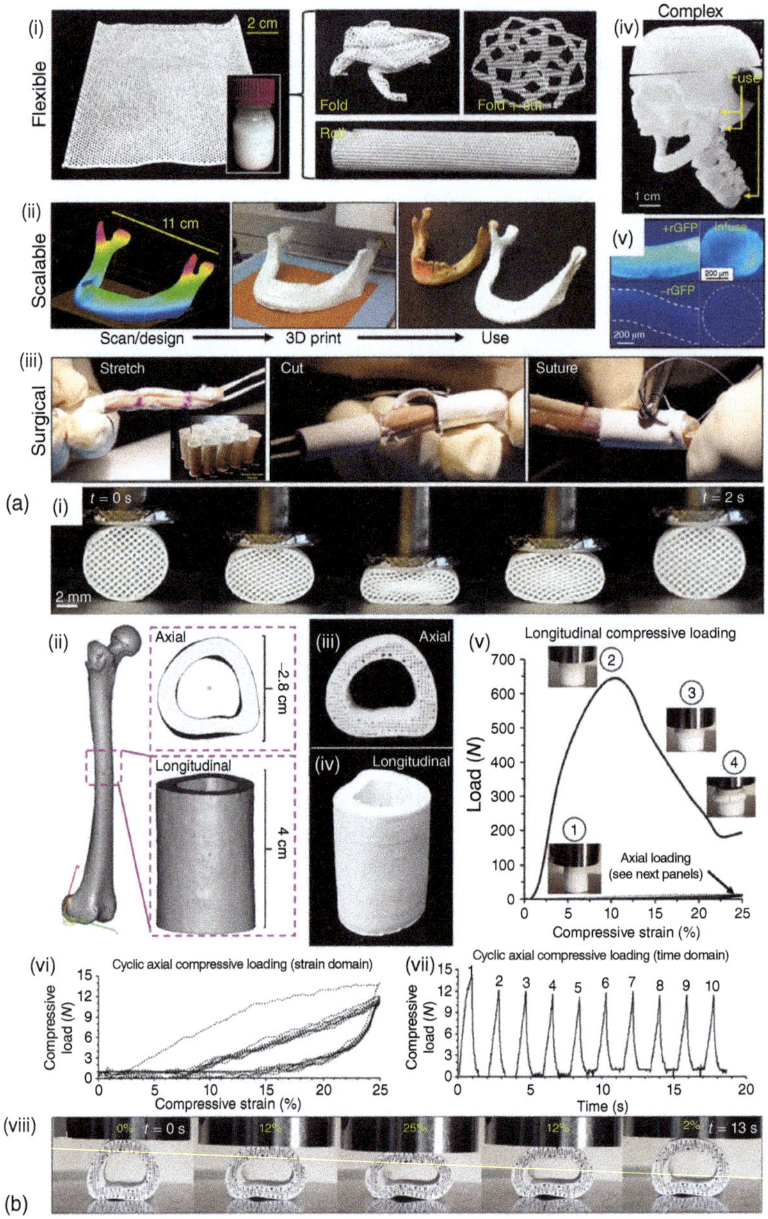
(i)
Flexible
2 cm
Fold
Fold + cut
Roll
Complex
(iv)
Fuse
1 cm
(ii)
Scalable
11 cm
Scan/design
3D print
Use
(v)
+rGFP
Infuse
200 μm
-rGFP
200 μm
(iii)
Surgical
Stretch
Cut
Suture
(a)
(i)
t = 0 s
t = 2 s
2 mm
(ii)
Axial
~2.8 cm
Longitudinal
4 cm
(iii)
Axial
(iv)
Longitudinal
(v)
Longitudinal compressive loading
Load (N)
Compressive strain (%)
Axial loading
(see next panels)
(vi)
Cyclic axial compressive loading (strain domain)
Compressive load (N)
Compressive strain (%)
(vii)
Cyclic axial compressive loading (time domain)
Compressive load (N)
Time (s)
(viii)
0% t = 0 s
12%
25%
12%
2% t = 13 s
(b)

Figure 3.7 (a) Fabrication of 3D-printed hyperelastic bone. (i) HA/PLGA liquid ink (inset) for 3D printing of a $12 \times 12\text{-cm}^2$ sheet, which could be rolled, folded, and cut. (ii) Human mandible could be 3D printed using the HA/PLGA liquid ink. (iii) HA/PLGA sleeves could be stretched, cut, and sutured to the human cadaveric tendon. (iv) The HA/PLGA liquid ink-enabled 3D printing of the human skull, skull cap, mandible, and upper thoracic. (v) Cross-sectional images of HA/PLGA fiber incorporated with (top) rGFP or (bottom) not, illuminated by black light. (b) Mechanical properties of hyperelastic bone. (i) A single compression and recovery cycle of the 3D-printed HA/PLGA. (ii) Adult human femur and femoral midshaft section longitudinal and axial views. (iii) Axial and (iv) longitudinal views of the 3D-printed hyperelastic bone femoral midshaft construct. (v) Longitudinal compressive loading profile of hyperelastic bone and (iv) femoral midshaft and photos at different percent strain points. (vi and vii) Cyclic compression loading profiles (10 cycles) of hyperelastic bone femoral midshaft loaded in the axial direction in (vi) the strain domain and (vii) the time domain. (viii) Photos of a single axial compression cycle displayed in (vi and vii) and corresponding percent strains. 3D, three-dimensional; HA, hydroxyapatite; PLGA, poly(lactic-*co*-glycolic acid); and rGFP, recombinant green fluorescent protein. Source: Jakus *et al.* [28] / with permission of Science publishing.

50 μm with a speed ranging from 1 to 10 000 droplets/s [11, 16b, 17]. One possible reason might be that the heat and mechanical stress utilized in inkjet-based (bio)printing can be harmful to cell survival [97]. Another reason might be the insufficient shape fidelity of hydrogel-based bioprinting constructs, and related crosslinking methods including chemical agents or pH utilized in the solidification of the deposited bioink droplets can be detrimental to the viability and functionality of the embedded cells [98].

In a study by Kim et al., the resolution at nanoscale for 3D-printed hybrid PCL grafts was improved by applying pin-modified electrified inkjets and the cell fate of osteoblasts that was controlled by the nano–microfibers (which were highly ordered or patterned) (Figure 3.11) [99]. The interlayered electrospun fibers enabled the 3D-printed hybrid constructs with high resolution and better initial adhesion, in the proliferation of MC3T3-E1 cells, which was benefitted from the surface-area-to-volume ratio and adequate physical properties. Besides, the nanoscale topography of the scaffold greatly influenced the attachment, proliferation, and differentiation of cells. Alternatively, Campbell and coworkers engineered a stem cell microenvironment based on inkjet-based printing for immobilizing spatially patterned bone morphogenetic protein-2 (BMP-2) on fibrin substrates [100]. The printing substrates could be relatively moved at x–y stages, and the inkjet printhead was fixed at 100 μm higher than the substrate. The piezoelectrically driven inkjet device (MicroJetTM) with a nozzle diameter of 30 μm produced 18 pl of drops with a diameter of ~25 μm and deposited spots with a diameter of 75 μm. When cultured under myogenic conditions, the osteogenic differentiation of mouse primary muscle-derived stem cells (MDSCs) on the printed pattern of BMP-2 was different from that of myogenic differentiation for the off-patterned cells. This study provided patterned growth factors to spatially control stem cell multidirectional differentiation.

To minimize the damage to cells and to prevent clogging of the printhead during 3D inkjet-based bioprinting, Cui and coworkers simultaneously co-bioprinted

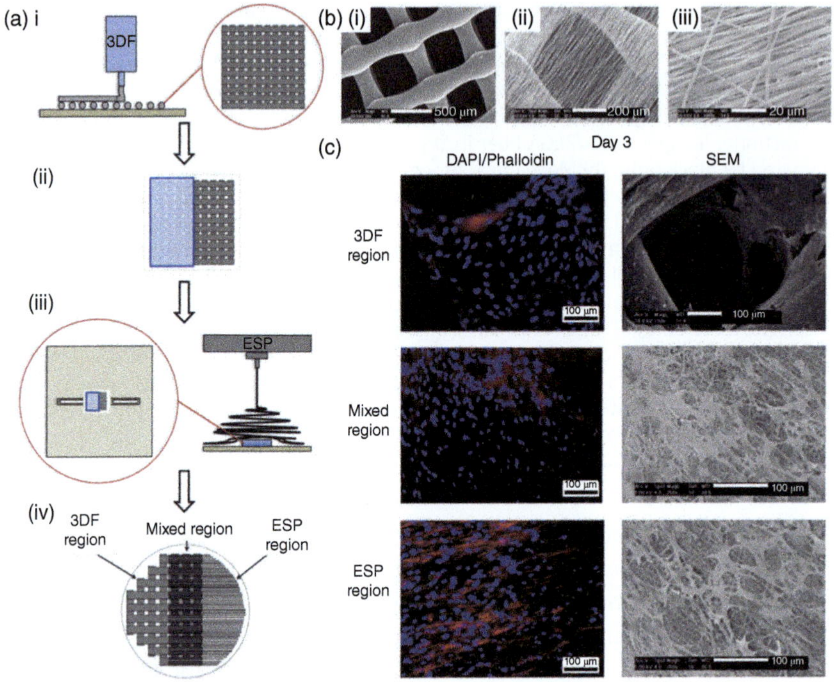

Figure 3.8 (a) Fabrication process of a triphasic scaffold: (i) generating a 3DF PCL scaffold using a Bioscaffolder device; (ii) partially covering ESP fibers on the 3DF PCL scaffold; (iii) covering of aligned ESP PLGA fibers on the uncovered area of the 3DF PCL scaffold; and (iv) obtaining the triphasic scaffold. (b) SEM images of the triphasic scaffold: (i) 3DF region; (ii) mixed region; and (iii) ESP region. (c) Cell staining and SEM morphology cultured for three days. 3DF, three-dimensional fiber deposited; DAPI, 4′,6-diamidino-2-phenylindole; ESP, electrospun; PCL, poly(ε-caprolactone); and SEM, scanning electron microscopy. Source: Criscenti *et al.* [95] / with permission of Institute of Physics (IOP) Publishing Group.

acrylated peptides with hBMSC-encapsulated acrylated PEG hydrogels followed by photopolymerization with minimal UV exposure [101]. A cell viability of $87.9 \pm 5.3\%$ was achieved by the resulted peptide-conjugated PEG scaffolds with minimal nozzle clogging benefitting from their low viscosity. The bioprinted cartilage and bone constructs exhibited better mechanical properties and excellent chondrogenic and osteogenic differentiation as evidenced by cartilage and mineral matrix deposition. Also, hBMSC hypertrophy was inhibited by the bioprinted PEG-peptide scaffolds during chondrogenic differentiation. Hence, the bioprinted hBMSC-laden PEG-peptide scaffolds could enhance chondrogenic and osteogenic differentiation for robust cartilage and bone formation with minimal clogging of the printhead. Additionally, the authors have also co-bioprinted a PEG-GelMA hybrid scaffold with better physical behavior and precise stem cell deposition by using an inkjet-based 3D bioprinting technique for cartilage and bone regeneration [102]. The firing frequency and drop volume were 3.6 kHz and 130 pl, respectively. Stem cells could be simultaneously assembled in a layer-by-layer manner within

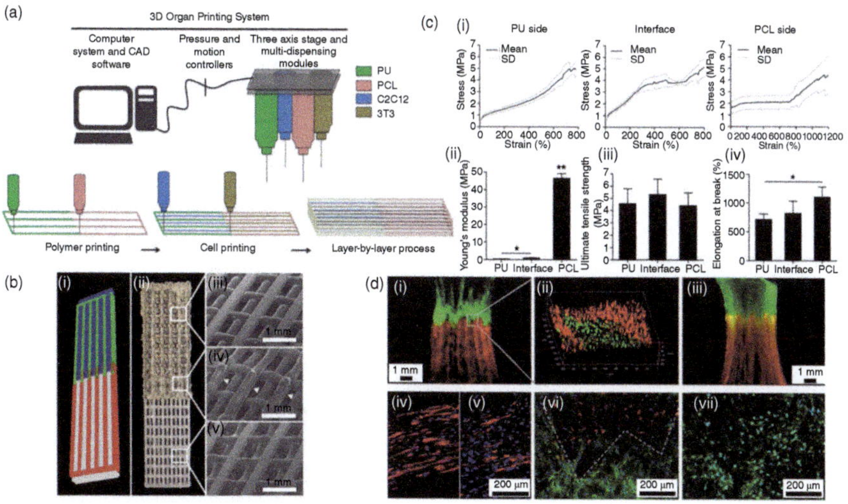

Figure 3.9 (a) Illustration of the IOP system. (b) 3D-bioprinted MTU construct: (i) CAD profile of cell-loaded MTU construct (green, PU; red, PCL; blue, C2C12 cells; and white, 3T3 cells); (ii) gross observation of MTU construct (top, PU side; bottom, PCL side; and interface, 10% of overlap area); and (iii–v) SEM observations of (iii) PU side, (iv) interface, and (v) PCL side. Arrowheads indicate the welding area of PU and PCL. (c) Tensile properties of MTU constructs. (i) The stress–strain curves of the MTU constructs consisting of three regions. (ii) Young's modulus, (iii) ultimate tensile strength, and (iv) elongation at break of the MTU constructs with three regions. $^{*}p < 0.05$, $^{**}p < 0.01$. (d) Fluorescence images of cell-loaded MTU constructs (green: C2C12 cells labeled with DiO; red: 3T3 cells labeled with DiI; and yellow: interface region): (i) images of cell interactions and movement cultured for (i) one day and (iii) seven days; (ii) 3D-reconstructed image of the interface region at one day; (iv–vii) immunofluorescence images of MTU constructs at seven days; and (iv and v) highly aligned C2C12 cells with multinucleated myotube on the PU side (red, (iv) desmin and (v) MHC; blue, DAPI); (vi) expressions of desmin (red) and Col-I (green) by C2C12 and 3T3 cells at the interface region (blue, DAPI); (vii) secretion of Col-I (green) by 3T3 cells on the PCL side (blue, DAPI). 3D, three-dimensional; 3T3, mouse embryonic fibroblasts; C2C12 cells, mouse myoblasts; CAD, computer-aided design; Col-I, collagen type I; DAPI, 4′,6-diamidino-2-phenylindole; DiI, 1,1′-dioctadecyl-3,3,3′,3′-tetramethylindocarbocyanine perchlorate; DiO, 3,3′-dioctadecyloxacarbocyanine perchlorate; IOP, integrated organ printing; MHC, myosin heavy chain; MTU, muscle-tendon unit; PCL, poly(ε-caprolactone); PU, polyurethane; SD, standard deviation; and SEM, scanning electron microscopy. Source: Merceron et al. [7a] / with permission of Institute of Physics (IOP) Publishing Group.

the scaffolds, avoiding additional cell encapsulation. This enabled inkjet-based bioprinting to be a prospective platform for generating biomimetic cartilage and bone tissues.

3.3.3 Light-Based (Bio)printing

These (bio)printers utilize laser or other types of light energy to eject or photopolymerize (bio)inks. These techniques are usually categorized based on the source of light energy used and the pattern formation method, such as vat polymerization (*e.g.* SLA, DLP, liquid-crystal display (LCD), volumetric additive manufacturing (VAM)),

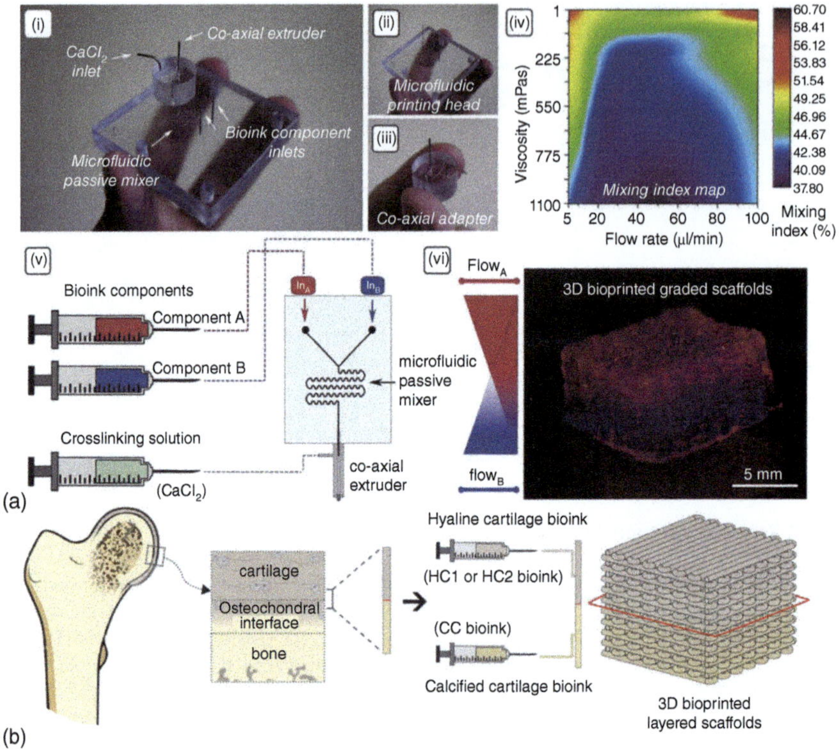

Figure 3.10 (a) (i) Microfluidic 3D bioprinting system. (ii) Microfluidic printhead and (iii) co-axial adapter. (iv) Mixing index heat map. (v and vi) A 3D-bioprinted graded scaffold. (b) Structural composition of the osteochondral interface and illustration of 3D-bioprinted multilayered scaffold for regenerating the osteochondral region. 3D, three-dimensional; $CaCl_2$, calcium chloride; CC, calcified cartilage; GelMA, gelatin methacryloyl; hAC, human articular chondrocyte; HC, hyaline cartilage; HC1 bioink, 4% w/v alginate + 6% w/v GelMA + 4% w/v CS-AEMA, containing only hMSCs; HC2 bioink, 4% w/v alginate + 6% w/v GelMA + 4% w/v CS-AEMA, containing a mixture of hMSCs and hACs at 3 : 1 ratio; CS-AEMA, 2-aminoethyl methacrylated chondroitin sulfate; and hMSC, human mesenchymal stem cell. Source: Idaszek *et al.* [96] / with permission of Institute of Physics (IOP) Publishing Group.

and laser-induced forward transfer (LIFT), among others. The light source is utilized to facilitate the formation of desired shapes either point-by-point (*e.g.* SLA and LIFT), layer-by-layer (e.g. DLP and LCD), or curing the entire volume (*e.g.* VAM) [6a, 15a, 83a, 103]. In most cases, the resolution of light-based (bio)printing is better than other (bio)printing techniques (up to 10 μm). Also, the cell viability is excellent (greater than 90%) as there is no mechanical/shear damage to the cells (no nozzles used), in particular in vat polymerization approaches, although photodamage is a possibility in turn [104]. Apart for the above advantages, the instrumentation used for this technique can be expensive, and multimaterial (bio)printing is oftentimes a complex and time-consuming process [15a, 105].

To improve the complexity of hierarchical structures, physical behavior, and bone-resident cell diversity, Zhang et al. developed hierarchical Haversian

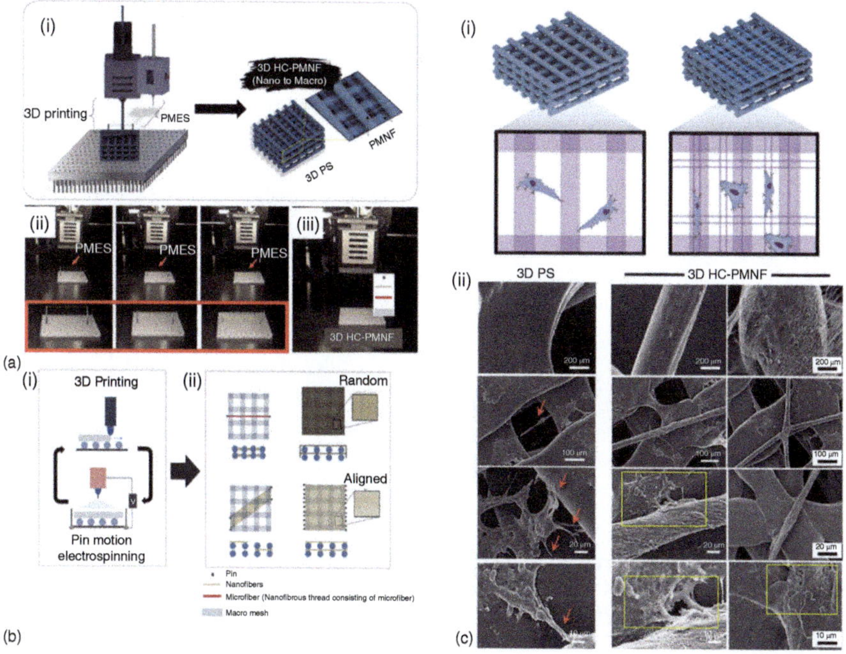

Figure 3.11 (a) (i) Schematic showing the fabrication process of 3D HC-PMNFs through combining 3D printing with PMES and (ii, iii) photos of (ii) PMES and (iii) 3D HC-PMNF through combining 3D printing with PMES. (b) (i) Illustration of generating 3D HC-PMNF and (ii) aligned or random fibers or mesh on the desired portion of the 3D-printed scaffold depending on the PINs. (c) (i) Illustration of attached cells on the macroscale 3D PS and nano- to macroscale 3D HC-PMNF and (ii) SEM images of attached MC3T3-E1 cells on 3D PS and 3D HC-PMNFs at three days. 3D, three-dimensional; HC-PMNF, hybrid constructs with pin-motion electrospun nanofiber; MC3T3-E1 cell, mouse embryonic osteoblast precursor cell; PMES, pin-motion electrospinning; PMNF, pin-motion electrospun nanofiber; PS, printed struts; and SEM, scanning electron microscopy. Source: Kim and Kim [99] / with permission of American Chemical Society.

bone-structured biomimetic scaffolds that consisted of $Ca_2MgSi_2O_7$ (AKT) and 45S5 BG using 3D DLP-based printing [7b]. The porosity and compressive property of the Haversian bone-mimicking scaffolds was controlled by regulating the microstructure. The rabbit BMSC and rabbit aortic endothelial cell (AEC) delivery in the biomimetic scaffolds promoted *in vitro* osteogenic, neurogenic, and angiogenic differentiation and enhanced the *in vivo* blood vessel ingrowth as well as the formation of new bone tissues. Findings from this work presented a novel technique for designing bone scaffolds with complex structures and diverse functionality by mimicking native bone tissues. In another study, Chen et al. the ECM/GelMA/exosome construct is 3D printed using the DLP-based printer (Figure 3.12) [106]. The fabrication system was composed of a DLP chip, a 405 nm wavelength visible light source with an output of 30 W/cm^2 and *x*–*y*–*z* stages. Digital masks generated by the DLP chip were used for transmitting visible light (11 mW/cm^2, 30 seconds) to the images on the ECM/GelMA/exosome ink to

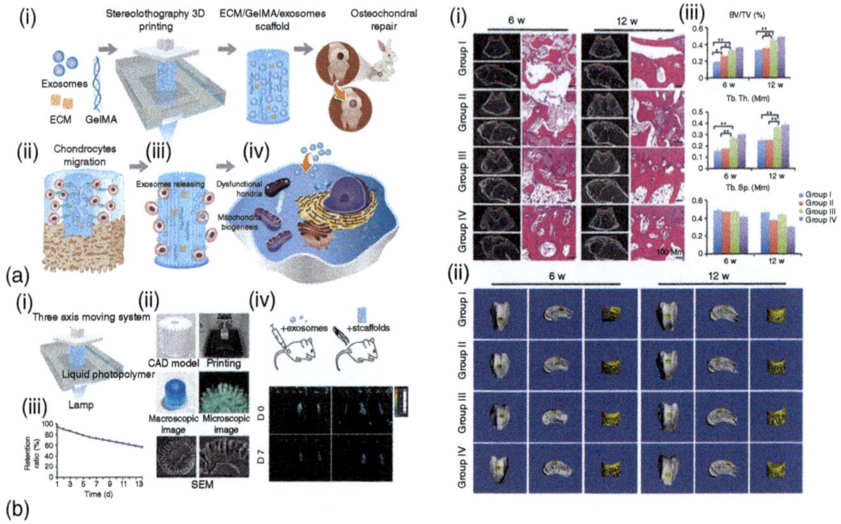

Figure 3.12 (a) Fabrication of ECM/GelMA/exosome scaffolds for osteochondral repair: (i) SLA 3D printing and implantation in an osteochondral defect; (ii) chondrocyte migration toward the defect regions; (iii) exosomes releasing from the 3D-printed scaffold; and (iv) the scaffold enhances the mitochondrial biogenesis of the chondrocytes. (b) 3D-printed ECM/GelMA/exosome scaffold with radially oriented structure: (i) process of SLA printing with a resolution of 50 μm; (ii) CAD file, printing process, and 3D-printed scaffold with methylene blue staining observed by microscopy and SEM; (iii) retention ratio of exosomes in the scaffold up to 14 days; and (iv) *in vivo* fluorescence imaging of exosomes labeled with Vybrant DiO (green) in PBS and the 3D-printed scaffold. (C) Subchondral bone repair using the 3D-printed ECM/GelMA/exosome scaffold with radially oriented structure in a rabbit model: (i) micro-CT reconstruction and histomorphological images of the osteochondral defect with different treatment; (ii) micro-CT reconstruction images of the repaired knees; (iii) BV/TV, Tb.Th, and TB.Sp values in all the groups. $^{*}p < 0.05$; $^{**}p < 0.01$. Group I, untreated defect; Group II, 3D-printed GelMA scaffold; Group III, 3D-printed ECM/GelMA scaffold; and Group IV, 3D-printed ECM/GelMA/exosome scaffold. 3D, three-dimensional; BV/TV, bone volume/total volume; CAD, computer-aided design; ECM, extracellular matrix; GelMA, gelatin methacryloyl; micro-CT, micro-computed tomography; PBS, phosphate-buffered saline; SEM, scanning electron microscopy; SLA, stereolithography apparatus; TB.Sp, trabecular spacing; and Tb.Th, trabecular thickness. Source: Chen *et al.* [106] / with permission of Ivyspring International Publisher.

crosslink the printed scaffolds with 4 × 4-mm^2 radially oriented 3D architectures. Exosomes could be sustainably released from the 3D-printed scaffolds for 14 days *in vitro* (>56%) and be retained for at least seven days *in vivo*. The 3D-printed scaffolds facilitated the regeneration of cartilage in the animal osteochondral defect model (Figure 3.12iii).

Besides, the ceramics powder (titanium dioxide [TiO_2]: <50 nm, aluminum oxide [Al_2O_3]: <15 nm, zirconium oxide [ZrO_2]: <20 nm, or HA: <60 nm) was incorporated in PEEK as a filling material for laser-assisted powder bed fusion (PBF) process via the selective laser sintering/selective laser melting (SLS/SLM) technique [107]. The 3D-printed composite samples were additionally heated during the SLS printing process to change the layer-by-layer manufacturing conditions at the glass transition temperatures of PEEK (~150 °C), but it also decreased the porosity of the 3D samples with a slight shrinkage. The strength properties of the 3D-printed ceramic–polymer compositions were thus inferior to the casted PEEK implants by almost 50%, indicating that the samples were suitable for heavy load applications, such as the hip joints prosthesis. In addition to that, Guder *et al.* reported the clinical application of 3D-printed (electron beam melting-enabled sintering) implants with highly cancellous titanium alloy (TiAl6V4) surfaces to enhance functionality and to reduce complications [108]. The authors evaluated four clinical cases treated with these 3D-printed prostheses. All the four patients acquired full weight bearing and limb salvage when continuously followed up for 21.25 months (average). It was clearly observed that soft tissues grew into the surfaces of the implants with improved bone functions and decreased complication rates.

3.4 Current Challenges and Future Directions

Based on the above discussions, great achievements have been achieved in 3D printing and bioprinting techniques, and there is also a strong potential that a common AM technique will be utilized for fabricating customized, patient-specific implants, and tissue models in orthopedics. However, some aspects of challenges and some barriers remain to be addressed for engineering clinically appropriate bone tissues with shape fidelity and stability, complex structure, and biological composition similar to the native bone tissues.

As potential implants for orthopedics, the shape fidelity and stability of the 3D-printed constructs in a bone defect are an extremely important aspect to fulfill the clinical requirements [109]. The development of cell-laden hydrogels, multimaterial bioprinting, and sequential crosslinking processes enable the self-supporting constructs to acquire mechanical properties for bioprinted cell-laden constructs [110]. A satisfied shape fidelity and preservation of original design structure can be achieved by incorporating nanomaterial fillers, such as high-consistency nanocellulose, silver nanoparticles, nSi, and nanoceramics, among others [38b, 89, 111]. Such attempts encourage us to systematically improve the printing parameters and geometrical compensation for structural integrity. In addition to the development of (bio)inks, other techniques such as electrospinning are also contributing by providing high shape stability and precise resolution for (bio)printers [84a, 112]. Lastly, the reversible crosslinked and stimuli-responsive materials or supramolecular hydrogels for biomimetic 4D (bio)printing are potential candidates for designing strengthened bioinks and acquiring the target shapes with complex 3D morphologies [29b, 79, 113].

To recapitulate the complex structure and the multicellular composition of the native bone, stem cells derived from adipose, cord blood, bone marrow, as well as endothelial cells have been co-bioprinted for producing vascularized bone grafts with perfusion patterns [7f, 114]. Furthermore, microfluidic channels in organ-on-chip models have been combined with multimaterial 3D (bio)printing for biomimetic tissue microarchitectures. These are fulfilling the single-cell encapsulation into microgels for sustained delivery of cells and bioactive factors *in vivo* and also for mimicking physiological conditions *in vitro* for drug screening and pathological studies [84b, 115]. The convergence of 3D printing with microfluidics facilitates to efficiently create complex organs or tissues with precisely patterned cells, heterogeneous biomaterials, and dynamic functions [116].

Even with the continuous developments of 3D (bio)printing technologies, it has already been clinically applied except a certain scenarios, such as the production of guide plates for orthopedic surgery [117]. To this end, the (bio)printed guide plates that is based on the CT-scanned images can be useful for assisting precise operations. The 3D-(bio)printed bone implants, such as prosthesis and fixtures, are the third-class medical devices according to the regulatory classifications [118]. Some clinical cases that have utilized the 3D-(bio)printed bone implants have been reported, including patients with trauma, fracture, deformity healing, osteonecrosis, and osteoporosis [118, 119]. Moreover, as the most promising technique, 3D (bio)printing is able to reconstruct structural and functional tissues [26]. To maximize the advantages of the technology, properly designed clinical trials of 3D-(bio)printed bone grafts would be valuable for the translation of numerous fundamental research toward pushing the final products being used for human applications.

Finally, large animal studies and long-term evaluations of the 3D-(bio)printed bone implants are still necessary to assess the efficacy of *in vivo* osseointegration with the surrounding musculoskeletal tissues [120]. Considering this aspect, the organ-on-chip technology based on microfluidics has been rapidly developing to mimic physiologically complex architectures and environments of native tissues, providing insights into normal human organ functions and disease pathophysiology, as well as more accurately predicting the safety and efficacy of investigational drugs in human beings [5b, 121], and can be possible alternatives to certain animal tests down the future.

Acknowledgments

This work was supported by the Brigham Research Institute.

References

1 Masaeli, R., Zandsalimi, K., Rasoulianboroujeni, M., and Tayebi, L. (2019). *Tissue Eng. Part B* 25: 387.

2 (a) Liu, Y., Luo, D., and Wang, T. (2016). *Small* 12: 4611. (b) Byambaa, B., Annabi, N., Yue, K. et al. (2017). *Adv. Healthc. Mater.* 6: 1700015.

3 Gao, G., Schilling, A.F., Yonezawa, T. et al. (2014). *Biotechnol. J.* 9: 1304.

4 Gibbs, D.M.R., Vaezi, M., Yang, S., and Oreffo, R.O.C. (2014). *Regener. Med.* 9: 535.

5 (a) Elder, B., Neupane, R., Tokita, E. et al. (2020). *Adv. Mater.* 32: e1907142. (b) Low, L.A., Mummery, C., Berridge, B.R. et al. (2020). *Nat. Rev. Drug Discovery* 20: 345. (c) Weisgrab, G., Ovsianikov, A., and Costa, P.F. (2019). *Adv. Mater. Technol.* 4: 1900275. (d) Smith, W.R., Hudson, P.W., Ponce, B.A. et al. (2018). *BMC Musculoskelet. Disord.* 19: 67.

6 (a) Ashammakhi, N., Hasan, A., Kaarela, O. et al. (2019). *Adv. Healthc. Mater.* 8: 1801048. (b) Faramarzi, N., Yazdi, I.K., Nabavinia, M. et al. (2018). *Adv. Healthc. Mater.* 7: e1701347.

7 (a) Merceron, T.K., Burt, M., Seol, Y.-J. et al. (2015). *Biofabrication* 7: 035003. (b) Zhang, M., Lin, R., Wang, X. et al. (2020). *Sci. Adv.* 6: eaaz6725. (c) Bittner, S.M., Smith, B.T., Diaz-Gomez, L. et al. (2019). *Acta Biomater.* 90: 37. (d) Moncal, K.K., Heo, D.N., Godzik, K.P. et al. (1972). *J. Mater. Res.* 2018: 33. (e) Urruela-Barrios, R., Ramirez-Cedillo, E., de Leon, A.D. et al. (2019). *Polymers* 11: 457. (f) Chiesa, I., Maria, C.D., Lapomarda, A. et al. (2020). *Biofabrication* 12: 025013.

8 du Chatinier, D.N., Figler, K.P., Agrawal, P. et al. (2021). *Biomicrofluidics* 15: 041304.

9 (a) Tamay, D.G., Usal, T.D., Alagoz, A.S. et al. (2019). *Front. Bioeng. Biotechnol.* 7: 164. (b) Qasim, M., Chae, D.S., and Lee, N.Y. (2019). *Int. J. Nanomed.* 14: 4333.

10 Zehnder, T., Sarker, B., Boccaccini, A.R., and Detsch, R. (2015). *Biofabrication* 7: 025001.

11 (a) Derby, B. (2010). *Annu. Rev. Mater. Res.* 40: 395. (b) Sears, N.A., Seshadri, D.R., Dhavalikar, P.S., and Cosgriff-Hernandez, E. (2016). *Tissue Eng. Part B* 22: 298. (c) Colosi, C., Costantini, M., Latini, R. et al. (2014). *J. Mater. Chem. B* 39: 6779. (d) Dababneh, A.B. and Ozbolat, I.T. (2014). *J. Manuf. Sci. Eng.* 136: 061016.

12 Li, Y., Zhang, F., Liu, Y., and Leng, J. (2020). *Sci. China Technol. Sci.* 63: 545.

13 Gopinathan, J. and Noh, I. (2018). *Biomater. Res.* 22: 11.

14 Li, Y., Li, Q., Li, H. et al. (2020). *J. Biomed. Mater. Res. Part B* 108: 2167.

15 (a) Ibrahim, A. (2018). *3D Bioprinting for Reconstructive Surgery Techniques and Applications*, vol. 245 (ed. D.J. Thomas, Z.M. Jessop and I.S. Whitaker). Woodhead Publishing. (b) Zheng, X., Huang, J., Lin, J. et al. (2019). *J. Biomater. Sci., Polym. Ed.* 30: 1172. (c) Lim, W., Kim, B., and Moon, Y.L. (2019). *Ann. Joint* 4: 7.

16 (a) Negro, A., Cherbuin, T., and Lutolf, M.P. (2018). *Sci. Rep.* 8: 17099. (b) Du, X., Fu, S., and Zhu, Y. (2018). *J. Mater. Chem. B* 6: 4397. (c) Obregon, F., Vaquette, C., Ivanovski, S., and Hutmacher, D.W. (2015). *J. Dent. Res.* 94: 143S.

17 Murphy, S.V. and Atala, A. (2014). *Nat. Biotechnol.* 32: 773.

18 (a) Ying, G., Jiang, N., Yu, C., and Zhang, Y.S. (2018). *Bio-Des. Manuf.* 1: 215. (b) Gungor-Ozkerim, P.S., Inci, I., Zhang, Y.S. et al. (2018). *Biomater. Sci.* 6: 915. (c) Zhang, Y.S. (2020). *Adv. Healthc. Mater.* 9: 2001043.

19 Groll, J., Burdick, J.A., Cho, D.-W. et al. (2019). *Biofabrication* 11: 013001.

20 Li, Q., Lei, X., Wang, X. et al. (2019). *Tissue Eng. Part A* 25: 1261.

21 Leucht, A., Volz, A.-C., Rogal, J. et al. (2020). *Sci. Rep.* 10: 5330.

22 Distler, T., Solisito, A.A., Schneidereit, D. et al. (2020). *Biofabrication* 12: 045005.

23 Antich, C., de Vicente, J., Jiménez, G. et al. (2020). *Acta Biomater.* 106: 114.

24 Fairag, R., Rosenzweig, D.H., Ramirez-Garcialuna, J.L. et al. (2019). *ACS Appl. Mater. Interfaces* 11: 15306.

25 Trombetta, R., Inzana, J.A., Schwarz, E.M. et al. (2017). *Ann. Biomed. Eng.* 45: 23.

26 Midha, S., Dalela, M., Sybil, D. et al. (2019). *J. Tissue Eng. Regener. Med.* 13: 925.

27 (a) Valino, A.D., Dizon, J.R.C., Espara, A.H. Jr., et al. (2019). *Prog. Polym. Sci.* 98: 101162. (b) Moroni, L., Boland, T., Burdick, J.A. et al. (2018). *Trends Biotechnol.* 36: 384.

28 Jakus, A.E., Rutz, A.L., Jordan, S.W. et al. (2016). *Sci. Transl. Med.* 8, 358ra127.

29 (a) Zhang, Z., Demir, K.G., and Gu, G.X. (2019). *Int. J. Smart Nano Mater.* 10: 205. (b) Gladman, A.S., Matsumoto, E.A., Nuzzo, R.G. et al. (2016). *Nat. Mater.* 15: 413. (c) Li, Y.-C., Zhang, Y.S., Akpek, A. et al. (2016). *Biofabrication* 9: 012001. (d) Sabzi, M., Babaahmadi, M., and Rahnama, M. (2017). *ACS Appl. Mater. Interfaces* 9: 24061.

30 Zhang, Y.S. and Khademhosseini, A. (2017). *Science* 356: eaaf3627.

31 Bai, X., Gao, M., Syed, S. et al. (2018). *Bioact. Mater.* 3: 401.

32 Chen, L., Wu, Z., Zhou, Y. et al. (2017). *J. Appl. Polym. Sci.* 134: 45271.

33 Ma, L., Wang, X., Zhao, N. et al. (2018). *ACS Appl. Mater. Interfaces* 10: 42146.

34 Diogo, G.S., Marques, C.F., Sotelo, C.G. et al. (2020). *ACS Biomater. Sci. Eng.* 6: 3664.

35 Nichol, J.W., Koshy, S.T., Bae, H. et al. (2010). *Biomaterials* 31: 5536.

36 (a) Gao, Q., Niu, X., Shao, L. et al. (2019). *Biofabrication* 11: 035006. (b) Castilho, M., Mouser, V., Chen, M. et al. (2019). *Acta Biomater.* 95: 297.

37 Xavier, J.R., Thakur, T., Desai, P. et al. (2015). *ACS Nano* 9: 3109.

38 (a) Chimene, D., Peak, C.W., Gentry, J.L. et al. (2018). *ACS Appl. Mater. Interfaces* 10: 9957. (b) Chimene, D., Miller, L., Cross, L. et al. (2020). *ACS Appl. Mater. Interfaces* 12: 15976.

39 Chimene, D., Kaunas, R., and Gaharwar, A.K. (2020). *Adv. Mater.* 32: 1902026.

40 Diloksumpan, P., de Ruijter, M., Castilho, M. et al. (2020). *Biofabrication* 12: 025014.

41 (a) Kilian, D., Ahlfeld, T., Akkineni, A.R. et al. (2020). *Sci. Rep.* 10: 8277. (b) Morgan, F.L.C., Moroni, L., and Baker, M.B. (2020). *Adv. Healthc. Mater.* 9: 1901798.

42 (a) Wang, C., Huang, W., Zhou, Y. et al. (2020). *Bioact. Mater.* 5: 82. (b) Li, J., Liu, X., Crook, J.M., and Wallace, G.G. (2020). *Front. Bioeng. Biotechnol.* 8: 824.

43 Schütz, K., Placht, A.-M., Paul, B. et al. (2017). *J. Tissue Eng. Regener. Med.* 11: 1574.

44 Zehnder, T., Boccaccini, A.R., and Detsch, R. (2017). *Biofabrication* 9: 025016.

45 Park, H. and Lee, K.Y. (2014). *J. Biomed. Mater. Res. Part A* 102: 4519.

46 (a) You, F., Wu, X., Kelly, M., and Chen, X. (2020). *Bio-Des. Manuf.* 3: 48. (b) Reakasame, S. and Boccaccini, A.R. (2018). *Biomacromolecules* 19: 3.

(c) Soltan, N., Ning, L., Mohabatpour, F. et al. (2019). *ACS Biomater. Sci. Eng.* 5: 2976.

47 Ahlfeld, T., Doberenz, F., Kilian, D. et al. (2018). *Biofabrication* 10: 045002.

48 (a) Zhai, P., Peng, X., Li, B. et al. (2002). *Int. J. Biol. Macromol.* 151: 1224. (b) Petta, D., D'Amora, U., Ambrosio, L., and Grijpma, D. (2020). *Biofabrication* 12: 032001.

49 Ozbolat, I.T. and Hospodiuk, M. (2016). *Biomaterials* 76: 321.

50 Shim, J.-H., Jang, K.-M., Hahn, S.K. et al. (2016). *Biofabrication* 8: 014102.

51 Zhai, X., Ruan, C., Ma, Y. et al. (2017). *Adv. Sci.* 5: 1700550.

52 (a) Abbadessa, A., Mouser, V.H.M., Blokzijl, M.M. et al. (2016). *Biomacromolecules* 17: 2137. (b) Camci-Unal, G., Cuttica, D., Annabi, N. et al. (2013). *Biomacromolecules* 14: 1085.

53 Yang, Y., Wang, M., Yang, S. et al. (2020). *Biofabrication* 12: 045013.

54 Lam, T., Dehne, T., Krüger, J.P. et al. (2019). *J. Biomed. Mater. Res. Part B* 107: 2649.

55 Jiang, Z., Diggle, B., Tan, M.L. et al. (2020). *Adv. Sci.* 7: 2001379.

56 Liu, F., Vyas, C., Poologasundarampillai, G. et al. (2017). *Macromol. Mater. Eng.* 303: 1700494.

57 (a) Guvendiren, M., Molde, J., Soares, R.M.D. et al. (2016). *Sci. Eng.* 2: 1679. (b) Cubo-Mateo, N. and Rodríguez-Lorenzo, L.M. (2020). *Polymers* 12: 1546.

58 Yeo, M., Lee, H., and Kim, G.H. (2016). *Biofabrication* 8: 035021.

59 Neufurth, M., Wang, X., Wang, S. et al. (2017). *Acta Biomater.* 64: 377.

60 Wang, Z., Wang, Y., Ito, Y. et al. (2016). *Sci. Rep.* 6: 20770.

61 Velioglua, Z.B., Pulata, D., Demirbakana, B. et al. (2019). *Connect. Tissue Res.* 60: 274.

62 Narayanan, L.K., Huebner, P., Fisher, M.B. et al. (2016). *ACS Biomater. Sci. Eng.* 2: 1732.

63 Wu, J.P., Li, L.L., Fu, C. et al. (2018). *Colloid Surf. B* 169: 233.

64 Tseng, J.W., Liu, C.Y., Yen, Y.K. et al. (2018). *Mater. Des.* 140: 209.

65 Li, Q.S., Zhao, W., Li, Y.X. et al. (2019). *Polymers* 11: 656.

66 Zheng, J.B., Kang, J.F., Sun, C.N. et al. (2021). *J. Mech. Behav. Biomed.* 118: 104475.

67 Chen, C., Yin, Y.R., Xu, H. et al. (2021). *Medicine* 100: 17.

68 Zhao, D.H., Liu, Y.D., Liu, B.H. et al. (2021). *ACS Appl. Mater. Interfaces* 13: 13714.

69 Mahmoudi, M., Burlison, S.R., Moreno, S. et al. (2021). *Mater. Interfaces* 13: 5529.

70 (a) Dziadek, M., Kudlackova, R., Zima, A. et al. (2019). *J. Biomed. Mater. Res. Part A* 107: 2479. (b) Carrodeguas, R.G. and Aza, S.D. (2011). *Acta Biomater.* 7: 3536.

71 Wang, C., Xue, Y., Lin, K. et al. (2012). *Acta Biomater.* 8: 350.

72 Bohner, M., Santoni, B.L.G., and Döbelin, N. (2020). *Acta Biomater.* 113: 23.

73 (a) Deng, Y., Jiang, C., Li, C. et al. (2017). *Sci. Rep.* 7: 5588. (b) Manchón, A., Alkhraisat, M., Rueda-Rodriguez, C. et al. (2015). *J. Biomed. Mater. Res. Part A* 103: 479.

74 Dong, Y., Duan, H., Zhao, N. et al. (2018). *Bio-Des. Manuf.* 1: 146.
75 (a) Mondal, S., Nguyen, T.P., Pham, V.H. et al. (2020). *Ceram. Int.* 46: 3443. (b) Shao, H., He, J., Lin, T. et al. (2019). *Ceram. Int.* 45: 1163.
76 Castilho, M., Moseke, C., Ewald, A. et al. (2014). *Biofabrication* 6: 015006.
77 (a) Tang, D., Tare, R.S., Yang, L.-Y. et al. (2016). *Biomaterials* 83: 363. (b) Pati, F., Ha, D.-H., Jang, J. et al. (2015). *Biomaterials* 62: 164. (c) Holloway, J.L., Ma, H., Rai, R., and Burdick, J.A. (2014). *J. Controlled Release* 191: 63.
78 (a) Li, J., Chen, M., Fan, X., and Zhou, H. (2016). *J. Transl. Med.* 14: 271. (b) Markstedt, K., Mantas, A., Tournier, I. et al. (2015). *Biomacromolecules* 16: 1489. (c) Dai, X., Zhang, Y., Gao, L. et al. (2015). *Adv. Mater.* 27: 3566.
79 Zhai, X., Ma, Y., Hou, C. et al. (2017). *ACS Biomater. Sci. Eng.* 3: 1109.
80 Wan, Z., Zhang, P., Liu, Y. et al. (2020). *Acta Biomater.* 101: 26.
81 Wang, C., Yue, H., Liu, J. et al. (2020). *Biofabrication* 12: 045025.
82 (a) Diaz-Gomez, L., Elizondo, M.E., Kontoyiannis, P.D. et al. (2020). *Tissue Eng. C* 26: 292. (b) Lia, J., Wu, C., Chu, P.K., and Gelinsky, M. (2020). *Mater. Sci. Eng., R* 140: 100543.
83 (a) Bagheri, A., Jin, J., and Appl, A.C.S. (2019). *Polym. Mater.* 1: 593. (b) Chen, X., Chen, G., Wang, G. et al. (2019). *Adv. Eng. Mater.* 22: 1901065.
84 (a) Aboubakr, S.H., Nery, S., Long, L. et al. (2017). *Univ. N. M. Orthop. Res. J.* 6: 110. (b) Richard, C., Neild, A., and Cadarso, V.J. (2020). *Lab Chip* 20: 2044.
85 Vaezi, M., Zhong, G., Kalami, H., and Yang, S. (2018). *Functional 3D Tissue Engineering Scaffolds Materials, Technologies and Applications*, vol. 235 (ed. Y. Deng and J. Kuiper). Woodhead Publishing.
86 (a) Vaezi, M. and Yang, S. (2014). *Rapid Prototyping of Biomaterials: Principles and Applications*, vol. 41 (ed. R. Narayan). Woodhead Publishing. (b) Boyle, B.M., Xiong, P.T., Mensch, T.E. et al. (2019). *Addit. Manuf.* 29: 100811.
87 (a) Wang, M., Favi, P., Cheng, X. et al. (2016). *Acta Biomater.* 46: 256. (b) Yan, Y., Chen, H., Zhang, H. et al. (2019). *Biomaterials* 190: 97.
88 (a) Tarafder, S., Brito, J.A., Minhas, S. et al. (2019). *Biofabrication* 12: 015008. (b) Huang, L., Du, X., Fan, S. et al. (2019). *Carbohydr. Polym.* 221: 146.
89 Klar, V., Pere, J., Turpeinen, T. et al. (2019). *Sci. Rep.* 9: 3822.
90 Lee, J.W., Kim, J.Y., and Cho, D.-W. (2010). *Int. J. Stem Cells* 3: 85.
91 Heo, D.N., Castro, N.J., Lee, S.J. et al. (2017). *Nanoscale* 9: 5055.
92 Seok, J.M., Rajangam, T., Jeong, J.E. et al. (2020). *J. Mater. Chem. B* 8: 951.
93 Zhang, J., Chen, Y., Xu, J. et al. (2018). *Regener. Biomater.* 5: 213.
94 Popov, A., Malferrari, S., Kalaskar, D.M., and J. (2017). *3D Print. Med.* 1: 191.
95 Criscenti, G., Longoni, A., Luca, A.D. et al. (2016). *Biofabrication* 8: 015009.
96 Idaszek, J., Costantini, M., Karlsen, T.A. et al. (2019). *Biofabrication* 11: 044101.
97 (a) Gao, G. and Cui, X. (2016). *Biotechnol. Lett.* 38: 203. (b) Kesti, M., Müller, M., Becher, J. et al. (2015). *Acta Biomater.* 11: 162.
98 (a) Murphy, S.V., Skardal, A., and Atala, A. (2013). *J. Biomed. Mater. Res. A* 101: 272. (b) Hennink, W.E. and van Nostrum, C. (2002). *Adv. Drug Delivery Rev.* 54: 13. (c) Wang, M.O., Vorwald, C.E., Dreher, M.L. et al. (2015). *Adv. Mater.* 27: 138. (d) Bose, S., Vahabzadeh, S., and Bandyopadhyay, A. (2013). *Mater. Today* 16: 496.

99 Kim, J.I. and Kim, C.S. (2018). *ACS Appl. Mater. Interfaces* 10: 12390.

100 (a) Phillippi, J.A., Miller, E., Weiss, L. et al. (2008). *Stem Cells* 26: 127. (b) Campbell, P.G., Miller, E.D., Fisher, G.W. et al. (2005). *Biomaterials* 26: 6762.

101 (a) Gao, G., Yonezawa, T., Hubbell, K. et al. (2015). *Biotechnol. J.* 10: 1568. (b) Cui, X. and Boland, T. (2009). *Biomaterials* 30: 6221.

102 Gao, G., Schilling, A.F., Hubbell, K. et al. (2015). *Biotechnol. Lett.* 37: 2349.

103 Maturi, M., Pulignani, C., Locatelli, E. et al. (2020). *Green Chem.* 22: 6212.

104 (a) Cidonio, G., Glinka, M., Dawson, J.I., and Oreffo, R.O.C. (2019). *Biomaterials* 209: 10. (b) Kim, S.H., Yeon, Y.K., Lee, J.M. et al. (2018). *Nat. Commun.* 9: 1620.

105 Castro, N.J., O'Brien, J., and Zhang, L.G. (2015). *Nanoscale* 7: 14010.

106 Chen, P., Zheng, L., Wang, Y. et al. (2019). *Theranostics* 9: 2439.

107 Shishkovsky, I., Sherbakov, V., Ibatullin, I. et al. (2018). *Compos. Struct.* 202: 651.

108 Guder, W.K., Hardes, J., Nottrott, M. et al. (2021). *J. Pers. Med.* 11: 351.

109 Ji, S. and Guvendiren, M. (2017). *Front. Bioeng. Biotechnol.* 5: 23.

110 Schwab, A., Levato, R., D'Este, M. et al. (2020). *Chem. Rev.* 120: 10850.

111 (a) Feng, X., Wu, Z., Xie, Y., and Wang, S. (2019). *BioResources* 14: 3701. (b) Chen, S., Yang, J., Jia, Y. et al. (2018). *Materials* 11: 2444. (c) Tetsuka, H. and Shin, S.R. (2020). *J. Mater. Chem. B* 15: 2930.

112 Vyas, C., Ates, G., Aslan, E. et al. (2020). *3D Print. Addit. Manuf.* 7: 105.

113 Rodell, C.B., MacArthur, J.W., Dorsey, S.M. et al. (2015). *Adv. Funct. Mater.* 25: 636.

114 (a) Ong, C.S., Yesantharao, P., Huang, C.Y. et al. (2018). *Pediatr. Res.* 83: 223. (b) Ni, T., Liu, M., Zhang, Y. et al. (1938). *Bioconjugate Chem.* 2020: 31.

115 (a) Mao, A.S., Shin, J.-W., Utech, S. et al. (2017). *Nat. Mater.* 16: 236. (b) Zhang, Y.S., Hou, X., de Santiago, G.T. et al. (2017). *Nat. Rev. Mater.* 2: 17016. (c) Miri, A.K., Mostafavi, E., Khorsandi, D. et al. (2019). *Biofabrication* 11: 042002.

116 Sun, H., Jia, Y., Dong, H. et al. (2020). *Curr. Opin. Chem. Eng.* 28: 1.

117 Gadia, A., Shah, K., and Nene, A. (2018). *Asian Spine J.* 12: 365.

118 Dong, X.P., Zhang, Y.W., Pei, Y.J. et al. (2020). *Bio-Des. Manuf.* 3: 122.

119 (a) Mesregah, M.K., Shams, A., Gamal, O., and Zaki, E.M. (2020). *Orthopedics* 43: 97. (b) Zhang, Y.Z., Cao, X.Y., Li, X.C. et al. (2018). *J. Orthop. Surg. Res.* 13: 167.

120 (a) Mavrogenis, A.F., Dimitriou, R., Parvizi, J., and Babis, G. (2009). *J. Musculoskelet. Neuronal Interact.* 9: 61. (b) Bollman, M., Malbrue, R., Li, C. et al. (2020). *Ann. N.Y. Acad. Sci.* 1463: 37.

121 Zhang, B., Montgomery, M., Chamberlain, M.D. et al. (2016). *Nat. Mater.* 15: 669.

4

Stem Cells and Their Application in Orthopedics

Xiao Li[1,2], Xiaobo Zhu[1,3], Jing Sun[1,2], Patrick S.H. Yung[1,3], Rocky S. Tuan[1,2], and Yangzi Jiang[1,2,3]

[1] *The Chinese University of Hong Kong, Institute for Tissue Engineering and Regenerative Medicine, Faculty of Medicine, Lo Kwee-Seong Integrated Biomedical Sciences Building, The Chinese University of Hong Kong, Hong Kong SAR 999077, P.R. China*
[2] *The Chinese University of Hong Kong, School of Biomedical Sciences, Faculty of Medicine, Lo Kwee-Seong Integrated Biomedical Sciences Building, The Chinese University of Hong Kong, Hong Kong SAR 999077, P.R. China*
[3] *The Chinese University of Hong Kong, Department of Orthopaedics and Traumatology, Faculty of Medicine, 9 Chak Cheung Street, Shatin, New Territories, Hong Kong SAR 999077, P.R. China*

4.1 Introduction

Stem cells are a group of undifferentiated cells with the characteristic of self-renewal ability (i.e. extensive proliferation) and differentiation ability. In general, stem cells can be divided into three types with varying potencies, including pluripotency, such as embryonic stem cells (ESCs) and induced-pluripotent stem cells (iPSCs), and multipotency (e.g. hematopoietic stem cells. HSCs; mesenchymal stem cells, MSCs). Another example of totipotent is the zygote, a fertilized egg, which can give rise to a whole individual animal or human. After numerous divisions, the zygote grows into the blastocyst. There are two main regions in a blastocyst, the inner cell mass and the trophoblast, and the inner cell mass is composed mainly by ESCs, which are regarded as pluripotent because they generate all cell types in an embryo.

The best studied multipotent stem cells are HSCs, which have been widely applied clinically as a therapy, such as in bone marrow transplants [1, 2]. In the 1960s, Ernest McCulloch and James Till firstly demonstrated the clonal nature of transplanted mouse marrow cells as spleen colonies [3, 4], which was the first identification of a stem cell, and then they established the properties of stem cells, i.e. self-renewal and differentiation, which still hold true today. These pioneering works on HSCs build up the ground of the application of bone marrow transplantation, which rescued many patients with leukemia and other blood cancers.

MSCs have been extensively studied on their multiple differentiation capabilities for decades. For example, MSCs are often used to replace and repair the damaged tissues as an important source for cell-based therapy in tissue engineering and regenerative medicine. They are subtypes of stromal cells and present in

Biofabrication for Orthopedics: Methods, Techniques, and Applications, First Edition.
Edited by Wenguo Cui, Xin Zhao, and Shen Liu.

adult bone marrow tissues. In 1970s, Friedenstein et al. reported a population of plastic-adherent cells from bone marrow [5]. These cells can form single-cell-derived colonies and then expand into colony-forming unit fibroblasts (CFU-fs) [5–8]. The proliferation and phenotype of CFU-f were studied in the following years, and these cells in CFU-f were able to differentiate into chondrocytes, osteoblasts, and adipocytes. In adult humans, most of the stem cells are derived from a single germ layer with multipotency, and the tissue-resident stem cells or progenitor cells are committed cells and exhibit oligopotency, which can form terminally differentiated cells that contribute to tissues and organs.

There are many studies reporting the favorable application of stem cells for bone and cartilage tissue regeneration. The clinical application of stem cells is investigated widely in tissue regeneration and musculoskeletal diseases as cell-based therapy. For example, autologous chondrocytes, cartilage tissue-specific stem/progenitor cells (TSPCs), and bone marrow-derived mesenchymal stem cells (BMSCs) were used to fill the osteochondral defects in patients and reported with favorable tissue regeneration results with long-term follow-ups [9–11]. Another example is the early report of MSCs. Clinical application in orthopedic diseases is for the treatment of three children with osteogenesis imperfecta (OI), and the patients showed increased total body bone mineral content and reduced frequencies of bone fracture after allogeneic bone marrow transplantation [12]. Recently, Hernigou et al. have reported a series of clinical trials to observe the effect of human bone marrow mesenchymal stem cell injection in subchondral lesions of knee osteoarthritis (OA) with 15 years of follow-up [13–15]. A case–controlled study revealed that BMSC injection enhanced the healing rate of rotator cuff repair as an adjunct therapy to reduce the tissue re-tear [16]. Therefore, the utility of stem cells are promising for the application in the field of orthopedic surgery in bone, cartilage, and tendon regeneration. This chapter is an introduction to the current and future potential applications of stem cells in orthopedics.

4.2 Mesenchymal Stem Cells (MSCs)

MSCs were firstly discovered and isolated from bone marrow based on their properties of adherence to plastic and formation of fibroblast-like colonies. Nowadays, MSCs have been identified in many other tissues throughout the body and are considered the most attractive cell type for tissue repair and potential treatments for orthopedic diseases (Figure 4.1).

4.2.1 Tissue Sources of MSCs

MSCs have been found in different tissues including but not limited to bone marrow, adipose [17], muscle [18], periosteum [19], umbilical cord blood [20], trabecular bone [21], synovial membrane, and skin [22, 23]. Although MSCs derived from different tissues show similar phenotypes by the definition of stem cell characterization criteria, such as cell surface markers and multilineage differentiation ability, they

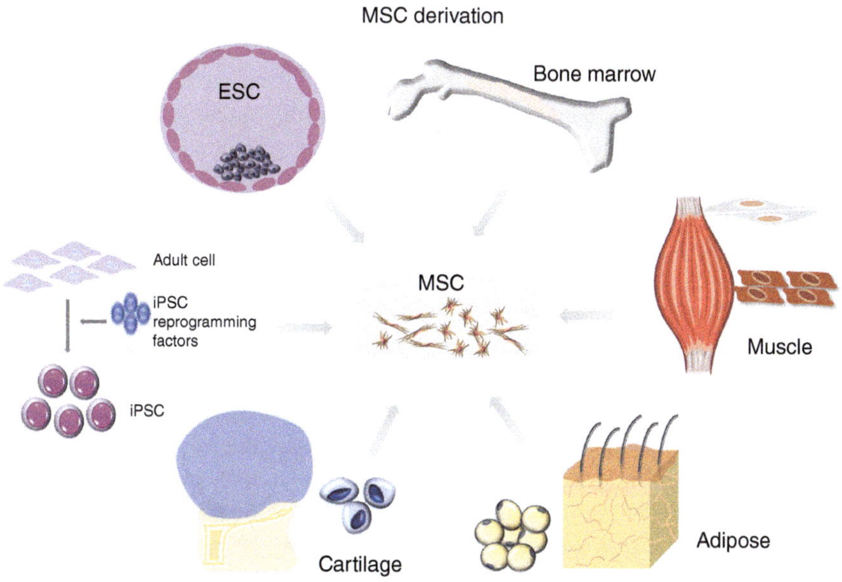

Figure 4.1 MSCs have been found in different tissues and organs such as bone marrow, adipose, muscle, and cartilage. In addition, ESCs can differentiate into MSCs because of totipotency. iPSCs are from reprogrammed somatic cells with pluripotency, and they also can generate MSCs *in vitro*.

still present diverse proliferation rate, colony size, differentiation capabilities, and immunosuppression abilities. BMSCs and adipose-derived mesenchymal stem cells (ADSCs) are most widely used in orthopedics because of their simple accessibility in clinics. However, the tissue sources make differences between stem cell types; for example, one comparative analysis found that ADSCs have a higher proliferation rate *in vitro* with a doubling time of 28 hours than those of 39 hours of BMSCs, and ADSCs have weaker capacity to suppress T-cell proliferation [24]. ADSCs also display higher colony frequency than BMSCs and umbilical cord blood-derived MSCs [20]. Apart from the tissue derivation, the age and health conditions of cell donors affect the phenotypes of MSCs. There are subtle differences in molecular expression between fetal MSCs and adult MSCs that indicates different proliferative capacities and committed lineages. For example, fetal MSCs expressed higher intercellular adhesion molecule 1 (CD54) and contained intracellular deposits of human leukocyte antigen (HLA), while adult MSCs showed increased expression of HLA class I and II molecules and vascular cell adhesion protein 1 (VCAM-1 and CD106) [25]. Besides, CD106 was positive in BMSCs while negative in ADSCs [24]. These differences in molecular signature expression levels may be related to their different derivation and migration ability.

MSCs and TSPCs, particularly mesenchyme-derived stem/progenitor cells, can differentiate into osteoblasts, adipocytes, chondroblasts, and other lineages [26]. These cells contribute to the formation of musculoskeletal system, and that is the basis of their application in orthopedics and traumatology [27, 28]. In addition, the immunomodulation ability of MSCs facilitates the therapeutic application in many scenarios. MSCs could affect the immune T-cell and B-cell response [29] by inhibiting T-cell proliferation, dendritic cells, and natural killer cell activation [30–32]. In the natural healing process, MSCs are recruited by inflammation or damage and modulate the inflammatory process [33, 34]. The cell recruitment and activation are regulated by bioactive cytokines; for example, monocyte chemoattractant protein-1 (MCP-1), macrophage inflammatory protein-1 alpha (MIP-1α, CCL3), and interleukin-8 (IL-8) are reported to enhance the migration of human bone marrow-derived mesenchymal stem cells (hBMSCs) [33]. Meanwhile, MSCs secrete a variety of cytokines, such as bone morphogenic proteins (BMPs), vascular endothelial growth factor (VEGF), and transforming growth factor-β (TGF-β), which promote vascularization, cell proliferation, cell differentiation, and endogenous tissue repair and regeneration [35]. Using MSCs in orthopedics as therapy is promising and practical because they involved in both tissue formation and immunomodulation. Here, we focus on several sources: MSCs that have been widely studied and used in orthopedic research and application, including BMSCs, ADSCs, TSPCs, ESC-derived MSCs and iPSC-derived MSCs, and the derivations of MSCs (e.g. exosomes).

4.2.2 Isolation, Identification, and Expansion of MSCs

Various protocols have been developed for MSC isolation from different species and tissue sources. Mechanical dissociation and enzymes (e.g. collagenase and trypsin) are generally employed for isolating MSCs from solid tissues [36]. MSCs from bone marrow tissue are usually isolated from native tissue using fast adhesion method followed by a density-gradient centrifugation over Ficoll or Percoll. During culture *in vitro*, the non-adherent hematopoietic cells were discarded with medium changes, and the fast adhesion cells (within ~36 hours) were collected, expanded, and characterized. MSCs generally appear as spindle-shaped fibroblast-like cells and adhere to culture plastic, and they form colonies and self-renew in the undifferentiated state. Scientists adopted multiple colony selection to amplify the MSC populations from mixture cultures in practice based on the fact that MSCs proliferate faster and form colonies; meanwhile, somatic cells are with limited proliferation capacities. For instance, when human MSCs were initially plated at 10 or 50 cells/cm^2, the cells expanded most rapidly on day 4 with a minimum average doubling time of about 10 hours [37]. After colony formation and cell expansion, the purity of MSCs can be further enhanced by surface marker-based selections. There is a great number of publications about the different cell surface markers of MSCs. The International Society for Cell Therapy (ISCT) proposed the basic characteristics of MSCs to unify the definition in 2006 because of the inconsistency on the defining characteristics of MSCs in early reports [38]. The criteria are (i) CD73, CD90, and CD105 positive; (ii) CD11b, CD14, CD19, CD34, CD45, and human leukocyte antigen (HLA-DR) negative; and (iii) possess the capacity to differentiate into osteoblasts, chondrocytes, and adipocytes *in vitro*.

After isolation, the *in vitro* culture conditions of MSCs are optimized by the tissue of cell derivation and their special requirements maintain the proliferation and differentiation capacity. The general used culture conditions are summarized in Table 4.1. Serum-supplemented Dulbecco's modified Eagle's medium (DMEM) or α-minimum essential medium (α-MEM) are widely applied basal medium for *in vitro* maintenance of MSCs. The *in vitro* culture and expansion are also important steps for the safety and efficacy of MSC-based therapy in clinical application, as they could influence the cell number, survival rate, and quality of therapeutic cells upon implantation. For instance, growth factors such as basic fibroblast growth factor (bFGF and FGF2), Wnt 3A, and interleukin-6 (IL-6) are known to increase the replicative capacity of MSCs [46–50]. Fetal calf serum (FCS) and fetal bovine serum (FBS) are frequently used to support MSCs growth; however, these xenogeneic products present disadvantages of potential immunogenicity and ethical concerns for clinical application. Human platelet lysate (hPL)-based biological factors were then explored as substitution for FCS or FBS [51]. hPL can stimulate the growth rate of ADSCs with the highest proliferation rate compared to cells in medium supplemented with FBS and bFGF [52]. In addition, bioreactors and physical signals such as mechanical signal [53], ultrasound [54], and electromagnetic fields [55] can also enhance the isolation and proliferation of stem cells, which are known for beneficial effects in stem cell cultures and have transient biological effects and easier synergy with other biochemical signals.

Further, long-term *in vitro* expansion may influence the biological functions of MSCs. BMSCs expressed lower levels of VCAM-1 and IL-6 in higher passages, which may be related to the immunomodulatory ability; meanwhile, other immunomodulatory cytokines such as indoleamine 2,3-dioxygenase (IDO), TGF-β, interleukin-10 (IL-10), and prostaglandin E2 (PGE2) had no significant changes after nine passages [56]. ADSCs displayed stable cytokine secretion functions with the increase of passages, exception of decreased IL-10 secretion in higher passages [56], and the level of cytokine IL-6 secreted from ADSCs was donor and site specific [52].

However, primary isolated MSCs cannot proliferate indefinitely *in vitro* because of both limited self-renewal ability and replicative cell senescence with decreased telomerase activity and telomere length [57, 58] and that is different from the MSCs generated from ESCs and iPSCs. ESCs are with the promise for cell therapy for they are in theory unlimited because of the embryonic derivation and can generate specialized cells [59]. However, the application of human ESCs is highly controversial because of ethical concerns [60, 61]. Allogenic ESCs from patients might inherit the oncogenic potential [62] and have the immunological rejection *in vivo*. iPSCs are reprogrammed somatic cells into a pluripotent ESC-like status by overexpressing a defined set of pluripotent transcription factors. These cells are firstly introduced by Takahashi and Yamanaka [63], and making autologous therapy with pluripotent and easily accessible cells becomes possible. The high proliferative capacity of iPSCs overcomes the limited availability of primary somatic cells such as chondrocytes. In addition, iPSCs can be induced to cells of all three germ layers, while tissue-specific stem cells or MSCs can only differentiate into several cell types. However, high rate of tumors occurred in mice generated from iPSCs that are currently far from being solved [64].

Table 4.1 Cell culture medium formulation, proliferation, and differentiation of mesenchymal stem cell *in vitro*.

Culture media (basic)	Proliferation (population doublings)	Differentiation (specific factors)			References
		Chondrogenic differentiation	Osteogenic differentiation	Adipogenic differentiation	
αMEM	10–30	FBS, TGF-β1, TGF-β3, insulin, and ascorbic acid	FBS, β-GP, dexamethasone, and ascorbic acid	FBS, insulin, dexamethasone, 3-isobutyl-1-methylxanthine, indomethacin, and hydrocortisone	[39–41]
DMEM	30–38	FBS, β-GP, dexamethasone, and ascorbic acid	FBS, TGF-βl, dexamethasone, ascorbic acid, and β-GP	FBS, hydrocortisone, 3-isobutyl-1-methylxanthine, and indomethacin	[40, 42]
Coon's modified Ham's F12 medium	22–23	FCS, ascorbic acid, and TGF-β1	FCS, ascorbic acid, β-GP, and dexamethasone	FCS, insulin, and dexamethasone	[43]
Duration		2–3 wk	2–3 wk	2–3 wk	[44]

DMEM, Dulbecco's modified Eagle's medium; αMEM, α-minimum essential medium; FCS, fetal calf serum; FBS, fetal bovine serum; β-GP, β-glycerophosphate; TGF-β1, transforming growth factor beta 1; and TGF-β3, transforming growth factor beta 3.

Source: Jiang et al. [45] With permission from Jiang Y, Wang D, Blocki A, Tuan RS. Mesenchymal stem cells in musculoskeletal tissue engineering, Table 49.2, In: Lanza R, Langer R, Vacanti J, Atala A, editors. Principle of Tissue Engineering (Fifth Edition). Elsevier, 2020, Chapter 49, pp. 883–915.

4.2.3 Current Cell Markers for MSC Identification

MSC cell surface marker profiles have stabilities and variations along with *in vitro* cell culture conditions and the *in vivo* tissue forming processes. For example, when MSCs are isolated freshly from bone marrow, the primary positive molecules to identify MSCs are CD105, CD73, and CD90 [65]. BMSCs gradually present a more uniform cell surface marker profile, while they were cultured for more than four passages *in vitro*. The colony-forming selected BMSCs present the same markers as their primary culture with CD105 (SH2), CD73 (SH3), and CD90; meanwhile, the other markers of CD29, CD44, CD71, CD106, CD120a, and CD124 are becoming positive, and CD14, CD34, and CD45 are negative [65]. Besides the ISCT general MSC cell markers [38], there are emerging additional cell surface markers to characterize the specific subtypes and subpopulations. This paragraph is a brief introduction of cell markers that had been used for identifying of MSCs from different tissues in the musculoskeletal system (Table 4.2).

Some surface markers of MSCs are inherent in native tissue, and some surface markers are obtained during culture. Among them, some are sensitive to culture conditions and passage number, and some can identify the subpopulations with specific functions. For example, CD106 expression declines in BMSCs after passages, especially after chondrogenic differentiation [41, 88]. hBMSCs cultured in FGF2-supplemented medium suppressed the positive rate of CD49, CD90, and CD146 [46]. $CD271^+$ hBMSCs involved in early osteogenesis and chondrogenesis, and they are able to repair chondral defects *in vivo* [89, 90]. Another subpopulation with higher proteoglycan production (an extracellular matrix [ECM] component generated by chondrocytes) can be identified by CD146, and $CD146^+$ hBMSCs present higher CFU-F and produce higher proteoglycan content than $CD146^-$ populations [91]. Meanwhile, $CD146^+$ BMSCs and $CD146^-$ BMSCs exhibited similar differentiation capacity to osteoblasts and adipocytes *in vitro* [92]. Additional surface markers were included apart from the minimum criteria of 2006 ISCT to further identify ADSCs [93]. For example, CD13, CD29, and CD44 are constitutively expressed at >80% of ADSCs, while CD31, CD45, and CD235a should be expressed less than 2% [93]. Compared to BMSCs, ADSCs display higher positive expression of CD34, CD36, and CD49d and negative expression of CD106 [57, 94]. However, it is difficult to distinguish ADSCs from normal fibroblasts, as both of them are positive for CD29, CD73, and CD105 and negative for CD14, CD31, and CD45 [95].

In terms of the tissue-specific stem/progenitor in joint tissues, infrapatellar fat pad-derived mesenchymal stem cells (IF-MSCs) share a similar profile with other MSC population for positive CD44, CD73, CD90, and CD105 and negative CD45, CD34, CD11b, CD19, and HLA-DR [96]. Studies also report unidentical phenotypes in some subpopulations of cultured IF-MSCs with relatively higher levels of CD34, CD45, CD106, and CD271 [66, 67, 97]. However, it remains unclear about the expression of CD34 [98], CD45 [67], and CD271 [99], while they are often reported to be negative on cultured IF-MSCs. Additionally, the surface markers are alterable based on the culture conditions, e.g. IF-MSCs exhibited high

Table 4.2 Summary of the markers expressed by human MSCs *in vitro*.

Cell types	Positive markers	Negative markers	Transition markers	References
BMSCs	CD9, CD13, CD29, CD44, CD49a, CD49b, CD49c, CD49d, CD49e, CD51, CD54, CD55, CD58, CD59, CD73, CD90, CD105, CD140b, CD146, CD147, CD151	CD3, CD11b, CD14, CD15, CD16, CD18, CD19, CD28, CD31, CD36, CD38, CD45, CD50, CD61, CD62e, CD62L, CD62P, CD102, CD104, CD133	CD10, CD31, CD34, CD49f, CD56, CD71, CD95, CD106, CD117, CD166, CD271	[66–72]
SM-MSCs	CD13, CD29, CD44, CD54, CD62, CD73, CD90, CD105, CD147, CD151	CD11a, CD14, CD31, CD49d, CD55, CD56, CD62, CD68, CD133	CD10, CD31, CD34, CD45, CD49a, CD106, CD117, CD146, CD166, CD271	[70–77]
SF-MSCs	CD13, CD44, CD49c, CD49d, CD49f, CD54, CD59, CD73, CD90, CD95, CD140b, CD147, CD151	CD31, CD36, CD45, CD117, CD271	CD10, CD34, CD71, CD105, CD106, CD166,	[66, 68, 78–80]
IF-MSCs	CD3, CD9, CD13, CD29, CD44, CD49c, CD49d, CD49e, CD49f, CD54, CD59 CD62e, CD73, CD90, CD105, CD140b, CD146, CD151, CD166	CD11a, CD11b, CD11c, CD14, CD18, CD31, CD133	CD31, CD34, CD36, CD45, CD50, CD56, CD106, CD117, CD271	[66, 68, 73, 81–83]
CDPCs	CD9, CD29, CD44, CD49c, CD49e CD49f, CD54, CD73, CD90, CD105, CD166	CD18, CD31, CD34, CD117, CD271	CD146	[84, 85]
TSPCs	CD44, CD90, CD133, CD146, Stro-l, Scx, Nestin, αSMA	CD34, CD45, CD106, CD117		[86]

BMSCs, bone marrow-derived MSCs; SM-MSCs, synovium-derived MSCs; SF-MSCs, synovial fluid-derived MSCs; IF-MSCs, infrapatellar fat pad-derived MSCs; CDPCs, chondrocyte-derived progenitor cells; and TSPCs, tendon-specific stem/progenitor cells.
Modified from Narcisi et al. 2017 [87].

CD10 level in culture medium added with either hPL or chemically reinforced media [99].

Tissue-specific stem/progenitors from cartilage (cartilage-derived MSCs, cartilage-derived stem/progenitor cells [CDPCs], CD-MSCs, and cartilage stem/progenitor cells [CSPCs]) can be isolated from primary isolated cells from human adult articular cartilage [9, 100] or menisci [101]. The articular cartilage-derived MSCs were positive for CD73, CD90, CD105 [38], CD29, CD49c, CD49d, CD49e, CD49f, CD106, CD146, CD9, and CD166 *in vitro* [102, 103]. Additionally, these cells display higher expression of CD44, which is a cell membrane receptor for hyaluronic acid (HA) [103]. A single-cell RNA-seq analysis identified the progenitor populations from meniscus tissue with a higher expression of $CD146^+$ cells [101]. CDPCs are found inherent in adult human articular cartilage, and they exhibited dynamic changes of phenotype from mature chondrocyte marker (COL2) to MSC markers (e.g. CD146 and many others) [9]. These cells were characterized by the ability to form colony and self-renew and possess multipotent differentiation capability and express positive stem cell markers such as STRO-1, CD9, CD29, CD44, CD54, CD73, CD90, CD105, and CD166 [84].

There are also TSPCs in tendon and ligament tissues. Distinct populations resident within tendon and ligament, including tendon fascicle-derived TSPCs, paratenon/epitenon-derived TSPCs, and perivascular TSPCs [104, 105]. Tendon fascicle-derived human tendon-derived stem/progenitor cells (hTSPCs) expressed specific transcriptional factor Scleraxis (Scx) as well as the MSC surface markers CD44, CD90, CD146, and Stro-l [86]. Animal model showed positive expression of Sca-1, Thy-1b (CD90.2), and CD44 in paratenon/epitenon-derived TSPCs, and higher expression levels of tenogenic marker for Scx and tenomodulin were exhibited in tendon-derived TSPCs compared to epitenon-derived TSPCs [106]. Human perivascular-derived TSPCs present Musashi-1, Scx, Nestin, alpha smooth muscle actin (αSMA), and CD133 [107], and a recent study identified Nestin as a novel marker for TSPCs in epitenon, paratenon, and perivascular niches [108].

Synovium-derived mesenchymal stem cells and synovial fluid-derived mesenchymal stem cells (SM-MSCs and SF-MSCs) can be isolated from and around synovium tissue. SM-MSCs express the mesenchymal markers CD29, CD44, CD16, and CD105 and present positive cell surface markers low-affinity nerve growth factor receptor (LNGFR) and THY-1 in an *ex vivo*-expanded population [109]. SF-MSCs are assumed to originate from SM-MSCs or BMSCs, and they express different levels of CD34, CD49c, CD54, CD71, CD106, CD117, CD166, and HLA-DR when compared to those of BMSCs [68, 78]. In addition to the mature tissues, the tissue-specific stem cell populations of human skeletal tissue were recently identified during tissue development. Human skeletal stem cells (hSSCs) have been recently identified from bone marrow cavity and articular cartilage in fetal and adult femoral heads [110], and they are positive of Podoplanin/CD146/CD73/CD164 within the growth plate and the diaphysis, which were confirmed by immunofluorescent staining. The commonly used cell surface markers to identify MSCs from different tissue sources in musculoskeletal system are summarized in Table 4.2.

In addition, human embryonic stem cell-induced mesenchymal stem cells (ESC-MSCs) exhibit a similar characteristic, spindle-shaped morphology as BMSCs. They express typical surface markers associated with adult MSCs, including CD13, CD15, CD105, CD106, CD271, and vascular endothelial growth factor A (VEGFA), but ESC-MSCs have lower expression level of these markers when compared to adult BMSCs. One of the characteristics of ESC-MSCs is the expression of hemangioblast cells and pericytes markers CD133 and FLK-1. The other is the expression of human telomerase reverse transcriptase (hTERT) and REX-1, which are typically associated with pluripotent stem cells. These unique expression profiles may mediate functional differences compared with MSCs from other sources [111]. Human-induced pluripotent stem MSCs (hiPS-MSCs) also produce colony-forming unit fibroblasts *in vitro*. They express a set of markers, including CD49d, CD73, CD90, CD105, CD140α, CD140β, CD271, STRO-1, and Vimentin, and these cells have no vascular endothelial CD34 and hematopoietic markers CD45. iPS-MSCs from mice exhibit specific surface markers for mice MSC such as CD105, CD140a, Sca-1, and CD44 [112, 113].

4.2.3.1 Differentiation of MSCs both *In Vitro* and *In Vivo*

MSCs can differentiate into osteoblasts, chondrocytes, fat cells, fibroblasts [114], neurons, glia cells [115], hepatocytes [116], and endothelial cells [117], which suggested that the differentiation potential of MSCs was not restricted to mesenchymal lineage, and they contribute to tissue development and formation in different ways. Therefore, many studies investigate the molecular regulation mechanism of MSC differentiation.

This paragraph will mainly introduce the musculoskeletogenic differentiation of MSCs. The research in development biology report valuable findings in embryonic skeletogenesis that clarify some signal pathways in MSC differentiation, such as WNT signaling, TGF-β/BMP signaling, and fibroblast growth factors (FGFs). The TGF-β family has been described widely in bone and cartilage development both in embryo and in adult [118, 119]. The modulation of TGF-β/BMPs in chondrogenesis and osteogenesis of MSCs has also been well established in mesenchymal cells [120–123]. The canonical Wnt signaling can stimulate osteogenic differentiation of MSCs and inhibit terminal differentiation of mature osteoblasts [124–127]. There are other signaling molecules that are participated in the MSC musculoskeletogenic differentiation pathways, such as FGF-2, epidermal growth factor (EGF)/epidermal growth factors receptor (EGFR) signaling, platelet-derived growth factor (PDGF), and Indian hedgehog (IHH)/Patched1 (PTCH1) signaling.

Chondrogenesis, osteogenesis, tenogenesis, and myogenesis differentiation of MSCs *in vitro* can be regulated by a number of factors, including growth factors, hormones, and extracellular matrix molecules (Table 4.1) [26]. For example, osteogenic induction media are usually supplemented with β-glycerophosphate, ascorbic acid, 1,25-dihydroxyvitamin D_3, and dexamethasone. The chondrogenic differentiation medium is DMEM that supplement with 1% insulin–transferrin–selenium premix, L-proline, TGF-β_3, and dexamethasone. In addition, a number of growth factors, including insulin-like growth factor 1 (IGF-1), EGF, VEGF, and PDGF, have also been reported to induce osteogenesis of MSCs [128]. Apart from chemical agents, physical stimuli, such as mechanical stimulation, pulsed electromagnetic field [129], and hydrostatic pressure, are also reported to promote MSC chondrogenesis and osteogenesis.

Osteogenesis and Bone Formation Osteogenesis and bone formation depend on the osteogenic differentiation of stem cells, and osteogenic differentiation is a complex process that is regulated by multiple endogenous and environmental factors [47, 130, 131]. In bone formation, the cells are derived from the MSCs and differentiated into osteoprogenitors, pre-osteoblast, osteoblast, and osteocytes (Figure 4.2, upper panel), which contribute to bone formation, and bone remodeling also depends on the activity of osteoclasts, which is believed to be differentiated from the myeloid precursors that are derived from HSCs (Figure 4.2, lower panel). TGF-β/BMPs signaling is an early regulator of MSC commitment to the osteogenic lineage [132]. Growth factors, such as IGF, FGF, and VEGF, are also involved in osteogenic differentiation [133–138]. WNT signaling is considered pivotal during osteoblast maturation, and osteoblast progenitors fail to pass through the terminal differentiation without Wnt proteins [139]. A Wnt antagonist was found to induce osteolytic lesions in multiple myeloma [140], while activation of the Wnt/β-catenin signaling is a frequent event in human osteosarcoma [141].

There are specific cellular markers to indicate the cell types during differentiation (Figure 4.2, upper panel); for instance, the MSC-derived osteoprogenitors express connective tissue growth factor (CTGF) [142]; RUNX2 and Osterix are osteogenic specific transcriptional factors during differentiation [143]; and Osterix is also required for the maturation of osteoblasts [144]. RUNX2 knockout in mice results in a complete lack of bone formation owing to maturational arrest of osteoblasts [145]. Osteopontin (OPN) and osteocalcin (OCN) are late markers in osteoblastic differentiation cascade [132]. In addition to osteoblasts and bone cells, osteoclastogenesis contribute to the bone resorption in bone tissue remodeling by their ability to degrade bone. The current understanding of osteoclast derivation is that (Figure 4.2, lower panel) the cells were originated from the monocyte/macrophage lineage cell through a stepwise process that is from HSCs [146–148]. First, the switch-on of the transcription factor named PU.1 drives the positive regulation of colony stimulating factor 1 (CSF-1) receptor/CSF1R (also known as macrophage colony-stimulating factor receptor [C-FMS]) in HSCs and activated the cell commitment toward myeloid progenitor cells. After the stimuli with macrophage colony-stimulating factor (M-CSF), these myeloid progenitor cells differentiated toward a monocyte–macrophage lineage (osteoclast precursors) following the expression of CSF1R. M-CSF and receptor activator of nuclear factor kappa-B ligand (RANKL) are two essential cytokines for osteoclastogenesis, and they trigger the differentiation of M-CSF-dependent progenitor cells to mononucleated pre-osteoclasts and further fuse to the mature multinucleated osteoclasts (Figure 4.2, lower panel) [147]. The activated osteoclasts secrete a cocktail of ions and degradative enzymes to break down the inorganic and organic components of

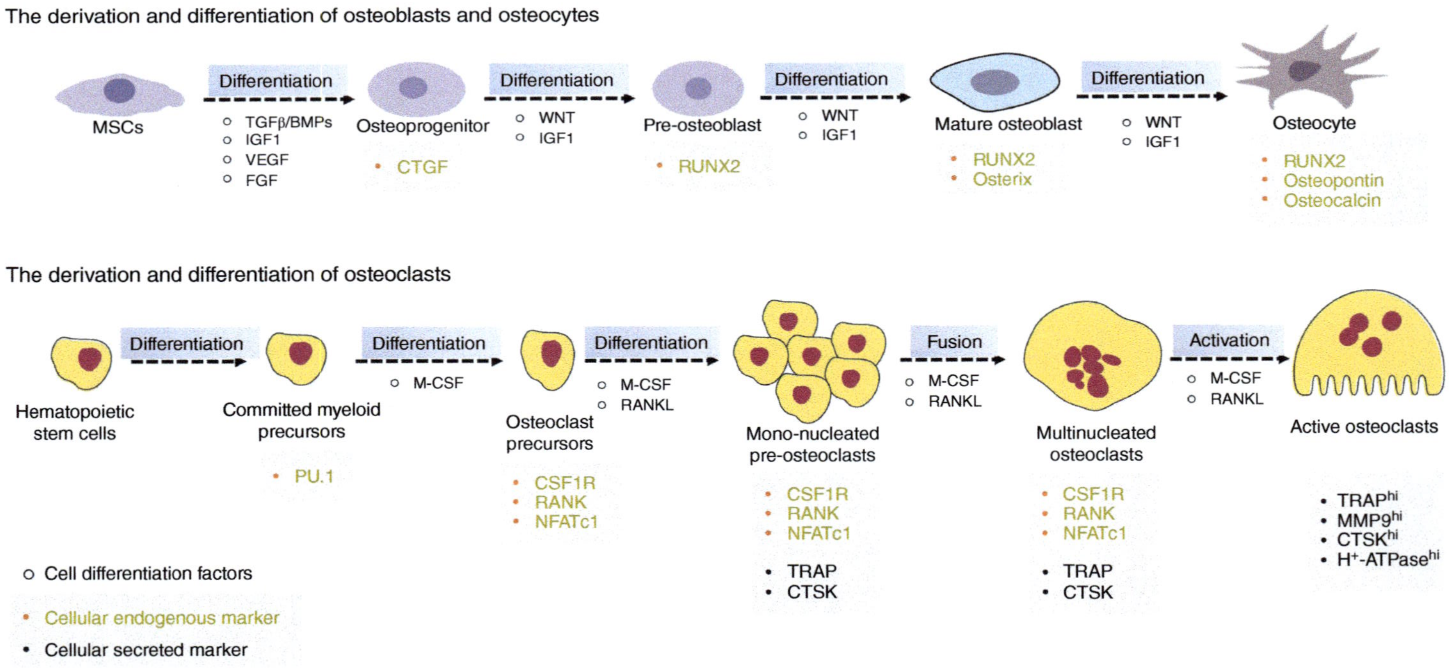

Figure 4.2 The derivation and differentiation of osteoblasts, osteocytes, and osteoclasts. (Upper panel) Cell derivation and major signal molecules involved in osteoblastogenesis and bone formation. (Lower panel) Cell derivation and major signal molecules involved in osteoclastogenesis. MSCs, mesenchymal stem cells; TGF-β, transforming growth factor beta; BMPs, bone morphogenic proteins; IGF1, insulin-like growth factor 1; VEGF, vascular endothelial growth factor; FGF, fibroblast growth factor; CTGF, connective tissue growth factor; WNT, wingless-type MMTV integration site; RUNX2, Runt-related transcription factor 2; M-CSF, macrophage colony-stimulating factor; RANKL, receptor activator of nuclear factor kappa-B ligand; CSF1R, colony stimulating factor 1 receptor; NFATc1, nuclear factor of activated T-cells, cytoplasmic 1; TRAP, TNF-receptor associated protein; CTSK, cathepsin K; and MMP9, matrix metallopeptidase 9.

bone, release minerals and factors embedded in the bone matrix, and regulate bone resorption and the levels of blood calcium.

In addition to the signaling factors, controlling the balance of several lineages of stem cell differentiation includes stimulations from microenvironment, such as chemical and physical factors. For instance, the stiffness of cell-adherent matrix could partially define the stem cell fate. Tensile strain and shear stress were demonstrated to enhance the osteogenic differentiation of hBMSC *in vitro* culture [149, 150]. Pulsed electromagnetic fields were tested to improve the osteoblastic differentiation of human mesenchymal stem cells (hMSCs) in the presence of BMP-2 *in vitro* [151–153]. As a load-bearing tissue, the *in vivo* bone formation and remodeling is also based on the mechanical loading, mechanical transduction, and related signaling, which have been recently investigated, and the focal adhesion kinase (FAK), Piezo1/2, Yes-associated protein (YAP1), are key factors involved [154–157].

The efficacy of transplanted BMSCs in cell therapy depends on their migration ability and capacity to differentiate into target tissues in injured sites such as in bone fractures [158]. Many factors were involved in osteogenesis and bone formation, such as stromal derived factor-1 (SDF-1), OPN, and extracellular matrix stiffness, which can regulate BMSCs homing [159]. The delivery methods of cells and bioactive molecules also influence the therapeutic efficacy of transplanted cells; direct injection of cells into joint cavities or injured tissues might reduce utilization of cells because of which the transplanted cells may spread. Cell sheet technique is a scaffold-free strategy to deliver cells in a unit that is superior to cell suspensions for 3D tissue formation [160, 161], which could be used in combination with scaffolds [162, 163]. Biomaterials, such as hydrogel [164], chitosan [165], hydroxyapatite [166], and collagen sponge-based scaffold [167], facilitate the expansion, differentiation, and delivery of BMSCs *in vivo*, with the advantages of supporting cells with a tissue mimetic microenvironment for bone tissue repair and regeneration. For instance, hydrogel substrates can improve hBMSC cell number with 25% increase in proliferation after seven days of culture compared to an increase of 10% in a conventional monolayer culture [164]. Gelatin-based structures, such as nanoparticles, microparticles, and 3D scaffold [168], were reported to induce MSC differentiation, and these constructs in combination with biochemical supplementation can induce osteogenic differentiation both *in vitro* and *in vivo*. A 3D collagen–glycosaminoglycan scaffold incorporated of calcium phosphate mineral improved new mineral synthesis of MSCs in osteogenic induction media [169]. Porous polylactic-*co*-glycolic acid (PLGA) scaffolds enhanced hBMSC expression of osteogenic genes OCN and OPN [170]. MSCs seeding on bio-scaffolds contributed to *in vivo* osteogenesis after long-term implantation [171]. However, these bio-scaffolds are still emphasized to provide an additional customized osteogenic induction environment for MSCs, and it is still a concern that the additional elements would cause more safety concerns in clinical applications. Induced pluripotent stem cell-derived mesenchymal stem cells (iMSCs) have also been proposed as an alternative to primary MSCs because if the decreased differentiation and proliferation capacity of primary MSCs with donor age and the duration of cell culture *in vitro* [172–174]. O. H. Jeon et al. co-cultured hiPSC-mesenchymal stem cells and macrophages on hydroxyapatite-coated PLGA/poly(L-lactic acid) (HA-PLGA/PLLA) scaffolds [175] and found mature bone-like tissue formation on such scaffolds during subcutaneous implantation in rodents.

Chondrogenesis and Cartilage Forming Growth factors, bio-scaffolds, and many other methods can enhance the chondrogenesis and cartilage forming abilities of stem cells. TGF-β and BMPs are mostly used for chondrogenesis; however, the requested subtypes of TGF-βs and BMPs are slightly different for BMSCs, ADSCs, and other derivation of MSCs [176]. For example, TGF-β1 and TGF-β3 are mostly used for BMSC chondrogenesis, and the chondrogenesis of ESC-MSCs is enhanced by adding BMP7 to the TGF-β1-supplemented induction medium [177]. Other growth factors and bio-factors have synergetic effect for MSC chondrogenesis and cartilage forming; for example, injections of TGF-β and ghrelin together with MSCs into rat osteochondral defect enhance the tissue regeneration [178]. Implantation of a composite hydrogel scaffold with a MSC homing peptide could increase the recruitment of endogenous MSCs, and repair the chondral defects in the rabbit model [179]. Subcutaneous implantation of pre-chondrogenic differentiated MSCs with scaffold in rats induced high levels of chondrogenic markers and osteogenic markers in the implanted MSC constructs [180]. Natural cartilage ECM-derived particles (CEDPs) can also be used as MSC carriers for cartilage formation [181]; BMSCs proliferated rapidly on the surface of CEDPs and differentiated into mature chondrocytes *in vitro*. These BMSC-laden CEDPs were used to repair trochlear cartilage defects of rats, and the joint function fully recovered after surgery [181].

CSPCs are TSPCs with significant chondrogenic and cartilage forming ability and have the ability to migrate to injury sites [182, 183], indicating the function to maintain tissue homeostasis [84]. CSPCs were found in adult human hyaline cartilage tissue (also known as cartilage/chondrocyte-derived progenitor cells, CDPCs) and have been demonstrated to form cartilage [9]. A low cell density (500 cells/cm^2, <50% confluence) and low glucose medium can promote CDPC proliferation *in vitro*, and they exhibited greater chondrogenic potential than BMSCs and can form hyaline-like cartilage tissue in nude mice without additional growth factor [9]. The CDPCs can be used in clinical therapy safely, and they have been applied to repair large-sized cartilage defects (6–13 cm^2) in young patients; the defects are fully filled with cartilage and joint function returned to normal after one year follow-up [9].

Tenogenesis Growth factors such as EGF, bFGF, and TGF-β1 and platelet-derived growth factor BB (PDGF-BB) were demonstrated to involve in tendon development and healing [184]; however, there are no standard induction protocols to induce MSCs into the tenogenic lineage. TGF-β3 induced an early expression of tendon marker Scx; meanwhile, BMP12, bFGF, and ascorbic acid could also induce the expression of tendon markers Decorin at later stage of differentiation [185]. Bioactive scaffolds consisting of silk and bFGF-releasing PLGA can stimulate MSCs to generate tendon/ligament both *in vitro* and *in vivo* [186]. In addition to bioactive factors, tensional and torsional mechanical stimulations are important for MSC tenogenesis, e.g. mechanical stimulation promoted MSCs to tenogenic differentiation in *in vitro* 3D-multilayered composite scaffolds and collagen gels [187, 188].

Similar to cartilage, TSPCs were identified [106] to reside within a unique niche that is composed of an extracellular matrix [86]. A variety of growth factors have been reported to influence TSPC proliferation and differentiation, including CTGF, PDGF-BB, bFGF, IGF1, and other members from TGF-β superfamily [106]. Recent studies have shown that platelet-rich plasma (PRP) treatments and BMSC exosomes can promote the proliferation of TSPCs both *in vivo* and *in vitro* [189–192]. TSPCs regulate inflammation during healing of acute tendon injuries via Jun N-terminal kinase (JNK) and STAT3 signaling [193]. In addition, TSPCs isolated from human patellar tendon can be differentiated toward fibrochondrogenic lineage, which implicated the potential application of TSPCs in bone–tendon junction repair [194]. To enhance tenogenesis *in vivo*, BMSCs were co-cultured with TSPC sheets, and the co-cultures showed a superior tendon healing effect compared with the single-cell source of BMSCs or TSPCs in a rat patellar tendon window injury model [190]. Similar regenerating effects of BMSCs and TSPCs were found in Achilles tendon rupture rats [195]. ADSCs mixed with PRP gels were also shown to enhance tendon healing and tensile strengths in rabbit models [196]. A randomized controlled trial on the therapy of tendonitis in equines revealed that ADSCs mixed with platelet concentration resulted in a greater organization of collagen fibers and decreased inflammation [197]. However, controversial results were reported in another study, which observed no superior effects in ADSC group compared to autologous serum group in tendonitis models [198]. The controversial results suggested that there are different effects for specific cell types for tenogenesis in disease conditions. In addition, clinical studies revealed that application of MSCs with HA preserved joint space and reduced degradation of cartilage volume in patients underwent anterior cruciate ligament (ACL) reconstruction [199]; therefore, more studies are needed for developing the clinical applications of MSCs for tendon diseases.

Myogenesis Satellite cells are skeletal muscle-specific unipotent stem cells that connect to the myofibers normally and remain in a quiescent state. In muscle injuries, various cytokines and growth factors are released by damaged tissue and activate satellite cells to proliferate and generate myogenic progenitors or myoblasts. Expression of C-X-C chemokine receptor type 4 (CXCR-4 or CD184) and muscle-derived fibroblast SDF-1 indicate the presence of muscle progenitor cells in muscles [200]. SRY-Box transcription factor 7 (SOX7) is a quiescent satellite cell marker [201], while desmin, myogenic factor 5 (Myf-5), myoblast determination protein (MyoD), and proliferating cell nuclear antigen (PCNA) are markers of proliferating satellite cells, indicating an active stage of the resident satellite cells [202–204]. The injured muscle could induce the release of growth factors such as FGF and hepatocyte growth factor and stimulate satellite cell proliferation and myogenesis [205]. However, it is still difficult to directly apply satellite cells as cell therapy in orthopedics because they are with high heterogeneity, small quantity, and limited myogenic differentiation ability.

The myogenesis of BMSCs are under investigation; for example, overexpression of notch intracellular domain (NICD), β-catenin, and Pax3 can induce the differentiation of MSCs toward the myogenic lineage [206–208]. Application of cyclic strain could accelerate skeletal myogenic differentiation of mouse BMSCs [209]. BMSCs are able to support skeletal muscle reconstruction by paracrine effect; IGF-1, IL-4, IL-6, and SDF-1 are known cytokines for BMSCs to interact with myoblasts [210]. BMSCs were also reported to facilitate muscle regeneration in Duchenne muscular dystrophy model mice by targeting C-X-C motif chemokine 12 (CXCL12) on muscle satellite cells [211]. *In vivo* experiment demonstrated that $CXCR4^{+}$ BMSCs promoted new myofiber formation in injured muscles [212].

In summary, with the accumulation of published data, the differentiation capabilities of MSCs are found to be affected by the source and status of MSCs. When comparing MSCs from fetal bone marrow, liver, lung, and spleen, the liver-derived MSCs showed less osteogenic differentiation and MSCs derived from spleen displayed inferior adipogenic differentiation ability [213]. Furthermore, the target tissues or neighboring tissues may influence the differentiation capacity of MSCs *in vivo*. For example, TSPCs are reported to reside within a unique niche, and depletion of components of the niche, such as biglycan and fibromodulin, affects the differentiation of TSPCs and impairs tendon formation *in vivo* [86]. These key elements need to be taken into consideration when designing the therapeutic strategies for applying stem cells for specific disease.

4.3 MSC-Derived Extracellular Vesicles (MSC-EVs) and Exosomes

Numerous studies revealed that the immunomodulatory effects of MSCs were attributed to MSC-sourced secretomes, such as soluble components and encapsulated extracellular vesicles (MSC-EVs) and exosomes [214]. MSC-derived exosomes are relatively small with a size of 30–200 nm [215]. These MSC-EVs and exosomes exert biological effects in a paracrine or endocrine manner. They are enriched with MSC-sourced bioactive molecules, including messenger RNAs, microRNAs, enzymes, cytokines, chemokines, immunomodulatory, and growth factors [216]. They express adhesion molecules (CD29, CD44, and CD73), which can guide them to the damaged tissues [215]. The paracrine secretion capacity is also one of the principal mechanisms that MSCs contribute to tissue regeneration via mediating tissue repair [217, 218]. As a cell-free product, the advantage of using MSC-EVs and exosomes instead of MSCs is that they can avoid the potential *in situ* clots of using cells and concerns about the long-term survival of engrafted cells, such as cell differentiation and possible malignant alteration [216]. ADSC-EVs and exosomes promote C2C12 cell proliferation [219], and intravenous injections of ADSC-EVs modulate different aspects of skeletal muscle regeneration in a cardiotoxin-induced acute skeletal muscle injury model [220]. ADSC-EVs could affect M1/M2 balance of infiltrating macrophages in muscle injury [221]. There are also other applications in orthopedics, MSC-derived exosomes can repair temporomandibular joint osteoarthritis in a rat model by reducing inflammation and improving matrix expression and subchondral bone architecture [222]. EVs and exosomes isolated from MSC-conditioned medium stimulated chondrocyte proliferation and bone regeneration in osteogenesis imperfecta animal model [223, 224]. Another advantage of using MSC-EVs and exosomes is that they can avoid the potential tumorigenesis in ESCs and iPSC-derived cells [225, 226], and both induced-pluripotent stem cell-mesenchymal stem cell-derived exosomes (iPSC-MSCs-Exos) and synovium-mesenchymal stem cell-derived exosomes (SM-MSC-Exos) can stimulate chondrocyte migration and proliferation in mouse OA model, and iPSC-MSCs-Exos have a superior therapeutic effect compared with SM-MSC-Exos [225]. In addition, exosomes from ESC-MSCs show a therapeutic effect on OA with decreased cartilage destruction and matrix degradation in the destabilization of the medial meniscus (DMM) model [226].

However, the clinical application of MSC-EVs needs to overcome several problems. It is difficult to produce sustained EVs and exosomes in large quantities [227], and the delivery method of this biological product is another big challenge. Recent research suggested integration of stem cell-derived exosomes with *in situ* hydrogel glue as a tissue patch is a solution for articular cartilage regeneration in a rabbit model [158].

4.4 Clinical Application of Stem Cells in Orthopedics

Since significant progress has been made in the past few decades and clinical protocols for the use of stem cells in orthopedics have been gradually established, in this paragraph, we listed and summarized some interesting clinical results of stem cell therapies in the treatment of bone, cartilage, tendon ligament, and spinal diseases.

4.4.1 Bone Defects, Non-union, and Osteogenesis Imperfecta

MSCs were utilized to enhance bone regeneration and union in the treatment of non-union, bone defect and to improve bone quality in osteogenesis imperfecta.

4.4.1.1 Bone Non-union and Defects

Delayed bone union or bone non-union after fracture usually needs revision surgeries. Percutaneous bone marrow grafting has shown promise to promote bone union in patients in a simple and minimally invasive way. Le Nail et al. reported that the percutaneous grafting with bone marrow autologous concentrate for open tibia fractures with non-union or delayed union could provide 53.5% (23/43) successful rate in union [228]. Wang et al. have recently employed an autologous BMSC-enriched β-tricalcium phosphate scaffold as a substitute for autologous bone graft to treat diaphyseal bone non-union, which can overcome the adverse events such as local hematoma, infection, and chronic pain because of autograft harvesting and achieve tissue regeneration. After nine months of follow-up, 27 patients (27/30; 90%) in the BMSC-enriched scaffold group acquired clinical union compared with

18 patients (18/20; 90%) in the autologous bone graft group, and the postoperative radiographic union score at nine months post-operation was similar between the two groups [229]. Thus, the BMSC-enriched scaffold represents a safe and effective treatment for bone non-union. In addition, there are clinical case reports of using MSCs for bone defects. Quarto et al. employed autologous MSCs in a macroporous hydroxyapatite scaffold to treat segmental bone defects in three patients with large bone defects. All the three patients showed abundant callus formation along the implants and good integration at the interfaces with the host bones by the second month after surgery [230].

4.4.1.2 Osteogenesis Imperfecta (OI)

OI is a genetic or heritable disorder characterized by osteopenia, giving rise to frequent fractures without obvious cause or injury, bone fragility, and deformities [231]. OI is mainly caused by mutations or defects in the gene expression of type I collagen, which led to the insufficiency or incorrect expression of type I collagen and the subsequent weakness of bone. Horwitz et al. reported the clinical case of three OI patients receiving transplantation of unmanipulated bone marrow donated from HLA-identical or single-antigen-mismatched siblings [12]. Significant improvement was found in all the three patients after 216 days of transplantation, a 13-year-old female OI patient showed approximately four times higher number of osteoblasts per high-power field (16.0 ± 3.0 versus 4.6 ± 1.8) from the biopsy of iliac wing, and reduced bone fracture occurrence rate was found in other patients (Patient 1: 3 versus 37 times; Patient 2: 2 versus 20 times). Three more patients were then enrolled into another clinical study organized by the same group, in which the allogeneic bone marrow transplantation was used as a treatment. The body bone mineral content of these patients increased at levels of 21.6–65.3 g above the baseline after three months of transplantation, and the OI patients had a median 7.5 cm increase in body length (range, 6.5–8.0 cm) compared with 1.25 cm (range, 1.0–1.5 cm) for age-matched untreated OI patients (as control group) after six months [232].

4.4.2 Osteochondral Defect and Osteoarthritis (OA)

Microfracture is a convenient and economic clinical solution for small- to medium-sized cartilage/osteochondral defects by creating channels on the subchondral bone plate, and allowing the bone marrow to flow and to fill, and the cartilage repair/resurfacing resulting from this type of marrow stimulation techniques is thought to be partially because of the recruitment and local differentiation of BMSCs. Stem cell-based therapy has been widely used to regenerate osteochondral defects before it progresses to OA. The most frequently used cells in clinical application and clinical studies for tissue formation include autologous chondrocytes [11, 233], adult stem cells such as cartilage-derived progenitors [9], and BMSCs [10, 234, 235]. Details of these clinical trials and studies are summarized in Table 4.3 [154]. Stem cell-based therapies are also widely investigated as potential therapy for degenerative diseases such as OA because of their

Table 4.3 Stem cell-based therapy for OA and cartilage defect.

Disease condition	Cell sources	Sample size and group	Placebo/control	Follow-up	Route of administration	Main results	Study group
Cartilage defect	CDPCs	15 patients	N/A	12 mo	MACI, mini-arthrotomy	The cartilage defects (6–13 cm^2) were completely repaired. Knee function (IKDC scores and Lysholm scores) was significantly improved	Jiang et al. [9]
	BMSCs	24 patients; BMSCs ($n = 12$), Placebo ($n = 12$)	Cell-free collagen gel sheet	10.5 mo (28–95 wk)	Implantation at the time of HTO	Arthroscopic and histological grading scores were significantly improved, but no significant clinical improvement	Wakitani et al. [10]
		3 patients	N/A	27 mo	Transplantation surgery	The cartilage defects were repaired, and clinical symptoms were improved	Wakitani et al. [236]
		1 patient	N/A	12 mo	Arthroscopic surgery	The defect was recovered with hyaline-like cartilage tissue, and clinical symptoms were improved	Kuroda et al. [235]
	SM-MSCs	10 patients (symptomatic single cartilage lesion in femoral condyle)	N/A	52 mo (37–80 mo)	Arthroscopic surgery	The MRI score and Lysholm score were significantly improved compared to pre-surgery	Sekiya et al. [237]
		5 patients (symptomatic knee chondral lesions; 1.5–3.0 cm^2)	N/A	24 mo	Mini-arthrotomy	The defects were successfully repaired. In addition, the VAS, Lysholm, and KOOS scores were significantly improved compared to pre-surgery	Shimomura et al. [238]

(*continued*)

Table 4.3 (Continued)

Diseasep condition	Cell sources	Sample sizep and group	Placebo/control	Follow-up	Route of administration	Main results	Study group
Osteoarthritis	Bone marrow aspirate concentrate (BMAC)	25 patients (bilateral knee OA); BMACs into one knee joint and placebo into the contralateral knee	Sterile saline	6 mo	Intra-articular injection	The pain relief (VAS scores and ICOAP scores) was significant compared to the baseline, but was non-significant compared to the saline-treated contralateral knee	Shapiro et al. [239] Phase I
	BMSCs	60 patients; Cohort1: BMSCs dose1(2.5×10^7 cells; $n = 10$), Placebo ($n = 5$); Cohort2: BMSCs dose2 (5×10^7 cells; $n = 10$), Placebo ($n = 5$); Cohort3: BMSCs dose3 (7.5×10^7 cells; $n = 10$), Placebo ($n = 5$); Cohort4: BMSCs dose4 (1.5×10^8 cells; $n = 10$), Placebo ($n = 5$)	PLASMA-LYTE A (a multiple electrolytes injection)	12 mo	Intra-articular injection	The lowest dose cohort (Cohort1; 2.5×10^7 cells) showed a trend toward improvement in all parameters including VAS, WOMAC, and ICOAP scores (though non-significant compared to the placebo group), while other higher doses cohorts neither showed significant symptomatic relief, nor showed acceptable safety	Gupta et al. [240] Phase II
	ADSCs	52 patients; ADSCs group ($n = 26$); Placebo ($n = 26$)	1% sodium hyaluronic acid	12 mo	Intra-articular injection	ADSCs group showed significant improvement in joint (WOMAC score) and cartilage volume (MRI quantification) in comparison with both baseline and placebo group	Lu et al. [241] Phase IIb

		18 patients; ADSCs low-dose: (1×10^7 cells; $n = 6$); ADSCs mid-dose: (2×10^7 cells; $n = 6$); ADSCs high-dose: (5×10^7 cells; $n = 6$)	N/A	24 mo	Intra-articular injection	ADSCs group with different dosage all showed significant improvement in joint function (WOMAC scores), life quality (SF-36 scores) and cartilage volume (MRI quantification). The high-dose group (5×10^7 cells) exhibited the highest improvement	Song et al. [242] Phase I/II
		24 patients; ADSCs group (1×10^8 cells; $n = 12$); Placebo ($n = 12$)	Sterile saline	6 mo	Intra-articular injection	ADSCs group showed a significant functional improvement and pain relief in comparison with baseline, without cartilage volume changes at 6 mo follow-up	Lee et al. [243] Phase IIb
		30 patients; ADSCs single-injection group: (1×10^8 cells; $n = 10$); ADSCs twice-injection group: (1×10^8 cells, injection at baseline and 6 mo; $n = 10$); Control group: ($n = 10$)	Conservative management	6 mo	Intra-articular injection	The pain, function, and life quality of both treatment groups were significantly improved compared to the control group	Freitag et al. [244] Phase II
	UC-MSCs	26 patients; single-dose UC-MSCs (20×10^6 cells, $n = 9$); repeated-dose UC-MSCs (20×10^6 cells, twice a year, $n = 9$); Placebo ($n = 8$)	Hyaluronic acid	12 mo	Intra-articular injection	Pain (WOMAC-A, pain scores) and function (WOMAC–C, function scores) were significantly improved in comparison with placebo group. Repeated-dose UC-MSCs (20×10^6 cells, twice a year) exhibited a superior effect to placebo group and single-dose group	Matas et al. [245] Phase I/II

(*continued*)

Table 4.3 (Continued)

Diseasep condition	Cell sources	Sample sizep and group	Placebo/control	Follow-up	Route of administration	Main results	Study group
		6 patients; Low-dose group ($1.15–1.25 \times 10^7$ cells; $n = 3$) High-dose group ($1.65–2.0 \times 10^7$ cells; $n = 3$)	N/A	7 yr	Intra-articular injection	Maturing repair tissue was observed at the 12 wk arthroscopic evaluation. The VAS and IKDC scores were improved at 24 wk. The improved clinical outcomes were stable over 7 yr of follow-up The histological findings at 1 yr showed hyaline-like cartilage MRI at 3 yr showed persistence of the regenerated cartilage No cases of osteogenesis or tumorigenesis was observed over 7 yr	Park et al. [246] Phase I/II

Abbreviations: VAS, visual analog scale; WOMAC, Western Ontario and Mc Master Universities Arthritis Index; ICOAP, intermittent and constant osteoarthritis pain; BMSCs, bone marrow-derived mesenchymal stromal cells; ADSCs, adipose-derived mesenchymal stromal cells; SM-MSCs, synovium-derived mesenchymal stromal cells; UC-MSCs, umbilical cord-derived mesenchymal stromal cells; MACI, matrix induced autologous chondrocyte implantation; MRI, magnetic resonance imaging; and HTO, high tibial osteotomy.
Source: Zhu et al. [154] / Frontiers Media / CC BY-4.0.

immunomodulatory ability. There are emerging phase I/II clinical trials employing BMSCs [239, 240], ADSCs [241, 242], and other allogenically derived MSCs such as umbilical cord-derived mesenchymal stromal cells [245] to treat OA knees. For instance, two phase II clinical trials showed that intra-articular injections of human ADSCs (5×10^7 cells) improved both pain condition and cartilage volume [241, 242]. The detailed information of these clinical trials is summarized in Table 4.3 [154].

Recently, clinical studies also show exciting results of using MSCs to strengthen subchondral bone to prevent OA progression, and the cell delivery method is different from the conventional intra-articular injection. Hernigou et al. released a series of clinical trial results of using BMSCs to treat subchondral bone marrow lesions (BMLs) in OA joints (Table 4.4) [15, 247, 248], with a clinical data of up to 15 years of follow-ups. Patients with different OA conditions (moderate to advanced OA, 60 cases; advanced OA, 140 cases; and OA secondary to knee osteonecrosis related to corticosteroids, 30 cases) were recruited, and these patients have comparable clinical symptom in bilateral knees and were randomized to accept subchondral bone MSC therapy or other surgical treatments. The patients were treated with subchondral bone MSC therapy on one knee, and the other knee was intra-articularly injected with MSCs ($n = 60$) or treated with total knee arthroplasty (TKA, $n = 140$) (Table 4.4 [154], Study 1 and Study 2). The clinical research suggested that (i) patients with subchondral bone MSC therapy postponed TKA for more than 10 years; (ii) patients with precise delivery of MSCs to subchondral bone shared similar clinical results (Knee score) as patients with TKA in both short- and long-term follow ups; and (iii) patients showed reduced size of BMLs after subchondral bone MSC therapy, which postponed or prevented the first TKA surgery in young patients with secondary osteonecrosis related to corticosteroids (Table 4.4, Study 3). It is quite possible that the local delivery of BMSCs enhanced the subchondral bone regeneration of the BMLs in OA joints and thus successfully prevented the further joint degeneration and delayed the time point of TKA in patients.

4.4.3 Tendon and Ligament Injury

Tendon and ligament injuries are common sports injury with limited regenerative ability. Once injured, tendon and ligament fail to retain the same structure, composition, and function as before. Application of biological grafts such as autografts, allografts, and resorbable biomaterials have the risk of various complications such as donor site morbidity, scar formation, infection, and tissue rejection. Therefore, enhancing the endogenous tissue repair ability with MSCs and TSPCs is under investigation.

4.4.3.1 Rotator Cuff Tear

Hernigou et al. conducted a case–control clinical study to evaluate the efficacy of biological augmentation of rotator cuff repair with autologous iliac crest BMSCs under arthroscopy. According to the 10-year follow-up study with ultrasound and magnetic resonance imaging (MRI) results, they found that the BMSC injection at

Table 4.4 Clinical results for the subchondral bone MSC therapy for OA patients.

Bilateral clinical study	OA condition before treatment	Number of knees	Age/gender	Treatment details	Clinical results		References
					2 yr follow-up	Long-term follow-up	
1 MSC SCB injection	Bilateral knee osteoarthritis	60	61 yr (48–72); 25M/35F	5727 BMSCs/ml; 20 ml (10 ml in medial tibial plateau; 10 ml in medial femoral condyle)	Knee score: 79.3 ± 12 Elevated K-L score: 11/60	Yearly arthroplasty incidence: 1.3% TKA incidence: 20% (12/60)	[247]
MSC IA injection		60		5727 BMSCs/ml; 20 ml into joint	Knee score: 64 ± 21 Elevated K–L score: 2/60	Yearly arthroplasty incidence 4.6%; TKA incidence: 70% (42/60)	
2 SCB MSC therapy	Bilateral osteoarthritis; secondary to knee ON related to corticosteroids	30	28 yr (18–40); 12M/18F	6500 BMSCs/ml; 40 ml (10 ml in medial tibial plateau; 10 ml in medial femoral condyle; 10 ml in lateral tibial plateau; 10 ml in lateral femoral condyle)	Knee score (3 mo): 81.3 ± 12	Knee score: 80.3 ± 11 TKA incidence: 10% (3/30)	[15]
TKA		30		TKA	Knee scores (3 mo): 79 ± 21	Knee score: 78.3 ± 23 TKA revision rate: 20% (6/30); re-revision: (2/6)	
3 SCB MSC therapy	Planned for staged-bilateral (TKA) for medial osteoarthritis, "comparable" pain in both knees	140	75.4 yr (65–90); 53M/87F	7800 BMSCs/mL; 20 ml, 10 ml in medial tibia plateau and 10 in. femoral condyle	Knee score: 81.3 ± 12	TKA incidence: 25% (25/140)	[248]
TKA		140		TKA	Knee score: 79 ± 21	TKA revision rate:15% (21/140)	

SCB, subchondral bone; MSC, mesenchymal stem cells; IA, intra-articular injection; and TKA, total knee arthroplasty.
Source: Zhu et al. [154] / Frontiers Media / CC BY-4.0.

a cell number level of 51 000 ± 25 000 could sufficiently enhance the healing rate of rotator cuff repair. In addition, BMSC injection prevented further ruptures (39 intact in 45 patients in the BMSC-treated group versus 20 intact in 45 patients in the control group) in the 10 years of follow-up period [16]. Interestingly, the same group found that patients with symptomatic rotator cuff tear have less endogenous MSCs at the tendon–bone interface tuberosity in comparison with those without rotator cuff injury [249], which could support the therapeutic effects of BMSC therapy.

4.4.3.2 Anterior Cruciate Ligament (ACL) Reconstruction

ACL rupture is another common sports injury and is regarded as an important risk factor for the development of OA, which accounts for an approximately a quarter of the overall knee OA population. To prevent OA after ACL injury, Wang et al. performed a double-blinded, randomized controlled trial regarding a single intra-articular injection of 75 million allogeneic immature mesenchymal precursor cells (MPCs) in a HA vehicle (MPCs + HA group, $n = 11$) for patients undergoing anterior cruciate ligament reconstruction (ACLR) [250], and HA injection as control (HA group, $n = 6$). MPC + HA-treated patients showed improvement on joint space width and less tibial bone expansion compared with the control group after 24 months with significant improvements in knee injury and osteoarthritis outcome score (KOOS) pain scores, symptom, activities of daily living, and SF-36 pain scores. These findings suggest that MPCs have the possibility to become a beneficial step for ACLR in ACL rupture patients and could prevent the progression of post-traumatic OA.

4.4.4 Spine

4.4.4.1 Spinal Fusion

Gan et al. reported the clinical use of enriched autologous bone marrow stem cells combined with porous β-tricalcium phosphate in posterior spinal fusion in patients with thoracolumbar fracture ($n = 19$) and degenerative disc disease ($n = 22$). After 34.5 months of follow-up, 95.1% patients had satisfying fusion results. Only four patients had a little exudation or swelling in wounds and recovered with conservative treatment [251]. A recent phase I/II prospective clinical trial also showed that the implantation of autologous BMSCs embedded with tricalcium phosphate in posterolateral spinal fusion is a safe and feasible procedure for patients with degenerative disc diseases. The clinical symptoms for back and radicular pain were relieved, and all patients except one resumed their work activities in a five-year follow-up period [252].

4.4.4.2 Spinal Cord Injuries (SCI)

A spinal cord injury (SCI) is the damage to the spinal cord caused by vehicle accidents, falls, violence, or sports-related injuries. SCI causes sensory loss and motor paralysis. They are normally treated with physical therapy, but many patients fail to recover because of the limited neural regeneration. Stem cell-based therapy may provide a potential treatment for intractable SCI.

Spinal cord and peripheral nerve injuries have a strong impact on affected patients. Recently, a single-center phase I/II clinical trial in which chronic SCI patients received four subarachnoid transplantations of allogenic hUC (human umbilical cord mesenchymal stem cells)-MSCs (monthly; 1×10^6 cells/kg) and were followed up for four times (1, 3, 6, and 12 months after final administration) was conducted. In total, 102 patients performed the safety assessment and finally 41 patients completed the efficacy assessment. Totally, 383 intra-thecal transplantations of hUC-MSCs were performed. Primary efficacy outcomes (IANR-SCIFRS total scores and American Spinal Injury Association [ASIA] scores, Table 4.5 [253] and Table 4.6 [254]) improved when compared with the baselines at different time points during the follow-up. Secondary efficacy outcomes evaluated by bladder and bowel functions, residual urine volume, and MRI also showed significant improvement at a final follow-up on muscle spasm in comparison with the baseline. There were no serious adverse effects (AEs) during the study. Mild AEs were reported after transplantation, such as fever (54/383; 14.1%), headache (16/383; 4.2%), transient increase in muscle tension (6/383; 1.6%), and dizziness (5/383; 1.3%), and they were resolved under conservative therapy [255].

In the clinical trial conducted by El-Kheir et al., similar improvements on ASIA scores and functional measurements were found after intrathecally administration of autologous BMSCs. Of note, they also found that thoracic SCI patients with shorter durations of injury and smaller cord lesions tended to achieve a higher rate of functional improvement than those with cervical injuries [256].

4.5 Considerations of Stem Cells and Derivations for Clinical Usage

To ensure the safety and effectiveness of cell/stem cell-based therapy, there are many concerns that need to be solved from bench side to clinical use. The donor (i.e. autologous, allogenic, and xenogeneic), origin of tissue, cell isolation and expansion procedure yield rate [257], quality control, cell number, dosage, frequency, and application methods will influence the therapeutic efficacy [258]. For instance, it was highly inconsistent in the preparation and bone marrow aspirate concentrate in published clinical studies about the application of bone marrow aspirate in musculoskeletal care [259]. In addition, because of the variation of PRP preparation and composition in current reports, it is difficult to compare the effects of PRP products between patients [260]. There has also been worked to explore the optimal protocol of stem cell culture and applications such as cell seeding densities, which showed different differentiation abilities *in vitro* and tissue formation abilities when applied *in vivo* [261, 262]. Therefore, standardized and stepwise descriptions of cell therapy protocol should be included in future studies to make the application of cells reproductive.

Storage of cell products by cryopreservation is crucial for clinical therapeutic purposes. Cryopreservation was reported to have no effect on BMSC morphology, proliferation, and differentiation potential, although there were still variabilities about the

Table 4.5 International association of neurorestoratology spinal cord injury functional rating scale (IANR-SCIFRS).

A. Upper limb movement	
(1) Eating and drinking	(3) Dressing
3 Normal	3 Normal
2 Finish independently with difficulty	2 Finish independently with difficulty
1 Some assistance	1 Some assistance
0 Total dependence	0 Total dependence
(2) Grooming	E. Sphincter control
3 Normal	(1) Bladder control
2 Finish independently with difficulty	3 Normal
1 Some assistance	2 Reflex voiding with partial sensation or control
0 Total dependence	1 Reflex voiding without sensation or control
	0 Total incontinence or urethral catheterization/cystostomy needed
	(2) Bowel control
	3 Normal
	2 Partial control with sensation
	1 Partial control without sensation or no control with
	0 Total incontinence (manual stimulation or laxative
	3 Normal required)
(3) Writing	F. Muscular tension (refers to the amount of tension or resistance to movement in a muscle)
3 Normal	3 Normal
2 Slow or sloppy; most words legible 3 Normal	2 Slight increase/decrease or mild spasm
1 Many words not legible 2 Partial control with sensation	1 Large increase/decrease or significant spasm
0 Unable to grip pen	0 Extreme stiffness or spasticity
B. Lower limb movement	G. Sweating
(1) Standing without brace	3 Normal
3 Normal	2 Slight decrease
2 Stand independently but unsteady	1 Significant decrease
1 Some assistance	0 Absent sweating
0 Cannot do	
(2) Walking without brace	
3 Normal	
2 Walk independently but slow or unsteady	
1 Some assistance	
0 Cannot do	

(*continued*)

Table 4.5 (Continued)

C. Trunk movement	H. Skin condition
(1) Sitting	3 Normal
3 Normal	2 Partial breakdown
2 Stable when still, but unstable when moving	1 Significant breakdown, often associated with edema
1 Unstable when still	0 Enduring bedsore or skin damage; severe edema
0 Cannot do	I. Pain
(2) Turning body over	3 No pain
3 Normal	2 Mild pain, ordinary pain killer effective
2 Finish independently with difficulty	1 Severe pain, narcotics required
1 Some assistance	0 Extreme pain, uncontrolled
0 Total dependence	
D. General movement	Sexual function (Rate for males only; do not include in overall score)
(1) Transfers: bed to chair/wheelchair	3 Normal
3 Normal	2 Finish independently with difficulty session or ejaculation
2 Can achieve erection and sexual penetration, but problems with sensation or ejaculation	1 Can achieve erection, but no sexual penetration, sensation, or ejaculation
1 Some assistance	0 Unable to achieve erection
0 Total dependence	
(2) Bathing	
3 Normal	
2 Finish independently with difficulty	
1 Some assistance	
0 Total dependence	

IANR-SCIFRS: SCI Functional Rating Scale of the International Association of Neurorestoratology [253].
Source: Huang et al. [253] / With Permission of SAGE Publications.

cryopreservation process [263]. There are different protocols reported in the literature with various cryoprotective agents, such as dimethyl sulfoxide (DMSO), bovine serum, glycerol, and antioxidants [264]. Preparation of therapeutic cells requires a non-animal origin protocol and a facility of good manufacturing practice (GMP) level for clinical use. Therefore, many industrial guidelines were developed for specific cell type and applications [258].

Clinical trials using human iPSC-derived cell products have emerged and are currently being evaluated for their efficacy and safety [9, 19]. However, effective translation of ESCs and iPSCs into clinical settings with uncompromised safety is still being investigated. Tumorigenicity during cellular transplantation is the major obstacle of using iPSCs and ESCs because undifferentiated ESCs/iPSCs can generate both teratomas and malignant tumors *in vivo* [265]. There might be risks of immune rejection of iPSC and iPSC derivations. There are reports of undifferentiated cells may be more immunogenic compared to differentiated iPSCs [266–268], and

Table 4.6 ASIA score.

Grade	
A	The impairment is complete. There is no motor or sensory function left below the level of injury
B	The impairment is incomplete. Sensory function, but not motor function, is preserved below the neurologic level (the first normal level above the level of injury), and some sensation is preserved in the sacral segments S4 and S5
C	The impairment is incomplete. Motor function is preserved below the neurologic level, but more than half of the key muscles below the neurologic level have a muscle grade less than 3 (i.e. they are not strong enough to move against gravity)
D	The impairment is incomplete. Motor function is preserved below the neurologic level, and at least half of the key muscles below the neurologic level have a muscle grade of 3 or more (i.e. the joints can be moved against gravity)
E	The patient's functions are normal. All motor and sensory functions are unhindered

American Spinal Injury Association (ASIA) [254] is a standardized neurological assessment for sensory and motor impairments of spinal cord injury.
Source: Roberts et al. [254] / Wolters Kluwer.

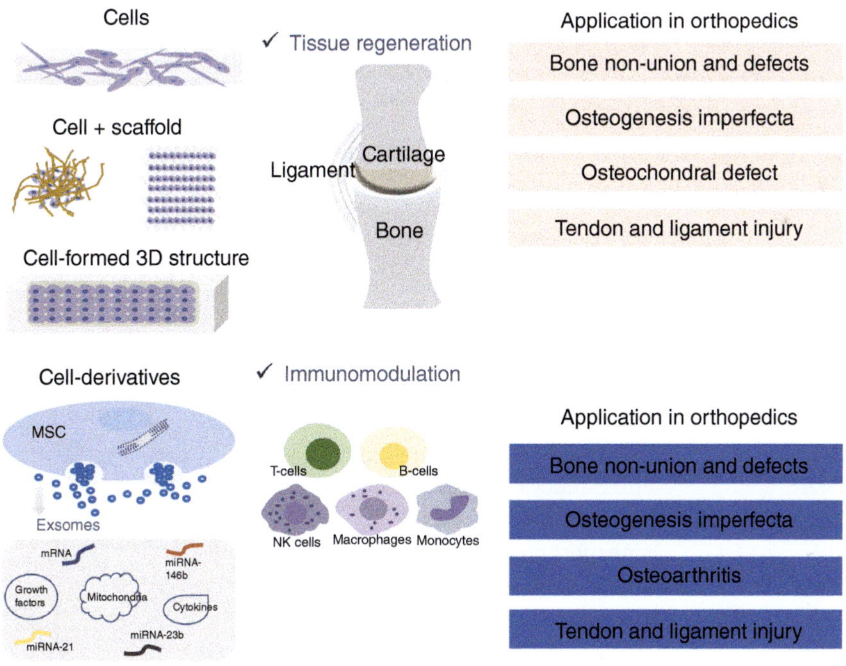

Figure 4.3 Summary of the derivation and application of MSCs in orthopedics.

iPSC-derived smooth muscle cells in mice initiated immune rejection while retinal epithelial cells did not [266–268]. There are many other potential challenges about iPSC application, such as genetic instability, chromosomal abnormalities, and loss of heterozygosity *in vitro*. In the future, technological advances such as gene editing and clarification of complex gene regulatory networks and cellular behaviors may propel iPSCs applications (Figure 4.3).

4.6 Conclusion

In summary, MSCs have different derivations, such as ESC, iPSCs, and tissue derived. The phenotype and characteristics are based on the derivation, isolation, and application methods. In orthopedic application, the widely used cell types are BMSCs, ADSCs, CDPCs, and TSPCs because of their accessibility and biosafety. The two major grounds for clinical application of MSCs are (i) stem cells contribute to tissue regeneration and (ii) MSCs have immunomodulatory ability. MSCs can be delivered to damaged tissues via different ways, including cells suspension, cells in scaffold, or in 3D constructs. The applications of biomaterials could improve the efficacy of MSCs in tissue repair. MSC-derived exosomes are important bio-factors that can modulate the inflammation. The clinical results of using stem cells as therapies were reported in bone non-union and defects, osteogenesis imperfecta, osteochondral defect, tendon and ligament, osteoarthritis, and so on.

Acknowledgments

The authors thank to the support of the National Key R&D Program of China (MOST, 2019YFA0111900 to YJ).

Abbreviations

ADSCs	adipose-derived MSCs
AEs	adverse effects
ACL	anterior cruciate ligament
α-MEM	α-minimum essential medium
αSMA	alpha smooth muscle actin
BMSCs	bone marrow-derived MSCs
BMP	bone morphogenic proteins
BMLs	bone marrow lesions
CFU-f	colony-forming unit fibroblasts
CD-MSCs	cartilage-derived MSCs
CDPCs	cartilage-derived stem/progenitor cells
CSPCs	cartilage stem/progenitor cells
CEDPs	cartilage ECM-derived particles

CXCR-4	C-X-C chemokine receptor type 4
CXCL12	C-X-C motif chemokine 12
CTGF	connective tissue growth factor
CSF1R	colony stimulating factor 1 receptor
CTSK	cathepsin K
DMM	destabilization of the medial meniscus
DMSO	dimethyl sulfoxide
DMEM	Dulbecco's modified Eagle's medium
ESCs	embryonic stem cells
ESC-MSCs	embryonic stem cell-induced MSCs
EGF	epidermal growth factors
FGF	fibroblast growth factor
FCS	fetal calf serum
FBS	fetal bovine serum
GMP	good manufacturing practice
HA	hyaluronic acid
hPL	human platelet lysate
hSSCs	human skeletal stem cells
hiPS-MSCs	human-induced pluripotent stem-MSCs
hUC-MSCs,	human umbilical cord mesenchymal stem cell-derived MSCs
IGF-1	insulin-like growth factor 1
IDO	indoleamine 2,3-dioxygenase
IL-6	interleukin-6
IL-8	interleukin-8
IL-10	interleukin-10
iPSCs	induced pluripotent stem cells
iPSC-MSCs-Exos	iPSC-MSC-derived exosomes
MSCs	mesenchymal stem cells
Myf-5	myogenic factor 5
MyoD	myoblast determination protein
MSC-EVs	MSC-derived extracellular vesicles
MPCs	mesenchymal precursor cells
M-CSF	macrophage colony-stimulating factor
MMP9	matrix metallopeptidase 9
NICD	notch intracellular domain
NFATc1	nuclear factor of activated T-cells, cytoplasmic 1
OI	osteogenesis imperfecta
OA	osteoarthritis
OPN	osteopontin
PGE2	prostaglandin E2
PDGF	platelet-derived growth factor
PLGA	polylactic-*co*-glycolic acid
PRP	platelet-rich plasma
PDGF-BB	platelet-derived growth factor BB
PCNA	proliferating cell nuclear antigen

RUNX2	Runt-related transcription factor 2
RANKL	receptor activator of nuclear factor kappa-B ligand
Scx	Scleraxis
SM-MSCs	synovium-derived MSCs
SF-MSCs	synovial fluid-derived MSCs
SDF-1	stromal derived factor-1
SOX7	SRY-Box transcription factor 7
SM-MSC-Exos	SM-MSC-derived exosomes
SCI	spinal cord injuries
TSPCs	tendon-specific stem/progenitor cells
TKA	total knee arthroplasty
TGF-β	transforming growth factor-β
TRAP	TNF receptor associated protein
VEGF	vascular endothelial growth factor
WNT	wingless-type MMTV integration site
YAP1	yes-associated protein

References

1 Brunstein, C.G., Fuchs, E.J., Carter, S.L. et al. (2011). *Blood* 118: 282.
2 Barrett, A.J., Horowitz, M.M., Pollock, B.H. et al. (1994). *N. Engl. J. Med.* 331: 1253.
3 Becker, A.J., McCulloch, E.A., and Till, J.E. (1963). *Nature* 197: 452.
4 Siminovitch, L., McCulloch, E.A., and Till, J.E. (1963). *J. Cell Comp. Physiol.* 62: 327.
5 Friedenstein, A.J., Chailakhjan, R.K., and Lalykina, K.S. (1970). *Cell Tissue Kinet.* 3: 393.
6 Friedenstein, A.J., Gorskaja, J.F., and Kulagina, N.N. (1976). *Exp. Hematol.* 4: 267.
7 Castro-Malaspina, H., Gay, R.E., Resnick, G. et al. (1980). *Blood* 56: 289.
8 Caplan, A.I. and Bruder, S.P. (2001). *Trends Mol. Med.* 7: 259.
9 Jiang, Y., Cai, Y., Zhang, W. et al. (2016). *Stem Cells Transl. Med.* 5: 733.
10 Wakitani, S., Imoto, K., Yamamoto, T. et al. (2002). *Osteoarthr. Cartil.* 10: 199.
11 Peterson, L., Vasiliadis, H.S., Brittberg, M., and Lindahl, A. (2010). *Am. J. Sports Med.* 38: 1117.
12 Horwitz, E.M., Prockop, D.J., Fitzpatrick, L.A. et al. (1999). *Nat. Med.* 5: 309.
13 Hernigou, P., Delambre, J., Quiennec, S., and Poignard, A. (2021). *Int. Orthop.* 45: 365.
14 Hernigou, P., Bouthors, C., Bastard, C. et al. (2021). *Int. Orthop.* 45: 391.
15 Hernigou, P., Auregan, J.C., Dubory, A. et al. (2018). *Int. Orthop.* 42: 2563.
16 Hernigou, P., Flouzat Lachaniette, C.H., Delambre, J. et al. (2014). *Int. Orthop.* 38: 1811.
17 Erickson, G.R., Gimble, J.M., Franklin, D.M. et al. (2002). *Biochem. Biophys. Res. Commun.* 290: 763.

18 Williams, J.T., Southerland, S.S., Souza, J. et al. (1999). *Am. Surg.* 65: 22.
19 De Bari, C., Dell'Accio, F., and Luyten, F.P. (2001). *Arthritis Rheumatol.* 44: 85.
20 Kern, S., Eichler, H., Stoeve, J. et al. (2006). *Stem Cells* 24: 1294.
21 Tuli, R., Seghatoleslami, M.R., Tuli, S. et al. (2003). *Mol. Biotechnol.* 23: 37.
22 Nishimura, K., Solchaga, L.A., Caplan, A.I. et al. (1999). *Arthritis Rheumatol.* 42: 2631.
23 Toma, J.G., Akhavan, M., Fernandes, K.J. et al. (2001). *Nat. Cell Biol.* 3: 778.
24 Zhu, X., Du, J., and Liu, G. (2012). *Clin. Lab.* 58: 897.
25 Gotherstrom, C., West, A., Liden, J. et al. (2005). *Haematologica* 90: 1017.
26 Jackson, W.M., Lozito, T.P., Djouad, F. et al. (2011). *J. Cell. Mol. Med.* 15: 2377.
27 Castillo-Cardiel, G., Lopez-Echaury, A.C., Saucedo-Ortiz, J.A. et al. (2017). *Dent. Traumatol.* 33: 38.
28 Winkler, T., Perka, C., von Roth, P. et al. (2018). *J. Cachexia Sarcopenia Muscle* 9: 880.
29 Corcione, A., Benvenuto, F., Ferretti, E. et al. (2006). *Blood* 107: 367.
30 Burchell, J.T., Strickland, D.H., and Stumbles, P.A. (2010). *Pharmacol. Ther.* 125: 1.
31 Spaggiari, G.M., Capobianco, A., Becchetti, S. et al. (2006). *Blood* 107: 1484.
32 Puissant, B., Barreau, C., Bourin, P. et al. (2005). *Br. J. Haematol.* 129: 118.
33 Wang, L., Li, Y., Chen, X. et al. (2002). *Hematology* 7: 113.
34 Longo, U.G., Berton, A., Alexander, S. et al. (2011). *Sports Med. Arthrosc. Rev.* 19: 380.
35 Porada, C.D. and Almeida-Porada, G. (2010). *Adv. Drug Delivery Rev.* 62: 1156.
36 Mastrolia, I., Foppiani, E.M., Murgia, A. et al. (2019). *Stem Cells Transl. Med.* 8: 1135.
37 Sekiya, I., Larson, B.L., Smith, J.R. et al. (2002). *Stem Cells* 20: 530.
38 Dominici, M., Le Blanc, K., Mueller, I. et al. (2006). *Cytotherapy* 8: 315.
39 Lee, R.H., Kim, B., Choi, I. et al. (2004). *Cell. Physiol. Biochem.* 14: 311.
40 Bruder, S.P., Jaiswal, N., and Haynesworth, S.E. (1997). *J. Cell. Biochem.* 64: 278.
41 Gronthos, S., Zannettino, A.C., Hay, S.J. et al. (2003). *J. Cell Sci.* 116: 1827.
42 Colter, D.C., Class, R., DiGirolamo, C.M., and Prockop, D.J. (2000). *Proc. Natl. Acad. Sci. U.S.A.* 97: 3213.
43 Muraglia, A., Cancedda, R., and Quarto, R. (2000). *J. Cell Sci.* 113 (Pt. 7): 1161.
44 Bonab, M.M., Alimoghaddam, K., Talebian, F. et al. (2006). *BMC Cell Biol.* 7: 14.
45 Jiang, Y., Wang, D., Blocki, A., and Tuan, R.S. (2020, Chapter 49). Mesenchymal stem cells in musculoskeletal tissue engineering, Table 49.2. In: *Principle of Tissue Engineering*, 5e (ed. R. Lanza, R. Langer, J. Vacanti and A. Atala), 883–915. Elsevier.
46 Hagmann, S., Moradi, B., Frank, S. et al. (2013). *Cell Prolif.* 46: 396.
47 Kolf, C.M., Cho, E., and Tuan, R.S. (2007). *Arthritis Res. Ther.* 9: 204.
48 Boland, G.M., Perkins, G., Hall, D.J., and Tuan, R.S. (2004). *J. Cell. Biochem.* 93: 1210.

49 Pricola, K.L., Kuhn, N.Z., Haleem-Smith, H. et al. (2009). *J. Cell. Biochem.* 108: 577.

50 Rider, D.A., Dombrowski, C., Sawyer, A.A. et al. (2008). *Stem Cells* 26: 1598.

51 Karadjian, M., Senger, A.S., Essers, C. et al. (2020). *Cells* 9.

52 Czapla, J., Matuszczak, S., Kulik, K. et al. (2019). *Stem Cell Res. Ther.* 10: 235.

53 Tang, X., Fan, L., Pei, M. et al. (2015). *Eur. Cell. Mater.* 30: 12.

54 Kusuyama, J., Bandow, K., Shamoto, M. et al. (2014). *J. Biolumin. Chemilumin.* 289: 10330.

55 Baek, S., Quan, X., Kim, S. et al. (2014). *ACS Nano* 8: 10125.

56 Moghadam, M., Tokhanbigli, S., Baghaei, K. et al. (2020). *Mol. Biol. Rep.* 47: 1723.

57 Bernardo, M.E., Zaffaroni, N., Novara, F. et al. (2007). *Cancer Res.* 67: 9142.

58 Liu, L., DiGirolamo, C.M., Navarro, P.A. et al. (2004). *Exp. Cell. Res.* 294: 1.

59 Ehnert, S., Glanemann, M., Schmitt, A. et al. (2009). *Langenbecks Arch. Surg.* 394: 985.

60 Berebichez-Fridman, R., Gomez-Garcia, R., Granados-Montiel, J. et al. (2017). *Stem Cells Int.* 2017: 2638305.

61 Gomez-Barrena, E., Sola, C.A., and Bunu, C.P. (2014). *Int. Orthop.* 38: 1803.

62 Schmitt, A., van Griensven, M., Imhoff, A.B., and Buchmann, S. (2012). *Stem Cells Int.* 2012: 394962.

63 Takahashi, K. and Yamanaka, S. (2006). *Cell* 126: 663.

64 Rodolfa, K., Di Giorgio, F.P., and Sullivan, S. (2007). *Differentiation* 75: 577.

65 Pittenger, M.F., Mackay, A.M., Beck, S.C. et al. (1999). *Science* 284: 143.

66 Ding, D.C., Wu, K.C., Chou, H.L. et al. (2015). *Cell Transplant.* 24: 1221.

67 English, A., Jones, E.A., Corscadden, D. et al. (2007). *Rheumatology* 46: 1676.

68 Alegre-Aguaron, E., Desportes, P., Garcia-Alvarez, F. et al. (2012). *Cells Tissues Organs* 196: 231.

69 de la Fuente, R., Abad, J.L., Garcia-Castro, J. et al. (2004). *Exp. Cell. Res.* 297: 313.

70 Shirasawa, S., Sekiya, I., Sakaguchi, Y. et al. (2006). *J. Cell. Biochem.* 97: 84.

71 Sakaguchi, Y., Sekiya, I., Yagishita, K., and Muneta, T. (2005). *Arthritis Rheumatol.* 52: 2521.

72 Han, S.A., Lee, S., Seong, S.C., and Lee, M.C. (2014). *Tissue Eng. Part A* 20: 2680.

73 Lopez-Ruiz, E., Peran, M., Cobo-Molinos, J. et al. (2013). *Osteoarthr. Cartil.* 21: 246.

74 Jo, C.H., Ahn, H.J., Kim, H.J. et al. (2007). *Cytotherapy* 9: 316.

75 Karystinou, A., Dell'Accio, F., Kurth, T.B. et al. (2009). *Rheumatology* 48: 1057.

76 Arufe, M.C., De la Fuente, A., Fuentes, I. et al. (2010). *J. Cell. Biochem.* 111: 834.

77 Jones, E., Churchman, S.M., English, A. et al. (2010). *Ann. Rheum. Dis.* 69: 450.

78 Morito, T., Muneta, T., Hara, K. et al. (2008). *Rheumatology* 47: 1137.

79 Lee, D.H., Sonn, C.H., Han, S.B. et al. (2012). *Osteoarthr. Cartil.* 20: 106.

80 Matsukura, Y., Muneta, T., Tsuji, K. et al. (2014). *Clin. Orthop. Relat. Res.* 472: 1357.

81 Jones, E.A., English, A., Henshaw, K. et al. (2004). *Arthritis Rheumatol.* 50: 817.

82 Wickham, M.Q., Erickson, G.R., Gimble, J.M. et al. (2003). *Clin. Orthop. Relat. Res.* 196.

83 Lopa, S., Colombini, A., Stanco, D. et al. (2014). *Eur. Cell. Mater.* 27: 298.

84 Jiang, Y. and Tuan, R.S. (2015). *Nat. Rev. Rheumatol.* 11: 206.

85 Ozbey, O., Sahin, Z., Acar, N. et al. (2014). *Acta Histochem.* 116: 763.

86 Bi, Y., Ehirchiou, D., Kilts, T.M. et al. (2007). *Nat. Med.* 13: 1219.

87 Narcisi, R., Cleary, M.A., Sivasubramaniyan, K. et al. (2017). MSC Populations for Cartilage Regeneration. In: *Cartilage Volume 3: Repair Strategies and Regeneration* (ed. S. Grässel and A. Aszódi), 35. Springer.

88 Fukiage, K., Aoyama, T., Shibata, K.R. et al. (2008). *Biochem. Biophys. Res. Commun.* 365: 406.

89 Mifune, Y., Matsumoto, T., Murasawa, S. et al. (2013). *Cell Transplant.* 22: 1201.

90 Kuci, S., Kuci, Z., Schafer, R. et al. (2019). *Sci. Rep.* 9: 1774.

91 Hagmann, S., Frank, S., Gotterbarm, T. et al. (2014). *BMC Musculoskelet. Disord.* 15: 322.

92 Harkness, L., Zaher, W., Ditzel, N. et al. (2016). *Stem Cell Res. Ther.* 7: 4.

93 Bourin, P., Bunnell, B.A., Casteilla, L. et al. (2013). *Cytotherapy* 15: 641.

94 Mohamed-Ahmed, S., Fristad, I., Lie, S.A. et al. (2018). *Stem Cell Res. Ther.* 9: 168.

95 Sachs, P.C., Francis, M.P., Zhao, M. et al. (2012). *Cell Tissue Res.* 349: 505.

96 Fu, Z., Song, X., Guo, L. et al. (2019). *Am. J. Sports Med.* 47: 2927.

97 Mochizuki, T., Muneta, T., Sakaguchi, Y. et al. (2006). *Arthritis Rheumatol.* 54: 843.

98 Khan, W.S., Adesida, A.B., Tew, S.R. et al. (2012). *Cell Prolif.* 45: 111.

99 Kouroupis, D., Bowles, A.C., Best, T.M. et al. (2020). *Am. J. Sports Med.* 48: 2013.

100 Williams, R., Khan, I.M., Richardson, K. et al. (2010). *PLoS One* 5: e13246.

101 Sun, H., Wen, X., Li, H. et al. (2020). *Ann. Rheum. Dis.* 79: 408.

102 Grogan, S.P., Barbero, A., Diaz-Romero, J. et al. (2007). *Arthritis Rheumatol.* 56: 586.

103 Levato, R., Webb, W.R., Otto, I.A. et al. (2017). *Acta Biomater.* 61: 41.

104 Dyment, N.A., Hagiwara, Y., Matthews, B.G. et al. (2014). *PLoS One* 9: e96113.

105 Yin, Z., Hu, J.J., Yang, L. et al. (2016). *Sci. Adv.* 2: e1600874.

106 Walia, B. and Huang, A.H. (2019). *J. Orthop. Res.* 37: 1270.

107 Tempfer, H., Wagner, A., Gehwolf, R. et al. (2009). *Histochem. Cell Biol.* 131: 733.

108 Mienaltowski, M.J., Adams, S.M., and Birk, D.E. (2013). *Tissue Eng. Part A* 19: 199.

109 Ogata, Y., Mabuchi, Y., Yoshida, M. et al. (2015). *PLoS One* 10: e0129096.

110 Chan, C.K.F., Gulati, G.S., Sinha, R. et al. (2018). *Cell* 175: 43.

111 Kopher, R.A., Penchev, V.R., Islam, M.S. et al. (2010). *Bone* 47: 718.
112 Himeno, T., Kamiya, H., Naruse, K. et al. (2013). *Biomed. Res. Int.* 2013: 259187.
113 Houlihan, D.D., Mabuchi, Y., Morikawa, S. et al. (2012). *Nat. Protoc.* 7: 2103.
114 Jiang, Y., Jahagirdar, B.N., Reinhardt, R.L. et al. (2002). *Nature* 418: 41.
115 Salehi, H., Amirpour, N., Niapour, A., and Razavi, S. (2016). *Stem Cell Rev. Rep.* 12: 26.
116 Schwartz, R.E., Reyes, M., Koodie, L. et al. (2002). *J. Clin. Invest.* 109: 1291.
117 Cao, Y., Sun, Z., Liao, L. et al. (2005). *Biochem. Biophys. Res. Commun.* 332: 370.
118 Hogan, B.L. (1996). *Genes Dev.* 10: 1580.
119 Horner, A., Kemp, P., Summers, C. et al. (1998). *Bone* 23: 95.
120 Johnstone, B., Hering, T.M., Caplan, A.I. et al. (1998). *Exp. Cell. Res.* 238: 265.
121 Mackay, A.M., Beck, S.C., Murphy, J.M. et al. (1998). *Tissue Eng.* 4: 415.
122 Denker, A.E., Haas, A.R., Nicoll, S.B., and Tuan, R.S. (1999). *Differentiation* 64: 67.
123 Barry, F., Boynton, R.E., Liu, B., and Murphy, J.M. (2001). *Exp. Cell. Res.* 268: 189.
124 Quarto, N., Behr, B., and Longaker, M.T. (2010). *Tissue Eng. Part A* 16: 3185.
125 Kahler, R.A. and Westendorf, J.J. (2003). *J. Biolumin. Chemilumin.* 278: 11937.
126 Kahler, R.A., Yingst, S.M., Hoeppner, L.H. et al. (2008). *Matrix Biol.* 27: 330.
127 Eijken, M., Meijer, I.M., Westbroek, I. et al. (2008). *J. Cell. Biochem.* 104: 568.
128 James, A.W. (2013). *Scientifica* 2013: 684736.
129 Tsai, M.T., Li, W.J., Tuan, R.S., and Chang, W.H. (2009). *J. Orthop. Res.* 27: 1169.
130 Kang, Q., Song, W.X., Luo, Q. et al. (2009). *Stem Cells Dev.* 18: 545.
131 Tang, Z., Li, X., Tan, Y. et al. (2018). *Regener. Biomater.* 5: 43.
132 Luu, H.H., Song, W.X., Luo, X. et al. (2007). *J. Orthop. Res.* 25: 665.
133 Deng, M., Mei, T., Hou, T. et al. (2017). *Stem Cell Res. Ther.* 8: 258.
134 Colenci, R., da Silva Assuncao, L.R., Mogami Bomfim, S.R. et al. (2014). *Arch. Oral Biol.* 59: 268.
135 Crane, J.L., Zhao, L., Frye, J.S. et al. (2013). *Bone Res.* 1: 186.
136 Deng, M., Luo, K., Hou, T. et al. (2018). *J. Cell. Physiol.* 233: 5792.
137 Cals, F.L., Hellingman, C.A., Koevoet, W. et al. (2012). *J. Tissue Eng. Regener. Med.* 6: 68.
138 Park, S.Y., Kim, K.H., Shin, S.Y. et al. (2013). *Tissue Eng. Part A* 19: 2495.
139 Rodda, S.J. and McMahon, A.P. (2006). *Development* 133: 3231.
140 Tian, E., Zhan, F., Walker, R. et al. (2003). *N. Engl. J. Med.* 349: 2483.
141 Haydon, R.C., Deyrup, A., Ishikawa, A. et al. (2002). *Int. J. Cancer* 102: 338.
142 Luo, Q., Kang, Q., Si, W. et al. (2004). *J. Biolumin. Chemilumin.* 279: 55958.
143 Augello, A. and De Bari, C. (2010). *Hum. Gene Ther.* 21: 1226.
144 Nakashima, K., Zhou, X., Kunkel, G. et al. (2002). *Cell* 108: 17.
145 Komori, T., Yagi, H., Nomura, S. et al. (1997). *Cell* 89: 755.
146 Lademann, F., Hofbauer, L.C., and Rauner, M. (2020). *Front. Cell Dev. Biol.* 8: 586031.

147 Boyle, W.J., Simonet, W.S., and Lacey, D.L. (2003). *Nature* 423: 337.

148 Zhao, B. and Ivashkiv, L.B. (2011). *Arthritis Res. Ther.* 13: 234.

149 Sumanasinghe, R.D., Bernacki, S.H., and Loboa, E.G. (2006). *Tissue Eng.* 12: 3459.

150 Yourek, G., McCormick, S.M., Mao, J.J., and Reilly, G.C. (2010). *Regener. Med.* 5: 713.

151 Schwartz, Z., Simon, B.J., Duran, M.A. et al. (2008). *J. Orthop. Res.* 26: 1250.

152 Bloise, N., Petecchia, L., Ceccarelli, G. et al. (2018). *PLoS One* 13: e0199046.

153 Ongaro, A., Pellati, A., Bagheri, L. et al. (2014). *Bioelectromagnetics* 35: 426.

154 Zhu, X., Chan, Y.T., Yung, P.S.H. et al. (2020). *Front. Cell Dev. Biol.* 8: 607764.

155 Zhang, X., Cai, D., Zhou, F. et al. (2020). *Biomaterials* 232: 119724.

156 Wu, H., Xu, T., Chen, Z. et al. (2020). *J. Cell. Physiol.* 235: 8653.

157 Wang, L., You, X., Lotinun, S. et al. (2020). *Nat. Commun.* 11: 282.

158 Lin, W., Xu, L., Zwingenberger, S. et al. (2017). *J. Orthop. Translat.* 9: 19.

159 Fu, X., Liu, G., Halim, A. et al. (2019). *Cells* 8.

160 Chen, G., Qi, Y., Niu, L. et al. (2015). *Biomed. Rep.* 3: 749.

161 Tian, Y., Xu, Y., Xue, T. et al. (2017). *Cell Death Dis.* 8: e2595.

162 Zou, X.H., Cai, H.X., Yin, Z. et al. (2009). *Cell Transplant.* 18: 433.

163 Chen, F., Zhou, Y., Barnabas, S.T. et al. (2007). *J. Biomech.* 40 (Suppl 1): S73.

164 Rana, D. and Ramalingam, M. (2017). *Mater. Sci. Eng., C* 76: 1057.

165 Xue, Y., Wu, M., Liu, Z. et al. (2019). *J. Mater. Sci. – Mater. Med.* 30: 47.

166 Wang, T., Yang, X., Qi, X., and Jiang, C. (2015). *J. Transl. Med.* 13: 152.

167 Zhang, B.Y., Xu, P., Luo, Q., and Song, G.B. (2021). *World J. Stem Cells* 13: 115.

168 Echave, M.C., Saenz del Burgo, L., Pedraz, J.L., and Orive, G. (2017). *Curr. Pharm. Des.* 23: 3567.

169 Caliari, S.R. and Harley, B.A. (2014). *Adv. Healthc. Mater.* 3: 1086.

170 Moser, C., Bardsley, K., El Haj, A.J. et al. (2018). *Front. Bioeng. Biotechnol.* 6: 161.

171 Duan, W., Chen, C., Haque, M. et al. (2018). *Stem Cell Res. Ther.* 9: 60.

172 Duscher, D., Rennert, R.C., Januszyk, M. et al. (2014). *Sci. Rep.* 4: 7144.

173 Yang, Y.K., Ogando, C.R., Wang See, C. et al. (2018). *Stem Cell Res. Ther.* 9: 131.

174 Palumbo, S., Tsai, T.L., and Li, W.J. (2014). *Stem Cells Dev.* 23: 852.

175 Jeon, O.H., Panicker, L.M., Lu, Q. et al. (2016). *Sci. Rep.* 6: 26761.

176 Jiang, Y., Lin, H., and Tuan, R.S. (2017). Overview: state of the art and future prospectives for cartilage repair. In: *Cartilage Volume 3: Repair Strategies and Regeneration* (ed. S. Grässel and A. Aszódi), 1. Cham: Springer https://doi.org/10.1007/978-3-319-53316-2_1.

177 Lee, P.T. and Li, W.J. (2017). *J. Cell. Biochem.* 118: 172.

178 Fan, L., Chen, J., Tao, Y. et al. (2019). *J. Orthop. Res.* 37: 1387.

179 Lu, J., Shen, X., Sun, X. et al. (2018). *Theranostics* 8: 5039.

180 Larson, B.L., Yu, S.N., Park, H. et al. (2019). *J. Tissue Eng. Regener. Med.* 13: 1453.

181 Yin, H., Wang, Y., Sun, Z. et al. (2016). *Acta Biomater.* 33: 96.

182 Seol, D., McCabe, D.J., Choe, H. et al. (2012). *Arthritis Rheumatol.* 64: 3626.

183 Koelling, S., Kruegel, J., Irmer, M. et al. (2009). *Cell Stem Cell* 4: 324.
184 Goncalves, A.I., Rodrigues, M.T., Lee, S.J. et al. (2013). *PLoS One* 8: e83734.
185 Perucca Orfei, C., Vigano, M., Pearson, J.R. et al. (2019). *Int. J. Mol. Sci.* 20.
186 Sahoo, S., Toh, S.L., and Goh, J.C. (2010). *Biomaterials* 31: 2990.
187 Altman, G.H., Horan, R.L., Martin, I. et al. (2002). *FASEB J.* 16: 270.
188 Rinoldi, C., Fallahi, A., Yazdi, I.K. et al. (2019). *ACS Biomater. Sci. Eng.* 5: 2953.
189 Zhang, J. and Wang, J.H. (2010). *Am. J. Sports Med.* 38: 2477.
190 Wu, T., Liu, Y., Wang, B. et al. (2016). *Tissue Eng. Part A* 22: 1229.
191 Zhou, Y., Zhang, J., Wu, H. et al. (2015). *Stem Cell Res. Ther.* 6: 173.
192 Yu, H., Cheng, J., Shi, W. et al. (2020). *Acta Biomater.* 106: 328.
193 Tarafder, S., Chen, E., Jun, Y. et al. (2017). *FASEB J.* 31: 3991.
194 Qin, S., Wang, W., Liu, Z. et al. (2020). *J. Orthop. Translat.* 22: 101.
195 Al-Ani, M., Xu, K., Sun, Y. et al. (2015). *Stem Cells Int.* 2015: 984146.
196 Uysal, C.A., Tobita, M., Hyakusoku, H., and Mizuno, H. (2012). *J. Plast. Reconstr. Aesthet. Surg.* 65: 1712.
197 Carvalho Ade, M., Badial, P.R., Alvarez, L.E. et al. (2013). *Stem Cell Res. Ther.* 4: 85.
198 Ahrberg, A.B., Horstmeier, C., Berner, D. et al. (2018). *BMC Musculoskelet. Disord.* 19: 230.
199 Trebinjac, S. and Gharairi, M. (2020). *Med. Arch.* 74: 387.
200 Ratajczak, M.Z., Majka, M., Kucia, M. et al. (2003). *Stem Cells* 21: 363.
201 Rajgara, R.F., Lala-Tabbert, N., Marchildon, F. et al. (2017). *Stem Cell Rep.* 9: 1139.
202 Scharner, J. and Zammit, P.S. (2011). *Skelet. Muscle* 1: 28.
203 Creuzet, S., Lescaudron, L., Li, Z., and Fontaine-Perus, J. (1998). *Exp. Cell. Res.* 243: 241.
204 Yablonka-Reuveni, Z. and Rivera, A.J. (1994). *Dev. Biol.* 164: 588.
205 Shea, K.L., Xiang, W., LaPorta, V.S. et al. (2010). *Cell Stem Cell* 6: 117.
206 Dezawa, M., Ishikawa, H., Itokazu, Y. et al. (2005). *Science* 309: 314.
207 Shang, Y., Zhang, C., Wang, S. et al. (2007). *Cytotherapy* 9: 667.
208 Gang, E.J., Bosnakovski, D., Simsek, T. et al. (2008). *Exp. Cell. Res.* 314: 1721.
209 Egusa, H., Kobayashi, M., Matsumoto, T. et al. (2013). *Tissue Eng. Part A* 19: 770.
210 Kasprzycka, P., Archacka, K., Kowalski, K. et al. (2019). *Stem Cell Res. Ther.* 10: 343.
211 Maeda, Y., Yonemochi, Y., Nakajyo, Y. et al. (2017). *Sci. Rep.* 7: 3305.
212 Kowalski, K., Dos Santos, M., Maire, P. et al. (2018). *Stem Cell Res. Ther.* 9: 258.
213 in 't Anker, P.S., Noort, W.A., Scherjon, S.A. et al. (2003). *Haematologica* 88: 845.
214 Gazdic, M., Volarevic, V., Arsenijevic, N., and Stojkovic, M. (2015). *Stem Cell Rev. Rep.* 11: 280.
215 Harrell, C.R., Jovicic, N., Djonov, V. et al. (2019). *Cells* 8.
216 Harrell, C.R., Fellabaum, C., Jovicic, N. et al. (2019). *Cells* 8.

217 Liang, X., Ding, Y., Zhang, Y. et al. (2014). *Cell Transplant.* 23: 1045.
218 Baglio, S.R., Pegtel, D.M., and Baldini, N. (2012). *Front. Physiol.* 3: 359.
219 Hong, P., Yang, H., Wu, Y. et al. (2019). *Stem Cell Res. Ther.* 10: 242.
220 Mitchell, R., Mellows, B., Sheard, J. et al. (2019). *Stem Cell Res. Ther.* 10: 116.
221 Lo Sicco, C., Reverberi, D., Balbi, C. et al. (2017). *Stem Cells Transl. Med.* 6: 1018.
222 Zhang, S., Teo, K.Y.W., Chuah, S.J. et al. (2019). *Biomaterials* 200: 35.
223 Takeuchi, R., Katagiri, W., Endo, S., and Kobayashi, T. (2019). *PLoS One* 14: e0225472.
224 Otsuru, S., Desbourdes, L., Guess, A.J. et al. (2018). *Cytotherapy* 20: 62.
225 Zhu, Y., Wang, Y., Zhao, B. et al. (2017). *Stem Cell Res. Ther.* 8: 64.
226 Wang, Y., Yu, D., Liu, Z. et al. (2017). *Stem Cell Res. Ther.* 8: 189.
227 Yamashita, T., Takahashi, Y., and Takakura, Y. (2018). *Biol. Pharm. Bull.* 41: 835.
228 Le Nail, L.R., Stanovici, J., Fournier, J. et al. (2014). *Int. Orthop.* 38: 1845.
229 Wang, X., Chu, W., Zhuang, Y. et al. (2019). *Cell Transplant.* 28: 212.
230 Quarto, R., Mastrogiacomo, M., Cancedda, R. et al. (2001). *N. Engl. J. Med.* 344: 385.
231 Wirth, T. (2012). *Orthopade* 41: 773.
232 Horwitz, E.M., Prockop, D.J., Gordon, P.L. et al. (2001). *Blood* 97: 1227.
233 Brittberg, M., Lindahl, A., Nilsson, A. et al. (1994). *N. Engl. J. Med.* 331: 889.
234 Bernhard, J.C. and Vunjak-Novakovic, G. (2016). *Stem Cell Res. Ther.* 7: 56.
235 Kuroda, R., Ishida, K., Matsumoto, T. et al. (2007). *Osteoarthr. Cartil.* 15: 226.
236 Wakitani, S., Nawata, M., Tensho, K. et al. (2007). *J. Tissue Eng. Regener. Med.* 1: 74.
237 Sekiya, I., Muneta, T., Horie, M., and Koga, H. (2015). *Clin. Orthop. Relat. Res.* 473: 2316.
238 Shimomura, K., Yasui, Y., Koizumi, K. et al. (2018). *Am. J. Sports Med.* 46: 2384.
239 Shapiro, S.A., Kazmerchak, S.E., Heckman, M.G. et al. (2017). *Am. J. Sports Med.* 45: 82.
240 Gupta, P.K., Chullikana, A., Rengasamy, M. et al. (2016). *Arthritis Res. Ther.* 18: 301.
241 Lu, L., Dai, C., Zhang, Z. et al. (2019). *Stem Cell Res. Ther.* 10: 143.
242 Song, Y., Du, H., Dai, C. et al. (2018). *Regener. Med.* 13: 295.
243 Lee, W.S., Kim, H.J., Kim, K.I. et al. (2019). *Stem Cells Transl. Med.* 8: 504.
244 Freitag, J., Bates, D., Wickham, J. et al. (2019). *Regener. Med.* 14: 213.
245 Matas, J., Orrego, M., Amenabar, D. et al. (2019). *Stem Cells Transl. Med.* 8: 215.
246 Park, Y.B., Ha, C.W., Lee, C.H. et al. (2017). *Stem Cells Transl. Med.* 6: 613.
247 Hernigou, P., Bouthors, C., Bastard, C. et al. (2020). *Int. Orthop.* 45: 365–373.
248 Hernigou, P., Delambre, J., Quiennec, S., and Poignard, A. (2020). *Int. Orthop.* 45: 391–399.
249 Hernigou, P., Merouse, G., Duffiet, P. et al. (2015). *Int. Orthop.* 39: 1219.
250 Wang, Y., Shimmin, A., Ghosh, P. et al. (2017). *Arthritis Res. Ther.* 19: 180.

251 Gan, Y., Dai, K., Zhang, P. et al. (2008). *Biomaterials* 29: 3973.
252 Blanco, J.F., Villaron, E.M., Pescador, D. et al. (2019). *Stem Cell Res. Ther.* 10: 63.
253 Huang, H., Xi, H., Chen, L. et al. (2012). *Cell Transplant.* 21 (Suppl 1): S23.
254 Roberts, T.T., Leonard, G.R., and Cepela, D.J. (2017). *Clin. Orthop. Relat. Res.* 475: 1499.
255 Yang, Y., Pang, M., Du, C. et al. (2021). *Cytotherapy* 23: 57.
256 El-Kheir, W.A., Gabr, H., Awad, M.R. et al. (2014). *Cell Transplant.* 23: 729.
257 Murray, I.R., Chahla, J., Safran, M.R. et al. (2019). *J. Bone Joint Surg. Am.* 101: 904.
258 Zhang, W., Heng, B.C., Jiang, Y.-Z., and Ouyang, H.-W. (2014). *J. Orthop. Translat.* 2: 56.
259 Piuzzi, N.S., Hussain, Z.B., Chahla, J. et al. (2018). *J. Bone Joint Surg. Am.* 100: 517.
260 Chahla, J., Cinque, M.E., Piuzzi, N.S. et al. (2017). *J. Bone Joint Surg. Am.* 99: 1769.
261 Bornes, T.D., Jomha, N.M., Mulet-Sierra, A., and Adesida, A.B. (2016). *Tissue Eng. Part C* 22: 208.
262 Cao, C., Zhang, Y., Ye, Y., and Sun, T. (2020). *J. Orthop. Surg. Res.* 15: 120.
263 Bahsoun, S., Coopman, K., and Akam, E.C. (2019). *J. Transl. Med.* 17: 397.
264 Zanata, F., Shaik, S., Devireddy, R.V. et al. (2016). *Adv. Exp. Med. Biol.* 951: 137.
265 Okita, K., Ichisaka, T., and Yamanaka, S. (2007). *Nature* 448: 313.
266 Garreta, E., Sanchez, S., Lajara, J. et al. (2018). *Curr. Transplant. Rep.* 5: 14.
267 de Almeida, P.E., Meyer, E.H., Kooreman, N.G. et al. (2014). *Nat. Commun.* 5: 3903.
268 Zhao, T., Zhang, Z.N., Westenskow, P.D. et al. (2015). *Cell Stem Cell* 17: 353.

5

Electrospinning Techniques

Tong Wu[1], Yuanfei Wang[2], Ziyi Zhou[1], Xiaopei Zhang[1], Na Liu[1], Zhendong Feng[1], Xiumei Mo[3], and Wenguo Cui[4]

[1]*Qingdao University, Institute of Neuroregeneration and Neurorehabilitation, Qingdao Medical College, 308 Ningxia Road, Qingdao 266071, P.R. China*
[2]*Qingdao Stomatological Hospital Affiliated to Qingdao University, Central Laboratory, 17 Dexian Road, Qingdao 266001, P.R. China*
[3]*Donghua University, College of Chemistry, Chemical Engineering and Biotechnology, State Key Lab for Modification of Chemical Fibers and Polymer Materials, 2999 North Renmin Road, Shanghai 201620, P.R. China*
[4]*Shanghai Institute of Traumatology and Orthopaedics, Ruijin Hospital, Shanghai Jiao Tong University School of Medicine, Department of Orthopaedics, Shanghai Key Laboratory for Prevention and Treatment of Bone and Joint Diseases, 197 Ruijin Second Road, Shanghai 200025, P.R. China*

5.1 Introduction

The development of electrospinning techniques involves a long history, which can be traced back to 1600, when William Gilbert brought the concept of electrospinning to the world [1]. He found that an electric field could make water droplets into a conical shape. Then, in 1882, Lord Rayleigh announced that a high voltage could make a liquid droplet full of electric charges [2]. In 1934 and 1944, Anton Formhals firstly filed patents about the manufacture of electrospinning setups, boosting the commercialization of electrospinning to produce textile yarns [3, 4]. About two decades later, between 1964 and 1969, Geoffrey Taylor described the formation of a Taylor cone during electrospinning [5]. A typical electrospinning setup includes a spinneret, a high voltage power supply, and a grounded collector. During the electrospinning process, the droplets are first charged by a high-voltage power supply. When the strength of the electric field exceeds the surface tension of the liquid droplet, a Taylor cone can be formed. Then, the liquid jet will be ejected from the spinneret, following which the solidified fibers can be collected using a conductive collector [6]. In recent years, the development of electrospinning technology has roughly gone through four stages: (i) screening the spinnability of different polymers and studying the influence of process parameters on fiber diameter and properties; (ii) studying the compositional diversification of electrospun nanofibers and the precise control of their fine structures; (iii) developing a wide range of applications of electrospun nanofibers in

Biofabrication for Orthopedics: Methods, Techniques, and Applications, First Edition.
Edited by Wenguo Cui, Xin Zhao, and Shen Liu.

the fields of energy, environment, biomedicine, and optoelectronics; and (iv) accelerating the mass production of electrospun fibers. Until now, electrospinning has been widely used to synthesize a variety of nano- or micro-scale fibers, attracting extensive attention from the researchers in the field of materials science and biomedicine [7]. As electrospun nanofibers are capable of mimicking the lateral structure and components of a native extracellular matrix (ECM), they boost the advanced development of electrospinning techniques.

The electrospinning techniques have been developing with the basic steps becoming diverse and simple. In particular, electrospinning techniques are no longer limited to traditional setups. A range of updated electrospinning techniques have been developed, each of which has its own advantages. For example, the coaxial electrospinning and emulsion electrospinning techniques are widely used in transporting bioactive effectors such as proteins, growth factors, antibiotics, etc. [8]. All these components can be incorporated into the core of nanofibers and thus released in a controlled manner [9]. Previous studies have found that bone regeneration and osteogenic differentiation could be guided using scaffolds fabricated by these techniques [10–12]. In addition, if one wants to replicate the complicated structures such as fiber bundles, it will be better to consider conjugated electrospinning or dynamic fluid electrospinning methods to generate fascicles by twisting the fibers together. All these electrospinning techniques have shown potential to suit for use in tissue regeneration, such as restoring impaired bone, cartilage, skin, muscle, rotator cuff, tendon, etc. [13–16]. In this chapter, we aim to review the recent progress of different electrospinning techniques and their wide range of applications in biomedicine (Figure 5.1). We will first introduce how to fabricate materials by different electrospinning techniques. Then, the unique applications in biomedicine, especially in orthopedics, will be demonstrated.

5.2 Different Types of Electrospinning Techniques

5.2.1 Traditional Electrospinning

A traditional electrospinning device mainly consists of three parts: a high voltage, an injection pump, and a collector (Figure 5.2a). During the electrospinning process, a high-voltage electrostatic field was established between the syringe needle and the collector to prepare nanofibers. During the electrospinning process, the high-voltage electrostatic field was used to stimulate the polymer-charged jet and then to obtain the polymer nanofibers by charged jet curing [18]. The nanofibrous structure produced by electrospinning is highly similar to that of ECM. In particular, their large specific surface area and interconnected pores make nanofibers excellent candidates to serve as tissue-repaired scaffolds and drug carriers. A rich variety of raw materials can be used for electrospinning, for example, natural materials, synthetic polymers, and inorganic materials. Natural materials such as gelatin, collagen, hyaluronic acid, and chitosan are of rich sources, good biocompatibility, and biodegradability [19].

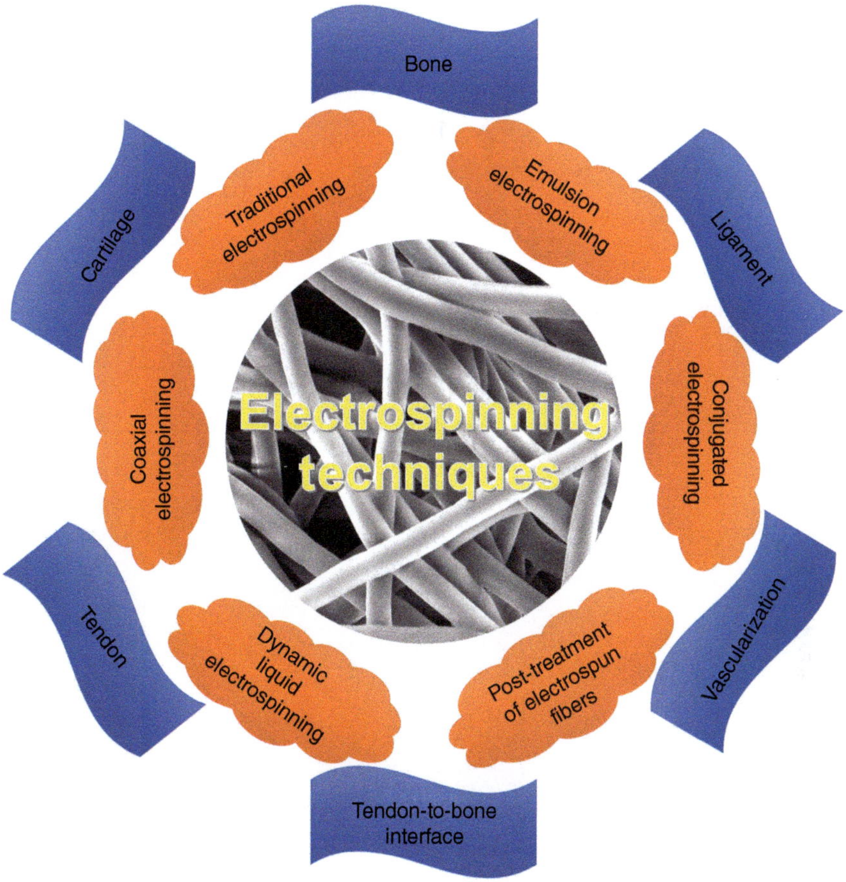

Figure 5.1 Typical scanning electron microscopy (SEM) image of electrospun nanofibers and the different electrospinning techniques used for a variety of applications involved orthopedic tissue repair.

For example, Ho et al. prepared nanofibrous scaffolds made of chitosan and demonstrated that mouse osteoblasts grew much better when cultured on nonwoven mats of chitosan nanofibers relative to the case of chitosan films. In addition, human osteoblasts were able to adhere and grow on the chitosan nanofibers, promoting the expression of osteopontin, osteocalcin, and alkaline phosphatase to stimulate osteoblast proliferation and maturation [20]. In another study, Behera et al. prepared nonwoven mats of nanofibers by electrospinning a blend of chitosan and TiO_2, which showed high porosity, high mechanical strength, and good flexibility [21]. In comparison to natural materials, synthetic polymers are more flexible in processing and manufacture. Synthetic polymers such as polylactic acid (PLA), polyglycolic acid (PGA), poly(lactic-*co*-glycolic acid) (PLGA), and polycaprolactone (PCL) have been widely used in tissue engineering. Among others, electrospinning a blend of

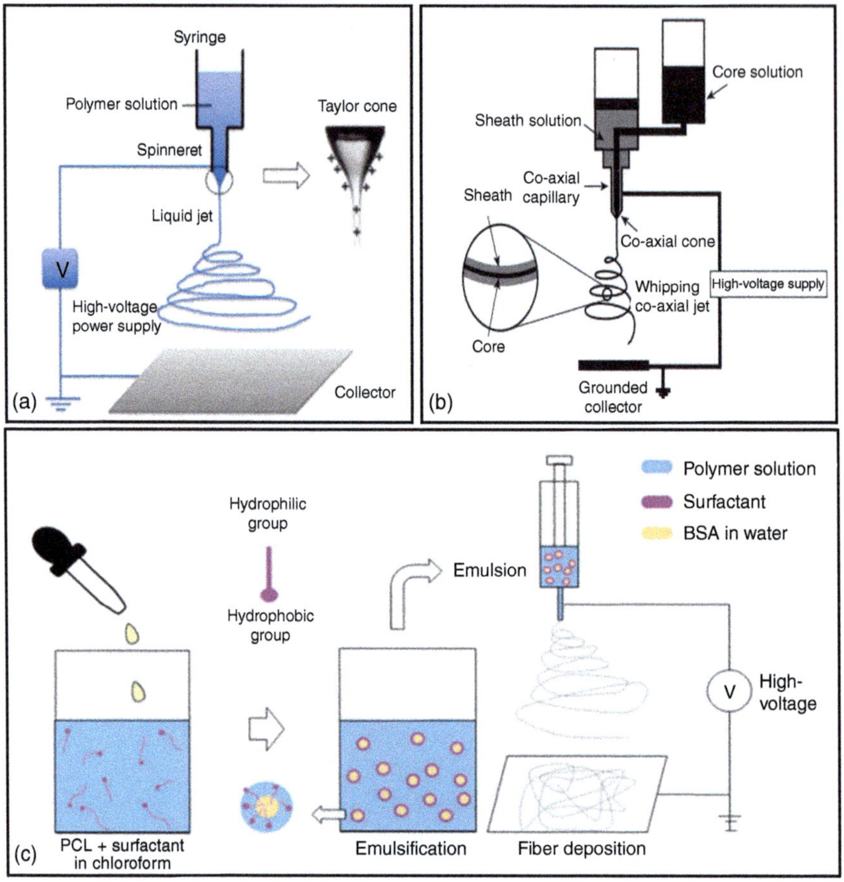

Figure 5.2 Schematic illustration showing the setups of (a) traditional electrospinning, (b) coaxial electrospinning, and (c) emulsion electrospinning. Source: Yoon et al. [17]/with permission of John Wiley & Sons.

natural materials and synthetic polymers is the most commonly used way. By loading bioactive substances into synthetic polymers, the advantageous properties of the two components can be combined simultaneously to prepare multifunctional scaffolds [22]. In a typical study, Kobayashi et al. implanted scaffolds made by electrospinning a blend of PGA and collagen underneath the rat fascia lata [23]. They showed that the composite nanofibers promoted the recruitment of host cells and accelerated the formation of peripheral blood vessels without the use of any bio-derived substance such as growth factors. As mentioned above, traditional electrospinning has its advantages as well as limitations and disadvantages. First of all, the traditional electrospinning technology is difficult to control the direction of electrospinning accurately and get a specific three-dimensional (3D) structure. Secondly, because of the rough surface of nanofibers, it is difficult to be directly applied in medical engineering [18].

5.2.2 Coaxial Electrospinning

Coaxial electrospinning was first introduced in 2002, which is an updated technique to prepare nanofibers with "core–sheath" structure based on the principle of a traditional electrospinning technique [24]. Coaxial electrospinning makes it possible to synthesize more nanofibers with different morphologies and applications from two or more polymer solutions. Importantly, nonpolymer materials (such as ceramics, metal oxides, and semiconductors) without filament forming properties can be processed into nanofibers using the coaxial electrospinning technique [24]. Compared with traditional electrospinning, the nozzle used for coaxial electrospinning has a coaxial and concentrical structure, which provides different paths for the shell and core solutions. With the development of coaxial electrospinning technology, it has become one of the most commonly used technologies to develop new drug carriers. In most cases, the drug molecules are integrated into the core and encapsulated by the sheath, which enables the sustained release of drugs from the core. In the later stage of development, two or more drugs are integrated into the core layer or two layers to produce synergistic effect. With the increasing variety of materials and the manufacturing of various structures, the application of coaxial electrospinning has been extended to the field of biomedicine (antibacterial and drug delivery), electrochemistry (batteries, supercapacitors, and energy collection), and environment (oil filters and molecular filters) [24]. A typical coaxial electrospinning device is presented in Figure 5.2b [25].

For example, bioactive proteins, metal ions, and drugs can be incorporated into the core of polymeric fibers. With the diffusion action and/or the degradation of shell materials, the encapsulated payloads can be gradually released [26]. The coaxial electrospinning process will be influenced by multiple parameters, such as environmental temperature and humidity, solution viscosity and concentration, flow rates of shell solution versus core solution, and the high voltage. When using the coaxial electrospinning to generate scaffolds for tissue repair and regeneration, one should usually try a range of electrospinning parameters to optimize the structure and components of the core–shell nanofibers [17].

In one typical demonstration, a cell-free scaffold consisting of core–shell fibers with micro-/submicro-scale structures was fabricated for bone regeneration. The composite fibers were prepared by coaxial electrospinning with cellulose acetate as the core solution, while a blend of silk fibroin, polyethylene oxide, nano-hydroxyapatite, and bone morphogenetic protein-2 (BMP-2) as the shell solution. The composite scaffold was verified to have strong mechanical property and biocompatibility, and BMP-2 could be released within a period of up to three weeks. Such a scaffold showed improvement in adhesion, proliferation, and osteogenic differentiation when cultured with the bone marrow-derived stem cells. Taken together, coaxial electrospinning represents a promising method for producing protein-incorporated nanofibrous scaffolds, showing good preservation of biological activity for tissue repair and regeneration.

5.2.3 Emulsion Electrospinning

Emulsion electrospinning is another normally used technique to generate core–shell nanofibers that encapsulate payloads. It has the same setup with traditional electrospinning except that the electrospinning solution is different (Figure 5.2c) [27]. The success of emulsion electrospinning mainly relies on the formation of a water-in-oil or oil-in-water emulsion to serve as the electrospinning solution. During electrospinning, the emulsion droplets could be stretched, electrospun, and subjected to solidification because of solvent evaporation to produce fibers. The emulsion electrospinning technique shows significance in preparing fibers containing active substances, special drugs, and other functional components [28–30]. The as-obtained core–shell nanofibers were proved to promote cell migration, proliferation, and gene expression by releasing bioactive agents in a controlled and sustainable manner and maintaining their biological activity. In one typical study, nanofibrous scaffolds made of a blend of PLGA and type I collagen, together with the incorporation of fluorescein isothiocyanate dextran, were fabricated by emulsion electrospinning [31]. From the measurement of the release behavior, it is shown that the release duration of the payload could be sustained as long as seven weeks. When cultured with osteoblastic-like MC3T3-E1 cells, such scaffolds showed improvement of osteoblastic activity and cytocompatibility.

Nowadays, emulsion electrospinning has still been widely used to generate functional nanofibers loading a wide range of bioactive components for use in controlled release and biomedicine. However, one should still pay more attention to the precise control of the placement of therapeutic agents when they are encapsulated into the core of nanofibers [32]. In addition, compared to the nanofibers fabricated by direct blend electrospinning and coaxial electrospinning from the same materials (i.e. PCL and L-ascorbate 2-phosphate), it has been reported that the nanofibers produced by emulsion electrospinning were of low mechanical properties [33]. However, the emulsion-electrospun nanofibers showed better continuous release of the encapsulated L-ascorbate 2-phosphate to promote osteogenic differentiation of suspension mononuclear cells in bone tissue engineering [34]. In this case, one should balance the physical properties of the nanofibers and the release behavior of the payload to boost their effect for use in tissue regeneration.

5.2.4 Conjugated Electrospinning

Conjugated electrospinning is a special technique used for producing nanoyarns by gathering a large number of nanofibers. As such, it has been usually used to increase the fiber diameter and improve the mechanical properties of nanofibers. The success of conjugated electrospinning relies on the attraction and twisting between positive-charged and negative-charged nanofibers generated from two opposite nozzles [35]. As shown in Figure 5.3a, two high power supplies were used to provide positive and negative charges, respectively, when the electrospinning

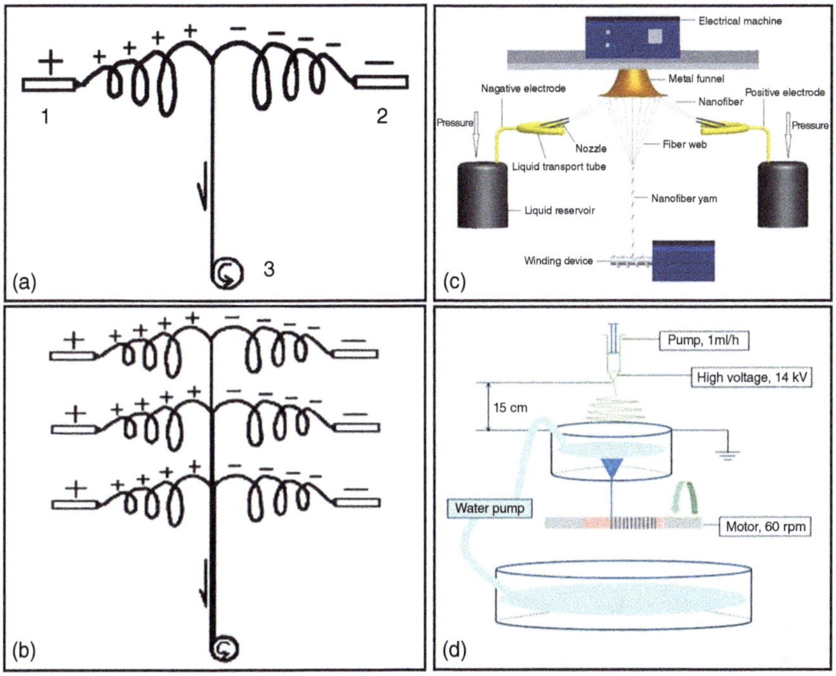

Figure 5.3 Schematic illustration of the setups of (a–c) conjugated electrospinning (a) using one pair of spinnerets; (b) using three pairs of spinnerets; (c) using one pair of spinnerets together with a rotating funnel; and (d) dynamic liquid electrospinning to generate nanofiber yarns. Source: (a and b) Li et al. [35]/with permission of John Wiley & Sons. (c) He et al. [36]/with permission of Springer Nature. (d) Wu et al. [37]/with permission of Royal Society of Chemistry.

process was performed from the two opposite sides. Three pairs of spinnerets can also be used to generate nanofiber yarns (Figure 5.3b). Because of the mutual attraction generated from the opposite charges, electrospun nanofibers would be entangled with each other before final collection by a rotating drum, thus forming circumferentially aligned nanofiber yarns. A rotating funnel can also be used to twist and collect the nanofibers produced from the two sides to further facilitate the formation of nanofiber yarns (Figure 5.3c) [36].

5.2.5 Dynamic Liquid Electrospinning

Dynamic liquid electrospinning is often employed to make porous scaffolds made of nanofiber yarns to resist the failure of 3D tissue regeneration when using planar nonwoven mats made of electrospun fibers [38]. As shown in Figure 5.3d, the success of dynamic liquid electrospinning mainly relies on the employment of a water basin with a hole at the center of the bottom. The water was recycled by a pump to maintain the water level, and thus the basin was continuously filled with water to create a vortex to twist the nanofibers deposited on the water surface to generate nanoyarns [37]. The nanoyarns were then pulled out through the hole and collected by a rotating drum to generate porous scaffolds. Such scaffolds have well-ordered microstructure, high porosity. In particular, the topological structure of electrospun nanoyarns is similar to that of natural collagen fibers. Compared with other porous scaffolds, the 3D structure of a scaffold composed of nanoyarns made it specifically appropriate to mimic the natural collagen fibers for bone or tendon regeneration.

5.2.6 Multi-fluid Electrospinning

Based on the demand of advanced functional nanomaterials with complex nanostructures, several multi-fluid electrospinning methods have been proposed, such as triaxial electrospinning and coaxial electrospinning with side-by-side cores (Figure 5.4) [39, 40]. These multi-fluid processes can be considered as some specific examples of coaxial electrospinning, which greatly expand the capability of electrospinning to produce novel nanostructures with different internal structures. The key factors of multi-fluid electrospinning are the design of a special spinneret,

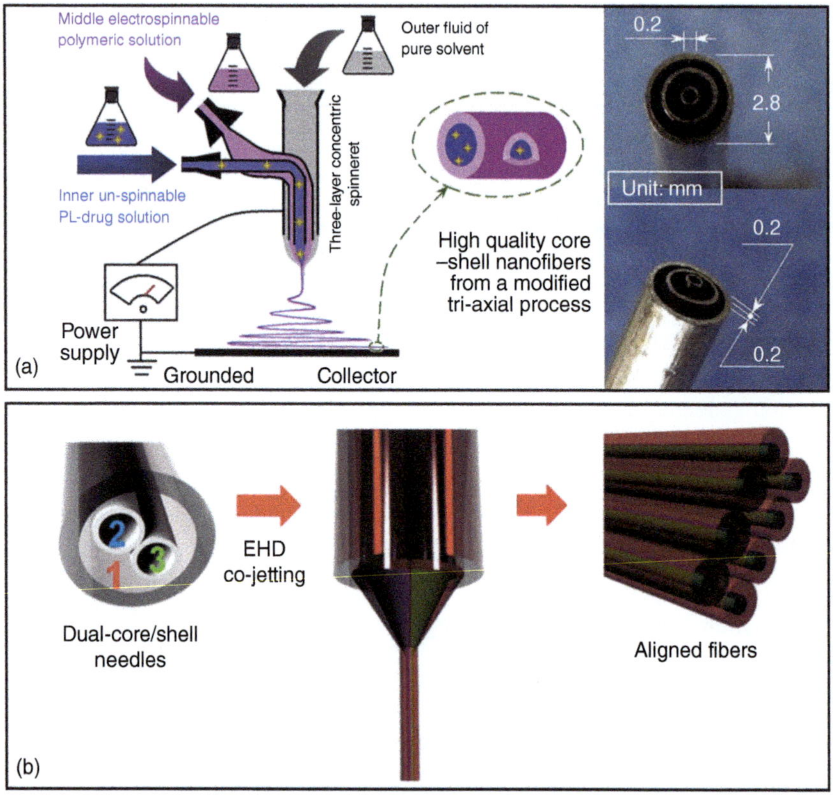

Figure 5.4 Schematic illustration showing special spinnerets: (a) triaxial (three-layer concentric) spinneret and (b) coaxial spinneret with a side-by-side core. Source: (a) Yang et al. [39]/with permission of Acta Materialia Inc. Published by Elsevier (b) Lee et al. [40]/with permission of American Chemical Society.

the compatibility of working solution, and special operating parameters. In these cases, complex nanostructures can be realized by direct electrospinning of various fluids with variable components and spatial distributions.

5.2.7 Electrospun Fibers Integrated with Post-treatments

A rich variety of electrospun fibers can be used with appropriate post-treatments for a target-specific application, for example, coating physical components or depositing electrosprayed microparticles on the fibers, welding of cross points among adjacent fibers, and 3D remolding. In this way, the physical, chemical, and/or biological properties of the fibrous scaffolds can be further improved.

5.2.7.1 Coating Physical Components on the Fibers

Most of the coating technologies make full use of the high specific surface area of nonwoven mats of nanofibers. In one study, FeOOH/g-C_3N_4 submicron particles were synthetized and incorporated in electrospun polyacrylonitrile nanofibers to obtain a nonwoven mat [41]. Then, a dense, smooth, and nondestructive layer of chitosan was coated on the surface of the mat to gain hydrophilicity. In other studies, proteins or growth factors were usually used as bioactive effectors to coat on the surfaces of nanofibers to provide biological cues [42, 43]. For example, fibronectin, collagen, and selectin were separately or together coated on electrospun PLA nanofibers to facilitate the viability, clonogenicity, and attachment of fucosyltransferase VI-treated umbilical cord blood $CD133^+$ cells (Figure 5.5a) [44]. Similarly, hydrogels are widely used as coating materials, such as calcium alginate, gelatin, and hyaluronic acid [50]. In this way, the advantages of hydrogels (e.g. lubricity, biocompatibility, and anti-biofouling properties) and the nanofibrous substrates (e.g. stiffness, toughness, and strength) can be combined.

5.2.7.2 Deposition of Electrosprayed Microparticles on the Fibers

Electrospray has been considered as a similar technique to electrospinning, which is usually used to deposit microparticles made of polymers or biomacromolecules on nonwoven mats of nanofibers to gain topographical cues and/or biological guidance. For the application in tissue regeneration, such composite scaffolds are specifically suitable for tuning the physical properties such as roughness and stiffness or manipulating cell behaviors, for example, cell attachment, migration, and differentiation. In one typical study, electrosprayed microparticles made of fatty acid were deposited on uniaxially aligned PCL microfibers (Figure 5.5b) [45]. By providing contact guidance from the roughness as well as the biological cue of the fatty acids, such scaffolds showed significant improvement in promoting the outgrowth of neurites. For the regeneration of bone tissues, simultaneous electrospraying of nanohydroxyapatite on electrospun polymeric nanofibrous scaffolds provided a better microenvironment for cell growth, holding great potential in healing bone defects and guiding bone tissue regeneration [51].

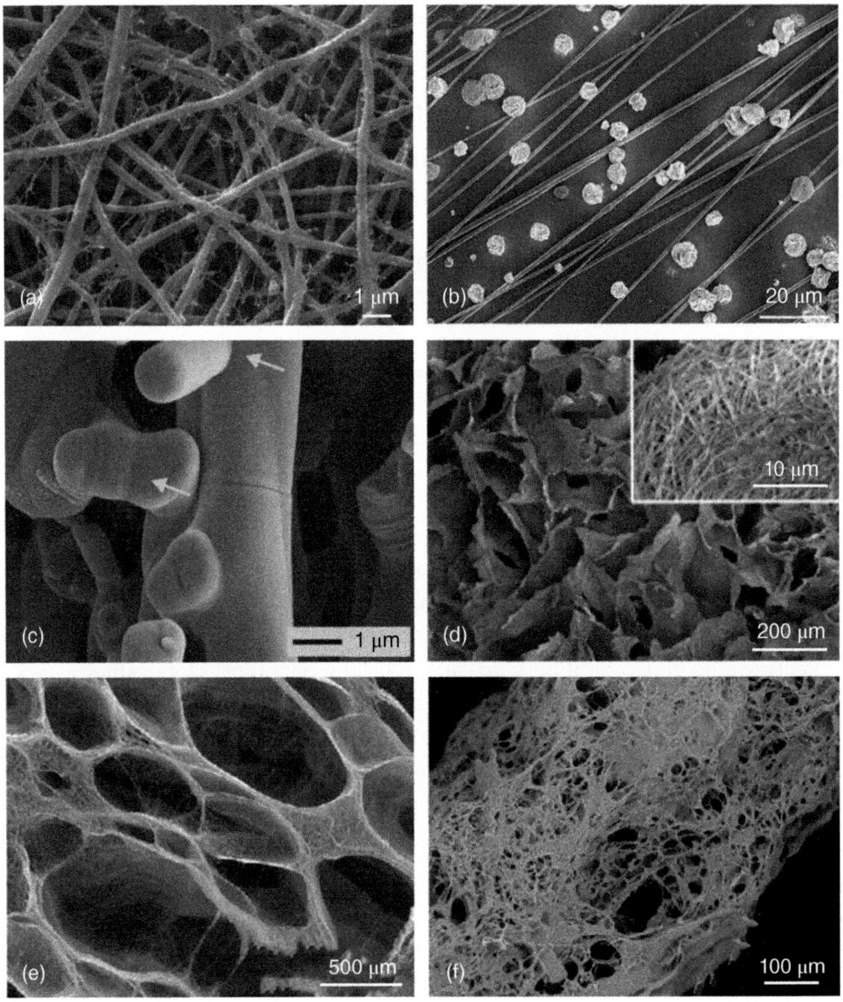

Figure 5.5 SEM images showing (a) the PLA nanofibers coated with fibronectin and collagen, (b) the deposition of electrosprayed microparticles made of fatty acid on aligned PCL nanofibers, (c) the welding of the cross points between electrospun PCL nanofibers after exposure to the vapor generated from 25 μl of dichloromethane in a closed vial for 60 minutes, (d) 3D fibrous scaffold made of bioactive glass (SiO_2–CaO) nanofibers with the use of chitosan as linkers, (e) expanded 3D nanofibrous scaffold made from planar PCL nanofiber mat, (f) 3D printed scaffolds with decellularized cow scapular cartilage matrix and gelatin/PLGA nanofibers as inks. Source: (a) Atashi et al. [44]/with permission of Springer Science Business Media, LLC, part of Springer Nature. (b) Xue et al. [45]/with permission of Wiley-VCH Verlag GmbH & Co. KGaA, Weinheim. (c) Li et al. [46]/with permission of WILEY-VCH Verlag GmbH & Co. KGaA, Weinheim. (d) Wang et al. [47]/with permission of American Chemical Society. (e) Chen et al. [48]/with permission of Wiley-VCH GmbH. (f) Chen et al. [49]/with permission of 2019 Elsevier B.V.

5.2.7.3 Welding of Nanofibers

In normal cases, nanofibers are collected layer-by-layer through the physical stacking to form the mats, which are slightly weak in mechanical strength when used as scaffolds for bone repair. A possible way is to enhance the interaction between neighboring nanofibers by welding the cross points of the nanofibers. Several strategies have been used to weld electrospun nanofibers. Among others, heat treatment, solvent vapor exposure, and light irradiation are most commonly used to get mechanically reinforced scaffolds without obvious structural changes. Heat treatment is often carried out by locally welding the nanofibers at a sufficiently high temperature. For example, when the temperature reached 70 °C, mesoporous carbonized nanofibers were welded to a network structure, which effectively improved the mechanical strength of the self-supporting carbon nanofiber membranes [52]. Another effective method is to expose the nanofibers to their suitable solvent in a sealed container. By controlling the volume of the solvent and the treatment duration, the welding extent can be controlled to weld only at the cross points of adjacent nanofibers. Using this method, PCL and PLGA nanofibers have been successfully welded at their cross points by dichloromethane and ethanol vapor, respectively (Figure 5.5c) [46]. In addition to solvent vapor exposure, photothermal irradiation can also be used to weld nanofibers simply by incorporating photothermal agents (e.g. organic agents and plasmonic nanostructures) into the nanofibers during electrospinning. For example, when indocyanine green, a light-absorbing dye, was blended with PCL to produce PCL/indocyanine green nanofibers, the welding of nanofibers can be triggered through the use of a near-infrared laser [53]. The welding extent can be simply tuned by changing the concentration of indocyanine green, power density, and irradiation duration. With the use of a photomask, one can also control the regions needed to be irradiated to obtain expected surface patterns.

5.2.7.4 Remolding Electrospun Nanofibers to 3D Scaffolds

The remolding of 2D electrospun nanofibers to construct 3D scaffolds is of great importance for bulk tissue regeneration. Among others, the nanofiber-based, 3D scaffolds made by freeze-drying, gas forming, or 3D printing show great potential for use in tissue regeneration [54]. In one demonstration, bioactive glass (SiO_2–CaO) nanofibers were homogenized and then reconstructed to 3D fibrous scaffolds with the use of chitosan as linkers by freeze-drying (Figure 5.5d) [47]. Such scaffolds showed feasibility in fitting the irregularly shaped bone defects and substantial promotion of bone regrowth and vascularization in a rat osteoporotic calvarial defect model. Using a similar method, the gelatin/PLA nanofibers were cut to short fibers, homogenized, and remolded to 3D porous scaffolds by freeze-drying [55]. In this case, the 3D scaffolds were cross-linked either by heating or chemical reaction, which showed excellent superabsorbent property, elasticity, and biocompatibility for cartilage regeneration. As a typical example of expanding a 2D nanofiber mat to a 3D nanofibrous scaffold by gas forming (Figure 5.5e), Xie and coworkers reported the fabrication of a class of 3D hierarchical scaffolds with gradations in pore size [48]. They showed that the 3D scaffold with a smaller pore size enhanced chondrogenic differentiation of bone mesenchymal stem cells compared with that with a larger pore size. Another emerging method to achieve 3D nanofibrous scaffolds is carried out with a 3D printing technique, together with the use of electrospun fiber-based inks. For example, decellularized cow scapular cartilage has been used to produce the decellularized matrix, which was then blended with gelatin/PLGA nanofibers to serve as the inks to generate 3D-printed scaffolds (Figure 5.5f) [49]. The as-obtained 3D scaffolds showed enhanced mechanical properties and facilitated the regeneration of cartilage tissues.

5.3 Typical Applications of Electrospun Fibers in Orthopedics

The musculoskeletal system represents the most important organs for human beings (Figure 5.6) [56]. Bone is the rigid body tissue with cells embedded in an abundant hard intercellular material, which contains two principal components: collagen and calcium phosphate. Collagen components can also be found in other musculoskeletal tissues. The tendon and ligament tissues are composed of collagen fiber bundles that are uniaxially aligned and wavy. At the tendon-to-bone interface, collagen fibers exhibit the gradient transition from aligned to random moving from the tendon site to bone. While the collagen fibers are circumferentially aligned in meniscus and annulus fibrosus, they are arranged by layered arrays in cartilage [56]. Bone diseases are the most common clinical diseases that affect the skeletal system, blood vessels, nerves, and soft tissues. Because of the lack of effective treatment, the prognosis of patients with bone defect, nerve fiber rupture, recurrent spasm, and vascular tendon adhesion is not sufficient. At present, surgical materials are less effective in treating bone defect, nerve fiber rupture, and repeated tendon spasm/adhesion after repair of high-energy trauma. Therefore, it is necessary to explore new and effective treatment methods [57]. Owing to the advanced development of electrospinning techniques, a wide range of scaffolds made of electrospun nanofibers have been applied to tissue repair and regeneration, including the reconstruction of musculoskeletal tissues. Because of its 3D structure, high aspect ratio, dense fiber network, high tissue affinity, and enhanced cell adhesion, drug-loaded electrospun fibers can be further combined with 3D printing technology and liposomes to prepare different tissues with multimodal peptides, such as bone. Therefore, electrospinning has a strong structural adjustability. In addition, drug-loaded electrospun fibers have become a new polymer material for the treatment of orthopedic diseases, which allows the drugs being controlled released at the local site of injured tissues [57].

5.3.1 Bone Tissue Repair

Because of the high incidence of bone injury caused by infection, tumor, and trauma, there is a great demand for bone transplantation in clinics [58]. Because

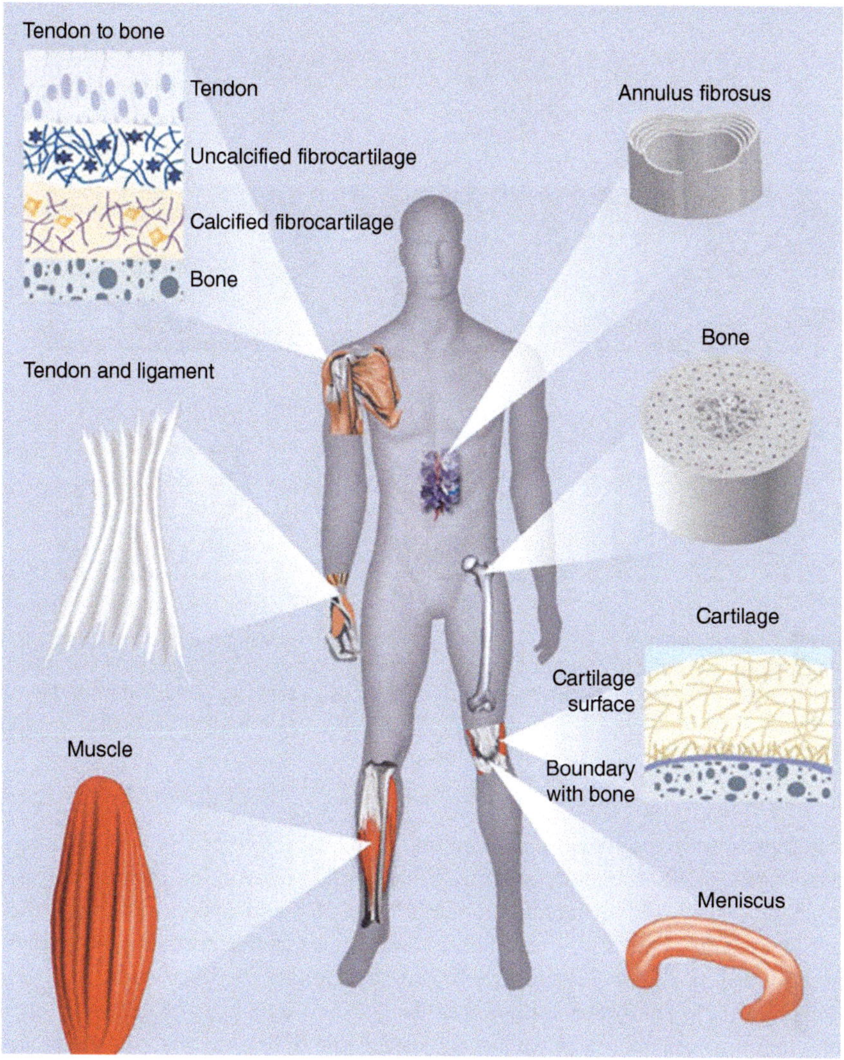

Figure 5.6 Schematic illustration of the musculoskeletal system. Source: Ma et al. [56]/with permission of Future Medicine.

of the limitation of the body self-healing capacity, the repair and reconstruction of bone defects are always difficult problems in orthopedics. Autogenous bone graft is normally considered as the gold standard for repairing severe bone defects because of its excellent capability of osteogenic induction and bone conductivity, together with the low immune reaction. However, the clinical application of autogenous bone or allogeneic bone is often limited by the shortage of donors [59]. As an emerging class of alternatives, tissue-engineered scaffolds have been designed and fabricated. In particular, a rich variety of electrospun nanofibers have been used as scaffolding materials for bone regeneration [60, 61]. Other additives can also be integrated

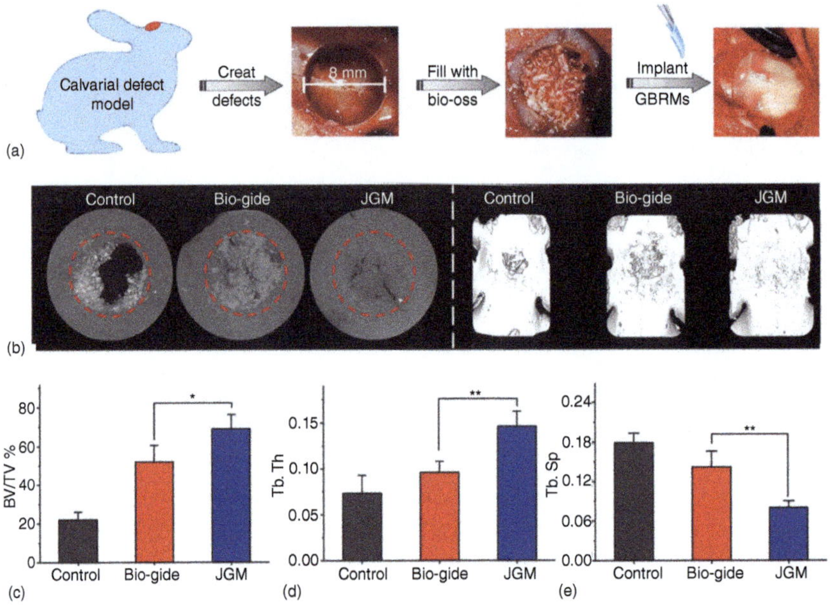

Figure 5.7 Evaluation of *in vivo* bone repair. (a) Schematic showing the operation. (b) Micro-CT images of bone formation post treatment by different scaffolds. (c) Bone volume fraction, (d) trabecular thickness, and (e) trabecular separation analyzed from the micro-CT images. *$P < 0.05$, **$P < 0.01$. Source: Wang et al. [62]/with permission of Wiley-VCH GmbH.

with electrospun nanofibers to increase the biological activities and thus enhance the osteogenesis. In one study, a Janus-guided bone regeneration membrane was constructed with random nanofibers made of gelatin and hydroxyapatite as the inner layer while aligned PCL nanofibers loading poly(methacryloxyethyltrimethyl ammonium chloride-*co*-2-Aminoethyl 2-methylacrylate hydrochloride) as the outer layer [62]. The inner surface was designed to enhance the adhesion and proliferation of osteoblasts as well as the osteogenic differentiation. On the other hand, the outer layer could help inhibit epithelial invasion and bacterial infection. When compared with a commercial product (e.g. Bio-Gide, tissue-derived collagen, and AlloDerm), such a Janus scaffold showed better performance for *in-vivo* bone tissue regeneration (Figure 5.7).

In addition, scaffolds made of electrospun nanofibers are capable of encapsulating related growth factors and controlling their release by simply blending together or incorporating them into the core of nanofibers [63]. For example, core–shell nanofibers with a blend of silk fibroin, PCL, and poly(vinyl alcohol) as the shell and BMP-2 as the core were fabricated by coaxial electrospinning [64]. Then, connective tissue growth factors were attached on the surface of the nanofiber mat by the layer-by-layer technique. In this way, the connective tissue growth factors could be fast released, while BMP-2 was released in a sustainable manner to improve the performance of bone regeneration.

Electrospun nanofibers also play an important role in the treatment of chronic osteomyelitis. In clinical surgery, the removal of necrotic bone will inevitably lead to the loss of periosteum and bone matrix, and the necrotic bone provides an ideal environment for the growth of bacteria [65]. To address this issue, electrospun nanofibers show possibility to treat chronic osteomyelitis by releasing both antibacterial and osteogenic drugs, together with the mechanical support by the nanofiber scaffolds [66, 67]. As a typical example, core–shell nanofibers with PCL as the core and gelatin as the shell were fabricated for bone formation [68]. Particularly, a traditional Chinese medicine (i.e. icariin) was added to the core, while a broad-spectrum antibacterial drug (i.e. moxifloxacin hydrochloride) was blended with gelatin. As such, an osteogenic and antibacterial periosteum was developed. Because of the dissimilar degradation rates of PCL and gelatin, time-dependent release of the two payloads was realized, showing effectiveness for *in vivo* bone formation in rabbit radius defects.

5.3.2 Vascularization

In addition to osteogenesis, angiogenesis represents another important aspect for bone defect repair. Actually, bone repair is mainly relied on the formation of blood vessels to ensure the supply of oxygen and nutrients and restore bone function. Although a rich variety of bone repair scaffolds have been investigated for bone tissue regeneration, vascularization remains an ongoing bottleneck issue that needs to be solved. It is a leading direction to construct a biomimetic microenvironment with porous structures to promote angiogenesis during bone tissue reconstruction, which has important scientific significance and clinical value.

Electrospun nanofibers highly simulate the 3D fibrous structures of native ECM, showing possibility to promote angiogenesis together with nutrient transportation during osteogenesis [69]. Shruthy et al. exploited a nanocomposite scaffold made of PLA fibrous yarns and silica-coated nanohydroxyapatite/gelatin, which could differentially release vascular endothelial growth factors within one week and BMP-2 and fibroblast growth factor-2 within three weeks [70]. In this way, neovascularization was promoted along with new bone formation in critical sized bone defects. Another possibility to reconstruct the bone defects with expected vascularization is enhancing the intimate cross talk between bony cells and endothelia cells [71]. By immobilization, anti-BMP-2 antibodies and antivascular endothelial growth factor antibodies with a parallel pattern on the surface of a scaffold made of PCL nanofibers, endogenous BMP-2, and vascular endothelial growth factors could be together bound to the scaffolds, thus contributing to the osteogenic and angiogenic differentiation of mesenchymal stem cells. Through the chick chorioallantoic membrane assay, the scaffold comprising bound endogenous bio-effectors induced an increased angiogenesis (Figure 5.8). Such a design for the scaffold enables the development of an effective approach to achieve vascularized bone tissue regeneration.

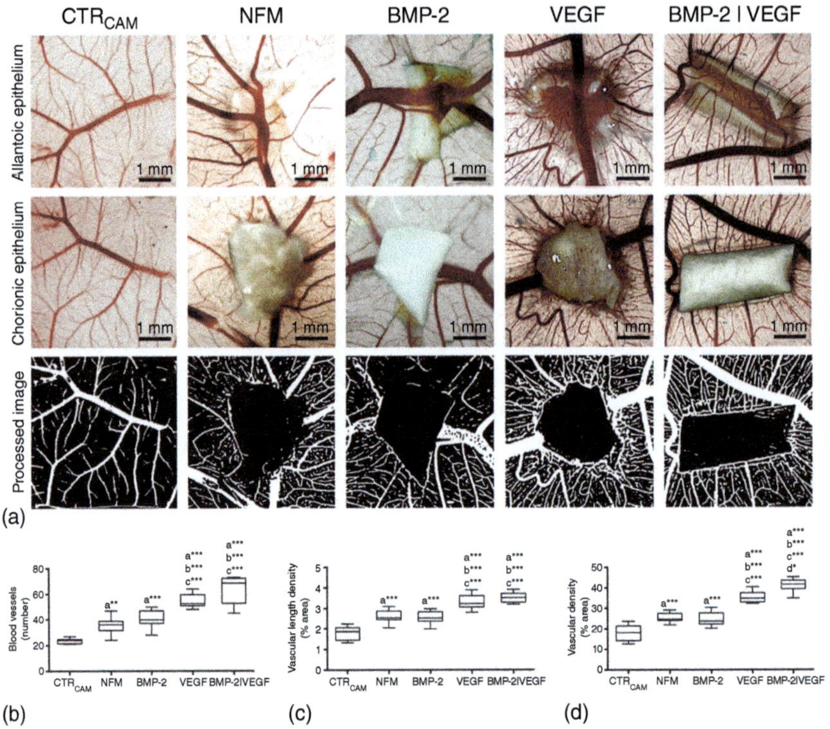

Figure 5.8 (a) Stereomicroscope photographs of *ex vivo* incubation with the nanofiber scaffolds immobilized with or without the growth factors on the chorioallantoic membrane of chick embryo. Quantification of (b) blood vessels, (c) vascular length density (i.e. ratio of skeletonized vasculature area to total area), and (d) vascular density (i.e. ratio of vasculature area to total selection area) of the different groups. $^{*}P < 0.01$, $^{**}P < 0.001$, $^{***}P < 0.0001$. Source: Casanova et al. [71]/with permission of Royal Society of Chemistry.

5.3.3 Cartilage Repair

Cartilage is composed of fibrous connective tissues constituting an important part of the organism, which is considered as an important supporting organ with certain hardness and elasticity. Because of its poor self-repair ability, the regeneration of cartilage tissues is still an unsolved problem in clinical treatment [72]. For example, the meniscus tear is one of the common cartilage injuries clinically. The current treatments include meniscus suture, meniscus partial resection, and other surgical strategies. However, the incidence of postoperative sequelae is also high, often causing secondary injury to patients. Tissue-engineered strategies based on growth factors and/or mesenchymal stem cells have shown good potential in the repair of cartilage defects. The local release of growth factors has exhibited effectiveness in creating a repairable environment for tissue regeneration. Mesenchymal stem cells have also been reported to prevent arthritis by reducing the inflammatory reaction and remold articular cartilage through the directional differentiation to chondrocytes, thus capable of delaying or even avoiding the joint replacement surgery [73].

Electrospun nanofibers provide a class of useful substrates for controlling the release of growth factors and manipulating the cell behaviors for cartilage repair. In addition, the nanofibrous scaffold can simulate the structure of natural ECM, and its high porosity and surface-to-volume ratio endow effective cell adhesion and nutrient transport [55, 74]. For example, it has been documented that either short-term delivery of basic fibroblast growth factors or continuous delivery of transforming growth factor-β3 from electrospun PCL nanofibers could increase the integration intensity for meniscus tissues of both juvenile and adult bovine defects [75]. In another study, a tissue-engineered scaffold was designed by integrating mesenchymal stem cells with electrospun PCL nanofibers, whose fiber direction matched with that of the meniscal circumferential fibers (Figure 5.9a) [76]. It has been proved that such a composite scaffold has good potential in maintaining the structural function of meniscus hoop and repairing the defected meniscus tissues with the regeneration of fibrocartilaginous tissue in a rabbit meniscus injury model (Figure 5.9b).

In addition to wrapping the 2D nanofiber mats surrounding the defect site, 3D nanofibrous scaffolds have also shown the capability of meeting the biochemical properties that are required for cartilage repair. In a typical demonstration, Chen et al. fabricated a porous 3D scaffold made of gelatin/PLA nanofibers and cross-linked with hyaluronic acid, which is a main component of ECM that lubricates joints and promotes wound healing [55]. Such a scaffold showed high absorbability and good cytocompatibility and promoted the repair of cartilage tissues using a rabbit articular cartilage loss model. In another study, Li et al. prepared a 3D composite scaffold made of gelatin/PCL nanofibers and decellularized cartilage ECM [77]. In this case, the nanofibers were reinforced by the biomimetic structures of natural cartilage to improve the mechanical properties. In particular, the decellularized cartilage components were functioned as bioactive effectors to enhance cellular responses and promote the secretion of collagen and glycosaminoglycan, thus improving the repair and regeneration of cartilage tissues. Although scaffolds made of electrospun nanofibers show promising potential in repairing cartilage tissues, they are only limited to animal experiments and have not been applied to clinical practice. Future works need to be carried out from clinical view and should be applied to more cases of different cartilage injuries.

5.3.4 Tendon and Ligament Repair

Tendons are composed of parallel collagenous bundles, whose function is to transfer muscle strength to bone, thus enabling normal movement. Similar to the structure of tendons, ligaments consist of dense strips of collagen, which are arranged along the length of the joint and inserted into the ends of the bone. In human body, ligaments exist with different sizes and directions. Lesions of tendons and ligaments represent over 40% of the musculoskeletal lesions [78]. The repair of tendon or ligament

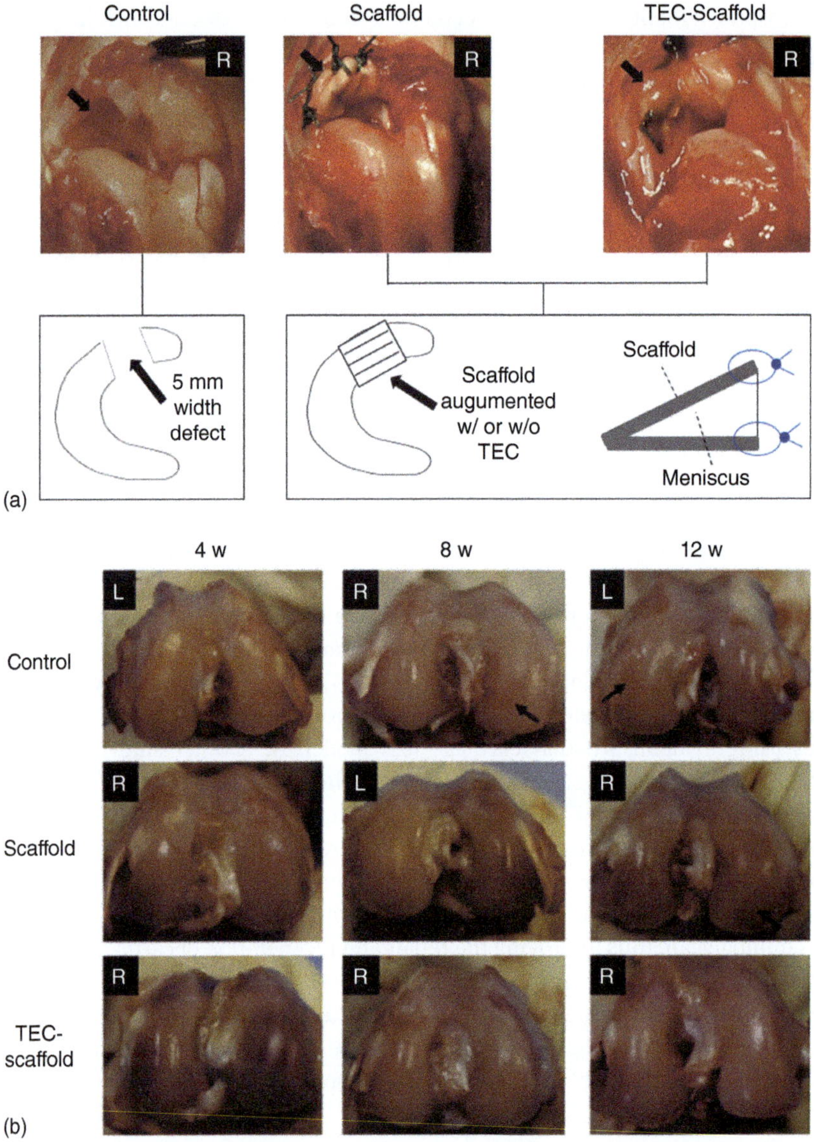

Figure 5.9 (a) Schematic illustration showing the surgical procedure to induce a meniscal injury and the treatment with the use of a scaffold (electrospun PCL nanofibers only) or a tissue-engineered scaffold (electrospun PCL nanofibers integrated with a sheet-like complex generated from the cultured mesenchymal stem cells, and the ECM synthesized by the cells). In the control group, no treatment was applied to the meniscal defect.
(b) Macroscopic evaluation of articular cartilage integrity on the medial femoral condyle after treatment with the use of the scaffold or the tissue-engineered scaffold up to 12 weeks. Source: Shimomura et al. [76]/with permission of Elsevier Ltd.

tissues usually requires a scaffold with sufficient mechanical properties and biological functions. Although there are a great majority of grafting materials emerged for repair tendon or ligament tissues, the recovery of normal mechanical strength and biofunctions is still not efficient.

Scaffolds made of electrospun nanofibers have been developed for use in tendon or ligament regeneration. Because of the low mechanical strength and stiffness of electrospun nanofibers relative to the normal tissues, an ongoing challenge is how to meet the rigid loading conditions of the native tendon or ligament. This is also the physiologically mechanical criterion that must be considered in designing the next generation of tendon/ligament scaffolds [79]. Besides, the biological functions should also be considered when designing the scaffolds. A perfect scaffold should be able to stimulate or regulate cell differentiation and growth to achieve the expected phenotype and structures, so as to restoring the native function. For example, electrospun nanofibrous scaffolds have been demonstrated to promote the phenotypic differentiation of tendon progenitor cells and mesenchymal stem cells to tendon/ligament-like cells, as well as the production of ECM [80–82].

Taking tendon as an example, electrospinning nanofibers have still been widely used for the repair of tendon. In one demonstration, Zhang et al. prepared a scaffold of aligned PLA nanofibers containing a small-molecule Trichostatin A, an inhibitor to histone deacetylases [81]. Such a scaffold could continuously release the Trichostatin A molecules to reduce the expression of histone deacetylases in the tendon stem/progenitor cells when cultured on the PLA nanofibers. In this way, the tenogenic differentiation of tendon stem/progenitor cells was enhanced, and *in situ* implantation of the scaffold in a rat model showed the restore of structural and mechanical properties of the regenerated Achilles tendon. In another study, hyaluronic acid hydrosols containing mitomycin-C was mixed with PLA solution and the surfactants to obtain the water-in-oil emulsions including micro-sol particles for electrospinning [83]. Through the use of such a scaffold, defect tendons were healed with up-regulation of apoptotic protein Bax expression and down-regulation of proteins Bcl2, collage I, collagen III, and α-smooth muscle actin (SMA) (Figure 5.10). In particular, the adhesion of tendon with surrounding tissues could also be prevented. Another study has also shown that the PLGA nanofibrous scaffolds loaded with doxycycline help promote the healing of Achilles tendon rupture [84]. We can conclude that electrospun nanofibers hold great potential to serve as substrates for use in tendon repair.

In addition to loading bioactive effectors into the nanofibrous scaffolds, the replication of the complex multiscale hierarchical structure of native tendons and ligaments also attracts the attention of researchers. Sensini et al. fabricated a scaffold consisting of multiple bundles of resorbable PLA nanofibers, which showed tailored dimensions and could be wrapped in a sheath of nanofibers to compact the construct [78]. The bundles were in turn composed of the nanofibers with a preferential direction. By measuring the tensile stiffness, strength, and toughness of the scaffold, it showed that the mechanical properties were in the required scope to repair tendons

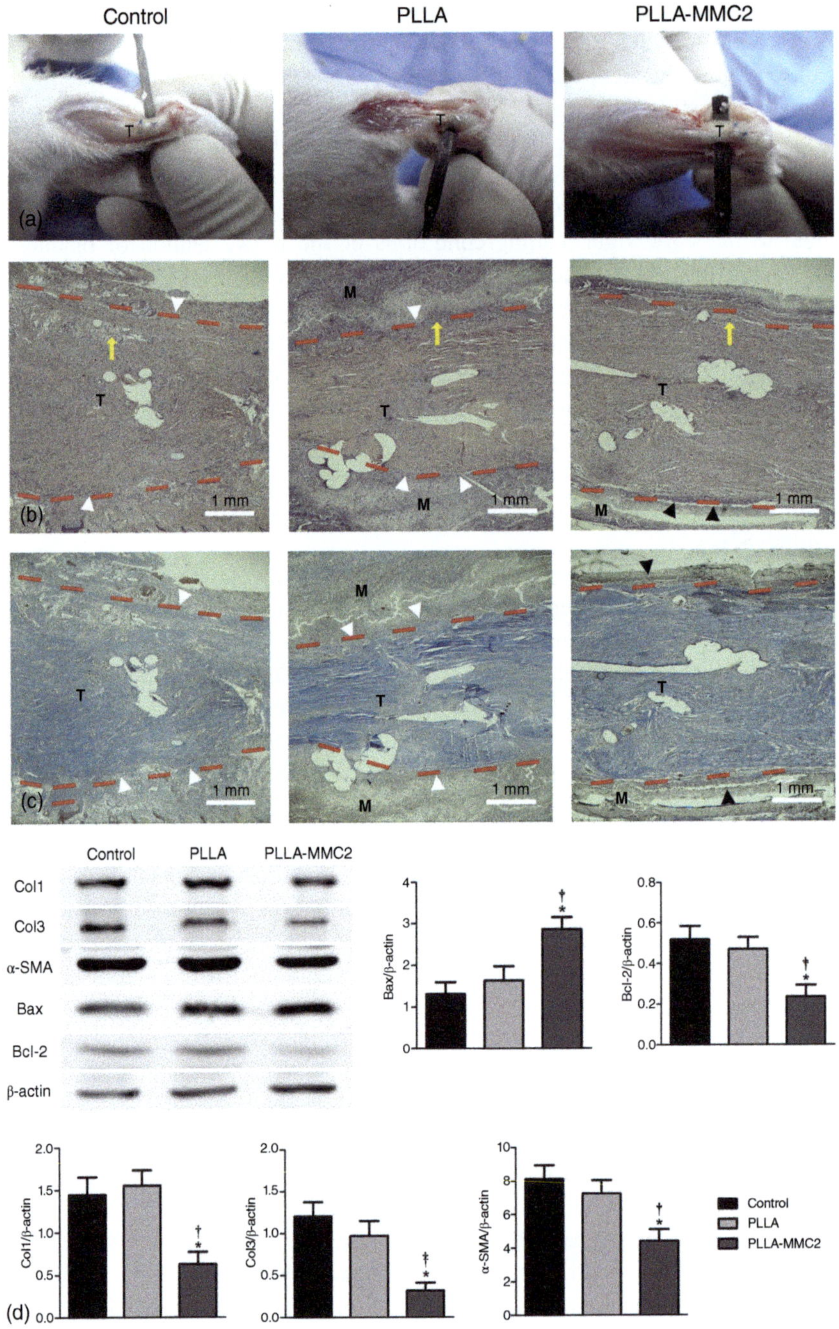

Figure 5.10 (a) Gross evaluation, (b) H&E, and (c) Masson staining of the rat Achilles tendon models without treatment (control) or with the use of PLA and PLA-Mitomycin-C (MMC2) nanofiber scaffolds. (d) Western blot assay for Bcl-2, Bax, collagen I, collagen III, and α-SMA expression in peritendinous adhesion tissues of the rat Achilles tendon post treatment (control) or treatment by PLA and PLA-mitomycin-C (MMC2) nanofiber scaffolds for three weeks. *$P < 0.05$ compared with control group, †$P < 0.05$ compared with the PLLA membrane group. Source: Zhao et al. [83]/with permission of Elsevier Ltd.

and ligaments. In addition, human fibroblasts were adhered and proliferated on such scaffolds. In another typical demonstration, a knitted scaffold made of core–sheath yarns with PCL microfibers as the core and PCL/silk fibroin nanofibers as the sheath for tendon repair [85]. When culturing rabbit bone marrow stem cells on the scaffold, the tenogenic differentiation of the stem cells was induced by up-regulating the expression levels of tendon-related genes. When implanting the scaffold into a rabbit patellar defect model, both the biomechanical reinforcement and tendon tissue regeneration were achieved. These approaches pave the way for designing the novel scaffolding materials with improved biomechanical properties for tendon and ligament regeneration.

From the current progress, there are still numerous obstacles for employing electrospun nanofibers to clinically repair tendon or ligament tissues. The relatively weak mechanical strength of electrospun nanofibers, especially the long-term mechanical stability during cyclic loading and temporal mechanics during biodegradation, are the most important issues that have not been solved [79]. Other obstacles include the precise replication of the complicated 3D structures of collagen bundles for native tendon or ligament tissues as well as the outcome of volumetric cell penetration. Scaffolds based on oriented nanofiber yarns may be further considered as a promising alternative. Besides, one should learn more about the changes of internal microenvironment and conditions after tendon/ligament injury so as to fabricate the ideal scaffolds.

5.3.5 Repair of Tendon-to-Bone Interface

The tendon–bone junction interface is a unique structure with gradations in components and alignment, which is composed of tendon, fibrocartilage, calcified fibrocartilage, and bone. From the histopathological results, the tendon-to-bone interface is difficult to be self-healed or surgical repair because of the lack of complete fibrocartilage regeneration, which will result in a weak mechanical strength as well as the possibility of secondary injury [86, 87]. Previous studies have shown that electrospun-aligned nanofibers can promote the proliferation of tendon cells and tenogenesis, which indicates the possibility to be further used for tendon-to-bone repair after additional modification. For example, uniaxially aligned PCL nanofibers loaded with kartogenin were used for rotator cuff tendon–bone healing [88]. Such a scaffold loaded with 100 mM kartogenin was demonstrated to encourage the differentiation of rat bone marrow stromal cells to chondrogenic and tenogenic cells. When applied to repair a rat rotator cuff tear, it also improved the formation of fibrocartilages and the organization of type-I and II collagen at the tendon–bone enthesis, together with an increased ultimate load to failure. In another demonstration, a dual-layer bipolar scaffold made of PLA nanofibers and nanohydroxyapatite-incorporated PLA nanofibers was designed to successfully generate nonmineralized fibrocartilage and mineralized fibrocartilage at the tendon-to-bone interface [89].

Precisely, to restore the function of native tendon-to-bone interface, a better way is repeating gradations of both the collagen fiber orientation and/or the

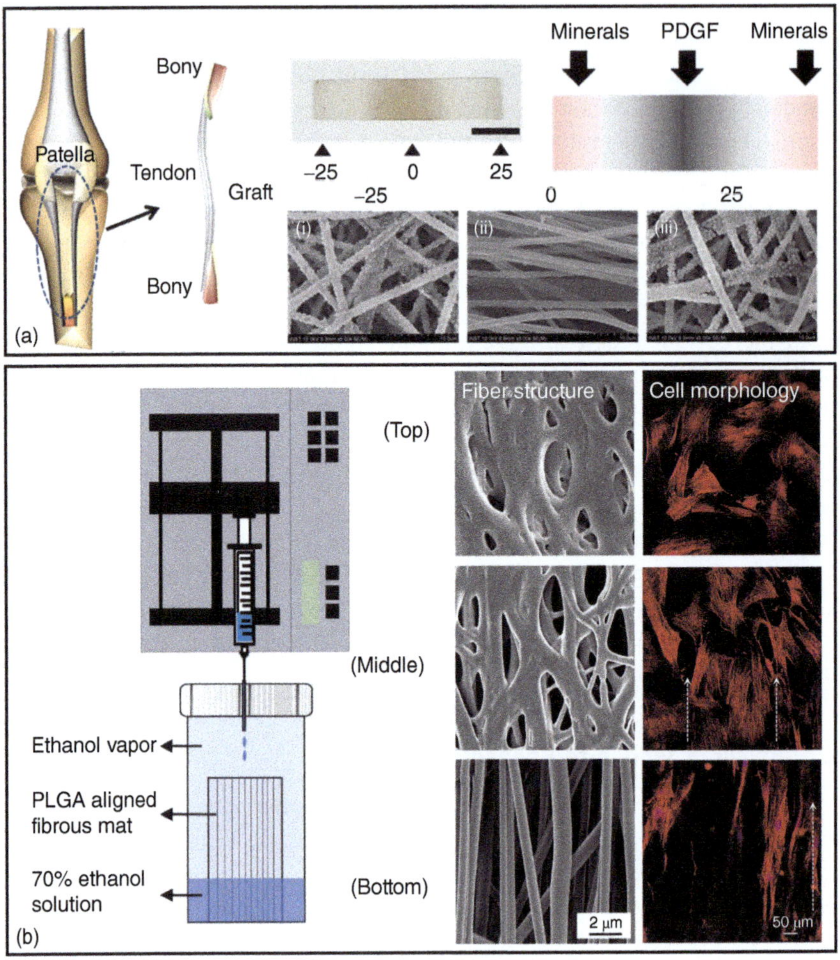

Figure 5.11 (a) Fabrication of a scaffold with random-aligned-random structure to mimic the tendon-to-bone interface. Photograph, schematic, and SEM images show the gradients of platelet-derived growth factors and minerals. (b) Schematic illustration of the strategy to weld aligned PLGA nanofibers to generate an aligned-to-random structure. When culturing with bone marrow stem cells, a transition for the cell morphology was obtained along with the fiber alignment. Source: (a) Madhurakkat Perikamana et al. [90]/with permission of Elsevier Ltd. (b) Li et al. [91]/with permission of WILEY-VCH Verlag GmbH & Co. KGaA, Weinheim.

mineral content from tendon to the bone. In a typical study, by harnessing the biochemical cues provided by immobilized platelet-derived growth factors and structural guidance generated from aligned PLA nanofibers, the tenogenic differentiation of adipose derived stem cells was significantly promoted [90]. By further generating a symmetrical gradient immobilization of platelet-derived growth factors and controlled mineralization on the nanofiber surface, an *in vitro* tissue interface graft mimicking the tendon–bone insertion was obtained (Figure 5.11a). In another demonstration, a gradient welding strategy was developed to generate the aligned-to-random structure with gradations in between using electrospun PLGA nanofibers, and bone marrow stem cells showed a transition from aligned to random for the morphology on the scaffold (Figure 5.11b) [91]. Such a scaffold can be further modified with gradient mineral content increasing from tendon to the bone region simply by a dipping method [92]. In addition, *in vivo* animal studies and investigation of clinical cases should be carried out to determine the therapeutic effects.

5.4 Conclusion and Future Outlook

Because of the constant development of materials and invention of setups, electrospun techniques exhibited great potential in producing scaffolds with variable components, structures, and biofunctions for use in orthopedic tissue repair and regeneration. At the current stage, advanced nanofiber scaffolds used for orthopedic tissue engineering mainly include those specifically designed with fine structures, incorporating functional components such as proteins, drugs, and growth factors, integrated with stem cells, as well as reconstructed to the 3D bulk matrix. However, because of the gap with the native tissues, the reconstruction and restore of impaired tissues are still difficult to meet all the requirements. To make these scaffolds more effective and specifically use for precision and translational medicine, one should precisely reproduce both macroscopic and microscopic structures of the tissues. The biocompatibility and biosafety of the raw materials used for fabricating scaffolds should also be taken into consideration, as the ultimate goal is clinical application. Although bioactive effectors have shown good performance in repairing almost all tissues in orthopedics, one should pay more attention to accurately control the releasing duration, half-time period, and localized delivery of the payloads. As the most concerned issue of the scientific community, if the scaffolds were designed with the use of stem cells, ethics is the major issue for pushing the products into the market. In general, as the physical properties and biological activities are obviously different for the different organs in orthopedics, one should still consider biomimetics as the first principle for the development of future alternatives. To this end, the design and fabrication of 3D scaffolds are of promising potential to restore the complex structures and function of the bulk tissues in orthopedics. In addition, for almost all of the scaffolds designed till now, how to realize the mass production is a common problem that needs to be solved.

References

1 Gilbert, W. (1958). *De Magnete*. Dover Publications.

2 Rayleigh, L. (2009). *Lond. Edinb. Dublin Philos. Mag. J. Sci.* 14: 184.

3 Formhals, A. (1934). US Patent 1, 975,504, filed 05 December 1930.

4 Formhals, A. (1944). US Patent 2, 349,950, filed 16 August 1938.

5 Xue, J., Wu, T., Dai, Y., and Xia, Y. (2019). *Chem. Rev.* 119: 5298.
6 Li, D. and Xia, Y. (2004). *Adv. Mater.* 16: 1151.
7 Wu, T., Mo, X., and Xia, Y. (2020). *Adv. Healthc. Mater.* 9: e1901761.
8 Chakraborty, S., Liao, I.C., Adler, A., and Leong, K.W. (2009). *Adv. Drug Delivery Rev.* 61: 1043.
9 Yarin, A.L. (2010). *Polym. Adv. Technol.* 22: 310.
10 da Silva, T., Gonçalves, R., Rocha, C. et al. (2019). *Mater. Sci. Eng., C* 97: 602.
11 Su, Y., Su, Q., Liu, W. et al. (2012). *Acta Biomater.* 8: 763.
12 Man, Z., Yin, L., Shao, Z. et al. (2014). *Biomaterials* 35: 5250.
13 Holzwarth, J. and Ma, P. (2011). *Biomaterials* 32: 9622.
14 Gluais, M., Clouet, J., Fusellier, M. et al. (2019). *Biomaterials* 205: 81.
15 Kim, H., Mandakhbayar, N., Kim, H. et al. (2020). *Biomaterials* 269: 120214.
16 Zhang, C., Yuan, H., Liu, H. et al. (2015). *Biomaterials* 53: 716.
17 Yoon, J., Yang, H.S., Lee, B.S., and Yu, W.R. (2018). *Adv. Mater.* 30: e1704765.
18 Tan, R., Yang, X., and Shen, Y. (2017). *Robot. Biomim.* 4: 17.
19 Soares, R.M.D., Siqueira, N.M., Prabhakaram, M.P., and Ramakrishna, S. (2018). *Mater. Sci. Eng., C* 92: 969.
20 Ho, M., Liao, M., Lin, Y. et al. (2014). *Int. J. Nanomed.* 9: 4293.
21 Behera, S.S., Das, U., Kumar, A. et al. (2017). *Int. J. Biol. Macromol.* 98: 329.
22 Liu, X., Holzwarth, J.M., and Ma, P.X. (2012). *Macromol. Biosci.* 12: 911.
23 Kobayashi, H., Terada, D., Yokoyama, Y. et al. (2013). *J. Biomed. Nanotechnol.* 9: 1318.
24 Han, D. and Steckl, A.J. (2019). *Chempluschem* 84: 1453.
25 Lu, Y., Huang, J., Yu, G. et al. (2016). *Wiley Interdiscip. Rev. Nanomed. Nanobiotechnol.* 8: 654.
26 Khalf, A. and Madihally, S.V. (2017). *Eur. J. Pharm. Biopharm.* 112: 1.
27 Hu, J., Prabhakaran, M.P., Ding, X., and Ramakrishna, S. (2015). *J. Biomater. Sci. Polym. Ed.* 26: 57.
28 Frizzell, H., Ohlsen, T.J., and Woodrow, K.A. (2017). *Int. J. Pharm.* 533: 99.
29 Maretschek, S., Greiner, A., and Kissel, T. (2008). *J. Controlled Release* 127: 180.
30 Bhattarai, R.S., Bachu, R.D., Boddu, S.H.S., and Bhaduri, S. (2018). *Pharmaceutics* 11: 11010005.
31 Wei, K., Li, Y., Mugishima, H. et al. (2012). *Biotechnol. J.* 7: 677.
32 Zhao, Y., Cao, X., and Jiang, L. (2007). *J. Am. Chem. Soc.* 129: 764.
33 Zhao, X., Lui, Y.S., Toh, P.W.J., and Loo, S.C.J. (2014). *Materials* 7: 7398.
34 Hadzir, S.N., Ibrahim, S.N., Abdul Wahab, R.M. et al. (2014). *Cytotherapy* 16: 674.
35 Li, X., Yao, C., Sun, F. et al. (2008). *J. Appl. Polym. Sci.* 107: 3756.
36 He, J., Zhou, Y., Qi, K. et al. (2013). *Fibers Polym.* 14: 1857.
37 Wu, T., Zheng, H., Chen, J. et al. (2017). *J. Mater. Chem. B* 5: 139.
38 Wu, T., Zhang, J., Wang, Y. et al. (2017). *J. Biomed. Nanotechnol.* 13: 303.
39 Yang, C., Yu, D.G., Pan, D. et al. (2016). *Acta Biomater.* 35: 77.
40 Lee, K.J., Park, T.H., Hwang, S. et al. (2013). *Langmuir* 29: 6181.
41 Zheng, S., Chen, H., Tong, X. et al. (2021). *Appl. Catal. B* 281: 119519.
42 Wu, T., Xue, J., Li, H. et al. (2018). *ACS Appl. Mater. Interfaces* 10: 8536.

43 Tanes, M.L., Xue, J., and Xia, Y. (2017). *J. Mater. Chem. B* 5: 5580.
44 Atashi, A., Islami, M., Mortazavi, Y., and Soleimani, M. (2018). *Mol. Biotechnol.* 60: 455.
45 Xue, J., Wu, T., Li, J. et al. (2019). *Angew. Chem. Int. Ed.* 58: 3948.
46 Li, H., Zhu, C., Xue, J. et al. (2017). *Macromol. Rapid Commun.* 38: 201600723.
47 Wang, L., Qiu, Y., Guo, Y. et al. (2019). *Nano Lett.* 19: 9112.
48 Chen, S., McCarthy, A., John, J.V. et al. (2020). *Adv. Mater.* 32: e2003754.
49 Chen, W., Xu, Y., Li, Y. et al. (2019). *Chem. Eng. J.* 382: 122986.
50 Liu, J., Qu, S., Suo, Z., and Yang, W. (2020). *Natl. Sci. Rev.* nwaa254.
51 Francis, L., Venugopal, J., Prabhakaran, M. et al. (2010). *Acta Biomater.* 6: 4100.
52 Shi, Z., Jin, G., Wang, J., and Zhang, J. (2017). *J. Electroanal. Chem.* 795: 26.
53 Wu, T., Li, H., Xue, J. et al. (2019). *Angew. Chem. Int. Ed.* 58: 16416.
54 Chen, Y., Shafiq, M., Liu, M. et al. (2020). *Bioact. Mater.* 5: 963.
55 Chen, W., Chen, S., Morsi, Y. et al. (2016). *ACS Appl. Mater. Interfaces* 8: 24415.
56 Ma, B., Xie, J., Jiang, J. et al. (2013). *Nanomedicine (Lond)* 8: 1459.
57 Mao, Y., Zhao, Y., Guan, J. et al. (2020). *Expert Opin. Drug Deliv.* 17: 993.
58 Berner, A., Boerckel, J.D., Saifzadeh, S. et al. (2012). *Cell Tissue Res.* 347: 603.
59 Zhang, J., Liu, W., Schnitzler, V. et al. (2014). *Acta Biomater.* 10: 1035.
60 Chen, J., Chu, B., and Hsiao, B.S. (2006). *J. Biomed. Mater. Res. Part A* 79: 307.
61 Yoshimoto, H., Shin, Y.M., Terai, H., and Vacanti, J.P. (2003). *Biomaterials* 24: 2077.
62 Wang, Q., Feng, Y., He, M. et al. (2020). *Adv. Funct. Mater.* 31: 2008906.
63 Seog-Jin, S., Hae-Won, K., and Jung-Hwan, L. (2016). *J. Nanomater.* 2016: 1.
64 Cheng, G., Yin, C., Tu, H. et al. (2019). *ACS Nano* 13: 6372.
65 Lew, D.P. and Waldvogel, F.A. (2004). *Lancet* 364: 369.
66 Di Prisco, I. (2010). *Biomaterials* 31: 9293.
67 Wei, S., Jian, C., Xu, F. et al. (2018). *J. Biomater. Appl.* 32: 1187.
68 Gong, M., Huang, C., Huang, Y. et al. (2019). *Nanomed. Nanotechnol. Biol. Med.* 17: 124.
69 Gizaw, M., Faglie, A., Pieper, M. et al. (2019). *Med One* 4: e190002.
70 Kuttappan, S., Mathew, D., Jo, J.I. et al. (2018). *Acta Biomater.* 78: 36.
71 Casanova, M.R., Oliveira, C., Fernandes, E.M. et al. (2020). *Biomater. Sci.* 8: 2577.
72 Hu, H., Dong, L., Bu, Z. et al. (2020). *J. Extracell. Vesicles* 9: 1778883.
73 Go, G., Jeong, S., Yoo, A. et al. (2020). *Sci. Robot.* 5: eaay6626.
74 Xia, H., Zhao, D., Zhu, H. et al. (2018). *ACS Appl. Mater. Interfaces* 10: 31704.
75 Ionescu, L., Lee, G., Huang, K., and Mauck, R.L. (2012). *Acta Biomater.* 8: 3687.
76 Shimomura, K., Rothrauff, B., Hart, D. et al. (2019). *Biomaterials* 192: 346.
77 Li, Y., Liu, Y., Xun, X. et al. (2019). *ACS Appl. Mater. Interfaces* 11: 36359.
78 Sensini, A., Gualandi, C., Focarete, M. et al. (2019). *Biofabrication* 11: 035026.
79 Brennan, D., Conte, A., Kanski, G. et al. (2018). *Adv. Healthc. Mater.* 7: e1701277.
80 Rothrauff, B.B., Lauro, B.B., Yang, G. et al. (2017). *Tissue Eng. Part A* 23: 378.
81 Zhang, C., Wang, X., Zhang, E. et al. (2017). *Acta Biomater.* S1742706117306086.
82 Sahoo, S., Ang, L.T., Goh, C.H., and Toh, S.L. (2010). *Differentiation* 79: 102.

83 Zhao, X., Jiang, S., Liu, S. et al. (2015). *Biomaterials* 61: 61.
84 Weng, C., Lee, D., Ho, J., and Liu, S. (2020). *Int. J. Nanomed.* 15: 125.
85 Cai, J., Xie, X., Li, D. et al. (2020). *Biomater. Sci.* 8: 4413.
86 Wong, M.W., Qin, L., Lee, K.M. et al. (2003). *Clin. Orthop. Relat. Res.* 413: 291.
87 Leung, K.S., Qin, L., Fu, L.K., and Chan, C.W. (2002). *Clin. Biomech.* 17: 594.
88 Zhu, Q., Ma, Z., Li, H. et al. (2019). *RSC Adv.* 9: 15582.
89 Li, X., Cheng, R., Sun, Z. et al. (2017). *Acta Biomater.* 61: 204.
90 Madhurakkat Perikamana, S.K., Lee, J., Ahmad, T. et al. (2018). *Biomaterials* 165: 79.
91 Li, H., Wu, T., Xue, J. et al. (2020). *Macromol. Rapid Commun.* 41: 1900579.
92 Liu, W., Lipner, J., Xie, J. et al. (2014). *ACS Appl. Mater. Interfaces* 6: 2842.

6

Joint Lubrication and Wear

Yi Wang, Keming Hu, Weiwei Zhao, and Hongyu Zhang

Tsinghua University, Department of Mechanical Engineering, State Key Laboratory of Tribology, No.30 Shuangqing Road, Haidian District, Beijing 100084, P.R. China

6.1 Introduction

Friction in biological systems of the human body, such as the joint, is normally low (cartilage secretes synovial fluid), but under conditions of contingency or disease, friction can become very high. Even though the lubrication properties of cartilage are excellent, friction and wear can occur with periodic mechanical loading of the human body, resulting in a gradual failure in the lubrication of the joint system. For adults, the damaged articular cartilage has little or no ability to repair itself (chondrocytes have a low intrinsic density), and the breakdown of lubrication leads to a number of diseases such as osteoarthritis. Currently, the widely used pharmacological injections have certain limitations because of the lack of long-term relief. Artificial joint replacement is considered as the last resort for the treatment of osteoarthritis. Although ultra-high-molecular-weight polyethylene (UHMWPE), the most commonly used polymer material in joint replacement, has high mechanical wear resistance and good biocompatibility, it still has defects such as attenuated lubrication, wear debris generation, and mismatched joint surfaces because of low creep resistance, which can result in unsatisfactory implantation outcomes or even more serious pathologies. It is therefore important to understand the lubrication mechanisms that lead to such low friction property of healthy joints, which has profound implications for both clinical strategies and biomedical applications [1].

Over the centuries, many pioneering scholars have explored this issue in different dimensions and depths, discussed the composition and mechanism of articular cartilage in natural healthy joint, and contributed to tribological theories such as hydrodynamic lubrication, boundary lubrication, two-phase lubrication, as well as, perhaps the most importantly, hydration lubrication, which are sequentially highlighted in this chapter. These tribological theories have gradually unraveled the mystery of joint lubrication mechanisms and are considered as the foundation for subsequent research.

Wang and Hu contributed equally.

Biofabrication for Orthopedics: Methods, Techniques, and Applications, First Edition.
Edited by Wenguo Cui, Xin Zhao, and Shen Liu.

6.2 Natural Joint Structure

6.2.1 Role of Articular Cartilage

Friction and wear are the main forms of energy dissipation and component failure in moving machine assemblies. It is estimated that about one third of the world's energy is consumed by various forms of friction. Consequently, the study of joint lubrication mechanisms *in vivo* is important for understanding the mechanisms of cartilage friction and wear, extending cartilage life, and reducing energy dissipation [2].

There are three types of joints in the human body, i.e. fibrous joints, cartilaginous joints, and synovial joints, and this chapter focuses on synovial joints. Major synovial joints, such as hip and knee joints, are unique and effective biotribological systems that exhibit extremely low friction coefficient ($\mu \approx 0.001$) between the two sliding cartilage surfaces under a wide range of shear rates and pressures in excess of 100 atm, which cannot be reproduced with any artificial biomaterials currently available.

The natural synovial joint is an extremely important weight-bearing tissue of the human body that sustains human daily activities. As shown in Figure 6.1a [3], the natural synovial joint is enclosed by a firm fibrous capsule that forms the joint capsule. The inner surface of joint capsule is covered with a highly vascularized layer of synovial membranes that secrete synovial fluid and at the same time provide nutrients to the joint. A thin layer of articular cartilage locates at the bony end of the joint. The synovial layer and articular cartilage tissues enclose the joint cavity, forming a closed biomechanical joint system.

The articular cartilage discussed in this section belongs to hyaline cartilage, which can be classified into three typical categories according to the nature and

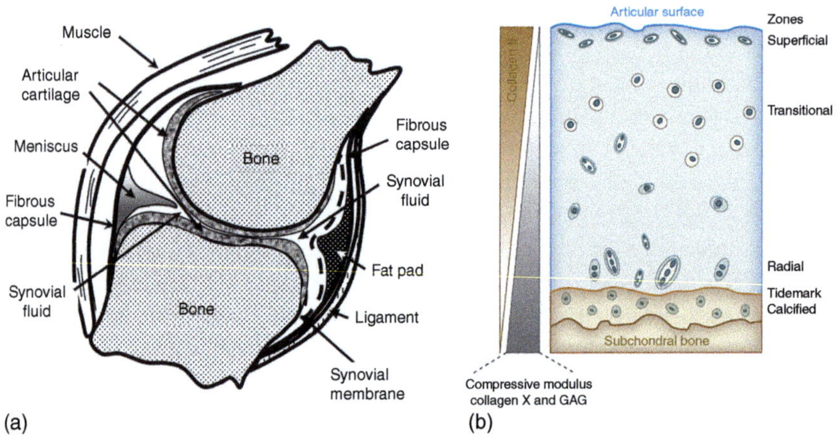

Figure 6.1 (a) The overall structure of synovial joint system. Source: Dowson [3]/with permission of Royal Society of Chemistry. (b) The regional division and the arrangement of articular cartilage including the superficial zone, transitional zone, radial zone, and calcified zone. GAG; glycosaminoglycans. Source: Jahn and Klein [4]/with permission of AIP Publishing.

arrangement of the fibers contained in the cartilage matrix. The natural articular cartilage is a water-rich, porous, and viscoelastic tissue free of blood vessels, lymphatic vessels, and nerves. It contains approximately 80% of interstitial water and 20% of extracellular matrix and is permeated with a large amount of biomacromolecules, proteins, chondrocytes, and other components [5]. In its healthy state, articular cartilage is a self-lubricating and self-healing system with a unique solid-phase structure. This structure is filled with interstitial fluid that is driven by pressure gradients generated by repeated joint friction, resulting in interstitial flow within the joint cavity. When the articular cartilage is loaded by joint motion, the instantaneous load is borne by the interstitial fluid, which is known as interstitial hydraulics. Because of the extremely low permeability of biomacromolecules such as proteoglycans, which limits interstitial flow, the water in the cartilage does not have time to escape from the load-bearing zone, thus creating hydraulics that shares the load generated by motion. Interstitial hydraulics is able to share 90–99% of the load on the cartilage for an extended period of time, preventing significant deformation of the cartilage and destruction of its tissues. When the hydraulics fails, the load is gradually taken up by the solid phase. This physiological mechanism greatly reduces the friction coefficient between the articular cartilage surfaces and ensures that the joint physiology functions properly for a long time.

6.2.2 Regional Classification of Articular Cartilage

Articular cartilage has a hierarchical structure with the fibrous reinforcement and a smooth surface, and the maturation of articular cartilage is marked by the appearance of tidemark [6]. According to the morphological changes of the cells and matrix, the layers can be divided into superficial zone, transitional zone, radial zone, and calcified zone from the cartilage surface downward, as shown in Figure 6.1b [4].

The superficial zone is about 40 μm thick and can be subdivided into two layers. The first layer is devoid of cells, and it consists of fine fibers and a small amount of proteoglycan. In the second layer, the cells are arranged in a flat disk-like dispersion with their long axis parallel to the articular surface. Compared to the other layers, the superficial region has a higher content of collagen fibers and water and a lower content of proteoglycans. The collagen fibers are arranged parallel to the joint surface, and the collagen fibers in the superficial layer have greater tension and strength than those in the deeper layer, which assists to resist shear forces during joint motion. At the same time, the dense collagen fibers form a mesh-like structure that ingests synovial fluid molecules and restricts the outflow of large molecules from the cartilage and the entry of other large molecules such as antibodies.

The transitional zone is about 500 μm from the cartilage surface, and it is generally several times larger than the superficial zone. The cells in this zone are metabolically active, and the collagen fibers in the matrix are thicker but less abundant than those in the superficial zone, which are arranged obliquely to the articular surface. The content of proteoglycan is higher and easier to extract than in the superficial region. The radial zone is the thickest part of the articular cartilage, where the chondrocytes are arranged in a columnar pattern and the cells begin to show signs of degeneration, which becomes more pronounced deeper into the layer. The collagen fibers are the thickest in this layer, aligned vertically, and part of them pass through the tidal marker and calcified zone into the subchondral bone, anchoring the cartilage tightly to the bone. The transitional zone and radial zone have good resistance to extrusion. The calcified zone is located between the tidemark and the subchondral bone. Generally, before bone maturation, this region is endochondral osteogenesis with chondrocyte degeneration, and after bone maturation, the matrix calcifies and the chondrocytes become smaller.

6.3 Joint Lubrication Mechanism

6.3.1 Tribological Theory

The superlow friction property of healthy synovial joint is well recognized, but the lubrication mechanism has been controversial, and with years of intensive research, the scholars have proposed different explanations. Nowadays, the prevailing view tends to agree that in natural joint, there is no dependence on a single lubrication mode, but rather an adaptive regulation of the articular cartilage to provide excellent lubrication behavior of the joint under different loads and sliding speeds [7, 8].

Inspired by Reynolds' lubrication theory, MacConaill proposed the hydrodynamic lubrication theory, which states that because the joint motion brings the synovial fluid between the joint surfaces, the structure of the joint cavity promotes the formation of a wedge-shaped fluid film, and friction is generated because of the internal interaction of the synovial fluid [9]. Charnley argued against this theory, who pointed out that as the relative motion between the joint surfaces is small and the applied load is large, and the motion is in the form of relative reciprocating motion, therefore, the fluid is bound to be squeezed out of the joint surfaces. Consequently, the fluid film cannot be produced, and the lubrication mechanism should be boundary lubrication [10]. Boundary lubrication involves a thin film of material molecules attached to the sliding substrates that move against each other, so that when the two surfaces slide against each other, the movement occurs at the interface between these boundary layers [11].

Although fluid film lubrication may be present in human joint under various states of motion, it is unlikely that the fluid film can form when the joint is at rest for a period of time in a load-bearing state. However, the joint is still well lubricated in this case, indicating that the synovial joint is in a boundary lubrication state. Charnley designed a pendulum test rig using the ankle joint and found that the amplitude of the pendulum decayed linearly with time, leading to the conclusion that friction was independent of velocity. Linn and Radin conducted experiments with dog ankle joints and observed that after adding hyaluronidase to the synovial fluid to reduce its viscosity, the friction coefficient of the joint remained almost unchanged [12]. When the added enzyme, such as trypsin, destroyed the proteins in the synovial fluid, the friction coefficient of the joint increased significantly without affecting the viscosity, indicating that the friction coefficient was independent of viscosity, and the joint was still in a boundary lubrication state.

With the progressive development of research in articular cartilage and synovial fluid, it is discovered that the synovial fluid is a non-Newtonian fluid with the viscosity inversely proportional to the shear rate and the elasticity exhibiting transient expansion. The elastohydrodynamic theory suggests that at high loads, the synovial fluid will not be pushed out of the joint contact surface, while the elastic deformation of the articular cartilage allows the synovial fluid to maintain a certain thickness even at high shear rates [13]. In 1966, Tanner measured the minimum lubricating film thickness of joint surface based on the elastofluid theory, and the result of 0.1 μm was greatly improved compared to the results calculated from hydrodynamic lubrication theory [14]. In 1978, Hamrock and Dowson also calculated the minimum lubricating film thickness of low elastic modulus materials to be 1.02 μm for ball contact with a flat surface [15]. These values are quite close to the roughness of the joint surface, indicating that it is entirely possible for the human joint to be in a state of elastohydrodynamic lubrication, even at low loads. The calculations of hydrodynamic lubrication and elastohydrodynamic lubrication are relatively well established, but in these calculations, the articular cartilage is still treated as a single-phase elastomer, ignoring the porous mechanical properties of the cartilage and the effects of interstitial flow. Therefore, the two-phase lubrication mechanism is proposed, where there is a solid phase dominated by collagen–protein polysaccharides and a liquid phase dominated by tissue fluids and ions in articular cartilage [16]. During the deformation of cartilage, the synovial fluid flows out of the solid-phase structure to bear the load and lubricate the system. However, the load-bearing capacity of the system will decrease with the increase in time and load, and eventually, the load will be transferred to the biomacromolecules present on the articular cartilage surface. In a previous study on the two-phase lubrication mechanism, Mansour and Mow reported that normal healthy articular cartilage was capable of generating a fluid-lubricated film at moderate sliding velocities of 25.4–76.2 mm/s [17].

Nowadays, boundary lubrication is considered as the main lubrication mechanism for human joint, especially at low speeds and high loads. Many scholars, such as Klein [18, 19], Daniel [20], and Hills and Butler [21], have been investigating boundary lubrication in the synovial joint, especially the role of various components of synovial fluid secreted by articular cartilage and the synergistic mechanism between them.

6.3.2 Hydration Lubrication

The phenomenon of superlubrication of natural joint is related to hydration of the biomacromolecules in articular cartilage. Hydration lubrication mechanism was first proposed by Klein and has been recognized as the primary reason of the superlubricity property of articular cartilage under high loads. Water, the natural medium in the biological lubrication system, is itself a low-viscosity lubricant and, as most biomolecular parts are ionized or polar, it is to some extent hydrated [1]. Therefore, in joint lubrication, the polar water molecules are strongly adsorbed surrounding the charged head groups of biomacromolecules, forming a hydrated layer mainly because of the interaction of the large dipole of water with the confined charges. This hydrated layer can bear large normal loads between the sliding surfaces because of the difficulty of extruding the hydrated water. Meanwhile, the hydration layer responds in a fluid manner owing to its rapid relaxation under shear action [22]. During the loading and relaxation process, the water molecules in the hydrated layer will be continuously exchanged with those in the human environment. In general, the hydrated layer formed surrounding the charged biomacromolecules in the articular cartilage has excellent load-bearing capacity and extremely low friction coefficient.

Various polymer brushes used to simulate articular cartilage biomacromolecules are also proposed based on this mechanism. Raviv et al. found that brushing on aqueous media with a charged polymer attached to the surface provided better lubrication than the surfactants [23]. Additionally, Briscoe et al. introduced phospholipids into boundary lubrication, which reduced friction coefficient by several orders of magnitude [24]. The phospholipid bases are prevalent in biological systems, for example, in cell membranes. Particularly, the phosphorylcholine head groups are typical zwitterionic structures that contain both the positively charged ($N^+(CH_3)_3$) and negatively charged (PO_4^-) domains. Thus, the hydration lubrication mechanism ensures that the phosphatidylcholine lipids in the articular cartilage matrix have strong boundary lubrication performance. Furthermore, Ma et al. recently proposed the energy dissipation mechanism for hydration lubrication corresponding to different loads, which contributed significantly to the development of boundary lubrication theory [25].

6.4 Joint Lubrication Behavior

6.4.1 Component

In boundary lubrication, the role of the biomacromolecules in joint synovial fluid is very important [26]. They form a cross-linked structure on articular cartilage surface, as shown in Figure 6.2a [27]. The synovial fluid in the human body's natural joint contains various biomacromolecules such as hyaluronic acid, glycoproteins, and phospholipids. Among them, hyaluronic acid and part of glycoproteins are secreted by synovial cells, while phospholipids exist in synovial fluid and articular cartilage surface. The synovial fluid and articular cartilage work synergistically to maintain an extremely low friction coefficient of the joint system.

6.4.2 Lubrication Performance

Hyaluronic acid, which is widely present in natural synovial fluid, vitreous humor, etc., is a linear polysaccharide polymer, and its chemical structure is shown in Figure 6.2b. Hyaluronic acid was once considered to be the primary lubricant of synovial joint for a long time owing to its relatively high viscosity, which could increase the thickness of the fluid film [28]. However, recently, it has been found that the load-bearing capacity of hyaluronic acid is very low (tens of kPa), which is

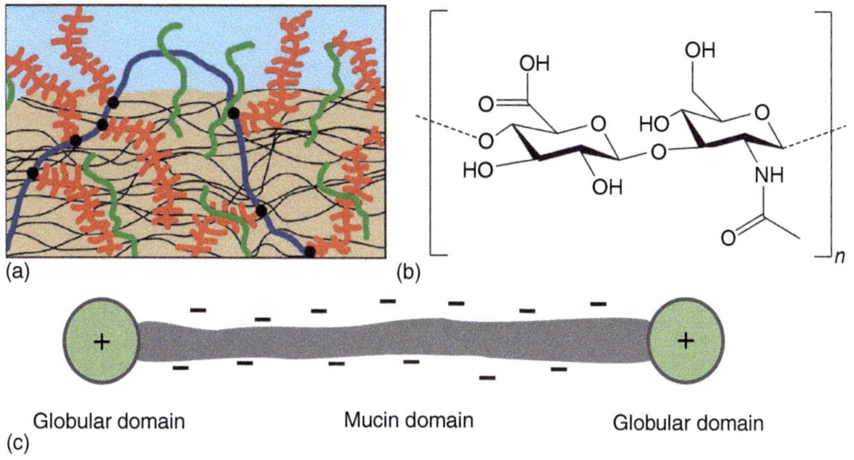

Figure 6.2 (a) The schematic structure formed by the biomacromolecules on articular cartilage. Blue: hyaluronan; red: aggrecan; and green: lubricin. Source: Seror et al. [27]/with permission of John Wiley & Sons. (b) The chemical structure of hyaluronic acid. (c) The schematic structure of lubricin with a negatively charged mucin domain in the middle and a positively charged globular domain at both the ends.

greatly different from the actual load applied to the human joint [29]. For example, Hills and Monds observed that after adding hyaluronidase to the synovial fluid, the degradation of hyaluronic acid did not affect its lubrication performance, indicating that hyaluronic acid might not occupy a dominant function in lubrication [30]. Additionally, the friction coefficient ($\mu \approx 0.2$–0.5) measured by surface force apparatus demonstrated that the lubrication performance of hyaluronic acid was very poor under high pressures [4]. At the same time, it has been shown that compared to physical (electrostatic) adsorption of hyaluronic acid on the mica surface, chemical adsorption (e.g. connection of hyaluronic acid with proteins or phospholipids) can achieve improved immobilization and correspondingly better load-bearing capacity [31]. Furthermore, it is indicated that hyaluronic acid can prevent phospholipase from decomposing phospholipids, thus increasing the gap between the phospholipid bilayers [32], and help the chondrocytes resist the damage from free radicals and promote cell proliferation [33]. Nowadays, in spite of the shear-thinning effect, hyaluronic acid is still employed in clinics for the treatment of osteoarthritis via intra-articular injection into the joint.

The second type of biomacromolecule is a glycoprotein named lubricin, which is present on the articular cartilage surface. Lubricin has a typical zwitterionic molecular structure, with negatively charged mucin domain in the middle and positively charged globular domain in both ends, as shown in Figure 6.2c. Owing to hydration of the charged groups in lubricin, efficient boundary lubrication can be achieved. Jay et al. used atomic force microscopy (AFM) to find that adhesion could be reduced when lubricin was adsorbed on two hydrophobic surfaces, as shown in Figure 6.3a [34]. In addition, they compared the rheological properties of synovial fluid with or without adding lubricin. They found that the diffusibility and elasticity of synovial

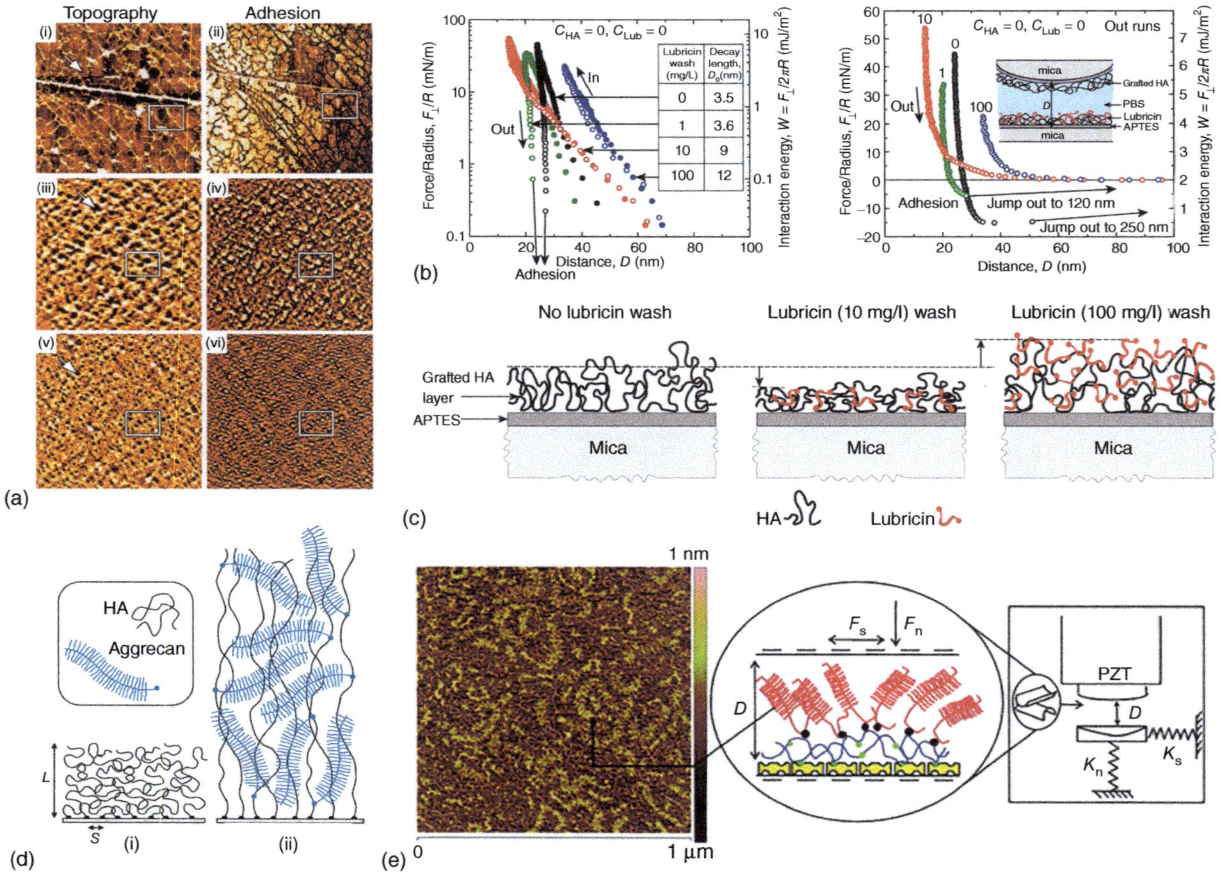
Topography
Adhesion
(i)
(ii)
(iii)
(iv)
(v)
(vi)
(a)
$C_{HA} = 0, C_{Lub} = 0$
Lubricin wash (mg/L)
Decay length, D_0(nm)
0
3.5
1
3.6
10
9
100
12
In
Out
Adhesion
Force/Radius, $F_\perp/R$ (mN/m)
Interaction energy, $W = F_\perp/2\pi R$ (mJ/m^2)
Distance, D (nm)
(b)
$C_{HA} = 0, C_{Lub} = 0$ Out runs
mica
Grafted HA
PBS
Lubricin
APTES
Jump out to 120 nm
Jump out to 250 nm
No lubricin wash
Lubricin (10 mg/l) wash
Lubricin (100 mg/l) wash
Grafted HA layer
APTES
Mica
(c)
HA
Lubricin
HA
Aggrecan
L
S
(i)
(ii)
(d)
1 nm
0
1 μm
(e)
F_s
F_n
D
PZT
K_n
K_s

Figure 6.3 (a) Atomic force microscopy (AFM) of lubricin coated onto highly ordered pyrolytic graphite. The topographic images and adhesion images of lubricin networks are formed at a concentration of 10 μg/ml (i and ii), 50 μg/ml (iii and iv), and 300 μg/ml (v and vi). The lubricin molecules are pseudo-colored to appear bright. Source: Jay et al. [34]/with permission of American College of Rheumatology. (b) The normal forces $F_{\perp}$ normalized by the surface radius of curvature R between two grafted hyaluronic acid layers on the mica surface with lubricin as a function of the mica–mica separation distance D. Source: Seror et al. [37]/with permission of American Chemical Society. (c) The thickness of lubricin-hyaluronic acid film on the surface influenced by lubricin concentration. Source: Das et al. [35]/with permission of American Chemical Society. (d) Putative morphology of the hyaluronan brushes, alone (i) and with aggrecan (ii). Source: Attili et al. [36]/with permission of Royal Society of Chemistry. (e) The AFM height image of a mica surface coated with aggrecan molecules extracted from bovine hip cartilage tissue, and the corresponding force–distance measurements between two mica surfaces at different test conditions using the surface force balance. HA: hyaluronic acid; APTES: (3-aminopropyl)triethoxysilane; PZT: piezoelectric tube. Source: Seror et al. [37]/with permission of American Chemical Society.

fluid without lubricin were worse, indicating that lubricin could eliminate the strain energy generated by movement in the synovial fluid and thus protect articular cartilage [38]. Das et al. found that lubricin could enter between hyaluronic acid to form a layer of viscoelastic colloid and achieve the friction coefficient of 0.09 under a load of 4 MPa, as shown in Figure 6.3b [35]. In this study, it was indicated that increasing the concentration of lubricin could increase the thickness of the adsorbed film, thus further reducing the friction coefficient. When the concentration of lubricin was low, the positively charged end of lubricin was electrostatically connected to hyaluronic acid, and the hyaluronic acid layer collapsed. On the contrary, when the concentration of lubricin was higher, the interaction between lubricin and hyaluronic acid reached saturation and excess lubricin accumulated on the surface, as shown in Figure 6.3c [35].

The third type of biomacromolecule is a polysaccharide named aggrecan. Similar to lubricin, the main chain of aggrecan is a polypeptide with a bottle-brush-like shape, which can bind to hyaluronic acid through connexin [39]. Attili and Richter employed a quartz crystal microbalance to find that aggrecan could be easily inserted between hyaluronic acid molecular brushes, and owing to electrostatic repulsion, hyaluronic acid molecular brushes expanded in the length direction, forming a self-assembled film with excellent elasticity and a thickness in micrometer level, as shown in Figure 6.3d [36]. However, this self-assembled film could only bear a low load of 0.5 kPa. In addition, Seror et al. used a surface force apparatus to investigate the possible boundary biolubricants in articular cartilage and observed that the lubrication performance of the hyaluronic acid–aggrecan complex was far better than that of hyaluronic acid itself, as shown in Figure 6.3e [37]. The friction coefficient of the complex reached a value as low as 0.01, but the lubrication layer could only bear a maximum load of 1.2 MPa.

The load-bearing capacity of the biomacromolecules such as hyaluronic acid and glycoproteins is very low, which is quite different from the actual load applied onto the natural joint (5–20 MPa), and the lubrication performance is also not that excellent as the articular cartilage. There is another type of biomacromolecules in the synovial fluid of joint, that is, phospholipids, which are typical amphiphilic molecules. Generally, the hydrophilic head group, which can attract water molecules through the charge–dipole interaction, is composed of choline and phosphoric acid. Based on the hydration lubrication mechanism, it has a high load-bearing capacity and a very low friction coefficient even under high pressures. The hydrophobic tail of phospholipids consists of two fatty acid chains. Phospholipids can form into spherical/columnar micelles, bilayers, vesicles, and other structures in the aqueous solution depending on the conditions such as molecular fatty acid chain length, chain volume, and optimal head area. As the choline quaternary ammonium salt at the head of the phospholipids is positively charged, the phospholipids can be easily adsorbed on negatively charged substances, for example, hyaluronic acid or glycoprotein. Phospholipids can reduce the interface energy when adsorbed on the surface, meanwhile achieving low friction coefficient. In a previous study, phospholipids have been proven to be excellent boundary lubricants [40, 41].

Kawano et al. used an animal model in their study to show that the combination of dipalmitoylphosphatidylcholine (DPPC) liposomes and hyaluronic acid with a high molecular weight improved the lubrication property than hyaluronic acid, as shown in Figure 6.4a [42]. Goldberg et al. studied the normal force and shear force of hydrogenated soy phosphatidylcholine (HSPC) small unilamellar vesicles (SUVs) adsorbed on the surface of mica at physiologically high salt concentrations (0.15 M $NaNO_3$). They found that the surface of mica was covered by a close-packed HSPC–SUVs layer with an overlayer of liposomes on top, and the friction coefficient of 0.0006 to 0.008 was obtained under the condition of contact pressure up to 6 MPa, as shown in Figure 6.4b, c [43]. The low friction at the interface is attributed to the hydration lubrication mechanism of the highly hydrated phosphorylcholine head groups exposed at the outer surface of the liposome. Duan et al. used AFM to measure the friction force on the titanium alloy/polymer surface. In particular, when DPPC was used as the boundary lubricant, a friction coefficient of 0.007 was obtained under a pressure of up to 15 MPa, as shown in Figure 6.4d [44]. In addition, it was noted that when the probe of AFM was hydrophilically modified and preadsorbed with liposomes, the friction coefficient and load-bearing capacity of the system could be further improved. These studies indicate that the phospholipids, as a component of the synovial fluid in the joint, contribute significantly to forming effective joint lubrication. Compared with hyaluronic acid, lubricin, and other biomacromolecules, the phospholipids have better lubrication performance owing to hydration lubrication, but the other biomacromolecules also play an indispensable and synergistic role in improving load-bearing and surface-anchoring property of the joint.

6.5 Artificial Biolubricants

Artificial biolubricants are very important to enhance the lubrication performance for both artificial joint with a replacement surgery and natural joint with progressive damage of articular cartilage. Researchers first investigated the tribological behavior of natural biomolecules on the surface of artificial joint materials.

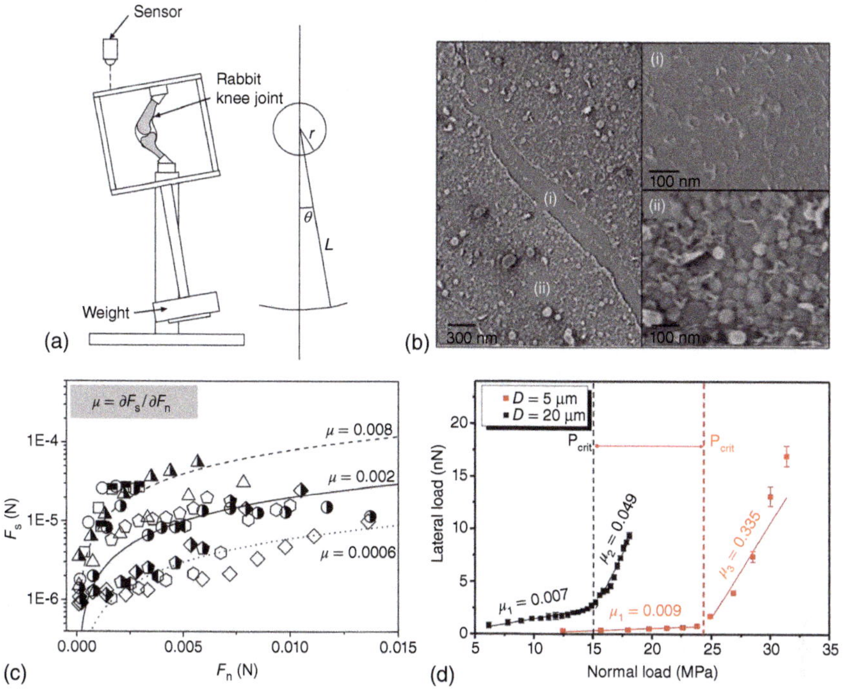

Figure 6.4 (a) The schematic diagram showing the pendulum friction tester employed for the measurement of friction coefficient. Source: Kawano et al. [42]. Reprinted with permission of American College of Rheumatology. (b) The cryo-scanning electron microscopy of a mica surface coated with HSPC–SUVs in 0.15 M $NaNO_3$ salt solution. (i) The traces of the original quasi-hexagonal close packing of the adsorbed liposome layer on the mica. (ii) The surface topography at a higher magnification, revealing that it is coated with an overlayer of larger and smaller liposome clusters. Source: Goldberg et al. [43]/with permission of Biophysical Society. (c) Shear force (F_s) as a function of normal force (F_n) compressing two surfaces. Open symbols: first entry to contact point; half-solid symbols: second entry (different symbols refer to different contact points). Source: Goldberg et al. [43]. Reprinted with permission of Biophysical Society. (d) The effect of the diameter of the polystyrene (PS) particle on the tribological behavior of the PS/Ti6Al4V surface. Source: Duan et al. [44]/with permission of American Chemical Society.

Serro et al. investigated the role of albumin and sodium hyaluronate (NaHA) in the biolubrication process, and they found that albumin could adsorb strongly and extensively on the surface of UHMWPE, but NaHA only resulted in a very low adsorption film with different morphologies and porosities. In addition, adding NaHA to protein solution showed improved adsorption of protein molecules on the surface of UHMWPE, as shown in Figure 6.5a [45]. With the rapid development of nanotechnology, an alternative approach is proposed to improve the lubrication of joint prostheses with a slow-release lubrication mechanism. Specifically, Lu et al. successfully prepared the carbon dots/poly(ethylene glycol)/chitosan (CDs/PEG/CS) composite gel, which was stored in laser-textured surfaces, and studied the lubrication and antifriction behaviors, as shown in Figure 6.5b [46]. The results demonstrated that the CDs associated with PEG through hydrogen bond

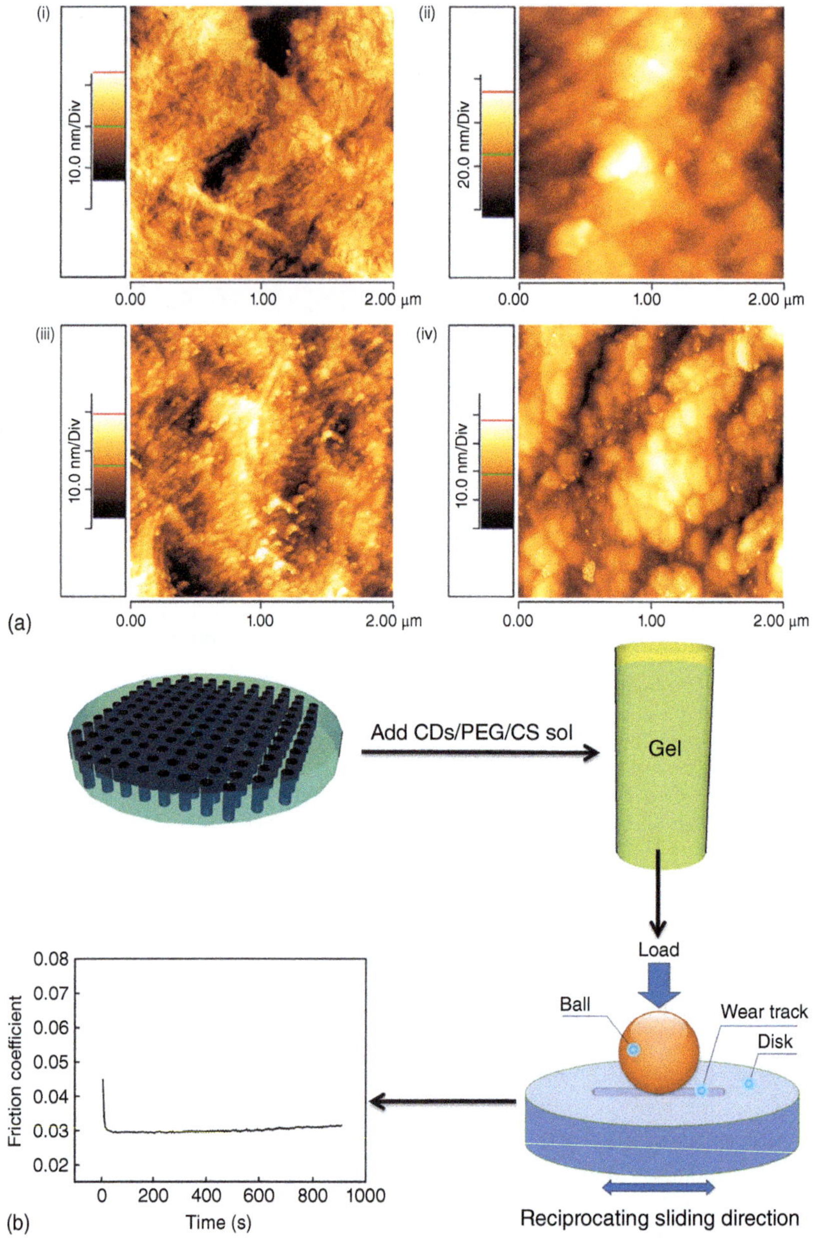

Figure 6.5 (a) Topographic AFM images of the surface of (i) bare UHMWPE and after incubation in solutions containing (ii) (bovine serum albumin [BSA]) = 4 mg/ml, (iii) [NaHA] = 0.5 mg/ml, and (iv) [BSA] = 4 mg/ml + [NaHA] = 0.5 mg/ml. Source: Serro et al. [45]/with permission of Elsevier. (b) The friction tests on textured surfaces with the storing of the CDs/PEG/CS composite gel as the lubricant, which has a slow-release effect. Source: Lu et al. [46]/with permission of Royal Society of Chemistry.

behaved as an excellent lubricant, and the texture-stored CDs/PEG/CS composite gel had a good slow-release effect, which was considered to be a promising method to improve the tribological performance of artificial joint.

Recently, phospholipid-coated mesoporous silica nanoparticles (MSNs@lip) were developed as an effective lubricant in order to treat osteoarthritis at an early stage, as shown in Figure 6.6a [47]. A series of tribological tests were performed under different experimental conditions, and the results indicated that MSNs@lip with multi-layers of phospholipids greatly reduced friction coefficient compared with MSNs. Additionally, the combination of polymer brushes and functional nanoparticles further expands the design rationale for bionic lubricants. Using the polymer brushes as the outer shell and the nanoparticles as the drug nanocarrier, functionalized nanoparticles modified by the polymer brushes have been prepared, which exert an excellent lubrication performance. Inspired by the structure of fresh *Euryale ferox* seed that possesses a slippery aril and a hard coat containing starchy kernel, Yan et al. biomimicked and synthesized a new kind of superlubricated nanoparticle, namely, poly(3-sulfopropyl methacrylate potassium salt) grafted mesoporous silica nanoparticles (MSNs-NH_2@PSPMK), employing a one-step photopolymerization method, as shown in Figure 6.6b [48]. The developed nanoparticles were endowed with enhanced lubrication by the grafted polyelectrolyte polymer because of the formation of a tenacious hydration layer surrounding the negative charges (SO_3^-) in PSPMK and were simultaneously featured with drug loading and release behavior as a result of the mesoporous channels in the MSNs. When encapsulated with a typical anti-inflammatory drug (diclofenac sodium), the lubrication property of the polymer-grafted nanoparticles was improved, while the drug release rate was sustained by increasing the thickness of the PSPMK layer, which could be simply achieved by the adjustment of the precursor monomer concentration in the photopolymerization process. As expected, the *in vitro* and *in vivo* experimental results demonstrated that the lubricated drug-loaded MSNs-NH_2@PSPMK nanoparticles could effectively protect the chondrocytes from oxidative stress-induced degeneration and thus inhibit the progressive development of osteoarthritis. In order to achieve controlled drug release, a type of supramolecular host–guest-functionalized MSNs activated by energy conversion from the joint movement was synthesized, as shown in Figure 6.6c [49]. The high friction coefficient from damaged joint caused more energy conversion, therefore weakening surface host–guest interaction and resulting in local cargo release. The hydrated layer on the nanoparticle surface played an important role in lubrication enhancement.

Additionally, Ji et al. employed a boundary lubricant known as 1,2-distearoyl-*sn*-glycero-3-phosphocholine (DSPC) liposomes for loading of D-glucosamine sulfate (GAS) to construct a synergistic strategy in the treatment of osteoarthritis, which allows for enhanced lubrication and sustained anti-inflammation properties, as shown in Figure 6.6d [50]. This kind of nanocarrier integrated the excellent anti-inflammatory effect of GAS and the lubrication performance of DSPC liposomes without the involvement of complicated synthesis processes, leading to suitable popularization. The experimental results showed that the GAS-loaded

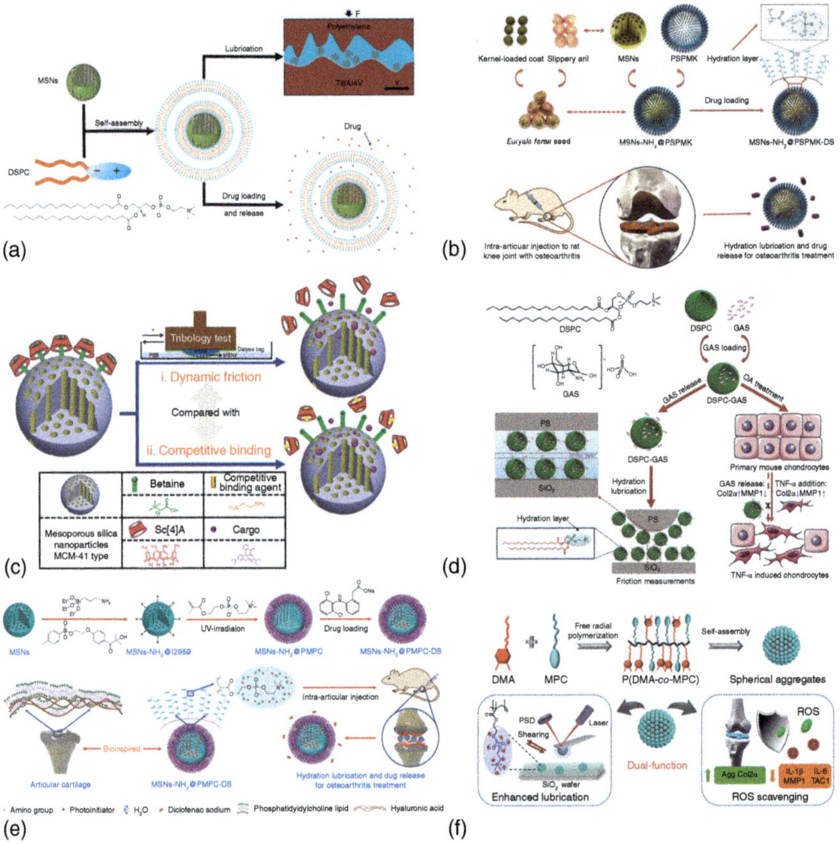

Figure 6.6 (a) The design of MSNs@lip nanoparticles with enhanced lubrication and drug release. Source: Sun et al. [47]/MDPI/CC BY-4.0. (b) The schematic illustration showing the design of superlubricated drug-loaded nanoparticles, i.e. MSNs-NH_2@PSPMK-DS (diclofenac sodium), which is inspired by the structure of fresh *Euryale ferox* seed, and the *in vitro* and *in vivo* experiments, showing the treatment of osteoarthritis via intra-articular injection by the synergetic effect of enhanced lubrication and sustained drug release. Source: Yan et al. [48]/with permission of John Wiley & Sons. (c) The schematic illustration of the friction-controlled drug delivery system. Two types of activation, (i) dynamic friction and (ii) competitive binding, are applied in this system. Source: Tan et al. [49]/with permission of Royal Society of Chemistry. (d) The simplified model of the preparation and characterization of the DSPC-GAS liposomes with sustained drug release and enhanced lubrication properties. Source: Ji et al. [50]/with permission of Royal Society of Chemistry. (e) The schematic illustration showing the synthesis of the superlubricated drug-loaded nanospheres via photopolymerization and the design of the biomimetic nanospheres, which is bioinspired by hydration lubrication of the phosphatidylcholine lipid in articular cartilage, for the synergetic treatment of osteoarthritis based on enhanced lubrication and sustained drug release. Source: Chen et al. [51]/with permission of Elsevier. (f) The schematic illustration of P(DMA-*co*-MPC) spherical aggregates and their dual functions as efficient lubricant and reactive oxygen species scavenger for the treatment of osteoarthritis. Source: Zhao et al. [52]/with permission of American Chemical Society.

DSPC liposomes could release GAS in a sustained manner while providing optimal lubrication in pure water and phosphate-buffered saline (PBS). The superiority of drug release and lubrication ability achieved with the GAS-loaded DSPC liposome in PBS was elucidated on the basis of salt-induced enhancement in liposomal stability and hydration lubrication of the hydrated salt ions. Inspired by the structure of phosphatidylcholine lipid with the presence of zwitterionic charges in articular cartilage, another type of lubricated nanospheres was successfully developed via photopolymerization, i.e. poly(2-methacryloyloxyethyl phosphorylcholine) grafted mesoporous silica nanospheres, as shown in Figure 6.6e [51]. The tribological and drug release tests showed improved lubrication and sustained drug release properties of the nanospheres. The *in vitro* and *in vivo* tests revealed that the superlubricated drug-loaded nanospheres effectively inhibited the development of osteoarthritis by the mechanism of up-regulating cartilage anabolic components and down-regulating cartilage catabolic proteases and pain-related gene. Therefore, it is considered that the nanospheres, with an integrated feature of both enhanced lubrication and sustained drug delivery, can be an efficient intra-articular nanomedicine for the treatment of osteoarthritis. Furthermore, Zhao et al. successfully designed and synthesized an efficient nanosphere system using the copolymer P(DMA-*co*-MPC), which was obtained by free radical copolymerization of the *N*-(3,4-dihydroxyphenethyl) methacrylamide (DMA) and 2-methacryloxyethyl phosphorylcholine (MPC). The test results demonstrated that this nanosphere system not only achieved enhanced lubrication performance but also simultaneously provided excellent anti-inflammatory and antioxidant functions toward oxidative stress-induced inflammation, therefore representing a promising drug-free therapy for the treatment of osteoarthritis, as shown in Figure 6.6f [52].

Similarly, Liu et al. also prepared charged polymer brush-grafted hollow silica nanoparticles (poly(3-sulfopropyl methacrylate potassium salt) [PSPMA]-g-HSNPs) with a core/shell structure, as shown in Figure 6.7a [53]. Because of the excellent hydration capability of the shell consisting of charged polymer brushes, the functional nanoparticles could achieve enhanced lubrication in the aqueous media based on hydration lubrication mechanism. Meanwhile, the hollow silica cores endowed the lubricated nanoparticles with drug loading and sustained release property. Additionally, Liu et al. employed surface-initiated atom transfer radical polymerization to graft polyelectrolyte brushes on the surface of thermo-sensitive microgels containing initiators and successfully synthesized a new kind of composite microgels modified by polyelectrolyte brushes, as shown in Figure 6.7b [54]. The poly(*N*-isopropylacrylamide) (PNIPAAm) microgels were used as the core to provide thermos-sensitive drug release, and the polyelectrolyte brushes grafted to the PNIPAAm core endowed the microgels with good hydration lubrication. It is considered that the synergetic dual-functions of the lubricated drug-loaded nanocarriers provide a promising strategy for the treatment of osteoarthritis.

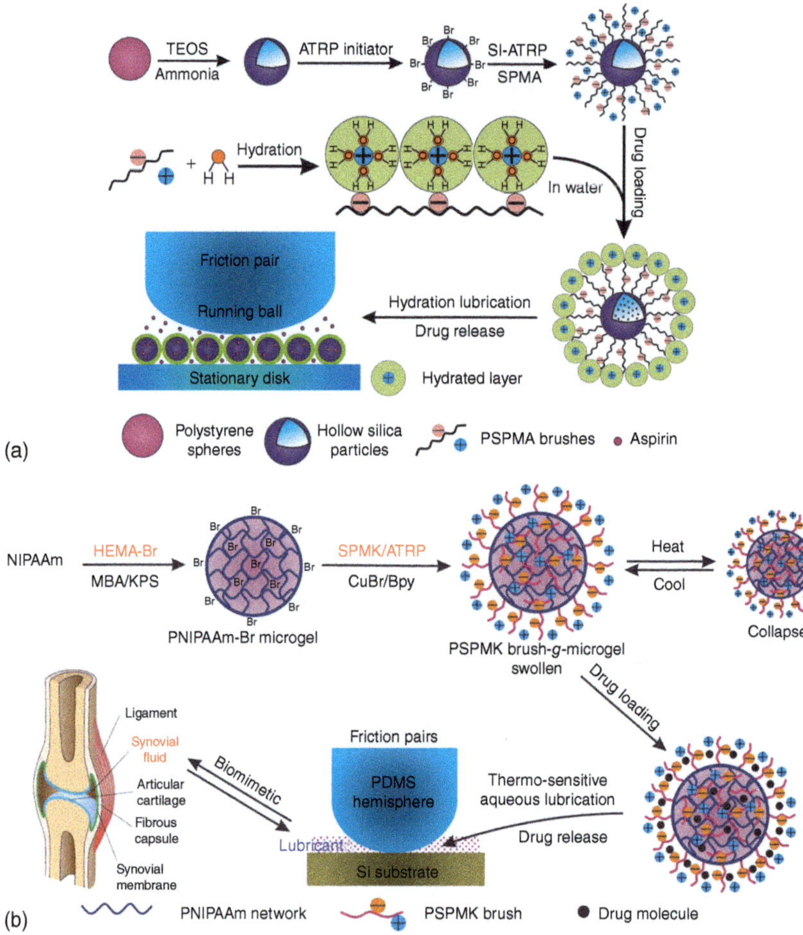

Figure 6.7 (a) The preparation of the PSPMA-*g*-HSNPs nanoparticles with a core/shell structure and the hydration lubrication mechanism of the negatively charged polymer brushes grafted on the surface of the drug-loaded nanoparticles for enhanced lubrication. Source: Liu et al. [53]/with permission of American Chemical Society. (b) Schematic illustration showing the fabrication of polymer brush-grafted PNIPAAm microgels, the drug loading and release of the spherical microgels, and the tribological process of the polyelectrolyte microgels as a biomimetic synovial fluid. HEMA: 2-Hydroxyethyl methacrylate; SPMK: 3-sulfopropyl methacrylate potassium salt; MBA: *N*,*N*′-methylenebis(acrylamide); KPS: potassium persulfate; TEOS: tetraethyl orthosilicate; ATRP: atom transfer radical polymerization. Source: Liu et al. [54]/with permission of American Chemical Society.

6.6 Artificial Joint Prosthesis

Artificial joint replacement is an important orthopedic surgery which is performed worldwide for the treatment of osteoarthritis at the end stage. Basically, the biomaterials that have been widely used in clinics for manufacturing artificial joints mainly include metal material, polymer material, and ceramic material.

6.6.1 Metal Material

Metal material has the advantage of high strength, high toughness, and easy processing, which makes them commonly used to fabricate artificial joint with complicated structure and high load-bearing capacity. In detail, these kinds of materials include stainless steel, cobalt–chromium–molybdenum alloy, and titanium alloy [55, 56]. Stainless steel has the advantage of good processing performance and high compressive strength. However, in the literature, it is indicated that stainless steel can generate corrosion and abrasive wear during the *in vivo* service in the body, which will produce metallic ions and eventually result in aseptic loosening and failure of the joint prosthesis [57]. Cobalt–chromium–molybdenum alloy has good wear resistance, corrosion resistance, mechanical property, and biocompatibility, which makes it an optimal biomaterial for joint prosthesis. However, the long-term use of cobalt–chromium–molybdenum alloy can increase the concentration of cobalt ions in the tissues around the joint, leading to allergic reaction. Titanium alloy has the advantage of excellent biocompatibility, low density, high strength, elastic modulus similar to the human bone, and strong biological interface bonding. By employing stainless steel and titanium alloy disks in contact with alumina pins to conduct friction corrosion tests in seawater, Chen et al. found that the friction coefficient of stainless steel was significantly higher than that of titanium alloy [58]. In addition, it was noted that corrosion of the metal materials was accelerated by friction and wear loss was accelerated by corrosion. Stainless steel was less resistant to sliding damage than titanium alloy.

6.6.2 Polymer Material

In the 1960s, Charnley first combined polytetrafluoroethylene acetabulum in contact with metal femoral head to establish artificial hip prosthesis for joint replacement surgery. At present, the commonly used polymer materials in artificial joint include UHMWPE and highly cross-linked polyethylene. UHMWPE has been widely used as an artificial joint material because of its excellent wear resistance and low friction coefficient. However, the production of polymeric wear debris caused by abrasion can generate adverse biological reaction and result in failure of the joint prosthesis because of osteolysis and aseptic loosening. Therefore, researchers have been continuously improving the antiwear properties of polyethylene [59]. The highly cross-linked polyethylene, which was developed in the late 1970s, has better wear resistance than UHMWPE and the wear resistance generally increases with the degree of cross-linking. It gradually becomes the common material to fabricate femoral component in artificial joint prosthesis. However, free radicals can generate during the cross-linking process and cause oxidative degradation, which has a significant adverse effect on toughness, ductility, and fatigue resistance of the joint prosthesis. In order to solve this issue, the second generation of highly cross-linked polyethylene is proposed, which adopts continuous multiple cycle annealing technology and therefore can reduce the formation of free radicals and maintain the mechanical properties of polyethylene. Additionally, in the third

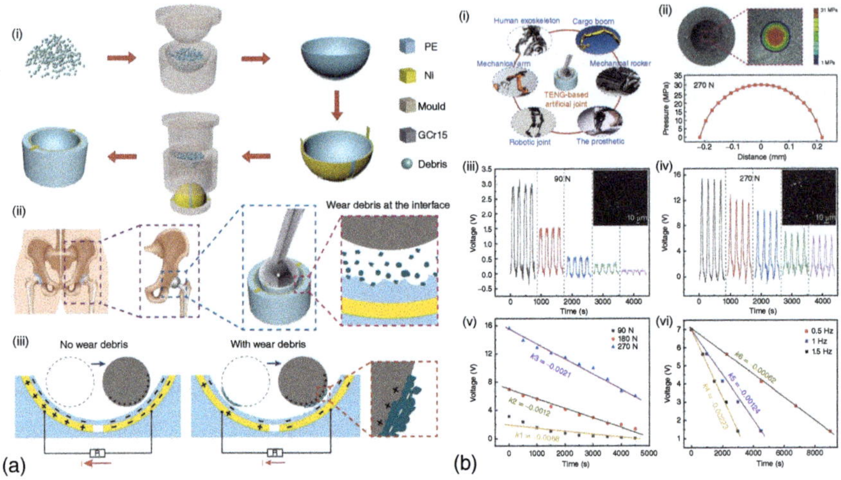

Figure 6.8 (a) The development of a self-powered triboelectric nanogenerator (TENG) sensor for detecting wear debris in artificial hip joint. (i) Fabrication process of TENG-based wear debris detection sensor. (ii) Schematic illustration of the wear debris sensor. (iii) Working mechanism of the fabricated TENG. (b) The test results showing the dependence of output open-circuit voltage of the TENG on external load and frequency from the joint wear simulator. The design of TENG sensor is a promising technique for monitoring wear debris in biomedical healthcare and mechanical industry. (i) Photograph of the setup simulating joint movement. (ii) The pressure distribution simulation of the ball-on-disk contact areas. The output open-circuit voltage of the TENG at load 90 N (iii) and 270 N (iv). The wear debris collected after movement are shown in the insert images. (v) Dependence of the output open-circuit voltage of the TENG on the external load. (vi) Dependence of the output open-circuit voltage of the TENG on the moving frequency. Source: Liu et al. [60]/with permission of Elsevier.

generation of highly cross-linked polyethylene, vitamin E has been added during the production process to further improve the antiwear properties of this material.

The generation of abrasive/fretting wear and correspondingly the wear debris is of great significance to prolong the lifetime of artificial joint prosthesis *in vivo*. Recently, a self-powered sensor based on triboelectric nanogenerator (TENG) was successfully designed and developed by Liu et al. for detecting polymeric wear debris generated in artificial joint prosthesis, as shown in Figure 6.8 [60]. Basically, the TENG with a steel ball and a polyethylene cup was fabricated by thermo-compression technique and designed into a ball-on-cup configuration of artificial hip joint, which could *in situ* detect the production of polyethylene wear debris. The dependence of electrical output signals on the amount and size of the debris was investigated, which indicated an obvious decrease in voltage with the increase in the amount and size of the particles. Furthermore, the fabricated sensor was tested in a custom-made joint wear simulator, and it achieved the real-time monitoring of polyethylene wear debris.

6.6.3 Ceramic Material

Ceramic material is a kind of biologically inert biomaterial with excellent hardness, wear resistance, corrosion resistance, and biocompatibility, which can further reduce the possibility of *in vivo* inflammatory reaction. Generally, alumina and zirconia are typical ceramic materials used in artificial joint prosthesis. However, ceramic material also has disadvantages such as brittleness, low yield strength, and being fragile under impact force and sensitive to stress concentration. Once cracks are formed, it is very easy to expand and therefore results in the fracture failure of ceramic material. These shortcomings have greatly limited the application of ceramic material in the field of artificial joint prosthesis. Currently, the ceramic artificial joint used in clinics is the fourth generation of ceramic material, which is manufactured by adding zirconia and strontium aluminum chromate to enhance the toughness and wear resistance properties and reduce the risk of fracture of the artificial joint prosthesis.

Acknowledgments

Y. Wang and K. Hu contributed equally to this work. This work was supported in part by the National Natural Science Foundation of China (52022043), the Tsinghua University-Peking Union Medical College Hospital Initiative Scientific Research Program of China (20191080593), and the Precision Medicine Foundation, Tsinghua University, China (10001020107).

References

1 Klein, J. (2013). *Friction* 1: 1.
2 Meng, Y.G., Xu, J., Jin, Z.M. et al. (2020). *Friction* 8: 221.
3 Dowson, D. (2012). *Faraday Discuss.* 156: 9.
4 Jahn, S. and Klein, J. (2018). *Phys. Today* 71: 48.
5 Linn, F.C. and Sokoloff, L. (1965). *Arthritis Rheum.* 8: 481.
6 Ratcliffe, A., Woo, S.L.-Y., and Mow, V.C. (1990). *Biomechanics of Diarthrodial Joints*. New York: Springer.
7 Katta, J., Jin, Z., Ingham, E., and Fisher, J. (2008). *Med. Eng. Phys.* 30: 1349.
8 Chang, D.P., Abu-Lail, N.I., Coles, J.M. et al. (2009). *Soft Matter* 5: 3438.
9 MacConaill, M.A. (1951). *Bone Joint J.* 33: 251.
10 Dowson, D. and Wright, V. (1981). *Introduction to the Biomechanics of Joints and Joint Replacement*. Mechanical Engineering Publications.
11 Hardy, W. and Bircumshaw, I. (1925). *Proc. R. Soc. London, Ser. A* 108: 1.
12 Linn, F.C. and Radin, E.L. (1968). *Arthritis Rheum.* 11: 674.
13 Dintenfass, L. (1963). *J. Bone Joint Surg. Am.* 45: 1241.
14 Tanner, R.L. (1966). *Phys. Med. Biol.* 11: 119.
15 Hamrock, B.J. and Dowson, D. (1981). *Ball Bearing Lubrication: The Elastohydrodynamics of Elliptical Contacts*. Wiley.
16 Mow, V.C., Kuei, S.C., Lai, W.M., and Armstrong, C.G. (1980). *J. Biomech. Eng.* 102: 73.
17 Mansour, J.M. and Mow, V.C. (1977). *J. Lubr. Technol.* 99: 163.

18 Raviv, U. and Klein, J. (2002). *Science* 297: 1540.
19 Klein, J. (2009). *Science* 323: 47.
20 Daniel, M. (2014). *Wien. Med. Wochenschr.* 164: 88.
21 Hills, B.A. and Butler, B.D. (1984). *Ann. Rheum. Dis.* 43: 641.
22 Lin, W.F. and Klein, J. (2021). *Adv. Mater.* 33: 2005513.
23 Raviv, U., Giasson, S., Kampf, N. et al. (2003). *Nature* 425: 163.
24 Briscoe, W.H., Titmuss, S., Tiberg, F. et al. (2006). *Nature* 444: 191.
25 Ma, L., Gaisinskaya-Kipnis, A., Kampf, N., and Klein, J. (2015). *Nat. Commun.* 6: 6060.
26 Jahn, S. and Klein, J. (2015). *Macromolecules* 48: 5059.
27 Seror, J., Sorkin, R., and Klein, J. (2014). *Polym. Adv. Technol.* 25: 468.
28 Ogston, A.G. and Stanier, J.E. (1953). *J. Physiol.* 119: 244.
29 Zhang, H.F., Wang, C.G., Li, H. et al. (2018). *Drug Des. Dev. Ther.* 12: 445.
30 Hills, B.A. and Monds, M.K. (1998). *Br. J. Rheumatol.* 37: 137.
31 Sotres, J. and Arnebrant, T. (2013). *Lubricants* 1: 102.
32 Nitzan, D.W., Nitzan, U., Dan, P., and Yedgar, S. (2001). *Rheumatology* 40: 336.
33 Singh, A., Corvelli, M., Unterman, S.A. et al. (2014). *Nat. Mater.* 13: 988.
34 Jay, G.D., Torres, J.R., Rhee, D.K. et al. (2007). *Arthritis Rheum.* 56: 3662.
35 Das, S., Banquy, X., Zappone, B. et al. (2013). *Biomacromolecules* 14: 1669.
36 Attili, S. and Richter, R.P. (2013). *Soft Matter* 9: 10473.
37 Seror, J., Merkher, Y., Kampf, N. et al. (2011). *Biomacromolecules* 12: 3432.
38 Jay, G.D., Torres, J.R., Warman, M.L. et al. (2007). *Proc. Natl. Acad. Sci. U.S.A.* 104: 6194.
39 Seror, J., Merkher, Y., Kampf, N. et al. (2012). *Biomacromolecules* 13: 3823.
40 Oguchi, T., Sakai, K., Sakai, H., and Abe, M. (2010). *Colloids Surf., B* 79: 205.
41 Corneci, M.C., Dekkiche, F., Trunfio-Sfarghiu, A.M. et al. (2011). *Tribol. Int.* 44: 1959.
42 Kawano, T., Miura, H., Mawatari, T. et al. (2003). *Arthritis Rheum.* 48: 1923.
43 Goldberg, R., Schroeder, A., Barenholz, Y., and Klein, J. (2011). *Biophys. J.* 100: 2403.
44 Duan, Y., Liu, Y., Li, J. et al. (2019). *Biomacromolecules* 20: 1522.
45 Serro, A.P., Degiampietro, K., Colaco, R., and Saramago, B. (2010). *Colloids Surf., B* 78: 1.
46 Lu, H.L., Ren, S.S., Zhang, P.P. et al. (2017). *RSC Adv.* 7: 21600.
47 Sun, T., Sun, Y.L., and Zhang, H.Y. (2018). *Polymers* 10: 513.
48 Yan, Y.F., Sun, T., Zhang, H.B. et al. (2019). *Adv. Funct. Mater.* 29: 1807559.
49 Tan, X.L., Sun, Y.L., Sun, T., and Zhang, H.Y. (2019). *Chem. Commun.* 55: 2593.
50 Ji, X.L., Yan, Y.F., Sun, T. et al. (2019). *Biomater. Sci.* 7: 2716.
51 Chen, H., Sun, T., Yan, Y.F. et al. (2020). *Biomaterials* 242: 119931.
52 Zhao, W.W., Wang, H., Han, Y. et al. (2020). *ACS Appl. Mater. Interfaces* 12: 51236.
53 Liu, G.Q., Cai, M.R., Zhou, F., and Liu, W.M. (2014). *J. Phys. Chem. B* 118: 4920.
54 Liu, G., Liu, Z., Li, N. et al. (2014). *ACS Appl. Mater. Interfaces* 6: 20452.
55 Wang, Z.Q. and Gao, D.R. (2014). *Mater. Des.* 53: 881.
56 Alemon, B., Flores, M., Ramirez, W. et al. (2015). *Tribol. Int.* 81: 159.

57 Diomidis, N., Mischler, S., More, N.S., and Roy, M. (2012). *Acta Biomater.* 8: 852.
58 Chen, J., Zhang, Q., Li, Q.A. et al. (2014). *Trans. Nonferrous Met. Soc. China* 24: 1022.
59 Moro, T., Takatori, Y., Ishihara, K. et al. (2004). *Nat. Mater.* 3: 829.
60 Liu, Y.Y., Zhao, W.W., Liu, G.X. et al. (2021). *Nano Energy* 85: 105967.

7

Microfluidic Biotechnology for "Bone-on-a-Chip"

Chaenyung Cha

Ulsan National Institute of Science and Technology (UNIST), Center for Multidimensional Programmable Matter, Department of Materials Science and Engineering, 50 UNIST-gil, Ulsan 44919, South Korea

7.1 Introduction

One of the most remarkable technological innovations in the past century has been microfabrication. The ability to make structures with micrometer-scale resolution using various materials has led to the creation of electronics industry that has defined this era and is likely to shape the future society. Various microfabrication techniques, such as lithography, deposition, and chemical and mechanical etching, which have now become commonplace and mostly automated, allow efficient, cost-effective, and high-volume production of miniaturized microdevices, such as integrated circuits ("microchips"), energy generators, sensors, and actuators [1, 2].

For each microfabrication process, various techniques are used in combination to ultimately develop a microdevice. For example, lithography, which is the most commonly utilized technology for integrated circuits and microelectromechanical systems (MEMSs), is generally defined as the overall methodology to fabricate sophisticated patterns on a substrate [3]. In particular, photolithography using photocurable materials is most commonly used for high precision and facile photoreaction [4]. It usually involves the deposition of curable material in defined space and thickness, followed by photocuring using a photomask (or a maskless method using laser or electron beam) to fabricate the material in a desired pattern. Afterward, etching is needed to remove unwanted, uncured materials for final development.

Even though the microfabrication technology has been developed and advanced mostly around the electronics industry, the possibility of fabricating miniaturized devices has attracted tremendous attention in the field of biomedicine [5, 6]. Most of the diagnostic tools in medicine involve the use of large, expensive equipment and require a significant amount of biological samples (e.g. blood and urine), which are cumbersome, cost-prohibitive, and time-consuming. Therefore, there is a great

Biofabrication for Orthopedics: Methods, Techniques, and Applications, First Edition.
Edited by Wenguo Cui, Xin Zhao, and Shen Liu.

need to scale down and produce smaller diagnostic devices in large quantities for cost-effectiveness, efficiency, and minimal invasiveness for patients. These biomedical microdevices has emerged as the key technology for "point-of-care" diagnostics, where simple and easily accessible and operable devices are used to test the patients and deliver fast and accurate diagnoses (Figure 7.1) [7]. The best-known examples that have become commonplace for general public are blood glucose monitors and pregnancy test kits.

More recently, there is an immense research trend in biomedical engineering to further scale down to a similar level of electronic microchips and integrate more sophisticated functions. For this objective, microfluidic technology is being widely and actively explored [8, 9]. Microfluidics refers to both the technology to control

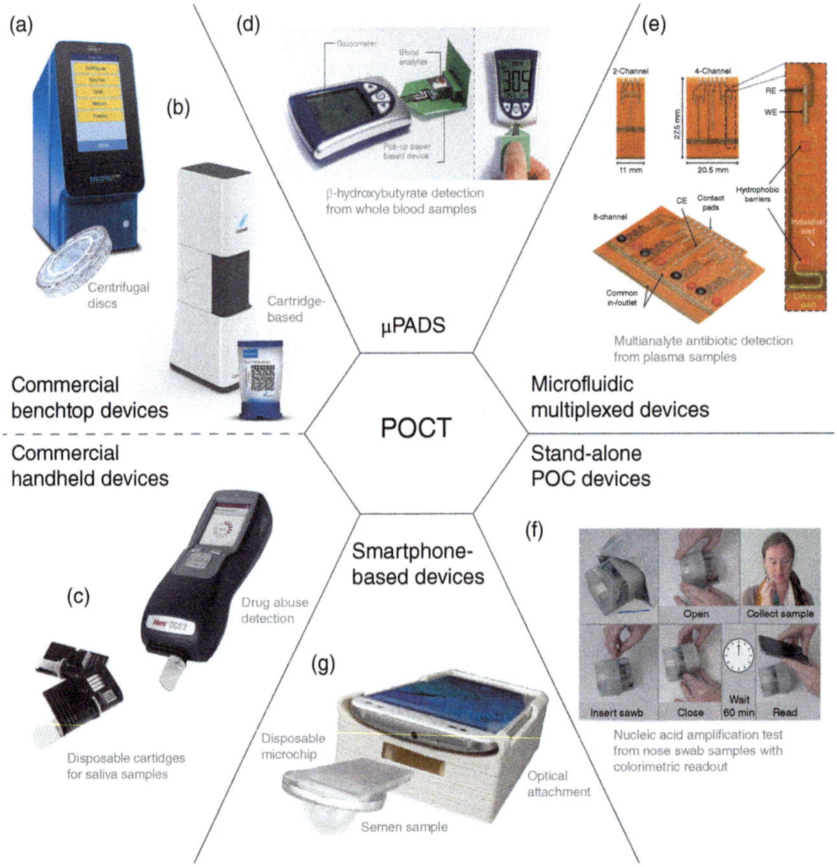

Figure 7.1 Representative examples of point-of-care technology (POCT): (a) Piccolo Xpress by Abbott, (b) GeneXpert Omni by Cepheid, (c) Alere DDS2 by Abbott, (d) a microfluidic paper-based analytical device (μPAD), (e) an electrochemical microfluidic multiplexed biosensor for multianalyte antibiotic detection, (f) a rapid, instrument-free, sample-to-result nucleic acid amplification test, (g) a smartphone-based semen analysis. Source: Dincer et al. [7] / John Wiley & Sons / CC BY 4.0.

and manipulate various fluids in microscale using microdevices and the scientific investigation of the fluid dynamics. It was first conceptualized and put into practice for inkjet printers that require a small amount of ink deposition. However, it has evolved into a bona fide subject of intense research focus in various fields, including physics and mechanical engineering that involves the design and fabrication of microchannels and the mathematical and physical interpretation of dynamic fluid motions. In chemical engineering, microfluidic devices are being developed and optimized in order to efficiently separate and analyze chemicals using small quantities of samples.

One of the first and important biomedical applications of microfluidics has been *in situ* preparation and analysis of biological molecules, especially DNA (Figure 7.2) [10, 11]. With the rise of genome research that requires the acquisition and analyses of massive amounts of genomic data, there is a great need for efficient and high-throughput methodology. Microfluidics has been considered ideal for this purpose as a microfluidic device can integrate various functions that can perform cell sorting, separation, lysis, and analysis, realizing a true micro total analysis system (μ-TAS). Cells can be sorted and separated either actively via physical forces (e.g. electric and magnetic field, acoustics, and optics) or passively via controlling the channel geometry [12]. DNA separation and analysis by electrophoresis and polymerase chain reaction (PCR) can also be accomplished on chip.

As the next step toward microfluidic technology in biomedicine, the growth of microfluidics in biomedical field has led to "organ-on-a-chip," which is considered the culmination of microfluidics-based tissue engineering [13–16]. Most of the

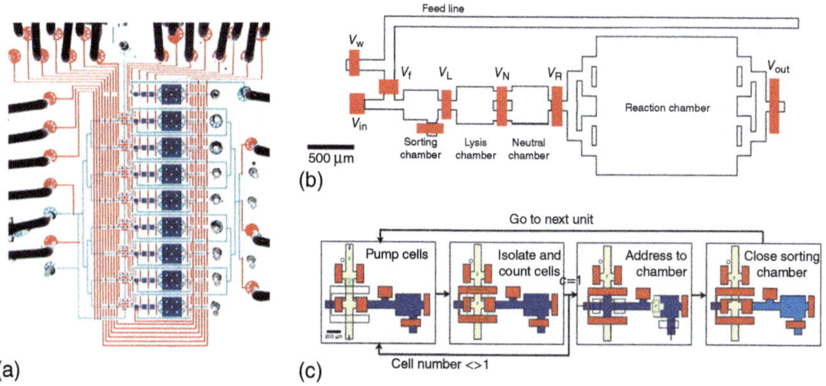

Figure 7.2 (a) The overall view of a microfluidic device for DNA analysis. Multiple amplification units are integrated in parallel. (b) Schematics of a single amplification unit for cell sorting, DNA extraction, and analysis. (c) Schematics of the experimental procedure of the cell sorting. Closed valves are shown in red, open valves are transparent, Cells are drawn in green. If a fluorescent cell is counted (c = 1), the sorting valve is opened and the cell moves to the sorting chamber. If the cell number is not equal to one (cell number<>1), the algorithm goes back to the previous step.Source: (a and b) Marcy et al. [10] / Plos One / CC BY-4.0.

biological research has been geared toward analyzing biomolecules (e.g. DNA, RNA, and proteins) or cellular phenotypes. As a result, there is a critical need to understand the physiology at a tissue level. The goal of tissue engineering is to engineer highly functional tissues using cells, supporting scaffolds, and soluble factors. Although the ultimate goal of creating a whole organ is still yet to come, various tissue constructs engineered thus far at a smaller scale have been shown to emulate tissue functions close to physiological conditions. These engineered tissues are slowly becoming a valuable tool for fundamental biological investigation that has been largely restricted to cells culture on hard, flat surfaces. Therefore, it could be easily envisioned that engineered tissues cultured within a microfluidic device can provide a sufficiently viable microenvironment, including media supply via microchannels, that can support tissue culture and incorporate tools necessary for biological analyses (Figure 7.3) [17]. The miniaturization also allows multiple types of tissues integrated in the same chip to emulate whole-body interactions, often termed "body-on-a-chip."

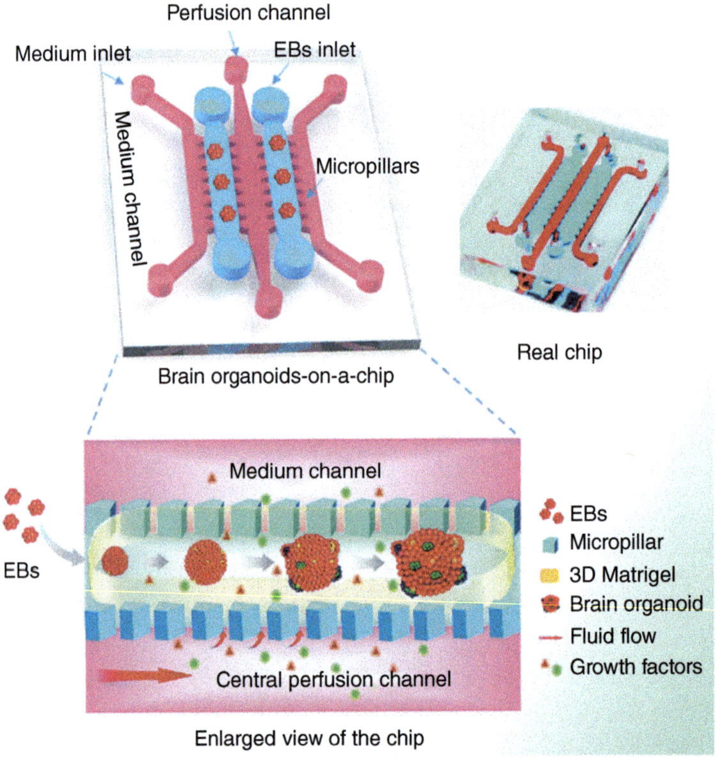

Figure 7.3 A representative example of a microfluidic organ-on-a-chip. It consists of separate channels for cell spheroids (embryonic bodies, EB) and culture media. The channels are interfaced via open pores between pillars to allow media exchange. EBs are induced to form brain-like organoids via media perfusion. Source: Wang et al. [17] / Royal Society of Chemistry.

In addition to fundamental biological investigation, the organ-on-a-chip has emerged as the next-generation drug screening platform, with the possibility to perform complex analyses on tissue function within [18–20]. In the process of drug discovery, screening is routinely performed mostly by identifying drug candidates that show increased binding affinity toward a target molecule and *in vitro* cell cultures to explore their physiological effects, before top candidates are preclinical and clinical trials. However, there has been a recent shift toward eliminating or significantly reducing the number of animal testing because of ethical concerns. This global trend has now forced the pharmaceutical community to adopt *in vitro* models that can closely emulate *in vivo* physiological conditions. The organ-on-a-chip technology is therefore widely viewed as the next paradigm of tissue-based drug screening platform that can produce more extensive biological information than conventional target-binding scheme, while covering most of the in-body, physiological responses expected for *in vivo* systems.

In this chapter, the basic principles of microfluidics are first introduced, including fabrication methods and important features. Then, its application for "organ-on-a-chip" emulating various aspects of bone tissues, such as mineralized tissue, bone marrow, and cancer metastasis, are further discussed in detail.

7.2 Basic Principles and Properties of Microfluidics

7.2.1 Fabrication of Microfluidic Devices via Soft Lithography

This growth in microfluidics has been largely aided by the development of soft lithography that allows the fabrication of microfluidic devices with desired channel geometries [21–23]. Unlike traditional photolithography performed on hard materials, such as rigid plastics, silicon/silicon oxide, and glass, soft lithography is designed to produce patterns on elastic materials. Having flexibility can afford a greater avenue for engineering various design features on a chip. Also, it becomes more pliable to attach functional modalities such as tubes for fluid injection. In addition, cells and tissues are more likely to survive and retain their functions within mechanically tunable and biocompatible polymeric substrates.

The most widely used elastic material for soft lithography is cross-linked polydimethylsiloxane (PDMS), a widely popular silicone elastomer commercially distributed by Dow Corning (under the brand, SYLGARD®). The PDMS elastomer has characteristics that are highly attractive for this purpose. It is easily cured by simply mixing the PDMS base (PDMS macromers) with a cross-linker (vinyl-capped PDMS) and the Pt catalyst, and the liquid form allows easy casting (Figure 7.4). It is mechanically tough and flexible while maintaining tunable elasticity based on cross-linking density. It is optically transparent, allowing easy visualization with conventional optical microscopy. It is gas permeable that allows oxygen intake for cell survival, while resisting a wide range of chemical reactions that could potentially undermine the device integrity or cause toxicity.

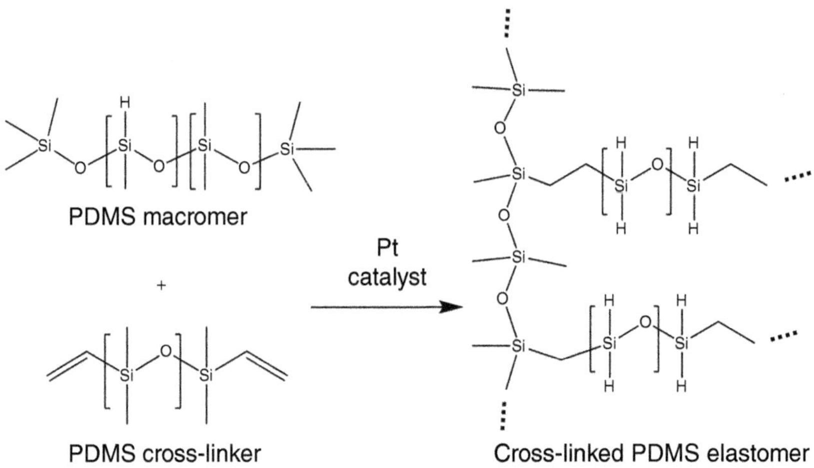

Figure 7.4 A cross-linking reaction to develop PDMS elastomer.

Even though soft lithography is often separately categorized, it is not completely devoid of conventional photolithography. Micropatterning device structures on elastomer are accomplished by casting precursor solution on a master, which is generally fabricated by photolithography. A general procedure for fabricating a microfluidic device is as follows (Figure 7.5) [24]:

- Fabrication of the master mold: A standard photolithography is applied on a silicon wafer to generate the desired channel structure. A photoresist is first spin-coated on a silicon wafer (SU-8 series are commonly used). UV is applied onto the wafer through a photomask that only allows a patterned UV to pass through.
- PDMS molding on the master: PDMS base and curing agent are mixed thoroughly, and the air bubbles generated during the mixing are removed by vacuum. The mixture is cast onto the master and baked for several hours at a high temperature (usually 80–90 °C) to develop the PDMS elastomer with a defined pattern ("stamp").
- PDMS–glass bonding: The PDMS stamp is detached from the master, and the holes for fluid inlets and outlets are made. The stamp and a cover glass are plasma treated and irreversibly bonded to each other to seal off the patterned area that becomes the channels for the microfluidic device. It should be noted that in addition to the PDMS stamp, other features, such as separation membranes, micropumps, and microvalves, are often fabricated separately, and assembled with the PDMS stamp, before the glass bonding.

After the device fabrication is complete, the inlets and outlets are connected with tubing to transport fluids in and out of the device. The precise amount and rate of fluid injection are generally controlled by an electronic pump, with a rate as low as a few picoliters per minute.

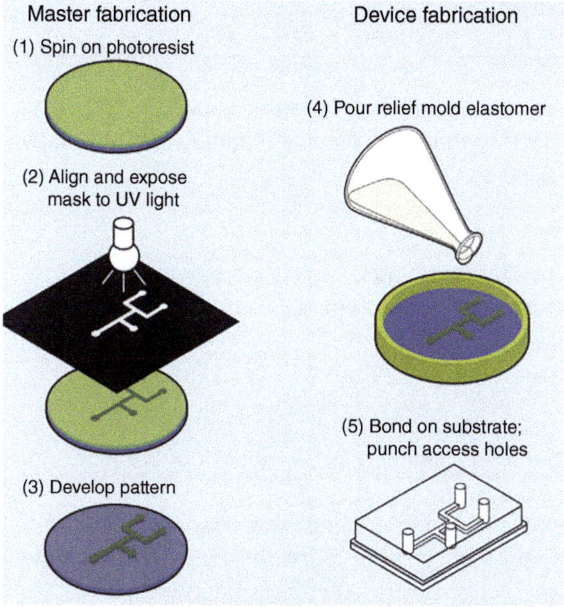

Figure 7.5 Schematics of soft lithography to fabricate a PDMS-based microfluidic device. Source: Walsh et al. [24] / With permission of Elsevier.

7.2.2 Fluid Controlling Components of Microfluidic Devices

The most basic function of a channel within the microfluidic device obviously is to transport the fluid to a desired location. Rather than a simple monotonous transport from the external pump, it is more advantageous to precisely control the fluid transport, in terms of rate, location, and frequency, within the microfluidic device. With a greater degree of complexity of microfluidic channel design, it becomes more difficult to either maintain or expedite the fluid transport. In addition, it is also desirable to independently control the fluid transport at different regions within a device. Therefore, various MEMS components that serve to control the fluids, namely, microvalves, micropumps, and micromixers, are often integrated [8].

- Reynolds number (Re)

  Re is a dimensionless value that is commonly used to determine the laminar or turbulent nature of a flow. It is defined as the ratio of inertial force to viscous force,

$$\text{Re} = \frac{\rho l v}{\mu}$$

where ρ is the density of the fluid, l is the diameter of the fluid channel, v is the velocity of the fluid, and μ is the viscosity of the fluid.

(Continued)

(Continued)

In most microfluidic devices, because of the small size of the fluid channels, the fluids are operated with Re much less than 100, often below 1, in which the flow is completely laminar. The transition from laminar to turbulent flow usually occurs when Re is between 2000 and 4000.

- Péclet number (Pe)

Pe is a dimensionless value that represents the relative proportions of convection and diffusion in a fluid transport system. It is defined as the ratio of convection rate to diffusion rate,

$$\mathrm{Pe} = \frac{l^2/D}{l/v} = \frac{lv}{D}$$

where l is the diameter of the fluid channel, v is the velocity of the fluid, and D is the diffusivity.

In a microfluidic system where the fluids are mostly laminar with low turbulence, the transport is mostly governed by diffusion, with Pe less than 1. On the other hand, Pe is greater than 1 for convection-dominant transport.

7.2.2.1 Microvalves

Microvalves are the simplest component to control the fluid transport by either opening or closing the channel. In addition, microvalves are employed to sort particles and cells to intended channels [8]. A typical valve consists of a flexible diaphragm that can be deformed to close off the flow channel, either completely or partially, and an actuator that controls the diaphragm. The valves are usually categorized by actuation mechanisms.

Pneumatic microvalves utilize externally applied air pressure that causes a diaphragm to be pushed close the channel (Figure 7.6a). The response time is very fast, and it can be independently controlled from the fluid transport as the air is separately supplied. The pneumatic pressure can also be generated by increasing the local temperature (i.e. thermopneumatic), which leads to increased pressure within a chamber and causes the diaphragm to deform. Compared to pneumatic valve, thermopneumatic valve is slow to respond, but it offers a greater extent of control by adjusting the temperature with a precision temperature controller.

Electromagnetic microvalves are controlled by an external electromagnet to actuate a magnetically active diaphragm (e.g. NiFe and NdFeB) (Figure 7.6b). It covers a larger distance than electrostatic microvalves, but the size becomes larger, and there is greater power consumption, as electricity is required to induce magnetic field. Electrostatic microvalves are controlled by an externally applied electric field through the embedded electrodes to a conductive diaphragm, which becomes deformed by an electrostatic force. Similarly, piezoelectric actuation, in which the electric field directly results in mechanical displacement, is also utilized. It offers the fastest and most precise actuation, although at a

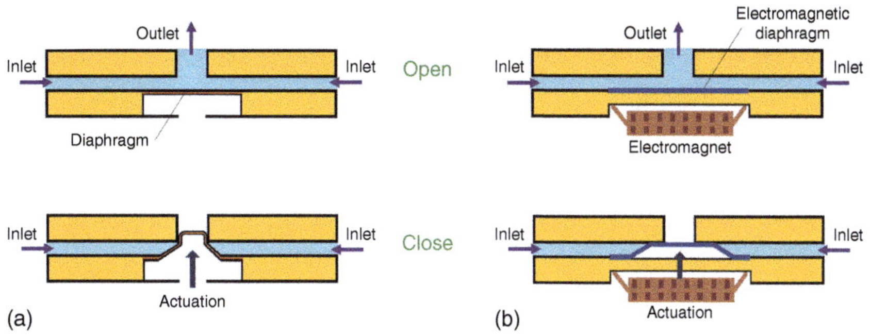

Figure 7.6 Schematics of a microvalve operation by (a) pneumatic and (b) electromagnetic actuations.

much reduced displacement (stroke). Unlike electromagnetic and piezoelectric microvalves, electrostatic microvalves do not require electromagnetically conductive materials that are difficult to integrate into elastomer-based microfluidic devices, and the elastomers with high flexibility and dielectric constant such as PDMS itself can be used as a movable diaphragm, but the degree of actuation is smaller.

7.2.2.2 Micropumps

A microfluidic pump is incorporated to locally supply a small amount of fluid to the region that is otherwise difficult to reach by simple injection. The micropumps are not completely dissimilar to the microvalves, in that the same actuation mechanisms and the resulting valve actions are utilized to generate the physical force required for pumping actions. Therefore, micropumps are similarly categorized based on the type of pumping actions.

The check-valve micropump consists of a chamber connected by inlet and outlet valves (Figure 7.7a) [25, 26]. When a membrane of the chamber is deflected to create low pressure, it allows the fluid injection into the chamber by opening the inlet valve. With the membrane back to its original position, there is increased pressure in the chamber that, in turn, forces the fluid out through the outlet. These "check" valves allow the pumping of a small and designated amount of fluids only through the action of membrane displacement, which is controlled by any actuation mechanism, and remain closed otherwise.

Diffuser micropumps, on the other hand, are "valve-less," indicating that the pumping of the fluid is controlled by the geometry of the inlet and outlet (Figure 7.7b) [26, 27]. Here, the opening that connects the inlet channel to the chamber ("inlet diffuser") becomes wider and that connects the outlet channel to the chamber ("outlet diffuser") becomes narrower, so there is a preferential flow from the chamber to the outlet. When the chamber is inflated, there is greater flow from the inlet to the chamber than from the outlet, thereby accumulating the fluid in the chamber, which is called the supply mode. When the chamber is compressed, a larger amount of fluid is pumped out to the outlet than to the inlet, called the

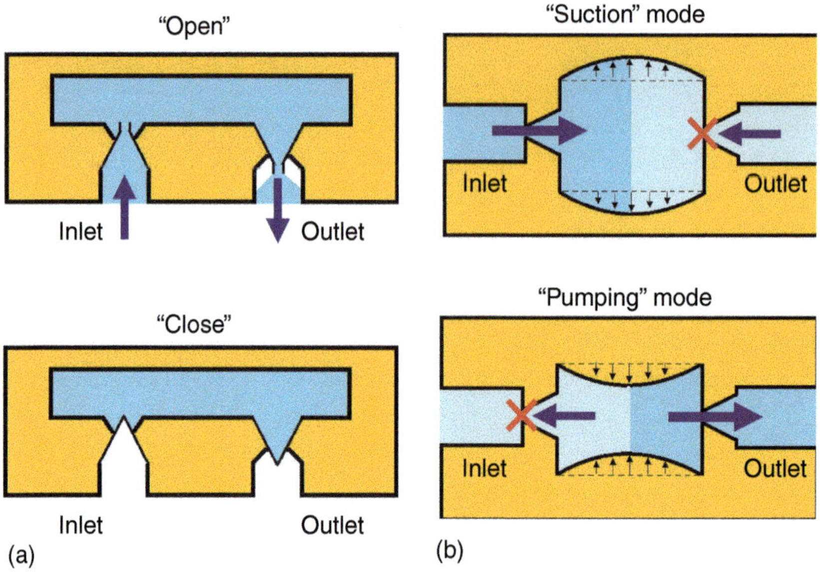

Figure 7.7 Schematics of micropump operations: (a) check-valve micropump and (b) diffuser micropump.

pumping mode. The diffuser micropumps are simple and easy to fabricate and do not require additional components, but there is a small inevitable back-flow during operation.

7.2.2.3 Micromixers

Unlike mixing fluids in macroscale that can be easily accomplished by inducing turbulent flow via shaking and stirring, most fluids within a microchannel are laminar in nature, so it is challenging to generate enough force to induce sufficient mixing. For this purpose, micromixers are commonly incorporated into a microfluidic device, especially for those that require mixing of multiple fluids, such as cell culture media.

Active micromixing relies on the same actuation schemes and operations of microvalves and micropumps to induce turbulent flow. On the other hand, mixing can be enhanced passively by introducing a winding and convoluted channel to increase advection along its path to allow sufficient diffusion (Figure 7.8) [28]. Unlike active micromixers that can induce efficient and immediate mixing by actuation, passive mixing can only be accomplished by a long path length and low flow rates. Nevertheless, this passive micromixing scheme is widely adopted because of its simplicity.

7.2.2.4 Concentration Gradient

The effect of drugs on biological systems is concentration dependent. One of the most efficient methods of exploring the concentration effect is by generating a chemical gradient. Within a microfluidic device, the concentration gradient is easily

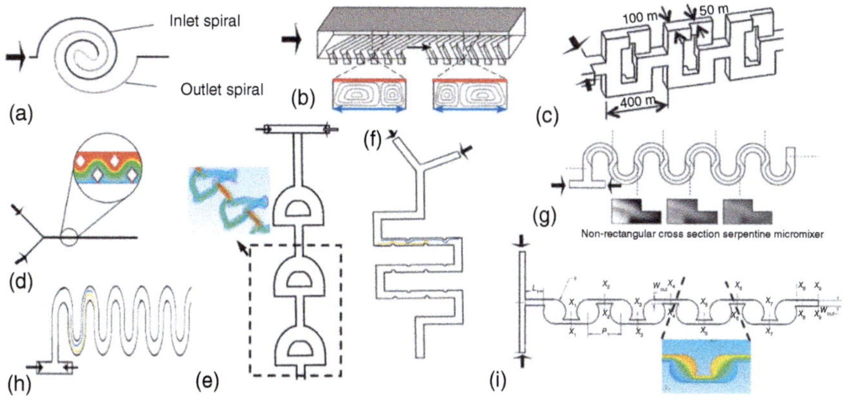

Figure 7.8 Schematics of various channel designs for passive micromixing: (a) spiral microchannels, (b) staggered herringbone, (c) Microfluidic baker's transformation, (d) obstacles placed in the fluid path, (e) a 3D "fishbone" design, (f–h) various "serpentine" channels, (i) a custom mixing channel for $Re = 1$. Source: Shanko et al. [28] / MDPI / CC BY-4.0.

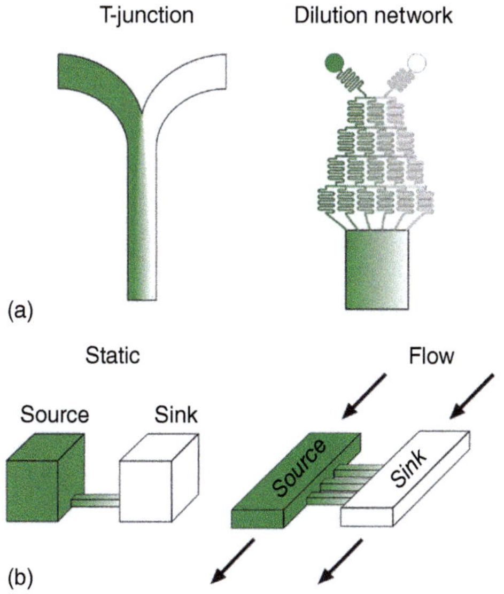

Figure 7.9 Examples of (a) flow-based and (b) diffusion-based microfluidic gradient generators. Source: Lin and Levchenko [29] / Frontiers Media / CC-BY 4.0.

generated by two or more compartments connected by channels or open channels aligned in parallel (Figure 7.9) [29]. Since the fluids are mostly in laminar flow, there is gradual diffusion between the fluids without turbulent mixing.

The passive micromixing via a convoluted channel can also be used to generate concentration gradients along its path (Figure 7.9) [29]. As two fluids flow through a series of passive micromixers, the reduced flow rates by the convoluted path result in slowed mixing via diffusion and convection and generation of chemical gradient.

7.2.3 Cell/Tissue Culture in Microfluidic Devices

For a microfluidic device to induce stable and long-term adhesion and phenotypical characteristics of cells and tissue, the device surface itself must be first modified to acquire cell-adhesive properties. Most elastomers and hard plastics are made of polymers that lack specific cell-responsive moieties. In addition, they are generally hydrophobic, especially for PDMS that is considered superhydrophobic, thereby resisting cell adhesion and wetting of aqueous media in general. To overcome this limitation, various surface modification techniques have been adopted to allow cell adhesion and on these platforms in order to fabricate significantly improve the quality of organ-on-a-chip technology [30].

The most common procedure to improve the hydrophilicity of elastomeric surface is to apply oxygen plasma, which results in the generation of hydrophilic functional groups, such as hydroxyl, epoxy, and ether groups, on the surface. However, the increase in hydrophilicity is often insufficient by oxygen plasma, and it diminished over time, because of the functionalized surface polymers becoming embedded via dynamic chain movement. To overcome this issue, more elaborate chemical modification schemes involving several wet-chemical reactions are often utilized. For example, silanization is often used to functionalize hydroxyl groups generated by oxygen plasma to attach additional functional groups, which are then further chemically modified to conjugate functional polymers. Alternatively, polymers can be directly grafted on the oxygen plasma-treated surface by radical polymerization, with hydrophilic functional groups acting as the initiator.

Chemical vapor deposition (CVD) provides a more uniform and direct method of generating thin films, which is able to completely cover the surface area, compared to the aforementioned chemical reaction schemes that only results in partial change in surface characteristics. The use of high energy in CVD allows deposition of both organic and inorganic materials in the gaseous state. In order to coat several layers, several cycles of film deposition are employed, often called as the layer-by-layer (LBL) approach. It should be noted that while wet-chemical reactions can be applied whenever PDMS molding has been completed, even after device fabrication via injection of reagents through channels, oxygen plasma and CVD that require high-energy treatment should be performed before the glass bonding step to finalize the device fabrication.

Having hydrophilic surface is not sufficient for promoting cell activities and tissue formation. It requires specific biological cues to derive specific cell adhesion and functions. Therefore, the surface is often coated with extracellular matrix (ECM) proteins, such as collagen and fibronectin, or cell-adhesive peptide sequences (e.g. Arg-Gly-Asp (RGD) peptide) for general cell adhesion [31]. DNA, RNA, and proteins (e.g. antibodies and receptors) are also often conjugated on the surface for more specific interactions.

Rather than simply culturing cells on the surface (2D), it is also possible to culture cells in 3D environment. A simple 3D tissue model can be created by inducing cell spheroid formation [32]. Highly proliferative cells with tight cell–cell junctions (and less affinity toward the cell–ECM interaction), such as stem cells and tumor cells, are well known to develop spherical cell aggregates, called spheroids, that are widely considered a miniature 3D tissue.

7.3 Microfluidic "Organ-on-a-Chip" Technology

7.3.1 Organ-on-a-Chip: Overview

Even though there have been significant advances in microfluidics for biomedical applications, especially in creating Bio-MEMS and μ-TAS for biosensing, the utilization of microfluidic technology in tissue engineering is still in early stages of development in comparison. This is mostly because of the fact that tissue engineering itself is still a burgeoning field, with several aspects such as cell source, scaffold materials, and culture methods still being heavily researched. As a result, the technology to integrate living biological tissues into a microfluidic system is continuously being developed and updated. Nevertheless, this "organ-on-a-chip" technology is recognized as the major and inevitable direction of this field.

Organ-on-a-chip has garnered this widespread attention mainly because of the critical limitations of the current screening tools for investigating biological responses in drug discovery. Thus far, *in vitro* cell culture and *in vivo* animal studies are the standard methods for screening drug candidates before clinical trials. However, the results obtained from these methods have often shown to be unreliable when predicting the human outcome. Conventional cell culture on a stiff 2D surface, universally practiced for its simplicity, does not accurately portray a 3D tissue environment and often rely on cells from established, commercially available, allogenic, and xenogeneic sources rather than from those directly derived from specific target patients. *In vivo* animal models can provide systemic physiological responses, such as pharmacokinetics and immune responses, which are not possible for *in vitro* studies. However, because of the fundamental difference between human and other species, those obtained from animal studies often do not translate well to human cases. In addition, with the growing ethical concern over animal treatment and sacrifice, there is a collective effort among the scientific community to eliminate or at least significantly reduce the animal testing.

Organ-on-a-chip technology is a promising biomedical μ-TAS technology that can overcome the limitations of both *in vitro* and *in vivo* investigations, by taking advantage of the practical level of tissue engineering. Although creating a whole organ is out of reach at this point, the ability to create a much smaller functional tissue has been well demonstrated in recent years, with the advancements in scaffold materials and cell culture methods, especially for patient-derived cells. Therefore, by integrating miniaturized tissues into a microfluidic device that can adequately support their viability and perform analyses could provide more complex biological analyses in a 3D tissue environment.

7.3.2 Biological Features of Organ-on-a-Chip Devices

7.3.2.1 Tissue Barrier Functions

Obviously, it is of paramount importance that the microfluidic device for organ-on-a-chip must have a compartment for cell/tissue culture to successfully maintain the viability and metabolic functions [16]. The simplest chamber would be an open space large enough for a desired number of cells connected by inlet and outlet channels. To take advantage of simplicity, multiple smaller chambers can be easily fabricated in series and parallel to acquire high-throughput capability.

One of the more challenging aspects of tissue structures that need to be emulated for organ-on-a-chip is the tissue barrier functions. The key difference between individual cells and tissues is the presence of membrane structures that separate different tissues and tightly control the mass transport to and fro, which results in varying phenotypical changes of cells in response to drugs according to the chemical gradient generated by the tissue depth and type of barrier. In addition, these membranes also act as the conduit for communications between different tissues. To fully take advantage of the co-culture capability of organ-on-a-chip, the simplest barrier structure is to place microchannels between two chambers in order to critically limit the fluid flow by diffusion, which is the same working principle of gradient generators (Figure 7.9). Depending on the channel geometry, the diffusion from one side to the other can be controlled, while generating the gradient in between. More precisely controlled transport can be attained by placing permeable or semipermeable membranes in between cell culture chambers (Figure 7.10) [33]. Porosity, thickness, and surface properties of these membranes are often controlled. Two different cells are cultured on the opposite sides of the membrane, and their mutual influence can be explored in the context of diffusion and mass transport. The type of materials used for membranes include polymer films and hydrogels. More active and physiologically relevant barrier is created by generating endothelial or epithelial layers on the permeable membranes. In addition to concentration, selective transport of particular chemicals across the membrane by the active transport mechanism of cells has also been investigated, such as the blood–brain barrier [34].

More recently, hydrogels have been increasingly adopted to provide a more suitable 3D microenvironment for tissue generation, while also acting as a biological barrier between different compartments (Figure 7.11) [35, 36]. The hydrogels possess tunable elasticity in a physiological range, high water content, and bioactive properties (e.g. cell adhesion molecules), which closely resemble the natural ECM and are a clear advantage over 2D membranes made from synthetic polymers.

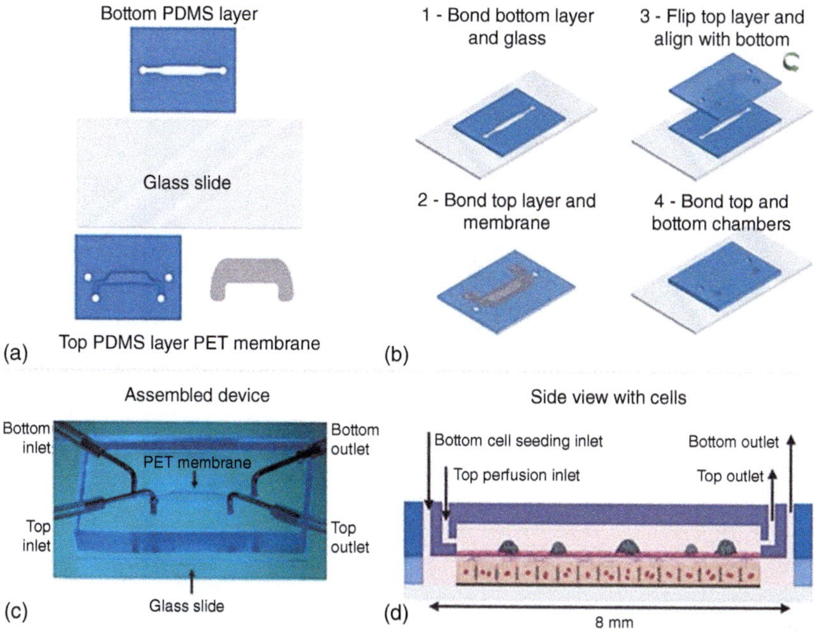

Figure 7.10 Schematics of fabricating a microfluidic device with a tissue barrier function; a membrane is placed within a cell culture chamber. (a) Schematic of the four parts used to assemble the device, (b-1) PDMS-glass bond bottom chamber assembled firmly by using an air plasma treatment, (b-2) PDMS-PET membrane top chamber prepared at 70 °C using a PDMS coating, and (b-3,4) assembly of the bottom and upper chamber, (c) the assembled microfluidic two-chamber device, (d) a side view of the device with the four cell types. Source: Prodanov et al. [33] / with permission of John Wiley & Sons.

7.3.2.2 External Stimulation

Another benefit of using a membrane is that it can be used to provide external stimuli, such as mechanical strain and electrical stimulation. Utilizing the same actuation mechanisms of microvalves and micropumps, as well as external machines, these membranes can be actuated to provide various physical forces (e.g. cyclic stretching, pulsatile, and turbulent flow) to stimulate the cells (Figure 7.12) [37]. These external stimulations can recreate the actual dynamic physiological conditions, such as mechanical loading on bone, lung aeration, and cardiac pumping.

7.3.2.3 Microvascular Network

The greatest attribute of a microfluidic device to recreate a tissue environment is the fact that the channels themselves can replicate the fluid transport function of the vascular system. However, the channel itself does not provide any biological functions imparted by endothelium and smooth muscle layer of the blood vessel. Therefore, endothelial cells (ECs) are often cultured along the inner lining of the channels to recreate a microvascular system within a microfluidic device (Figure 7.13) [38]. To help improve the adhesion and subsequent activities of ECs, various ECM proteins,

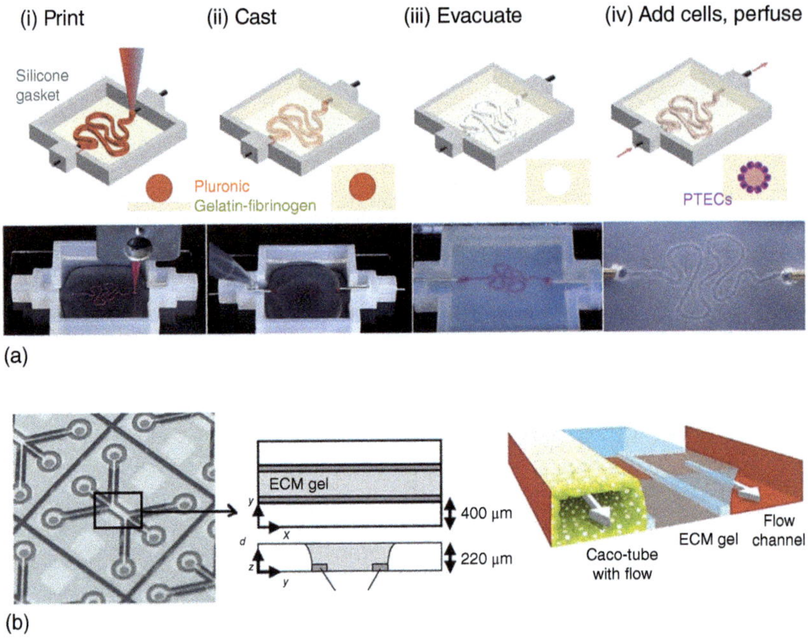

Figure 7.11 Hydrogel-integrated organ-on-a-chip. (a) A hydrogel-based vascularized tissue consisting of proximal tubule epithelial cells (PTECs) is fabricated inside a microfluidic chamber. (b) Hydrogel as a biological barrier is placed between an epithelial tube and a flow channel. Source: (a) Homan et al. [35] / Springer Nature / CC BY-4.0 and (b) Trietsch et al. [36] / Springer Nature / CC BY-4.0.

such as collagen, fibronectin, and laminin, are conjugated to the device surface [39]. In addition to ECM proteins, collagen, or fibronectin for adhesion, EC-specific junction proteins such as platelet endothelial cell adhesion molecule (PECAM) and vascular endothelial (VE)-cadherin have been conjugated to further stimulate EC functions. The controlled flow within these channels allows the investigation of the effect of fluid mechanics (e.g. shear stress and cyclic pressure) on the ECs.

To induce sufficient attachment of ECs to fully cover a channel is a significant challenge. Even with ECM-coated surfaces, it is difficult to induce enough proliferation and migration of ECs within a channel, especially at places difficult to access such as corners and upper side of the channel. Alternatively, ECs capable of sprouting and tubular formation embedded in ECM scaffolds, such as collagen and fibrin hydrogels, are placed in the channels. These ECs are in a microenvironment more conducive for more extensive vascular network formation [40].

The angiogenic properties of ECs cultured within a microfluidic device are often investigated in response to soluble factors (e.g. growth factors, cytokines, and synthetic chemicals) and cell–cell interactions. In a typical setup, ECs are cultured in a chamber connected to another chamber or a flow channel via microchannels. The extent of angiogenesis through these microchannels is explored by altering

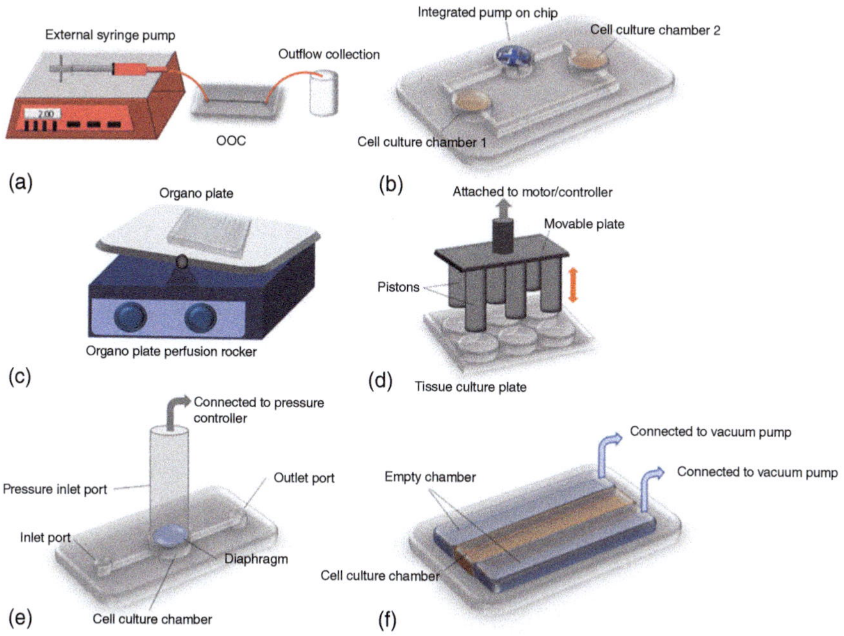

Figure 7.12 Various methods of applying external mechanical stimuli to organ-on-a-chip (OOC). (a) external syringe pump is delivering shear flow, (b) micro-pump is integrated within a microfluidic OOC, (c) OrganoPlate (Mimetas) perfusion rocker is used to generate the shear flow by tilting the stage, (d) compressive force is delivered to the cells on a tissue culture plate by a moveable piston, (e) compressive force is delivered by the pressure controller to the diaphragm placed on top of the cells, (f) a periodic pressure is generated by applying vacuum to the peripheral chambers. Source: (a–f) Reproduced with permission from Kaarj el al. [37]/ MDPI / CC-BY-4.0.

experimental parameters such as flow dynamics, chemical gradients, co-cultured cells, and mechanical properties (Figure 7.14) [41].

7.3.3 Recent Advances in "Bone-on-a-Chip" Technology

With continued development in organ-on-a-chip technology, microfluidic systems incorporating cells and tissues that represent various organs are continuously being developed and optimized. Developing "bone-on-a-chip," in particular, has been a significant challenge because, unlike most soft tissues, bone is made up of highly complex and heterogeneous tissue structures. Cortical bone, also called compact bone, accounts for the majority of bone tissue and is made of highly mineralized, dense columns containing osteoblasts and osteocytes, called osteons. Cancellous bone, also called spongy bone, on the other hand, is found on the end of long bone and highly porous and contains bone marrow that houses stromal cells. These cells become either hematopoietic stem cells (HSCs) giving rise to blood cells or mesenchymal stem cells (MSCs) that differentiate into various cell types, such as

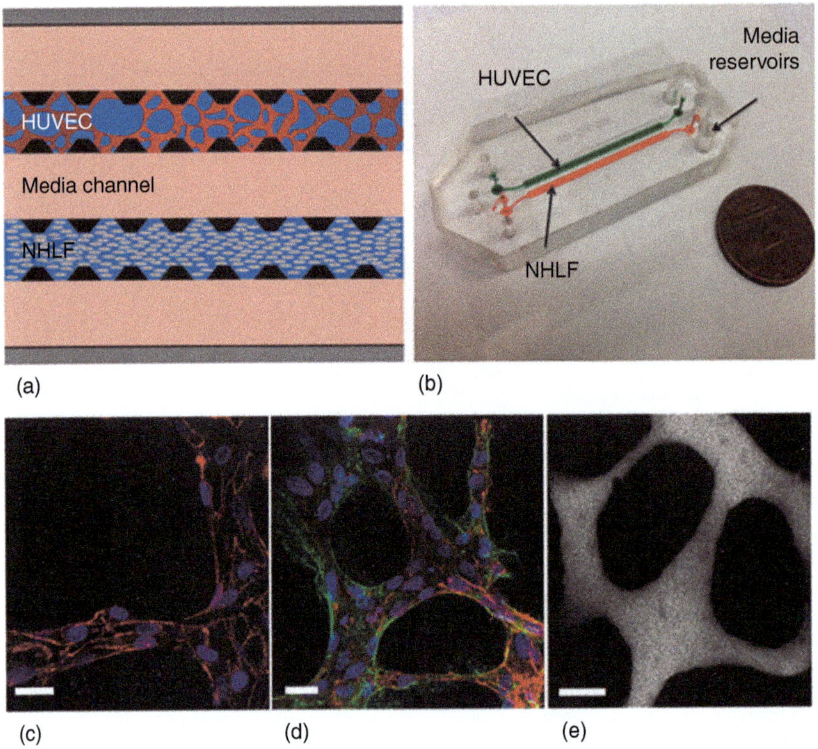

Figure 7.13 Microfluidic microvascular network. (a, b) Two channels are laden with human umbilical vein endothelial cells (HUVECs) and normal human lung fibroblasts (NHLFs). Characteristic markers of endothelial cells are determined by fluorescent imaging of (c) VE-cadherin (red) and (d) type IV collagen (green) and lumen (red). (e) Perfusion through the vascular formation is determined by the fluorescent probe. Scale bar: 20 μm. Source: Chen et al. [38] / with permission of Oxford University Press.

adipocytes, osteocytes, and chondrocytes. Therefore, extensive research efforts are underway to recreate these different aspects of bone tissue within organ-on-a-chip platforms.

7.3.4 Mineralization and Osteogenesis

Mineralization is the hallmark that separates bone tissue from other soft tissues. However, it is difficult to induce mineral formation within engineered tissues, as high water content within the tissues is not a favorable environment for ion precipitation of calcium and phosphate to develop hydroxyapatite (HA), the major bone mineral. In bone tissue engineering, there are two different approaches to induce and investigate mineralization within bone tissues: chemical and biological.

For chemical approaches, the mineralization can be induced by spontaneous precipitation of calcium and phosphate dissolved in an aqueous media onto the tissue

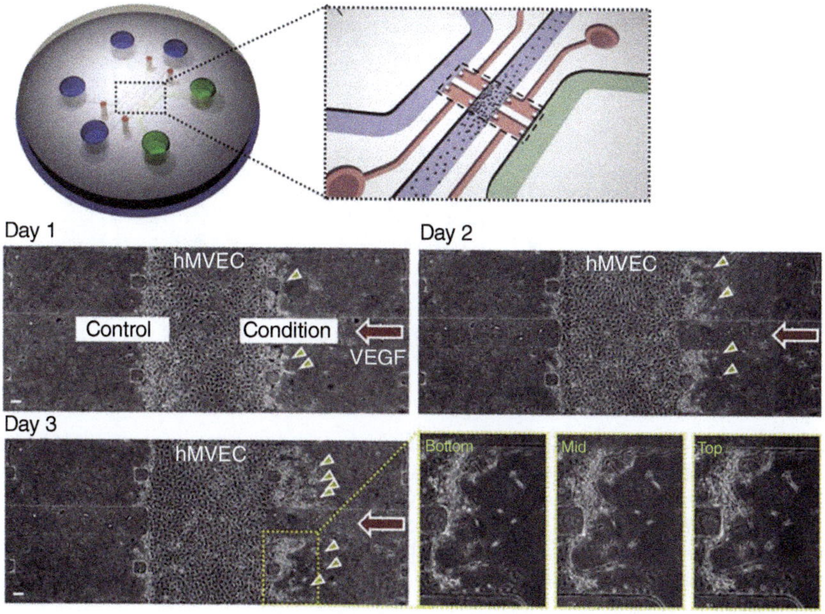

Figure 7.14 A microfluidic angiogenesis chip. Human microvascular endothelial cells (hMVECs) cultured in the middle channel are connected to a vascular endothelial growth factor (VEGF)-flowing channel (green) on one side and a control channel (purple) on the other side. Angiogenic response of hMVEC toward VEGF is demonstrated. Source: Chung et al. [41] / with permission of Springer Nature.

scaffold. The simplest method is to immerse tissue scaffolds in aqueous biological buffered media containing small amounts of calcium and phosphates, such as simulating body fluid, for an extended period of time, usually several days up to a few weeks [42]. The scaffolds are usually presented with negatively charged functional groups (e.g. carboxylate, sulfate, and phosphate) so that the initial nucleation with calcium ions can be facilitated. This method has the advantage of maintaining biocompatible environment during mineralization but is also time-consuming and insufficient to induce mineralization throughout the scaffold because of diffusion limit. This is especially problematic for a microfluidic device with embedded living cells and tissues. Alternatively, the mineral particles that are separately fabricated can be encapsulated into the tissue scaffold. In addition to HA, other types of calcium phosphate minerals, such as mono-, di-, and tri-calcium phosphate, also have been explored.

Numerous previous studies reported thus far have demonstrated that these minerals are highly beneficial in promoting osteoblast activities, angiogenesis, and osteogenic differentiation of MSCs. For example, Jusoh et al. developed a bone-on-a-chip device that incorporated fibrin hydrogel-containing HA particles designed to mimic osteon structures [43]. Angiogenesis into the fibrin hydrogel was highly promoted by the presence of HA particles. More recently, a microfluidic chip was fabricated using sintered HA-based ceramics rather than conventional PDMS

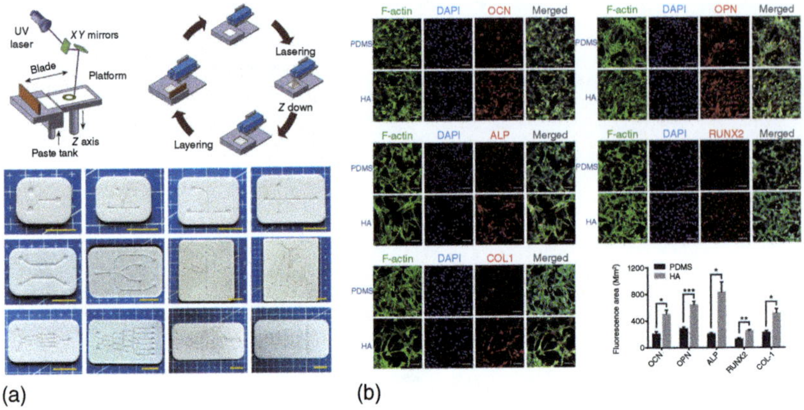

Figure 7.15 Hydroxyapatite (HA)-based ceramic bone-on-a-chip. (a) HA-ceramic microfluidic mold is fabricated by stereolithography performed on a HA-containing resin. (b) Osteoblasts are cultured in the HA-ceramic microfluidic device, and the increase in osteogenic properties, compared to PDMS surface, is examined by visualizing characteristic markers (F-actin and alkaline phosphatase [ALP]). $^{*}p < 0.05$, $^{**}p < 0.01$, $^{***}p < 0.001$. Source: Tang et al. [44] / with permission of Elsevier.

elastomers or hard plastics (Figure 7.15) [44]. In this study by Tang et al., a HA green body with a microfluidic channel pattern was first created by stereolithography performed on a HA-containing photocurable resin, which was then further sintered to obtain the HA-ceramic microfluidic device. Because of the presence of HA on the device surface, osteoblasts showed much greater osteogenic activities, as compared with those cultured on the PDMS elastomer. This study demonstrated that HA-based ceramic materials can provide effective osteogenic environment for a bone-on-a-chip platform.

For biological approaches, osteogenic cells are induced to produce calcium mineral deposits to their surroundings. Osteoblasts have the natural ability to produce calcium minerals. However, it can be further promoted by providing osteogenic growth factors or external (mechanical and electrical) stimulations [45, 46]. In addition to osteoblasts, biological mineralization could be realized via osteogenic differentiation of MSCs, which results in calcium mineral deposits around the cells. Compared to chemical approaches, the cellular mineralization is more physiologically relevant, but it is difficult to generate extensive mineralization throughout the engineered tissue constructs. Sheyn et al. cultured MSCs in a microfluidic device with an osteogenic medium and monitored their osteogenic differentiation by quantitative analyses of characteristic protein markers (e.g. osteocalcin, osteopontin, collagen I, and bone sialoprotein) and fluorescent imaging of calcium and phosphate ions [47]. It was demonstrated that the osteogenic activities were significantly modulated by the flow condition compared with the static culture. Bahmaee et al. also similarly introduced an "osteogenesis-on-a-chip." Here, a highly porous polymeric network, developed by a polymerized high internal phase emulsion (polyHIPE) technique, was used as the 3D cell culture platform

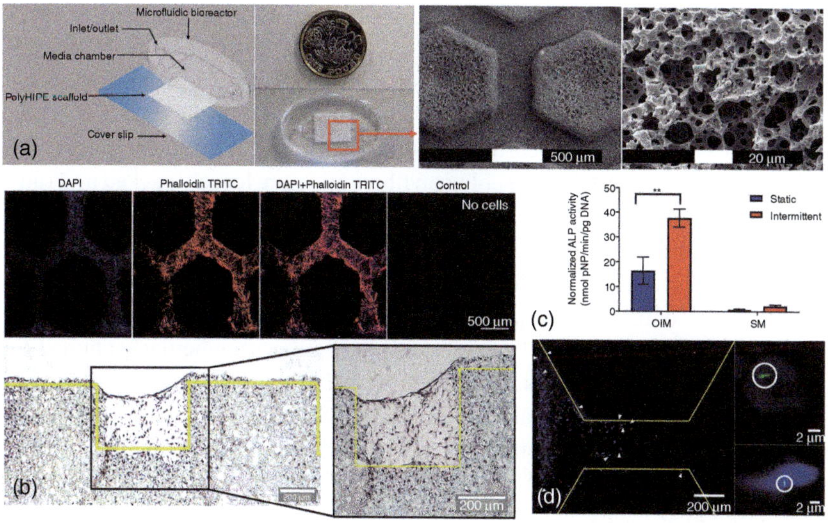

Figure 7.16 (a) A highly porous scaffold, developed by a polymerized high internal phase emulsion (polyHIPE) technique, encapsulated with MSCs is integrated into a microfluidic device. (b) MSCs show significant proliferation and spreading into the scaffold. (c) Osteogenic differentiation is enhanced in osteogenesis induction medium (OIM) in intermittent flow condition, as identified by alkaline phosphatase (ALP) expression. $^{**}p < 0.01$. (d) The presence of primary cilia (green) indicates mechanosensing of fluid shear stress by the cells. Source: Bahmaee et al. [48] / Frontiers Media / CC BY 4.0.

(Figure 7.16) [48]. The MSCs could infiltrate this porous polyHIPE substrate to several hundreds of micrometers in depth, forming 3D tissue-like networks, and eventually covering the entire channel area. In addition, the osteogenic differentiation by the induction media could be promoted in intermittent shear stress compared to static condition. This is another example of material innovation guiding the advancement in microfluidics, promoting cellular activities by adopting a bioactive material as the microfluidic platform.

The effect of co-culture on osteogenic differentiation of MSCs has also been explored. For example, MSCs demonstrated immunomodulatory effect on the co-cultured macrophages by suppressing M1 (pro-inflammatory) polarization and enhancing M2 (anti-inflammatory) polarization [49]. However, under an osteogenic environment, the macrophages were shown to increase the osteogenic differentiation of MSCs by acquiring the M1 phenotype.

7.3.5 Osteochondral Tissue

It is also a point of great biomedical interest to emulate osteochondral (OC) tissue, a highly heterogeneous interstitial area between articular cartilage and mineralized bone, as debilitating osteoarthritic diseases on OC tissue affecting millions of people worldwide. The OC tissue consists of hierarchical layers of superficial zone, middle zone, deep zone, calcified cartilage, and subchondral bone [50, 51].

- Superficial zone (10–20% of total cartilage): Collagen fibers are densely packed and align parallel to the articular surface. Within this layer, there are flattened chondrocytes residing along the direction of aligned collagen fibers. Because of its density and alignment, it is capable of resisting shear and tensile forces.
- Middle zone (40–60% of total cartilage): It consists of thicker collagen fibers and higher proteoglycan content. The collagen fibers are more loosely packed and randomly oriented. There is a lower density of chondrocytes than the superficial zone. Because of its large volume and elasticity, it is mainly responsible for absorbing and withstanding compressive forces.
- Deep zone (30% of total cartilage): It contains the thickest collagen fibers and the highest proteoglycan content. The collagen fibers are perpendicularly aligned to the articular surface. It provides the greatest protection against compressive forces.
- Calcified cartilage: HA minerals are deposited within the organic matrix consisting of collagen fibers and proteoglycans. The chondrocyte population is very low and mostly hypertrophic. Its role is to anchor and make a gradual transition between softer cartilage and subchondral bone.
- Subchondral bone: It consists of a subchondral bone plate and a trabecular bone, underneath the calcified cartilage. It contains various bone cells: osteoblasts and osteoclasts. It is heavily vascularized for both the bone cells within itself and the chondrocytes in the neighboring articular cartilage.

It is obviously a significant challenge to recreate this complex and the heterogeneous nature of the OC tissue. The organ-on-a-chip is viewed as an ideal platform for this purpose, as it can be designed to contain different compartments to develop sub-tissues separately while interconnected. For example, Lin et al. developed a microfluidic platform having a tissue culture chamber with two distinct fluid inlets (Figure 7.17) [52]. In this study, mesenchymal progenitor cells derived from induced pluripotent cells (iMPCs) embedded in a gelatin hydrogel was placed in the chamber. Chondrogenic medium (CM) and osteogenic medium (OM) were separately injected to induce "bi-phasic" cartilage and bone tissue formation, respectively, emulating the OC tissue.

Pirosa et al. similarly prepared a HA-containing polycaprolactone (PCL) scaffold, which was laden with gelatin hydrogel containing MSCs [53]. CM, OM, and

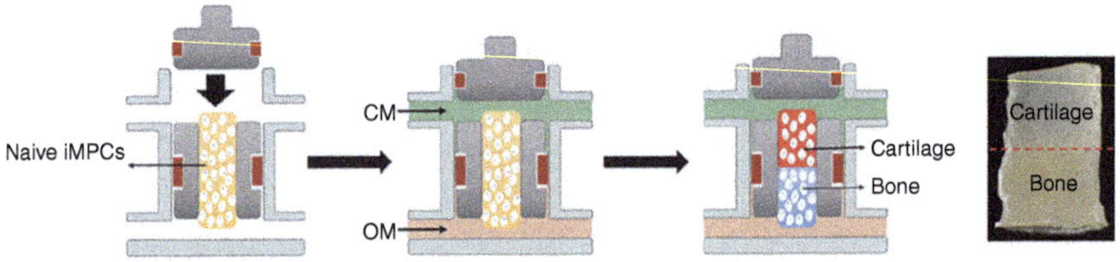

Figure 7.17 A schematic illustration of an organ-on-a-chip platform to develop osteochondral (OC) tissue. Mesenchymal progenitor cells (iMPCs) embedded in gelatin hydrogel were placed in a tissue culture chamber. Chondrogenic medium (CM) and osteogenic medium (OM) were injected into the chamber at opposite sides to generate cartilage and bone tissues, respectively. Source: Lin et al. [52] / Frontiers Media / CC BY-4.0.

endothelial growth medium (EGM) were provided separately in order to develop the vascularized OC tissue.

7.3.6 Bone Marrow

Bone marrow has been a hot topic of interest to biologists and bioengineers, as it provides the intricate microenvironment for HSCs and MSCs. Their growth, maintenance, activation, and differentiation are tightly regulated by the complex interplay between cells and matrices in bone marrow. Also, bone marrow has been implicated in cancer growth and metastasis. However, because of the complexity of the tissue structure, it has been a significant challenge to recreate "bone marrow-on-a-chip."

The most important requisite for creating a successful bone marrow-on-a-chip is the procurement of HSCs. HSCs used to develop tissue constructs to emulate bone marrow are generally obtained by harvesting from donors (both animals and humans) because of the lack of established cell lines of HSCs. In general, from harvested bone marrow, a small subset of cells expressing HSC markers, most notably CD34, are isolated and expanded *in vitro* before integrating into a microfluidic platform [54].

Unlike other adult stem cells, the maintenance, self-renewal, and differentiation of HSCs have been a tremendous challenge [55, 56]. HSCs are highly heterogeneous, less lineage-committed, and under the complex influences of soluble factors (e.g. cytokines and growth factors) and cell–cell and cell–ECM interactions. This challenge is not surprising, considering that HSCs residing in bone marrow are tightly regulated. Thus, extensive research efforts are underway to develop and optimize culture methods for HSCs, controlling media formulations and physicomechanical properties of scaffolds. Taking advantage of the physiological function of MSCs residing in bone marrow, co-culture with MSCs has also been utilized to support HSC expansion and maintenance, as they produce various growth factors and cytokines beneficial to HSCs [57].

HSCs, like other cell types, are cultured on a substrate coated with ECM proteins. Most previous studies have utilized scaffolds containing fibronectin, collagen, and laminin as the *de facto* substrates for stable HSC adhesion and *in vitro* expansion [58–60]. Microfluidic platforms are also commonly coated with these ECM proteins for HSC culture [61]. To further enhance the HSC activities through substrate engineering, other types of materials have been adopted to control the surface properties. For example, Mehta et al. incorporated poly(diallydimethylammonium) chloride, a polyelectrolyte, and clay, a group of silicate minerals, along with ECM proteins onto the PDMS substrate of a microfluidic device to significantly enhance the proliferation of bone marrow stromal cells [62]. Even without ECM proteins, the mineralized scaffold has also been shown to improve the HSC culture. Sieber et al. developed a microfluidic device that incorporated a highly porous HA-coated zirconia scaffold embedded with both MSCs and HSCs (Figure 7.18) [63]. The morphology of this ceramic scaffold closely resembled the human bone marrow. The initial MSC culture in this scaffold produced bone marrow-related

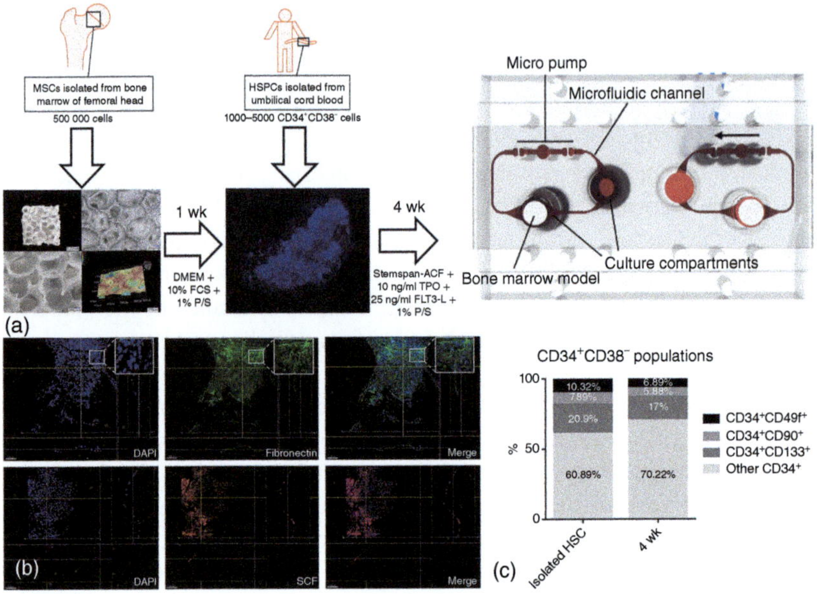

Figure 7.18 (a) MSCs are first cultured within a HA-coated zirconia scaffold to generate a bone marrow niche for the subsequent HSC culture. The scaffold is then integrated into a microfluidic device. (b) Bone marrow niche created by MSCs, identified by fibronectin and stem cell factor (SCF), is well maintained after four weeks. (c) The compositions of HSCs having different cell surface markers are largely maintained. Source: Sieber et al. [63] / with permission of John Wiley & Sons.

proteins, such as angiopoietin, fibronectin, intercellular adhesion molecule 1 (ICAM-1), pleiotrophin, and osteopontin. These proteins built up within the scaffold provided a suitable bone marrow-like microenvironment for subsequent HSC culture.

Using a highly unique approach, Torisawa et al. first developed bone marrow tissue by implanting a collagen scaffold containing demineralized bone powder and bone morphogenic proteins into a mouse [64]. The long-term *in vivo* incubation, up to eight weeks, resulted in both HSCs and differentiated blood cells occupying the scaffold, similar to the actual bone marrow. This engineered bone marrow tissue was then harvested and integrated into a microfluidic device, and the feasibility of this bone marrow-on-a-chip as a screening platform was demonstrated via radiation toxicity studies.

7.3.7 Cancer

Bone tissue has been a unique focal point in cancer research because of its propensity for metastasis [65]. Almost all types of solid tumors that reach advanced stages metastasize to other organs, and bone metastasis is one of the most frequent outcomes, next to lung and liver. For example, over 60% of metastatic breast cancer and prostate cancer cases lead to bone metastasis. This is due to several contributing factors. First, bone is heavily vascularized, with microvasculature extending throughout osteons. Metastatic circulating cancer cells and their aggregates become lodged into the capillaries and extravasated into bone tissue. Second, bone marrow, which is highly porous and a site of continuous exchange of blood cells and plasma, has also been implicated as the site of metastasis. Third, bone is continuously undergoing deposition and resorption by osteoblasts and osteoclasts, respectively. Infiltrated cancer cells have been shown to disrupt their activities, leading to expedited osteoblastic or osteolytic destruction of bone tissue.

Despite its significance, bone cancer metastasis has been difficult to diagnose and examine, as conventional imaging-based tools are not adept at densely mineralized conditions, especially at early stages. Consequently, a major research focus of bone-on-a-chip has been creating a suitable model for cancer metastasis to bone, in order to understand the underlying mechanisms and ultimately develop diagnostic and treatment options. The simplest bone cancer-on-a-chip involves the co-culture of osteoblasts and/or osteoclasts with cancer cells. For example, Hao et al. developed a microfluidic device containing osteoblasts in a chamber (Figure 7.19) [66]. The chamber was covered by a semipermeable membrane designed to efficiently remove metabolic products while retaining important proteins to ensure safe long-term

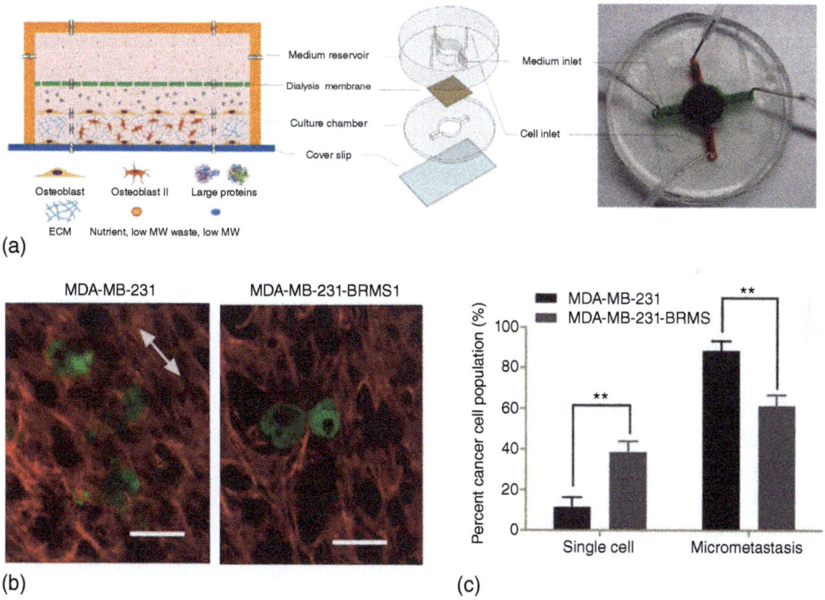

Figure 7.19 (a) Osteoblasts are cultured in a microfluidic chamber with a dialysis membrane, so the metabolic waste is removed and important proteins remain in the chamber in order to efficiently generate bone tissue. (b) Malignant breast cancer (MDA-MB-231) cells show greater extravasation into the bone tissue than non-malignant (MDA-MB-231-BRMS1) cells. (c) Malignant MDA-MB-231 cells undergo greater metastasis than non-malignant MDA-MB-231-BRMS1 cells. **$p < 0.01$. Source: Hao et al. [66] / with permission of John Wiley & Sons.

culture in order to induce osteoblast maturation, as identified by characteristic protein expression and mineral deposition. Breast cancer cells were then delivered via a microchannel, and their extravasation into the bone tissue and subsequent proliferation were studied.

The bone-on-a-chip incorporating vasculature is also widely investigated to study bone metastasis via extravasation of circulating cancer cells. According to a report by Jeon et al., they developed a complex tissue structure consisting of vasculature formed by ECs and bone mimicking microenvironment by osteoblasts and MSCs in a microfluidic chamber, which was flanked by two open parallel channels [67]. Breast cancer cells flown through these channels were shown to extravasate into the vascularized bone tissue, more predominantly than those without the MSCs or osteoblasts. In addition, the number of extravasating cancer cells was in fact lower under flow condition than static condition, which was resulted from lower vascular permeability by the shear stress on ECs. However, under the same flow condition, the cancer cells were able to infiltrate deeper into the bone tissue, highlighting the complex role of shear stress on vasculature and cancer metastasis.

7.4 Conclusion and Future Perspectives

The organ-on-a-chip has certainly been established as the future paradigm of cell and tissue culture platform. This technology is expected to continue to play a major role in bone tissue engineering, as conventional methods of engineering large bone tissue constructs are not capable of recreating the shear complexity of bone tissue physiology. With the continued development of bone-on-a-chip technology, there will be a greater emphasis placed on the physiological maturation of engineered tissues. For this, a deeper understanding of fundamental biology and tissue physiology, as well as developing advanced cell culture methods, must be proceeded in parallel.

The ultimate biomedical potential of organ-on-a-chip lies in drug screening. Conventional drug screening involves identifying drug candidates interacting with target molecules. However, this does not take into account the tissue and organ responses as well as variable immune responses in a whole-body environment. The organ-on-a-chip that can recreate the whole-body experience, including immune response, could mean the future paradigm in drug screening. At the current stage of bone-on-a-chip, there are several key issues that need to be resolved in order to reach the ultimate goal. First, target bone diseases should be successfully recreated within the engineered bone tissue. As normal and abnormal tissues require different culture methods and demonstrate different phenotypical characteristics, it is imperative that the chip platform be able to accommodate both types of tissues. Second, the immune system should be integrated within the chip. Unlike the bone cells within the tissue, immune cells, such as macrophages, not only reside within the tissue but also circulate around the vasculature. Thus, maintenance of circulating immune cells that also interact with bone tissue within a chip should be a major focus. Third, *in situ* biological evaluation methods should be integrated with the chip for increased usability and convenience, a crucial prerequisite for high-throughput analyses.

Acknowledgments

This work was supported by the Basic Science Research Program and Leading Foreign Research Institute Recruitment Program through the National Research Foundation of Korea (NRF) funded by the Ministry of Science and ICT (2018R1D1A1B07048522 and 2018K1A4A3A01063890).

References

1 Hierlemann, A., Brand, O., Hagleitner, C., and Baltes, H. (2003). *Proc. IEEE* 91: 839.
2 Wise, K. and Najafi, K. (1991). *Science* 254: 1335.
3 Mishra, M.K., Dubey, V., Mishra, P.M., and Khan, I. (2019). *J. Eng. Res. Rep.* 4: 1.
4 del Barrio, J. and Sánchez-Somolinos, C. (2019). *Adv. Opt. Mater.* 7: 1900598.
5 Lee, S.J. and Lee, S.Y. (2004). *Appl. Microbiol. Biotechnol.* 64: 289.
6 Park, T.H. and Shuler, M.L. (2003). *Biotechnol. Prog.* 19: 243.
7 Dincer, C., Bruch, R., Costa-Rama, E. et al. (2019). *Adv. Mater.* 31: 1806739.
8 Ashraf, M.W., Tayyaba, S., and Afzulpurkar, N. (2011). *Int. J. Mol. Sci.* 12: 3648.
9 Zhong, Q., Ding, H., Gao, B. et al. (2019). *Adv. Mater. Technol.* 4: 1800663.
10 Marcy, Y., Ishoey, T., Lasken, R.S. et al. (2007). *PLoS Genet.* 3: e155.
11 Ma, S., Murphy, T.W., and Lu, C. (2017). *Biomicrofluidics* 11: 021501.
12 Wyatt Shields IV, C., Reyes, C.D., and López, G.P. (2015). *Lab Chip* 15: 1230.
13 Ahadian, S., Civitarese, R., Bannerman, D. et al. (2018). *Adv. Healthcare Mater.* 7: 1700506.
14 Lee, S.H. and Sung, J.H. (2018). *Adv. Healthcare Mater.* 7: 1700419.
15 Coluccio, M.L., Perozziello, G., Malara, N. et al. (2019). *Microelectron. Eng.* 208: 14.
16 Zhang, B., Korolj, A., Lai, B.F.L., and Radisic, M. (2018). *Nat. Rev. Mater.* 3: 257.
17 Wang, Y., Wang, L., Guo, Y. et al. (2018). *RSC Adv.* 8: 1677.
18 Kimura, H., Sakai, Y., and Fujii, T. (2018). *Drug Metab. Pharmacokinet.* 33: 43.
19 Skardal, A., Shupe, T., and Atala, A. (2016). *Drug Discovery Today* 21: 1399.
20 Lee, J.B. and Sung, J.H. (2013). *Biotechnol. J.* 8: 1258.
21 Kim, P., Kwon, K.W., Park, M.C. et al. (2008). *Biochip J.* 2: 1.
22 Xia, Y. and Whitesides, G.M. (1998). *Annu. Rev. Mater. Sci.* 28: 153.
23 Kiran Raj, M. and Chakraborty, S. (2020). *J. Appl. Polym. Sci.* 137: 48958.
24 Walsh, D.I., Kong, D.S., Murthy, S.K., and Carr, P.A. (2017). *Trends Biotechnol.* 35: 383.
25 Andersson, H., van der Wijngaart, W., Nilsson, P. et al. (2001). *Sens. Actuators, B* 72: 259.

26 Laser, D.J. and Santiago, J.G. (2004). *J. Micromech. Microeng.* 14: R35.

27 Au, A.K., Lai, H., Utela, B.R., and Folch, A. (2011). *Micromachines* 2: 179.

28 Shanko, E.-S., van de Burgt, Y., Anderson, P.D., and den Toonder, J.M. (2019). *Micromachines* 10: 731.

29 Lin, B. and Levchenko, A. (2015). *Front. Bioeng. Biotechnol.* 3: 39.

30 Zhou, J., Ellis, A.V., and Voelcker, N.H. (2010). *Electrophoresis* 31: 2.

31 Shakeri, A., Jarad, N.A., Leung, A. et al. (2019). *Adv. Mater. Interfaces* 6: 1900940.

32 Moshksayan, K., Kashaninejad, N., Warkiani, M.E. et al. (2018). *Sens. Actuators, B* 263: 151.

33 Prodanov, L., Jindal, R., Bale, S.S. et al. (2016). *Biotechnol. Bioeng.* 113: 241.

34 Booth, R. and Kim, H. (2012). *Lab Chip* 12: 1784.

35 Homan, K.A., Kolesky, D.B., Skylar-Scott, M.A. et al. (2016). *Sci. Rep.* 6: 34845.

36 Trietsch, S.J., Naumovska, E., Kurek, D. et al. (2017). *Nat. Commun.* 8: 262.

37 Kaarj, K. and Yoon, J.-Y. (2019). *Micromachines* 10: 700.

38 Chen, M.B., Whisler, J.A., Jeon, J.S., and Kamm, R.D. (2013). *Int. Biol.* 5: 1262.

39 Akther, F., Yakob, S.B., Nguyen, N.-T., and Ta, H.T. (2020). *Biosensors* 10: 182.

40 Buchanan, C.F., Voigt, E.E., Szot, C.S. et al. (2014). *Tissue Eng. Part C* 20: 64.

41 Chung, S., Sudo, R., Vickerman, V. et al. (2010). *Ann. Biomed. Eng.* 38: 1164.

42 Cha, C., Kim, E.-S., Kim, I.W., and Kong, H. (2011). *Biomaterials* 32: 2695.

43 Jusoh, N., Oh, S., Kim, S. et al. (2015). *Lab Chip* 15: 3984.

44 Tang, Q., Li, X., Lai, C. et al. (2021). *Bioact. Mater.* 6: 169.

45 Meng, S., Zhang, Z., and Rouabhia, M. (2011). *J. Bone Miner. Metab.* 29: 535.

46 Chatakun, P., Núñez-Toldrà, R., Díaz López, E.J. et al. (2014). *Cell. Mol. Life Sci.* 71: 113.

47 Sheyn, D., Cohn-Yakubovich, D., Ben-David, S. et al. (2019). *Microfluid. Nanofluid.* 23: 99.

48 Bahmaee, H., Owen, R., Boyle, L. et al. (2020). *Front. Bioeng. Biotechnol.* 8: 1042.

49 Romero-López, M., Li, Z., Rhee, C. et al. (2020). *Tissue Eng. Part A* 26: 1099.

50 Sophia Fox, A.J., Bedi, A., and Rodeo, S.A. (2009). *Sports Health* 1: 461.

51 Nukavarapu, S.P. and Dorcemus, D.L. (2013). *Biotechnol. Adv.* 31: 706.

52 Lin, Z., Li, Z., Li, E.N. et al. (2019). *Front. Bioeng. Biotechnol.* 7: 411.

53 Pirosa, A., Gottardi, R., Alexander, P.G. et al. (2021). *Biomaterials* 272: 120773.

54 Sidney, L.E., Branch, M.J., Dunphy, S.E. et al. (2014). *Stem Cells* 32: 1380.

55 Seita, J. and Weissman, I.L. (2010). *WIREs Syst. Biol. Med.* 2: 640.

56 Walasek, M.A., van Os, R., and de Haan, G. (2012). *Ann. N. Y. Acad. Sci.* 1266: 138.

57 Mehrasa, R., Vaziri, H., Oodi, A. et al. (2014). *Int. J. Mol. Cell. Med.* 3: 1.

58 Choi, J.S. and Harley, B.A.C. (2017). *Sci. Adv.* 3: e1600455.

59 Susek, K.H., Korpos, E., Huppert, J. et al. (2018). *Matrix Biol.* 67: 47.

60 Nelson, M.R. and Roy, K. (2016). *J. Mater. Chem. B* 4: 3490.

61 Santos Rosalem, G., Gonzáles Torres, L.A., de Las Casas, E.B. et al. (2020). *PLoS One* 15: e0243840.

62 Mehta, G., Kiel, M.J., Lee, J.W. et al. (2007). *Adv. Funct. Mater.* 17: 2701.

63 Sieber, S., Wirth, L., Cavak, N. et al. (2018). *J. Tissue Eng. Regener. Med.* 12: 479.
64 Torisawa, Y.-s., Spina, C.S., Mammoto, T. et al. (2014). *Nat. Methods* 11: 663.
65 Macedo, F., Ladeira, K., Pinho, F. et al. (2017). *Oncol. Rev.* 11: 321.
66 Hao, S., Ha, L., Cheng, G. et al. (2018). *Small* 14: 1702787.
67 Jeon, J.S., Bersini, S., Gilardi, M. et al. (2015). *Proc. Natl. Acad. Sci. U.S.A.* 112: 214.

8

Bioactive Glasses in Orthopedics

Amel Houaoui and Jonathan Massera

Tampere University, Faculty of Medicine and Health Technology, Korkeakoulunkatu 3, Tampere 33720, Finland

8.1 First Bioactive Glass

The first bioactive glass (BAG) has been developed in 1969 by Prof. L.L. Hench [1]. The glass developed was called 45S5 and was made of $45SiO_2$–$24.5Na_2O$–24.5CaO–$6P_2O_5$ (in wt%). The large amount of alkali and alkaline earth ions in the glass composition, along with the rather low amount of glass former (i.e. SiO_2 and P_2O_5), leads to a glass prone to dissolve in any aqueous solution. The mechanism of dissolution has been widely reported and can be summarized as follows [2]:

1) Alkali-H^+ ion exchange. Alkali ions are released into the solution and substituted with H^+ ions in the glass structure, thus leading to a rise in the solution pH. This stage is controlled by diffusion and consequently slows down with time.
2) The silica network (backbone of the glass structure) starts to partly dissolve releasing Si(OH4) in the solution. This has been associated with the breaking of Si—O—Si bonds and the formation of SiOH end units. This event occurs at the glass/solution interface, and therefore, its kinetic is surface controlled.
3) The insoluble silica then re-polymerizes to form a silica gel at the surface of the reactive glass.
4) The BAG surface, covered with silanol, acts as nucleation point for the precipitation and growth of a CaP layer at the glass surface. The CaP reactive layers are formed due to the migration of Ca and P ions from the core of the glass to the surface and as a consequence of the supersaturation of the surrounding with those ions.
5) Initially, the CaP-reactive layer is amorphous and with time crystallizes into a hydroxyapatite (HA) layer.

Later, the ability of BAGs to support the precipitation of reactive HA layer was considered a first sign of bioactivity. However, as such, it is not sufficient to claim the bioactive nature of a biomaterial. BAG 45S5 was found to bond to living tissue, and

Houaoui and Massera contributed equally.

Biofabrication for Orthopedics: Methods, Techniques, and Applications, First Edition.
Edited by Wenguo Cui, Xin Zhao, and Shen Liu.

such bonding was associated with the precipitation of the hydroxyapatite [3]. *In vivo*, following the precipitation of HA comes the adsorption of biological moieties at the HA layer interface and the action of macrophages. Next, the stem cells attach to the material surface and initiate to differentiate into the needed cell lineage. Finally, the matrix will form and mineralize.

The BAG 45S5 has found significant space in clinics and especially in orthopedics with the establishment of Perioglass® in 1995 and NovaBone® in 2002. The role of Perioglass and NovaBone was to harness the natural healing process. In short, 45S5 BAG demonstrated the ability to recruit stem cells and, through signaling pathways, to induce differentiation of cells into osteoblastic lineage. The 45S5 glass was then marketed as an osteostimulative biomaterial [4].

8.2 Bioactive Glass Versatility

The advantage of BAGs, over any other class of biomaterials, is its ability to tailor its composition [5]. Tailoring of the composition not only allows us to have control over the dissolution rate but is also beneficial in controlling the release of therapeutic ions. As an example, K. Karlson and Ö. Anderson developed the BAG S53P4 (composition in wt%: $53SiO_2$–$23Na_2O$–$20CaO$–$4P_2O_5$) soon after the discovery of 45S5. The glass S53P4 is nowadays sold as BonAlive®. S53P4 is indicated as a bone graft substitute in craniofacial procedures or as a bone substitute after bone tumor removal [6]. More recently, new clinical studies have shown the benefit of using S53P4 in the treatment of aneurysmal bone cyst in children [7] as a filler material in mastoid obliteration surgery [8] or as a filler in large mandibular defects [9]. The increasing interest in the BAG S53P4 is not only because of the inherent ability of BAGs to promote new bone formation but, in this case, to a surprisingly efficient antibacterial activity. Indeed, S53P4 BAG has been studied on over 50 patients in the treatment of chronic osteomyelitis [10]. The results showed that the antibacterial effect was not mediated by the antibiotics and is a promising solution in times where antibiotic resistance is growing.

Aside from the traditional BAGs already available in the market, new glass compositions are being developed in view of delivering ions or light with therapeutic interest. As can be seen in Figure 8.1, glasses can be produced as bulk, powder, fibers, or scaffolds. The main advantage of BAGs lies in their ability to be doped with ions that can participate in promoting biological functions.

8.2.1 Osteogenesis

The main class of bioactive glasses belongs to the silicate glass family. Si ions are known to contribute to metabolic processes and to the formation of new bone [11]. The mechanism of silicate glass dissolution, as said before, is associated with the formation of a Si-rich layer, which also promotes the precipitation of HA at the surface of the glass, thus providing a beneficial substrate for the attachment and proliferation of osteoblast cells but also stem cells [12]. Furthermore, Si intake is beneficial for the

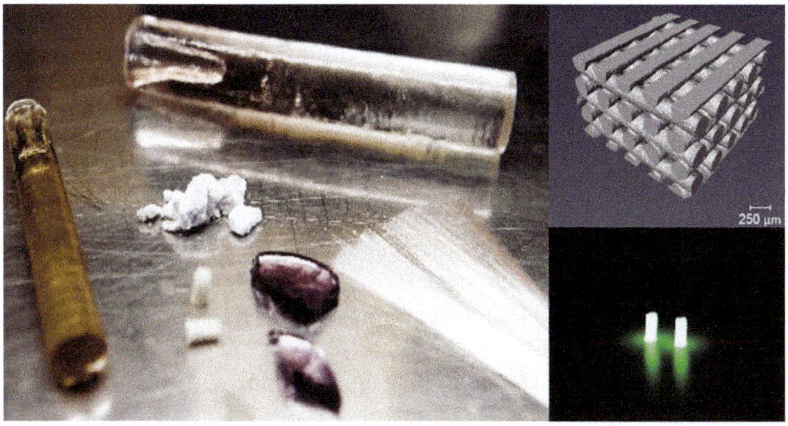

Figure 8.1 Bioactive glasses with various compositions processed as bulk, powder, fiber, or scaffold (top right). Bottom right image shows a bioactive glass exhibiting persistent luminescent properties.

production of a denser bone mineral [13] as well as to promote collagen formation and osteoblastic differentiation [14].

While Ca^{2+} is an ion that has been traditionally included in BAGs (see 45S5 and S53P4 BAGs) to promote HA precipitation as well as to favor osteoblast proliferation, differentiation, and mineralization [15], other divalent cations can substitute Ca^{2+} to further invoke the therapeutic effect. For example, Ca can be replaced by Mg and/or Sr. Mg, when substituted for Ca, is known to decrease the speed of dissolution of the glass as Mg does not solely enter the glass network as a modifier but more likely as an intermediate [16, 17]. However, Mg has been associated with improved new bone formation as well as improved bone-cell adhesions and stability [18, 19]. Sr can also replace Ca in the glass network. The use of Sr in BAG is linked to the use of such ions to treat osteoporosis. Generally, the substitution of Sr for Ca is associated with an increase in the glass dissolution because of the large ionic radius of Sr ions. The benefit of Sr lies in its ability to be incorporated in the HA layer to form a Sr-substituted HA reactive layer, which is more dense and less prone to chemical attack during the bone regeneration [20–22]. Sr ions have proven to be beneficial for the attachment and proliferation of human gingival fibroblasts [21]. *In vivo*, Sr ions have been linked with an increased formation of new and denser bone [23]. Lastly, Zn ions have also been incorporated in BAGs and successfully promoted bone formation through activation of protein synthesis in osteoblastic cells and by increasing adenosine triphosphate (ATP) activity [24].

8.2.2 Angiogenesis

While in orthopedics, it is widely accepted that the bioactive material must be osteoconductive, or even better osteoinductive, one must keep in mind that the primary reason for bone graft failure is the lack of vascularization. Angiogenesis is pivotal in the graft success as it provides a connection between the implant and the surrounding tissue. Furthermore, vascular networks are of utmost importance in the tissue regeneration process [25–27].

In BAGs, while the focus was, initially, on the osteogenesis, lately much effort has focused on developing new BAGs with enhanced angiogenic properties. For instance, while boron ions were known to stimulate bone formation [28], careful studies of the impact of boron release from BAGs have unraveled the ability of this ion to promote vascularization. For instance, when expose to a control boron release, bone marrow stromal cells overexpressed vascular endothelial growth factors (VEGFs) [29]. Durand et al. studied the impact of boron release from a boron containing 45S5-modified glass ($43SiO_2$–$24.5Na_2O$–$24.5CaO$–$6P_2O_5$–$2B_2O_3$ in wt%). The addition of boron to the traditional 45S5 BAG was linked with an increased expression of integrin $\alpha_v\beta_3$ and higher vasculature density in the embryonic quail chorioallantoic membrane (CAM). The angiogenic response obtained by the boron containing BAG ion release was equivalent to applying 10 μg/ml of basic fibroblast growth factor [30].

Aside from boron, copper has also been found to have potential for angiogenesis. Indeed, the use of $CuSO_4$ associated with growth factors demonstrated synergistic effect. The use of low concentration of Cu ions was then reported to be promising toward promoting vascularization of implants [31]. Later, such results were confirmed using Cu ions released from mesoporous BAGs in an *in vivo* zebrafish model. The study by Romero-Sánchez et al. demonstrates that while the Cu-free mesoporous BAG promotes vascularization, such event is enhanced when Cu is added to the mesoporous glass [32]. Furthermore, a study by Wang et al. demonstrated that Cu-doped BAGs (3 wt% of Cu doped into $18SiO_2$–$36B_2O_3$–$6Na_2O$–$8K_2O$–$8MgO$–$22CaO$–$2P_2O_5$) elicit a significantly higher capacity of the bone to regenerate alongside with increased angiogenesis stimulation, in a rat calvarial model [33].

8.2.3 Antimicrobial Activity

Despite the breakthrough in modern medicine and the acceptation of medical devices aiming at augmenting, treating, repairing, or even replacing tissues, surgeons are facing new challenges. Indeed, the implantation of foreign device is associated with risk such as, but not limited to, "biomaterials associated infections" [34]. Such challenge is even amplified by the growing bacteria drug resistance. Therefore, new strategies must be developed to combat potential infections, which may lead to implant failures. BAGs, such as S53P4, have already been found to evoke surprising antimicrobial activity [35, 36]. When bacteria were incubated with S53P4, cell shrinkage and damage to the membrane were evidenced by scanning electron microscopy (SEM) analysis [37]. The antimicrobial properties are believed to be a consequence of the rise in pH at the glass/liquid interface but also to the release of silica, calcium, and phosphorus ions, which was proven to disturb the bacteria's membrane as well as increasing the osmotic pressure [38, 39]. Indeed, the release of phosphorus and calcium ions was confirmed to play an important role in bacterial eradication as phosphate-based glasses exhibit even higher antimicrobial properties than traditional silicate BAGs [40]. Furthermore, BAGs by-products are not only efficient in eradicating planktonic bacteria but also efficient in preventing biofilm formation [41]. Preventing/suppressing biofilms are of paramount importance in a clinical context. Biofilm forming strains are believed to develop, more easily, multidrug resistance and therefore to prevail even after drug delivery. The advantage of BAGs comes from the assumption that bacteria cannot grow resistant to pH change or increased osmotic pressure.

In order to increase the efficiency for bacterial eradication, BAGs can be doped with ions such as Ag^+ or Cu^{2+}. Ag ions have been the most biocide ions studied in BAGs. It is suggested that Ag^+ disrupts the functions of the bacterial cell membrane and proteins by binding to DNA and thiol groups [42]. However, while many studies have evidenced the antimicrobial properties of Ag-doped BAGs, little work has investigated such effect *in vivo* [43]. One of the challenges when using Ag as an antimicrobial agent is to eradicate bacteria without compromising cytocompatibility. Cu ion has also been heavily studied as it is a well-known antimicrobial ion [44]. While Cu exhibits lower antimicrobial activities, it presents the advantage to elicit less cytotoxic reaction, when compared with Ag [40].

8.3 Alternative Bioactive Glasses

Keeping in mind the high versatility of glass in general, phosphate-based BAGs are an interesting alternative to silicate BAGs. Indeed, the main limitations of silicate BAGs are (i) uncontrolled and rapid crystallization, upon heating, preventing their drawing into fibers or sintering into scaffolds, without adverse crystallization [45, 46] and (ii) remnants of silica BAGs even 14 years post-surgery (while not being found to have any adverse effect). While borosilicate glasses may overcome, partly, the limitation of silicate glass, phosphate glasses show properties of particular interest [47–51]. As said earlier, the control release of therapeutic ions may invoke specific biological events beneficial to the patient. The main advantage lies in the dissolution mechanism of phosphate-BAGs [52]. As opposed to silicate BAGs that exhibit a non-congruent dissolution process, phosphate glasses dissolve in a congruent manner.

The dissolution mechanism and its implication on the reactive layer formation can be seen by SEM analysis of glass particles immersed in simulated body fluid, as shown in Figure 8.2.

As shown in Figure 8.2, in the case of silicate BAG, upon immersion, a thick silica-rich layer forms in between the parent glass and the precipitated HA. On the other hand, when phosphate BAGs are immersed in simulated body fluids, the reactive layer precipitated directly at the glass surface. It should be noted that in the case of metaphosphate glass, the reactive layer has been found to be closer in composition to a dicalcium phosphate dihydrate (DCPD) [53] while HA or octacalcium phosphate was found to precipitate at the surface of invert-phosphate

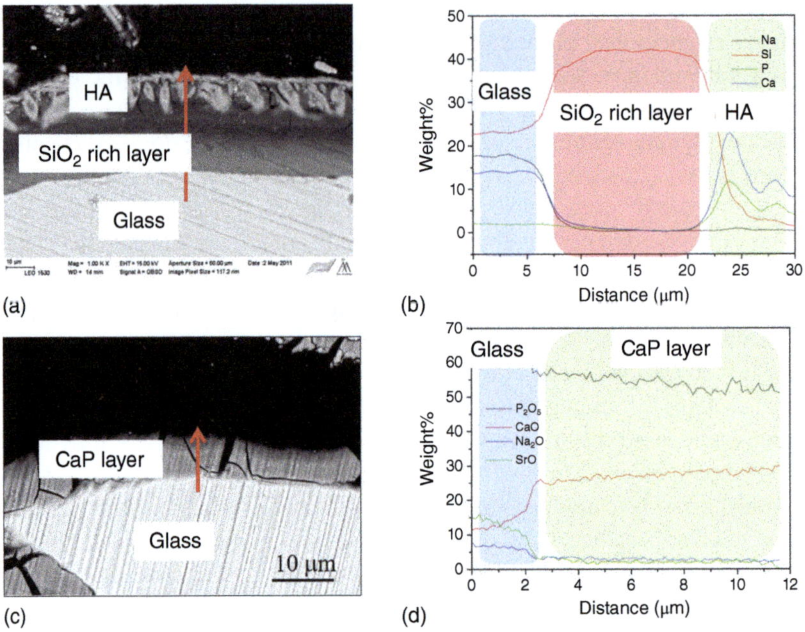

Figure 8.2 SEM images and energy-dispersive X-ray spectroscopy (EDX) line scan of a bioactive silicate glass (a and b) and a bioactive phosphate glass (c and d) immersed for 72 hours in simulated body fluid.

glasses [54, 55]. Furthermore, control over the glass composition enables to optimize the cell–material interactions. While the congruent dissolution may lead to a lack of cell attachment at early time points, tailoring the ions released (nature and concentration) revealed to yield cell–material interaction similar to those seen with traditional silicate BAGs [21, 56–58]. It should be pointed out that a significant amount of work is being conducted to control the phosphate glass surface physicochemical properties to favor early cell adhesion [59, 60].

Such unique properties enable to control and predict the release of ions over an extended period of time. Such dissolution mechanism has led to the development of Se-, Cu-, and Co-doped COSECURE glass boluses tested on animal to overcome acute as well as marginal deficiencies [61]. Furthermore, the lack of change of the parent glass surface composition during the dissolution enables a more complete dissolution of the biomaterial, over time, and could potentially prevent remnant glass at the surgical site. Such property is even of greater importance when considering the use of BAG as a filler in a composite or a hybrid material.

8.4 Bioactive Glasses in Composites and Hybrid Materials

To remedy its hardness, its brittleness, and the difficulties in shaping BAGs, they are usually combined with organic matrices. These matrices act as the organic matrix of the bone. They are easier to process into a 3D scaffold, which will support cell proliferation and differentiation. They are used for bone tissue engineering applications as they have a minimal immunologic response and they present a good biocompatibility, which will promote cell behavior [62, 63]. The development of organic/inorganic composites or hybrids allows to combine the properties of each phase to obtain a material having all the characteristics and properties needed to repair the bone [64].

8.4.1 Materials Used as Organic Phase in Composites or Hybrid Devices for Bone Repair

The most used organic matrix in composites and hybrids are polymers. These polymers can be natural or synthetic. Both exhibit advantages and drawbacks.

8.4.1.1 Natural Polymers

Natural polymers were the first biodegradable biomaterials used clinically because of their similarities with bone extracellular matrix [65]. They exhibit interesting properties, i.e. good biodegradability, and biological advantages (i.e. no toxicity, non-immunogenicity, and intrinsic osteogenic capacities), which favor cell attachment and promote chemotaxis [66]. Nowadays, the most used natural polymer in orthopedics is collagen as it is also part of the bone composition, allows the nucleation/deposition of mineral crystals and provides binding sites for osteogenic proteins [65–68]. Other natural polymers such as chitosan, alginate, and gelatin can be used. Gelatin is a natural denatured polymer, a derivative of collagen. It is composed of amino acids (hydroxyproline, proline, or sequences such as RGD – arginine–glycine–aspartic acid). Low cost, immunogenicity, aqueous solution solubility, and the different sources of collagen (fish, cattle bone, and pig skin) from which it can be extracted are a few examples of the advantages of this protein. Gelatin is used in tissue engineering studies because of its biocompatible and biodegradable properties. It is considered as a suitable biomaterial to mimic the extracellular matrix because of its functional groups and the possibility to form 3D scaffolds with a porous structure [69, 70]. However, the gelatin needs to be cross-linked in order to stay self-supported at biological temperatures.

Chitosan has received great attention because it is biocompatible, biodegradable, and osteoconductive. It can be used in different forms for orthopedic applications, for example, films, fibers, and sponges. Its processability is one of its advantage for bone tissue engineering. Moreover, the Food and Drug Administration (FDA) approved the use of chitosan in wound dressing. However, chitosan scaffolds have low osteoinductivity and low mechanical properties. In addition, other limitations of chitosan have been reported, such as its pH-dependent solubility and its rapid degradation *in vivo* [71–73].

Alginate is an anionic natural polymer known for its biocompatible, biodegradable, and hydrophilic properties. In physiological conditions, alginate can form porous gels. An important disadvantage of using alginate as a biomaterial for

Table 8.1 Summary of the most representative natural polymers used as building materials for bone repair.

Polymer	Source	Advantages	Drawbacks	Main formulations
Collagen	Ovine, porcine, equine, and bovine	Biocompatible, biodegradable, hydrophilic, cell-binding properties, and low antigenicity	Relatively weak mechanical stiffness, low antigenicity, transmission of pathogens, induces allergic reactions	Polymeric and mineralized scaffolds
Gelatin	Denaturation and hydrolysis of natural collagen	Biocompatible, biodegradable, cell-binding properties, low cost, and gelling properties	Low mechanical properties, and fast degradation rate	Hydrogel and micro-/nanosphere
Silk	Lepidoptera larvae, some arachnids, and various flies	Biocompatible, biodegradable, flexible processability, high mechanical strength, and thermally stable	Slow degradation rate (2–4 yr)	Polymeric and composite scaffolds
Alginate	Brown seaweed	Biocompatible, biodegradable, relatively low cost, easy gelation by ionic cross-linking, and easy chemical modification with adhesion RGD ligands	Poor cell–material interaction due to inherent lack of cell adhesivity and low protein adsorption	Hydrogel and micro-/nanosphere
Cellulose	Wood, plants, tunicates, and algae	Biodegradable, biocompatible, and high mechanical performance	Low cell-binding properties	Reinforcement of composite scaffolds
Chitosan	Exoskeleton of crustaceans and mollusks, insect cuticles, and fungi	Bioactive, biocompatible, biodegradable, antibacterial and nonimmunogenic properties, and ability for cell ingrowth	Relatively weak mechanical strength and stability	Sponges, composite scaffolds, hydrogels, and micro-/nanosphere
Starch	Corn, potato, wheat, and tapioca	Biocompatible, biodegradable, and low cost	Lack of processability and low surface area	Composite scaffolds

Source: Reprinted with permission from Rodríguez [76]; published by Elsevier.

tissue engineering is represented by the lack of sites for protein adsorption and cell attachment [74, 75].

An exhaustive list of natural polymers used for bone repair is presented in Table 8.1 [76].

Despite the benefit of natural polymers, they do have significant disadvantages such as their poor mechanical strength, their high speed of degradation, and their limited supply [77]. However, the main disadvantage is the lack of reproducibility [78]. Synthetic polymers are an alternative as their reproducibility, tailorable degradation rate, and mechanical strength can overcome the limitations of natural polymers.

8.4.1.2 Synthetic Polymers

Bone tissue engineering necessitates bioresorbable materials with degradation rate tailored to the repair strategy and tissue regeneration. Therefore, biodegradable synthetic polymers are favored. They should also present adapted mechanical properties, good stability, and promote cell adhesion, growth, migration, and differentiation. Polymer materials can have different characteristics such as molecular weight, polydispersity, and crystallinity. All those characteristics will have a significant impact on the degradation rate, which would strongly affect the polymer scaffold properties [62, 79–82]. They can be processed by different fabrication techniques such as extrusion, injection, or compression molding. Their properties and degradation can be affected by the type of processing technique, as high processing temeprature and mechanical stresses can lead to polymer degradation or polymer chain scission.

The most used polymers for bone tissue engineering are poly(glycolic) acid (PGA), poly(lactic) acid (PLA), and poly(lactic-*co*-glycolic) acid (PLGA) (Figure 8.3).

PGA is an aliphatic polyester produced by chain polymerization. PGA has a degradation time of 6–12 months *in vivo*; this is why it is used in bone tissue engineering and medical applications [80, 84].

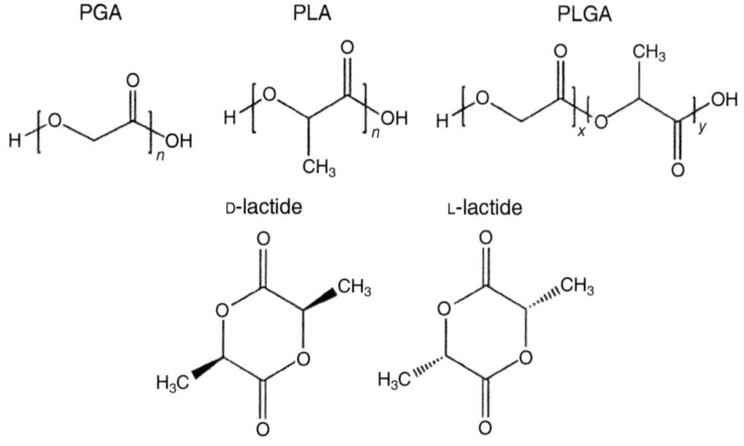

Figure 8.3 Chemical structure of PGA, PLA, PLGA, and the enantiomers D- and L-lactide. Source: Reprinted with permission from Miranda et al. [83]; © Creative Commons CC BY License, published by MDPI Antibiotics.

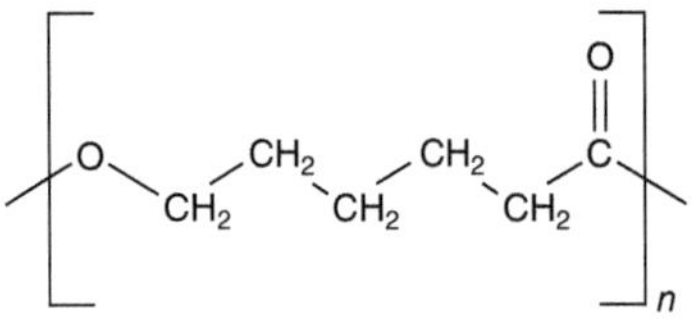

Figure 8.4 Structure of poly(ε-caprolactone) (PCL). Source: Reprinted with permission from McKeen [89]; published by Elsevier.

PLA is synthesized by the cyclic dimer of lactic acid, which exists as two optical isomers: D- and L-lactide are the naturally occurring isomer and DL-lactide is the synthetic blend of D-lactide and L-lactide. The molar mass of the polymer as well as the degree of crystallinity has a significant influence on the mechanical properties. This polymer is commonly used in biomedical devices because of its excellent biocompatibility with the human body, its biodegradability, thermal plasticity, and suitable mechanical properties. However, PLA has some drawbacks. For instance, it has an insufficient degradation rate and an inability to fully integrate with the bone [80, 85–88].

PLGA is a copolymer synthesized from the two homopolymers PLA and PGA. Synthesizing a copolymer allows us to combine the properties of the homopolymer used. For example, the faster biodegradation of the PGA allows us to optimize the degradation of PLA. Their biodegradation is optimized by controlling the amounts of both polymers [83, 85].

Poly(ε-caprolactone) (PCL) is also a polymer used for bone tissue engineering. PCL (Figure 8.4) is synthesized by the ring-opening polymerization of cyclic monomer ε-caprolactone in the presence of stannous octoate, serving as a catalyst. Its degradation can take between two and three years *in vivo* [62, 89, 90].

These different polymers, natural or synthetic, can be used in scaffolds comprising a mineral matrix as BAG for bone repair. The interaction between BAGs and the organic matrix can be different, leading to composites or hybrid devices.

8.4.2 Composite Materials

For bone repair, the implant should have the same properties as the tissue. Much current research is focused on the fabrication of bioactive composite materials, with the bioactive phase incorporated as a filler into the bioresorbable polymer matrix. In particular, composites based on biodegradable polymers are being increasingly studied because this combination does not require a revision surgery for their removal, as the newly formed bone gradually substitutes the implanted scaffold during degradation [91–95].

The use of a bioactive inorganic calcium phosphate filler in composite materials for bone biomaterials was pioneered by Bonfield et al. in the early 1980s with the production of particulate HA/polyethylene composites [96]. To date, the two inorganic matrices mostly used are the hydroxyapatite and the tricalcium phosphate because of their biocompatibility, their osteoconduction, and osteointegration [97, 98], but a new category of ceramics, BAGs, is increasingly studied for composite materials.

8.4.2.1 Bioactive Glasses Used in Composites

The BAG is a mineral material developed for bone repair. Few compositions are commercialized as 45S5 (Bioglass®) or S53P4 (BonAlive) [99]. As discussed earlier, these BAGs have shown their ability of osteointegration and bone repair. However, BAGs are too stiff, brittle, and fragile when used alone [100]. Thus, they are combined to organic matrices, which can contain them and allow them to keep their bioactivity. Polymer/ceramic composite scaffolds represent a convenient alternative because of the possibility to tailor their properties (e.g. mechanical and structural behavior, degradation kinetics, and bioactivity).

Different types of BAGs can be used in composites for bone tissue engineering, such as silicate, borosilicate, or phosphate glasses. The composition of the silicate and borosilicate glasses varies in function of SiO_2 and B_2O_3, respectively, while for the phosphate glasses, it depends on the P_2O_5 oxide concentration.

The BAGs most studied are the silicate glasses. The first one developed and commercialized for bone applications is 45S5. This is the standard to which are compared all the silicate glasses. Moreover, many glasses are developed based on the composition of 45S5, such as S53P4 or 13-93, which are also FDA approved (Table 8.2).

45S5 and S53P4 are two silicate-commercialized BAGs as they present the best bioactivity properties. However, when included in polymer-based composites obtained for instance by extrusion, their high reactivity can induce the degradation of the polymer while processing, as was shown by Vergnol et al. with PLA/45S5 BAG composites [101]. This polymer degradation in the composite can lead to less-suitable mechanical properties for bone repair. The degradation reaction is not well known; thus, this is an issue which researchers can focus on. Nevertheless, this phenomenon can be limited when using more stable glasses such as 13-93. 13-93 has a much lower degradation rate than traditional BAGs and, thus, limits the degradation of the organic matrix. Houaoui et al. as well as Lyyra et al. developed composites based on PLA and 13-93, and the results showed that the polymer is still degraded with this BAG but much less than when compared to the one processed with 45S5 or S53P4 [95, 102]. Indeed, the combination of PLA with 13-93 processed by twin-screw extrusion reduces the decrease in the molecular weight of the polymer during melt processing [95]. However, one drawback of this BAG is the dissolution rate, which can be too low compared to 45S5 or S53P4. One possibility to remediate the slow degradation rate would be to replace some silicate by borate, as the boron would increase the dissolution rate. This leads to the development of borosilicate glasses.

Table 8.2 Nominal glass composition (weight%).

Glass	SiO_2	Na_2O	CaO	P_2O_5	K_2O	MgO
45S5	45	24.5	24.5	6	—	—
S53P4	53	23	20	4	—	—
13-93	53	6	20	4	12	5

Borosilicate glasses are compositions including SiO_2 but also B_2O_3 in order to tailor the properties of silicate glass. Indeed, the addition of boron in its composition allows us to increase its dissolution rate and favor the conversion to apatite. The silicate BAG most used for that is 13-93. Indeed, in its composition, different amounts of SiO_2 can be replaced by B_2O_3, giving a higher reactivity of the glass with an increasing content of B_2O_3 [103]. Houaoui et al. demonstrated that 20% of silica replaced by boron oxide in 13-93 allows a faster dissolution of the glass in solution, a favored precipitation of calcium phosphate in simulated body fluid and commitment to osteoblastic lineage of C2C12 myoblastic cells. The borosilicate allowed a higher reactivity while keeping the same degradation of the polymer as when it is included with 13-93 [95]. However, while borosilicate presents an important reactivity and bioactivity, the content of boron has an influence on cell behavior. Indeed, a high amount of boron in the glass can be toxic for the cells [104, 105].

Phosphate glasses, based on the P_2O_5 glass-forming network, have been developed for biomedical use. These glasses are interesting for bone and musculoskeletal tissue engineering as resorbable materials as they present a faster degradation rate compared to silicate glasses and have the ability to degrade completely in aqueous solutions [103, 106]. This makes their great interest. Knowles et al. studied degradable composites based on polyhydroxy butyrate and phosphate glass particles, which showed that the glass releases Ca^{2+} and P^{5+} leading to enhanced bone formation. This release helps the osteoblasts to form the new hard tissue [107].

The interest of the BAG lies in its dissolution and bioactivity. Its bioactivity would bring the signals to the cells to differentiate into bone-forming cells; thus, its influence on the cells is also investigated.

8.4.2.2 Bioactive Glass Influence on Cell Behavior *In Vitro*

Every material aimed to be used for bone repair must undergo different tests, starting by *in vitro* experiments, before *in vivo* testing. The biocompatibility of a composite must be investigated. Biocompatibility refers to the ability of a material to have a good integration into the host without toxicity but also the ability for it to interact with the environment and support cellular activity.

To investigate the biocompatibility *in vitro*, cell viability, proliferation, and morphology is studied. To study the cytocompatibility of BAG/polymer composites, different types of cells can be used as mesenchymal stem cells (MSCs, from human, rats or mice), human osteosarcoma cell lines, mouse pre-osteoblastic cell line (MC3T3-E1), mouse myoblastic cells (C2C12), or human primary osteoblasts [108].

Disks based on P(L,DL)LA with 70% of 13-93 BAG (weight%), obtained by compression molding of composite rods, showed the ability to support the growth of C2C12 myoblastic cells. However, these same composites based on P(L,DL)LA with a borosilicate glass, the 13-93B20 (same glass as 13-93 where 20% of silicon dioxide was replaced by boric oxide) showed a lower proliferation but a differentiation into an osteoblastic lineage more accentuated. Indeed, boron is known to decrease the proliferation while promoting osteogenesis [47, 95].

The cytocompatibility of P(L,DL)LA/45S5 composites (20%, 30%, or 50% of BAG, weight%) studied with MG63 osteoblastic cells (isolated from human osteosarcoma) until 10 days of incubation showed a good cell viability on the composites [101]. Porous scaffolds based on P(DL)LA polymer and 0%, 5%, or 40% of 45S5 (weight%) were obtained by freeze-drying, and their biocompatibility was studied with human adipose-derived stem cells [109]. The composites with these three contents of BAG supported cell attachment and growth. The ALP activity was analyzed, showing an increased activity with a higher content of 45S5.

Yao et al. and Lu et al. studied composites based on PLGA/45S5 with different contents of BAG. Lu et al. showed that the behavior of SaOS-2 cells is glass content dependent. Indeed, 10% and 50% of glass content were studied, and with 10%, the proliferation was more important than with higher content [110]. These results are also demonstrated by Yao et al. using MSC [111]. Lu et al. also showed that the ALP activity was more important with 10% of glass content. These studies show that there is probably a threshold of glass content in composites allowing cell survival and proliferation.

Porous scaffolds based on PLGA microspheres containing 30 wt% 45S5 glass particles showed the ability to support the proliferation of rat bone marrow-derived MSCs and a higher ability to favor differentiation of MSCs into an osteoblastic lineage when compared to the PLGA scaffold alone [111]. Also studied by Lu et al., porous composites of PLGA/45S5 microspheres (25% wt% of glass) present a better support of collagen type I synthesis by SaOS-2, when compared to the scaffolds without glass or tissue control polystyrene [112].

Some studies were also done with composites based on PCL and BAG [108, 113, 114]. In Jo et al. study, proliferation of MC3T3-E1 cells was higher on the composites compared to the polymer alone. The same trend was observed when the composite materials were implanted in a rat calvarial defect [113]. Erdemli et al. worked on the same composite with a third component, bone demineralized matrix (BDM), or calcium sulfate (CS). SaOS-2 cell had a better viability on PCL/glass or PCL/glass/DBM (demineralized bone matrix) composites compared to PCL/glass/CS of PCL alone. Cell spreading was also better on PCL/glass or PCL/glass/DBM [114].

These different studies show that the biocompatibility of a material containing BAG depends on the glass composition and dissolution rate but also the glass loading in the final product.

It was shown that the composition of the glass can have an effect on the cellular behavior. For example, in some studies, it was noted that a borosilicate glass could slow down the cell proliferation when compared to a silicate glass [95]. The reason for such behavior is not well understood, but the faster dissolution of the borosilicate glass could be one reason. Therefore, the content of boric oxide in the composition needs to be controlled as when in a high amount, it could speed up the dissolution rate, which in turn can reveal to be toxic for the cells.

Another parameter to consider is the glass loading in the material. As written above, in many studies, it was shown that a high content of glass can have a negative effect on the cellular proliferation. Indeed, a high content of glass means an important ion release, which can disturb the cellular proliferation. Moreover, the size of the glass particles also needs to be controlled, as small particles present a high reactive surface, thus a high dissolution rate which can also negatively impact the cells behavior.

To summarize, for the development of a composite based on BAG, the type of glass must be chosen carefully in order to have a dissolution adapted to the result aimed. Its particle size, but also its content in the final material, has to be controlled to promote the best cellular response *in vitro* before *in vivo* experimentations.

8.4.2.3 Composites' Fabrication Methods

Different forms of composites can be obtained, and their shapes depend on the processing methods. Different techniques can be used to form composite materials based on polymer with BAG dispersed in the organic phase. They usually follow the methods used to shape the polymer, with the addition of the glass powder in the process. Processing these types of materials require high temperature, usually higher than the melting temperature of the polymer, as for example, the extrusion process or compression molding [95, 115]. These methods are used only for synthetic polymers, as natural polymers necessitate softer techniques to prevent degradation.

Ideally, scaffolds for bone repair need to be porous in order to allow cell colonization [116–118]. The pores should be interconnected for cellular and vascular growth, transport of nutrients, and metabolic wastes. The optimal size for bone tissue growth is between 250 and 350 μm. To obtain this porosity, many methods can be used as the solvent casting and particulate leaching, gas foaming, freeze-drying, electrospinning, or 3D printing [119–127].

Each technique has some specificity as its ease or cost, the shape and size of the obtained materials, its percentage of porosity, and pore size.

Solvent casting and particulate leaching require the use of a mixture of BAG powder in a polymer solution uniformly mixed with salt particles of a specific diameter, which will leave a porous structure after solvent evaporation and immersion in water. However, the pore size and shape are limited by the size of the salt particles [119].

Gas foaming is the technique where molded composites made of a biodegradable polymer and BAG are placed under high pressures with gas-foaming agents, such as carbon dioxide or nitrogen, until the polymer is saturated, resulting in nucleation and growth of gas bubbles with sizes ranging between 100 and 500 μm in the composites, creating the porosity [120–122].

Freeze-drying allows us to prepare porous composite scaffolds. In the freezing stage, the polymer solution from the composite is cooled down to a certain temperature at which all materials are in the frozen state and the solvent forms ice crystals, forcing the polymer molecules to aggregate into the interstitial spaces. In the second phase, the solvent is removed by applying a pressure lower than the equilibrium vapor pressure of the frozen solvent. When the solvent is completely sublimated, a dry polymer scaffold with an interconnected porous microstructure remains. The porosity of the scaffolds depends on the concentration of the polymer solution; pore size distribution is affected by the freezing temperatures [123, 124].

The electrospinning technique uses electrical charges to draw fine fibers up to the nanometer scale. It has been widely employed to fabricate porous scaffolds with nanofibrous architectures that can mimic the structure and biological functions of the natural extracellular matrix [26]. This technique is able to generate fibers with diameters ranging from 2 nm to several micrometers using solutions of both natural and synthetic polymers, with small pore sizes and high surface area-to-volume ratios [127, 128].

3D printing is a method that allows us to control the shape and the architecture of the scaffold, its porosity, and pore size. Different techniques of 3D printing exist as low-temperature deposition manufacturing (LDM), fused deposition modeling (FDM), stereolithography (SLA), selective laser sintering (SLS) [125, 127, 129]. FDM uses the extrusion of composite filaments based on thermoplastic polymer and deposes the semi-molten composites on a heated platform by layer-by-layer. FDM allows us to obtain scaffolds with interesting mechanical properties and structural integrity thanks to a fusion between layers [127, 129]. LDM requires materials prepared in a solution, and the scaffold is obtained thanks to the layer-by-layer technique on a platform maintained below 0 °C [129]. SLS is based on the sintering of a composite powder by a CO_2 laser beam, while heating it above the transition temperature. The process is repeated layer-by-layer until the entire 3D model is obtained. The model should be heated at the end to obtain the final density. This technique is used to obtain scaffolds with irregular shapes [129]. SLA is based on the photopolymerization of a resin using a UV laser. The model is emerged in the resin chamber, and the process is repeated layer-by-layer until the entire construct is produced [126, 127].

Effort is devoted to the development of porous, high-strength composite structures for the regeneration of human bone at load-bearing sites. Composites allow the tailoring of the mechanical properties for the specific application to closely match the stiffness of the natural tissue. At the same time, this alternative enables the BAG to be shaped and helps to mimic the bone thanks to its osteoproperties. Its bioactivity will allow the osteointegration of the implant in the defect, which will help bone growth and therefore bone repair. The final properties of the implant depend on the characteristics of the organic phase (such as its chemical composition, molecular weight, and crystallinity) but also on the properties of the glass used (such as their volume fraction, their size, their dispersion, and their reactivity). Finally, the synthesis conditions, as the high temperatures, could influence the final properties of the implant. Indeed, it is possible for them to influence the integrity of the compounds, which can affect the properties of the composite. Moreover, one drawback with many composites is that the inorganic phase is dispersed in the organic phase, and they interact on a micrometer scale, which can result in a gradient of resorption rates during dissolution. This inevitably leads to material instability *in vivo* [130]. Plus, it can be difficult to control properties such as the degradation of the material and surface chemistry characteristics because of the different resorption rate of the components and the masking of the bioactive component. To overcome those drawbacks, the alternative proposed is to develop hybrid materials [38, 130].

8.4.3 Hybrid Materials

Hybrid inorganic–organic materials are made, as composites, of an organic and an inorganic phase, which are mixed homogeneously or heterogeneously. However, unlike the composites, these two phases are bonded together [131]. Hybrid materials are increasingly developed as they require soft chemistry processes that allow us to preserve the organic phase and keep its mechanical properties adapted to bone repair [132]. They are commonly formed by the sol–gel method, which provides low processing temperatures. During this process, the constituent phases interact at a molecular level. That allows a better control of material parameters such as degradation rates or mechanical properties [133]. Hybrids obtained can have different forms such as gels, thin films, and fibers.

Different classes of hybrids exist depending on the links between the organic and inorganic phases. Class I hybrids present no covalent links between organic and inorganic components but rather the bonding occurs through van der Waals or hydrogen bonding or by electrostatic forces. Class II hybrid is the category in which the organic and inorganic phases of the material are linked by strong covalent links [38, 131]. The class II hybrids are the most studied for bone repair, as their covalent links cannot be broken at biological temperature, for example. This is the case of the gelatin, which cannot be dissolved at biological temperatures when cross-linked.

Hybrid scaffolds for bone tissue engineering were first studied by Julian R. Jones who investigated gelatin/BAG hybrids. The BAG was obtained by hydrolysis and condensation, under acidic conditions, of a silica precursor, i.etetraethyl orthosilicate (TEOS). [38, 130, 134]. The processed hybrid was composed of a single phase comprising the silica network and the gelatin.

Another type of hybrids exists, where two distinct phases, covently linked, coexist and is called hybrid nanocomposites (Figure 8.5) [131].

Houaoui et al. developed a new generation of hybrid nanocomposites based on gelatin and BAG particles [135]. Working with BAG particles is of great interest as it would allow us to add in its composition all the elements interesting for a specific therapeutic action as Si, known to promote osteogenesis, Ag for an antimicrobial activity, Li or B for angiogenesis, a property of paramount importance for successful outcome of a bone implant (Figure 8.6) [136–144]. Houaoui et al. added boron

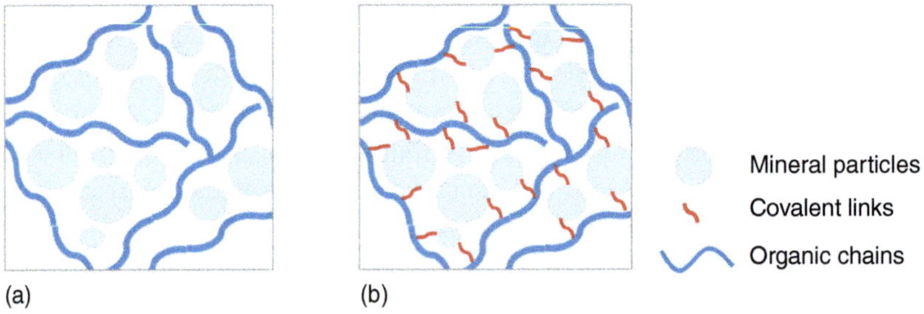

Figure 8.5 Schematic representation of a composite material (a) and a hybrid nanocomposite material (b). Source: Houaoui [64].

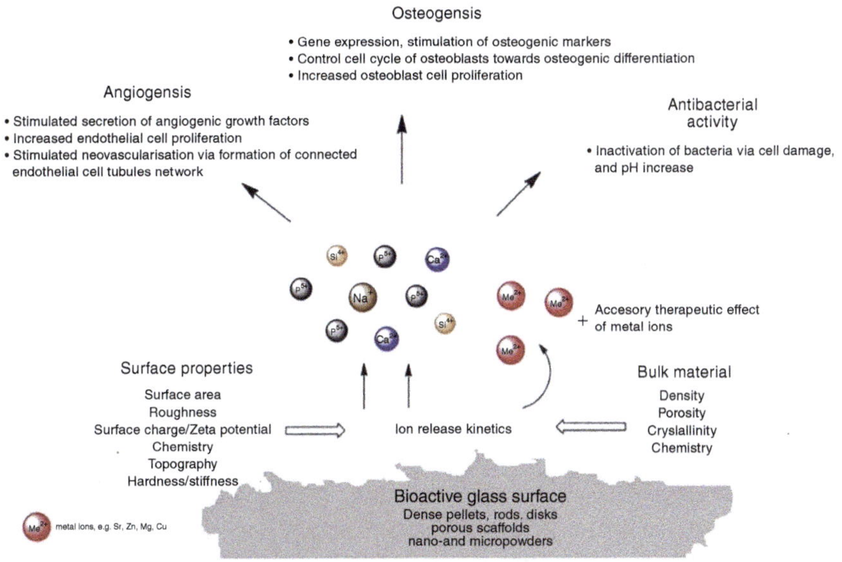

Figure 8.6 Overview of biological responses to ionic dissolution products of bioactive glasses. Source: Reprinted with permission from Hoppe et al. [136]; published by Elsevier.

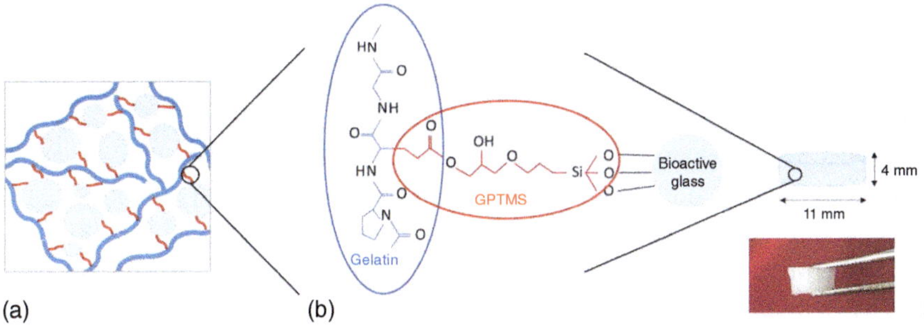

Figure 8.7 (a) Schematic representation (inspired from Mahony et al. [18]) and (b) chemical structure of the hybrid made of gelatin and bioactive glass (BAG) covalently linked by 3-glycidoxypropyltrimethoxysilane (GPTMS). Source: Reprinted with permission from Houaoui et al. [135] / MDPI / CC BY 4.0.

in the 13-93 glass in order to increase the dissolution rate and promote bioactivity [135].

Hybrid scaffolds are materials where the organic and inorganic phases interact chemically at the nanoscale level [130]. The covalent links between the organic and inorganic phases can be obtained by bifunctional and hybrid molecules, which can link inorganic and organic elements. Mahony et al. and Houaoui et al. used, as coupling agent between the gelatin and the BAG, the 3-glycidoxypropyltrimethoxysilane (GPTMS) (Figure 8.7).

GPTMS is a coupling agent that has been shown to be successful for the cross-linking of gelatin with inorganic phases [130, 134]. Moreover, in addition to the

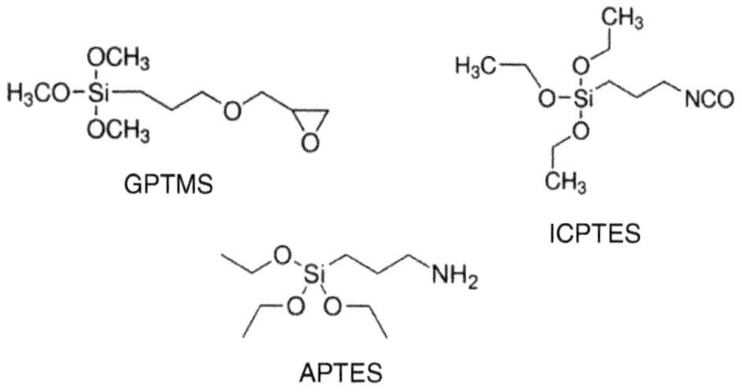

Figure 8.8 Structure of different sol–gel precursors used in hybrid materials. Source: Refs. [145–149].

GPTMS, other coupling agents can be used, such as 3-aminopropyltriethoxysilane (APTES) or 3-isocyanatopropyltriethoxysilane (ICPTES) (Figure 8.8) [145–149].

The use of these coupling agents allows to link organic and inorganic phases in hybrid materials. However, the content of coupling agent needs to be optimized in order to have enough links between both organic and inorganic phases, adapted mechanical properties, while avoiding toxicity from potential unreacted coupling agents. Indeed, if used in a high amount, these molecules can be toxic for the cells, which limit their use. Mahony et al. studied different *C*-factor (molar ratio of GPTMS/gelatin) and found that with higher *C*-factor (*C*-factor of 1500), the stiffness of the materials was higher, which promotes MSC adhesion and differentiation [130]. However, Houaoui et al. chose a *C*-factor of 500 for their developed hybrids after optimization as a higher *C*-factor was toxic for the culture of MC3T3-E1 cells [135]. Thus, the content of coupling agent must be controlled in order to find the best properties of the materials for bone tissue engineering while promoting cell viability, growth, and differentiation. Hybrid approaches are suitable for tissue engineering only if the employed coupling reagents can be completely removed before cell addition or implantation or is not toxic once released in medium [130, 150]. In addition, most of the covalently cross-linked hydrogels do not allow direct incorporation of cells inside the hydrogel, making necessary to seed the cells on its surface and, if there is a suitable open macroporosity, let them migrate inside the scaffold.

The development of hybrid materials is interesting, allowing to shape the materials according to the shape and size of the defect, allowing its use for bone filling with tunable properties by controlling the coupling agent amount, BAG content, as well as the BAG composition. Many parameters still need to be optimized and controlled in order to avoid cell toxicity, while promoting bioactivity and cell differentiation into the aimed lineage. For that, no hybrid material is commercialized nowadays, but they remain promising for this application, even if some issues still need to be solved.

The comparison of composite and hybrid materials is shown in Table 8.3.

Table 8.3 Comparison of composite and hybrid materials developed for bone tissue engineering.

	Composites	Hybrids
Requirements	• Custom-made materials adapted to the shape and size of the defects • Porous materials • Mechanical properties adapted to the targeted application • Biocompatible, biodegradable, and osteostimulative	
Difference in the structure	Inorganic matrix *dispersed* in the organic matrix	Inorganic matrix *covalently linked* to the organic matrix thanks to a coupling agent
Coupling agent	None	Bifunctional agents as the GPTMS, ICPTES, and APTES
Synthesis methods	Methods requiring high temperatures (extrusion, 3D-printing, compression molding, etc.), or solvents	Soft chemistry processes
Advantages	• Mechanical properties adapted to the aimed application	• Stability of the material • Control on the organic and inorganic phases dissolution thanks to the coupling agent • Mechanical properties adapted to the aimed application
Drawbacks	• Synthesis technique can have an influence on the integrity of the organic phase • Different degradation rate between the organic and inorganic phase	• Coupling agent in a high amount can be toxic
Applications	• Bone mechanical support (Interference screw) • Bone filling (dental putty, etc.)	Not used to date but application aimed: bone filling

8.4.4 Composites in Clinics

Bone tissue engineering is an area in which many researchers are focused. Some materials based on BAG promoting bone regeneration are already developed for clinical use. They can be pre-formed or injected into the defect. Many materials are used nowadays as metals, synthetic or natural polymers, and ceramics. Researchers found that the best option was to mimic the bone, which is composed of an organic and an inorganic phase. Thus, composites are widely developed and used in clinics, in order to combine the properties of both phases, as in the bone. These composites are used in dental and orthopedic regenerative medicine.

Injectable putty is available based on the polymer and BAG, Bioglass or BonAlive for dental surgery [151, 152]. Commercialized by NovaBone (NovaBone Products LLC, Jacksonville, FL) or Noraker (Noraker, Lyon, France), they are used in dental surgery. Wang et al. showed that a bone defect in the ovine model filled with the NovaBone putty was repaired at 42% with new bone, compared to 24% with NovaBone particulate and 1.2% for the defects that were not filled with a substitute [153]. NovaBone putty is a composite made of 45S5 glass of 32–710 μm, polyethylene glycol, and glycerin. Other putties are also used based on 45S5 but with different particle size. Indeed, Synergy Biomedical (Biostructures LLC, Newport Beach, CA) presents a putty with BAG spheres of 400 μm as they would leave interstices between each other for bone ingrowth [151].

Moldable composites based on collagen and Bioglass were also developed by NovaBone, NovaBone Macroform®. These substitutes would be used to fill bone defects and thanks to their malleability, they could be adapted to the size and shape of the defect. The use of collagen as the organic matrix allows us to have a material that mimics the bone, as having the same organic matrix. The collagen would bring sites for cell adhesion and protein binding, thus helping in cells recruitment for the repair of the defect. This collagen/BAG porous scaffold will facilitate bone growth as a consequence of its porosity but also its ease of vascularization, which is important for an efficient repair.

Noraker developed composites for bone mechanical support as LockActiv®. LockActiv is a bioactive and bioresorbable interference screw based on P(L,DL)LA and 15% 45S5 BAG (weight%). LockActiv is a tibial and femoral fixation of the tendon graft or autogenous tendon used in the reconstruction of the knee's ligaments. After the surgeon has created a hole into the patient bone, the screw will fix the detached ligament to the hole. This composite is interesting as it is fully bioresorbable and will allow the regeneration of the bone tunnel in tibia and femur. By combining the polymer and the glass, it would have the mechanical properties and the bioactivity adapted to osteointegrate and repair the defect.

8.5 Conclusion

BAGs are a unique type of biomaterials. Their versatility allows to produce biomaterials ranging from nearly inert to bioactive and/or able to release ions of therapeutic interest. Their surface chemistry can be tailored to improve/control protein adsorption. They can be produced as granules, blocks, fibers, or scaffolds. However, some limitations exist, which can be overcome by designing composites or hybrid biomaterials. Bioactive composites still need to be improved to be suitable to each type, size, and shape of bone defects. However, their mechanical, bioactive, biocompatible, and osteostimulative properties cannot be neglected even if their development remains a challenge. These innovative scaffolds still need improvement to be used in priority compared to autologous grafts, but their development is showing a promising avenue.

Hybrid materials are materials that show molecular interactions between the organic and inorganic matrices. Such interactions allow us to have a better control on the dissolution properties of the scaffolds and their mechanical properties. However, some investigations on these materials and their coupling agents are still needed in order to obtain scaffolds cyto- and biocompatible in the long term and thus allow their use in clinical applications.

Acknowledgments

Dr. Amel Houaoui and Prof. Jonathan Massera contributed equally to this work. This work was supported in part by the Academy of Finland and the Jane and Aatos Erkko foundation. AH would also like to thank CY Initiative of Excellence (CY-Cergy Paris Université, France) for supporting her PhD project which allowed to develop devices presented in this book chapter.

References

1 Hench, L.L. (2006). The story of Bioglass®. *J. Mater. Sci. - Mater. Med.* 17 (11): 967–978.
2 Hench, L.L. (1991). Bioceramics: from concept to clinic. *J. Am. Ceram. Soc.* 74 (7): 1487–1510.
3 Hench, L.L. and Paschall, H.A. (1973). Direct chemical bond of bioactive glass-ceramic materials to bone and muscle. *J. Biomed. Mater. Res.* 7 (3): 25–42.
4 Sollazzo, V., Palmieri, A., Scapoli, L. et al. (2010). PerioGlas® acts on human stem cells isolated from peripheral blood. *Dent. Res. J.* 7 (1): 28–34.
5 Arstila, H. (2008). Crystallization characteristics of bioactive glasses. PhD thesis submitted at Åbo Akademi, https://www.doria.fi/handle/10024/44023.
6 van Gestel, N.A.P., Geurts, J., Hulsen, D.J.W. et al. (2015). Clinical applications of S53P4 bioactive glass in bone healing and osteomyelitic treatment: a literature review. *Biomed Res. Int.* 2015: 1–12.
7 Syvänen, J., Nietosvaara, Y., Kohonen, I. et al. (2018). Treatment of aneurysmal bone cysts with bioactive glass in children. *Scand. J. Surg.* 107 (1): 76–81.
8 Vos, J., de Vey Mestdagh, P., Colnot, D. et al. (2017). Bioactive glass obliteration of the mastoid significantly improves surgical outcome in non-cholesteatomatous chronic otitis media patients. *Eur. Arch. Otorhinolaryngol.* 274 (12): 4121–4126.
9 Stoor, P. and Apajalahti, S. (2017). Osteotomy site grafting in bilateral sagittal split surgery with bioactive glass S53P4 for skeletal stability. *J. Craniofac. Surg.* 28 (7): 1709–1716.
10 Malat, T.A., Glombitza, M., Dahmen, J. et al. (2018). The use of bioactive glass S53P4 as bone graft substitute in the treatment of chronic osteomyelitis and infected non-unions – a retrospective study of 50 patients. *Z. Orthop. Unfall.* 156 (2): 152–159.
11 Carlisle, E.M. (1981). Silicon: a requirement in bone formation independent of vitamin D1. *Calcif. Tissue Int.* 33 (1): 27–34.
12 Damen, J.J.M. and Ten Cate, J.M. (1992). Silica-induced precipitation of calcium phosphate in the presence of inhibitors of hydroxyapatite formation. *J. Dent. Res.* 71 (3): 453–457.
13 Carlisle, E.M. (1970). Silicon: a possible factor in bone calcification. *Science* 167 (3916): 279–280.

14 Reffitt, D.M., Ogston, N., Jugdaohsingh, R. et al. (2003). Orthosilicic acid stimulates collagen type 1 synthesis and osteoblastic differentiation in human osteoblast-like cells in vitro. *Bone* 32 (2): 127–135.

15 Maeno, S., Niki, Y., Matsumoto, H. et al. (2005). The effect of calcium ion concentration on osteoblast viability, proliferation and differentiation in monolayer and 3D culture. *Biomaterials* 26 (23): 4847–4855.

16 Massera, J., Hupa, L., and Hupa, M. (2012). Influence of the partial substitution of CaO with MgO on the thermal properties and in vitro reactivity of the bioactive glass S53P4. *J. Non-Cryst. Solids* 358 (18, 19): 2701–2707.

17 Watts, S.J., Hill, R.G., O'Donnell, M.D., and Law, R.V. (2010). Influence of magnesia on the structure and properties of bioactive glasses. *J. Non-Cryst. Solids* 356 (9, 10): 517–524.

18 Zreiqat, H., Howlett, C.R., Zannettino, A. et al. (2002). Mechanisms of magnesium-stimulated adhesion of osteoblastic cells to commonly used orthopaedic implants. *J. Biomed. Mater. Res.* 62 (2): 175–184.

19 Yamasaki, Y., Yoshida, Y., Okazaki, M. et al. (2002). Synthesis of functionally graded $MgCO_3$ apatite accelerating osteoblast adhesion. *J. Biomed. Mater. Res.* 62 (1): 99–105.

20 Massera, J. and Hupa, L. (2014). Influence of SrO substitution for CaO on the properties of bioactive glass S53P4. *J. Mater. Sci. - Mater. Med.* 25 (3): 657–668.

21 Massera, J., Kokkari, A., Närhi, T., and Hupa, L. (2015). The influence of SrO and CaO in silicate and phosphate bioactive glasses on human gingival fibroblasts. *J. Mater. Sci. - Mater. Med.* 26 (6): 196–205.

22 Kargozar, S., Montazerian, M., Fiume, E., and Baino, F. (2019). Multiple and promising applications of strontium (Sr)-containing bioactive glasses in bone tissue engineering. *Front. Bioeng. Biotechnol.* 7: 161.

23 Marie, P.J., Ammann, P., Boivin, G., and Rey, C. (2001). Mechanisms of action and therapeutic potential of strontium in bone. *Calcif. Tissue Int.* 69 (3): 121–129.

24 Kwun, I.-S., Cho, Y.-E., Lomeda, R.-A.R. et al. (2010). Zinc deficiency suppresses matrix mineralization and retards osteogenesis transiently with catch-up possibly through Runx 2 modulation. *Bone* 46 (3): 732–741.

25 Grosso, A., Burger, M.G., Lunger, A. et al. (2017). It takes two to tango: coupling of angiogenesis and osteogenesis for bone regeneration. *Front. Bioeng. Biotechnol.* 5: 68.

26 Saghiri, M., Asatourian, A., Garcia-Godoy, F., and Sheibani, N. (2016). The role of angiogenesis in implant dentistry part II: the effect of bone-grafting and barrier membrane materials on angiogenesis. *Med. Oral Patol. Oral Cir. Bucal* 21 (4): e526–e537.

27 Bone & Joint Research Group, Centre for Human Development, Stem Cells and Regeneration, Developmental Origins of Health and Disease, Institute of Developmental Sciences, University of Southampton, Southampton, SO16 6YD, U.K., Kanczler, J., and Oreffo, R. (2008). Osteogenesis and angiogenesis: the potential for engineering bone. *Eur. Cell. Mater.* 15: 100–114.

28 Uysal, T., Ustdal, A., Sonmez, M.F., and Ozturk, F. (2009). Stimulation of bone formation by dietary boron in an orthopedically expanded suture in rabbits. *Angle Orthod.* 79 (5): 984–990.

29 Balasubramanian, P., Hupa, L., Jokic, B. et al. (2017). Angiogenic potential of boron-containing bioactive glasses: in vitro study. *J. Mater. Sci.* 52 (15): 8785–8792.

30 Haro Durand, L.A., Vargas, G.E., Romero, N.M. et al. (2015). Angiogenic effects of ionic dissolution products released from a boron-doped 45S5 bioactive glass. *J. Mater. Chem. B* 3 (6): 1142–1148.

31 Gérard, C., Bordeleau, L.-J., Barralet, J., and Doillon, C.J. (2010). The stimulation of angiogenesis and collagen deposition by copper. *Biomaterials* 31 (5): 824–831.

32 Romero-Sánchez, L.B., Marí-Beffa, M., Carrillo, P. et al. (2018). Copper-containing mesoporous bioactive glass promotes angiogenesis in an in vivo zebrafish model. *Acta Biomater.* 68: 272–285.

33 Wang, H., Zhao, S., Zhou, J. et al. (2014). Evaluation of borate bioactive glass scaffolds as a controlled delivery system for copper ions in stimulating osteogenesis and angiogenesis in bone healing. *J. Mater. Chem. B* 2 (48): 8547–8557.

34 Li, B. and Webster, T.J. (2018). Bacteria antibiotic resistance: new challenges and opportunities for implant-associated orthopedic infections. *J. Orthop. Res.* 36 (1): 22–32.

35 Zhang, D., Munukka, E., Leppäranta, O. et al. (2006). Comparison of antibacterial effect of three bioactive glasses. *Key Eng. Mater.* 309–311: 345–348.

36 Drago, L., Toscano, M., and Bottagisio, M. (2018). Recent evidence on bioactive glass antimicrobial and antibiofilm activity: a mini-review. *Materials* 11 (2): 326.

37 Drago, L., Vecchi, E.D., Bortolin, M. et al. (2015). Antimicrobial activity and resistance selection of different bioglass S53P4 formulations against multidrug resistant strains. *Future Microbiol.* 10 (8): 1293–1299.

38 Jones, J.R. (2013). Review of bioactive glass: from Hench to hybrids. *Acta Biomater.* 9 (1): 4457–4486.

39 Lindfors, N., Romano, C., Scarponi, S. et al. (2017). Bioactive glasses in infection treatment, chapter 14. In: *Bioactive Glasses: Fundamentals, Technology and Applications* (ed. A.R. Boccaccini, D.S. Brauer and L. Hupa), 316–335. The Royal Society of Chemistry.

40 Mishra, A., Petit, L., Pihl, M. et al. (2017). Thermal, structural and in vitro dissolution of antimicrobial copper-doped and slow resorbable iron-doped phosphate glasses. *J. Mater. Sci.* 52 (15): 8957–8972.

41 Badave, G.K. (2015). Biofilm producing multidrug resistant *Acinetobacter baumannii*: an emerging challenge. *J. Clin. Diagn. Res.* 9 (1): DC08–DC10.

42 Simchi, A., Tamjid, E., Pishbin, F., and Boccaccini, A.R. (2011). Recent progress in inorganic and composite coatings with bactericidal capability for orthopaedic applications. *Nanomed. Nanotechnol. Biol. Med.* 7 (1): 22–39.

43 Rivadeneira, J. and Gorustovich, A. (2017). Bioactive glasses as delivery systems for antimicrobial agents. *J. Appl. Microbiol.* 122 (6): 1424–1437.

44 Jaiswal, S., McHale, P., and Duffy, B. (2012). Preparation and rapid analysis of antibacterial silver, copper and zinc doped sol–gel surfaces. *Colloids Surf., B* 94: 170–176.

45 Massera, J., Fagerlund, S., Hupa, L., and Hupa, M. (2012). Crystallization mechanism of the bioactive glasses, 45S5 and S53P4. *J. Am. Ceram. Soc.* 95 (2): 607–613.

46 Filho, O.P., Latorre, G., and Hench, L. (1996). Effect of crystallization on apatite-layer formation of bioactive glass 45S5. *J. Biomed. Mater. Res.* 30 (4): 509–514.

47 Ojansivu, M., Mishra, A., Vanhatupa, S. et al. (2018). The effect of S53P4-based borosilicate glasses and glass dissolution products on the osteogenic commitment of human adipose stem cells. *PLoS One* 13 (8): 1–20.

48 Balasubramanian, P., Büttner, T., Miguez Pacheco, V., and Boccaccini, A.R. (2018). Boron-containing bioactive glasses in bone and soft tissue engineering. *J. Eur. Ceram. Soc.* 38 (3): 855–869.

49 Erasmus, E.P., Sule, R., Johnson, O.T. et al. (2018). In vitro evaluation of porous borosilicate, borophosphate and phosphate bioactive glasses scaffolds fabricated using foaming agent for bone regeneration. *Sci. Rep.* 8 (1): 3699.

50 Fabert, M., Ojha, N., Erasmus, E. et al. (2017). Crystallization and sintering of borosilicate bioactive glasses for application in tissue engineering. *J. Mater. Chem. B* 5 (23): 4514–4525.

51 Al-Hajjar, M., Fisher, J., Williams, S. et al. (2013). Effect of femoral head size on the wear of metal on metal bearings in total hip replacements under adverse edge-loading conditions. *J. Biomed. Mater. Res. Part B* 101B (2): 213–222.

52 Bunker, B.C., Arnold, G.W., and Wilder, J.A. (1984). Phosphate glass dissolution in aqueous solutions. *J. Non-Cryst. Solids* 64 (3): 291–316.

53 Massera, J., Petit, L., Cardinal, T. et al. (2013). Thermal properties and surface reactivity in simulated body fluid of new strontium ion-containing phosphate glasses. *J. Mater. Sci. - Mater. Med.* 24 (6): 1407–1416.

54 Kasuga, T., Hattori, T., and Niinomi, M. (2012). Phosphate glasses and glass-ceramics for biomedical applications. *Phosphorus Res. Bull.* 26: 8–15.

55 Nommeots-Nomm, A., Houaoui, A., Pradeepan Packiyanathar, A. et al. (2020). Phosphate/oxyfluorophosphate glass crystallization and its impact on dissolution and cytotoxicity. *Mater. Sci. Eng., C* 117: 111269.

56 Mishra, A., Ojansivu, M., Autio, R. et al. (2019). In-vitro dissolution characteristics and human adipose stem cell response to novel borophosphate glasses. *J. Biomed. Mater. Res. Part A* 107 (9): 2099–2114.

57 Kiani, A., Lakhkar, N.J., Salih, V. et al. (2012). Titanium-containing bioactive phosphate glasses. *Philos. Trans. R. Soc. Math. Phys. Eng. Sci.* 370 (1963): 1352–1375.

58 Islam, M.T., Felfel, R.M., Abou Neel, E.A. et al. (2017). Bioactive calcium phosphate-based glasses and ceramics and their biomedical applications: a review. *J. Tissue Eng.* 8: 204173141771917.

59 Hyunh, N.B., Palma, C.S.D., Rahikainen, R. et al. (2021). Surface modification of bioresorbable phosphate glasses for controlled protein adsorption. *ACS Biomater. Sci. Eng.* 7 (9): 4483–4493.

60 Azizi, L., Turkki, P., Huynh, N. et al. (2021). Surface modification of bioactive glass promotes cell attachment and spreading. *ACS Omega* 6 (35): 22635–22642.

61 Zervas, G., Telfer, S.B., Carlos, G., and Anderson, P. (1988). The effect of soluble-glass boluses containing copper, cobalt and selenium on the blood composition of ewes. *Anim. Feed Sci. Technol.* 21 (1): 23–29.

62 Song, R., Murphy, M., Li, C. et al. (2018). Current development of biodegradable polymeric materials for biomedical applications. *Drug Des. Dev. Ther.* 12: 3117–3145.

63 Zarif, M.-E. (2018). A review of chitosan-, alginate-, and gelatin-based biocomposites for bone tissue engineering. *Biomater. Tissue Eng. Bull.* 5 (3, 4): 97–109.

64 A. Houaoui (2021). Développement de matériaux composites et hybrides à base de verre bioactif pour la bio-ingénierie osseuse, PhD Thesis, CY Cergy Paris Université. https://hal.archives-ouvertes.fr/tel-03456688/(accessed 09 December 2021).

65 Wahl, D.A., Sachlos, E., Liu, C., and Czernuszka, J.T. (2007). Controlling the processing of collagen-hydroxyapatite scaffolds for bone tissue engineering. *J. Mater. Sci. - Mater. Med.* 18 (2): 201–209.

66 Davis, H. and Leach, J. (2008). Hybrid and composite biomaterials in tissue engineering. In: *Topics in Multifunctional Biomaterials and Devices* (ed. N. Ashammakhi), 1–26.

67 Salgado, A.J., Coutinho, O.P., and Reis, R.L. (2004). Bone tissue engineering: state of the art and future trends. *Macromol. Biosci.* 4 (8): 743–765.

68 Sachlos, E., Gotora, D., and Czernuszka, J.T. (2006). Collagen scaffolds reinforced with biomimetic composite nano-sized carbonate-substituted hydroxyapatite crystals and shaped by rapid prototyping to contain internal microchannels. *Tissue Eng.* 12 (9): 2479–2487.

69 Yang, G., Xiao, Z., Long, H. et al. (2018). Assessment of the characteristics and biocompatibility of gelatin sponge scaffolds prepared by various crosslinking methods. *Sci. Rep.* 8 (1): 1616.

70 Sharifi, E., Azami, M., Kajbafzadeh, A.-M. et al. (2016). Preparation of a biomimetic composite scaffold from gelatin/collagen and bioactive glass fibers for bone tissue engineering. *Mater. Sci. Eng., C* 59: 533–541.

71 LogithKumar, R., KeshavNarayan, A., Dhivya, S. et al. (2016). A review of chitosan and its derivatives in bone tissue engineering. *Carbohydr. Polym.* 151: 172–188.

72 Balagangadharan, K., Dhivya, S., and Selvamurugan, N. (2017). Chitosan based nanofibers in bone tissue engineering. *Int. J. Biol. Macromol.* 104: 1372–1382.

73 Preethi Soundarya, S., Haritha Menon, A., Viji Chandran, S., and Selvamurugan, N. (2018). Bone tissue engineering: scaffold preparation using chitosan and other biomaterials with different design and fabrication techniques. *Int. J. Biol. Macromol.* 119: 1228–1239.

74 Dalheim, M.Ø., Vanacker, J., Najmi, M.A. et al. (2016). Efficient functionalization of alginate biomaterials. *Biomaterials* 80: 146–156.

75 Shaheen, T.I., Montaser, A.S., and Li, S. (2019). Effect of cellulose nanocrystals on scaffolds comprising chitosan, alginate and hydroxyapatite for bone tissue engineering. *Int. J. Biol. Macromol.* 121: 814–821.

76 Rodríguez, G.R. (2019). Natural polymers for bone repair, chapter 8. In: *Bone Repair Biomaterials*, vol. 34 (ed. K.M. Pawelec and J.A. Planell), 199–232.

77 Salehi-Nik, N., Rezai Rad, M., Nazeman, P., and Khojasteh, A. (2017). Polymers for oral and dental tissue engineering. In: *Biomaterials for Oral and Dental Tissue Engineering* (ed. L. Tayebi and K. Moharamzadeh), 25–46. Elsevier.

78 Filippi, M., Born, G., Chaaban, M., and Scherberich, A. (2020). Natural polymeric scaffolds in bone regeneration. *Front. Bioeng. Biotechnol.* 8: 474.

79 Ashter, S.A. (2016). Types of biodegradable polymers. In: *Introduction to Bioplastics Engineering* (ed. S.A. Ashte), 81–151. Elsevier.

80 Benatti, A.C.B., Pattaro, A.F., Rodrigues, A.A. et al. (2019). Bioreabsorbable polymers for tissue engineering: PLA, PGA, and their copolymers. In: *Materials for Biomedical Engineering* (ed. A.-M. Holban and A.M. Grumezescu), 83–116. Elsevier.

81 Sabir, M.I., Xu, X., and Li, L. (2009). A review on biodegradable polymeric materials for bone tissue engineering applications. *J. Mater. Sci.* 44 (21): 5713–5724.

82 Santos, A.R. Jr., (2010). Bioresorbable polymers for tissue engineering. In: *Tissue Engineering* (ed. D. Eberli). InTech https://doi.org/10.5772/8580.

83 Miranda, C.S., Ribeiro, A.R.M., Homem, N.C., and Felgueiras, H.P. (2020). Spun biotextiles in tissue engineering and biomolecules delivery systems. *Antibiotics* 9 (4): 174.

84 Ikada, Y. and Tsuji, H. (2000). Biodegradable polyesters for medical and ecological applications. *Macromol. Rapid Commun.* 16, 21 (3): 117–132.

85 Ginsac, N. (2011). Caractérisation de matériaux composite polyacide lactique-bioverre pour application dans la réparation osseuse, Autre. INSA de Lyon, p. 150.

86 Cheng, Y., Deng, S., Chen, P., and Ruan, R. (2009). Polylactic acid (PLA) synthesis and modifications: a review. *Front. Chem. China* 4 (3): 259–264.

87 Schmidmaier, G., Baehr, K., Mohr, S. et al. (2006). Biodegradable polylactide membranes for bone defect coverage: biocompatibility testing, radiological and histological evaluation in a sheep model. *Clin. Oral Implants Res.* 17 (4): 439–444.

88 Gupta, B., Revagade, N., and Hilborn, J. (2007). Poly(lactic acid) fiber: an overview. *Prog. Polym. Sci.* 32 (4): 455–482.

89 McKeen, L.W. (2012). Environmentally friendly polymers. In: *Permeability Properties of Plastics and Elastomers* (ed. L.W. McKeen), 287–304. Elsevier.

90 Fang, R., Zhang, E., Xu, L., and Wei, S. (2010). Electrospun PCL/PLA/HA based nanofibers as scaffold for osteoblast-like cells. *J. Nanosci. Nanotechnol.* 10 (11): 7747–7751.

91 Niemelä, T. and Kellomäki, M. (2011). Bioactive glass and biodegradable polymer composites. In: *Bioactive Glasses* (ed. H.O. Ylänen), 227–245. Elsevier.

92 Navarro, M., Aparicio, C., Charles-Harris, M. et al. (2006). Development of a biodegradable composite scaffold for bone tissue engineering: physicochemical, topographical, mechanical, degradation, and biological properties. In: *Ordered Polymeric Nanostructures at Surfaces*, vol. 200 (ed. G.J. Vancso), 209–231. Berlin/Heidelberg: Springer.

93 Niemela, T., Niiranen, H., Kellomaki, M., and Tormala, P. (2005). Self-reinforced composites of bioabsorbable polymer and bioactive glass with different bioactive glass contents. Part I: initial mechanical properties and bioactivity. *Acta Biomater.* 1 (2): 235–242.

94 Niemelä, T., Niiranen, H., and Kellomäki, M. (2008). Self-reinforced composites of bioabsorbable polymer and bioactive glass with different bioactive glass contents. Part II: in vitro degradation. *Acta Biomater.* 4 (1): 156–164.

95 Houaoui, A., Lyyra, I., Agniel, R. et al. (2020). Dissolution, bioactivity and osteogenic properties of composites based on polymer and silicate or borosilicate bioactive glass. *Mater. Sci. Eng., C* 107: 110340.

96 Bonfield, W., Grynpas, M.D., Tully, A.E. et al. (1981). Hydroxyapatite reinforced polyethylene – a mechanically compatible implant material for bone replacement. *Biomaterials* 2 (3): 185–186.

97 Shimazaki, K. and Mooney, V. (1985). Comparative study of porous hydroxyapatite and tricalcium phosphate as bone substitute. *J. Orthop. Res.* 3 (3): 301–310.

98 Roberts, T.T. and Rosenbaum, A.J. (2012). Bone grafts, bone substitutes and orthobiologics: the bridge between basic science and clinical advancements in fracture healing. *Organogenesis* 8 (4): 114–124.

99 Fiume, E., Barberi, J., Verné, E., and Baino, F. (2018). Bioactive glasses: from parent 45S5 composition to scaffold-assisted tissue-healing therapies. *J. Funct. Biomater.* 9 (1): 24.

100 Wang, M. (2003). Developing bioactive composite materials for tissue replacement. *Biomaterials* 24 (13): 2133–2151.

101 Vergnol, G., Ginsac, N., Rivory, P. et al. (2016). In vitro and in vivo evaluation of a polylactic acid-bioactive glass composite for bone fixation devices: polylactic acid-bioactive glass composite for bone fixation devices. *J. Biomed. Mater. Res. Part B* 104 (1): 180–191.

102 Lyyra, I., Leino, K., Hukka, T. et al. (2021). Impact of glass composition on hydrolytic degradation of polylactide/bioactive glass composites. *Materials* 14 (3): 667.

103 Rahaman, M.N., Xiao, W., and Huang, W. (2018). Bioactive glass composites for bone and musculoskeletal tissue engineering, chapter 11. In: *Bioactive Glasses*, 2e (ed. H.O. Ylänen), 285–336. Woodhead Publishing.

104 Fu, Q., Rahaman, M.N., Bal, B.S. et al. (2010). Silicate, borosilicate, and borate bioactive glass scaffolds with controllable degradation rate for bone tissue engineering applications. II. In vitro and in vivo biological evaluation. *J. Biomed. Mater. Res. Part A* 95A (1): 172–179.

105 Fu, Q., Rahaman, M.N., Fu, H., and Liu, X. (2010). Silicate, borosilicate, and borate bioactive glass scaffolds with controllable degradation rate for bone tissue engineering applications. I. Preparation and in vitro degradation. *J. Biomed. Mater. Res. Part A* 95A (1): 164–171.

106 Knowles, J.C. (2003). Phosphate based glasses for biomedical applications. *J. Mater. Chem.* 13 (10): 2395.

107 Knowles, J.C. and Hastings, G.W. (1993). In vitro and in vivo investigation of a range of phosphate glass-reinforced polyhydroxybutyrate-based degradable composites. *J. Mater. Sci. - Mater. Med.* 4 (2): 102–106.

108 Verrier, S., Gough, J.E., and Boccaccini, A.R. (2011). Bioactive glass containing composites for bone and musculoskeletal tissue engineering scaffolds. In: *Bioactive Glasses* (ed. H.O. Ylänen), 162–188. Elsevier.

109 Lu, W., Ji, K., Kirkham, J. et al. (2014). Bone tissue engineering by using a combination of polymer/bioglass composites with human adipose-derived stem cells. *Cell Tissue Res.* 356 (1): 97–107.

110 Lu, H., Tang, A., Oh, S. et al. (2005). Compositional effects on the formation of a calcium phosphate layer and the response of osteoblast-like cells on polymer-bioactive glass composites. *Biomaterials* 26 (32): 6323–6334.

111 Yao, J., Radin, S., Leboy, P.S., and Ducheyne, P. (2005). The effect of bioactive glass content on synthesis and bioactivity of composite poly(lactic-*co*-glycolic acid)/bioactive glass substrate for tissue engineering. *Biomaterials* 26 (14): 1935–1943.

112 Lu, H.H., El-Amin, S.F., Scott, K.D., and Laurencin, C.T. (2003). Three-dimensional, bioactive, biodegradable, polymer-bioactive glass composite scaffolds with improved mechanical properties support collagen synthesis and mineralization of human osteoblast-like cells in vitro. *J. Biomed. Mater. Res.* 64A (3): 465–474.

113 Jo, J.-H., Lee, E.-J., Shin, D.-S. et al. (2009). In vitro/in vivo biocompatibility and mechanical properties of bioactive glass nanofiber and poly(ε-caprolactone) composite materials. *J. Biomed. Mater. Res. Part B* 91B (1): 213–220.

114 Erdemli, O., Çaptug, O., Bilgili, H. et al. (2010). In vitro and in vivo evaluation of the effects of demineralized bone matrix or calcium sulfate addition to polycaprolactone – bioglass composites. *J. Mater. Sci. - Mater. Med.* 21 (1): 295–308.

115 Naseri, S., Boccaccini, A.R., and Nazhat, S.N. (2017). Bioactive glass particulate-incorporated polymer composites, chapter 10. In: *Bioactive Glasses: Fundamentals, Technology and Applications* (ed. A.R. Boccaccini, D.S. Brauer and L. Hupa), 236–256. The Royal Society of Chemistry.

116 Hutmacher, D.W. (2001). Scaffold design and fabrication technologies for engineering tissues – state of the art and future perspectives. *J. Biomater. Sci., Polym. Ed.* 12 (1): 107–124.

117 Gerhardt, L.-C. and Boccaccini, A.R. (2010). Bioactive glass and glass-ceramic scaffolds for bone tissue engineering. *Materials* 3 (7): 3867–3910.

118 Rahaman, M.N., Bal, B.S., and Huang, W. (2014). Review: emerging developments in the use of bioactive glasses for treating infected prosthetic joints. *Mater. Sci. Eng., C* 41: 224–231.

119 Mikos, A.G., Thorsen, A.J., Czerwonka, L.A. et al. (1994). Preparation and characterization of poly(L-lactic acid) foams. *Polymer* 35 (5): 1068–1077.

120 Di Maio, E., Mensitieri, G., Iannace, S. et al. (2005). Structure optimization of polycaprolactone foams by using mixtures of CO_2 and N_2 as blowing agents. *Polym. Eng. Sci.* 45 (3): 432–441.

121 Haugen, H., Ried, V., Brunner, M. et al. (2004). Water as foaming agent for open cell polyurethane structures. *J. Mater. Sci. - Mater. Med.* 15 (4): 343–346.

122 Parks, K.L. and Beckman, E.J. (1996). Generation of microcellular polyurethane foams via polymerization in carbon dioxide. II: foam formation and characterization. *Polym. Eng. Sci.* 36 (19): 2417–2431.

123 Liapis, A.I. and Bruttini, R. (1994). A theory for the primary and secondary drying stages of the freeze-drying of pharmaceutical crystalline and amorphous solutes: comparison between experimental data and theory. *Sep. Technol.* 4 (3): 144–155.

124 Pikal, M., Shah, S., Roy, M., and Putman, R. (1990). The secondary drying stage of freeze drying: drying kinetics as a function of temperature and chamber pressure. *Int. J. Pharm.* 60 (3): 203–207.

125 Brunello, G., Sivolella, S., Meneghello, R. et al. (2016). Powder-based 3D printing for bone tissue engineering. *Biotechnol. Adv.* 34 (5): 740–753.

126 Guduric, V. (2017). 3D printing and characterization of PLA scaffolds for layer-by-layer bioassembly in Tissue Engineering, Human health and pathology. Université de Bordeaux. p. 264.

127 Tereshchenko, V., Larionov, P., Kirilova, I. et al. (2016). Materials and methods of bone tissue engineering. *Hir. Pozvonočnika* 13: 72–81.

128 Zhu, N. and Che, X. (2013). Biofabrication of tissue scaffolds. In: *Advances in Biomaterials Science and Biomedical Applications* (ed. R. Pignatello). InTech https://doi.org/10.5772/54125.

129 Babilotte, J., Guduric, V., Le Nihouannen, D. et al. (2019). 3D printed polymer-mineral composite biomaterials for bone tissue engineering: fabrication and characterization. *J. Biomed. Mater. Res. Part B* 107 (8): 2579–2595.

130 Mahony, O., Tsigkou, O., Ionescu, C. et al. (2010). Silica-gelatin hybrids with tailorable degradation and mechanical properties for tissue regeneration. *Adv. Funct. Mater.* 20 (22): 3835–3845.

131 Sanchez, C., Julián, B., Belleville, P., and Popall, M. (2005). Applications of hybrid organic–inorganic nanocomposites. *J. Mater. Chem.* 15 (35, 36): 3559.

132 Zha, J. and Roggendorf, H. (1991). Sol-gel science, the physics and chemistry of sol-gel processing, Ed. by C.J. Brinker and G.W. Scherer, Academic Press, Boston 1990, xiv, 908 pp., bound?ISBN 0-12-134970-5. *Adv. Mater.* 3 (10): 522.

133 Novak, B.M. (1993). Hybrid nanocomposite materials – between inorganic glasses and organic polymers. *Adv. Mater.* 5 (6): 422–433.

134 Mahony, O., Yue, S., Turdean-Ionescu, C. et al. (2014). Silica–gelatin hybrids for tissue regeneration: inter-relationships between the process variables. *J. Sol-Gel Sci. Technol.* 69 (2): 288–298.

135 Houaoui, A., Szczodra, A., Lallukka, M. et al. (2021, 2021). New generation of hybrid materials based on gelatin and bioactive glass particles for bone tissue regeneration. *Biomolecules* 11 (3): 444.

136 Hoppe, A., Güldal, N.S., and Boccaccini, A.R. (2011). A review of the biological response to ionic dissolution products from bioactive glasses and glass-ceramics. *Biomaterials* 32 (11): 2757–2774.

137 Xynos, I.D., Edgar, A.J., Buttery, L.D.K. et al. (2001). Gene-expression profiling of human osteoblasts following treatment with the ionic products of Bioglass® 45S5 dissolution. *J. Biomed. Mater. Res.* 55 (2): 151–157.

138 Day, R.M. (2005). Bioactive glass stimulates the secretion of angiogenic growth factors and angiogenesis in vitro. *Tissue Eng.* 11 (5, 6): 768–777.

139 Gorustovich, A.A., Roether, J.A., and Boccaccini, A.R. (2010). Effect of bioactive glasses on angiogenesis: a review of in vitro and in vivo evidences. *Tissue Eng. Part B* 16 (2): 199–207.

140 Leu, A. and Leach, J.K. (2008). Proangiogenic potential of a collagen/bioactive glass substrate. *Pharm. Res.* 25 (5): 1222–1229.

141 Allan, I., Newman, H., and Wilson, M. (2001). Antibacterial activity of particulate Bioglass® against supra- and subgingival bacteria. *Biomaterials* 22 (12): 1683–1687.

142 Gorriti, M.F., López, J.M.P., Boccaccini, A.R. et al. (2009). In vitro study of the antibacterial activity of bioactive glass-ceramic scaffolds. *Adv. Eng. Mater.* 11 (7): B67–B70.

143 Hu, S., Chang, J., Liu, M., and Ning, C. (2009). Study on antibacterial effect of 45S5 Bioglass®. *J. Mater. Sci. - Mater. Med.* 20 (1): 281–286.

144 Bi, L., Rahaman, M.N., Day, D.E. et al. (2013). Effect of bioactive borate glass microstructure on bone regeneration, angiogenesis, and hydroxyapatite conversion in a rat calvarial defect model. *Acta Biomater.* 9 (8): 8015–8026.

145 Chen, S., Osaka, A., Ikoma, T. et al. (2011). Fabrication, microstructure, and BMP-2 delivery of novel biodegradable and biocompatible silicate–collagen hybrid fibril sheets. *J. Mater. Chem.* 21 (29): 10942.

146 Liu, Y.-L., Su, Y.-H., and Lai, J.-Y. (2004). In situ crosslinking of chitosan and formation of chitosan–silica hybrid membranes with using γ-glycidoxypropyltrimethoxysilane as a crosslinking agent. *Polymer* 45 (20): 6831–6837.

147 Maeda, H., Kasuga, T., and Hench, L.L. (2006). Preparation of poly(L-lactic acid)-polysiloxane-calcium carbonate hybrid membranes for guided bone regeneration. *Biomaterials* 27 (8): 1216–1222.

148 Rhee, S.-H., Choi, J.-Y., and Kim, H.-M. (2002). Preparation of a bioactive and degradable poly(e-caprolactone)/silica hybrid through a sol–gel method. *Biomaterials* 7, 23 (24): 4915–4921.

149 Gurunathan, T., Mohanty, S., and Nayak, S.K. (2016). Hyperbranched polymers for coating applications: a review. *Polym. Plast. Technol. Eng.* 55 (1): 92–117.

150 Connell, L.S., Gabrielli, L., Mahony, O. et al. (2017). Functionalizing natural polymers with alkoxysilane coupling agents: reacting 3-glycidoxypropyl trimethoxysilane with poly(γ-glutamic acid) and gelatin. *Polym. Chem.* 8 (6): 1095–1103.

151 Jones, J.R., Brauer, D.S., Hupa, L., and Greenspan, D.C. (2016). Bioglass and bioactive glasses and their impact on healthcare. *Int. J. Appl. Glass Sci.* 7 (4): 423–434.

152 Schallenberger, M.A., Rossmeier, K., Lovick, H.M. et al. (2014). Comparison of the osteogenic potential of OsteoSelect demineralized bone matrix putty to NovaBone calcium-phosphosilicate synthetic putty in a cranial defect model. *J. Craniofac. Surg.* 25 (2): 657–661.

153 Wang, Z., Lu, B., Chen, L., and Chang, J. (2011). Evaluation of an osteostimulative putty in the sheep spine. *J. Mater. Sci. - Mater. Med.* 22 (1): 185–191.

Biofabrication for Orthopedics

Biofabrication for Orthopedics

Methods, Techniques, and Applications

Edited by
Wenguo Cui
Xin Zhao
Shen Liu

Volume 2

WILEY-VCH

Editors

Prof. Wenguo Cui
Shanghai Jiao Tong University School of Medicine
Shanghai Institute of Traumatology and Orthopaedics, Ruijin Hospital
197 Ruijin 2nd Road
200025 Shanghai
China

Prof. Xin Zhao
Hong Kong Polytechnic University
Department of Biomedical Engineering
11 Yok Choi Road, Hung Hom
999077 Hong Kong
Hong Kong

Prof. Shen Liu
Shanghai Jiao Tong University
Affiliated Sixth People's Hospital
600 Yishan Road
Department of Orthopaedics
200233 Shanghai
China

Cover Image: © SciePro/Adobe Stock Photos

Library of Congress Card No.: applied for

British Library Cataloguing-in-Publication Data
A catalogue record for this book is available from the British Library.

Bibliographic information published by the Deutsche Nationalbibliothek
The Deutsche Nationalbibliothek lists this publication in the Deutsche Nationalbibliografie; detailed bibliographic data are available on the Internet at <http://dnb.d-nb.de>.

Print ISBN: 978-3-527-35170-1
ePDF ISBN: 978-3-527-83136-4
ePub ISBN: 978-3-527-83135-7
oBook ISBN: 978-3-527-83137-1

Typesetting Straive, Chennai, India
Printing and Binding CPI Group (UK) Ltd, Croydon, CR0 4YY

C112261_030922

Contents

Volume 1

Foreword from Prof. Changsheng Liu *xi*
Foreword from Prof. Yingze Zhang *xiii*
Foreword from Prof. Lianfu Deng *xv*
Foreword from Prof. Cato T. Laurencin *xvii*
Preface *xix*

Part I Biofabrication Techniques *1*

1 **Current Progress and Technological Challenges in Translational 3D Bioprinting** *3*
Greeshma Ratheesh, Natividad G. Cerezo, Weidong Gao, Jayanti Mendhi, Prashant Sonar, and Yin Xiao

2 **Bioceramics for Promoting Bone Regeneration** *25*
Meng Zhang, Hongshi Ma, and Chengtie Wu

3 **3D Printing and Bioprinting Strategies Applied Toward Orthopedics** *55*
Zongliang Wang, Prajwal Agrawal, and Yu Shrike Zhang

4 **Stem Cells and Their Application in Orthopedics** *93*
Xiao Li, Xiaobo Zhu, Jing Sun, Patrick S.H. Yung, Rocky S. Tuan, and Yangzi Jiang

5 **Electrospinning Techniques** *133*
Yuanfei Wang, Ziyi Zhou, Xiaopei Zhang, Na Liu, Zhendong Feng, Xiumei Mo, Wenguo Cui, and Tong Wu

6 **Joint Lubrication and Wear** *159*
Yi Wang, Keming Hu, Weiwei Zhao, and Hongyu Zhang

7 **Microfluidic Biotechnology for "Bone-on-a-Chip"** *181*
Chaenyung Cha

8 **Bioactive Glasses in Orthopedics** *211*
Amel Houaoui and Jonathan Massera

Volume 2

Foreword from Prof. Changsheng Liu *xiii*
Foreword from Prof. Yingze Zhang *xv*
Foreword from Prof. Lianfu Deng *xvii*
Foreword from Prof. Cato T. Laurencin *xix*
Preface *xxi*

Part II Biomedical Applications in Orthopedics *243*

9 **3D Printing for Orthopedics** *245*
Lan Li and Qing Jiang
9.1 Overview of 3D Printing Technology *245*
9.2 Bone Tissue Engineering and 3D Printing *246*
9.3 Cartilage Tissue Engineering and 3D Printing *247*
9.4 Structural Requirements of 3D Printing *248*
9.4.1 Pore Size *248*
9.4.2 Porosity *249*
9.4.3 Pore Structure *249*
9.5 Biomaterials for 3D Printing *250*
9.5.1 Metal *251*
9.5.2 Bioceramics *252*
9.5.3 Polymer Materials *253*
9.6 Application of 3D Printing in Cell Printing and Orthopedic Tissue Engineering *254*
9.6.1 Inkjet 3D Printing *254*
9.6.2 Extrusion-Based 3D Bioprinting *255*
9.6.3 Laser-Assisted Printing *256*
9.7 Future Prospects *256*
Acknowledgments *258*
References *258*

10 **Bone Implants (Bone Regeneration and Bone Cancer Treatments)** *265*
Hangxiang Sun, Xingzhi Zhou, Yongxing Zhang, Lei Zhang, Xiaohua Yu, Zhaoming Ye, and Cato T. Laurencin
10.1 Bone Regeneration *265*
10.1.1 Introduction *265*

10.1.2 Biological Characteristics of Bone *266*
10.1.2.1 Bone Biology *266*
10.1.2.2 Osteogenesis and Bone Healing (Biomineralization) *267*
10.1.3 Role of Implants for Bone Regeneration *268*
10.1.3.1 Providing Structural Support *271*
10.1.3.2 Regulating Stem Cell Behaviors *273*
10.1.3.3 Activating Gene-Associated Therapy *274*
10.1.3.4 Application of Orthobiologics *275*
10.1.3.5 Generating Stimulus Responses for Bone Healing *276*
10.1.3.6 Converging Multiple Synergistic Strategies *279*
10.1.4 Concluding Remarks and Future Outlook *281*
10.2 Bone Cancer Treatments *281*
10.2.1 Clinical Approach to Bone Cancer *281*
10.2.1.1 Introduction *281*
10.2.1.2 Traditional Approaches *282*
10.2.1.3 Surgery *283*
10.2.1.4 Chemotherapy *283*
10.2.1.5 Radiotherapy *284*
10.2.1.6 Molecular Targeted Therapy *284*
10.2.1.7 Other Therapy *285*
10.2.2 State-of-the-Art Nanotechnology *285*
10.2.2.1 Nanotechnology for Bone Cancer Therapy *285*
10.2.2.2 Cytotoxic Agents *286*
10.2.2.3 Drug Delivery Platform *286*
10.2.2.4 Phototherapy *287*
10.2.2.5 Gene Therapy *289*
10.2.2.6 Summary and Future Direction *289*
10.2.3 Biofabrication for Bone Cancer Treatments *291*
10.2.3.1 Introduction *291*
10.2.3.2 Application in Bone Cancer Treatments *291*
10.2.3.3 Summary and Future Direction *294*
Acknowledgments *295*
References *295*

11 Bionic Fixation: Design, Biomechanics, and Clinical Application *323*
Wei Chen and Qi Zhang
11.1 Bionics and Medical Bionics *323*
11.2 Structural Bionics in the Field of Orthopedics and Traumatology *324*
11.2.1 Elastic Bionic Fixation Device for Distal Tibiofibular Syndesmosis *324*
11.2.2 Minimally Invasive Adjustable Plate for Pelvic Fractures *327*
11.2.3 Bionic Implants for Intertrochanteric Fracture of the Femur *328*
11.2.4 Bionic Fixation of Calcaneal Fracture *331*
11.3 Bionic Materials *335*

11.3.1 Application and Designation of New Composite Materials *335*
11.3.2 Bionic Design *337*
11.4 Future Perspectives and Current Limitations *337*
Acknowledgments *338*
References *338*

12 Cartilage Injury and Repair *341*
Yusheng Li, Ning Wang, Di Liu, and Zijun Cai
12.1 Introduction *341*
12.2 Pathology of Cartilage Injury *342*
12.3 Clinical Characteristics of Cartilage Injury *343*
12.4 Evaluation of Cartilage Injuries *343*
12.4.1 Arthroscopic Evaluation of Cartilage Injury *343*
12.4.2 Histopathology Evaluation of Cartilage Injury *344*
12.4.3 Magnetic Resonance Imaging (MRI) Assessment of Cartilage Injury *345*
12.5 Clinical Strategies of Cartilage Repair *347*
12.5.1 Debridement, Cartilage Shaving, and Joint Lavage *347*
12.5.2 Bone Marrow Stimulation and Augmentation *348*
12.5.3 Osteochondral Autografts and Allografts *349*
12.5.4 Autologous Chondrocyte Implantation (ACI) *351*
12.6 Potential and Advanced Strategies of Cartilage Repair *352*
12.6.1 Tissue Engineering Approaches *352*
12.6.2 Platelet-Rich Plasma (PRP) for Cartilage Repair *354*
12.6.3 Gene Editing for Cartilage Repair *356*
12.7 Conclusion *357*
References *358*

13 Biofabrication for Cartilage Regeneration *367*
Sien Lin, Boguang Yang, and Gang Li
13.1 Introduction *367*
13.2 Biological Responses of Cartilage Injury *369*
13.3 Biofabrication: Traditional Cellular Products *370*
13.4 Biofabrication: Bioprinting *374*
13.5 Biofabrication: Electrospinning *382*
13.6 Biofabrication: Microfluidic *386*
Acknowledgments *389*
References *389*

14 Intervertebral Disc Repair *401*
Yong Cao, Tianding Wu, Chunyue Duan, JinSong Li, Yuxin Jing, Jiaqi Xu, Daudi R. Manini, and Jianzhong Hu
14.1 Introduction *401*
14.2 Technique for Intervertebral Disc Therapy *405*
14.2.1 Cell-Based Treatment Techniques *405*

14.2.2 Patient Selection for Cell Therapy *408*
14.3 Gene Therapy for IVD *409*
14.4 Molecular Therapy for IVD Degeneration *411*
14.4.1 Anticatabolic Regulators *411*
14.4.2 Mitogens *412*
14.4.3 Chondrogenic Morphogens *412*
14.4.4 Intracellular Regulators *414*
14.5 Biomaterial Therapy *414*
14.6 Summary *417*
Authorship Contribution Statement *417*
References *417*

15 New Prospects in Skin Tissue Engineering and Fabrication *423*
Lei Wang, Xiaonan Liu, and Ke Shen
15.1 Skin Anatomy and Related Cell Types *423*
15.2 Types of Skin Wounds and Wound Healing Process *425*
15.3 Treatments for Wound Healing *427*
15.4 Skin Regeneration Using Tissue Engineering *429*
15.4.1 Local Stem Cells *429*
15.4.2 Induced Pluripotent Stem Cells (iPSC) *435*
15.4.3 Bone Marrow Stem Cells (BMSC) *437*
15.4.4 Summary *439*
15.5 Skin Substitutes *439*
15.5.1 3D Printing *440*
15.5.2 Acellular Skin Substitute *440*
15.5.3 Cell Allogeneic Skin Substitutes *441*
15.5.4 Cellular Autologous Skin Substitutes *441*
15.5.5 Summary *447*
Acknowledgments *447*
References *447*

16 Techniques for Biofabrication of Vascular and Vascularized Tissue *461*
Zhe Zhang, Qiang Yang, and Chunming Wang
16.1 Introduction *461*
16.2 3D Printing *462*
16.2.1 Inkjet Bioprinting *463*
16.2.2 Extrusion Bioprinting *463*
16.2.3 Laser-Assisted Bioprinting *465*
16.3 Microfabrication *468*
16.3.1 Micropatterning/Microfluidics *468*
16.3.2 Micromodule Assembly *469*
16.4 Electrospinning *471*
16.5 Decellularized Vascular Tissue *473*

16.6 Cell-Only Fabrication *473*
16.6.1 Cell-Only Bioprinting *474*
16.6.2 Cell Sheet *474*
16.7 Conclusion and Outlook *475*
Acknowledgments *476*
References *476*

17 Vascularization *481*
Xiao Lu, Jianxi Lu, Fangchun Jin, and Yajie Lu
17.1 Introduction *481*
17.2 Is Porous Structure Needed for Vascularization? *482*
17.3 Promotion of Vascularization in Biomaterials *486*
17.3.1 Vascular Graft *486*
17.3.2 Growth Factors *487*
17.3.2.1 Vascular Endothelial Growth Factor *488*
17.3.2.2 Bone Morphogenetic Protein *489*
17.3.2.3 Basic Fibroblast Growth Factor *490*
17.3.2.4 Multifactor Combination *490*
17.3.3 Stem Cells *491*
17.3.3.1 Endothelial Cells and Endothelial Progenitor Cells *491*
17.3.3.2 Mesenchymal Stem Cells *492*
17.3.3.3 Embryonic Stem Cells *493*
17.3.3.4 Induced Pluripotent Stem Cells *493*
17.3.4 Microfabricated Vasculature *494*
17.4 Vascularization in Porous Biomaterials *495*
17.4.1 Influence of the Pore Diameter *495*
17.4.2 Influence of the Interconnection Diameter *498*
17.5 Vascularization of Porous Biomaterials in Bones *500*
17.5.1 Influence of the Porous Structure *501*
17.5.2 Guided Vascular Growth in Porous Biomaterials *503*
17.5.3 Clinical Application of Guided Vascularization with Porous Biomaterials *505*
17.6 Conclusion *507*
References *509*

18 Muscle Repair *513*
Rong Bao, Wei Wang, Shen Liu, and Cunyi Fan
18.1 Introduction *513*
18.2 Anatomy of Skeletal Muscle *514*
18.2.1 Morphological Characteristics of Skeletal Muscle *514*
18.2.2 Composition of Skeletal Muscle *514*
18.3 Pathophysiology After Skeletal Muscle Injury *515*
18.3.1 Inflammatory Phase *515*
18.3.2 Repair Phase *515*
18.3.3 Remodeling Phase *516*

18.4 Biofabrication for Skeletal Muscle Regeneration *517*
18.4.1 Materials for Fabrication (Natural, Synthetic, and Hybrid) *517*
18.4.2 Biochemical Substances for Fabrication (Cytokines, Genetic Drugs, Micromolecules, Etc.) *518*
18.4.3 Acellular Scaffolds *519*
18.4.4 Cellular Scaffolds for Skeletal Muscle Regeneration *521*
18.4.5 Tissue Engineering for Muscle Repair *523*
18.5 Application and Prospect *525*
18.5.1 Promising Strategies for Skeletal Muscle Regeneration *525*
18.5.2 Challenges and Future Perspectives *527*
References *528*

19 Rotator Cuff Repair *535*
Chongyang Wang and Yaohua He
19.1 Introduction *535*
19.2 Histology of the Rotator Cuff Tendon-to-Bone Enthesis *537*
19.3 Risk Factors Associated with High Failure Rates of Rotator Cuff Repair *537*
19.3.1 Rotator Cuff Tear Size *539*
19.3.2 Fatty Infiltration of Rotator Cuff Muscles *540*
19.3.3 Failure to Regenerate the Microstructure of Normal Tendon-to-Bone Enthesis *541*
19.3.4 Low BMD of the Humeral Head *541*
19.4 Biofabrication for Rotator Cuff Repair *541*
19.4.1 Biofabrication Promotes the Regeneration of Enthesis Structure *542*
19.4.1.1 Biofabrication Interplate Between the Tendon and Bone *542*
19.4.1.2 Biofabrication Overlays the Repaired Rotator Cuff Tendon *543*
19.4.2 Biofabrication Bridges the Torn Tendon and Bone *544*
19.4.3 Biofabrication Inhibits Fat Infiltration of Rotator Cuff Muscles *546*
19.4.4 Biofabrication Increases Bone Mineral Density *547*
19.5 Biological Augmentation for Rotator Cuff Repair *547*
19.5.1 Cell-Based Strategies *547*
19.5.2 Cell-Derived Hybrid Biological Factors *549*
19.5.3 Growth Factors *550*
19.5.4 Drugs or Small Molecules *550*
19.6 Future Perspectives of Biofabrication for Rotator Cuff Repair *551*
Acknowledgments *553*
References *553*

20 Tendon Repair *559*
Xiao Wang
20.1 Introduction *559*
20.2 Tendon Biology *560*
20.3 Tendon Injury *562*

20.4 Tendon Healing *563*
20.5 Tendon Repair *564*
20.5.1 Physical Modalities *564*
20.5.2 Mechanical Loading and Mobilization *565*
20.5.3 Cytokines and Growth Factors *566*
20.5.4 Tissue Engineering *567*
20.5.5 Prevention of Adhesions *568*
20.6 Conclusion *568*
References *568*

21 Nerve Regeneration *581*
Zhiwen Yan, Wei Chen, Yun Qian, and Cunyi Fan
21.1 Peripheral Nerve Regeneration *581*
21.1.1 Introduction *581*
21.1.2 Microenvironmental Imbalance *583*
21.1.3 Microenvironmental Rebalance *585*
21.2 Spinal Cord Regeneration *587*
21.2.1 Introduction *587*
21.2.2 Pathophysiological Process *588*
21.2.3 Bio-fabrication for Spinal Cord Regeneration *590*
21.3 Clinical Application of Nerve Conduit *594*
21.3.1 Clinical Classification of Nerve Injuries *594*
21.3.2 Clinical Evaluation *595*
21.3.3 Surgical Repair of Peripheral Nerve Injuries *597*
21.3.4 Surgical Repair of Spinal Cord Injuries *600*
21.3.5 Future Development of NGCs *601*
21.4 Conclusion *601*
Acknowledgments *602*
References *602*

Index *607*

Foreword from Prof. Changsheng Liu

I would like to thank Profs Wenguo Cui, Xin Zhao, and Shen Liu for inviting me to write this foreword. I am honored to take part in the knowledge exchange of our scientific culture, as well as to promote training of our students and professionals.

With the rapid development of biomaterials and biofabrication technologies, a variety of biomaterials have been made for the repair and regeneration of bone, muscle, tendon, and other orthopedic tissues and organs. This book focuses on the current state of biofabrication technologies and materials in orthopedics biomedical applications. The various highlights of this book include a comprehensive analysis of different biofabrication methods for constructing relevant biomaterials in the orthopedics field, with an emphasis on the recent developments of technologies involving 3D printing, bioceramic, electrospinning, microfluidics, bioactive glass, etc. A detailed look into these biofabrication techniques using different biomaterials is also included in terms of functional performance, advantages, and disadvantages for repairing and regeneration in orthopedics, of which the choices of biomaterials are especially critical since their applications must be based on a thorough knowledge of the anatomy and physiology of the bone, muscle, and adjacent tissue. With the content being organized clearly, practically, rigorously, and most importantly, up to date with current knowledge, this review will therefore provide the research community a reference source for current approaches to biomaterials and scaffold preparation in orthopedics, as well as to reinforce our readers' understanding of materials' biomedical applications in orthopedics.

In essence, this book is clear and well written in a way that provides highly useful and relevant content to support trained orthopedic doctors and biomaterial researchers. This helps give the readers the basis of the preparation of a variety of biomaterials and thus enhance the understanding of their application in orthopedics. Through this book, the authors have faith that the readers and researchers can generate next-generation biomaterials for orthopedic applications.

Changsheng Liu
East China University of Science and Technology, P.R. China

Foreword from Prof. Yingze Zhang

In the past decades, traditional orthopedic metal implants such as plates and rivets have benefited many patients with orthopedic diseases. However, problems such as mechanical failure and poor efficacy always exist, which need to be solved through innovation. Nowadays, interdisciplinary convergence and cross-border integration of several technologies are becoming the norm and will continue to give birth to the forefront of new disciplines. Thanks to the development of tissue engineering and regenerative medicine, a large number of new materials and technologies with great application prospects in orthopedic medicine have emerged, providing new opportunities for the diagnosis and treatment of orthopedic diseases.

This book systematically demonstrates the widely popular and eye-catching biomaterials and biofabrication technologies, explains in detail their advantages and disadvantages in orthopedics, and lists their applications in the repair of bone, muscle, ligament, tendon, and other orthopedic organs.

Each chapter of this book emphasizes the impact of advanced materials and biofabrication technologies in the field of orthopedics, including how biofabrication affects all substances used in bone implants (from 3D printing to electrospinning and microfluidic technology). One of the highlights of this book is apparently to cover the most innovative biomaterial manufacturing technologies and emphasizes the safety of biofabrication that we have not fully understood and need to develop solutions. It also provides concise reasons on why we should consider using these materials to complete the regeneration and repair of orthopedic systems, discussing in-depth beyond the typical trial-and-error thinking of traditional orthopedics.

The editors and authors of this book hope to enrich readers and researchers in the field of orthopedics, orthopedic biomaterials, and their clinical applications, and in my opinion, this book is certainly a must-have for all those who want to create new solutions to the most stubborn problems in orthopedics, so that we can improve the quality of life for patients of implants with novel biofabrication technologies. We believe that new orthopedic biofabrication technologies for bone tissue regeneration and a better understanding of the interface between biomaterials and adjacent tissues will be an important part of orthopedic biomaterials and implants in the future for better orthopedic surgery.

Ying ze zhang

Yingze Zhang
Department of Orthopedic Surgery
Third Hospital of Hebei
Medical University Director
NHC Key Laboratory
of Intelligent Orthopaedic
Equipment, Shijiazhuang
P.R. China

Foreword from Prof. Lianfu Deng

With the rapid development of medical science and engineering technology, the field of orthopedics has also made great progress. In recent years, biomaterial implants have been widely used in the replacement treatment or adjunctive therapy of injured bones to replace, support localization, or repair human bones, joints, and soft tissues. The clinical demand for orthopedic implanted medical devices has been maintaining a rapid growth rate. However, due to the differences in disease or injury types, the demand for personalized design and fabrication of biomaterials and biomaterials with specific functions is also increasing. This also greatly promoted the development of orthopedic materials through biofabrication technology.

The continuous innovation and development of biomaterial fabrication technologies (3D (three-dimensional) bioprinting, electrospinning, and microfluidic) have improved the utilization of biomaterials, including the fabrication of complex 3D structures, the improvement of biomaterial interface and cell interaction, and the development of controlled drug-release systems. The development and biofabrication of advanced biomaterials in orthopedics field covers the interdisciplinary study of materialogy, tissue engineering and regenerative medicine, cell and molecular biology, and clinical medicine. The biofabrication techniques bridge basic research and clinical treatment, help to accelerate the orthopedic basic scientific research to the engineering application of the industrialization process, and promote the development of targeted treatment strategies for different orthopedics diseases or injuries. It is meaningful to advance the development progress of translational medicine.

It is timely to compile this book on advanced fabrication technology of biomaterials and its application in orthopedics. This book summarizes the latest scientific research progress and development trend of orthopedics worldwide, aiming to track the frontier of the discipline, introduce advanced theories and technologies, and promote the development of orthopedics and orthopedics technology. This book will help doctors, researchers, and students of medical specialty to have a good command of the design, fabrication, and application of biomaterials as well as the regulatory mechanism of bone regeneration process and better understand the development direction of biological fabrication and orthopedic technology.

Lianfu Deng
Shanghai Key Laboratory
for Prevention and Treatment
of Bone and Joint Diseases
Shanghai Institute of
Traumatology and Orthopaedics
Ruijin Hospital, Shanghai Jiao Tong
University School of Medicine
P.R. China

Foreword from Prof. Cato T. Laurencin

It is always a great privilege to be invited to write a foreword for a book. Moreover, when the book enlarges our scientific knowledge and viewpoints while promoting the training of our students and professionals, it is even more welcome. I am therefore more than happy to provide my insights regarding such books.

The convergence of the life, physical, and engineering sciences has led to the creation of a new multidisciplinary field, which I have termed as "regenerative engineering." Orthopedic surgery is a prominent field that stands to gain from the insights of regenerative engineering. In particular, biofabrication techniques developed in this field applying the principles of regenerative engineering have the potential of bringing revolutionary treatments to clinical practice. This book summarizes important aspects of biofabrication techniques for orthopedic surgery and offers valuable information of state-of-the-art technological advancements that can ultimately become a practical toolbox for physicians, scientists, and engineers. While traditional orthopedic surgery books cover conventional strategies, this book focuses on newer biological strategies such as bioprinting, stem-cell therapy, and organ-on-a-chip technology. The chapters encourage readers to examine and embrace novel technologies to address clinical challenges. The content is well organized while covering a wide range of related fields. The knowledge and information provided by all the contributors are up to date, covering comprehensively technological aspects as well as clinical applications involved in orthopedic surgery practice.

The editors of this book, Profs Wenguo Cui, Xin Zhao, and Shen Liu, have assembled an excellent team and have created an important contribution to our field. The contributors to this book are respected scholars and professionals that come from a broad background including orthopedic surgery, biomedical engineering, stem-cell biology, and physics. This allows the book to have insights regarding the current status and future trends of this field.

Scientific books must evolve to capture important aspects of the corresponding fields. This book is published at the right time when many exciting new biofabrication technologies have arisen, many of them inspired by the field of regenerative engineering. I congratulate again the editors of the book. This well-organized contribution to the world's literature will be relevant to scientists, engineers, and clinicians everywhere.

Cato T. Laurencin, MD, PhD
Member, National Academy of Science/
National Academy of Engineering
Albert and Wilda Van Dusen Distinguished
Professor of Orthopaedic Surgery
Professor of Chemical and
Biomolecular Engineering
Professor of Materials Science
and Engineering
Professor of Biomedical Engineering

Director, The Raymond and Beverly Sackler
Center for Biomedical, Biological, Physical
and Engineering Sciences

Chief Executive Officer
The Connecticut Convergence Institute
for Translation in Regenerative Engineering
The University of Connecticut

Preface

Wenguo Cui

Xin Zhao

Shen Liu

Bone, as one of the most essential organs in the body, not only supports the stability of body along with protection of other organs, but also provides hematopoietic function and stores minerals such as calcium and phosphorus. Unfortunately, when some damages/diseases, such as fracture, osteoarthritis, and osteoporosis, appear in the bones, these functions will be disrupted, leading to the systemic responses as seen clinically. Researchers are therefore proposing to develop techniques to remodel the damaged/diseased bone functions. Last decades, biofabrication techniques, including three-dimension (3D) bioprinting, 3D printing, electrospinning, microfluidics, and stem-cell therapy, have been widely applied in tissue regeneration. Some products of these biofabrication techniques were even commercialized, such as Absorb GT1 and XinSorb. *Biofabrication for Orthopedics – Special Topic* in Wiley mainly focuses on the development of biofabrication and applications in orthopedics, comprising 20 chapters in total, as detailed below.

Areas Covered in the Special Topic

The book starts with the two popular biofabrication techniques in the recent decades, namely, translational 3D bioprinting and 3D printing, which are comprehensively reviewed by Xiao and colleagues in the choice of bioink, cell types, the remodeling and maturation of bioprinting scaffolds, and by Zhang and colleagues in the choice of 3D printing polymers. Bioceramics, as a class of inorganic and non-metallic materials, is introduced by Wu and colleagues in the preparation of various bioceramics, along with the potential applications for the hard tissue repair. Additionally, Massera and colleagues provide an overview about the synthesis and application of different bioactive glasses. In addition to the bioactive scaffolds, the mesenchymal-stem-cells-derived extracellular vesicles and exosomes can also be regarded as biofabrications, which are thoroughly discussed by Jiang and colleagues. With the use of advanced biomaterials, two common techniques, electrospinning and microfluidics, are also reported by Wu and colleagues, and Cha and colleagues, respectively.

The application of 3D printing in orthopedics, including, but not limited to, bone engineering and cartilage engineering, is provided by Jiang and colleagues. Implantation and fixation are important processes in orthopedics. Laurencin and colleagues review implants that provide structural support and regulate stem-cell behaviors, while Chen and colleagues review the fixation of distal tibiofibular syndesmosis and intertrochanteric fracture of the femur. Besides bones, cartilages are also discussed starting with pathology cartilage injuries and clinical magnetic resonance imaging assessments that are introduced by Wang, followed by the review by Li and colleagues detailing scaffolds and cell-derived products for the cartilage regeneration. Additionally, intervertebral disc, as another special organ, is also reviewed by Cao and colleagues in terms of its current understanding and its repair using biofabrication-based treatments. Moreover, Zhang and colleagues also provide a comprehensive overview on joint lubrication.

In the process of bone regeneration, angiogenesis plays a critical role in accelerating regeneration. In this regard, Lu and colleagues thoroughly describe how biofabrication guided angiogenesis in controlling the pore diameter and interconnection diameter. Furthermore, Wang and colleagues review in detail the applications of different biofabrication techniques in vascularization such as bioprinting, electrospinning, or microfluidics. Since bone injuries usually are accompanied by skin, muscle, nerve, and tendon injuries, Wang and colleagues introduce various types of cells involved in the skin regeneration, and how the cell-derived products regulate the repairing. Fan and colleagues additionally review the functions of muscles, the influences of volumetric muscle loss (VML), and how it can be repaired via different biofabrication techniques. Nerve regeneration including peripheral nerve regeneration, spinal cord regeneration, and the clinical applications of nerve conduit are reviewed by Qian and colleagues. As a tendon complex, He and colleagues introduce the rotator cuff under clinical use and show how biofabrications regulate rotator cuff repairing.

In summary, this book provides a comprehensive review of novel biofabrication techniques including 3D printing, electrospinning, microfluidics, exosomes, and more for applications in bone and bone-related regeneration. We believe this book can serve as the definitive reference of the latest technologies that could inspire medical practitioners and translational researchers to improve the quality of life of patients, whether it is to transform clinical procedures in present or to propose novel procedures of improved biomaterials in the near future.

27 October 2021

Wenguo Cui
Department of Orthopaedics
Shanghai Key Laboratory
for Prevention and Treatment
of Bone and Joint Diseases
Shanghai Institute of Traumatology
and Orthopaedics, Ruijin Hospital
Shanghai Jiao Tong University School
of Medicine, 197 Ruijin 2nd Road
Shanghai 200025
P.R. China

Xin Zhao
Department of Biomedical Engineering
the Hong Kong Polytechnic University
Hung Hom, Hong Kong
P.R. China

Shen Liu
Department of Orthopaedics
Shanghai Jiao Tong University
Affiliated Sixth People's Hospital
600 Yishan Road, Shanghai 200233
P.R. China

Part II

Biomedical Applications in Orthopedics

9

3D Printing for Orthopedics

Lan Li[1,2] *and Qing Jiang*[1,2]

[1] *Drum Tower Hospital affiliated to Medical School of Nanjing University, State Key Laboratory of Pharmaceutical Biotechnology, Division of Sports Medicine and Adult Reconstructive Surgery, Department of Orthopedic Surgery, 321 Zhongshan Road, Nanjing 210000, P.R. China*
[2] *Jiangsu Engineering Research Center for 3D Bioprinting, 321 Zhongshan Road, Nanjing 210000, P.R. China*

9.1 Overview of 3D Printing Technology

In recent decades, regenerative medicine and tissue engineering research has focused on promoting the function and structural regeneration or replacement of damaged tissues and organs (such as bone, blood vessels, skin, nerve, and cartilage) [1–6]. The goal of promoting tissue regeneration could be achieved by creating porous three-dimensional (3D) scaffolds to provide an appropriate microenvironment for tissue and organ regeneration. These tissue engineering scaffolds need to fulfill the following requirements: (i) biocompatibility, (ii) biodegradability or absorbance after implantation, (iii) appropriate mechanical properties, and (iv) specific shape. Traditional porous scaffold preparation methods, such as the porogen method, template method, and foaming method, have their defects, such as uncontrollable internal pore structures, complex forming process, and simple external form, which cannot fully meet the needs of tissue engineering scaffolds [2, 7, 8].

The 3D printing technology based on computer-aided design (CAD) or computed tomography (CT) imaging 3D reconstruction has unique advantages in porous or customized structure fabrication [9]. Typically, the realization of 3D printing requires the cooperation of a 3D printer, computer, 3D design software, and bioink. Briefly, the 3D models are generated using CAD or other software, imported into the 3D printing equipment, and finally printed using bioink (Figure 9.1).

The 3D printing techniques include stereolithography (SLA) [10], digital light processing (DLP) [11], inkjet printing [12], fused deposition printing (FDM) [13], selective laser sintering (SLS) [14], and electron beam melting (EBM) [15]. SLA and DLP technologies use ultraviolet (UV) or laser to induce the polymerization of liquid materials containing photoinitiators to create 3D structures. The working principle of an inkjet 3D printer is similar to that of a traditional inkjet printer; both can spray droplets, but the former ejects bioink that can form stereostructures. The FDM

Biofabrication for Orthopedics: Methods, Techniques, and Applications, First Edition.
Edited by Wenguo Cui, Xin Zhao, and Shen Liu.

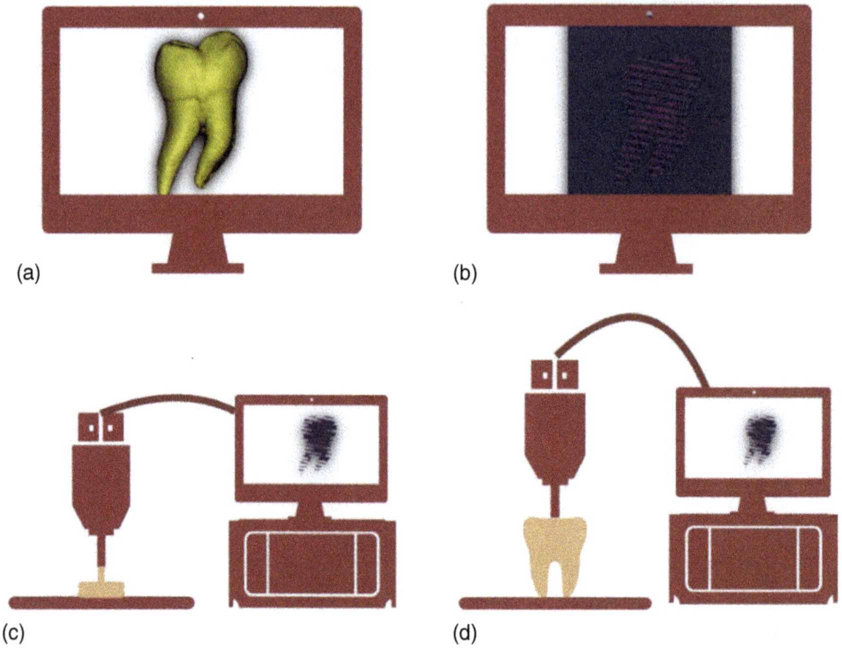

Figure 9.1 The process of 3D printing. (a) Importing the 3D model. (b) Slicing the model and planning the printing path. (c) Stacking the model layer by layer with bioink. (d) Finish printing.

technique uses thermoplastic materials that are heated, melted, and then extruded by pressure to construct 3D structures. SLS and EBM techniques use high-power laser or high-energy electron beam heating to melt the thermoplastic powder materials and fabricate 3D structures. These methods have been adapted in various fields, including architectural modeling, art, mechanical engineering, tissue engineering, and regenerative medicine.

9.2 Bone Tissue Engineering and 3D Printing

The treatment of large-scale bone defects caused by high-energy injury or pathological fracture is a major clinical challenge. Among all the treatment methods, internal fixation (intramedullary nail), external fixation (Ilizarov technique), autogenous, or allogeneic bone grafts are the most effective options, and autologous bone transplantation is considered as the gold standard for repairing large bone defects [16]. However, the limited range of donor sources, high morbidity of the donor site, and prolonged operation time are the common reasons for hindering the wide application of autologous bone graft [17]. The use of allogeneic or xenograft is also limited because of its shortcomings, such as immune rejection, disease transmission, and lack of osteogenic properties [18]. These risks might lead to bone absorption, nonunion, and failure of transplantation [19, 20].

In recent years, various biomaterials have been developed as substitutes for autogenous bone for tissue engineering. Hydroxyapatite (HA), β-tricalcium phosphate (β-TCP), calcium sulfate, and their composites are widely used in orthopedic reconstruction because of components similar to natural bone. Some commercial products based on these components, such as chronOS (β-TCP implants, Synthes, Switzerland), Stimulan Pellets/Dispenser (calcium sulfate, Biocomposites, England), Synatite (HA implants, S.B.M., France), Dragonbio (β-TCP/bioactive glass implants Mindray, China), Osteon bone substitute material (β-TCP/HA, Genoss, Korea), and NovaBone bone graft (bioactive glass, NovaBone Products, USA), have been used in clinics. However, because of varied bone defect types, the available commercial biomaterials cannot be bridged or filled with the missing bone tissue completely according to the anatomical shape and structure, thereby limiting their clinical application [21].

Tissue engineering based on 3D printing technology has overcome these difficulties and modified the traditional treatment of bone defects [22]. According to medical imaging data, 3D printing technology uses biomaterials to fabricate scaffolds with complex structures directly. These scaffolds can fulfill the specific needs of patients and realize the accurate design on macro- and microscale to provide an optimal structure for bone regeneration *in vivo* [23]. The continual development of 3D printing technology has identified a large number of tissue engineering substitutes with osteogenic potential in preclinical research.

9.3 Cartilage Tissue Engineering and 3D Printing

The predisposing factors of cartilage injury usually comprise osteoarthritis (OA) or sports trauma. Cartilage injury is characterized by progressive loss of hyaline cartilage, leading to pain, swelling, and stiffness in the joints. The treatment of this disease will cause a significant economic burden; the annual medical cost related to cartilage injury in the United States is between 3.4 and 13.2 billion USD [24]. Currently, the gold standard for treating cartilage injury is joint replacement, i.e. to replace the diseased cartilage and adjacent bone tissue with metal and polymer prosthesis. However, this method cannot guarantee the patients to return to the previous living standard, and the repeat surgery rate within 10 years is about 12% [25]. Therefore, developing a new method to improve the regeneration of cartilage is imperative.

The structure and component distribution of cartilage scaffold constructed by traditional tissue engineering methods is uniform. However, articular cartilage is a highly anisotropic tissue, and its composition and tissue alter gradually with depth. The tissue of natural cartilage can be divided into three regions from top to bottom: surface layer, middle layer, and deep layer; the biomechanical properties of each layer are different [26]. Therefore, one of the hotspot research directions of tissue-engineered scaffolds is the simulation of natural cartilage and subchondral bone structures. Another major challenge of cartilage tissue engineering is that the mechanical properties of cartilage scaffolds are often lower than those of natural cartilage and cannot bear a high level of compression and shear load for a prolonged period, which is the main reason for transplantation failure.

The layer-by-layer assembly of 3D printing technology allows accurate distribution of cells, bioactive factors, and biomaterials in the 3D space [27, 28]. It can realize the fabrication of cartilage scaffolds with bionic microstructure, components, and biomechanical properties similar to natural cartilage [29, 30]. In addition, because cartilage damage is often accompanied by subchondral bone damage, layered printing of cartilage and subchondral bone to prepare a biphasic structure scaffold is a major advantage of 3D printing technology.

9.4 Structural Requirements of 3D Printing

The microstructure of tissue-engineered scaffolds plays a critical role in maintaining cell viability and promoting tissue growth. Compared to the traditional manufacturing methods (such as porogen, template, and foaming methods), 3D printing technology can prepare tissue engineering scaffolds with controllable complex microstructure. The characteristics of scaffold microstructure are usually determined by pore size, porosity, stacking thickness, and pore configuration.

9.4.1 Pore Size

Pore size is a vital requirement of scaffolds for bone and cartilage regeneration. Among all the scaffold manufacturing techniques, 3D printing is the most suitable for fabricating porous structures as it can construct the internally connected porous structure [31]. A large number of studies have shown that the pore size of scaffolds is crucial for osteogenesis and chondrogenesis. When the pore size is between 100 and 700 μm, the cell migration and growth can be stimulated and is also conducive to the occurrence of neovascularization [32]. The ideal tissue scaffolds should meet the requirements of internal fluid transport and nutrient supply to promote the growth of external tissues. The pore size should not be large, which would otherwise result in the limited bridge formation between cells or reduce the stiffness of the scaffold structure. Therefore, pore size is required to balance the biological and mechanical characteristics with respect to specific tissue regeneration.

Li et al. prepared three kinds of porous Ti-6Al-4V scaffolds with different pore sizes (300–400, 400–500, and 500–700 μm) by EBM, and the elastic modulus was 3.7–1.7 GPa [33]. The scaffold with a pore size of 300–400 μm demonstrated the optimum cell behavior, and the osteogenic effect was verified in the following one year based on animal experiments. The study found that the scaffolds with aperture 300–400 μm could promote bone regeneration and integrate with the surrounding bone tissue.

Gao et al. have designed and fabricated a thermos-responsive supramolecular hydrogel biphasic scaffold based on acrylamide using the thermal-assisted extrusion 3D printing technique [34]. The pore size was 630–700 μm. In addition, transforming growth factor-beta 1 (TGF-β1) and -TCP were used for simultaneous regeneration of cartilage and bone in the upper and lower structures of the scaffolds. The gradient scaffold showed optimal osteogenic and chondrogenic ability in the *in vitro* and *in vivo* experiments.

9.4.2 Porosity

Porosity plays a momentous role in the migration of cells. Appropriate porosity improves the combination of cells and scaffolds and expands the interacting surface area of scaffolds with the surrounding tissues. Scaffolds with suitable porosity can promote bone and cartilage regeneration while maintaining mechanical properties [35].

Kim et al. printed magnesium phosphate ceramic scaffolds with various porosities using paste-extrusion-based 3D printing technique. When the porosity increased from 25% to ≥30%, the strength of the scaffolds decreased, and the degradation rate increased significantly. An *in vivo* study showed that when the porosity was 30–45%, the degradation rate and strength of this bioceramic scaffold were suitable for new bone regeneration [36]. Shim et al. manufactured 3D absorbable polycaprolactone (PCL) membranes with different porosities (30%, 50%, and 70%) using the thermal melting 3D printing technique. Then, the effect in guiding bone regeneration has been evaluated. Both *in vitro* and animal experiments showed that when the porosity was low (30%), the osteogenic differentiation ability and new bone formation volume were significantly higher than those of the other groups [37].

The porosity requirement of cartilage scaffolds is usually higher than that of bone scaffolds. Levorson et al. evaluated the cellular behavior of fibrin and PCL electrospun scaffolds (made by electrospinning 3D printing technique) with different porosities (86–92%) and found that the scaffolds with high porosity can maintain the cell function in serum-free condition, promote the deposition of glycosaminoglycan, and form a microenvironment similar to the natural extracellular matrix [38].

9.4.3 Pore Structure

The pore structure can guide the growth and remodeling of the tissues surrounding the scaffolds. The optimization of the internal scaffold geometry mainly focuses on the optimization design of the pore structure.

Some studies demonstrate that the curvature of the cell surface is a major factor in determining the regeneration rate of the tissue. The actin fibers and myosin have a higher density on the concave surface, indicating a high local cellular stress state near the concave. However, changes in the surface curvature might alter the cell attachment rate, cell migration speed, and cell morphology (including cell diffusion region) [39]. This phenomenon shows that the tissue regeneration rate changes with altered curvature. The regeneration speed of tissue on the concave surface is faster than that of convex and flat surfaces. Therefore, the design and manufacture of pore structures are crucial for the optimization of scaffold geometry [40].

Ferlin et al. found that stem cells were more likely to prefer adipogenic and chondrogenic differentiation in 3D-printed scaffolds made of ordered cubic pores. The osteogenic differentiation could be observed in the scaffolds of ordered circular pores [23]. Bobbert et al. used the SLS technique to manufacture titanium alloy scaffolds with different pore sizes and porosities based on triply periodic minimal surface (TPMS). They characterized the topological structure, quasi-static mechanical properties, fatigue resistance, and permeability [41]. The results showed that these scaffolds have morphological characteristics similar to trabeculae, including mean surface curvature close to zero. The biomechanical properties were also similar to the trabecular bone, thus providing sufficient mechanical support for bone regeneration and bone integration while avoiding stress occlusion. The porous scaffold also showed high fatigue resistance, which simulates the topological structure, mechanics, and physical properties of the bone trabeculae. These characteristics indicated that the TPMS method is feasible for bone tissue engineering scaffold design.

In addition, because the structure distribution of natural tissue is not uniform, the hierarchical porous structure is also a hotspot in the design of scaffold geometry. The changing characteristics from nanometer to millimeter scale facilitate the scaffolds to meet the requirements of bionics regarding the morphology, structure, and function [42]. Shi et al. constructed a porous gradient structure based on the octahedral unit by forward solid modeling method. They obtained a titanium alloy scaffold with a solid layer and two different porosity layers by SLS printing. The elastic modulus of this scaffold was close to that of the cortical bone and also showed biocompatibility in the cell culture [43]. Afshar et al. constructed porous scaffolds with linear gradient transformation based on P-shaped surface and D-shaped surface by TPMS method. The scaffolds were fabricated using the Polyjet 3D printing technique. Compared to the uniform porous structure, the linear gradient porous structure has better mechanical properties in tension and compression. Such a linear structure without obvious delamination can simulate the microstructure changing trend of natural bone and cartilage. Besides, this structure can improve the mechanical properties while maintaining the stiffness, which provides a novel idea for the design and preparation of the bionic structure [44].

However, studies on the exact effect of the porous structure on osteogenesis of 3D printing scaffolds are insufficient. The majority of the studies have focused on regulating the mechanical properties of scaffolds, and only pore size and porosity are considered in 3D-printed bone and cartilage scaffolds. The pore geometry related to these tissue engineering scaffolds needs to be explored further.

9.5 Biomaterials for 3D Printing

The biomaterials for 3D printing can be classified into natural or synthetic materials. According to the types of printing methods, the most commonly used materials are metal, ceramic, and polymers (Figure 9.2).

(a) (b) (c)

Figure 9.2 3D printing tissue engineering scaffolds made of different materials. (a) Titanium alloy scaffold. (b) Bioceramic scaffold; (c) Polymer scaffold.

9.5.1 Metal

The metal materials that can be used for 3D printing include alloys based on iron, cobalt, chromium, magnesium, titanium, and nickel, which have suitable modulus and strength for bone tissue engineering. The biological inertia of these metal materials also ensures the safety of their usage. Among all metal materials, titanium alloys have several advantages, such as good biocompatibility, high strength-to-weight ratio, low elastic modulus, and excellent corrosion resistance. Therefore, it is widely used in bone regeneration and remodeling of massive bone defects, for example, providing support in the form of a plate, mesh, or cage during bone healing [45–47]. Presently, a variety of porous titanium-based scaffolds prepared by 3D printing techniques have shown adequate results in several preclinical and clinical studies.

However, there are some difficulties in the application of metal materials: (i) the material requirements for 3D printing are higher than traditional methods, especially the particle size distribution with respect to the metal; (ii) the metal will corrode after long-term implantation *in vivo* and release in metal ions. These metal ions cannot be removed by the body, which might cause osteolysis around the prosthesis or some systemic symptoms [48, 49]; and (iii) metals are usually used as permanent implants because they cannot be degraded. However, the newborn functional tissues can only surround the implants but not replace them.

Thus, the researchers are currently focusing on biodegradable metal materials. Such metal is expected to be gradually degraded in the host body, and the released degradation products can stimulate the regeneration of the surrounding tissue, which would be completely degraded after tissue healing in the implantation area [50]. Iron-based and magnesium-based biodegradable metals meet the requirements mentioned above. A previous study has shown that Fe–Mn alloy-based bone scaffolds can be prepared by inkjet 3D printing technique using Fe–Mn powder as a powder bed and a water-based organic solvent as a binder [51]. Preosteoblasts showed adequate cell viability on the scaffold. The degradation rate was higher, and the deposition of calcium, phosphorus, and other bone mineral components was more obvious compared to the pure iron scaffold. The scaffold prepared by this kind of metal material is suitable for the repair of non-load bearing areas, such as the craniofacial region.

Mg-based biodegradable alloys have excellent bioactivity and osteogenic ability and can also be used to prepare scaffolds by inkjet 3D printing. Mostafa et al. developed a novel porous magnesium scaffold, which forms a connected structure with 43.2–54.2% porosity [52]. The elastic modulus of the scaffold was 0.2–1 GPa, and the compressive strength was 4.3–6.2 MPa, which was close to that of cancellous bone. In the animal study by Liu et al., the porous Mg scaffold was implanted into the femoral condyle of New Zealand rabbits and was found to be degraded from the outside to the inside after three months. In addition, the growth of the newly born bone tissue was observed following the degradation of the scaffolds [53]. Therefore, Mg-based biodegradable alloy is a promising biomaterial for segmental bone defect repair.

9.5.2 Bioceramics

Bioceramics contain metallic and non-metallic components, which have been widely used in 3D printing because of their optimal biocompatibility, high mechanical strength, good biodegradability, and osteoinductive ability [54]. HA and TCP are two of the most widely used bioceramics.

HA is commonly found in human teeth and bones, and scaffolds printed with this material have mechanical strength similar to natural bone. HA scaffolds with complex internal pore structures can be prepared by layer-by-layer 3D printing technology. These structures improve the internal microenvironment of scaffolds and provide sufficient oxygen and nutrients for cell proliferation [55]. According to a previous study, HA and TCP composite porous scaffolds were produced by 3D printing and sintering. This composite bioceramic scaffold showed excellent mechanical properties, high biocompatibility, and low cytotoxicity after sintering [56]. Some studies demonstrated that ceramic calcium silicate ($CaSiO_3$) with adequate osteogenic potential could be printed accurately and stimulate bone repair in animal models [57].

Correia et al. printed bioceramic scaffolds with 40–60% porosity by extrusion-based 3D printing using TCP/sodium alginate bioink and then loaded with functionalized silver nanoparticles. This scaffold structure is suitable for bone tissue regeneration because of its good mechanical properties, biocompatibility, and bactericidal ability [58]. Yang et al. designed tricalcium silicate bioceramic slurry without high-temperature sintering. After 3D printing, it can self-assemble in water to form a stable scaffold structure. Through surface modification of the scaffold and drug loading, this slurry enhances the adhesion of rat bone marrow mesenchymal stem cells and the expression of alkaline phosphatase and also shows remarkable osteogenic ability in animal models [59].

Fabrication of bioceramic material grafts with complex internal structures and customized shapes by traditional methods is rather challenging. However, the bionic extracellular matrix structure for cell attachment and proliferation can be designed and prepared by 3D printing technology [60]. The high-resolution design of pore connectivity renders biocompatibility and compressive strength to the bioceramic materials, thereby facilitating tissue regeneration and bone repair in non-load-bearing and load-bearing areas.

9.5.3 Polymer Materials

Polymer materials are widely used in fabricating various tissue engineering scaffolds. Compared to metals and ceramics, polymer materials have great design flexibility because their composition and structure can be customized according to specific needs. Based on the molecular design, polymer materials have the features of hydrolysis or enzymatic hydrolysis, which is conducive to its degradation in the body. Also, the bioactive factors, drugs, or proteins can be grafted by side-chain modification, such that the polymer scaffolds can promote tissue regeneration. Common polymers are divided into natural and synthetic polymers. Natural polymers include hyaluronic acid, gelatin, alginate, chitosan, collagen, and other polymers from natural tissues, while synthetic polymers include poly(lactic *co* glycolic acid) (PLGA), poly(methyl methacrylate) (PMMA), PCL, and polyethylene glycol (PEG) [61].

Typically, the polymer materials are constructed layer by layer in extrusion-based 3D printing. Natural polymer materials and some synthetic polymers can be easily made into hydrogel-type bioinks. Their hydration and biocompatibility are similar to those of the biological tissues, and their mechanical strength can be adjusted to form a stable three-dimensional structure. Some studies preferred photo-crosslinked hyaluronic acid hydrogel to encapsulate the chondrocytes to repair the injured cartilage. The advantage of this photo-crosslinked hyaluronic acid hydrogel is its injectable property and application in various irregular-shaped cartilage defects. Nettles et al. used photo-crosslinked hyaluronic acid hydrogel to wrap the chondrocytes for the repair of porcine cartilage defects. The photopolymerization hydrogel can be tightly bonded with the surrounding cartilage and can significantly promote the formation of a cartilage-related matrix [62].

Owing to its mechanical strength, the artificial polymer is suitable as the rigid network structure of the scaffold. Temple et al. developed a series of customized 3D printing PCL scaffolds, which have anatomically shaped external profiles and various internal porosities and can treat large craniomaxillofacial bone defects [63]. In this preclinical study, the external shape of the scaffold was designed according to the patients' imaging data, while the internal porosity of the scaffold promotes the differentiation of human adipose-derived stem cells into blood vessels and bone. In another study, macroporous and microporous structures were also observed in 3D-printed PLGA scaffolds with different pore sizes [64]. The modulus was similar to that of trabecular bone, and human osteoblast proliferation, alkaline phosphatase, and osteonectin expression were also observed. These results indicated that these scaffolds have excellent bone conduction properties that promote osteogenesis.

9.6 Application of 3D Printing in Cell Printing and Orthopedic Tissue Engineering

The commonly used 3D printing technologies for cell printing are inkjet 3D printing, extrusion-based 3D printing, and laser-assisted 3D printing. The different characteristics of these technologies need to be considered for various applications and purposes. The main technical characteristics are listed in Table 9.1.

9.6.1 Inkjet 3D Printing

Inkjet 3D printing is the most commonly used type of printing in biological applications. A controlled volume of liquid can be delivered to a predetermined location. Inkjet 3D printers usually use thermal or piezoelectric to spray droplets onto the substrate [64, 65]. The thermal inkjet printer generates a pressure pulse by heating the print head to force the nozzle to produce droplets. This local heating (200–300 °C) does not exert any significant effect on the viability or post-printing function of biomolecules and mammalian cells [66, 67]. The advantages of a thermal inkjet printer include high printing speed, low cost, and a wide application range. However, when cells and materials are exposed to thermal and mechanical stress, the defects such as low controllability of droplet direction, uneven droplet size, frequent nozzle clogging, and difficulty in cell encapsulation also limit the application of thermal inkjet 3D printing technology.

The piezoelectric inkjet printer generates sound waves in the printing head through piezoelectric crystal, breaks the liquid into droplets at regular intervals, and ejects the droplets out of the nozzle through the rapidly changing shape. This technique can produce uniform size droplets, control the spray direction of droplets accurately, and avoid the effect of high temperature and shear force on cells [68].

The inkjet 3D printing technology can repair the cartilage tissue without any additional damage. The implanted tissue should be integrated with the natural

Table 9.1 Characteristics of different 3D printing technologies.

Parameters	Types		
	Inkjet print	Microextrusion print	Laser-assisted print
Material viscosity	3.5–12 mPa/s	>30 mPa/s	1–300 mPa/s
Crosslink	Chemical/photo-polymerization	Chemical/photo-polymerization/temperature/shear stress	Chemical/photo-polymerization
Speed	1–10 000 drop/s	10–15 μm/s	200–1600 mm/s
Accuracy	<1 pl	>5 μm	Microscale
Cell viability	>85%	40–80%	>95%
Cell density	$<10^6$/ml	High, cell spheres is acceptable	10^8/ml

cartilage to repair the lesions of different sizes and thicknesses. Cui et al. improved the technology of hot inkjet printing and accurately printed human articular chondrocytes and PEG dimethyl acrylate layer by layer for cartilage repair [69]. The bioink can crosslink through photopolymerization during the 3D printing process, and compared to manual extrusion, the printing speed is faster, the risk of cell pollution is smaller, and cell vitality is higher [70].

Furthermore, the combination of electrospinning and inkjet printing technology can be applied to construct the scaffold of multilayer bionic cartilage structure *in vitro* [71], which maintains the cell function as the mechanical and structural characteristics. The inkjet printing technology can also be used to make bone-like structures [72]. This bionic structure can promote the continuous differentiation of stem cells in animal models and form highly mineralized tissues; its density is similar to that of endogenous bone tissue.

9.6.2 Extrusion-Based 3D Bioprinting

The principle of extrusion-based 3D printer is similar to the dispensing machine. Various types of bioinks, such as hydrogel, biocompatible polymer, and cell spheres, could be used in this technique [73]. The most common extruding process is generated by pneumatic or mechanical (piston or screw system) mechanism [74, 75]. The advantage of pneumatic printing is that it has a simpler driving mechanism, and the extrusion force is limited only by the air pressure. The mechanical driving mechanism has a small and complex component structure, providing a larger printing space than pneumatic printing equipment when the equipment volume is similar. However, its extrusion force cannot be increased because of the limitation in the equipment volume.

Fedorovich et al. used human osteoblast precursors, chondrocytes, and alginate hydrogel as the biological ink and printed the porous scaffold through the extrusion-based printing method and then implanted it into mice. The results showed that the cells could survive in the preset area and differentiate into bone or cartilage [76]. Esfahani et al. employed this technique to print bioceramic scaffolds with sufficient mechanical strength for the treatment of defects on load-bearing bone. Compared to the pore structure of other traditional patterns, the hexagonal porous geometry achieved a large contact area and enhanced the ability of load transfer between the print layers [77]. Hung et al. synthesized the aqueous dispersion of elastic biodegradable polyurethane (PU) nanoparticles and used polyethylene oxide as a tackifier for fabricating 3D printing scaffolds. Then, chondrocytes were implanted, and their potential in cartilage tissue engineering was evaluated. The results showed that chondrocytes had satisfactory migration, proliferation, and extracellular matrix expression in this 3D printing polyurethane scaffold [78]. Yang et al. used type I collagen or agarose mixed with sodium alginate as bioink, which was combined with chondrocytes to construct a 3D cartilage tissue by extrusion-based bioprinting. The effect of different bioinks in 3D printing cartilage scaffolds was analyzed. The results exhibited that the bioink composed of sodium alginate and collagen significantly promoted cell adhesion, accelerated cell proliferation, enhanced cartilage-specific gene expression, inhibited chondrocyte dedifferentiation, and maintained chondrocyte phenotype [79].

Bella et al. developed a handheld 3D printing pen, which directly printed the bioink and cultured cells into cartilage defects by coaxial extrusion. The study also verified the feasibility of this technique in a goat model; early cartilage regeneration was observed at the *in situ* printing site after eight weeks [80]. Lipskas et al. used a robot-assisted method to fill the bone and cartilage defects *in situ* in the *in vitro* experiment. The results showed that this technology could repair the defect area rather satisfactorily [81].

9.6.3 Laser-Assisted Printing

Laser-assisted bioprinting (LAB) is a technology of additive manufacturing based on the principle of laser-induced forward transfer [82]. It was originally designed to print metal materials and since then has been successfully applied to biomaterials, such as peptides, DNA, and cells [83]. Although the application scope of LAB is less than that of inkjet printing or extrusion printing, it is widely used in the manufacture of tissues and organs. The laser does not directly contact the cells or bioink, but the pulse is focused through the energy absorption layer to generate high-pressure bubbles that push the bioink and cells to the base. Therefore, this printing method can maintain the vitality and function of the cells in high-speed and high-resolution printing.

Keriquel et al. tried to repair the skull defect *in situ* using this technique [84]. A bioink composed of nano-HA, collagen, and preosteoblasts was utilized in the study, and the bone defect area was repaired after eight weeks. Olivia et al. used a similar method to print scaffold in the rat skull defect region and evaluated the vascularization of the regenerated tissue. The results showed obvious angiogenesis in the defect area, which is critical for bone formation and reconstruction [85].

9.7 Future Prospects

Currently, the mainstream application of 3D printing technology in bone and cartilage regeneration is the traditional tissue engineering strategy, i.e. fabricating scaffolds *in vitro* and co-culturing with seed cells to induce the cells to differentiate into the target tissue and implant these scaffolds *in vivo*. However, this method is a time-consuming, complex operation, and hence, it cannot be used in the treatment of all clinical diseases.

Therefore, some researchers have begun to focus on the *in situ* printing in the lesion area. Nonetheless, the implementation of *in situ* printing has several limitations. The handheld 3D printing pen mentioned above can only fulfill the extrusion function of bioink but cannot achieve the unique trajectory planning and porous structure molding of 3D printing. Limited by its forming principle, the laser-assisted 3D printing technology cannot stack high-layer thickness and fill a large volume of osteochondral lesions. Therefore, the *in situ* 3D printing technology based on

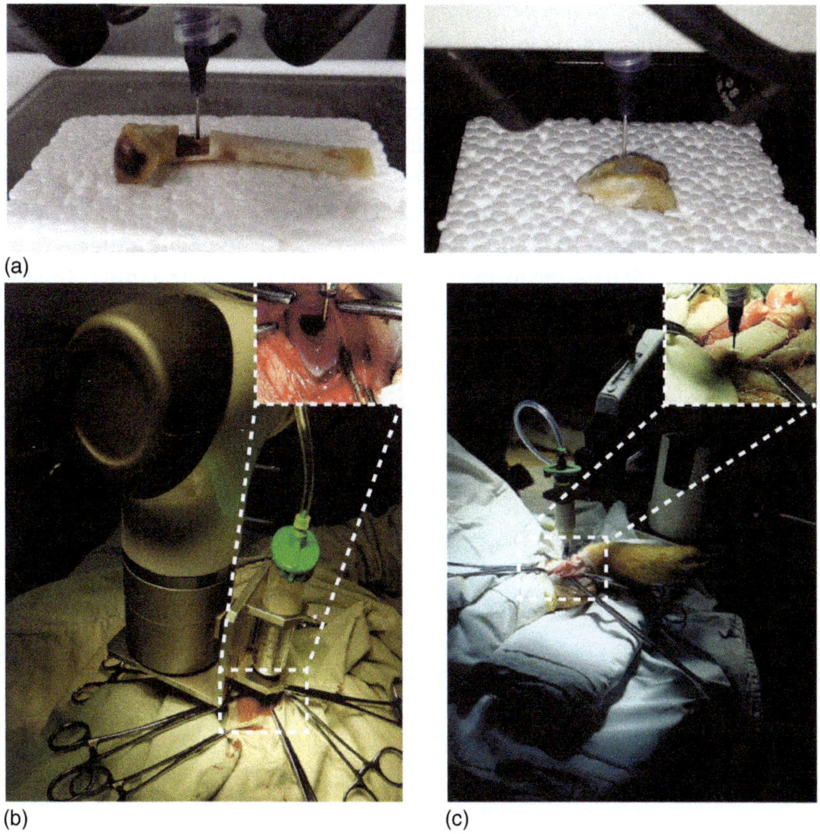

Figure 9.3 The *in situ* 3D print based on microextrusion print. (a) Repair the bone and osteochondral defects using a traditional 3D printer. (b) Repair the cartilage defects in rabbit model using a 6-dof robot. (c) Repair the bone defects in swine model using a 4-dof robot.

extrusion and robot-assisted technique, owing to the wide application of the biological ink and controllable molding process, is the most likely technical solution to achieve this goal. In a study conducted by Jiang et al. in 2017, the feasibility of using a 3D scanner and a traditional extrusion-based 3D printer was explored to fix the bone and cartilage lesions *in situ* (Figure 9.3a) [85, 86]. The findings demonstrated that the accuracy and crosslinking time were acceptable. In the subsequent studies, the 4-dof and 6-dof robots were employed to act as the 3D printer (Figure 9.3b,c). The extrusion spray nozzle was assembled at the end of the robot. The bone and cartilage injuries in living animal models were repaired by this technology [87, 88]. Therefore, achieving the 3D models of the target area using 3D scanning, CT, or magnetic resonance imaging (MRI), importing the data into the robotic 3D printer, and then completing the damage repair using *in situ* 3D printing technology may be the next-generation clinical solution.

However, treating lesions by fabricating implants *in situ* using 3D printing technology faces challenges, including synthesizing biomaterials, forming scaffolds with defect-like shapes, and developing an appropriate 3D bioprinter. Among these challenges, the fast/biocompatible crosslinking reaction of bioink is the foundation of *in situ* 3D printing. The commonly used crosslinking methods, such as high or low temperature, unphysiological ionic concentrations, or strong ultraviolet or blue light, are not suitable for *in vivo* 3D printing because of extremely slow reaction speed or undesirable side products . Moreover, the physiological environment is filled with body fluid, which might induce negative effects on the gelation of bioink and unfavorable printing shapes. Reducing the reaction time with mild reaction conditions *in situ* is vital for the realization of precision printing structures. Another crucial factor is the 3D bioprinter, the direct executor of the *in situ* printing process. Traditional 3D printers usually consist of a ball-screw drive system, which is large and heavy. The printing space is limited by the length of the ball-screw system, which hinders the convenience and maneuverability in direct 3D printing for tissue repair. A 3D bioprinter with small volume and large working space is the future direction in the development of such equipment.

Acknowledgments

This work was supported by the National Natural Science Foundation of China (81730067 and 81991514), the National Key Research and Development Project (2018YFF0301101), the Natural Science Foundation of Jiangsu Province (BK20200121), Jiangsu Provincial Key Medical Center Foundation, and the Jiangsu Provincial Medical Outstanding Talent Foundation.

References

1 Amini, A.R., Laurencin, C.T., and Nukavarapu, S.P. (2012). Bone tissue engineering: recent advances and challenges. *Crit. Rev. Biomed. Eng.* 40: 363–408.

2 Kim, J.E., Kim, S.H., and Jung, Y. (2016). Current status of three-dimensional printing inks for soft tissue regeneration. *Tissue Eng. Regener. Med.* 13: 636–646.

3 Gu, X., Ding, F., and Williams, D.F. (2014). Neural tissue engineering options for peripheral nerve regeneration. *Biomaterials* 35: 6143–6156.

4 Jung, J.P., Bhuiyan, D.B., and Ogle, B.M. (2016). Solid organ fabrication: comparison of decellularization to 3D bioprinting. *Biomater. Res.* 20: 27.

5 Song, Y., Su, X., Firouzian, K.F. et al. (2020). Engineering of brain-like tissue constructs via 3D cell-printing technology. *Biofabrication* 12 (3): 035016.

6 Li, X., Liu, L., Zhang, X., and Xu, T. (2018). Research and development of 3D printed vasculature constructs. *Biofabrication* 10 (3): 032002.

7 Chan, B.P. and Leong, K.W. (2008). Scaffolding in tissue engineering: general approaches and tissue-specific considerations. *Eur. Spine J.* 17: 467–479.

8 Gu, B.K., Park, S.J., Kim, M.S. et al. (2013). Fabrication of sonicated chitosan nanofiber mat with enlarged porosity for use as hemostatic materials. *Carbohydr. Polym.* 97: 65–73.

9 Bikas, H., Stavropoulos, P., and Chryssolouris, G. (2016). Additive manufacturing methods and modelling approaches: a critical review. *Int. J. Adv. Manuf. Technol.* 83: 389–405.

10 Quan, Z., Wu, A., Keefe, M. et al. (2015). Additive manufacturing of multi-directional preforms for composites: opportunities and challenges. *Mater. Today* 18: 503–512.

11 Holmes, L.R. and Riddick, J.C. (2014). Research summary of an additive manufacturing technology for the fabrication of 3D composites with tailored internal structure. *JOM* 66: 270–274.

12 Zhang, Y.S., Yue, K., Aleman, J. et al. (2017). 3D bioprinting for tissue and organ fabrication. *Ann. Biomed. Eng.* 45: 148–163.

13 Turner, B.N., Strong, R., and Gold, S.A. (2014). A review of melt extrusion additive manufacturing processes: I. Process design and modeling. *Rapid Prototyp. J.* 20: 192–204.

14 Ii, W.S.L., Zhang, B., Ziegert, J. et al. (2015). In-situ metrology system for laser powder bed fusion additive process. *Procedia Manuf.* 1: 393–403.

15 Sing, S.L., An, J., Yeong, W.Y., and Wiria, F.E. (2016). Laser and electron-beam powder-bed additive manufacturing of metallic implants: a review on processes, materials and designs. *J. Orthop. Res.* 34 (3): 369–385.

16 Cyril, M., Brian Thomas, B., and Wade, S. (2015). Management of segmental bone defects. *J. Am. Acad. Orthop. Surg.* 23: 143–153.

17 Loeffler, B.J., Kellam, J.F., Sims, S.H., and Bosse, M.J. (2012). Prospective observational study of donor-site morbidity following anterior iliac crest bone-grafting in orthopaedic trauma reconstruction patients. *J. Bone Joint Surg. Am.* 94: 1649–1654.

18 Delloye, C., Cornu, O., Druez, V., and Barbier, O. (2007). Bone allografts: what they can offer and what they cannot. *J. Bone Joint Surg. Br.* 89: 574–579.

19 Athanasiou, V.T. (2010). Histological comparison of autograft, allograft-DBM, xenograft, and synthetic grafts in a trabecular bone defect: an experimental study in rabbits. *Med. Sci. Monit.* 16: 24.

20 Dimitriou, R., Jones, E., Mcgonagle, D. et al. (2011). Bone regeneration: current concepts and future directions. *BMC Med.* 9: 66.

21 Jardini, A.L., Larosa, M.A., Filho, R.M. et al. (2014). Cranial reconstruction: 3D biomodel and custom-built implant created using additive manufacturing. *J. Craniomaxillofac. Surg.* 42: 1877–1884.

22 Jones, N. (2012). Science in three dimensions: the print revolution. *Nature* 487: 22.

23 Ferlin, K.M., Prendergast, M.E., Miller, M.L. et al. (2016). Influence of 3D printed porous architecture on mesenchymal stem cell enrichment and differentiation. *Acta Biomater.* 32: 161–169.

24 Buckwalter, J.A., Saltzman, C., and Brown, T. (2004). The impact of osteoarthritis: implications for research. *Clin. Orthop. Relat. Res.* 427: S6–S15.

25 Labek, G., Thaler, M., Janda, W. et al. (2011). Revision rates after total joint replacement. *J. Bone Joint Surg. Br.* 93B (3): 293–297.

26 Gannon, A.R., Nagel, T., and Kelly, D.J. (2012). The role of the superficial region in determining the dynamic properties of articular cartilage. *Osteoarthr. Cartil.* 20: 1417–1425.

27 Vhm, M., Levato, R., Bonassar, L.J. et al. (2017). Three-dimensional bioprinting and its potential in the field of articular cartilage regeneration. *Cartilage* 8: 327–340.

28 Murphy, S.V. and Atala, A. (2014). 3D bioprinting of tissues and organs. *Nat. Biotechnol.* 32: 773–785.

29 Schuurman, W., Khristov, V., Pot, M.W. et al. (2011). Bioprinting of hybrid tissue constructs with tailorable mechanical properties. *Biofabrication* 3: 021001.

30 Visser, J., Melchels, F.P., Jeon, J.E. et al. (2015). Reinforcement of hydrogels using three-dimensionally printed microfibres. *Nat. Commun.* 6: 6933.

31 Nagarajan, N., Dupret-Bories, A., Karabulut, E. et al. (2018). Enabling personalized implant and controllable biosystem development through 3D printing. *Biotechnol. Adv.* 36: 521–533.

32 Wang, X., Xu, S., Zhou, S. et al. (2016). Topological design and additive manufacturing of porous metals for bone scaffolds and orthopaedic implants: a review. *Biomaterials* 83: 127–141.

33 Li, G., Wang, L., Pan, W. et al. (2016). In vitro and in vivo study of additive manufactured porous Ti6Al4V scaffolds for repairing bone defects. *Sci. Rep.* 6: 34072.

34 Gao, F., Xu, Z.Y., Liang, Q.F. et al. (2018). Direct 3D printing of high strength biohybrid gradient hydrogel scaffolds for efficient repair of osteochondral defect. *Adv. Funct. Mater.* 28: 1706644.

35 Roy, T.D., Simon, J.L., Ricci, J.L. et al. (2010). Performance of degradable composite bone repair products made via three-dimensional fabrication techniques. *J. Biomed. Mater. Res. Part A* 66A: 283–291.

36 Kim, J.A., Lim, J., Naren, R. et al. (2016). Effect of the biodegradation rate controlled by pore structures in magnesium phosphate ceramic scaffolds on bone tissue regeneration in vivo. *Acta Biomater.* 44: 155–167.

37 Shim, J.H., Jeong, J.H., Won, J.Y. et al. (2017). Porosity effect of 3D-printed polycaprolactone membranes on calvarial defect model for guided bone regeneration. *Biomed. Mater.* 13: 015014.

38 Levorson, E.J., Sreerekha, P.R., Chennazhi, K.P. et al. (2016). Fabrication and characterization of multiscale electrospun scaffolds for cartilage regeneration. *Biomed. Mater.* 8: 014103.

39 Zadpoor, A.A. (2015). Bone tissue regeneration: the role of scaffold geometry. *Biomater. Sci.* 3: 231–245.

40 Bidan, C.M., Kommareddy, K.P., Rumpler, M. et al. (2013). Geometry as a factor for tissue growth: towards shape optimization of tissue engineering scaffolds. *Adv. Healthc. Mater.* 2: 186–194.

41 Bobbert, F.S.L., Lietaert, K., Eftekhari, A.A. et al. (2017). Additively manufactured metallic porous biomaterials based on minimal surfaces: a unique combination of topological, mechanical, and mass transport properties. *Acta Biomater.* 53: 572–584.

42 Kureshi, A. (2010). Alignment hierarchies: engineering architecture from the nanometre to the micrometre scale. *J. R. Soc. Interface* 7 (Suppl 6): S707–S716.
43 Shi, J., Yang, J., Li, Z. et al. (2017). Design and fabrication of graduated porous Ti-based alloy implants for biomedical applications. *J. Alloys Compd.* 728: 1043–1048.
44 Afshar, M., Anaraki, A.P., Montazerian, H. et al. (2016). Additive manufacturing and mechanical characterization of graded porosity scaffolds designed based on triply periodic minimal surface architectures. *J. Mech. Behav. Biomed. Mater.* 62: 481–494.
45 Spoerke, E.D., Murray, N.G., Li, H. et al. (2010). Titanium with aligned, elongated pores for orthopedic tissue engineering applications. *J. Biomed. Mater. Res. Part A* 84A: 402–412.
46 Sakat, M.S., Kilic, K., Altas, E. et al. (2016). Comminuted frontal sinus fracture reconstructed with titanium mesh. *J. Craniofac. Surg.* 27: e207.
47 Sebastiaan Souer, J., David, R., Stefan, M. et al. (2010). Comparison of functional outcome after volar plate fixation with 2.4-mm titanium versus 3.5-mm stainless-steel plate for extra-articular fracture of distal radius. *J. Hand Surg.* 35: 398–405.
48 Chung, W.O., Wataha, J.C., Hobbs, D.T. et al. (2011). Peroxotitanate- and monosodium metal-titanate compounds as inhibitors of bacterial growth. *J. Biomed. Mater. Res. Part A* 97A: 348–354.
49 Davis, R.R., Hobbs, D.T., Rania, K. et al. (2010). Titanate particles as agents to deliver gold compounds to fibroblasts and monocytes. *J. Biomed. Mater. Res. Part A* 93A: 864–869.
50 Zheng, Y.F., Gu, X.N., and Witte, F. (2014). Biodegradable metals. *Mater. Sci. Eng., R* 77: 1–34.
51 Chou, D.T., Wells, D., Hong, D. et al. (2013). Novel processing of iron–manganese alloy-based biomaterials by inkjet 3-D printing. *Acta Biomater.* 9: 8593–8603.
52 Mostafa, Y., Mehdi, R., Daryoosh, V. et al. (2017). Porous magnesium-based scaffolds for tissue engineering. *Mater. Sci. Eng., C* 71: 1253–1266.
53 Liu, Y.J., Yang, Z.Y., Tan, L.L. et al. (2014). An animal experimental study of porous magnesium scaffold degradation and osteogenesis. *Braz. J. Med. Biol. Res.* 47: 715–720.
54 Seitz, H., Rieder, W., Irsen, S. et al. (2005). Three-dimensional printing of porous ceramic scaffolds for bone tissue engineering. *J. Biomed. Mater. Res. Part B* 74B: 782–788.
55 Leukers, B., Gülkan, H., Irsen, S.H. et al. (2005). Hydroxyapatite scaffolds for bone tissue engineering made by 3D printing. *J. Mater. Sci. - Mater. Med.* 16: 1121–1124.
56 Warnke, P.H., Seitz, H., Warnke, F. et al. (2010). Ceramic scaffolds produced by computer-assisted 3D printing and sintering: characterization and biocompatibility investigations. *J. Biomed. Mater. Res. Part B* 93B: 212–217.

57 Wu, C., Wei, F., Zhou, Y. et al. (2012). 3D-printing of highly uniform $CaSiO_3$ ceramic scaffolds: preparation, characterization and in vivo osteogenesis. *J. Mater. Chem.* 22: 12288–12295.

58 Correia, T.R., Figueira, D.R., Sá, K.D.D. et al. (2016). 3D printed scaffolds with bactericidal activity aimed for bone tissue regeneration. *Int. J. Biol. Macromol.* 93: 1432–1445.

59 Yang, C., Wang, X., Ma, B. et al. (2017). 3D-printed bioactive Ca_3SiO_5 bone cement scaffolds with nano surface structure for bone regeneration. *ACS Appl. Mater. Interfaces* 9: 5757.

60 Castilho, M., Moseke, C., Ewald, A. et al. (2014). Direct 3D powder printing of biphasic calcium phosphate scaffolds for substitution of complex bone defects. *Biofabrication* 6: 015006.

61 Li, L., Yu, F., Zheng, L. et al. (2019). Natural hydrogels for cartilage regeneration: modification, preparation and application. *J. Orthop. Translat.* 17: 26–41.

62 Nettles, D.L., Parker Vail, T., Morgan, M.T. et al. (2004). Photocrosslinkable hyaluronan as a scaffold for articular cartilage repair. *Ann. Biomed. Eng.* 32: 391–397.

63 Temple, J.P., Hutton, D.L., Hung, B.P. et al. (2015). Engineering anatomically shaped vascularized bone grafts with hASCs and 3D-printed PCL scaffolds. *J. Biomed. Mater. Res. Part A* 102: 4317–4325.

64 Kai, L., Tai, W.F.V., Tong, C. et al. (2009). Proliferation and differentiation of human osteoblasts within 3D printed poly-lactic-*co*-glycolic acid scaffolds. *J. Biomater. Appl.* 23: 533.

65 Cui, X., Boland, T., D'Lima, D.D. et al. (2012). Thermal inkjet printing in tissue engineering and regenerative medicine. *Recent Pat. Drug Deliv. Formul.* 6: 149–155.

66 Xu, T., Jin, J., Gregory, C. et al. (2005). Inkjet printing of viable mammalian cells. *Biomaterials* 26: 93–99.

67 Xu, T., Gregory, C.A., Molnar, P. et al. (2006). Viability and electrophysiology of neural cell structures generated by the inkjet printing method. *Biomaterials* 27: 3580–3588.

68 Saunders, R.E., Gough, J.E., and Derby, B. (2008). Delivery of human fibroblast cells by piezoelectric drop-on-demand inkjet printing. *Biomaterials* 29: 193–203.

69 Cui, X., Breitenkamp, K., Finn, M.G. et al. (2012). Direct human cartilage repair using three-dimensional bioprinting technology. *Tissue Eng. Part A* 18: 1304.

70 Kim, T.K., Sharma, B., Williams, C.G. et al. (2003). Experimental model for cartilage tissue engineering to regenerate the zonal organization of articular cartilage. *Osteoarthr. Cartil.* 11: 653–664.

71 Xu, T., Binder, K.W., Albanna, M.Z. et al. (2013). Hybrid printing of mechanically and biologically improved constructs for cartilage tissue engineering applications. *Biofabrication* 5: 015001.

72 Paolo, D.C., Georg, B., Minhaj Siddiqui, M. et al. (2007). Isolation of amniotic stem cell lines with potential for therapy. *Nat. Biotechnol.* 25: 100–106.

73 Peltola, S., Melchels, F., Grijpma, D.W., and Kellomaki, M. (2008). A review of rapid prototyping techniques for tissue engineering purposes. *Ann. Med.* 40: 268–280.

74 Chang, C., Boland, E.D., Williams, S.K., and Hoying, J.B. (2011). Direct-write bioprinting three-dimensional biohybrid systems for future regenerative therapies. *J. Biomed. Mater. Res. Part B* 98B: 160–170.

75 Visser, J., Peters, B., Burger, T.J. et al. (2013). Biofabrication of multi-material anatomically shaped tissue constructs. *Biofabrication* 5: 035007.

76 Fedorovich, N.E., Schuurman, W., Wijnberg, H.M. et al. (2012). Biofabrication of osteochondral tissue equivalents by printing topologically defined, cell-laden hydrogel scaffolds. *Tissue Eng. Part C* 18: 33–44.

77 Roohani-Esfahani, S.I., Newman, P., and Zreiqat, H. (2016). Design and fabrication of 3D printed scaffolds with a mechanical strength comparable to cortical bone to repair large bone defects. *Sci. Rep.* 6: 19468.

78 Hung, K.C., Tseng, C.S., and Hsu, S.H. (2015). Synthesis and 3D printing of biodegradable polyurethane elastomer by a water-based process for cartilage tissue engineering applications. *Adv. Healthc. Mater.* 3: 1578–1587.

79 Yang, X., Lu, Z., Wu, H. et al. (2018). Collagen-alginate as bioink for three-dimensional (3D) cell printing based cartilage tissue engineering. *Mater. Sci. Eng., C* 83: 195.

80 Di Bella, C., Duchi, S., O'Connell, C.D. et al. (2017). In-situ handheld 3D bioprinting for cartilage regeneration. *J. Tissue Eng. Regener. Med.* 12: 611–621.

81 Lipskas, J., Deep, K., and Yao, W. (2019). Robotic-assisted 3D bio-printing for repairing bone and cartilage defects through a minimally invasive approach. *Sci. Rep.* 9: 3746.

82 Barron, J.A., Ringeisen, B.R., Kim, H. et al. (2004). Application of laser printing to mammalian cells. *Thin Solid Films* 453, 454: 383–387.

83 Dinca, V., Kasotakis, E., Catherine, J. et al. (2008). Directed three-dimensional patterning of self-assembled peptide fibrils. *Nano Lett.* 8: 538–543.

84 Keriquel, V., Oliveira, H., Rémy, M. et al. (2017). In situ printing of mesenchymal stromal cells, by laser-assisted bioprinting, for in vivo bone regeneration applications. *Sci. Rep.* 7: 1778.

85 Olivia, K., Davit, H., Murielle, R. et al. (2019). In situ prevascularization designed by laser-assisted bioprinting: effect on bone regeneration. *Biofabrication* 11: 045002.

86 Li, L., Yu, F., Shi, J. et al. (2017). In situ repair of bone and cartilage defects using 3D scanning and 3D printing. *Sci. Rep.* 7 (1): 9416.

87 Ma, K., Zhao, T., Yang, L. et al. (2020). Application of robotic-assisted in situ 3D printing in cartilage regeneration with HAMA hydrogel: an in vivo study. *J. Adv. Res.* 23: 123–132.

88 Li, L., Shi, J.P., Ma, K.W. et al. (2021). Robotic in situ 3D bio-printing technology for repairing large segmental bone defects. *J. Adv. Res.* 30: 75–84.

10

Bone Implants (Bone Regeneration and Bone Cancer Treatments)

Hangxiang Sun[1,2,3], Xingzhi Zhou[1,2,3], Yongxing Zhang[1,2,3], Lei Zhang[4,5], Xiaohua Yu[1,2,3], Zhaoming Ye[1,2,3], and Cato T. Laurencin[4,5,6]

[1] *Zhejiang University School of Medicine, Department of Orthopedic Surgery, the Second Affiliated Hospital, Hangzhou City, Zhejiang Province, P.R. China*
[2] *Orthopedics Research Institute of Zhejiang University, Hangzhou City, Zhejiang Province, P.R. China*
[3] *Key Laboratory of Motor System Disease Research and Precision Therapy of Zhejiang Province, Hangzhou City, Zhejiang Province, P.R. China*
[4] *University of Connecticut, Department of Biomedical Engineering, Storrs, CT, 06269, USA*
[5] *University of Connecticut, Department of Chemical and Biomolecular Engineering, 233 Glenbrook Rd, Storrs, CT, 06269, USA*
[6] *University of Connecticut Health, Department of Orthopedic Surgery, Farmington, CT, 06030, USA*

10.1 Bone Regeneration

10.1.1 Introduction

The skeletal system is one of the most important tissues in the human body; it not only provides structural support but also plays a pivotal role in various metabolic activities. Bone possesses considerable self-healing capacity; thus, it can recover from bone damage in most cases. However, bone defects resulting from complex bone fractures, resection of bone tumors, or other contexts may fail to heal by themselves [1]. These diseases will lead to an increase in medical costs, loss of productivity, and affecting society in various other aspects [2]. Apart from calculable expenditure, pain and loss of mobility are also huge spiritual burdens for patients.

Bone grafting is a commonly applied surgical intervention in bone defects. In general, bone grafts can be easily categorized into biological materials and synthetic substitutes. Biological grafts include autografts, allografts, and xenografts. Autograft is defined as harvesting bone fragments and transplanting them to another site in the same patient. Autograft is considered as the "gold standard" as autologous bone exhibited complete histocompatibility, excellent osteoconductivity, and osteoinductivity. However, it is severely suffered from limited supply and donor site morbidity,

Hangxiang Sun and Xingzhi Zhou contributed equally.

Biofabrication for Orthopedics: Methods, Techniques, and Applications, First Edition.
Edited by Wenguo Cui, Xin Zhao, and Shen Liu.

including donor site pain, wound complications, and increased operation duration [3]. The transplantation of tissues from one individual to another is termed as allografts. Allografts provide good osteoconductivity, but they have no osteogenic properties and thus may result in delayed union and non-union. Moreover, it can also introduce higher risks of immune rejection or carrying infective agents [4, 5]. Xenografts are taken from other species and are limited by a strong immune response and potential infection risk [6, 7].

Because of the drawbacks associated with autografts and allografts, the use of synthetic substitutes has been substantially increasing in recent years. Biomaterials for bone grafts have been rapidly evolving since the twentieth century. The first generation of biomaterials is represented by metals, some ceramics, and synthetic polymers. The common ground is that they are bioinert and provide mechanical support for bone grafts [8]. The second generation of biomaterials is considered bioactive as they can stimulate bone regeneration and form a strong bond between bone–materials interface [4, 8]. The third generation of biomaterials is aimed to induce desired cellular responses via modification at the molecular level to achieve tissue regeneration [9, 10]. With a focus on activating specific genes or regulation of target signaling pathways, the third generation of biomaterials is generally bioresponsive [11], which may better accommodate tissue healing processes. The maturation of nanotechnology and related fields at nanometer levels accelerates the development of the third generation of biomaterials.

The field of bone regeneration has gone through vigorous revolution in the past three decades. Regarding the paradigm shift of technological advances for bone implants, we focus on the combination of biomaterial design and tissue engineering in recent progress. Firstly, we reviewed the biological characteristics of bone and the natural process of fracture healing. Then, we summarized several strategies in designing biomaterials to promote bone regeneration and list some effective reports. Finally, we emphasized the importance and potential of manufacturing advanced composite materials.

10.1.2 Biological Characteristics of Bone

10.1.2.1 Bone Biology

The human skeletal system consists of more than 200 unique bones. According to the type, bones can be divided into four categories: long bones (e.g. femur bone), short bones (e.g. calcaneus), irregular bones (e.g. Maxilla), and flat bones (e.g. cranial bones). Macroscopically, bones have two distinct tissue types: cortical and trabecular. Cortical bone has high density and low porosity, leading to substantial strength and rigidity. In contrast, trabecular bone is highly porous with its pore space filled with bone marrow, vessels, etc., which provides partial mechanical support and plays a role in mineral homeostasis [12]. Although the function and shape are different, both trabecular and cortical bones have microstructure called bone structural units (osteons). Osteons are also called Haversian systems in cortical bone. Blood vessels and nerves travel through the central Haversian canal and are

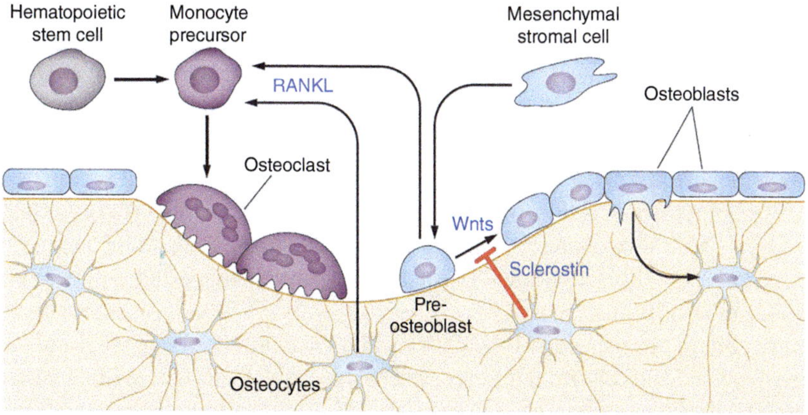

Figure 10.1 Osteoclasts and osteoblasts: osteoclasts are derived from hematopoietic stem cells. Under the stimulation of M-CSF and RANKL, monocyte precursors fuse into mature osteoclasts, which play a dominating role in bone resorption. Osteoblasts derived from mesenchymal stromal cells synthesize bone matrix and participate in mineralization. A few mature osteoblasts continue to differentiate into osteocytes. Osteocytes secrete RANKL and sclerostin, a Wnt signaling pathway inhibitors, to regulate bone formation. Source: Riddle and Clemens [10] Reprinted with permission of American Physiological Society.

surrounded by several layers of bone tissues [13]. Bone structural units in trabecular bone are saucer-shaped bone packets parted by cement lines [13].

From a materials perspective, bone is an ordered connective tissue composed of organic and inorganic bone ground substance and cells. Extracellular matrix makes up almost 90% of bone volume [9, 14–18]. Collagen-based organic matrix provides flexibility while the mineral elements provide stiffness [9, 13]. Hydroxyapatite (HA) [$Ca_{10}(PO4)_6(OH)_2$] is the main crystallization mode of inorganic matrix; in addition to good mechanical properties, the inorganic matrix serves as an ion reservoir (storing a great quantity of calcium, phosphorus, etc.). Collagen types I (about 90%), III, IV, X, and XII constitute the majority of the organic structural components; thus, pathological collagen deposition will result in brittle bones [9]. Non-collagenous proteins (such as fibronectin, osteopontin, osteocalcin, and bone sialoprotein [BSP]) also play a central role during cell adhesion, osteoblast/osteoclast differentiation, tissue mineralization, and bone remodeling [19, 20]. Bone cells (Figure 10.1) making up about 10% of bone volume are derived from either hematopoietic mesenchymal stem cells (HSCs) or mesenchymal stem cell lineage (MSCs) [19, 21]. HSCs give rise to osteoclasts, the principal cells that mediate bone resorption, and MSCs give rise to osteoprogenitor cells, osteoblasts, osteocytes, and bone lining cells [22].

10.1.2.2 Osteogenesis and Bone Healing (Biomineralization)

Physiological processes of bone biogenesis and fracture healing are the theoretical basis of modern bone implants research. In general, bone tissue is formed in two distinct manners: intramembranous or endochondral bone formation [23, 24].

In the process of intramembranous bone formation, mesenchymal cells condensate in the initial stage of osteogenesis. Condensate cells proliferate rapidly and migrate outward; then, some of them are differentiated into osteoblasts to mediate direct ossification. Moreover, the mesenchymal cells can become proliferating chondrocytes and hypertrophic chondrocytes, followed by the transformation of osteoblasts, matrix mineralization, blood vessel ingrowth, and other physiological processes to form new bones, which is known as endochondral bone formation. These complex processes are strictly regulated by growth factors and associated signaling pathways, such as Runx2, Ihh/PTHrP, Wnts, FGFs, IGF-1, and vascular endothelial growth factor (VEGF) [23–25].

Healthy bones have strong self-healing capability (Figure 10.2). When a fracture occurs, the blood and cells from the damaged blood vessels form a three-dimensional fibrin clot called hematoma. Hematoma not only plays the function of stop bleeding but also has great significance for fracture repair [22, 24]. Some evidence suggests that the release of transforming growth factor-β1 (TGF-β1) and platelet-derived growth factor (PDGF) by platelets in hematoma may promote early bone production [23, 24]. Meanwhile, VEGF mediates angiogenesis [23–25]. Bone morphogenic proteins (BMPs) and TGF-β are the main growth factors that orchestrate the proliferation and differentiation of mesenchymal cells into chondrocytes and osteoblasts. Then, the hematoma is replaced by granulation tissue and chondrocytes. With the production of cartilaginous matrix, granulation tissue becomes cartilaginous callus with stronger tensile strength [22, 24]. After that, osteocalcin can be detected, and osteoid gradually replaces the cartilage matrix [27]. With the mineralization proceeding, the callus changes into the woven bone in which the collagen fibers are randomly oriented [22, 27]. The woven bone still needs several months to remodeling, and this process follows Wolff's law. Remodeling transformes the disorganized calcified component with more ordered ones. Arriving here, the repair of the bone has been completed, and the new bone has a similar performance to that before the injury. Here, we list some factors that have been reported in the process of repairing bone injury (Table 10.1) [22, 23, 27, 28].

The process of fracture healing has laid down the foundation for bone regeneration strategies. With advances in basic science research, researchers have developed various approaches to recapitulate the fracture healing process for new bone generation. More specifically, these approaches usually involve one or more following events: sequential activation of osteogenic genes, the participation of periosteal cells and immune cells, secretion of regulatory factors, and regulation of signaling pathways [9, 18, 29].

10.1.3 Role of Implants for Bone Regeneration

In order to make biomaterials play a pre-designed role in the body, it is necessary to get a whole picture of mechanical properties, microstructure, and physiological effects of materials. Mechanical strength is the first consideration of orthopedic implants [30, 31]. An implant that replaces the damaged tissue should possess appropriate mechanical properties (such as tensile strength, compressive strength,

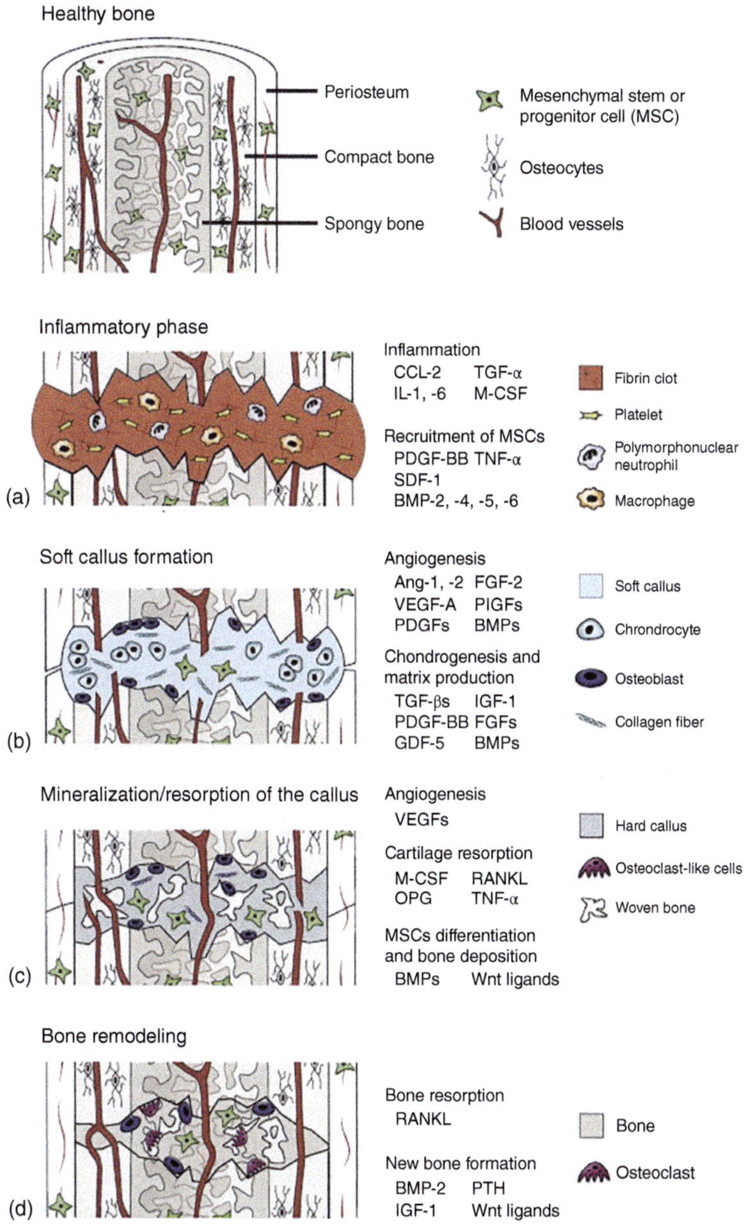

Figure 10.2 Healing process of bone fracture. (a) Inflammatory phase, (b) soft callus formation, (c) mineralization/resorption of the callus, and (d) bone remodeling. BMP, bone morphogenetic protein; FGF, fibroblast growth factor; GDF-5, growth/differentiation factor 5; IGF-1, insulin-like growth factor 1; M-CSF, macrophage colony-stimulating factor; OPG, osteoprotegerin; PDGF, platelet-derived growth factor; PlGF, placenta growth factor; PTH, parathyroid hormone; RANKL, receptor activator of nuclear factor κB ligand; SDF-1, stromal cell-derived factor 1; TGF-β, transforming growth factor β; TNF-α, tumor necrosis factor α; and VEGF, vascular endothelial growth factor. Source: Martino et al. [26] Reprinted with permission of Elsevier.

Table 10.1 Factors that have been reported in the process of repairing bone injury.

	Related signaling molecules in bone healing
Signaling molecules	**Function**
IL-1	Recruiting inflammatory cells
IL-6	Recruiting fibroblasts
VEGF	Neovascularization
FGFs	Angiogenesis
BMPs	Regulating differentiation of MSCs
TGF-β	Inducing release of other growth factors; inducing the proliferation of MSCs
IGF	Supporting osteoblast; enhancing the effect of PDGF
PDGF	Mediating proliferation of fibroblasts and endothelial cells; inducing MSC migration, activation, and proliferation
Noggin	Inhibiting the activity of BMP
Chordin	Providing negative feedback control

flexural strength, etc.). On the macro level, implants need to resist the load of megapascals range to maintain stability and prevent deformation. Mechanical mismatch of implants and bone tissue inevitably leads to mechanical shielding, which results in implant failure. For instance, bone resorption caused by stress shielding was observed in hip and knee prostheses [32]. These prostheses have quiet higher elastic modulus (220–230 GPa) than cortical bone (20–30 GPa), which makes them bear the main load [31]. Bone is a mechanically sensitive tissue. Without sufficient mechanical stimulation for a long time, bone resorption occurs and eventually leads to the loosening of implants. Microcosmically, studies have shown that the elasticity of biomaterials regulates the bone formation [33]. Mechanical stiffness between 11 and 60 kPa directs mesenchymal stem cells to osteogenic fate and expressing higher osteoblastic markers [33–35]. On the other hand, benefiting from the development of nanotechnology, researchers have found that architectural and microstructural properties (such as composition, porosity, grain size, etc.) have impacts on regeneration [36, 37]. As far as porosity is concerned, there is no clear view of the minimum size for bone ingrowth at present. Pore interconnects with more than 50 μm diameter have shown obvious bone ingrowth [38, 39]. Different materials may be slightly different, but generally speaking, the larger the pore, the more obvious the effect is. It should be noted that the alteration of porosity can significantly change the strength of materials; meanwhile, increased surface area caused by pores will accelerate the degradation of materials.

With the invention of bioactive materials in the 1960s and the introduction of tissue "engineering" in the 1980s, the physiological effects of biomaterials have been paid more attention [40, 41]. The role of bone implants during bone regeneration has been rapidly enriched. *In vivo* reactions of orthopedic implants such as corrosion, dissolution, wear, degradation, and cell interaction are essential indexes reflecting the outcome of implantation [42, 43]. The reactions were also used to establish a sustained release system [44]. It is worth mentioning that these reactions are totally different, depending on their constituted materials. Biological coatings and cell strategies lead to intense and controllable physiological effects. A combination of scaffold and cells can effectively lead to the formation of new bone. Besides, scaffolds carrying genes and bioactive proteins are also widely used in orthopedic implants by leveraging the biological function of these agents to promote bone tissue formation. Mechanical properties, microstructure, and physiological effects promote and restrict each other. Here, we list a part of different mainstream strategies in stimulating bone regeneration to present the current status of bone implants.

10.1.3.1 Providing Structural Support

Structural support is one of the most basic and important functions of implants. It includes serving as bone void fillers, allowing for the ingrowth of blood vessels and cells, allowing entry of nutrients and exit of metabolic by-products, etc. Pre-clinical and clinical research studies have found that defects beyond the critical size will not heal by themselves; however, implants such as demineralized bone powder may be used to provide structural support and enable healing of critical-sized bone defects [45, 46]. Porous material facilitates bone regeneration (Table 10.2). Porosity is not an isolated factor. It is closely related to connectivity and permeability. These factors jointly affect bone ingrowth. Some points of view propose that appropriate manufacturing processes and materials will stimulate bone regeneration [53, 54]. Computational topology design (CTD) and solid free-form fabrication (SFF) make these ideas come true [55]. In the case of hydroxyapatite, it can be designed to vary in size, porosity, and strength. Small-sized hydroxyapatite particles can adapt to different defects.

The filler provides contact surface and mechanical stability for bone growth. Bone growth arises from the edge of the defect. A compact bone–biomaterial interface can effectively promote osseointegration and positive communication between both

Table 10.2 Effects of pore size of different materials on bone regeneration.

Material	Pore size (μm)	Outcomes
Calcium aluminate	Larger than 75–100	Minimum pore size for significant ingrowth of natural bone [47]
Hydroxyapatite	300–400	The optimal size of BMP-induced osteogenesis [48, 49]
CaP-based scaffolds	5–50	Cell penetration [50–52]
Titanium	Larger than 50	More bone ingrowth [38]
poly(D,L-lactide-*co*-glycolide)	Larger than 16	Hematoma stabilization and osteopromotion

sides. Osseointegration was defined as the formation of bone tissues rather than fibrous tissues around implants [56]. Bragdon et al. found that an implant motion of 20 μm does not impair osseointegration. Still, when micromotion reaches 40 or 150 μm, it would inhibit bone ingrowth in a distal femur implantation model of dog [57]. Bone cements, represented by PMMA (polymethyl methacrylate), are common materials used to fill the defect for its high plasticity [58]. In terms of PMMA, it has a low degradation rate and hardly stimulates bone regeneration. Composite bone cement, such as HA-PMMA and electrospun poly(ε-caprolactone) (PCL)/PMMA fibrous scaffolds, can promote bone healing more effectively, although they cannot degrade considerably [59, 60]. An ideal implant should be able to dissolve or reabsorbed with the proceeding of bone regeneration as the healing process progresses and offer space for newly formed bone tissue [61]. Calcium phosphate bone cements (CPCs) based on one or more calcium orthophosphates, another kind of bone cement commonly used in clinics, possess better biodegradability. However, they still need approximately 78 weeks (about 1.5 years) to degrade [62]. Thus, the development of biodegradable materials has become a new demand. Hydrogels are highly hydrated 3D networks with controllable physical properties and good biodegradability [63]. Natural (such as gelatin and hyaluronic acid) and synthetic polymers (such as polyethylene glycol [PEG]) can be added as raw materials for hydrogels [64]. This kind of material can be completely degraded in just a few weeks with the repair. Stimuli-responsive hydrogel can even degrade in few hours [65]. Hydrogels with good plasticity and internal and surface structure are widely used as scaffolds, but most hydrogels are unsatisfactory in mechanical properties [64].

Optimizing structural properties of implants (e.g. porosity and permeability) is a common strategy to provide mechanical support. This strategy aims to provide a structure that is similar to the natural environment to promote bone regeneration [1]. Classic scaffold materials allow new bone formation on the surface or down inside pores, channels, or pipes [56, 66]. This characterization is termed osteoconductivity and is closely related to architecture [56, 67]. Porosity is an essential characteristic in biomaterials. Generally, metals have poor osteoconductivity, but porous titanium implants were found to allow bone ingrowth in rabbits [68]. Porous titanium and its alloys are widely investigated to get better implant outcomes [68, 69]. Porous hydroxyapatite also increases bone ingrowth and bone contact [70]. Interconnected pores and suitable pore size are more conducive for tissue ingrowth. It is worth mentioning that the existence of pores inevitably changes the mechanical strength, which should be paid attention to in the design. Apart from porosity, permeability plays an auxiliary role in evaluating the flow of nutrients, which is a vital factor that determines the cell survival [71].

Currently, additive manufacturing technology combined with algorithm has been able to produce materials that fully fit the defect and highly simulate the natural structure [72]. Even so, providing structural support alone cannot bring the regeneration potential into full play and other strategies need to be combined.

10.1.3.2 Regulating Stem Cell Behaviors

The difficulty of harvesting and expansion limits tissue-specific cells such as osteoblasts. Stem cells with high differentiation potential and low immunogenicity are applied extensively in bone repair therapy. Using engineered bone implants to carry stem cells or to mobilize the homing differentiation of endogenous stem cells is important for bone repair. Various types of stem cells have been demonstrated to differentiate into osteogenic lineage (e.g. human embryonic stem cells (hESCs), MSCs, and induced pluripotent stem cells [iPSCs]) [73–75]. On the other hand, in the process of revascularization, endothelial progenitor cells (EPCs) participate in the formation of some blood vessels [76]. Besides, periosteum-derived progenitor cells were involved in fracture healing [77]. Overall, it is a promising strategy to mobilize the regeneration potential of stem cells fully.

Osteoprogenitor Cells They are stem cells located in bone tissues. These cells can be found in bone marrow, endosteum, adipose, and other endogenous locations [78]. Osteoprogenitors are the precursors of highly differentiated osteocytes (e.g. osteoblast) and can self-proliferate and self-renew [79]. Some research reports that chondrocyte-like progenitor replantation after *in vitro* amplification can induce endochondral bone formation [80]. TGF-β can induce MSCs toward chondrogenesis and get a transient hypertrophic phenotype that accelerates bone formation *in vivo* after being implanted with agarose shell [81, 82].

Mesenchymal Stem Cells (MSCs) They are a group of stem cells with multipotency. MSCs can generate four lineages, bone, cartilage, fat, and muscle, and can also be isolated from these tissues. Compared with other stem cells, mesenchymal stem cells are more abundant, easier to extract, expand, and differentiate and have fewer ethical concerns. Adipose-derived stem cells (ADSCs) seeded in polylactic acid (PLA) scaffolds and coral scaffolds have been studied, respectively, and both experiments showed bone regeneration [83, 84]. In other studies, scaffold compounds with MSCs are also more effective than just scaffold [85–87]. In order to improve the repair efficiency, endothelial cells and BMSCs seeded together on a biodegradable implant demonstrated more significant bone formation in severe combined immunodeficient mice [88]. However, cell encapsulation strategies are limited by cell sources and uncertainties in amplification, such as production restriction in elderly patients and reduced regenerative ability during expansion. Some researchers turn to stimulate homing and differentiation of endogenous stem cells [89–92]. As mentioned before, the recruitment of MSCs is a crucial biological event in bone fracture healing. Stromal cell-derived factor 1 (SDF-1)/CXCR4 signing axis and SMAD signal pathway induced by TGF-β1 during bone resorption lead the process [93, 94]. Ji et al. used PCL/gelatin electrospun membranes to carry SDF-1α. SDF-1α-loaded membranes present significant recruitment of BMSCs *in vitro* and more bone formed *in vivo* [95]. In addition, incorporations of implants with statins, platelet-rich plasma (PRP), or self-assembling peptide (SAP) have been studied for MSCs homing [96–98]. Besides, a project mimicking the trabecular bone used a magnesium-doped hydroxyapatite/type I collagen scaffold to induce osteogenic differentiation of bone marrow-derived mesenchymal stem cells (BMSCs) *in vitro*; meanwhile, *in vivo* experiments showed osteoinductivity [99].

Embryonic Stem Cells (ESCs) and Induced Pluripotent Stem Cells (iPSCs) Embryonic stem cells (ESCs) were first isolated by Thomson et al. from human blastocysts in 1998 [100]. ESCs are totipotent stem cells capable of developing into all three embryonic germ layers [100]. *In vivo* and *in vitro* experiments have proved that it can differentiate into osteogenic lineage and any other cell types from human body [101, 102]. For instance, human ESCs derived MSCs seeded on macroporous calcium phosphate effectively differentiated into osteoblast and contributed to bone mineral synthesis and expression of osteogenic markers *in vitro* [101], which also demonstrated similar potency in healing cranial defects when compared to human BMSCs [102]. However, because of their human blastocyst origin, ESCs have inevitable ethical problems and the risks of immune rejection and teratoma formation [103, 104].

iPSCs were firstly reported by Takahashi and Yamanaka in 2006 where they successfully reprogrammed fibroblasts to iPSCs possessing pluripotency using a series of transcription factors [105], namely, Oct4, SOX2, c-Myc, and Klf4. In comparison with ESCs, iPSCs can be easily derived from human somatic cells, which circumvent the ethical concerns associated with ESCs. iPSCs also enable precision medicine as customized therapy can be realized using iPSCs [106]. Thus, investigators have been explored to treat bone defects combining with tissue engineering scaffolds. For example, Levi et al. utilized BMP-2 releasing poly-L-lactic acid scaffolds to carry hiPSCs and found that these iPSCs were induced into osteogenic progenitor and generated robust new bone tissue [107]. However, the efficiency and safety of ESCs and iPSCs are still under investigation and their clinical application in large scale is still unclear.

10.1.3.3 Activating Gene-Associated Therapy

Delivering specific genetic information to the defect and induce the bone regeneration have turned out to be effective [108]. Activating gene-associated therapies have the advantages of continuously producing nascent protein *in vivo* and lower cost compared to protein delivering [108]. These therapies avoid the possible side effects of high protein doses and lose the efficacy of protein. Homing factors (e.g. VEGF) and genes associated with osteogenesis (e.g. BMPs) are most commonly used in implants for bone regeneration [109, 110].

Injection is a direct implementation and presents promising outcomes in animal experiments, but expression of genes in non-target sites and low transfection efficiency urge to be solved [108, 111]. The incorporations of specific genes and implants significantly improve the efficiency of gene delivery. Gene-activated matrices (GAMs), defined as scaffold system carrying genes, provide a feasible method for locally maintaining a high concentration of DNA vector to enhance the transfection [112]. Fang et al. firstly used collagen sponge to absorb plasmid DNA containing BMP-4 or parathyroid hormone gene fragments in a simple manner, which have been shown to regulate bone formation *in vivo* [113]. Adeno-associated virus encoding BMP-2 immobilized to PCL scaffold indicated more mineral formation and increased mechanical strength at the defect site [114]. Hydrogel and other biodegradable implants encapsulate the genes and change the local releasing process. Research studies on silk-elastin-like protein polymer (SELP) hydrogels have found a sustaining release of adenoviruses for more than four weeks *in vitro* [115]. *In vivo* results demonstrate a more persistent local gene expression compared with direct injection [115]. The releasing rate is also determined by the structure (e.g. pore size) and degradation of the scaffold [116]. The responsive release is another way of design. Cui et al. synthesized poly(lactic-*co*-glycolic acid) (PLGA)/HA/poly(l-lactic acid)-block-aniline pentamer (PLA-AP)/phBMP-4 composite scaffold with responsive gene release and gene expression [117]. When electrical stimulation was induced, the scaffold can significantly increase the release of pSTAR-hBMP-4 plasmid, and the expression of the gene can be strictly controlled in the presence of doxycycline. On the other hand, encapsulation or immobilization of vectors in implants also reduced the host immune response and antivirus neutralization thus increased the efficiency and security [118, 119]. Previous studies have confirmed that viral vectors have higher transfection efficiency than non-viral vectors, but there is the risk of immune response *in vivo* [118, 120]. Sailaja et al. use alginate microspheres to encapsulate recombinant adenovirus reduced vector-specific immune response [121]. Similar conclusions were found in poly(lactic-glycolic) acid copolymer scaffold [122]. Several studies related to PEG and other polymers have shown protection on virus vectors from antivirus neutralization [123–125]. In addition to DNA fragments, mRNA is an emerging medium in activating gene-associated therapies [126, 127].

Besides, *ex vivo* delivery using expanded cells is an alternative way to activate relative genes. ADSCs transfected with VEGF were co-implanted with poly(lactide-*co*-glycolide) PLAGA, and the scaffold indicated noticeable vascular growth [128]. This approach has the advantages of precise control of gene transduction and expression level. Moreover, operators are allowed to sorting and scrutinizing modified cells, the further security is guaranteed [129]. *Ex vivo* delivery is also limited by surgical invasion, complex technology, and cost [129]. Although it still remains unclear which method is more suitable, the strategies to activate specific genes are no doubt becoming an important branch for bone regeneration.

10.1.3.4 Application of Orthobiologics

The biological process of bone regeneration is complex and delicate. It has been found that BMP-2 [130, 131], PDGF [131], IGF [132], and other growth factors play essential roles in fracture repair. Local injection is an application scenario of growth factor, while local retention of buffer delivery is short [133] and supraphysiological doses can easily lead to severe side effects [134]. Strategies of using implants to deliver growth factors attempt to release single or multiple inducerons to regulate and accelerate natural regeneration mechanisms while reducing side effects.

One of the most classic ways of delivery is to harness the interactions between target growth factors and implant [135]. Many teams have reported the effectiveness of using collagen sponges in which BMP-2 is physically adsorbed [133]. The isoelectric points of BMP-2 and collagen sponge are quite different [133]. When the pH is between the two isoelectric points, the electrostatic interaction between the material and the target cargo significantly increases the loading efficiency [136]. On the other hand, growth factors can be delivered by specific immobilization in scaffolds [137, 138]. Kuhl et al. reported that epidermal growth factor (EGF) tethered with scaffolds retained greater activity than physically adsorbed [137]. However, if the functional region of the protein is cross-linked with the scaffold, these growth factors are exposed to the risks of losing their activity. In response, Martino and his coworkers proposed an engineered fibrin scaffold with the 12th to 14th type III FN repeat (FN III12–14) [139]. This motif could bind growth factors and prolong the retention time of growth factors [139, 140].

Incorporation of growth factors into biomaterials, especially polymeric matrices, has been a promising approach to deliver growth factors. Since Langer and Folkman reported that polymers could release proteins continuously in 1976, this field has attracted much attention [141]. In general, growth factors are trapped in the scaffold and released with the degradation of polymer structure [138]. Tran et al. observed a slow and sustained protein release from PLGA-PEG-PLGA microspheres up to 36 days without a burst phase [142]. Hydrogels are widely used in these delivery systems because of their flexible yet firm structure, excellent biocompatibility, and biodegradability [143]. The degradation, swelling, and deformation of hydrogels can be used as leverages to control their release kinetics [143]. Moreover, microcapsules containing growth factors were loaded to form hybrid scaffolds. Patel et al. added gelatin particles carrying VEGF and BMP-2 into the scaffolds, thus accelerating early bone regeneration [144]. Yilgor et al. loaded two kinds of microcapsules consisting of PLGA and poly(3-hydroxybutyrate-*co*-3-hydroxyvalerate) (PHBV) on the chitosan scaffolds to realize the differential release of two factors [145]. Such hybrid systems provide the possibility of introducing precise control in the growth factor release process.

10.1.3.5 Generating Stimulus Responses for Bone Healing

Although utility of soluble factors and adhesive factors has been widely applied for bone regeneration, multiple stimuli, especially mechanical signals, also participate in the regulation of cell homeostasis and regeneration [146–148]. Bone cells sense the deformation and shear induced by fluid flow and transform them into biochemical reaction [149–151]. Piezoelectric effect and streaming potential are also considered to be perceived by bone cells [152, 153]. Another evidence supporting the effectiveness of implant-mediated stimuli in promoting bone regeneration is that these stimuli can affect the stem cell fate commitment [154].

Mechanical Stimuli Cells can sense the local mechanical niche at the nanometer to micron level; thus, a set of mechanical parameters of implants are manipulated to promote bone regeneration. Pelham and Wang designed an elastic polyacrylamide substrate with controllable flexibility and determined that the stiffness of extracellular matrix has effects on cells as early as 1997 [155]. In the next two decades, elastic polymer networks represented by hydrogels, such as PEG and alginate, were extensively explored because the elastic modulus of the hydrogel system could cover all human tissues by changing the cross-linking density [35, 156, 157]. The current perception is that integrin binds to ligand and mediates the exchange of mechanical signals and cell traction force [158, 159]. These stimuli have shown promising results in regulating the differentiation of stem cells. In work by Engler et al., MSCs cultured on rigid metrics (34 kPa) resulted in higher expression of osteocalcin and CBFα1 [160]. It is generally considered that stem cells cultured on a stiff substrate are more likely to differentiate into osteoblasts and chondrocytes [161, 162]. On the other hand, previous mechanical niche may also be involved. Yang et al. reported that human mesenchymal stem cells (hMSCs) with a longer previous culture period on supraphysiologically rigid tissue culture polystyrene ($E \approx 3$ GPa) are more inclined to osteogenic differentiation [154]. It is worth mentioning that the mechanical environment of three-dimensional culture is more complex and cannot be completely transformed from the 2D culture [163]. Luo et al. prepared alginate/mesoporous silica nanoparticles (MSNs) incorporated hydrogel carrying signal molecules. The MSCs encapsulated by this hybrid hydrogel showed higher expression of alkaline phosphatase (ALP), osteocalcin (OCN), Runx2, etc., than those seeded on the hydrogel [163]. It is probably that the aggregation of cells enhances intercellular communication [163]. In the three-dimensional environment, extracellular matrix (ECM) secreted by the cell itself is deposited and forms a *de novo* pericellular matrix, which may induce more efficient regulation in cell differentiation [164]. Loebel et al. found that when cut the link between mesenchymal stromal cell and nascent proteins, nuclear translocation of Yes-associated protein/transcriptional coactivator with PDZ-binding domain (YAP/TAZ) was reduced, and the cells showed adipogenic differentiation [165]. Although it is difficult to separate mechanical cues alone, the related research is of great significance for investigating bone regeneration.

Electrical Stimuli Since the discovery of the piezoelectric effect in tissues in 1950s, a large number of studies have tried to introduce electrical signals to promote bone repair [166–170]. External electrical stimuli, bioelectrical signals, and physiological electric fields have been reported to regulate cell behavior and affect bone regeneration [171–174]. Conducting polymers (CPs) are a set of biomaterials commonly used in neuroscience [175] and controlled drug release [176, 177]. The conductive polymer was simply mixed with the scaffold material to introduce electrical stimulation into the implant site. Shahini et al. found that the composite scaffold containing nano-poly(3,4-ethylenedioxythiophene) (PEDOT) enhanced the viability of mesenchymal stem cells by introducing electrical stimulation [178] (Figure 10.3). The characteristics of piezoelectric materials are suitable for orthopedic implants, converting physiological load into electrical signals for repair. Nevertheless, most of the research studies still focused on static state [179]. Piezoelectric ceramics are polarized during sintering, thus carrying surface charges. Chen et al. reported that the polarized surface of potassium sodium niobate (KNN) ceramics could effectively

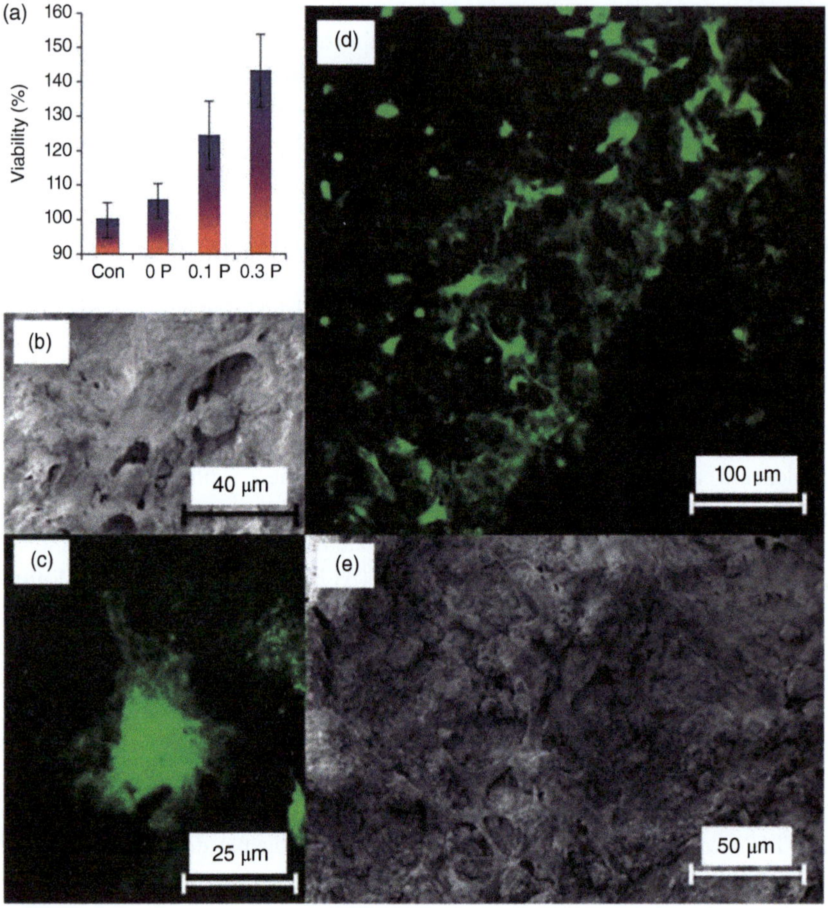

Figure 10.3 0 P, 0.1 P, and 0.3 P represent the addition from 0%, 0.1%, and 0.3% (w/w) poly(3,4-ethylenedioxythiophene) poly(4-styrene sulfonate) (PEDOT : PSS) in the mixture of 10% (w/v) gelatin and 30% (w/v) bioactive glass nanoparticles. Higher number of living cells on the scaffold with higher concentration of PEDOT : PSS. (b) Scanning electron microscope (SEM) and (c) confocal fluorescent microscopy of cells on 0.3 P scaffold. (d) Confocal fluorescent and (e) SEM of cell distribution on the 0.3 P scaffold. Source: Shahini et al. [178] / with permission of Dove Medical Press Ltd.

adsorb protein and promote cell proliferation [180]. It is necessary to control many parameters to give full play to the characteristics of such materials.

Other Stimuli In addition to the aforementioned stimuli, such as ultrasound [181], photothermal [182], etc., have also been reported for bone repair. Low-intensity pulsed ultrasound is considered to increase bone metabolism [183] and promote bone formation [184, 185]. Recently, Crasto et al. also reported a controlled release system using ultrasound for the delivery of BMP-2 [186]. Mild heat stimulation is considered to promote bone formation and the introduction of implants with photothermal effect provides a more efficient way. The composite scaffolds composed of black phosphorus nanosheets and PLGA (BPs@PLGA) possess excellent near-infrared (NIR) photothermal response [182]. Under the condition of periodic low-dose infrared irradiation, the composite implant can effectively promote bone formation, which may be achieved by up-regulating heat shock protein. The introduction of abiotic signals is one of the branches of bone regeneration research, and there are still many mechanisms that need to be solved.

10.1.3.6 Converging Multiple Synergistic Strategies

Although implant modification strategies have displayed promising results in bone regeneration, there is still an urgent need to integrat these solutions for the treatment of critical size bone defects. Scaffolds with suitable stiffness, growth factors, or well-controlled microstructure mimicing a more vivid natural extracellular environment [187–189].

Most of the composite scaffolds take multiple strategies to play a synergistic effect in bone regeneration. The combination of mechanical stimulation and biological signals is one common pattern. The rigidity of transglutaminase cross-linked gelatin (TG-Gel) can be simply adjusted by the concentration of gelatin [190]. The 3D culture in TG-Gel scaffolds presented rigidity-dependent switch from proliferation to osteogenic differentiation [190]. After adding biochemical signal BMP-2, higher ALP activity and more calcium nodules were detected *in vitro*, indicating that the biochemical and mechanical signal promotes osteogenic differentiation synergistically [190]. Kong and coworkers reported the effect of substrate rigidity on the transfection efficiency of non-viral gene vector [191]. When the substrate rigidity increased from 20 to 110 kPa, the uptake of pDNA-poly(ethyleneimine) (PEI) condensates increased significantly [191] (Figure 10.4). It is speculated that the cell proliferation and release of pDNA from the condensates induced by substrate improve the efficiency in common [191]. Besides, Nii et al. presented a co-regulation case of fibronectin and stiffness [192].

Adding biochemical signals or functional genes while regulating stem cells is another method. Implants are designed to carry multiple cargoes. Zhao et al. achieved synchronous delivery of BMSCs and growth factors in GelMA microspheres using microfluidic-assisted technology [193]. After four weeks of culture *in vitro*, it was found that BMSC continued to grow in GelMA microspheres and the expression of osteogenic markers increased, such as ALP, osteopontin, and osteocalcin [193]. The largest amount of new bone formation was observed in the GelMA/BMSC/BMP-2 group, after implantation in the rabbit femoral defect for four weeks [193].

Responsive scaffolds can also introduce multiple cues at the same time. Shuai et al. utilized PH-sensitive mPEG-bPEI-PAsp (DIP-BzA) cationic micelle carrying simvastatin and siRNA targeting the noggin gene [194]. Simvastatin enhanced the expression of BMP-2 while siRNA inhibited noggin gene, thus playing a synergistic role in regeneration [194]. Multiple synergistic strategies are inseparable from excellent mastery of materials and pathological processes; meanwhile, they are also the main direction of future research.

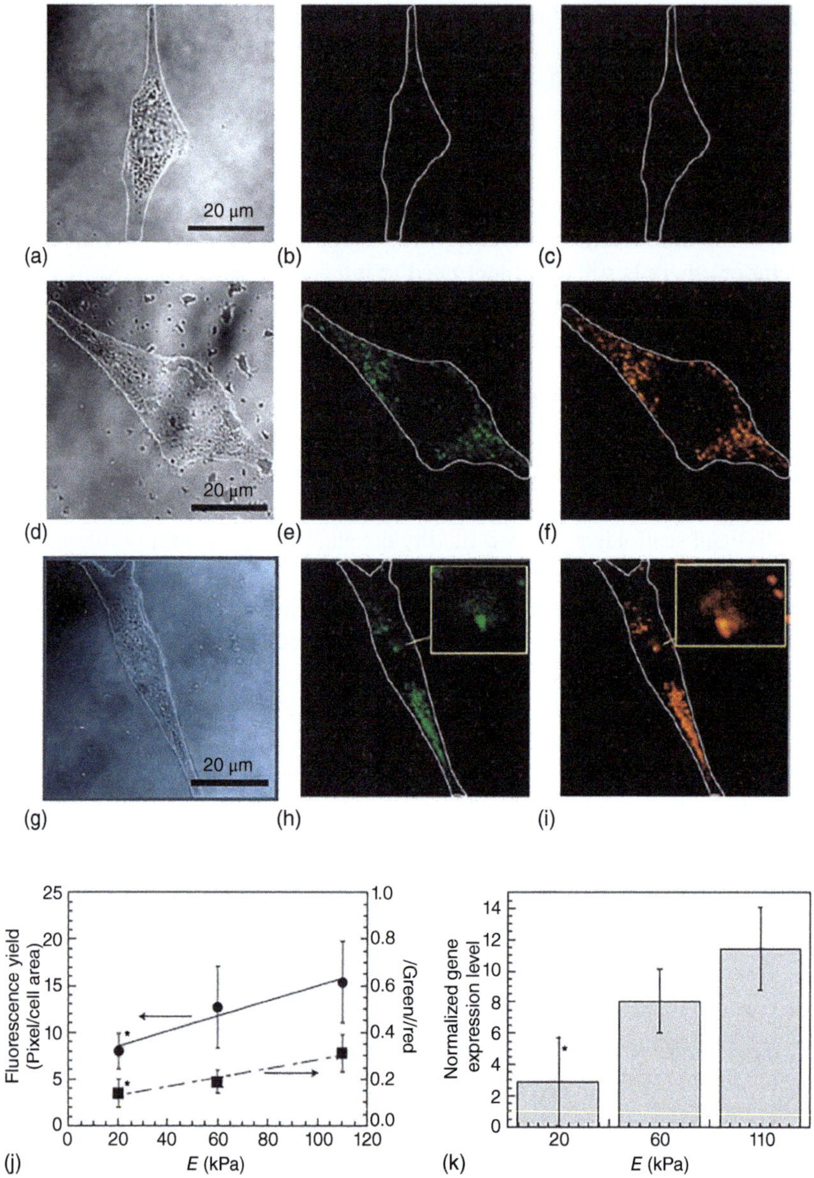

Figure 10.4 (a–c), (d–f), and (g–i) are the images obtained when the elastic moduli (*E*) of the hydrogel varies at 2 060 110 kPa and the white line outlines the cells observed. The amount of pDNA uptaken by cells increased in proportion to *E* (a–j). The expression of the marker gene β-galactosidase was detected. When *E* changed from 20 to 100 kPa, the expression of gene also increased significantly (k). In Figure 10.4k, the significance of "*" has been drawn in the first group (20 kPa). Source: Kong et al. [191] / with permission of Elsevier.

10.1.4 Concluding Remarks and Future Outlook

Currently, bone implants can partially meet the needs of bone regeneration. Throughout the development of bone implants, from improving biocompatibility to enhancing natural regeneration process, researchers have achieved substantial progress toward the clinical translation of these strategies. Research on stem cells and developmental biology has enhanced our understanding of bone regeneration. Stimulus-responsive scaffolds are designed to meet the changing requirements of bone tissue regeneration. Future research will continue to focus on the appropriate spatiotemporal presentation of exogenous clues. Nanotechnology and gene engineering provide more abundant tools for research. The scale of drug delivery carriers has been gradually evolving from micro to nano. The focus of next stage exploration is also shifting from cell population to intercellular interaction and we are increasingly emphasizing the importance of bone microenvironment. Moreover, small molecules that affect paracrine and autocrine secretion are also used to induce cell behavior. In some studies of gene modification, apart from using traditional viral and non-viral vectors, miRNA has become a new choice. It is worth mentioning that the wide application of 3D printing technology provides the basis for customized implants and personalized treatment. Unraveling the complex mechanism of the regeneration process and mimicking it through bioengineering technology, finally translating these efforts from bench side to clinics is the final destination of the bone healing research community.

10.2 Bone Cancer Treatments

10.2.1 Clinical Approach to Bone Cancer

10.2.1.1 Introduction

Bone cancer is a group of bone diseases involving uncontrolled cell growth with the potential of invasion and spread. The exact cause and mechanism of bone cancer are still unclear, but certain types of bone cancer are connected with hereditary factors, previous radiation exposure, and other bone conditions such as Paget's disease. It affects people in different ages and is slightly more common in males than in females. Skeletal bones can host numerous types of cancers, which are called as "primary bone cancer;" meanwhile, they also commonly receive metastases from other malignancies, including breast, lung, renal, prostate, and thyroid cancers, named as "secondary bone cancer" [195]. The five-year survival rate of bone cancer mainly depends on the tumor type, location, and stage.

Primary bone cancers include osteosarcoma, Ewing sarcoma, and chondrosarcoma [196]. They account for less than 1% of diagnosed cancers each year and are associated with significant morbidity and mortality [197]. Osteosarcoma, the most common bone cancer, most often occurs in children and adolescents. It typically develops in the metaphysis of long bones, specifically the distal femur, proximal tibia, and proximal humerus. Metastasis to the lungs is common. Specific diseases are associated with a high risk of developing osteosarcoma, including Li-Fraumeni syndrome characterized by a mutation in the TP53 gene [198, 199] and retinoblastoma characterized by a mutation in RB1 gene [200, 201]. Besides, osteosarcoma is also the most common radiation-induced bone sarcoma [202]. Ewing sarcoma is the second most common bone cancer. It is similar to osteosarcoma in terms of presenting symptoms, age at occurrence, and treatment and often occurs in the lower limbs and pelvis. Balanced chromosomal ewing sarcoma (EWS)/erythroblast transformation specific (ETS) translocations have been implicated in the pathogenesis of Ewing sarcoma [203–205]. Prognosis for osteosarcoma and Ewing sarcoma depends on the presence of metastasis, which lowers the five-year survival rate to 20–30%. Chondrosarcoma is the rarest bone cancer, primarily affecting adults older than 40 years and occurs in long bones of limbs and pelvis. More than half of patients with chondrosarcoma exhibit mutations in either IDH1 or IDH2 [206, 207]. The survival rates are higher because most of these tumors are low-grade lesions.

Timely diagnosis is challenging because of late patient presentation, non-specific symptoms that mimic common musculoskeletal injuries, and low suspicion by physicians. Diagnostic methods include chest imaging, imaging of the primary site (plain radiographs, computed tomography [CT], and magnetic resonance imaging [MRI]), and bone scan. Plain radiography is the preferred diagnostic test. Radiographic suspicion of a bone malignancy should prompt quick referral to a cancer center for multidisciplinary care. Biopsy should be performed for patients with a suspected primary bone cancer.

Metastatic tumors involving bone refer to secondary tumors caused by primary malignant tumors (carcinoma or sarcoma) in the organs and tissues outside the bone, metastasizing to the bones through the blood or lymphatic system. The incidence of bone metastases account for 15–20% of systemic metastatic tumors, second only to lung metastases and liver metastases, while supreme to primary malignant bone tumors [208]. The prevalence age is 40–60 years, and there are more men patients than women with a ratio of approximately 3 : 1. Bone metastases mostly occur in the axial bones, especially the thoracic spine, lumbar spine, pelvis, and the proximal end of the limbs. Malignant tumors originated from the breast, lung, prostate, kidney, and thyroid are the most common primary lesions, accounting for about 80% of all cases. The treatment of bone metastases is based on comprehensive treatment, and the prognosis depends on the source of the tumor and the location and number of metastases.

10.2.1.2 Traditional Approaches

The treatment options depend on the tumor type, stage, size, and location, as well as the patient's age and general health. Current curative strategies for bone cancer include surgery, chemotherapy, radiation therapy, targeted therapy, etc. The use of neoadjuvant and adjuvant chemotherapy, in combination with surgery, has improved survival rates to nearly 80% for patients with localized disease, and 90–95% of patients do not require limb amputation [209].

Table 10.3 Enneking's criteria for surgical margins in musculoskeletal tumors.

Margin	Dissection
Intralesional	Within the lesion
Marginal	Through the pseudocapsule or reactive tissue
Wide	Lesion (including biopsy scar), pseudocapsule and/or reactive zone, and an unviolated cuff of normal tissue completely surrounding the mass removed as a single block
Radical	Entire anatomic compartment containing the tumor removed as one block

Source: Enneking et al. [212]. Reprinted with permission of Wolters Kluwer.

10.2.1.3 Surgery

As the usual treatment for bone cancer, surgery aims to safely and completely remove the tumor and yet preserve as much function as possible [210, 211]. Surgical margins should be wide enough to minimize potential local recurrence and narrow enough to maximize function. According to Enneking's definition, surgery margins should be at least wide and margins can be classified into intralesional, marginal, wide, and radical [212]. Surgical procedures are classified by the type of the margin they achieve (Table 10.3).

Local tumor control can be achieved either through limb-sparing resection or amputation. Limb-sparing resection is preferred if reasonable functional outcomes can be achieved. Tremendous progress in surgical techniques and in biomedical engineering, as well as positive effects of chemotherapy, have led to a major shift away from amputation toward limb-sparing surgery, and limb-sparing surgery has become an essential approach to bone cancer [213–218]. The most basic requirement of limb-sparing is en bloc resection.

Most patients after limb-sparing surgery require reconstructive surgery to regain limb function [214, 219]. Reconstruction technology mainly includes allograft, autograft, inactivation replantation, and prothesis arthroplasty. Each technology has its strengths and weaknesses.

10.2.1.4 Chemotherapy

Chemotherapy is the use of non-specific intracellular poisons to inhibit mitosis and tumor cell division. Chemotherapeutic agents are mainly divided into alkylating agents, antimetabolites, anti-microtubule agents, topoisomerase inhibitors, and cytotoxic antibiotics. Despite being widely used in clinical practice, chemotherapy has a range of side effects, including immunosuppression, gastrointestinal distress, anemia, alopecia, infertility, peripheral neuropathy, and cognitive impairment [220–225].

According to the National Comprehensive Cancer Network (NCCN) guidelines, the first-line chemotherapy regimens include different combinations of cisplatin, doxorubicin, high-dose methotrexate, and ifosfamide [196]. Multiagent chemotherapy regimens, including ifosfamide and/or cyclophosphamide, etoposide, doxorubicin and/or dactinomycin, and vincristine have been shown to be effective in patients with localized Ewing's sarcoma. The first-line chemotherapy regimens include VAC/IE (vincristine, doxorubicin, and cyclophosphamide alternating with ifosfamide and etoposide), VAI (vincristine, doxorubicin, and ifosfamide), and VIDE (vincristine, ifosfamide, doxorubicin, and etoposide) [226–229].

Patients who have Ewing sarcoma or newly diagnosed osteosarcoma usually receive chemotherapy. Chemotherapy is not typically used to treat chondrosarcoma or chondroma. Conventional chondrosarcoma (grades 1–3) and dedifferentiated chondrosarcomas have no standard chemotherapy options [210]. Although surgery remains an essential part of osteosarcoma treatment, the addition of adjuvant and neoadjuvant chemotherapy has improved clinical outcomes in patients with localized osteosarcoma. Neoadjuvant chemotherapy has become standard practice in the treatment of osteosarcoma. Neoadjuvant chemotherapy was first introduced using the combination of methotrexate, doxorubicin, and vincristine, with or without cyclophosphamide [205]. One potential advantage of neoadjuvant chemotherapy is immediate treatment of the primary tumor and micrometastases. Response in the primary tumor may also permit safer resection procedures. Preoperative chemotherapy is preferred for high-grade osteosarcoma and periosteal lesions before excision, and postoperative chemotherapy is recommended for patients with pathologically high-grade diagnosis.

10.2.1.5 Radiotherapy

Radiotherapy (RT) involves ionizing radiation to control or kill cancer cells by damaging cancerous DNA by energy, photon, or charged particle. There are three main types of radiation therapy, external beam radiation therapy, brachytherapy, and systemic radioisotope therapy. In order to deliver high-radiation doses and maximize normal tissue sparing, specialized techniques of external beam radiation therapy such as intensity-modulated radiotherapy (IMRT) or stereotactic radiosurgery (SRS) should be considered as clinically indicated [230, 231]. Radiotherapy is used either as a curative therapy or as part of adjuvant therapy after surgery to prevent tumor recurrence. Although radiation therapy is painless, high doses can cause acute, long term, or cumulative side effects.

Osteosarcoma was long considered a radio-resistant tumor, but research studies suggested that radiotherapy might have a positive influence on patients who were unable to undergo complete surgical resection [216]. RT alone should also be applied for Ewing sarcoma that cannot be completely removed. Definitive RT alone can also be regarded as an alternative if en bloc resection is not feasible in cases of chondroma. Moreover, RT can provide a satisfactory local control in giant cell tumor of bone.

10.2.1.6 Molecular Targeted Therapy

Targeted therapy aims to use the drug interfering with specific molecules involved in tumor growth and spread of cancer cells for targeting tumor removal, with lower toxicity and better drug tolerability compared to conventional chemotherapy.

Denosumab (a RANKL-specific inhibitor), approved by Food and Drug Administration (FDA) in 2013, has demonstrated significant activity in patients with unresectable or recurrent giant cell tumor of bone (GCTB) [232–234]. Meanwhile, the increased knowledge of osteosarcoma pathogenesis has prompted the advent of numerous small-molecule kinase inhibitors such as insulin-like growth factor receptor (IGF-R), PDGF/platelet-derived growth factor receptor (PDGFR), VEGF/vascular endothelial growth factor receptor (VEGFR), and human epidermal growth factor receptor (HER-2). Among these targeted drugs, the VEGFR inhibitor sorafenib demonstrates its effects and advantages in patients with relapsed and unresectable high-grade osteosarcoma even after the failure of standard multimodal therapy [235]. Moreover, regorafenib demonstrated clinically meaningful antitumor activity with recurrent, progressive, metastatic osteosarcoma after the failure of conventional chemotherapy [236].

10.2.1.7 Other Therapy

Immunotherapy has emerged as an effective alternative to traditional therapies because of the potential to distinguish tumor and healthy cells. Immunotherapy harnesses the body's immune system to attack tumor cells. There are various cancer immunotherapeutic approaches, including monoclonal antibodies, adoptive cell therapies, immune checkpoint inhibitors, and non-specific cancer immunotherapies [237–242]. Over the past few decades, immunotherapy has gained prominence as a viable therapeutic option for bone tumor. The addition of muramyl tripeptide phosphatidylethanolamine (MTP-PE) to chemotherapy significantly improves the overall survival in newly diagnosed patients with osteosarcoma [243]. It is worth mentioning that anti-PD-1 antibodies or PD-1 inhibitors may be a potential immunotherapeutic target for metastatic osteosarcoma based on the previous research [244, 245].

In addition, high-dose therapy followed by stem cell transplant (HDT/SCT) has been evaluated in patients with osteosarcoma [246]. Serial arterial embolization is effective in patients with giant cell tumors and may be used alone or in conjunction with other therapy [247, 248].

10.2.2 State-of-the-Art Nanotechnology

10.2.2.1 Nanotechnology for Bone Cancer Therapy

Despite much progress in recent years, traditional therapeutic strategies for bone cancers are suboptimal. Only patients with early-stage tumor could be cured by surgery. Chemotherapy and radiotherapy for bone cancer always suffer from poor outcomes because of several reasons. Firstly, lack of bone targeting fails the anticancer agents to accumulate at the sites of cancer and thus causes adverse effects because of the non-specific biodistribution of drugs. Secondly, the extracellular matrix of bone may obstruct the penetration and accumulation of anticancer agents [249]. Hence, there is an urgent need to develop new therapeutic strategies for bone cancer.

Nanotechnology has shown great potential in improving the sensitivity of diagnosis and therapeutic effect in treating a variety of cancers [250–253]. Nanotherapeutics overcomes many inherent problems of current cancer treatment and offers new solutions for developing cancer therapeutics [254]. Nanoparticles display unique biological and physicochemical properties. Compared to conventional chemotherapeutic agents or biomolecule, the size of nanoparticles enables the combination of multiple therapeutic functions into a single platform, which can facilitate solubilization, protection from degradation, sustained release, targeting, and triggered activation. Specifically, nanoparticles display unique hemodynamic properties and preferentially accumulate in tumors because of the enhanced permeability and retention (EPR) effect [255, 256]. For bone cancer, advanced nanotechnology allows a controlled release of agents and displays improved efficacy and safety. Furthermore, nanoparticles can be functionalized with targeting moieties, and the bone-targeted nanoparticles could preferentially deliver the drugs to the sites of tumor.

10.2.2.2 Cytotoxic Agents

Cytotoxic agents are substances that kill cancer cells directly. These agents may stop cancer cells from growing and dividing and may cause shrinking of tumor size.

Gold nanoparticles and silver nanoparticles can decrease osteosarcoma cell viability [256–258]. Metal oxide nanoparticles can also exhibit anticancer activity. The cytotoxicity of TiO_2 and Al_2O_3 nanoparticles in osteosarcoma cells has been studied and confirmed [259]. ZnO and CeO_2 nanoparticles are effective against osteosarcoma by triggering reactive oxygen species (ROS) production [260, 261]. In particular, ZnO nanoparticles can trigger the death of osteosarcoma cells through mediating autophagy and apoptosis in a ROS-dependent manner [262].

Hydroxyapatite and other inorganic nanoparticles have also been shown to have selective cytotoxicity to bone cancer cells [263]. Selenium-doped hydroxyapatite nanoparticles (Se-HANs), which can fill the bone defect generated from bone cancer removal, induce apoptosis of bone tumor cells by an inherent caspase-dependent apoptosis pathway synergistically orchestrated with the generation of ROS (Figure 10.5) [264]. In a separate work, fullerene C60 nanocrystals (nano-C60) elicit a direct cytotoxic effect in osteosarcoma cells via ROS production [265].

10.2.2.3 Drug Delivery Platform

Drug delivery refers to approaches, formulations, technologies, and systems for delivering a therapeutic agent/drug to its site in the body to achieve its desired therapeutic effect and improve its efficacy and safety by controlling the rate, time, and place of release [266].

Nanotechnology used as drug delivery platforms demonstrate several advantages, including improved drug solubilization, controlled and sustained release, enhanced safety, protection from degradation and unique hemodynamic properties, and biodistribution profiles [255, 267]. These capabilities can potentially reduce off-target effects and lower the amount of drugs that must be administered [268]. Many novel nanotechnology-based drug delivery platforms have been developed

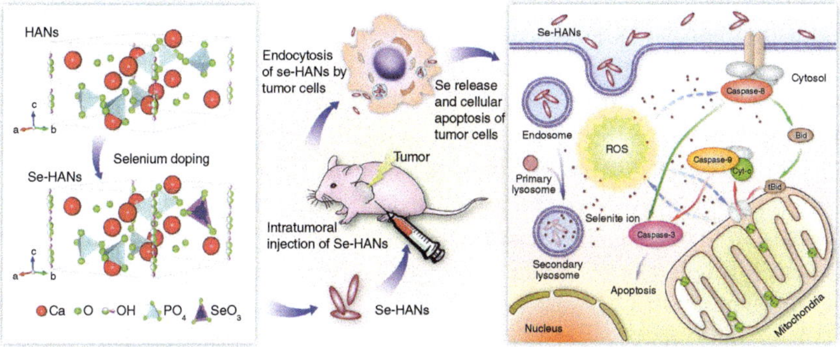

Figure 10.5 Preparation and working principle of antitumor Se-HANs nanoparticles. Source: Wang et al. [264] Reprinted with permission of American Chemical Society.

over the past few decades, including polymer–drug conjugates [267], liposomes [269], microspheres [270], nanoparticles [271], dendrimers [272], extracellular vesicles [273], and proteins [274, 275]. Several platforms have served for the treatment of bone cancer now.

There have been various chemotherapeutics involved in nanotechnology for bone tumor. For instance, bisphosphonates (BPs), with a high affinity to the hydroxyapatite matrix of the bone, are widely used in the treatment of bone resorption disorders [276]. Bisphosphonate-bound polymeric nanoparticles have been developed to deliver chemotherapeutics for bone cancer treatment and demonstrate malignant bone-targeting capability [277–279]. Besides polymeric nanoparticles, hydroxyapatite (HAp) nanoparticles functionalized with bisphosphonate as a bone targeting moiety have also been designed as a drug delivery vehicle for chemotherapeutics in the treatment of osteosarcoma [280]. As a matter of fact, bisphosphonates are the most widely used bone targeting ligands [281, 282].

Another type of nanotechnology that could be useful in chemotherapeutic delivery for bone cancer is represented by microspheres. Calcium phosphate-phosphorylated adenosine hybrid microspheres show a high doxorubicin loading capacity and pH-responsive release property [283]. In addition, calcium carbonate has also been designed for fabricating doxorubicin-loading hybrid microspheres [284]. Moreover, hyaluronate nanogel exhibits the enhanced antitumor effect by delivery of doxorubicin and cisplatin compared to one drug alone and their combination [285].

10.2.2.4 Phototherapy

Collateral damage to normal cells caused by surgery, chemotherapy, radiotherapy, and other existing cancer therapies prompts studies on non-invasive and selective therapies. Phototherapy, which involves the use of exogenous photosensitizing agents to increase the efficiency of light radiation, is becoming one of the most promising therapies for cancer, including photothermal therapy (PTT) and photodynamic therapy (PDT) [286].

PTT utilizes nanomaterials called ablation agents to absorb particular light irradiation and convert the energy into heat to kill cancer cells. Compared to normal cells, cancer cells are more susceptible to elevated heat [287]. At tissue temperatures >60 °C, which are usually achieved with PTT, cancer cells suffer from irreversible damage owing to protein denaturation and membrane destruction [288–290]. Thus, it is possible to provide heat only to the tumor, making PTT selective and non-invasive to surrounding normal tissues. To generate heat internally, the light irradiation should be able to penetrate healthy tissues with minimal side effects. NIR light can penetrate tissues with relatively mild damage, making NIR advantageous to PTT [291, 292].

For decades, nanomaterials for photothermal cancer treatment have included gold-based nanomaterials [293–296], copper-based nanomaterials [297, 298], carbon-based nanomaterials [299, 300], CPs [301, 302], and so on. The aforementioned nanomaterials have been applied in the clinical treatment of osteosarcoma, such as gold nanorods (GNRs) [303], carbon aerogel (CA) [304], and transition metal carbides, and nitrides (MXenes) [305]. In addition, bismuth-doped nanomaterials have attracted considerable interest in bone tumor research [306–308]. The high photothermal efficiency of these bismuth-doped agents indicates that they are promising candidates for bone cancer treatment. To prepare bone-targeted nanoparticles with anticancer activity, phytic acid-capped platinum nanoparticles were designed and showed high affinity to hydroxyapatite [309], which is the matrix of human bone.

PDT is a therapeutic modality that uses a light-activatable chemical substance termed as photosensitizer (PS) and a light source with a specific wavelength to induce cytotoxic effects through the generation of ROS via type I or type II photochemical reactions [310, 311]. Antitumor effects of PDT derive from three interrelated mechanisms: direct cytotoxic effects on tumor cells, anti-vascular effects, and induction of the inflammatory reaction that can activate the immune system [312]. PDT can destroy the tumor cells in a safe and effective manner as the photosensitizer is not individually toxic without the presence of external light.

Photodynamic agents for potential anticancer applications can be classified into various types: organic molecules and nanoparticles (such as cyanine-based and porphyrin-based agents) [313], noble metal materials (such as gold nanoparticles) [295], carbon-based materials (such as graphene oxide and carbon nanotubes) [314, 411], silica and silicon nanoparticles [315], quantum dots (QDs) [316], upconversion nanoparticles (UCNPs) [317], two-dimensional nanomaterials (such as MnO_2 nanosheet, MoS_2 nanosheets, and black phosphorus) [318–320], and other materials [321, 322]. PDT has been licensed for the treatment of several cancers, including skin cancer, head and neck cancer, lung cancer, and esophageal carcinoma [323]. Studies indicate that certain photodynamic agents have promising clinical applications in PDT of bone cancer. For instance, zinc phthalocyanine (ZnPc) as a second-generation photosensitizer has been used to treat osteosarcoma [324, 325]. Lack of tumor targeting efficiency is one of the current challenges of PDT for bone cancer. Mesenchymal stem cells (MSCs) as delivery vehicles have proven to be a promising tool for targeted delivery [326, 327]. MSCs loaded with PMMA nanoparticles (FNPs) decorated with the photosensitizer porphyrin represent a targeted delivery strategy and exert a ROS-mediated cytotoxic effect on osteosarcoma cells [328, 329].

10.2.2.5 Gene Therapy

Gene therapy is a promising option for cancer treatment involving the introduction of target gene into cells to reduce or restore gene expression. Nucleic acid-based therapeutics such as plasmid DNA, mRNA, siRNA, and miRNA could be delivered into tumor by nanotechnology. To date, nanomaterials including dendrimers, liposomes, proteins, and inorganic nanoparticles have been used as nanocarriers for gene therapy [330–333]. Binding nucleic acids to nanomaterials will increase their serum stability and nuclease resistance and promote cellular internalization and endosomal escape [249].

A bone-targeted alendronate-functionalized gelatin nanoparticle was used as gene carriers, and the targeted nanoparticles showed high efficiency in the delivery of genes into osteosarcoma cells [334]. Polymers are widely used to deliver therapeutic genes. An aminated polyglycerol dendritic nanocarrier has been synthesized to deliver osteosarcoma dormancy-associated miRNAs [335]. Moreover, a novel polysaccharide derivative was prepared for gene therapy, and siAEG-1 was chosen as the target gene for osteosarcoma therapy with good effectiveness [336].

10.2.2.6 Summary and Future Direction

Nanotechnology has emerged as a highly promising tool against cancer. The advantage of nanotechnology over conventional bone cancer treatment includes reduced systemic toxicity and augmented therapeutic efficacy because of controlled release and improved pharmacokinetics and pharmacodynamics. Although nanotechnology shows promising advantages in treating bone cancers, there are still multiple challenges to be addressed.

First, synthesizing nanomaterials is a complicated process. It is crucial to determine the optimal physicochemical parameters precisely. Nevertheless, systematic parallel screening of multifarious nanomaterial properties remains challenging because of the difficulties of precise and reproducible synthesis of nanoparticle libraries with distinct characteristics [337, 338]. Compared with traditional bulk techniques, microfluidic technology enables nanomaterials with a narrower size distribution, tunable chemical and physical characteristics, and greater batch-to-batch reproducibility [339–342]. Particle replication in nonwetting templates (PRINT) is another promising technology with precise control of size, shape, chemical composition, drug loading, and surface properties [343–345]. Second, nanomaterials can get into cells and be transported throughout the circulation, making them hard to trace [346]. Moreover, nanomaterials exhibit high surface activity because of surface effects [347], making them susceptible to modification by toxic compounds. Third, it should also be noted that the vast majority of the current understanding of nanomaterial behavior *in vivo* is based on animal data, and its pre-clinical and clinical data remain largely unexplored.

The power of light may decrease as it penetrates tissue; thus, the curative effect of PTT may be reduced in practical use [348, 349]. For this reason, for a complete cure, it is reasonable to combine PTT with other therapeutic modalities. PTT combining with several therapies such as PDT [349], immunotherapy [350], and chemotherapy [351] have shown synergistic anti-tumor effect in certain solid tumors. Such multimodal synergistic therapies should also have great potential in the treatment of bone cancers. As confirmed previously, some PTT agents possess the potential to work as nanocarriers [352]. Therefore, the future work should also focus on multifunctional ablation agents. The limited penetration depth of light is also challenging in PDT. For deep PDT, X-ray, and internal lights can be optional light source [353], and there are increasing interests in novel photosensitizing nanosystems and photosensitizer-delivering nanosystems, such as UCNPs [354]. Moreover, tumor hypoxia, lack of tumor targeting efficiency, and side effects from undesirable white light activation and self-catalyzation are the other unresolved challenges of PDT that need to be addressed in the future [355].

Nanotechnology plays indispensable roles in current immunotherapy in improving the efficacy and developing novel therapeutic modalities. Nanoimmunotherapy can be realized by developing nanomedicines that target (i) cancer cells, (ii) immune microenvironment, and (iii) the peripheral immune system [356]. Nanoimmunotherapy targeting cancer cells can promote immunogenic cell death (ICD), which is induced by certain types of [357] chemotherapies [358] as well as PDT [359, 360] and other stimuli [361, 362]. Also, nanoimmunotherapy targeting the tumor immune microenvironment (TIME) is the second strategy for anticancer immunotherapy. Nanomedicines can modulate the TIME in two different ways, including alleviating immunosuppression [363–365] and promoting immune activation [366–368]. Nanoimmunotherapy targeting the peripheral immune system such as lymph nodes and the spleen receives increasing attention. The mechanism gives improvement in antigen presentation [369, 370], generation of cytotoxic T cells, as well as enhancement in the viability and activity of T cells [371, 372]. Until now, no study on bone-targeted immunotherapy is reported. The bone-targeted nanoparticles may be used to specifically deliver checkpoint inhibitors and PD-1/PD-L1 antibodies to bone tumor microenvironments for immunotherapy.

In recent years, many novel therapeutic modalities based on nanotechnology have received ever-increasing attention. For instance, sonodynamic therapy (SDT), which utilizes low-intensity ultrasound combined with sonosensitizers to generate ROS, becomes a promising alternative in the treatment of deep tumors. Compared with phototherapy, SDT provides many advantages, such as no phototoxicity, deeper penetration depth, and fewer side effects [373]. In addition, chemodynamic therapy (CDT), an ROS-based strategy that involves exploiting the acidity and endogenous H_2O_2 through Fenton or Fenton-like reactions, has demonstrated new possibilities for selective cancer intervention involving specific manipulation of dysregulated redox homeostasis [374] and good effectiveness in oncology [375, 376]. Besides, the use of magnetic hyperthermia for drug release in cancer therapy has been widely studied [377]. In a nutshell, all the aforementioned approaches provide new and encouraging directions for bone cancer treatment.

10.2.3 Biofabrication for Bone Cancer Treatments

10.2.3.1 Introduction

Surgical resection of bone cancers can result in large bone defects that require clinical intervention. Bone implants, including autografts, allografts, and bio-inert metallic devices, are currently the standard method of treatment for bone defects caused by bone cancer. However, these bone implants come with their own drawbacks. For autografts, a major limitation is the volume of bone required, as well as expensive procedures, deformities, possible donor site injury, and surgical risks. Compared to autografts, allografts exhibit reduced osteoconductivity because of the lack of a cellular component, and the use of bone allografts leads to a risk of disease transmission from the donor material and immunogenic reactions [378]. Metal implants can cause stress shielding, a phenomenon in which the implants take up most of the load and the surrounding bone tissues begin to resorb, leading to bone embrittlement or even bone refracture. Moreover, the lack of vascularization is the reason for the failure of many of these standard clinical methods. Poor vascularity impairs the exchange of nutrients and waste products between tissue and blood, as well as the cell transport to the injury site, resulting in graft necrosis [379].

At present, bone graft substitutes using biological or synthetic materials typically suffer from poor osseointegration and impaired bone tissue formation, thus driving the need for enhanced growth rate. Biofabrication techniques may provide a method for controlling the uniformity of cell distribution and localization on a scaffold to enhance cellular function [380, 381]. Unique multilayered structure would be needed for different stages of cell proliferation, maturation, and differentiation in bone formation and remodeling, and biofabrication can also enable timed release of growth factors or other compounds to promote bone formation. Sophisticated and user-controlled biofabrication can generate customized 3D structures closer to the anatomical bone construction. Biofabricated bone will hopefully substitute the need for donor bone grafts, allowing patients to recover from surgery sooner and reducing the physical and psychological risk of morbidity. Consequently, it provides a more affordable approach for bone cancer patients, as well as a novel therapeutic option for individuals who cannot be cured by traditional therapy [382].

10.2.3.2 Application in Bone Cancer Treatments

3D printing is a biofabrication approach to building biocompatible scaffolds and patient-specific bone implants using 3D multilayered constructs with outstanding mechanical and osteoconductive characteristics [383]. Compared to the scaffolds prepared by traditional methods, 3D-printed scaffolds allow precise manipulation of scaffold geometry, interconnecting hole structure, and porosity, especially for the latter, which has significant implications for the biological behavior of scaffolds, including nutrient transportation, cell spreading, vascularization, waste product removal, and ingrowth of new bone from the periphery to the interior of the scaffolds [384, 385]. 3D printing technology has evolved into a versatile technology capable of producing complex functional 3D structures from a variety of materials such as ceramics, polymer, and metal, giving a strong boost in bone cancer therapy.

Bioactive glass (BG) is a representative biomaterial for the regeneration of bone tissue with proven biocompatibility, osteoconductivity, osteoinductivity, and osteointegrativity [386, 387]. For this reason, BG scaffolds with interconnected macropores and precisely controlled internal structures fabricated through 3D printing are widely considered to be ideal for bone tissue regeneration [388, 389]. Research showed that 3D-printing BG scaffolds integrated with 2D Ti_3C_2 MXenes were fabricated for the versatile scaffold construction for bone cancer treatment. 2D Ti_3C_2 MXenes kill the bone cancer cells based on their unique photothermal conversion property, and the 3D BG component promotes osteogenesis by assisting the differentiation of mesenchymal stem cells (MSCs) into osteoblasts. Especially, the titanium-based species, as the biodegradation product of Ti_3C_2 MXenes, also accelerate the formation of new bone [390] (Figure 10.6). Similarly, another 2D

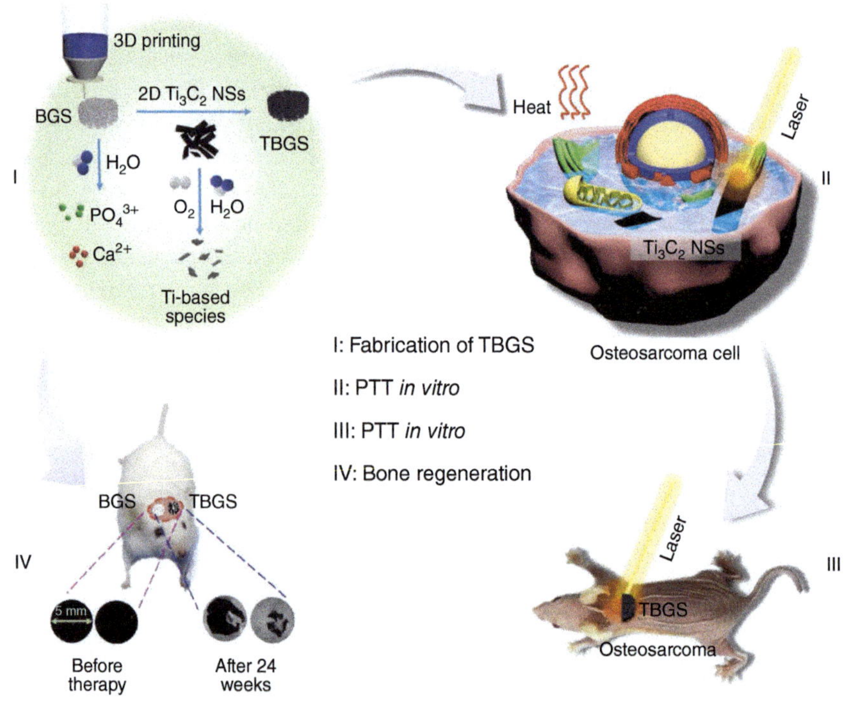

Figure 10.6 Schematic illustration of the fabrication of Ti_3C_2-BG scaffold, ablation of bone cancer, and regeneration of bone tissue. Source: Pan et al. [390] / with permission from John Wiley & Sons, Ltd.

Nb2C MXene loaded with nitric oxide (NO) donor was integrated into the large macropores of 3D-printing BG scaffolds. On the one hand, the high photothermal conversion efficiency of 2D MXene is attributed to the anti-tumor activity and the subsequent release of high concentrations of NO. On the other hand, the low concentration of NO can enhance angiogenesis and osteogenesis based on BG scaffolds. The anti-tumor effect is enhanced by the combined effect of NO-based gas therapy and photonic hyperthermia, and the generation of local NO circulation stimulates bone regeneration based on the coordination of NO with 3D-printing BG scaffold [391] as well. In another study, 3D-printed BG scaffolds were functionalized by semiconductor $CuFeSe_2$ nanocrystals. This bifunctional scaffold can effectively inhibit the growth of bone cancer *in vivo* through the photothermal effect induced by $CuFeSe_2$ and induce bone regeneration by the combination of released Ca, Si, P, Fe, Cu, and Se ions [392].

Tricalcium phosphate (TCP), a group of calcium phosphate ceramics, is another common 3D printing scaffold material because of its excellent ability to stimulate osteogenesis. It is reported that curcumin, a type of hydrophobic biomolecule with anti-osteosarcoma activity, was encapsulated within liposomes, which was incorporated into a 3D printing TCP scaffold. The 3D scaffold offers mechanical support for cell adhesion, and the prolonged release of the drug from the liposome facilitates the achievement of increased bioavailability and reduced toxicity. This work demonstrated a novel bifunctional scaffold not only inhibiting bone cancer cells but also promoting the formation of healthy bone [393] (Figure 10.7). Research also showed that metal–organic framework Cu-TCPP nanosheets were incorporated onto the surface of β-TCP scaffolds to fabricate composite scaffolds [394]. The scaffolds can kill osteosarcoma by exploiting the photothermal properties of Cu-TCPP

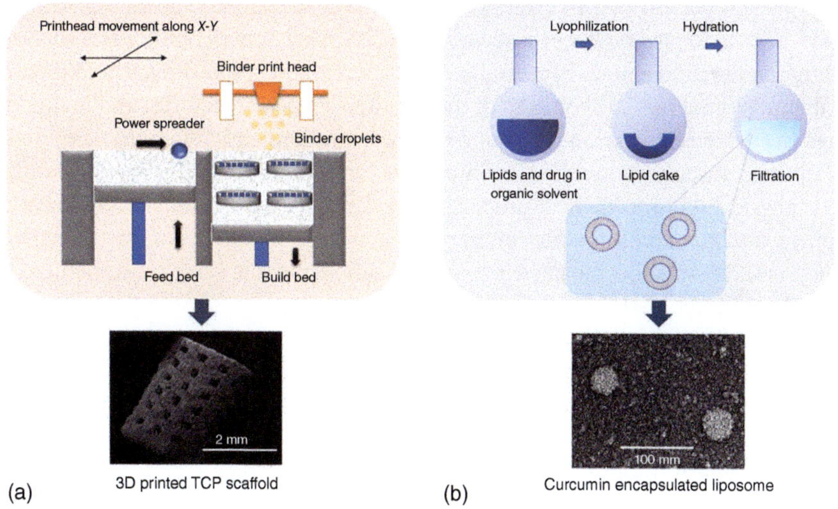

Figure 10.7 Schematic illustration of fabricating a bifunctional scaffold. (a) 3D printing to synthesize a TCP scaffold with a designed porosity and (b) formation of the liposome. Source: Sarkar and Bose [393] / with permission from Elsevier.

nanosheets. In addition, Cu ions released from Cu-TCPP nanosheets can act synergistically to stimulate new bone formation. Similar bifunctional bio-ceramic scaffolds were prepared in combination of 3D printing and dopamine therapy as photothermal agent for tumor therapy [395]. Researchers also reported a bifunctional graphene oxide-modified TCP composite scaffold with efficient PTT and improved osteogenic differentiation performance through micro-extrusion-based 3D printing [396].

Apart from bioceramics, 3D printing of titanium alloy has also been used to treat bone cancers [397]. For example, selective laser sintering (SLS) was performed to fabricate 3D-printed titanium alloy implants with micro-scale rough topography. The as-printed scaffolds were then electrochemically anodized to generate nanosurface topography for the loading of chemotherapeutics. The dual topography implants are shown to perform as a multifunctional chemotherapeutic agent release system for the treatment of bone cancer together with bone support.

Apart from scaffolds featuring a single antitumor mechanism, a versatile porous scaffold for delivery of 2D black phosphorus nanosheets, doxorubicin, and osteogenic peptide was developed through a micro-extrusion-based 3D printing [398]. The synergistic effect of PTT and chemotherapy made early tumor ablation and long-term prevention of tumor recurrence achieved. Fe-$CaSiO_3$ composite scaffolds were also designed via the 3D printing technique, demonstrating the synergistic effects of photothermal and ROS tumor-killing therapies. Moreover and impressively, the scaffolds also possess distinct mechanical properties for cortical bone defects associated with bone cancer [399].

10.2.3.3 Summary and Future Direction

Large bone defects caused by bone cancer excision remained a considerable challenge in orthopedics. As the current bone graft materials have their own limitations, many new strategies have been developed to engineer functional bone tissues as alternatives to autogenous bone grafts like 3D scaffolding [400, 401]. Scaffolds can be manufactured into 3D volumes of the required shapes, structures, and architectures to enable cell attachment, proliferation, migration, and differentiation [402]. An optimal scaffold should be biocompatible, biodegradable, and possessing sufficient physical and mechanical properties. In addition, the scaffold should have a sophisticated hierarchical structure in order to mimic that processed by the native bone tissue. Meanwhile, the need for vascularization in tissue substitutes should not be neglected [403, 404].

Biofabrication, especially 3D printing technology, can build pre-shaped, porous biocompatible scaffolds with improved mechanical and osteoconductive properties using 3D multilayered constructs [405]. Synthetic materials, in particular, bioceramics, which can be used to produce 3D-printed bone scaffolds, are thought to be particularly promising solutions for bone defect repair and bone cancer treatment. There is a broad range of bioceramics for 3D-printed scaffolds, including hydroxyapatite (HA), TCP, biphasic calcium phosphate (BCP), calcium silicate (CS), bioactive glasses (BGs), and bioceramics doped with different elements [406]. Novel functional bioceramic scaffolds have been applied to repair the bone defects caused by surgical excision and kill the possibly remaining tumor cells by means of PTT, chemotherapy, and ROS tumor therapy.

Nevertheless, there are still some limitations about the present biofabrication and 3D printing technology. Printing resolution limits porosity and complex-layered structure [407]. Moreover, internal structural alignment manners lead to poor mechanical strength of 3D-printed scaffolds [407]. Mechanical properties, degradation rates, accuracy, and other parameters of the 3D-printed scaffolds require constant optimization. Advances in 3D printing and biofabrication technologies require improved material resolution, complexity, reproducibility, and synthesis rate [408]. However, as an innovative technology, 3D printers with high performance could be very highly priced. Therefore, more technological developments and cross-disciplinary cooperation are necessary in order to reduce costs [405].

In recent years, a promising and innovative biofabrication strategy has now been added to the 3D printing industry: bioprinting. Bioprinting allows the selective distribution of biomaterials, cells, growth factors, or combinations thereof to create artificial multicellular tissues/organs in three dimensions [409]. A number of studies have attempted to produce bone tissue through bio-printing [410]. Although there is still a long way to fabricate fully high-simulated functional bone tissue and engineered 3D scaffolds, we are the strong believer that biofabrication technology will continue to bring new development and breakthrough in the field of bone cancer treatment.

Acknowledgments

H.S. and Z.Z. contributed equally to this work. This work was financially supported by the National Key Research and Development Program of China (grant no. 2018YFC1105401 and 2018YFC1105404), the National Natural Science Foundation of China (81872173, 82072959, and 31870959), the General Horizontal Project of Zhejiang University (KH20192916), and the Zhejiang Undergraduate Talent Project (grant no. 2020R401218). A Project Supported by the Scientific Research Fund of Zhejiang Provincial Education Department.

References

1 Ho-Shui-Ling, A., Bolander, J., Rustom, L.E. et al. (2018). Bone regeneration strategies: engineered scaffolds, bioactive molecules and stem cells current stage and future perspectives. *Biomaterials* 180: 143–162.

2 Lubeck, D.P. (2003). The costs of musculoskeletal disease: health needs assessment and health economics. *Best Pract. Res. Clin. Rheumatol.* 17: 529–539.

3 Dohmen, P.M., Ozaki, S., Verbeken, E. et al. (2002). Tissue engineering of an auto-xenograft pulmonary heart valve. *Asian Cardiovasc. Thorac. Ann.* 10: 25–30.

4 Yu, X., Tang, X., Gohil, S.V., and Laurencin, C.T. (2015). Biomaterials for bone regenerative engineering. *Adv. Healthc. Mater.* 4: 1268–1285.
5 Hench, L.L. (1980). Biomaterials. *Science* 208: 826–831.
6 Kulkarni, R.K., Pani, K.C., Neuman, C. et al. (1966). Polylactic acid for surgical implants. *Arch. Surg.* 93: 839–843.
7 Shin, H., Jo, S., and Mikos, A.G. (2003). Biomimetic materials for tissue engineering. *Biomaterials* 24: 4353–4364.
8 Hench, L.L. and Polak, J.M. (2002). Third-generation biomedical materials. *Science* 295: 1014–1017.
9 Fuchs, R.K., Thompson, W.R., and Warden, S.J. (2019). Bone biology. In: *Bone Repair Biomaterials*, Woodhead Publishing Series in Biomaterials, 2e (ed. K.M. Pawelec and J.A. Planell), 15–52. Woodhead Publishing https://doi.org/10.1016/b978-0-08-102451-5.00002-0.
10 Riddle, R.C. and Clemens, T.L. (2017). Bone cell bioenergetics and skeletal energy homeostasis. *Physiol. Rev.* 97: 667–698.
11 Revuelta, J., Fraile, I., Monterrey, D.T. et al. (2021). Heparanized chitosans: towards the third generation of chitinous biomaterials. *Mater. Horiz.* 8: 2596–2614.
12 Skedros, J.G., Holmes, J.L., Vajda, E.G., and Bloebaum, R.O.Y.D. (2005). Cement lines of secondary osteons in human bone are not mineral-deficient: new data in a historical perspective. *Anat. Rec. A Discov. Mol. Cell Evol. Biol.* 803: 781–803.
13 Buck, D.W. and Dumanian, G.A. (2012). Bone biology and physiology: part I. The fundamentals. *Plast. Reconstr. Surg.* 129: 1314–1320.
14 Martin, E. and Shapiro, J.R. (2007). Osteogenesis imperfecta: epidemiology and pathophysiology. *Curr. Osteoporos. Rep.* 5: 91–97.
15 Gordon, J.A.R., Diamant, M., Rieneck, K. et al. (2007). Bone sialoprotein expression enhances osteoblast differentiation and matrix mineralization in vitro. *Bone* 41: 462–473.
16 Roach, H.I. (1994). Why does bone matrix contain non-collagenous proteins? The possible roles of osteocalcin, osteonectin, osteopontin and bone sialoprotein in bone mineralisation and resorption. *Cell Biol. Int.* 18: 617–628.
17 Hunter, G.K. and Goldberg, H.A. (1993). Nucleation of hydroxyapatite by bone sialoprotein. *Proc. Natl. Acad. Sci. U.S.A.* 90: 8562–8565.
18 Clarke, B. (2008). Normal bone anatomy and physiology. *Clin. J. Am. Soc. Nephrol.* 3 (Suppl 3): 131–139.
19 Olsen, B.R. and Berendsen, A.D. (2017). Bone development. *Physiol. Behav.* 176: 139–148.
20 Percival, C.J. and Richtsmeier, J.T. (2013). Angiogenesis and intramembranous osteogenesis. *Dev. Dyn.* 242: 909–922.
21 Kronenberg, H.M. (2003). Developmental regulation of the growth plate. *Nature* 423: 332–336.
22 Shiu, H.T., Leung, P.C., and Ko, C.H. (2018). The roles of cellular and molecular components of a hematoma at early stage of bone healing. *J. Tissue Eng. Regener. Med.* 12: e1911–e1925.

23 Lieberman, J.R., Daluiski, A., and Einhorn, T.A. (2002). The role of growth factors in the repair of bone biology and clinical applications. *J. Bone Joint Surg. Ser. A* 84: 1032–1044.

24 Bigham-Sadegh, A. and Oryan, A. (2015). Basic concepts regarding fracture healing and the current options and future directions in managing bone fractures. *Int. Wound J.* 12: 238–247.

25 Street, J., Winter, D., Wang, J.H. et al. (2000). Is human fracture hematoma inherently angiogenic? *Clin. Orthop. Relat. Res.* 224–237.

26 Martino, M.M., Briquez, P.S., Maruyama, K., and Hubbell, J.A. (2015). Extracellular matrix-inspired growth factor delivery systems for bone regeneration. *Adv. Drug Delivery Rev.* 94: 41–52.

27 Harwood, P.J., Newman, J.B., and Michael, A.L.R. (2010). (ii) An update on fracture healing and non-union. *Orthop. Trauma* 24: 9–23.

28 Dimitriou, R., Tsiridis, E., and Giannoudis, P.V. (2005). Current concepts of molecular aspects of bone healing. *Injury* 36: 1392–1404.

29 Kloen, P. et al. (2003). BMP signaling components are expressed in human fracture callus. *Bone* 33: 362–371.

30 Hollister, S.J. (2005). Porous scaffold design for tissue engineering. *Nat. Mater.* 4: 518–524.

31 Gotman, I. (1997). Characteristics of metals used in implants. *J. Endourol.* 11: 383–389.

32 Bauer, T.W. and Schils, J. (1999). The pathology of total joint arthroplasty. II. Mechanisms of implant failure. *Skeletal Radiol.* 28: 483–497.

33 Huebsch, N., Lippens, E., Lee, K. et al. (2015). Matrix elasticity of void-forming hydrogels controls transplanted-stem-cell-mediated bone formation. *Nat. Mater.* 14: 1269–1277.

34 Kim, T.H., An, D.B., Oh, S.H. et al. (2015). Creating stiffness gradient polyvinyl alcohol hydrogel using a simple gradual freezing-thawing method to investigate stem cell differentiation behaviors. *Biomaterials* 40: 51–60.

35 Huebsch, N., Arany, P.R., Mao, A.S. et al. (2010). Harnessing traction-mediated manipulation of the cell/matrix interface to control stem-cell fate. *Nat. Mater.* 9: 518–526.

36 Wang, X., Padture, N.P., Tanaka, H., and Ortiz, A.L. (2005). Wear-resistant ultra-fine-grained ceramics. *Acta Materialia* 53: 271–277.

37 Bignon, A., Chouteau, J., Chevalier, J. et al. (2003). Effect of micro- and macroporosity of bone substitutes on their mechanical properties and cellular response. *J. Mater. Sci. - Mater. Med.* 14: 1089–1097.

38 Otsuki, B., Takemoto, M., Fujibayashi, S. et al. (2006). Pore throat size and connectivity determine bone and tissue ingrowth into porous implants: three-dimensional micro-CT based structural analyses of porous bioactive titanium implants. *Biomaterials* 27: 5892–5900.

39 Lu, J.X., Flautre, B., Anselme, K. et al. (1999). Role of interconnections in porous bioceramics on bone recolonization in vitro and in vivo. *J. Mater. Sci. - Mater. Med.* 10: 111–120.

40 Hench, L.L. (2006). The story of bioglass. *J. Mater. Sci. - Mater. Med.* 17: 967–978.
41 Langer, R. and Vacanti, J.P. (1993). Tissue engineering. *Science* 260: 920–926.
42 Klopfleisch, R. and Jung, F. (2017). The pathology of the foreign body reaction against biomaterials. *J. Biomed. Mater. Res. Part A* 105: 927–940.
43 Mouthuy, P.-A., Snelling, S.J.B., Dakin, S.G. et al. (2016). Biocompatibility of implantable materials: an oxidative stress viewpoint. *Biomaterials* 109: 55–68.
44 Yu, X., Khalil, A., Dang, P.N. et al. (2014). Multilayered inorganic microparticles for tunable dual growth factor delivery. *Adv. Funct. Mater.* 24: 3082–3093.
45 El-Rashidy, A.A., Roether, J.A., Harhaus, L. et al. (2017). Regenerating bone with bioactive glass scaffolds: a review of in vivo studies in bone defect models. *Acta Biomater.* 62: 1–28.
46 Schmitz, J.P. and Hollinger, J.O. (1986). The critical size defect as an experimental model for craniomandibulofacial nonunions. *Clin. Orthop. Relat. Res.* 299–308.
47 Hulbert, S.F., Young, F.A., Mathews, R.S. et al. (1970). Potential of ceramic materials as permanently implantable skeletal prostheses. *J. Biomed. Mater. Res. Part A* 4: 433–456.
48 Tsuruga, E., Takita, H., Itoh, H. et al. (1997). Pore size of porous hydroxyapatite as the cell-substratum controls BMP-induced osteogenesis. *J. Biochem.* 121: 317–324.
49 Kuboki, Y., Jin, Q., and Takita, H. (2001). Geometry of carriers controlling phenotypic expression in BMP-induced osteogenesis and chondrogenesis. *J. Bone Joint Surg. Am.* 83-A Suppl: S105–S115.
50 Polak, S.J., Rustom, L.E., Genin, G.M. et al. (2013). A mechanism for effective cell-seeding in rigid, microporous substrates. *Acta Biomater.* 9: 7977–7986.
51 Lan Levengood, S.K., Polak, S.J., Wheeler, M.B. et al. (2010). Multiscale osteointegration as a new paradigm for the design of calcium phosphate scaffolds for bone regeneration. *Biomaterials* 31: 3552–3563.
52 Bernstein, A., Niemeyer, P., Salzmann, G. et al. (2013). Microporous calcium phosphate ceramics as tissue engineering scaffolds for the repair of osteochondral defects: histological results. *Acta Biomater.* 9: 7490–7505.
53 Tour, G., Wendel, M., and Tcacencu, I. (2011). Cell-derived matrix enhances osteogenic properties of hydroxyapatite. *Tissue Eng. Part A* 17: 127–137.
54 Szpalski, C., Wetterau, M., Barr, J., and Warren, S.M. (2012). Bone tissue engineering: current strategies and techniques-part I: scaffolds. *Tissue Eng. Part B* 18: 246–257.
55 Hollister, S.J. (2006). Porous scaffold design for tissue engineering. *Nat. Mater.* 5: 590.
56 Albrektsson, T. and Johansson, C. (2001). Osteoinduction, osteoconduction and osseointegration. *Eur. Spine J.* 10: S96–S101.
57 Bragdon, C.R., Burke, D., Lowenstein, J.D. et al. (1996). Differences in stiffness of the interface between a cementless porous implant and cancellous bone in

vivo in dogs due to varying amounts of implant motion. *J. Arthroplasty* 11: 945–951.

58 Vaishya, R., Chauhan, M., and Vaish, A. (2013). Bone cement. *J. Clin. Orthop. Trauma* 4: 157–163.

59 Son, S.R., Linh, N.T.B., Yang, H.M., and Lee, B.T. (2013). In vitro and in vivo evaluation of electrospun PCL/PMMA fibrous scaffolds for bone regeneration. *Sci. Technol. Adv. Mater.* 14.

60 Itokawa, H., Hiraide, T., Moriya, M. et al. (2007). A 12 month in vivo study on the response of bone to a hydroxyapatite-polymethylmethacrylate cranioplasty composite. *Biomaterials* 28: 4922–4927.

61 Chen, K.Y., Shyu, P.-C., Dong, G.-C. et al. (2009). Reconstruction of calvarial defect using a tricalcium phosphate-oligomeric proanthocyanidins cross-linked gelatin composite. *Biomaterials* 30: 1682–1688.

62 Frankenburg, E.P., Goldstein, S.A., Bauer, T.W. et al. (1998). Biomechanical and histological evaluation of a calcium phosphate cement. *J. Bone Joint Surg. Am.* 80: 1112–1124.

63 Pina, S., Oliveira, J.M., and Reis, R.L. (2015). Natural-based nanocomposites for bone tissue engineering and regenerative medicine: a review. *Adv. Mater.* 27: 1143–1169.

64 Gibbs, D.M.R., Black, C.R.M., Dawson, J.I., and Oreffo, R.O.C. (2016). A review of hydrogel use in fracture healing and bone regeneration. *J. Tissue Eng. Regener. Med.* 10: 187–198.

65 Oliva, N., Shitreet, S., Abraham, E. et al. (2012). Natural tissue microenvironmental conditions modulate adhesive material performance. *Langmuir* 28: 15402–15409.

66 Rezwan, K., Chen, Q.Z., Blaker, J.J., and Boccaccini, A.R. (2006). Biodegradable and bioactive porous polymer/inorganic composite scaffolds for bone tissue engineering. *Biomaterials* 27: 3413–3431.

67 LeGeros, R.Z. (2008). Calcium phosphate-based osteoinductive materials. *Chem. Rev.* 108: 4742–4753.

68 Lopez-Heredia, M.A., Goyenvalle, E., Aguado, E. et al. (2008). Bone growth in rapid prototyped porous titanium implants. *J. Biomed. Mater. Res. Part A* 85: 664–673.

69 Li, Y., Yang, C., Zhao, H. et al. (2014). New developments of Ti-based alloys for biomedical applications. *Materials* 7: 1709–1800.

70 Gariboldi, M.I. and Best, S.M. (2015). Effect of ceramic scaffold architectural parameters on biological response. *Front. Bioeng. Biotechnol.* 3: 1–11.

71 Silva, M.M.C.G., Cyster, L.A., Barry, J.J.A. et al. (2006). The effect of anisotropic architecture on cell and tissue infiltration into tissue engineering scaffolds. *Biomaterials* 27: 5909–5917.

72 Schwab, A., Levato, R., D'Este, M. et al. (2020). Printability and shape fidelity of bioinks in 3D bioprinting. *Chem. Rev.* 120: 11028–11055.

73 Jones, E.A., Giannoudis, P.V., and Kouroupis, D. (2016). Bone repair with skeletal stem cells: rationale, progress to date and clinical application. *Ther. Adv. Musculoskelet. Dis.* 8: 57–71.

74 Li, F., Bronson, S., and Niyibizi, C. (2010). Derivation of murine induced pluripotent stem cells (iPS) and assessment of their differentiation toward osteogenic lineage. *J. Cell. Biochem.* 109: 643–652.

75 Tang, M., Chen, W., Weir, M.D. et al. (2012). Human embryonic stem cell encapsulation in alginate microbeads in macroporous calcium phosphate cement for bone tissue engineering. *Acta Biomater.* 8: 3436–3445.

76 Ma, X.L., Sun, X.L., Wan, C.Y. et al. (2012). Significance of circulating endothelial progenitor cells in patients with fracture healing process. *J. Orthop. Res.* 30: 1860–1866.

77 Roberts, S.J., van Gastel, N., Carmeliet, G., and Luyten, F.P. (2015). Uncovering the periosteum for skeletal regeneration: the stem cell that lies beneath. *Bone* 70: 10–18.

78 Ibrahim, A., Bulstrode, N.W., Whitaker, I.S. et al. (2017). Nanotechnology for stimulating osteoprogenitor differentiation. *Open Orthop. J.* 10: 849–861.

79 Nahian, A. and Davis, D.D. (2022). *Histology, Osteoprogenitor Cells*. Treasure Island (FL): StatPearls. Available from: https://www.ncbi.nlm.nih.gov/books/NBK559160/.

80 Weiss, H.E., Roberts, S.J., Schrooten, J., and Luyten, F.P. (2012). A semi-autonomous model of endochondral ossification for developmental tissue engineering. *Tissue Eng. Part A* 18: 1334–1343.

81 Narcisi, R., Quarto, R., Ulivi, V. et al. (2012). TGF β-1 administration during ex vivo expansion of human articular chondrocytes in a serum-free medium redirects the cell phenotype toward hypertrophy. *J. Cell. Physiol.* 227: 3282–3290.

82 Vinardell, T., Sheehy, E.J., Buckley, C.T., and Kelly, D.J. (2012). A comparison of the functionality and in vivo phenotypic stability of cartilaginous tissues engineered from different stem cell sources. *Tissue Eng. Part A* 18: 1161–1170.

83 Di Bella, C., Farlie, P., and Penington, A.J. (2008). Bone regeneration in a rabbit critical-sized skull defect using autologous adipose-derived cells. *Tissue Eng. Part A* 14: 483–490.

84 Cui, L., Liu, B., Liu, G. et al. (2007). Repair of cranial bone defects with adipose derived stem cells and coral scaffold in a canine model. *Biomaterials* 28: 5477–5486.

85 Petite, H., Viateau, V., Bensaïd, W. et al. (2000). Tissue-engineered bone regeneration. *Nat. Biotechnol.* 18: 959–963.

86 Yousefifard, M., Maleki, S.N., Askarian-Amiri, S. et al. (2019). A combination of mesenchymal stem cells and scaffolds promotes motor functional recovery in spinal cord injury: a systematic review and meta-analysis. *J. Neurosurg. Spine* 32: 269–284.

87 Viateau, V., Guillemin, G., Bousson, V. et al. (2007). Long-bone critical-size defects treated with tissue-engineered grafts: a study on sheep. *J. Orthop. Res.* 25: 741–749.

88 Kaigler, D., Krebsbach, P.H., West, E.R. et al. (2005). Endothelial cell modulation of bone marrow stromal cell osteogenic potential. *FASEB J.* 19: 665–667.

89 Both, S.K., van Apeldoorn, A.A., Jukes, J.M. et al. (2011). Differential bone-forming capacity of osteogenic cells from either embryonic stem cells or bone marrow-derived mesenchymal stem cells. *J. Tissue Eng. Regener. Med.* 5: 180–190.

90 Mareschi, K., Ferrero, I., Rustichelli, D. et al. (2006). Expansion of mesenchymal stem cells isolated from pediatric and adult donor bone marrow. *J. Cell. Biochem.* 97: 744–754.

91 Zhou, S., Greenberger, J.S., Epperly, M.W. et al. (2008). Age-related intrinsic changes in human bone-marrow-derived mesenchymal stem cells and their differentiation to osteoblasts. *Aging Cell* 7: 335–343.

92 Siegel, G., Kluba, T., Hermanutz-Klein, U. et al. (2013). Phenotype, donor age and gender affect function of human bone marrow-derived mesenchymal stromal cells. *BMC Med.* 11: 146.

93 Kitaori, T., Ito, H., Schwarz, E.M. et al. (2009). Stromal cell-derived factor 1/CXCR4 signaling is critical for the recruitment of mesenchymal stem cells to the fracture site during skeletal repair in a mouse model. *Arthritis Rheum.* 60: 813–823.

94 Tang, Y., Wu, X., Lei, W. et al. (2009). TGF-β1-induced migration of bone mesenchymal stem cells couples bone resorption with formation. *Nat. Med.* 15: 757–765.

95 Ji, W., Yang, F., Ma, J. et al. (2013). Incorporation of stromal cell-derived factor-1α in PCL/gelatin electrospun membranes for guided bone regeneration. *Biomaterials* 34: 735–745.

96 Yueyi, C., Xiaoguang, H., Jingying, W. et al. (2013). Calvarial defect healing by recruitment of autogenous osteogenic stem cells using locally applied simvastatin. *Biomaterials* 34: 9373–9380.

97 Lu, J., Shen, X., Sun, X. et al. (2018). Increased recruitment of endogenous stem cells and chondrogenic differentiation by a composite scaffold containing bone marrow homing peptide for cartilage regeneration. *Theranostics* 8: 5039–5058.

98 Malhotra, A., Pelletier, M.H., Yu, Y., and Walsh, W.R. (2013). Can platelet-rich plasma (PRP) improve bone healing? A comparison between the theory and experimental outcomes. *Arch. Orthop. Trauma Surg.* 133: 153–165.

99 Minardi, S., Corradetti, B., Taraballi, F. et al. (2015). Evaluation of the osteoinductive potential of a bio-inspired scaffold mimicking the osteogenic niche for bone augmentation. *Biomaterials* 62: 128–137.

100 Thomson, J.A., Itskovitz-Eldor, J., Shapiro, S.S. et al. (1998). Embryonic stem cell lines derived from human blastocysts. *Science* 282: 1145–1147.

101 Chen, W., Liu, J., Manuchehrabadi, N. et al. (2013). Umbilical cord and bone marrow mesenchymal stem cell seeding on macroporous calcium phosphate for bone regeneration in rat cranial defects. *Biomaterials* 34: 9917–9925.

102 Taiani, J.T., Buie, H.R., Campbell, G.M. et al. (2014). Embryonic stem cell therapy improves bone quality in a model of impaired fracture healing in the mouse; tracked temporally using in vivo micro-CT. *Bone* 64: 263–272.

103 de Miguel-Beriain, I. (2015). The ethics of stem cells revisited. *Adv. Drug Delivery Rev.* 82, 83: 176–180.

104 Bahadur, G., Morrison, M., and Machin, L. (2010). Beyond the 'embryo question': human embryonic stem cell ethics in the context of biomaterial donation in the UK. *Reprod. Biomed. Online* 21: 868–874.

105 Takahashi, K. and Yamanaka, S. (2006). Induction of pluripotent stem cells from mouse embryonic and adult fibroblast cultures by defined factors. *Cell* 126: 663–676.

106 Perez, J.R., Kouroupis, D., Li, D.J. et al. (2018). Tissue engineering and cell-based therapies for fractures and bone defects. *Front. Bioeng. Biotechnol.* 6: 1–23.

107 Levi, B., Hyun, J.S., Montoro, D.T. et al. (2012). In vivo directed differentiation of pluripotent stem cells for skeletal regeneration. *Proc. Natl. Acad. Sci. U.S.A.* 109: 20379–20384.

108 Evans, C.H. (2012). Gene delivery to bone. *Adv. Drug Delivery Rev.* 64: 1331–1340.

109 Endo, M., Kuroda, S., Kondo, H. et al. (2006). Bone regeneration by modified gene-activated matrix: effectiveness in segmental tibial defects in rats. *Tissue Eng.* 12: 489–497.

110 Geiger, F., Bertram, H., Berger, I. et al. (2005). Vascular endothelial growth factor gene-activated matrix (VEGF165-GAM) enhances osteogenesis and angiogenesis in large segmental bone defects. *J. Bone Miner. Res.* 20: 2028–2035.

111 Baltzer, A.W., Lattermann, C., Whalen, J.D. et al. (2000). Genetic enhancement of fracture repair: healing of an experimental segmental defect by adenoviral transfer of the BMP-2 gene. *Gene Ther.* 7: 734–739.

112 Luo, D. and Saltzman, W.M. (2000). Enhancement of transfection by physical concentration of DNA at the cell surface. *Nat. Biotechnol.* 18: 893–895.

113 Fang, J., Zhu, Y.Y., Smiley, E. et al. (1996). Stimulation of new bone formation by direct transfer of osteogenic plasmid genes. *Proc. Natl. Acad. Sci. U.S.A.* 93: 5753–5758.

114 Dupont, K.M., Boerckel, J.D., Stevens, H.Y. et al. (2012). Synthetic scaffold coating with adeno-associated virus encoding BMP2 to promote endogenous bone repair. *Cell Tissue Res.* 347: 575–588.

115 Hatefi, A., Cappello, J., and Ghandehari, H. (2007). Adenoviral gene delivery to solid tumors by recombinant silk-elastin like protein polymers. *Pharm. Res.* 24: 773–779.

116 von Cresce, A.W., Dandu, R., Burger, A. et al. (2008). Characterization and real-time imaging of gene expression of adenovirus embedded silk-elastin like protein polymer hydrogels. *Mol. Pharmaceutics* 5: 891–897.

117 Cui, L., Zhang, J., Zou, J. et al. (2020). Electroactive composite scaffold with locally expressed osteoinductive factor for synergistic bone repair upon electrical stimulation. *Biomaterials* 230: 119617.

118 Jang, J.-H., Schaffer, D.V., and Shea, L.D. (2011). Engineering biomaterial systems to enhance viral vector gene delivery. *Mol. Ther.* 19: 1407–1415.

119 Kreppel, F. and Kochanek, S. (2008). Modification of adenovirus gene transfer vectors with synthetic polymers: a scientific review and technical guide. *Mol. Ther.* 16: 16–29.

120 Bez, M., Pelled, G., and Gazit, D. (2020). BMP gene delivery for skeletal tissue regeneration. *Bone* 137: 115449.

121 Sailaja, G., HogenEsch, H., North, A. et al. (2002). Encapsulation of recombinant adenovirus into alginate microspheres circumvents vector-specific immune response. *Gene Ther.* 9: 1722–1729.

122 Beer, S.J., Matthews, C.B., Stein, C.S. et al. (1998). Poly(lactic-glycolic) acid copolymer encapsulation of recombinant adenovirus reduces immunogenicity in vivo. *Gene Ther.* 5: 740–746.

123 Kim, P.-H., Kim, T.-I., Yockman, J.W. et al. (2010). The effect of surface modification of adenovirus with an arginine-grafted bioreducible polymer on transduction efficiency and immunogenicity in cancer gene therapy. *Biomaterials* 31: 1865–1874.

124 Eto, Y., Gao, J.-Q., Sekiguchi, F. et al. (2005). PEGylated adenovirus vectors containing RGD peptides on the tip of PEG show high transduction efficiency and antibody evasion ability. *J. Gene Med.* 7: 604–612.

125 Fisher, K.D., Stallwood, Y., Green, N.K. et al. (2001). Polymer-coated adenovirus permits efficient retargeting and evades neutralising antibodies. *Gene Ther.* 8: 341–348.

126 Elangovan, S., Khorsand, B., Do, A.-V. et al. (2015). Chemically modified RNA activated matrices enhance bone regeneration. *J. Controlled Release* 218: 22–28.

127 Balmayor, E.R., Geiger, J.P., Aneja, M.K. et al. (2016). Chemically modified RNA induces osteogenesis of stem cells and human tissue explants as well as accelerates bone healing in rats. *Biomaterials* 87: 131–146.

128 Jabbarzadeh, E., Starnes, T., Khan, Y.M. et al. (2008). Induction of angiogenesis in tissue-engineered scaffolds designed for bone repair: a combined gene therapy-cell transplantation approach. *Proc. Natl. Acad. Sci. U.S.A.* 105: 11099–11104.

129 D'Mello, S., Atluri, K., Geary, S.M. et al. (2017). Bone regeneration using gene-activated matrices. *AAPS J.* 19: 43–53.

130 Salazar, V.S., Gamer, L.W., and Rosen, V. (2016). BMP signalling in skeletal development, disease and repair. *Nat. Rev. Endocrinol.* 12: 203–221.

131 Richardson, T.P., Peters, M.C., Ennett, A.B., and Mooney, D.J. (2001). Polymeric system for dual growth factor delivery. *Nat. Biotechnol.* 19: 1029–1034.

132 Xian, L., Wu, X., Pang, L. et al. (2012). Matrix IGF-1 maintains bone mass by activation of mTOR in mesenchymal stem cells. *Nat. Med.* 18: 1095–1101.

133 Geiger, M., Li, R.H., and Friess, W. (2003). Collagen sponges for bone regeneration with rhBMP-2. *Adv. Drug Delivery Rev.* 55: 1613–1629.

134 Carragee, E.J., Hurwitz, E.L., and Weiner, B.K. (2011). A critical review of recombinant human bone morphogenetic protein-2 trials in spinal surgery: emerging safety concerns and lessons learned. *Spine J.* 11: 471–491.

135 Yu, X., Suárez-González, D., Khalil, A.S., and Murphy, W.L. (2015). How does the pathophysiological context influence delivery of bone growth factors? *Adv. Drug Delivery Rev.* 84: 68–84.

136 Friess, W., Uludag, H., Foskett, S. et al. (1999). Characterization of absorbable collagen sponges as rhBMP-2 carriers. *Int. J. Pharm.* 187: 91–99.

137 Kuhl, P.R. and Griffith-Cima, L.G. (1996). Tethered epidermal growth factor as a paradigm for growth factor-induced stimulation from the solid phase. *Nat. Med.* 2: 1022–1027.

138 Boerckel, J.D., Kolambkar, Y.M., Dupont, K.M. et al. (2011). Effects of protein dose and delivery system on BMP-mediated bone regeneration. *Biomaterials* 32: 5241–5251.

139 Martino, M.M., Tortelli, F., Mochizuki, M. et al. (2011). Engineering the growth factor microenvironment with fibronectin domains to promote wound and bone tissue healing. *Sci. Transl. Med.* 3: 100ra89.

140 Martino, M.M. and Hubbell, J.A. (2010). The 12th–14th type III repeats of fibronectin function as a highly promiscuous growth factor-binding domain. *FASEB J.* 24: 4711–4721.

141 Langer, R. and Folkman, J. (1976). Polymers for the sustained release of proteins and other macromolecules. *Nature* 263: 797–800.

142 Tran, V.-T., Karam, J.-P., Garric, X. et al. (2012). Protein-loaded PLGA-PEG-PLGA microspheres: a tool for cell therapy. *Eur. J. Pharm. Sci.* 45: 128–137.

143 Li, J. and Mooney, D.J. (2016). Designing hydrogels for controlled drug delivery. *Nat. Rev. Mater.* 1.

144 Patel, Z.S., Young, S., Tabata, Y. et al. (2008). Dual delivery of an angiogenic and an osteogenic growth factor for bone regeneration in a critical size defect model. *Bone* 43: 931–940.

145 Yilgor, P., Tuzlakoglu, K., Reis, R.L. et al. (2009). Incorporation of a sequential BMP-2/BMP-7 delivery system into chitosan-based scaffolds for bone tissue engineering. *Biomaterials* 30: 3551–3559.

146 Fu, J., Wang, Y.-K., Yang, M.T. et al. (2010). Mechanical regulation of cell function with geometrically modulated elastomeric substrates. *Nat. Methods* 7: 733–736.

147 Discher, D.E., Janmey, P., and Wang, Y.-L. (2005). Tissue cells feel and respond to the stiffness of their substrate. *Science* 310: 1139–1143.

148 Vogel, V. and Sheetz, M. (2006). Local force and geometry sensing regulate cell functions. *Nat. Rev. Mol. Cell Biol.* 7: 265–275.

149 Klein-Nulend, J., Bakker, A.D., Bacabac, R.G. et al. (2013). Mechanosensation and transduction in osteocytes. *Bone* 54: 182–190.

150 Duncan, R.L. and Turner, C.H. (1995). Mechanotransduction and the functional response of bone to mechanical strain. *Calcif. Tissue Int.* 57: 344–358.

151 Bacabac, R.G., Smit, T.H., Mullender, M.G. et al. (2004). Nitric oxide production by bone cells is fluid shear stress rate dependent. *Biochem. Biophys. Res. Commun.* 315: 823–829.

152 Ahn, A.C. and Grodzinsky, A.J. (2009). Relevance of collagen piezoelectricity to 'Wolff's law': a critical review. *Med. Eng. Phys.* 31: 733–741.

153 Hastings, G.W. and Mahmud, F.A. (1988). Electrical effects in bone. *J. Biomed. Eng.* 10: 515–521.

154 Yang, C., Tibbitt, M.W., Basta, L., and Anseth, K.S. (2014). Mechanical memory and dosing influence stem cell fate. *Nat. Mater.* 13: 645–652.

155 Pelham, R.J. Jr., and Wang, Y.l. (1997). Cell locomotion and focal adhesions are regulated by substrate flexibility. *Proc. Natl. Acad. Sci. U.S.A.* 94: 13661–13665.

156 Kloxin, A.M., Kasko, A.M., Salinas, C.N., and Anseth, K.S. (2009). Photodegradable hydrogels for dynamic tuning of physical and chemical properties. *Science* 324: 59–63.

157 Mitragotri, S. and Lahann, J. (2009). Physical approaches to biomaterial design. *Nat. Mater.* 8: 15–23.

158 Laakso, J.M., Lewis, J.H., Shuman, H., and Ostap, E.M. (2008). Myosin I can act as a molecular force sensor. *Science* 321: 133–136.

159 Schiller, H.B., Friedel, C.C., Boulegue, C., and Fässler, R. (2011). Quantitative proteomics of the integrin adhesome show a myosin II-dependent recruitment of LIM domain proteins. *EMBO Rep.* 12: 259–266.

160 Engler, A.J., Sen, S., Sweeney, H.L., and Discher, D.E. (2006). Matrix elasticity directs stem cell lineage specification. *Cell* 126: 677–689.

161 Olivares-Navarrete, R., Lee, E.M., Smith, K. et al. (2017). Substrate stiffness controls osteoblastic and chondrocytic differentiation of mesenchymal stem cells without exogenous stimuli. *PLoS One* 12: e0170312.

162 Zouani, O.F., Kalisky, J., Ibarboure, E., and Durrieu, M.-C. (2013). Effect of BMP-2 from matrices of different stiffnesses for the modulation of stem cell fate. *Biomaterials* 34: 2157–2166.

163 Luo, Z., Zhang, S., Pan, J. et al. (2018). Time-responsive osteogenic niche of stem cells: a sequentially triggered, dual-peptide loaded, alginate hybrid system for promoting cell activity and osteo-differentiation. *Biomaterials* 163: 25–42.

164 Maia, F.R., Fonseca, K.B., Rodrigues, G. et al. (2014). Matrix-driven formation of mesenchymal stem cell-extracellular matrix microtissues on soft alginate hydrogels. *Acta Biomater.* 10: 3197–3208.

165 Loebel, C., Mauck, R.L., and Burdick, J.A. (2019). Local nascent protein deposition and remodelling guide mesenchymal stromal cell mechanosensing and fate in three-dimensional hydrogels. *Nat. Mater.* 18: 883–891.

166 Shamos, M.H. and Lavine, L.S. (1967). Piezoelectricity as a fundamental property of biological tissues. *Nature* 213: 267–269.

167 Gjelsvik, A. (1973). Bone remodeling and piezoelectricity – I. *J. Biomech.* 6: 69–77.

168 Vasquez-Sancho, F., Abdollahi, A., Damjanovic, D., and Catalan, G. (2018). Flexoelectricity in bones. *Adv. Mater.* 30.

169 Fukada, E. and Yasuda, I. (1964). Piezoelectric effects in collagen. *Jpn. J. Appl. Phys.* 3: 117–121.

170 Fukada, E. and Yasuda, I. (1957). On the piezoelectric effect of bone. *J. Phys. Soc. Jpn.* 12: 1158–1162.

171 Funk, R.H.W. (2015). Endogenous electric fields as guiding cue for cell migration. *Front. Physiol.* 6: 143.

172 McCaig, C.D. and Zhao, M. (1997). Physiological electrical fields modify cell behaviour. *Bioessays* 19: 819–826.

173 Reid, B. and Zhao, M. (2014). The electrical response to injury: molecular mechanisms and wound healing. *Adv. Wound Care* 3: 184–201.

174 Balint, R., Cassidy, N.J., and Cartmell, S.H. (2013). Electrical stimulation: a novel tool for tissue engineering. *Tissue Eng. Part B* 19: 48–57.

175 Fattahi, P., Yang, G., Kim, G., and Abidian, M.R. (2014). A review of organic and inorganic biomaterials for neural interfaces. *Adv. Mater.* 26: 1846–1885.

176 Guo, B., Glavas, L., and Albertsson, A.-C. (2013). Biodegradable and electrically conducting polymers for biomedical applications. *Prog. Polym. Sci.* 38: 1263–1286.

177 Abidian, M.R., Kim, D.-H., and Martin, D.C. (2006). Conducting-polymer nanotubes for controlled drug release. *Adv. Mater.* 18: 405–409.

178 Shahini, A., Yazdimamaghani, M., Walker, K.J. et al. (2013). 3D conductive nanocomposite scaffold for bone tissue engineering. *Int. J. Nanomed.* 9: 167–181.

179 Tandon, B., Blaker, J.J., and Cartmell, S.H. (2018). Piezoelectric materials as stimulatory biomedical materials and scaffolds for bone repair. *Acta Biomater.* 73: 1–20.

180 Chen, W., Yu, Z., Pang, J. et al. (2017). Fabrication of biocompatible potassium sodium niobate piezoelectric ceramic as an electroactive implant. *Materials* 10: 345.

181 Tanzer, M., Harvey, E., Kay, A. et al. (1996). Effect of noninvasive low intensity ultrasound on bone growth into porous-coated implants. *J. Orthop. Res.* 14: 901–906.

182 Tong, L., Liao, Q., Zhao, Y. et al. (2019). Near-infrared light control of bone regeneration with biodegradable photothermal osteoimplant. *Biomaterials* 193: 1–11.

183 Sato, W., Matsushita, T., and Nakamura, K. (1999). Acceleration of increase in bone mineral content by low-intensity ultrasound energy in leg lengthening. *J. Ultrasound Med.* 18: 699–702.

184 Warden, S.J., Favaloro, J.M., Bennell, K.L. et al. (2001). Low-intensity pulsed ultrasound stimulates a bone-forming response in UMR-106 cells. *Biochem. Biophys. Res. Commun.* 286: 443–450.

185 Sun, J.S., Hong, R.C., Chang, W.H. et al. (2001). In vitro effects of low-intensity ultrasound stimulation on the bone cells. *J. Biomed. Mater. Res. Part A* 57: 449–456.

186 Crasto, G.J., Kartner, N., Reznik, N. et al. (2016). Controlled bone formation using ultrasound-triggered release of BMP-2 from liposomes. *J. Controlled Release* 243: 99–108.

187 Lin, D., Chai, Y., Ma, Y. et al. (2019). Rapid initiation of guided bone regeneration driven by spatiotemporal delivery of IL-8 and BMP-2 from hierarchical MBG-based scaffold. *Biomaterials* 196: 122–137.

188 Lee, K.Y., Alsberg, E., Hsiong, S. et al. (2004). Nanoscale adhesion ligand organization regulates osteoblast proliferation and differentiation. *Nano Lett.* 4: 1501–1506.

189 Cipitria, A., Boettcher, K., Schoenhals, S. et al. (2017). In-situ tissue regeneration through SDF-1α driven cell recruitment and stiffness-mediated bone regeneration in a critical-sized segmental femoral defect. *Acta Biomater.* 60: 50–63.

190 Tan, S.J., Fang, J.Y., Yang, Z. et al. (2014). The synergetic effect of hydrogel stiffness and growth factor on osteogenic differentiation. *Biomaterials* 35: 5294–5306.

191 Kong, H.J., Liu, J., Riddle, K. et al. (2005). Non-viral gene delivery regulated by stiffness of cell adhesion substrates. *Nat. Mater.* 4: 460–464.

192 Nii, M., Lai, J.H., Keeney, M. et al. (2013). The effects of interactive mechanical and biochemical niche signaling on osteogenic differentiation of adipose-derived stem cells using combinatorial hydrogels. *Acta Biomater.* 9: 5475–5483.

193 Zhao, X., Liu, S., Yildirimer, L. et al. (2016). Injectable stem cell-laden photocrosslinkable microspheres fabricated using microfluidics for rapid generation of osteogenic tissue constructs. *Adv. Funct. Mater.* 26: 2809–2819.

194 Huang, J., Lin, C., Fang, J. et al. (2018). pH-sensitive nanocarrier-mediated codelivery of simvastatin and noggin siRNA for synergistic enhancement of osteogenesis. *ACS Appl. Mater. Interfaces* 10: 28471–28482.

195 Ferguson, J.L. and Turner, S.P. (2018). Bone cancer: diagnosis and treatment principles. *Am. Fam. Physician* 98: 205–213.

196 Sybil Biermann, J., Adkins, D.R., Agulnik, M. et al. (2013). Bone cancer: clinical practice guidelines in oncology. *JNCCN J. Natl. Compr. Cancer Netw.* 11: 688–723.

197 Stiller, C.A., Trama, A., Serraino, D. et al. (2013). Descriptive epidemiology of sarcomas in Europe: report from the RARECARE project. *Eur. J. Cancer* 49: 684–695.

198 Malkin, D., Li, F.P., Strong, L.C. et al. (1990). Germ line p53 mutations in a familial syndrome of breast cancer, sarcomas, and other neoplasms. *Science* 250: 1233–1238.

199 McIntyre, J.F., Smith-Sorensen, B., Friend, S.H. et al. (1994). Germline mutations of the p53 tumor suppressor gene in children with osteosarcoma. *J. Clin. Oncol.* 12: 925–930.

200 Feugeas, O., Guriec, N., Babin-Boilletot, A. et al. (1996). Loss of heterozygosity of the RB gene is a poor prognostic factor in patients with osteosarcoma. *J. Clin. Oncol.* 14: 467–472.

201 Weichselbaum, R.R., Beckett, M., and Diamond, A. (1988). Some retinoblastomas, osteosarcomas, and soft tissue sarcomas may share a common etiology. *Proc. Natl. Acad. Sci. U.S.A.* 85: 2106–2109.

202 Kalra, S., Grimer, R.J., Spooner, D. et al. (2007). Radiation-induced sarcomas of bone. *J. Bone Joint Surg. Ser. B* 89: 808–813.

203 Delattre, O., Zucman, J., Melot, T. et al. (1991). The Ewing family of tumors – a subgroup of small-round-cell tumors defined by specific chimeric transcripts. *N. Engl. J. Med.* 325: 1412–1417.

204 De Alava, E. and Gerald, W.L. (2000). Molecular biology of the Ewing's sarcoma/primitive neuroectodermal tumor family. *J. Clin. Oncol.* 18: 204–213.

205 Grünewald, T.G.P., Cidre-Aranaz, F., Surdez, D. et al. (2018). Ewing sarcoma. *Nat. Rev. Dis. Prim.* 4: –5.

206 Amary, M.F., Bacsi, K., Maggiani, F. et al. (2011). IDH1 and IDH2 mutations are frequent events in central chondrosarcoma and central and periosteal chondromas but not in other mesenchymal tumours. *J. Pathol.* 224: 334–343.

207 Lu, C., Venneti, S., Akalin, A. et al. (2013). Induction of sarcomas by mutant IDH2. *Genes Dev.* 27: 1986–1998.

208 Fornetti, J., Welm, A.L., and Stewart, S.A. (2018). Understanding the bone in cancer metastasis. *J. Bone Miner. Res.* 33: 2099–2113.

209 Bernthal, N.M., Federman, N., Eilber, F.R. et al. (2012). Long-term results (>25 years) of a randomized, prospective clinical trial evaluating chemotherapy in patients with high-grade, operable osteosarcoma. *Cancer* 118: 5888–5893.

210 Bielack, S., Carrle, D., and Jost, L. (2008). Osteosarcoma: ESMO clinical recommendations for diagnosis, treatment and follow-up. *Ann. Oncol.* 19: 2007–2009.

211 Ritter, J. and Bielack, S.S. (2010). Osteosarcoma. *Ann. Oncol.* 21: 320–325.

212 Enneking, W.F., Spanier, S.S., and Goodman, M.A. (2003). A system for the surgical staging of musculoskeletal sarcoma. *Clin. Orthop. Relat. Res.* 4–18.

213 Arndt, C.A. and Crist, W.M. (1999). Common musculoskeletal tumors of childhood and adolescence. *N. Engl. J. Med.* 341: 342–352.

214 Gosheger, G., Gebert, C., Ahrens, H. et al. (2006). Endoprosthetic reconstruction in 250 patients with sarcoma. *Clin. Orthop. Relat. Res.* 450: 164–171.

215 Gherlinizoni, F., Picci, P., Bacci, G., and Campanacci, D. (1992). Limb sparing versus amputation in osteosarcoma. Correlation between local control, surgical margins and tumor necrosis: Istituto Rizzoli experience. *Ann. Oncol.* 3: 23–27.

216 Ozaki, T., Flege, S., Kevric, M. et al. (2003). Osteosarcoma of the pelvis: experience of the cooperative osteosarcoma study group. *J. Clin. Oncol.* 21: 334–341.

217 Grimer, R.J. (2005). Surgical options for children with osteosarcoma. *Lancet Oncol.* 6: 85–92.

218 Andreou, D., Bielack, S.S., Carrle, D. et al. (2011). The influence of tumor- and treatment-related factors on the development of local recurrence in osteosarcoma after adequate surgery. An analysis of 1355 patients treated on neoadjuvant Cooperative Osteosarcoma Study Group protocols. *Ann. Oncol.* 22: 1228–1235.

219 Malawer, M.M. and Chou, L.B. (1995). Prosthetic survival and clinical results with use of large-segment replacements in the treatment of high-grade bone sarcomas. *J. Bone Joint Surg. Ser. A* 77: 1154–1165.

220 Koppelmans, V., Breteler, M.M.B., Boogerd, W. et al. (2012). Neuropsychological performance in survivors of breast cancer more than 20 years after adjuvant chemotherapy. *J. Clin. Oncol.* 30: 1080–1086.

221 Rodgers, G.M., Becker, P.S., Blinde, M. et al. (2012). Cancer- and chemotherapy-induced anemia. *J. Natl. Compr. Cancer Netw.* 10: 628–653.

222 Paus, R., Haslam, I.S., Sharov, A.A., and Botchkarev, V.A. (2013). Pathobiology of chemotherapy-induced hair loss. *Lancet Oncol.* 14: e50–e59.

223 Baghdadi, M., Wada, H., Nakanishi, S. et al. (2016). Chemotherapy-induced IL34 enhances immunosuppression by tumor-associated macrophages and mediates survival of chemoresistant lung cancer cells. *Cancer Res.* 76: 6030–6042.

224 Argyriou, A.A., Bruna, J., Genazzani, A.A., and Cavaletti, G. (2017). Chemotherapy-induced peripheral neurotoxicity: management informed by pharmacogenetics. *Nat. Rev. Neurol.* 13: 492–504.

225 Spears, N., Lopes, F., Stefansdottir, A. et al. (2019). Ovarian damage from chemotherapy and current approaches to its protection. *Hum. Reprod. Update* 25: 673–693.

226 Grier, H.E., Krailo, M.D., Tarbell, N.J. et al. (2003). Addition of ifosfamide and etoposide to standard chemotherapy for Ewing's sarcoma and primitive neuroectodermal tumor of bone. *N. Engl. J. Med.* 348: 694–701.

227 Paulussen, M., Craft, A.W., Lewis, I. et al. (2008). Results of the EICESS-92 study: two randomized trials of Ewing's sarcoma treatment – cyclophosphamide compared with ifosfamide in standard-risk patients and assessment of benefit of etoposide added to standard treatment in high-risk patients. *J. Clin. Oncol.* 26: 4385–4393.

228 Paulussen, M., Ahrens, S., Dunst, J. et al. (2001). Localized Ewing tumor of bone: final results of the Cooperative Ewing's Sarcoma Study CESS 86. *J. Clin. Oncol.* 19: 1818–1829.

229 Hawkins, M.M., Lancashire, E.R., Winter, D.L. et al. (2008). The British Childhood cancer Survivor Study: objectives, methods, population structure, response rates and initial descriptive information. *Pediatr. Blood Cancer* 50: 1018–1025.

230 Patel, S. and DeLaney, T.F. (2008). Advanced-technology radiation therapy for bone sarcomas. *Cancer Control* 15: 21–37.

231 McGovern, S.L. and Mahajan, A. (2012). Progress in radiotherapy for pediatric sarcomas. *Curr. Oncol. Rep.* 14: 320–326.

232 Chawla, S., Henshaw, R., Seeger, L. et al. (2013). Safety and efficacy of denosumab for adults and skeletally mature adolescents with giant cell tumour of bone: interim analysis of an open-label, parallel-group, phase 2 study. *Lancet Oncol.* 14: 901–908.

233 Thomas, D., Henshaw, R., Skubitz, K. et al. (2010). Denosumab in patients with giant-cell tumour of bone: an open-label, phase 2 study. *Lancet Oncol.* 11: 275–280.

234 Branstetter, D.G., Nelson, S.D., Manivel, J.C. et al. (2012). Denosumab induces tumor reduction and bone formation in patients with giant-cell tumor of bone. *Clin. Cancer Res.* 18: 4415–4424.

235 Grignani, G., Palmerini, E., Dileo, P. et al. (2012). A phase II trial of sorafenib in relapsed and unresectable high-grade osteosarcoma after failure of standard multimodal therapy: an Italian sarcoma group study. *Ann. Oncol.* 23: 508–516.

236 Duffaud, F., Mir, O., Boudou-Rouquette, P. et al. (2019). Efficacy and safety of regorafenib in adult patients with metastatic osteosarcoma: a non-comparative, randomised, double-blind, placebo-controlled, phase 2 study. *Lancet Oncol.* 20: 120–133.

237 Rosenberg, S.A., Yang, J.C., and Restifo, N.P. (2004). Cancer immunotherapy: moving beyond current vaccines. *Nat. Med.* 10: 909–915.

238 Weiner, L.M., Dhodapkar, M.V., and Ferrone, S. (2009). Monoclonal antibodies for cancer immunotherapy. *Lancet* 373: 1033–1040.

239 Rosenberg, S.A. and Restifo, N.P. (2015). Adoptive cell transfer as personalized immunotherapy for human cancer. *Science* 348: 62–68.

240 June, C.H., O'Connor, R.S., Kawalekar, O.U. et al. (2018). CAR T cell immunotherapy for human cancer. *Science* 359: 1361–1365.

241 Ribas, A. and Wolchok, J.D. (2018). Cancer immunotherapy using checkpoint blockade. *Science* 359: 1350–1355.

242 Schumacher, T.N., Scheper, W., and Kvistborg, P. (2019). Cancer neoantigens. *Annu. Rev. Immunol.* 37: 173–200.

243 Meyers, P.A., Schwartz, C.L., Krailo, M.D. et al. (2008). Osteosarcoma: the addition of muramyl tripeptide to chemotherapy improves overall survival – a report from the children's oncology group. *J. Clin. Oncol.* 26: 633–638.

244 Lussier, D.M., O'Neill, L., Nieves, L.M. et al. (2015). Enhanced T-cell immunity to osteosarcoma through antibody blockade of PD-1/PD-L1 interactions. *J. Immunother.* 38: 96–106.

245 Zheng, B., Ren, T., Huang, Y. et al. (2018). PD-1 axis expression in musculoskeletal tumors and antitumor effect of nivolumab in osteosarcoma model of humanized mouse. *J. Hematol. Oncol.* 11: 1–13.

246 Fagioli, B.F., Aglietta, M., Tienghi, A. et al. (2015). High-dose chemotherapy in the treatment of relapsed osteosarcoma: an Italian Sarcoma Group Study. *J. Clin. Oncol.* 20: 2150–2156.

247 Lin, P.P., Guzel, V.B., Moura, M.F. et al. (2002). Long-term follow-up of patients with giant cell tumor of the sacrum treated with selective arterial embolization. *Cancer* 95: 1317–1325.

248 Hosalkar, H.S., Jones, K.J., King, J.J. et al. (2007). Serial arterial embolization for large sacral giant-cell tumors: mid- to long-term results. *Spine* 32: 1107–1115.

249 Gao, X., Li, L., Cai, X. et al. (2021). Targeting nanoparticles for diagnosis and therapy of bone tumors: opportunities and challenges. *Biomaterials* 265: 120404.

250 Yang, K., Feng, L., Shi, X., and Liu, Z. (2013). Nano-graphene in biomedicine: theranostic applications. *Chem. Soc. Rev.* 42: 530–547.

251 Lamprecht, A. (2015). Nanomedicines in gastroenterology and hepatology. *Nat. Rev. Gastroenterol. Hepatol.* 12: 195–204.

252 Rodríguez-Nogales, C., González-Fernández, Y., Aldaz, A. et al. (2018). Nanomedicines for pediatric cancers. *ACS Nano* 12: 7482–7496.

253 van der Meel, R., Sulheim, E., Shi, Y. et al. (2019). Smart cancer nanomedicine. *Nat. Nanotechnol.* 14: 1007–1017.

254 Haag, R. and Kratz, F. (2006). Polymer therapeutics: concepts and applications. *Angew. Chem. Int. Ed.* 45: 1198–1215.

255 Wolfram, J. and Ferrari, M. (2019). Clinical cancer nanomedicine. *Nano Today* 25: 85–98.

256 Matsumura, Y. and Maeda, H. (1986). A new concept for macromolecular therapeutics in cancer chemotherapy: mechanism of tumoritropic accumulation of proteins and the antitumor agent smancs. *Cancer Res.* 46: 6387–6392.

257 Rahim, M., Khan, M.S., Khan, M.S. et al. (2014). Glycation-assisted synthesized gold nanoparticles inhibit growth of bone cancer cells. *Colloids Surf., B* 117: 473–479.

258 Nayak, D., Ashe, S., Rauta, P.R. et al. (2016). Bark extract mediated green synthesis of silver nanoparticles: evaluation of antimicrobial activity and antiproliferative response against osteosarcoma. *Mater. Sci. Eng., C* 58: 44–52.

259 Di Virgilio, A.L., Reigosa, M., and Lorenzo De Mele, M.F. (2010). Response of UMR 106 cells exposed to titanium oxide and aluminum oxide nanoparticles. *J. Biomed. Mater. Res. Part A* 92: 80–86.

260 Alpaslan, E., Yazici, H., Golshan, N.H. et al. (2015). pH-dependent activity of dextran-coated cerium oxide nanoparticles on prohibiting osteosarcoma cell proliferation. *ACS Biomater. Sci. Eng.* 1: 1096–1103.

261 Sisubalan, N., Ramkumar, V.S., Pugazhendhi, A. et al. (2018). ROS-mediated cytotoxic activity of ZnO and CeO_2 nanoparticles synthesized using the *Rubia cordifolia* L. leaf extract on MG-63 human osteosarcoma cell lines. *Environ. Sci. Pollut. Res.* 25: 10482–10492.

262 He, G., Ma, Y., Zhu, Y. et al. (2018). Cross talk between autophagy and apoptosis contributes to ZnO nanoparticle-induced human osteosarcoma cell death. *Adv. Healthc. Mater.* 7: 1–13.

263 Qing, F., Wang, Z., Hong, Y. et al. (2012). Selective effects of hydroxyapatite nanoparticles on osteosarcoma cells and osteoblasts. *J. Mater. Sci. - Mater. Med.* 23: 2245–2251.

264 Wang, Y., Wang, J., Hao, H. et al. (2016). In vitro and in vivo mechanism of bone tumor inhibition by selenium-doped bone mineral nanoparticles. *ACS Nano* 10: 9927–9937.

265 Xu, J., Wang, H., Hu, Y. et al. (2019). Inhibition of CaMKIIα activity enhances antitumor effect of fullerene C60 nanocrystals by suppression of autophagic degradation. *Adv. Sci.* 68, 1801233.

266 Jain, K.K. (2014). Current status and future prospects of drug delivery systems. *Methods Mol. Biol.* 1141: 1–56.

267 Ekladious, I., Colson, Y.L., and Grinstaff, M.W. (2019). Polymer–drug conjugate therapeutics: advances, insights and prospects. *Nat. Rev. Drug Discovery* 18: 273–294.

268 Safra, T., Muggia, F., Jeffers, S. et al. (2000). Pegylated liposomal doxorubicin (doxil): reduced clinical cardiotoxicity in patients reaching or exceeding cumulative doses of 500 mg/m^2. *Ann. Oncol.* 11: 1029–1033.

269 Torchilin, V.P. (2005). Recent advances with liposomes as pharmaceutical carriers. *Nat. Rev. Drug Discovery* 4: 145–160.

270 Tahara, Y., Mukai, S.A., Sawada, S.I. et al. (2015). Nanocarrier-integrated microspheres: nanogel tectonic engineering for advanced drug-delivery systems. *Adv. Mater.* 27: 5080–5088.

271 Irvine, D.J. (2011). Drug delivery: one nanoparticle, one kill. *Nat. Mater.* 10: 342–343.

272 Mignani, S., Rodrigues, J., Tomas, H. et al. (2018). Dendrimers in combination with natural products and analogues as anti-cancer agents. *Chem. Soc. Rev.* 47: 514–532.

273 El Andaloussi, S., Mäger, I., Breakefield, X.O., and Wood, M.J.A. (2013). Extracellular vesicles: biology and emerging therapeutic opportunities. *Nat. Rev. Drug Discovery* 12: 347–357.

274 Spicer, C.D., Jumeaux, C., Gupta, B., and Stevens, M.M. (2018). Peptide and protein nanoparticle conjugates: versatile platforms for biomedical applications. *Chem. Soc. Rev.* 47: 3574–3620.

275 Wang, A.Z., Langer, R., and Farokhzad, O.C. (2012). Nanoparticle delivery of cancer drugs. *Annu. Rev. Med.* 63: 185–198.

276 Low, S.A. and Kopeček, J. (2012). Targeting polymer therapeutics to bone. *Adv. Drug Delivery Rev.* 64: 1189–1204.

277 Morton, S.W., Shah, N.J., Quadir, M.A. et al. (2014). Osteotropic therapy via targeted layer-by-layer nanoparticles. *Adv. Healthc. Mater.* 3: 867–875.

278 Yin, Q., Tang, L., Cai, K. et al. (2016). Pamidronate functionalized nanoconjugates for targeted therapy of focal skeletal malignant osteolysis. *Proc. Natl. Acad. Sci. U.S.A.* 113: E4601–E4609.

279 Rudnick-Glick, S., Corem-Salkmon, E., Grinberg, I., and Margel, S. (2016). Targeted drug delivery of near IR fluorescent doxorubicin-conjugated poly(ethylene glycol) bisphosphonate nanoparticles for diagnosis and therapy of primary and metastatic bone cancer in a mouse model. *J. Nanobiotechnol.* 14: 1–11.

280 Wu, V.M., Mickens, J., and Uskoković, V. (2017). Bisphosphonate-functionalized hydroxyapatite nanoparticles for the delivery of the bromodomain inhibitor JQ1 in the treatment of osteosarcoma. *ACS Appl. Mater. Interfaces* 9: 25887–25904.

281 Cole, L.E., Vargo-Gogola, T., and Roeder, R.K. (2016). Targeted delivery to bone and mineral deposits using bisphosphonate ligands. *Adv. Drug Delivery Rev.* 99: 12–27.

282 Giger, E.V., Castagner, B., and Leroux, J.C. (2013). Biomedical applications of bisphosphonates. *J. Controlled Release* 167: 175–188.

283 Zhou, Z.F., Sun, T.-W., Chen, F. et al. (2017). Calcium phosphate-phosphorylated adenosine hybrid microspheres for anti-osteosarcoma drug delivery and osteogenic differentiation. *Biomaterials* 121: 1–14.

284 Li, J., Jiang, H., Ouyang, X. et al. (2016). $CaCO_3$/ttraethylenepentamine-graphene hollow microspheres as biocompatible bone drug carriers for controlled release. *ACS Appl. Mater. Interfaces* 8: 30027–30036.

285 Zhang, Y., Wang, F., Li, M. et al. (2018). Self-stabilized hyaluronate nanogel for intracellular codelivery of doxorubicin and cisplatin to osteosarcoma. *Adv. Sci.* 5: 1700821.

286 Li, X., Lovell, J.F., Yoon, J., and Chen, X. (2020). Clinical development and potential of photothermal and photodynamic therapies for cancer. *Nat. Rev. Clin. Oncol.* 17: 657–674.

287 Ito, A., Shinkai, M., Honda, H. et al. (2003). Heat shock protein 70 expression induces antitumor immunity during intracellular hyperthermia using magnetite nanoparticles. *Cancer Immunol. Immunother.* 52: 80–88.

288 Hildebrandt, B., Ahlers, O., Dieing, A. et al. (2002). The cellular and molecular basis of hyperthermia. *Crit. Rev. Oncol. Hematol.* 43: 33–56.

289 Pérez-Hernández, M., Del Pino, P., Mitchell, S.G. et al. (2015). Dissecting the molecular mechanism of apoptosis during photothermal therapy using gold nanoprisms. *ACS Nano* 9: 52–61.

290 Melamed, J.R., Edelstein, R.S., and Day, E.S. (2015). Elucidating the fundamental mechanisms of cell death triggered by photothermal therapy. *ACS Nano* 9: 6–11.

291 Weissleder, R. (2001). A clearer vision for in vivo imaging: progress continues in the development of smaller, more penetrable probes for biological imaging. *Nat. Biotechnol.* 19: 316–317.

292 Vogel, A. and Venugopalan, V. (2003). Mechanisms of pulsed laser ablation of biological tissues. *Chem. Rev.* 103: 577–644.

293 Huang, X., El-Sayed, I.H., Qian, W., and El-Sayed, M.A. (2006). Cancer cell imaging and photothermal therapy in the near-infrared region by using gold nanorods. *J. Am. Chem. Soc.* 128: 2115–2120.

294 Wang, Y., Black, K.C.L., Luehmann, H. et al. (2013). Comparison study of gold for photothermal cancer treatment. *ACS Nano* 7: 2068–2077.

295 Vankayala, R., Huang, Y.K., Kalluru, P. et al. (2014). First demonstration of gold nanorods-mediated photodynamic therapeutic destruction of tumors via near infra-red-light activation. *Small* 10: 1612–1622.

296 Vankayala, R., Lin, C.C., Kalluru, P. et al. (2014). Gold nanoshells-mediated bimodal photodynamic and photothermal cancer treatment using ultra-low doses of near infra-red light. *Biomaterials* 35: 5527–5538.

297 Tian, Q., Tang, M., Sun, Y. et al. (2011). Hydrophilic flower-like CuS superstructures as an efficient 980 nm laser-driven photothermal agent for ablation of cancer cells. *Adv. Mater.* 23: 3542–3547.

298 Zhou, M., Zhang, R., Huang, M. et al. (2010). A chelator-free multifunctional [64Cu]CuS nanoparticle platform for simultaneous micro-PET/CT imaging and photothermal ablation therapy. *J. Am. Chem. Soc.* 132: 15351–15358.

299 Dai, C., Chen, Y., Jing, X. et al. (2017). Two-dimensional tantalum carbide (MXenes) composite nanosheets for multiple imaging-guided photothermal tumor ablation. *ACS Nano* 11: 12696–12712.

300 Panwar, N., Soehartono, A.M., Chan, K.K. et al. (2019). Nanocarbons for biology and medicine: sensing, imaging, and drug delivery. *Chem. Rev.* 119: 9559–9656.

301 Cheng, L., Yang, K., Chen, Q., and Liu, Z. (2012). Organic stealth nanoparticles for highly effective in vivo near-infrared photothermal therapy of cancer. *ACS Nano* 6: 5605–5613.

302 Yang, J., Choi, J., Bang, D. et al. (2011). Convertible organic nanoparticles for near-infrared photothermal ablation of cancer cells. *Angew. Chem. Int. Ed.* 50: 441–444.

303 Li Volsi, A., Scialabba, C., Vetri, V. et al. (2017). Near-infrared light responsive folate targeted gold nanorods for combined photothermal-chemotherapy of osteosarcoma. *ACS Appl. Mater. Interfaces* 9: 14453–14469.

304 Dong, S., Zhang, Y.-N., Wan, J. et al. (2020). A novel multifunctional carbon aerogel-coated platform for osteosarcoma therapy and enhanced bone regeneration. *J. Mater. Chem. B* 8: 368–379.

305 Yin, J., Han, Q., Zhang, J. et al. (2020). MXene-based hydrogels endow polyetheretherketone with effective osteogenicity and combined treatment of osteosarcoma and bacterial infection. *ACS Appl. Mater. Interfaces* 12: 45891–45903.

306 Lu, Y., Li, L., Lin, Z. et al. (2018). Enhancing osteosarcoma killing and CT imaging using ultrahigh drug loading and NIR-responsive bismuth sulfide@mesoporous silica nanoparticles. *Adv. Healthc. Mater.* 7: 1–12.

307 Cheng, J., Wang, W., Xu, X. et al. (2020). $AgBiS_2$ nanoparticles with synergistic photodynamic and bioimaging properties for enhanced malignant tumor phototherapy. *Mater. Sci. Eng., C* 107, 110324.

308 Wang, L., Long, N.J., Li, L. et al. (2018). Multi-functional bismuth-doped bioglasses: combining bioactivity and photothermal response for bone tumor treatment and tissue repair article. *Light Sci. Appl.* 7: 1–13.

309 Zhou, Z., Fan, T., Yan, Y. et al. (2019). One stone with two birds: phytic acid-capped platinum nanoparticles for targeted combination therapy of bone tumors. *Biomaterials* 194: 130–138.

310 Celli, J.P., Rizvi, I., Evans, C.L. et al. (2010). Imaging and photodynamic therapy: mechanisms, monitoring and optimization. *Chem. Rev.* 110: 2795–2838.

311 Dolmans, D.E.J.G.J., Fukumara, D., and Jain, R.K. (2003). Photodynamic therapy for cancer. *Nat. Rev. Cancer* 3: 375–380.

312 Agostinis, P., Berg, K., Cengel, K.A. et al. (2011). Photodynamic therapy of cancer: an update. *CA Cancer J. Clin.* 61: 250–281.

313 Jung, H.S., Verwilst, P., Sharma, A. et al. (2018). Organic molecule-based photothermal agents: an expanding photothermal therapy universe. *Chem. Soc. Rev.* 47: 2280–2297.

314 Chung, S., Revia, R.A., and Zhang, M. (2019). Graphene quantum dots and their applications in bioimaging, biosensing, and therapy. *Adv. Mater.* 1904362: 1–26.

315 Piao, Y., Burns, A., Kim, J. et al. (2008). Designed fabrication of silica-based nanostructured particle systems for nanomedicine applications. *Adv. Funct. Mater.* 18: 3745–3758.

316 Shen, Y., Sun, Y., Yan, R. et al. (2017). Rational engineering of semiconductor QDs enabling remarkable $^1O_2$ production for tumor-targeted photodynamic therapy. *Biomaterials* 148: 31–40.

317 Zhang, P., Steelant, W., Kumar, M., and Scholfield, M. (2007). Versatile photosensitizers for photodynamic therapy at infrared excitation. *J. Am. Chem. Soc.* 129: 4526–4527.

318 Ji, D.K., Zhang, Y., Zang, Y. et al. (2016). Targeted intracellular production of reactive oxygen species by a 2D molybdenum disulfide glycosheet. *Adv. Mater.* 28: 9356–9363.

319 Chen, Y., Ye, D., Wu, M. et al. (2014). Break-up of two-dimensional MnO_2 nanosheets promotes ultrasensitive pH-triggered theranostics of cancer. *Adv. Mater.* 26: 7019–7026.

320 Wang, H., Yang, X., Shao, W. et al. (2015). Ultrathin black phosphorus nanosheets for efficient singlet oxygen generation. *J. Am. Chem. Soc.* 137: 11376–11382.

321 Li, X., Kim, C.-Y., Lee, S. et al. (2017). Nanostructured phthalocyanine assemblies with protein-driven switchable photoactivities for biophotonic imaging and therapy. *J. Am. Chem. Soc.* 139: 10880–10886.

322 Nam, C., Jang, Y., Lee, K.S. et al. (2006). Insertion of a specular reflective and transmissive nano-oxide layer into giant magnetoresistance spin-valve structure. *J. Nanosci. Nanotechnol.* 6: 3483–3486.

323 Brown, S.B., Brown, E.A., and Walker, I. (2004). The present and future role of photodynamic therapy in cancer treatment. *Lancet Oncol.* 5: 497–508.

324 Yu, W., Ye, M., Zhu, J. et al. (2018). Zinc phthalocyanine encapsulated in polymer micelles as a potent photosensitizer for the photodynamic therapy of osteosarcoma. *Nanomed. Nanotechnol. Biol. Med.* 14: 1099–1110.

325 Yu, W., Wang, Y., Zhu, J. et al. (2019). Autophagy inhibitor enhance ZnPc/BSA nanoparticle induced photodynamic therapy by suppressing PD-L1 expression in osteosarcoma immunotherapy. *Biomaterials* 02: 192.

326 Gao, Z., Zhang, L., Hu, J., and Sun, Y. (2013). Mesenchymal stem cells: a potential targeted-delivery vehicle for anti-cancer drug, loaded nanoparticles. *Nanomed. Nanotechnol. Biol. Med.* 9: 174–184.

327 Xu, C., Feng, Q., Yang, H. et al. (2018). A light-triggered mesenchymal stem cell delivery system for photoacoustic imaging and chemo-photothermal therapy of triple negative breast cancer. *Adv. Sci.* 5, 1800382.

328 Duchi, S., Sotgiu, G., Lucarelli, E. et al. (2013). Mesenchymal stem cells as delivery vehicle of porphyrin loaded nanoparticles: effective photoinduced in vitro killing of osteosarcoma. *J. Controlled Release* 1–13: https://doi.org/10.1016/j.jconrel.2013.03.012.

329 Lenna, S., Bellotti, C., Duchi, S. et al. Mesenchymal stromal cells mediated delivery of photoactive nanoparticles inhibits osteosarcoma growth in vitro and in a murine in vivo ectopic model. *J. Exp. Clin.* 39: 40.

330 Yang, K., Miron, R.J., Bian, Z., and Zhang, Y.F. (2018). A bone-targeting drug-delivery system based on Semaphorin 3A gene therapy ameliorates bone loss in osteoporotic ovariectomized mice. *Bone* 114: 40–49.

331 Zhang, G., Guo, B., Wu, H. et al. (2012). A delivery system targeting bone formation surfaces to facilitate RNAi-based anabolic therapy. *Nat. Med.* 18: 309–316.

332 Wang, M., Liu, H., Li, L., and Cheng, Y. (2014). A fluorinated dendrimer achieves excellent gene transfection efficacy at extremely low nitrogen to phosphorus ratios. *Nat. Commun.* 5: 1–8.

333 Xin, Y., Huang, M., Guo, W.W. et al. (2017). Nano-based delivery of RNAi in cancer therapy. *Mol. Cancer* 16: 134.

334 Mekhail, G.M., Kamel, A.O., Awad, G.A. et al. (2016). Synthesis and evaluation of alendronate-modified gelatin biopolymer as a novel osteotropic nanocarrier for gene therapy. *Nanomedicine* 11: 2251–2273.

335 Tiram, G., Segal, E., Krivitsky, A. et al. (2016). Identification of dormancy-associated microRNAs for the design of osteosarcoma-targeted dendritic polyglycerol nanopolyplexes. *ACS Nano* 10: 2028–2045.

336 Wang, F., Pang, J.-D., Huang, L.-L. et al. (2018). Nanoscale polysaccharide derivative as an AEG-1 siRNA carrier for effective osteosarcoma therapy. *Int. J. Nanomed.* 13: 857–875.

337 Nel, A.E., Mädler, L., Velegol, D. et al. (2009). Understanding biophysicochemical interactions at the nano-bio interface. *Nat. Mater.* 8: 543–557.

338 Shi, J., Kantoff, P.W., Wooster, R., and Farokhzad, O.C. (2017). Cancer nanomedicine: progress, challenges and opportunities. *Nat. Rev. Cancer* 17: 20–37.

339 Valencia, P.M., Basto, P.A., Zhang, L. et al. (2010). Single-step assembly of homogenous lipid-polymeric and lipid-quantum dot nanoparticles enabled by microfluidic rapid mixing. *ACS Nano* 4: 1671–1679.

340 Rhee, M., Valencia, P.M., Rodriguez, M.I. et al. (2011). Synthesis of size-tunable polymeric nanoparticles enabled by 3D hydrodynamic flow focusing in single-layer microchannels. *Adv. Mater.* 23: 79–83.

341 Chen, D., Love, K.T., Chen, Y. et al. (2012). Rapid discovery of potent siRNA-containing lipid nanoparticles enabled by controlled microfluidic formulation. *J. Am. Chem. Soc.* 134: 6948–6951.

342 Kim, Y., Chung, B.L., Ma, M. et al. (2012). Mass production and size control of lipid-polymer hybrid nanoparticles through controlled microvortices. *Nano Lett.* 12: 3587–3591.

343 Rolland, J.P., Maynor, B.W., Euliss, L.E. et al. (2005). Direct fabrication and harvesting of monodisperse, shape-specific nanobiomaterials. *J. Am. Chem. Soc.* 127: 10096–10100.

344 Perry, J.L., Herlihy, K.P., Napier, M.E., and Desimone, J.M. (2011). PRINT: a novel platform toward shape and size specific nanoparticle theranostics. *Acc. Chem. Res.* 44: 990–998.

345 Xu, J., Wong, D.H.C., Byrne, J.D. et al. (2013). Future of the particle replication in nonwetting templates (PRINT) technology. *Angew. Chem. Int. Ed.* 52: 6580–6589.

346 Wang, B., He, X., Zhang, Z. et al. (2013). Metabolism of nanomaterials in vivo: blood circulation and organ clearance. *Acc. Chem. Res.* 46: 761–769.

347 Nel, A. (2005). Air pollution–related illness: effects of particles. *Science* 308: 804–806.

348 Fomina, N., Sankaranarayanan, J., and Almutairi, A. (2012). Photochemical mechanisms of light-triggered release from nanocarriers. *Adv. Drug Delivery Rev.* 64: 1005–1020.

349 Guan, Q., Zhou, L.-L., Li, Y.-A. et al. (2019). Nanoscale covalent organic framework for combinatorial antitumor photodynamic and photothermal therapy. *ACS Nano* https://doi.org/10.1021/acsnano.9b06467.

350 Zhang, D., Zheng, Y., Lin, Z. et al. (2019). Artificial engineered natural killer cells combined with antiheat endurance as a powerful strategy for enhancing photothermal-immunotherapy efficiency of solid tumors. *Small* 15: 1–13.

351 Zhu, X., Gong, Y., Liu, Y. et al. (2020). Ru@CeO_2 yolk shell nanozymes: oxygen supply in situ enhanced dual chemotherapy combined with photothermal therapy for orthotopic/subcutaneous colorectal cancer. *Biomaterials* 242: 119923.

352 Chen, Y.W., Chen, P.J., Hu, S.H. et al. (2014). NIR-triggered synergic photo-chemothermal therapy delivered by reduced graphene oxide/carbon/mesoporous silica nanocookies. *Adv. Funct. Mater.* 24: 451–459.

353 Hu, J., Tang, Y.'a., Elmenoufy, A.H. et al. (2015). Nanocomposite-based photodynamic therapy strategies for deep tumor treatment. *Small* 11: 5860–5887.

354 Yang, M., Yang, T., and Mao, C. (2019). Enhancement of photodynamic cancer therapy by physical and chemical factors. *Angew. Chem. Int. Ed.* 58: 14066–14080.

355 Zhou, Z., Song, J., Nie, L., and Chen, X. (2016). Reactive oxygen species generating systems meeting challenges of photodynamic cancer therapy. *Chem. Soc. Rev.* 45: 6597–6626.

356 Shi, Y. and Lammers, T. (2019). Combining nanomedicine and immunotherapy. *Acc. Chem. Res.* 52: 1543–1554.

357 Zhang, Q., Wei, W., Wang, P. et al. (2017). Biomimetic magnetosomes as versatile artificial antigen-presenting cells to potentiate T-cell-based anticancer therapy. *ACS Nano* 11: 10724–10732.

358 Zhao, X., Yang, K., Zhao, R. et al. (2016). Inducing enhanced immunogenic cell death with nanocarrier-based drug delivery systems for pancreatic cancer therapy. *Biomaterials* 102: 187–197.

359 Duan, X., Chan, C., Guo, N. et al. (2016). Photodynamic therapy mediated by nontoxic core-shell nanoparticles synergizes with immune checkpoint blockade to elicit antitumor immunity and antimetastatic effect on breast cancer. *J. Am. Chem. Soc.* 138: 16686–16695.

360 He, C., Duan, X., Guo, N. et al. (2016). Core-shell nanoscale coordination polymers combine chemotherapy and photodynamic therapy to potentiate checkpoint blockade cancer immunotherapy. *Nat. Commun.* 7: 1–12.

361 Duan, X., Chan, C., and Lin, W. (2019). Nanoparticle-mediated immunogenic cell death enables and potentiates cancer immunotherapy. *Angew. Chem. Int. Ed.* 58: 670–680.

362 Galluzzi, L., Buqué, A., Kepp, O. et al. (2017). Immunogenic cell death in cancer and infectious disease. *Nat. Rev. Immunol.* 17: 97–111.

363 Lu, K., He, C., Guo, N. et al. (2016). Chlorin-based nanoscale metal-organic framework systemically rejects colorectal cancers via synergistic photodynamic therapy and checkpoint blockade immunotherapy. *J. Am. Chem. Soc.* 138: 12502–12510.

364 Zanganeh, S., Hutter, G., Spitler, R. et al. (2016). Iron oxide nanoparticles inhibit tumour growth by inducing pro-inflammatory macrophage polarization in tumour tissues. *Nat. Nanotechnol.* 11: 986–994.

365 Zheng, Y., Tang, L., Mabardi, L. et al. (2017). Enhancing adoptive cell therapy of cancer through targeted delivery of small-molecule immunomodulators to internalizing or noninternalizing receptors. *ACS Nano* 11: 3089–3100.

366 Shae, D., Becker, K.W., Christov, P. et al. (2019). Endosomolytic polymersomes increase the activity of cyclic dinucleotide STING agonists to enhance cancer immunotherapy. *Nat. Nanotechnol.* 14: 269–278.

367 Yuan, H., Jiang, W., von Roemeling, C.A. et al. (2017). Multivalent bi-specific nanobioconjugate engager for targeted cancer immunotherapy. *Nat. Nanotechnol.* 12: 763–769.

368 Zhang, Y., Li, N., Suh, H., and Irvine, D.J. (2018). Nanoparticle anchoring targets immune agonists to tumors enabling anti-cancer immunity without systemic toxicity. *Nat. Commun.* 9: 6.

369 Kuai, R., Ochyl, L.J., Bahjat, K.S. et al. (2017). Designer vaccine nanodiscs for personalized cancer immunotherapy. *Nat. Mater.* 16: 489–498.

370 Min, Y., Roche, K.C., Tian, S. et al. (2017). Antigen-capturing nanoparticles improve the abscopal effect and cancer immunotherapy. *Nat. Nanotechnol.* 12: 877–882.

371 Smith, T.T., Stephan, S.B., Moffett, H.F. et al. (2017). In situ programming of leukaemia-specific T cells using synthetic DNA nanocarriers. *Nat. Nanotechnol.* 12: 813–822.

372 Stephan, M.T., Moon, J.J., Um, S.H. et al. (2010). Therapeutic cell engineering with surface-conjugated synthetic nanoparticles. *Nat. Med.* 16: 1035–1041.

373 Liang, S., Deng, X., Ma, P. et al. (2020). Recent advances in nanomaterial-assisted combinational sonodynamic cancer therapy. *Adv. Mater.* 2003214: 1–35.

374 Tang, Z., Liu, Y., He, M., and Bu, W. (2019). Chemodynamic therapy: tumour microenvironment-mediated Fenton and Fenton-like reactions. *Angew. Chem. Int. Ed.* 58: 946–956.

375 Ren, Z., Sun, S., Sun, R. et al. (2020). A metal–polyphenol-coordinated nanomedicine for synergistic cascade cancer chemotherapy and chemodynamic therapy. *Adv. Mater.* 32: 1–10.
376 Chen, Y., Huang, Y., Zhou, S. et al. (2020). Tailored chemodynamic nanomedicine improves pancreatic cancer treatment via controllable damaging neoplastic cells and reprogramming tumor microenvironment. *Nano Lett.* 20: 6780–6790.
377 Moros, M., Idiago-López, J., Asín, L. et al. (2019). Triggering antitumoural drug release and gene expression by magnetic hyperthermia. *Adv. Drug Delivery Rev.* 138: 326–343.
378 Cornu, O., Druez, V., and Barbier, O. (2007). Bone allografts: what they can offer and what they cannot. *J. Bone Joint Surg. Br.* 89: 574–580.
379 Mercado-Pagán, Á.E., Shanjani, Y., Yang, Y., and Stahl, A.M. (2015). Vascularization in bone tissue engineering constructs. *Ann. Biomed. Eng.* 433: 718–729.
380 Guillotin, B. and Guillemot, F. (2011). Cell patterning technologies for organotypic tissue fabrication. *Trends Biotechnol.* 29: 183–190.
381 Guillotin, B., Souquet, A., Catros, S. et al. (2010). Laser assisted bioprinting of engineered tissue with high cell density and microscale organization. *Biomaterials* 31: 7250–7256.
382 Tang, D., Tare, R.S., Yang, L.-Y. et al. (2016). Biofabrication of bone tissue: approaches, challenges and translation for bone regeneration. *Biomaterials* 83: 363–382.
383 Brunello, G., Sivolella, S., Meneghello, R. et al. (2016). Powder-based 3D printing for bone tissue engineering. *Biotechnol. Adv.* 34: 740–753.
384 Bose, S., Vahabzadeh, S., and Bandyopadhyay, A. (2013). Bone tissue engineering using 3D printing. *Mater. Today* 16: 496–504.
385 Zhang, L., Yang, G., Johnson, B.N., and Jia, X. (2019). Three-dimensional (3D) printed scaffold and material selection for bone repair. *Acta Biomater.* 84: 16–33.
386 Brauer, D.S. (2015). Bioactive glasses – structure and properties. *Angew. Chem. Int. Ed.* 54: 4160–4181.
387 Kargozar, S., Baino, F., Hamzehlou, S. et al. (2018). Bioactive glasses: sprouting angiogenesis in tissue engineering. *Trends Biotechnol.* 36: 430–444.
388 Wu, C., Zhou, Y., Fan, W. et al. (2012). Hypoxia-mimicking mesoporous bioactive glass scaffolds with controllable cobalt ion release for bone tissue engineering. *Biomaterials* 33: 2076–2085.
389 Do, A.V., Khorsand, B., Geary, S.M., and Salem, A.K. (2015). 3D printing of scaffolds for tissue regeneration applications. *Adv. Healthc. Mater.* 4: 1742–1762.
390 Pan, S., Yin, J., Yu, L. et al. (2020). 2D MXene-integrated 3D-printing scaffolds for augmented osteosarcoma phototherapy and accelerated tissue reconstruction. *Adv. Sci.* 7, 1901511.

391 Yang, Q., Yin, H., Xu, T. et al. (2020). Engineering 2D mesoporous silica@MXene-integrated 3D-printing scaffolds for combinatory osteosarcoma therapy and NO-augmented bone regeneration. *Small* 1906814: 1–15.

392 Dang, W., Li, T., Li, B. et al. (2018). A bifunctional scaffold with $CuFeSe_2$ nanocrystals for tumor therapy and bone reconstruction. *Biomaterials* 160: 92–106.

393 Sarkar, N. and Bose, S. (2019). Liposome-encapsulated curcumin-loaded 3D printed scaffold for bone tissue engineering. *ACS Appl. Mater. Interfaces* 11: 17184–17192.

394 Dang, W., Ma, B., Li, B. et al. (2019). 3D printing of metal-organic framework nanosheets-structured scaffolds with tumor therapy and bone construction. *Biofabrication* 4: 465705.

395 Ma, H., Luo, J., Sun, Z. et al. (2016). 3D printing of biomaterials with mussel-inspired nanostructures for tumor therapy and tissue regeneration. *Biomaterials* 111: 138–148.

396 Ma, H., Jiang, C., Zhai, D. et al. (2016). A bifunctional biomaterial with photothermal effect for tumor therapy and bone regeneration. *Adv. Funct. Mater.* 26: 1197–1208.

397 Maher, S., Lima-Marques, L., Evdokiou, A. et al. (2017). Engineering of micro to nano structured 3D printed drug-releasing titanium implants for enhanced osseointegration and localized delivery of anticancer drugs. *ACS Appl. Mater. Interfaces* 935: 29562–29570.

398 Wang, C., Ye, X., Zhao, Y. et al. (2020). Cryogenic 3D printing of porous scaffolds for in situ delivery of 2D black phosphorus nanosheets, doxorubicin hydrochloride and osteogenic peptide for treating tumor resection-induced bone defects. *Biofabrication* 12: 35004.

399 Ma, H., Li, T., Huan, Z. et al. (2018). 3D printing of high-strength bioscaffolds for the synergistic treatment of bone cancer. *NPG Asia Mater.* 10: 31–44.

400 Wade, R.J. and Burdick, J.A. (2014). Advances in nanofibrous scaffolds for biomedical applications: from electrospinning to self-assembly. *Nano Today* 9: 722–742.

401 Zhang, Y.S., Zhu, C., and Xia, Y. (2017). Inverse opal scaffolds and their biomedical applications. *Adv. Mater.* 29.

402 Zhang, Y.S. and Xia, Y. (2015). Multiple facets for extracellular matrix mimicking in regenerative medicine. *Nanomedicine* 10: 689–692.

403 Karageorgiou, V. and Kaplan, D. (2005). Porosity of 3D biomaterial scaffolds and osteogenesis. *Biomaterials* 26: 5474–5491.

404 Almubarak, S., Nethercott, H., Freeberg, M. et al. (2016). Tissue engineering strategies for promoting vascularized bone regeneration. *Bone* 83: 197–209.

405 Vaezi, M., Seitz, H., and Yang, S. (2013). A review on 3D micro-additive manufacturing technologies. *Int. J. Adv. Manuf. Tech.* 67: 1721–1754.

406 Wu, C. (2018). 3D-printed bioceramic scaffolds: from bone tissue engineering to tumor therapy. *Acta Biomater.* 79: 37–59.

407 Zhou, X., Zhou, G., Junka, R. et al. (2021). Fabrication of polylactic acid (PLA)-based porous scaffold through the combination of traditional bio-fabrication and 3D printing technology for bone regeneration. *Colloids Surf., B* 197: 111420.

408 Prendergast, M.E. and Burdick, J.A. (2020). Recent advances in enabling technologies in 3D printing for precision medicine. *Adv. Mater.* 32: 1–14.

409 Groll, J., Boland, T., Blunk, T. et al. (2016). Biofabrication: reappraising the definition of an evolving field. *Biofabrication* 8, 013001.

410 Mota, C., Camarero-Espinosa, S., Baker, M.B. et al. (2020). Bioprinting: from tissue and organ development to in vitro models. *Chem. Rev.* 120: 10547–10607.

411 Murakami, T., Nakatsuji, H., Inada, M. et al. (2012). Photodynamic and photothermal effects of semiconducting and metallic-enriched single-walled carbon nanotubes. *J. Am. Chem. Soc.* 134: 17862–17865.

11

Bionic Fixation: Design, Biomechanics, and Clinical Application

Wei Chen and Qi Zhang

The Third Hospital of Hebei Medical University, Department of Orthopedic Surgery, No. 139 Ziqiang Road, Shijiazhuang 050051, China

11.1 Bionics and Medical Bionics

As an independent discipline, the term bionics was officially introduced in September 1960. The term bionics was developed by Major Steele, an officer of the United States Air Force, based on the Latin word "bios" and the suffix "nic" at the first symposium on bionics, which was held in the United States. He defined that "Bionics is the study of systems that work in a way that mimics biological systems, or in a way that has the properties or features of biological systems, or in a manner similar to biological systems."

Modern bionics has expanded to many fields, including architecture, aircraft manufacturing, dress making, robot, sports, and medicine. Bionics has been promoted and developed in these fields and industries. Medical bionics is the application and development of the theories, concepts, and techniques of biomimicry in the medical field, mainly through the imitation of both macroscopic and microcosmic structures, material composition, function and control of human body to design, develop, and use bionic instruments, tools, and techniques to repair or replace part of the human body and restore the impaired functions. Medical bionics has become a hot branch of biomimicry and attracts scientists from both basic research and physicians. The technologies in the medical bionics can be classified into structural bionics, functional bionics, material bionics, mechanical bionics, and control bionics. The definition of bionic fixation in orthopedics contains material bionic and structural bionic. The former includes artificial bone and ligaments or joint capsule sutures (polymer materials) that simulate the structure of human fibrous tissue. The latter includes new types of internal fixation that simulate the structure and function of human amphiarthroses. The main trend in the development of "bionic fixation" is to improve the design and biocompatibility of materials and to develop novel "bionic fixation" that are highly biocompatible and more responsive to human bone trabecular ingrowth.

Biofabrication for Orthopedics: Methods, Techniques, and Applications, First Edition.
Edited by Wenguo Cui, Xin Zhao, and Shen Liu.

In the field of traumatology and orthopedics, bionic concepts, techniques, implants, and tools, which imitate the structure, morphology, material composition, biomechanical features, and functions of human bone, joints, ligament, and other tissues, have been developed and applied in the management of bone defects, infections, fractures, and dislocations (such as posterior pelvic ring disruptions and separation of tibiofibular syndesmosis). Structural bionics and material bionics are the most common types of biomimicry applied in the field of orthopedics and traumatology.

11.2 Structural Bionics in the Field of Orthopedics and Traumatology

Metallic implants are widely used in the management of traumatic fractures and dislocations. The most commonly used implants are made of stainless steel and titanium alloy, which can provide rigid fixation for fractures and dislocations. However, the modulus of elasticity of titanium alloy and stainless steel are much higher than that of the human bone, which may lead to local stress concentration of the fractured bone using metallic implants. In addition, the commercial implants cannot match the irregular shape and biomechanical features of human bone and joints. For example, the reconstruction plates cannot match the irregular shape of the posterior pelvic ring and cannot imitate the structure and function of the posterior sacroiliac ligament complex. Another example is the unstable syndesmotic injuries, which are usually fixed using metallic screw. The rigid screw fixation has some drawbacks: the prolonged immobilization and related complications, late syndesmotic widening after screw removal, screw loosening, or screw breakage. To avoid the drawbacks of traditional rigid fixation devices and improve the property of the implants, bionic fixation devices, which can match the structure, shape, biomechanical features, and the material composition of human bone and joint, have been developed, tested, and even applied in the management of orthopedic patients.

11.2.1 Elastic Bionic Fixation Device for Distal Tibiofibular Syndesmosis

The distal tibiofibular syndesmosis (DTS) injury is the primary injury type of the ankle trauma, accounting for 1–11% of all ankle lesions [1]. In clinical practice, the DTS as a microdynamic system is responsible for weight transmission and walking as well as the movement of the fibula combined with the tibia in all directions [2, 3]. However, a traditional rigid internal fixation system has considered the limitation of the physiological micromovement of the DTS, which may be against the early rehabilitative training. It can also lead to a series of complications including implant loosening or breakage, chronic ankle instability, or even post-traumatic ankle arthritis. An ideal implant to stabilize the tibiofibular syndesmosis should allow early mobilization for weight-bearing and be strong enough to maintain reduction of the

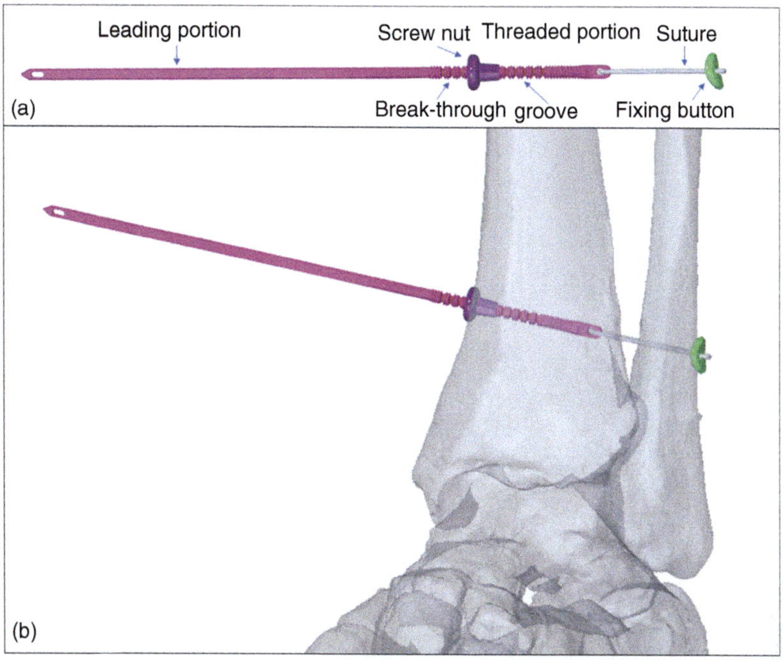

Figure 11.1 The elastic bionic device for distal tibiofibular syndesmosis (a) and the schematic diagram of the fixation model (b).

syndesmosis. Therefore, we have designed the elastic bionic device, which combines the features of both rigid and flexible fixations (Figure 11.1) [4]. The rigid part of this device is composed of a screw (with an easily break-off groove) and a screw nut. The posterior portion of the screw body is slippery and the screw tail is round, connecting to the fixing button of the screw tail by a fixed length of high strength non-absorbable suture. The rigid part can afford an improved rigidity and stability, and the high strength non-absorbable suture located between the tibia and fibula may retain the motion of the syndesmosis to the maximum degree, based on the data obtained through the biomechanical study.

Wang et al. [5] treated 17 patients with DTS injury with the use of the elastic bionic device. During the operation, a hole from the posterolateral side of the fibula to the anteromedial part of the tibia was drilled. The length of the hole was measured, and then a screw of matched length was passed through the hole with the fixing button of the screw tail tightly attaching to the fibula. The screw nut was installed on the tibial side to make the device tightened properly. Finally, the exposed leading portion of the screw was broken off (Figure 11.2). Postoperative radiographs were taken to demonstrate the reduction and fixation of DTS (Figure 11.3). Excellent clinical outcomes as determined using the American Orthopaedic Foot and Ankle Society (AOFAS) scores were reported in all patients. The mean AOFAS scores at 12 months postoperatively were 96.50 ± 5.60, with 16 cases rated as excellent, one as good, zero as fair, and zero as poor. The authors concluded that the use of the "bionic fixation" device

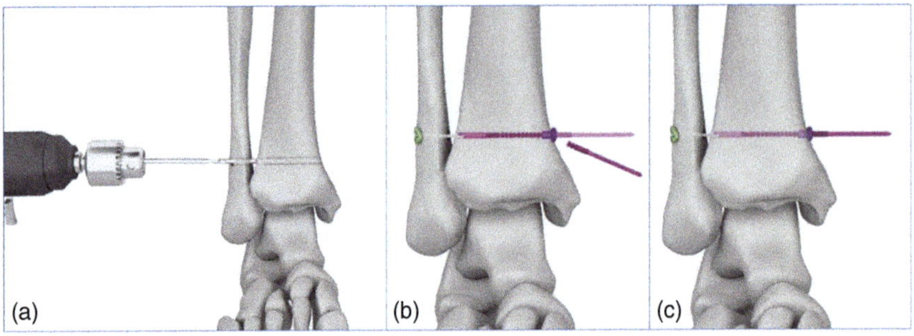

Figure 11.2 Schematic diagrams showing the main procedures for fixation of the tibiofibular syndesmosis using the elastic device. (a) A hole was drilled from the posterolateral side of fibula to anteromedial tibia. (b) A main screw was passed through the hole from the fibular side and the fixing button of the screw tail was tightly attached to the fibula; then, the screw nut was installed and adjusted on the tibial side to make the construct tightened properly. (c) The exposed leading portion of the screw was broken off.

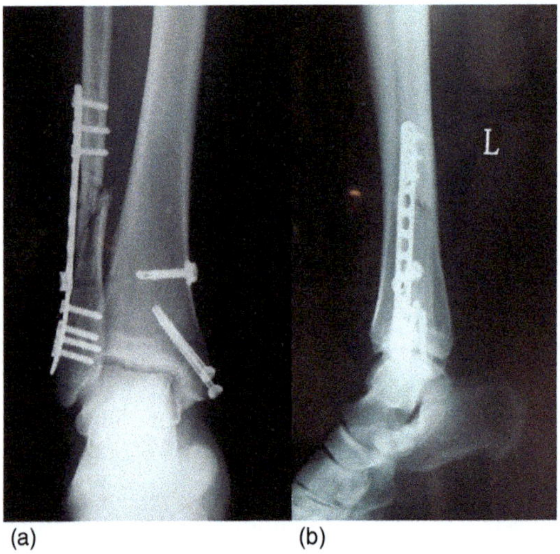

Figure 11.3 A male patient sustained left DTS, who were treated using the elastic bionic device. Postoperative radiographs showed that the DTS was anatomically reduced and the internal implants remained in a proper position. (a) Antero-posterior view and (b) lateral view.

for treating DTS injury appears to be a promising option because it can provide stable and reliable fixation, rapid functional recovery, and relatively fewer complications. The use of screw fixation for DTS separation not only makes the joint lose its micromotion characteristics but also predisposes to complications such as screw breakage and loosening; the use of TightRope flexible fixation has a higher risk of complications such as loosening and osteolysis at the contact surface. We combine the screws and TightRope flexible fixation. In clinical practice, the DTS as a micro-dynamic system is responsible for weight transmission and walking as well as the movement of the fibula combined with the tibia in all directions. The non-absorbable suture located between the tibia and fibula can avoid the rigid fixation of screws for DTS. The non-absorbable suture elasticity can adapt to the micromotion of the DTS.

11.2.2 Minimally Invasive Adjustable Plate for Pelvic Fractures

Posterior pelvic ring disruptions are serious injuries, often caused by high-energy trauma and accompanied by other fractures or injuries of the body [6, 7]. The posterior pelvic ring is composed of the posterior part of the ilium, sacrum, and the posterior sacroiliac ligament complex, which is similar to the structure of the suspended bridge (Figure 11.4). The posterior pelvic ring plays an important role in supporting the upper body weight, stabilizing the pelvic ring, and transmitting the upper body weight from spine to the lower extremities [8]. Anatomical reduction and rigid fixation is a key procedure to restore the anatomical structure and biomechanical functional of the posterior pelvic ring.

An iliosacral screw (ISS) and a tension band plate (TBP) are the two commonly used internal fixators in the treatment of posterior pelvic ring disruptions [9–11]. However, they are technique-demanding and lack the ability to restore the anatomical structure and biomechanical features.

Chen et al. invented a minimally invasive adjustable plate (MIAP) that consists of two Z-shaped brackets at each end of an adjustable connection bar (Figure 11.5). The structure of the MIAP simulates the suspended bridge construct of the posterior pelvic ring and can restore the biomechanical characteristics of the normal pelvic ring in the conduction of weight-load bearing. The results of finite element analysis have verified that there was no significant difference in biomechanical stability among MIAP, ISS, and TBP fixation of posterior pelvic ring disruptions. However, the fatigue damage, screw loosening risk, and stress shielding of the ISS system and the MIAP system are lower than that of the TBP system. However, MIAP and ISS fixation are more conducive to the healing process than TBP fixation.

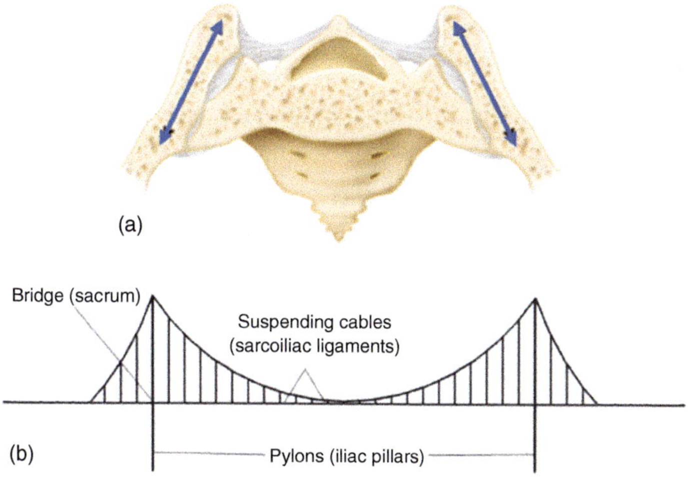

Figure 11.4 The suspended bridge structure of the posterior pelvic ring: (a) the schematic diagram of the posterior pelvic ring and (b) the structure of the suspended bridge.

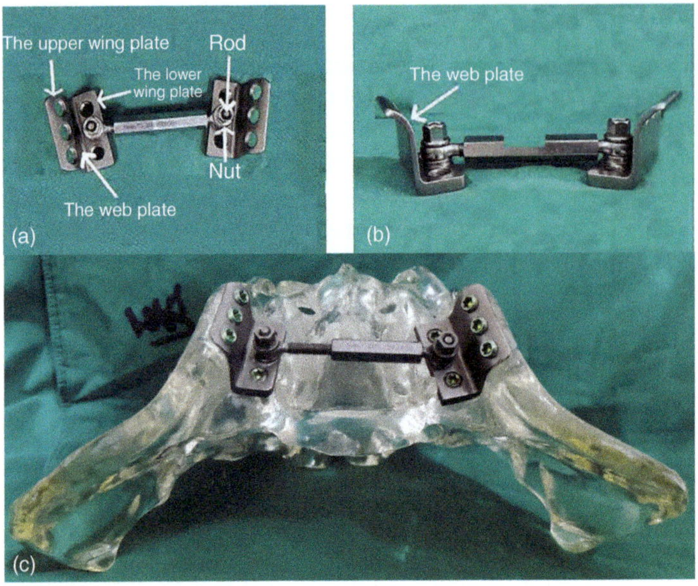

Figure 11.5 The photograph of the minimally invasive adjustable plate for the posterior pelvic ring. (a) Anteroposterior position of entire plate, (b) lateral position of the entire plate, and (c) the pelvic model fixed using the minimally invasive adjustable plate.

The MIAP can be used to fix the posterior pelvic ring disruptions without prebending, which could reduce the compressed or separated fractures/dislocations by adjusting the length of the connection bar (Figure 11.6). The MIAP has been applied in the management of posterior pelvic ring injury in a minimally invasive fashion. Rigid internal fixation and excellent functional recovery have been achieved in such patients (Figure 11.7).

11.2.3 Bionic Implants for Intertrochanteric Fracture of the Femur

Intertrochanteric fracture is the common lower extremity trauma, especially in the elderly [13–15], which damages the structure of compressive and tensile trabecular bone of the proximal femur. Bone trabeculae can provide essential mechanical support and stress conduction, which bear 40–70% of the load at the proximal end of the femur; meanwhile, they can transform the bending moment generated by the body weight that load down the femoral head, having it equally distributed to the bone cortex of the proximal femur. Studies have shown that when intertrochanteric fractures occur, tissue-level damages begin with trabeculae. Local violence during fall in the side initially leads to the microfractures of approximately 1.5–6.4% bone trabeculae of the proximal femur, subsequently progressing to the rupture of the bone trabecular meshwork, and eventually, the bone cortex is involved, which means a fracture. However, the significance of the reconstruction of the cracked trabecular in the treatment of fractures is often overlooked.

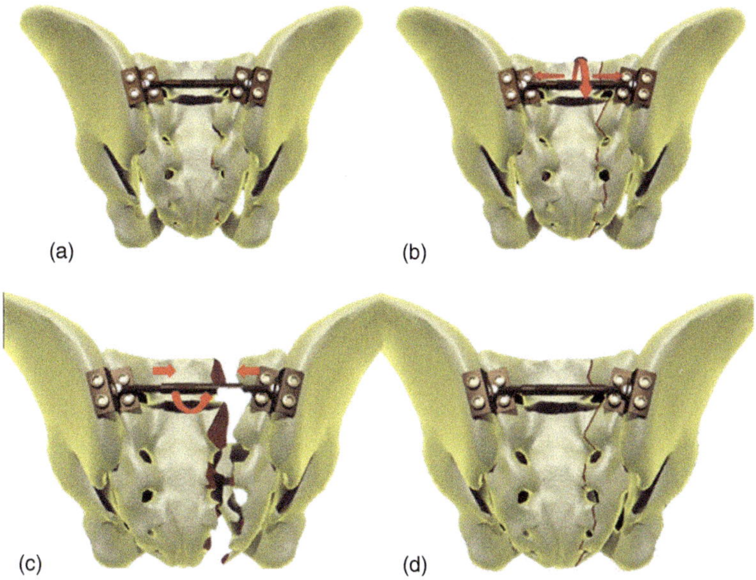

Figure 11.6 The schematic diagrams showing the reduction ability of MIAP in the treatment of compressed and separated sacral fractures. (a) The mode charts of compressed sacral fracture in Denis zone II. (b) The compressed fractures were distracted by rotating the hexagonal tube to elongate the connection bar of MIAP. (c) The mode charts of separated longitudinal fracture of the sacrum. (d) The separated fractures were compressed by rotating the hexagonal tube to shorten the connection bar. Source: Chen et al. [12] Reprinted with permission of Elsevier.

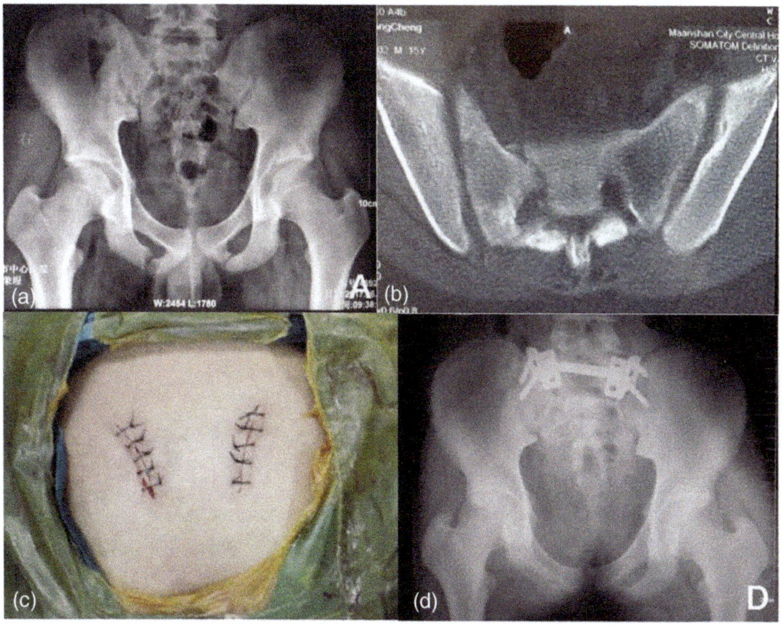

Figure 11.7 A 15 year-old boy sustained compressed sacral fractures on the right side with neural injury: (a) anteroposterior view and (b) axial computed tomography (CT) image. The MIAP was used to reduce and fix the fracture in a minimally invasive fashion: (c) the photograph of the incision and (d) the anteroposterior view of pelvis taken postoperatively.

Intertrochanteric fractures of the femur can be fixed with the use of a dynamic hip screw (DHS), Gamma nail, and proximal femoral nail anti-rotation (PFNA). However, the implants will further damage the structure of compressive and tensile bone trabecular. Implant failure, such as implant loosing, breakage, or backward, is disastrous complication encountered with the traditional internal fixation system. In addition, the difference between the elastic modulus of the bone and that of the implant can cause the so-called "stress shielding," which can subsequently inhibit the transfer of physiological stress of bone tissues and impair the process of fracture healing.

To solve this problem, Zhang et al. proposed the concept of bionic fixation of traumatic fracture and designed the bionic implants with a porous structure, which allowed the bone trabeculae to grow in (Figures 11.8 and 11.9). It is speculated that the porous structure of the bionic implant could help the reconstruction of the trabeculae and therefore improve the fixing force of the implant, promote the cortical-cancellous biphasic healing of fractures, and restore the mechanical support and conduction role of the fractured bone. The above hypothesis has been preliminarily confirmed by finite element analysis.

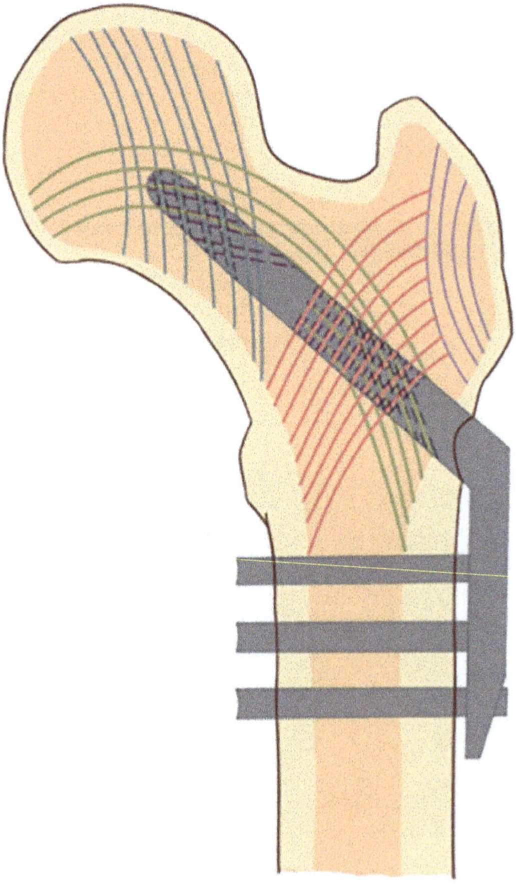

Figure 11.8 The schematic diagrams of a bionic dynamic hip screw for intertrochanteric fractures. Source: Cun et al. [16] Reprinted with permission of Springer Nature.

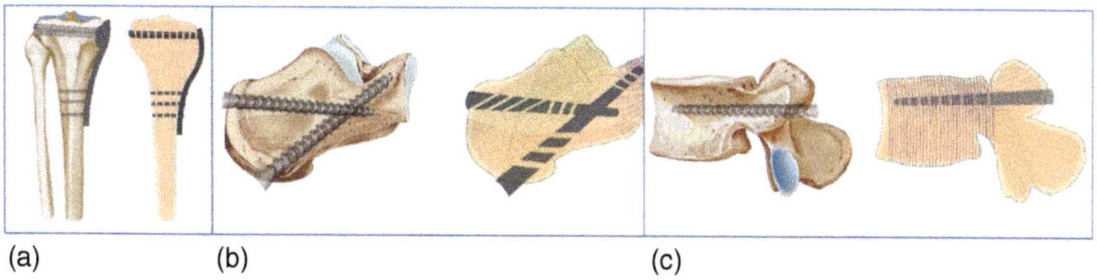

Figure 11.9 The schematic diagrams of bionic implants for tibial plateau fracture (a), calcaneal fracture (b), and fracture of lumbar vertebra (c).

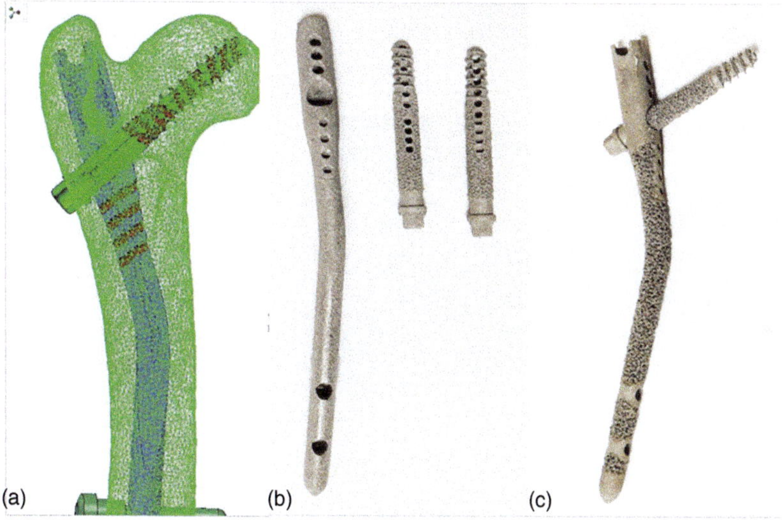

Figure 11.10 The bionic Gamma nail for intertrochanteric fracture of the femur of the sheep: (a) the diagram of the nail inserting into the femur of the sheep, (b) the 3D-printed nail and lag screw, and (c) the assembled nail.

Finite element analysis was used to imitate the trabecular orientation of the human proximal femur to improve the biocompatibility of the fixation. The result shows that the bionic DHS and Gamma nail can make the stress distribution of the fractured bone fixed using the bionic implants closer to the stress distribution of the intact bone than that fixed using traditional implants for fixating the intertrochanteric fracture, thereby aiding in protecting the healing bone and restoring the bone structure [16]. Based on the results from finite element analysis, bionic Gamma nail was designed to fix the intertrochanteric fractures of the femur of sheep (Figures 11.10–11.14).

11.2.4 Bionic Fixation of Calcaneal Fracture

The treatment of displaced intra-articular calcaneal fractures remains challenging for orthopedic surgeons. The aim of surgery in the treatment of displaced intra-articular calcaneal fractures is to restore the normal anatomy, repair soft tissue

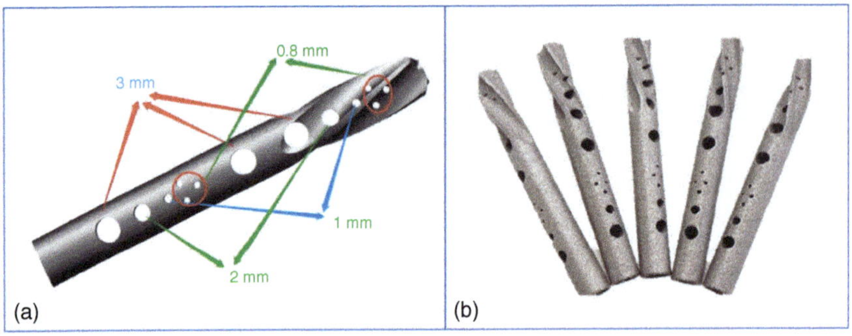

Figure 11.11 The bionic screw with holes of different size: (a) the design diagram of the bionic screw. (b) 3D-printed screw.

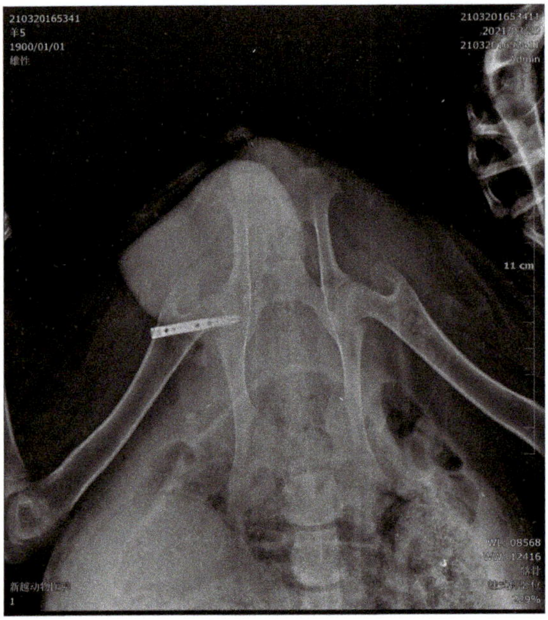

Figure 11.12 The bionic screw was inserted into the femoral neck of the sheep to investigate whether the trabecular bone grows into the holes.

injuries, and facilitate the return to normal physiological function. At present, a large number of studies reported the outcomes of calcaneal fractures fixed using different internal fixation systems, such as locking or anatomical plate and screws, hollow screws, external fixator, and Kirschner wires. However, there is a lack of studies regarding "bionic fixation" in the management of calcaneal fracture. Kinner et al. [17] designed a novel bionic plate, which imitates the trabecular orientation of the human calcaneus and the trajectory architecture of the heel. This novel plate has a robust construction because of the frame design; the locking holes are arranged in a way that the screws can easily be placed in areas with best bone quality. The alignment of the plate lugs is based on the direction of force, imitating the anatomic trajectories of the calcaneus to provide a high degree of strength. Hence, this is referred to as "bionic" plate design.

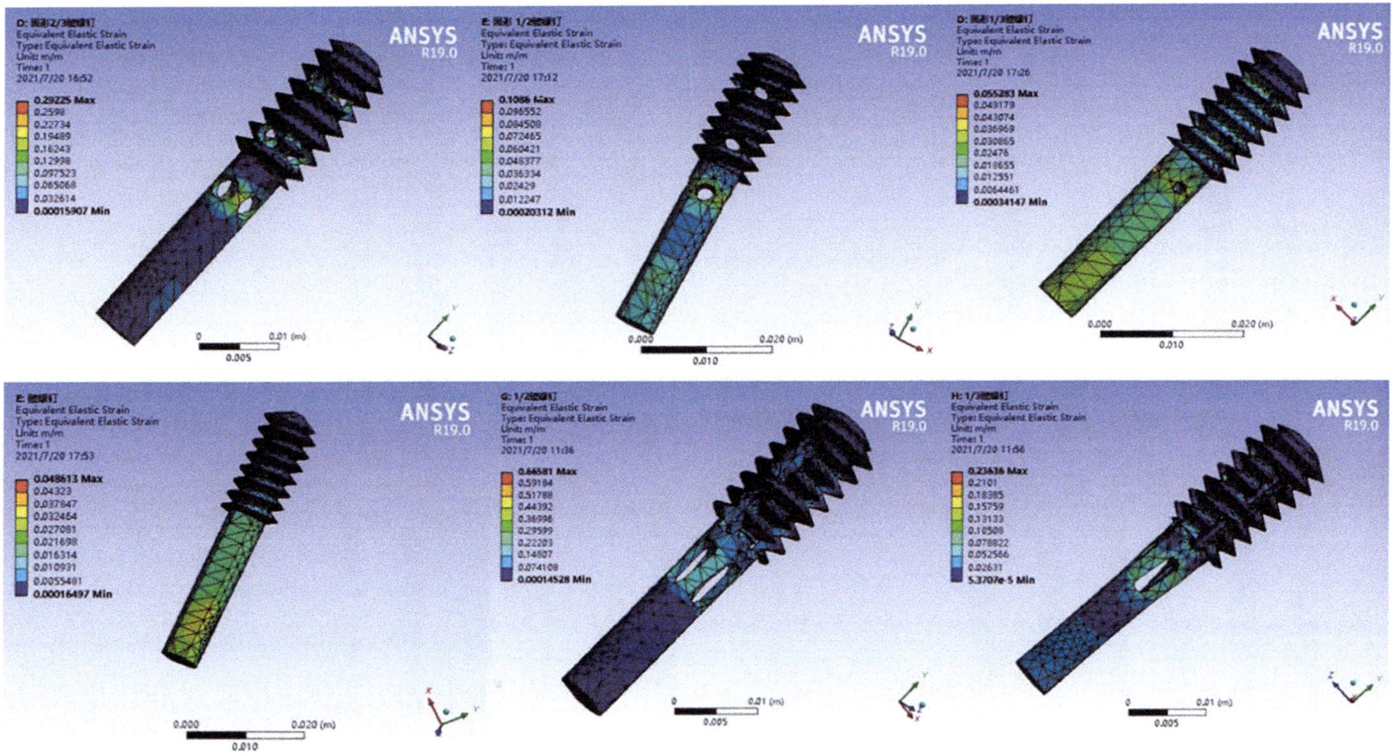

Figure 11.13 Comparing biomechanical characters among six types of bionic implants using finite element analysis.

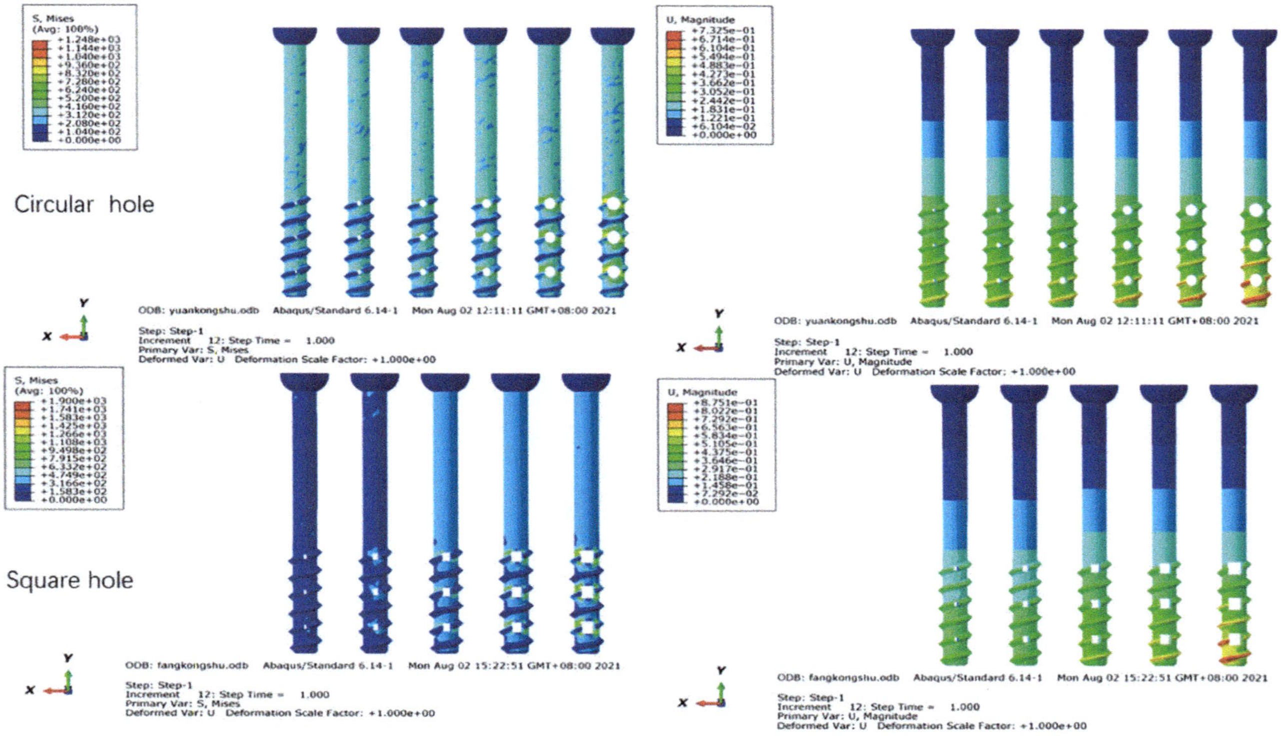

Figure 11.14 Comparing biomechanical effect of different-sized circular hole and square hole on screw using finite element analysis.

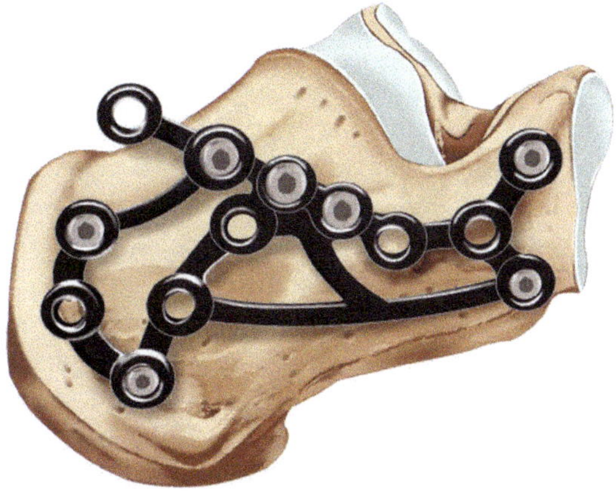

Figure 11.15 The bionic calcaneal plate and the calcaneal model fixed with this plate.

Biomechanical test shows that the bionic fixation offers higher stability and load tolerance while being more resistant to fatigue when compared to a traditional calcaneal locking plate (Figure 11.15). Ten patients with displaced intra-articular calcaneal fractures were treated using the novel bionic plate and promising outcomes without neither implant failure nor loss of reduction were seen at 12-month follow ups.

11.3 Bionic Materials

Bionic materials are characterized by the composition, structure, and biomechanics close to human organ and tissue, high biocompatibility, and good resistance to wear, fatigue, and deformation, with or without the feature of bioabsorbability. Currently, most of the bionic materials used for research and clinical application are biodegradable or a combination of biodegradable and non-biodegradable materials. Zhang et al. [18] and Kikuchi [19] have successfully synthesized a hydroxyapatite (HA)/collagen composite, which can be successfully involved in bone reconstruction, as it can be replaced by new bone.

11.3.1 Application and Designation of New Composite Materials

Different materials have different mechanical and biological characteristics, and each have their own strengths and limitations. As for metal materials, magnesium has good biocompatibility, biosafety, and bioabsorbability. Its biomechanical properties of strong stiffness and high strength make it minimize the stress shielding effect of metal in the fixation of human bone [20]. Zinc also plays an important role in many physiological and cellular processes. They both have good biocompatibility as implants. However, as bionic metal materials applied in the field of orthopedics and traumatology, both magnesium and zinc have insufficient ductility and instable degradation rate [21]. For polymer materials, collagen could promote bone formation and initiate and propagate mineralization. Chitin/Chitosan (CS) derivatives are also widely used in various biomedical applications because of their unique properties such as biocompatibility, mucoadhesion, non-toxicity, and capability to form gels. Polylactic acid (PLA), polycaprolactone (PCL), etc. are non-toxic, have good mechanical strength and high biocompatibility, and are easy to process. However, low mechanical strength and hydrophilicity of bionic polymer materials cannot be ignored [22, 23]. For inorganic materials, HA has the ability of bone induction, which may be due to HA itself being one of the components of bone. Bioglass (BG) is similar to the inorganic components in bone tissue and widely used in hard and soft tissue repair because of its good osteogenic induction and biocompatibility. Because of the low strength and brittleness of most inorganic materials, their application as load-bearing is limited. During the operation or long-term maintenance, it is easy to be fragmented.

Any single material cannot achieve the best bone repairing effect. In order to make up for the deficiency of single material in biological, physical, and chemical properties, a variety of materials have been synthesized into composite materials through a certain process to meet the needs.

Zhang et al. [18] designed and produced bionic mineralized collagen (MC) nanofibrils. This bionic MC was similar to the natural bone in terms of both the composition and the nanostructure. Bionic MC is based on the principle of *in situ* bionic mineralization and molecular self-assembly technology, which uses the ordered arrangement of collagen to carry out multi-level bionic from the nanosize. During the preparation of the composite, HA crystals grew on the surface of the collagen fibrils with triple helices and their crystallographic *c*-axes oriented along the longitudinal axes of the fibrils. The MC fibrils aligned parallel to each other to form MC fibers. For the first time, the new hierarchical self-assembly structure of the collagen–HA composite was verified by conventional and high-resolution transmission electron microscopy (HR-TEM), as shown in Figure 11.16 [17].

Natural bone tissue is a complex biomineralization system with layered structure. The porous structure of MC is similar to cancellous bone, which is conducive to

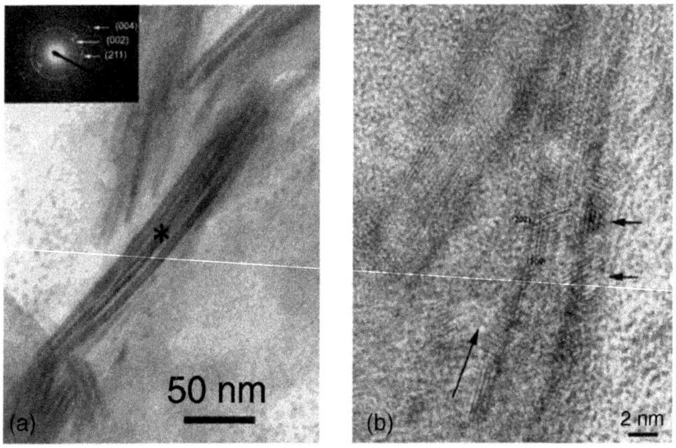

Figure 11.16 (a) Higher magnification of the MC fibrils. The inset shows the selected area electron diffraction pattern of the MC fibrils. The asterisk is the center of the area, and the diameter of the area is about 200 nm. (b) High-resolution transmission electron microscopy (HR-TEM) image of MC fibrils. The long arrow indicates the longitude direction of the collagen fibril. Two short arrows indicate two hydroxyapatite (HA) nanocrystals.

the growth of cells, new bone tissue, and microvessels, and provides an interconnected channel to create the best environment for bone growth [24]. This similarity with natural bone tissue makes the MC material not only have good biocompatibility and bone conduction but also have mechanical properties close to natural bone. Besides, the MC scaffold has a good ability to promote bone regeneration [25, 26]. When the MC scaffold is used, the growth and osteogenic differentiation of human mesenchymal stromal cells (hMSCs) have been increased; the markers of osteogenesis such as runt-related transcription factor 2 (RUNX2) and osteocalcin (OCN) were upregulated [27].

11.3.2 Bionic Design

Bionic design plays an important role in the modification of materials. To improve the characteristics of composite materials as well as the mechanical and biological performance requirements, bionic design and surface modification of materials can be conducted, such as cartilage repair bracket and artificial intervertebral disc. Artificial bone and ligaments or joint capsule sutures (polymer materials) are also a kind of typical bionic materials that simulate the structure of human fibrous tissue.

The combination of MC and high molecular polymer has also become a research hotspot in recent years. A series of porous MC/PCL scaffolds with different pore features were fabricated via freeze-drying, and Cui and coworkers [28] applied the scaffolds in a Sprague-Dawley rat cranial bone calvarial defect model. The mechanical properties, osteogenic induction, and biocompatibility of MC scaffolds have been improved by doping different components.

Microspheres (MSs) are organic or inorganic, spherical, free-flowing particles ranging from 1 to 1000 μm in diameter that may encapsulate drugs or bioactive molecules. MS serves as a part of bionic composite materials to improve the biological functions of scaffolds. In an MS-incorporated scaffold, MSs can regulate scaffold functions through its own characteristics, such as mechanical strength and material components, or by carrying biomolecules [29]. Poly-L-lactic acid (PLLA) microspheres with an injectable, porous structure have proved to be an ideal drug-carrying material capable of uniformly and efficiently loading drugs and able to promote the sustained and effective release of drugs using a special structure [30]. Liu et al. [31] constructed artificial bionic cartilage microspheres by encapsulating mesenchymal stromal cells (MSCs) in collagen microspheres and cultured in a chondrogenic-inducing medium. Then, bionic cartilage acellular matrix microsphere (BCAMM) scaffolds were fabricated from the cultured microspheres. As a result, the BCAMM successfully induced MSC chondrogenesis without any additional fetal bovine serum or induction components.

11.4 Future Perspectives and Current Limitations

In conclusion, the main research purpose for the future is to develop composite materials with the following bionic design with stable performance, easy access, and with similar structure and properties to those of the human bone. The advantages observed in the use of "bionic fixation" are as follows: the restoration of joint stability, rapid functional recovery, improved quality of reduction, and better clinical outcome. For future application to "bionic fixation," we envision the possibility of achieving better bone trabecular ingrowth, which will decrease the number of implant failure. "Bionic fixation" in orthopedics promises to be available as a fully implantable device for the fracture, controlled via an implanted device to promote bone ingrowth. In the future, it would be interesting to implement such bionic fixation to the fracture site as they hold promise for tremendous health improvements and an overall increase in quality of life. Despite several promising aspects offered by innovative "bionic" design, there are still several limitations, which must be faced before the widespread use of similar devices. The major limitation of bionic fixation presented in this article is that they were mainly short-term follow-up applied to clinic; therefore, longer follow-up time regarding bionic fixation in the treatment of fracture must be performed in view of their safety and functionality.

Acknowledgments

This work was supported in part by the National Natural Science Foundation of China (82072447), Hebei Province fund for Distinguished Young Scholars (H2021206329), and Hebei Province Outstanding Youth Foundation Project (H2017206104).

References

1 Grass, R., Herzmann, K., Biewener, A., and Zwipp, H. (2000). *Unfallchirurg* 103: 520.
2 Close, J.R. (1956). *J. Bone Joint Surg. Am.* 38: 761.
3 Rasmussen, O., Tovborg-Jensen, I., and Boe, S. (1982). Distal tibiofibular ligaments. *Acta Orthop. Scand.* 53: 681.
4 Xu, G., Chen, W., Zhang, Q. et al. (2013). *Scand. J. Trauma Resusc. Emerg. Med.* 21: 71.
5 Wang, L., Zhang, Y., Song, Z. et al. (2018). *Int. Orthop.* 42: 221.
6 Tiziani, S., Dienstknecht, T., Osterhoff, G. et al. (2019). *Int. Orthop.* 43: 1779.
7 Gordon, W.T., Fleming, M.E., Johnson, A.E. et al. (2018). *Mil. Med.* 183: 115.
8 van Berkel, D., Ong, T., Drummond, A. et al. (2019). *BMJ Open* 9: e032111.
9 Moed, B.R. and Geer, B.L. (2006). *J. Orthop. Trauma* 20: 378.
10 Sagi, H.C., Ordway, N.R., and DiPasquale, T. (2004). *J. Orthop. Trauma* 18 (3): 138.
11 Griffin, D.R., Starr, A.J., Reinert, C.M. et al. (2006). *J. Orthop. Trauma* 20 (1): S30–S36.
12 Chen, W., Hou, Z., Su, Y. et al. (2013). *Injury* 44: 975.
13 Civinini, R., Paoli, T., Cianferotti, L. et al. (2019). *Int. Orthop.* 43: 187.

14 Chen, W., Lv, H., Liu, S. et al. (2017). *Lancet Global Health* 5: e807.

15 Kumar, P., Rajnish, R.K., Sharma, S., and Dhillon, M.S. (2020). *Int. Orthop.* 44: 623.

16 Cun, Y., Dou, C., Tian, S. et al. (2020). *Int. Orthop.* 44: 551.

17 Kinner, B., Kerschbaum, M., Bley, C. et al. (2015). *Int. Orthop.* 39: 111.

18 Zhang, W., Liao, S.S., and Cui, F.Z. (2003). *Chem. Mater.* 15: 3221.

19 Kikuchi, M. (2013). *Biol. Pharm. Bull.* 36 (11): 1666.

20 Kamrani, S. and Fleck, C. (2019). *Biometals* 32 (2): 185.

21 Hernández-Escobar, D., Champagne, S., Yilmazer, H. et al. (2019). *Acta Biomater.* 97: 1.

22 Riaz, S., Fatima, N., Rasheed, A. et al. (2018). *Int. J. Biomater.* 2018: 13.

23 Han, G., Liu, S., Pan, Z. et al. (2020). *Carbohydr. Polym.* 229: 115.

24 Wubneh, A., Tsekoura, E.K., Ayranci, C., and Uludağ, H. (2018). *Acta Biomater.* 80: 1.

25 Trivedi, S., Srivastava, K., Saluja, T.S. et al. (2020). *Odontology* 108: 251.

26 Zhang, Z., Li, Z., Zhang, C. et al. (2018). *Int. J. Nanomed.* 13: 7503.

27 de Melo Pereira, D., Eischen-Loges, M., Birgani, Z.T., and Habibovic, P. (2020). *Front. Bioeng. Biotechnol.* 23: 8.

28 Wang, S., Yang, Y., Koons, G.L. et al. (2020). *Mater. Sci. Eng., C* 106: 110186.

29 He, Q., Zhang, J., Liao, Y. et al. (2020). *Biotechnol. Adv.* 39: 107459.

30 Xu, Y., Gu, Y., Cai, F. et al. (2020). *Adv. Funct. Mater.* 30: 2006333.

31 Liu, J., Wang, X., Lu, G. et al. (2019). *J. Mater. Chem. B* 7 (4): 640.

12

Cartilage Injury and Repair

Yusheng Li, Ning Wang, Di Liu, and Zijun Cai

Xiangya Hospital of Central South University, Department of Orthopedics, Xiangya Road 87, Hunan 410008, China

12.1 Introduction

Cartilage is a supportive connective tissue with greater toughness, mainly composed of chondrocytes and extracellular matrix (ECM). Because of the lack of blood vessels and lymphatic vessels, nutrients from joint fluid or perichondrium only have to gradually penetrate into the cartilage matrix to nourish chondrocytes. Cartilage can be mainly divided into three categories: hyaline cartilage composed of collagen fibers, fibrils, and the surrounding amorphous matrix (mainly distributed in the trachea and bronchial walls, articular surfaces, etc.); fibrocartilage composed of cross-arranged collagen fibers (mainly distributed in intervertebral discs, including knee meniscus); and elastic cartilage composed of collagen fibers and elastic fibers (mainly distributed in the auricle, epiglottis, etc.) [1].

Articular hyaline cartilage is most vulnerable to injury because it needs to bear and absorb the mechanical loading from joint movement. Although articular cartilage is thin, it can be divided into four layers according to the matrix arrangement and chondrocyte state (Figure 12.1): (i) superficial layer or resting zone (fibers are parallel to the cartilage surface, and chondrocytes are in round shape and may function as a reservoir of chondrocyte progenitors once cartilage injury occurred [2]); (ii) middle layer or proliferating zone (fibers are randomly distributed, and chondrocytes are divided and flattened into organized columns); (iii) deep layer or prehypertrophic zone (fibers are perpendicular to the surface, and chondrocytes continue cell cycle and will exit in order to enter hypertrophic differentiation); and (iv) calcified layer or hypertrophic zone (acting as the transition between cartilage and the underlying subchondral bone and hypertrophic chondrocytes with enlarged cell size and enhanced mineralized matrix generation) [3–5].

Biofabrication for Orthopedics: Methods, Techniques, and Applications, First Edition.
Edited by Wenguo Cui, Xin Zhao, and Shen Liu.

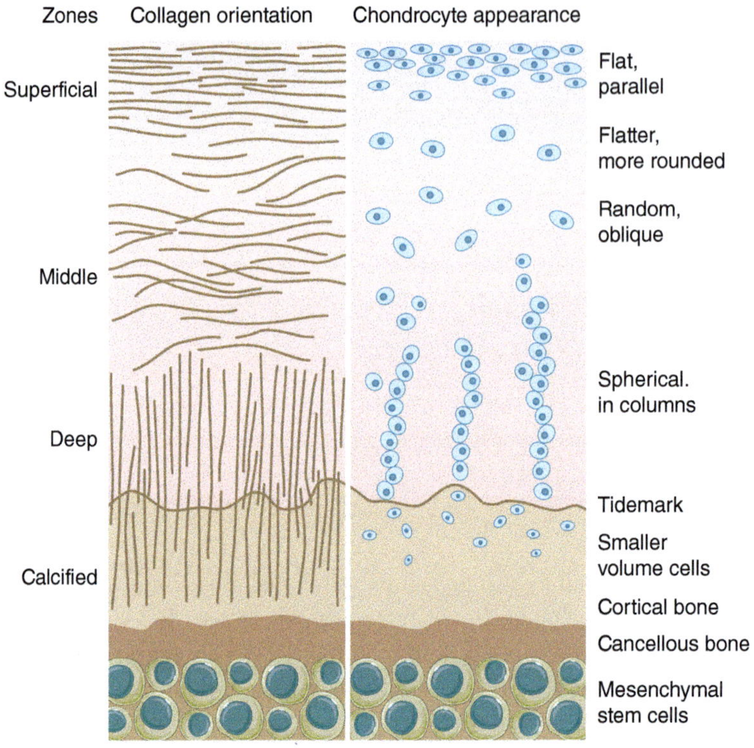

Figure 12.1 Anatomy schematic and histology of articular cartilage.

12.2 Pathology of Cartilage Injury

Cartilage injury usually occurs in knee, hip, ankle, and elbow joints. According to the etiology, articular cartilage injury could be divided into acute injury (usually caused by severe high-kinetic energy trauma of the joint) and chronic injury (usually caused by aging, obesity, etc.). The pathogenic process following articular cartilage injury contains four stages: immediate reaction phase, acute phase, symptomatic phase, and chronic phase (posttraumatic osteoarthritis [PTOA]) [6]. In the immediate reaction phase, cartilage experienced swelling, collagen fibers rupture, glycosaminoglycan (GAG) is lost, and chondrocytes located at the injury site were necrosed. In the acute phase, proinflammatory cytokines were released into synovial fluid, leukocytes infiltrated into articular joint, the balance of ECM anabolic and catabolic was broken, and chondrocytes started to apoptosis. In the asymptomatic phase, inflammation as well as several clinical symptoms including synovial effusion and joint pain could still persist in this stage and last for months. If the impact of cartilage injury could not be self-limited over months, then articular joint would enter the chronic damage phase and finally end up with PTOA [7].

12.3 Clinical Characteristics of Cartilage Injury

Cartilage injury is quite common especially among athletes with a suggested overall prevalence as high as up to 36% just in knee joint [8, 9]. The clinical symptoms of cartilage injury will vary depending on the injury type, extent, chronicity, accompanied with or without other joint compartments injury, etc., while common symptoms include pain, swelling and stiffness (especially worse during sports activities), joint instability (patients may also experience symptoms of their knee joint giving away or may have a grinding sensation in daily activities), joint locking (cartilage broken pieces lie freely in the joint known as "loose body," it might cause the patient unable to move their joint from a particular motion in a particular time), decreased range of motion of the affected joint, and sometimes even joint deformity. Once a cartilage injury happened, it cannot heal optimally on its own because of the lack of blood supply; cartilage injuries could lead to early degeneration of joint and finally cause secondary osteoarthritis (OA).

12.4 Evaluation of Cartilage Injuries

There are many methods applied for evaluating cartilage injuries from different joints [10–14]. In this part, we mainly introduce the scoring system for assessing knee joint cartilage injury or degeneration because knee joint is the predominant joint for osteochondral injuries and osteoarthritis development [13]. Meanwhile, this part also restricts itself to arthroscopy and histopathology assessment since they are well accepted as golden standard in clinics and research.

12.4.1 Arthroscopic Evaluation of Cartilage Injury

The Outerbridge score was first described in 1961, which is based on macroappearance or arthroscopic inspection of the cartilage surface. An Outerbridge score contains five levels from grade 0 to 4; grade 0 is a normal and smooth cartilage, grade 1 describes cartilage softening and dynamic probing, grade 2 describes partial thickness lesions less than 1.5 cm in diameter, grade 3 lesions are greater than 1.5 cm in diameter or have a full thickness fissure, and grade 4 lesions involve complete cartilage loss with exposed subchondral bone [15]. Although Outerbridge score has been widely used both for clinical and scientific purpose, it still has some defects. The most common criticisms of this scoring system are lack of precise description of cartilage lesion depth and its inconsistent and poor reproducibility among orthopedic surgeons [15]. Therefore, Tim Dwyer et al. proved a high correlation between the International Cartilage Repair Society (ICRS) scoring system and histological grading of depth; they also showed that the arthroscopic ICRS classification system has good interobserver and intraobserver reliability [16]. This means that the ICRS scoring system could be a good supplement for Outerbridge score to some extent.

The ICRS committee was founded in 1997 and has been focusing on developing a standardized cartilage injury and repair evaluating system [17]. The ICRS scoring system is used to determine the severity of cartilage lesions and the quality of cartilage repairment. It also contains five grades from grade 0 to 4: grade 0 represents normal intact cartilage; grade 1 means chondral softening and blistering, superficial lesions, and lacerations; grade 2 are lesions and fissures extending down to <50% of cartilage depth; grade 3 also means partial loss of cartilage thickness with cartilage lesions extending down >50% of cartilage depth (from cartilage surface down to the calcified layer); and grade 4 describes full-thickness cartilage loss with the exposure of subchondral bone [16].

12.4.2 Histopathology Evaluation of Cartilage Injury

As for histopathological assessment of cartilage degeneration, Mankin et al. proposed the histologic grading system (Mankin scoring system) since 1971 [18]. Because Mankin score was established on end-stage cartilage degeneration, it has some significant limitations when assessing mild and moderate cartilage degeneration [19, 20]. In order to overcome those limitations, the Osteoarthritis Research Society International (OARSI) group developed a more useful method to assess cartilage degeneration histopathology at 2000. Notably, the cartilage of experimental animal models is different from that of a human being. For example, mouse knee joint cartilage is only 30 μm thick, which is approximately 50-fold thinner than that of a human [21], so when it comes to assess mouse knee joint cartilage and human knee joint cartilage, even the principles and main content of OARSI score are similar, there are still some detailed differences. The OARSI score applied on humans contains two parts: the OARSI grade and stage. The OARSI grade evaluation system is based on depth progression into cartilage. Grade 0 represents normal cartilage; grade 1 is an uneven but intact cartilage surface, sometimes with superficial fibrillation; grade 2 appears with cartilage surface discontinuity; grade 3 has vertical fissures or clefts; grade 4 represents cartilage erosion including mid zone excavation; grade 5 means cartilage total denudation; and grade 6 means joint deformation and osteophytes formation (Figure 12.2) [22]. The OARSI group also developed a stage system to assess the invasive range of cartilage degeneration in whole knee joint. According to the extent of joint involvement, cartilage degeneration could be divided into five stages, from stage 0 to 4: no cartilage degeneration sign seen is stage 0, while <10% joint extent is stage 1, stage 2 is 10–25%, stage 3 is 25–50%, and stage 4 is >50% joint extent involvement. In conclusion, grade is an index of the cartilage degeneration severity, stage is defined as the horizontal extent of cartilage involvement irrespective of the underlying grade, and the OARSI score is a grade and a stage component, which represents a combined assessment of cartilage degeneration within the joint [20, 22].

As mentioned above, the OARSI score system applied in animal studies is different from that applied in humans; histologists developed various OARSI standards according to different animal species and their knee joint characteristics [23–25]. Since mouse model is the most frequently used animal model in

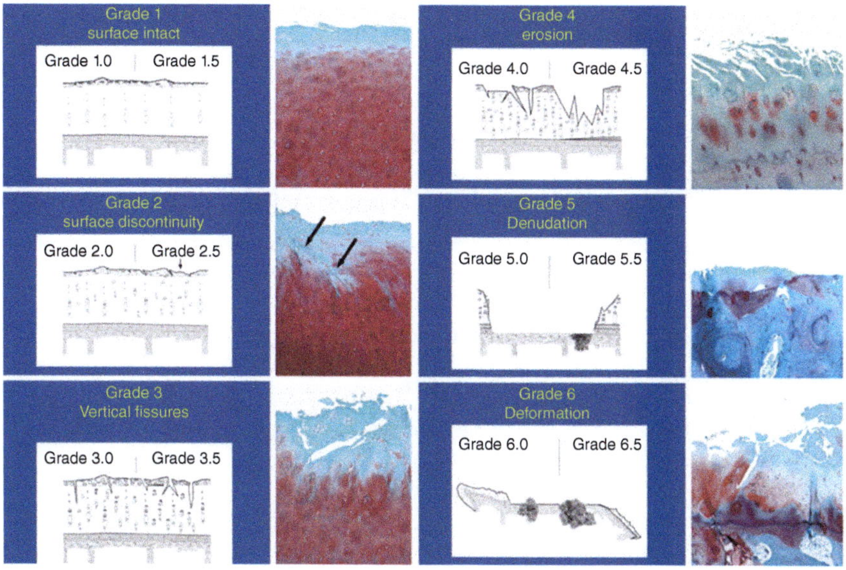

Figure 12.2 OARSI grade system for evaluating osteochondral samples from human knee joint cartilage: graphs from upper left to downright showed schematic panel and histopathology examples of OARSI grade 1–6. (Grade 0 means normal cartilage that is not shown here). Source: Pritzker et al. [22]/with permission of Elsevier.

cartilage degeneration research field, we mainly focused on introducing OARSI score in assessing mouse knee joint cartilage. Mouse knee joint is small and can be harvested intact, and it is easier to section the entire joint rather than just an osteochondral part. In order to have enough representative sections and slices to convenient histologist to evaluate the entire joint cartilage degeneration pathology, the histologist usually obtains 13–16 slides from the whole joint sectioning, where each slice is harvested at approximately 80 μm intervals and with three 4–6 μm sections on it. The OARSI grade system for mouse knee joint cartilage assessment is also different from that of humans; it directly integrates the extent assessment of articular cartilage into its grading system (Figure 12.3) [23].

12.4.3 Magnetic Resonance Imaging (MRI) Assessment of Cartilage Injury

Magnetic resonance imaging (MRI) is a powerful noninvasive tool for detecting joint compartment injury and repair, including muscle, tendon, ligament, cartilage, etc. [26, 27]. Currently, there are many MRI techniques applied in cartilage evaluation from research perspective, such as standard spin-echo (SE) and gradient-recalled echo (GRE) sequences for investigating the structure of articular cartilage: T2 mapping, T1ρ imaging, sodium imaging, and diffusion-weighted imaging for evaluating the collagen network and proteoglycan content in cartilage [28]. However, the most commonly used MRI techniques in musculoskeletal system are intermediate-weighted, fast spin-echo (FSE) with or without fat

Grade	Osteoarthritic damage
0	Normal
0.5	Loss of Safranin-O without structural changes
1	Small fibrillations without loss of cartilage
2	Vertical clefts down to the layer immediately below the superficial layer and some loss of surface lamina
3	Vertical clefts/erosion to the calcified cartilage extending to <25% of the articular surface
4	Vertical clefts/erosion to the calcified cartilage extending to 25–50% of the articular surface
5	Vertical clefts/erosion to the calcified cartilage extending to 50–75% of the articular surface
6	Vertical clefts/erosion to the calcified cartilage extending >75% of the articular surface

Figure 12.3 OARSI scoring system for evaluating mouse knee joint. Source: Glasson et al. [23]/with permission of Elsevier.

suppression and three-dimensional (3D) fat-suppressed gradient-recalled echo (GRE) acquisition [29, 30].

Normal cartilage contains approximately 75–80% water by weight and therefore appears less intense to joint fluid on T2-weighted images [31]. FSE combines heavy T2 weighting and exhibits a high signal intensity of bone and dark cartilage against bright synovial fluid. However, cartilage signal in fat-suppressed 3D gradient echo sequences is relatively high compared with adjacent fat-suppressed joint tissue [32]. Similar to arthroscopy and histology evaluation, MRI assessment for cartilage injury and defects also developed various classification systems. The most commonly used MRI assessment system of cartilage defects was developed by Yulish et al. at 1987, and it contains four grades: grade 1 means abnormal intrachondral signal with a normal chondral surface; grade 2 appeared with irregularity of cartilage surface with/without focal loss of less than 50% of the cartilage thickness; grade 3 has severe surface irregularity with a cartilage loss of 50–100% of the thickness; and grade 4 represents complete loss of articular cartilage with exposure of subchondral bone [17, 32, 33]. If a solitary cartilage injury or lesion site is presented in MRI, a detailed evaluation should also be considered regarding the MRI grade, including the anatomic location, lesion size, margins, bone marrow edema, and subchondral cysts. All the conditions are important for choosing the potential treatment options and could affect the patients' prognosis [34].

12.5 Clinical Strategies of Cartilage Repair

Under physiological conditions, articular cartilage exerts inherent limited capacity for the repair of chondral defects, and it remains an essential clinical challenge for orthopedic surgeons [35]. In cartilage, highly differentiated chondrocyte is the major resident cell and blood vessel is absent, compromising the formation of fibrin clot, transfer of repair-related factors, proliferation and migration of chondrocytes, and biogenesis of cellular matrix. Consequently, paucity of vascular supply and differentiated chondrocytes mainly contributes to the disability of cartilage for restoration. Meanwhile, many individuals suffer from various symptoms related to degenerative articular cartilage, including chronic pain, swelling of joints, locking or catching, and loss of locomotion. In a prospective study, 61% patients who have undergone knee arthroscopy were concomitant with chondral or osteochondral lesions [36]. Therefore, orthopedic surgeons continue to explore methods to restore focal lesions of cartilage. Currently, effective pharmaceutical interventions for cartilage injury is still not available, while surgical strategies for cartilage lesions have evolved over the past several decades, including bone marrow stimulation techniques, chondral and osteochondral grafts, and autologous chondrocyte implantation (ACI) [37] (Figure 12.4).

12.5.1 Debridement, Cartilage Shaving, and Joint Lavage

Arthroscopic debridement and cartilage shaving, as palliative treatments, removes loose bodies, fragments, debris and hypertrophic synovium, trims blocking osteophytes and meniscal tears, and shaves impaired or degenerative joint tissues, which mitigates mechanical symptoms of patients to some extent [38]. However, this surgical operation aims to ameliorate the pathophysiological alterations in mechanical structure rather than promoting articular cartilage repair. In a randomized and controlled trial, the arthroscopic surgery (arthroscopic debridement and cartilage shaving together) for OA did not show any clinical superiority than optimized physical and medical therapy [39]. It is consistent with the study conducted by Laupattarakasem et al. [40], which also indicated that arthroscopic debridement may had no therapeutic value for patients with knee OA. However, a study concluded that arthroscopic debridement improves the short-and-medium-term postoperative outcomes through disposing of focal articular cartilage damage [41].

Joint lavage, either with or without arthroscopy, is a conventional treatment modality to improve local symptoms and joint move function in the short term [42]. Theoretically, joint lavage could eliminate inflammatory cytokines, biological enzymes, and debris and extract intra-articular free bodies through injecting and withdrawing fluid, which potentially result in synovitis, pain, and arthropathy [42, 43]. Most of the studies have demonstrated that joint lavage neither alleviates pain nor mitigates physical disability in patients with knee OA compared to control groups. In a controlled trial, a total of 180 patients with knee osteoarthritis (KOA) were enrolled and randomly assigned to receive arthroscopic lavage, arthroscopic debridement, or placebo surgery [44]. Therapeutic effects were evaluated after

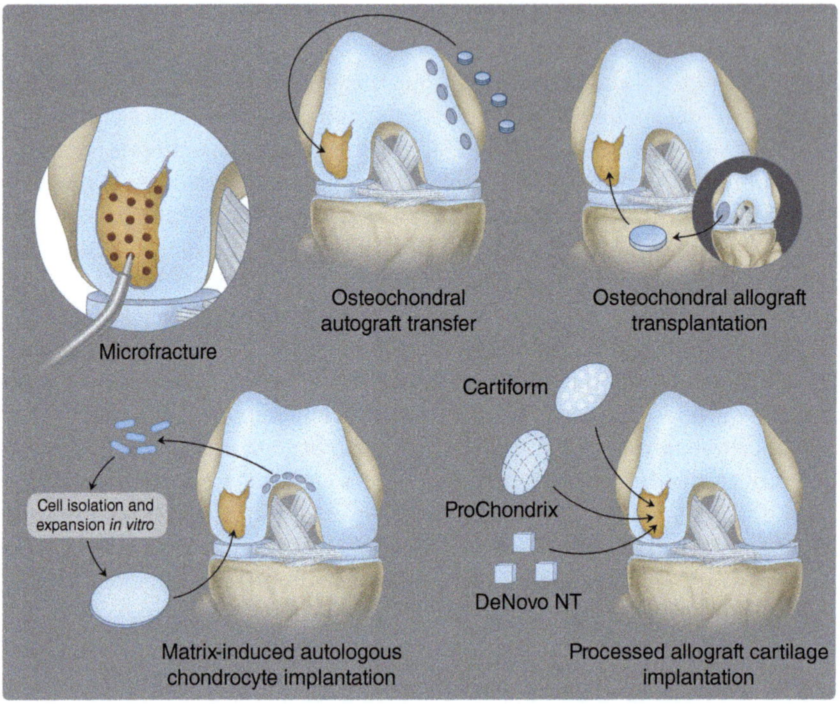

Figure 12.4 Repair strategies for cartilage defects. The currently used repair strategies for cartilage defects include microfracture; osteochondral autograft transfer; osteochondral allograft transplantation; implantation of processed allograft cartilage such as DeNovo NT, ProChondrix, and Cartiform; and matrix-induced autologous chondrocyte implantation. The choice of treatment method depends on the size and type of the defect, the expertise and preferences of the surgeon, and patient-specific factors such as age and activity level.

24 months' follow-up with self-reported scores for pain and function and one objective test. Intriguingly, the statistical and clinically meaningful difference was not observed between surgical intervention groups and placebo groups. However, present clinical trials are limited in sample size, inconsistent evaluation indexes, and unreasonable methodology [38, 45]. According to reported research studies, joint lavage, arthroscopic debridement, and cartilage shaving provide transient pain relief and little additional benefits for patients.

12.5.2 Bone Marrow Stimulation and Augmentation

Bone marrow stimulation technique is a reparative strategy for cartilage lesions and predominantly contains drilling (penetration of the subchondral plate using a drill), abrasion arthroplasty (arthroscopic debridement of the base of the lesion down to bleeding bone), and microfracture [46]. Microfracture is an often-utilized orthopedic method characterized by safety, cost-effectiveness, and technical simplicity and thus is recommended as the first-line treatment modality for small-sized focal chondral defects. During the surgery, the calcified layer is removed by debridement, and a surgical awl is used to make several perforations in the subchondral bone plate of exposed cartilage defects [47]. Subsequently, marrow bleeding is released from small holes and fills defective sites of articular cartilage, resulting in the formation of fibrocartilaginous repair tissue [47, 48]. However, fibrocartilage is not as durable as hyaline articular cartilage, and the quality of fibrous repair tissue has sub-optimal biomechanical and biochemical properties, which is easily declined because of its enrichment of type I collagen. Consistent with the inferior properties of fibrocartilage, accumulating evidence has demonstrated that microfracture only improves clinical outcomes at short-term follow-up and does not show better therapeutic effects in the long run [49–51]. With regard to postoperative clinical outcomes, most of the studies have reported that it markedly ameliorated dysfunctions of knee during the first 24 months, while the higher failure rate of surgery can be expected regardless of the size of defects after the five-year period of initial operation [49, 52]. However, in a recent randomized controlled trial, Ibarra et al. [53] reported that after the microfracture surgery, there are significant clinical improvements in a mean of six years compared with patients' preoperative status. There are still some intrinsic shortcomings in microfracture, such as vulnerable fibrocartilage, reduced cartilage volume, and adverse biomechanical changes. To counteract the potential disadvantages of microfracture and create native microenvironments for bone marrow stem cells, microfracture-based techniques, biological augmentation of microfracture, has been introduced and practiced in clinic. The additional augmentation mainly contains all kinds of bioactive molecules (e.g. hyaluronic acid [HA] and growth factors), acellular scaffolds, and micronized acellular cartilage ECM from allografts [54, 55]. In a multicenter prospective study, patients underwent microfracture augmented by cartilage allograft ECM, obtained improved patient-reported and clinically significant outcomes, and decreased failure and complication rates at a two-year follow-up [56]. In addition, microfracture augmentation with trypsin pretreatment and growth factor-functionalized self-assembling peptide hydrogel scaffold was shown to improve functional outcomes and histologic parameters and elevate proteoglycan content in cartilage defects of equine models [57].

Although articular cartilage has intrinsic limited capacity of repair, bone marrow stimulation is still a ubiquitous orthopedic approach to facilitate the restoration and regeneration of chondral lesions. The majority of existing studies demonstrated that microfracture-induced fibrocartilage plays a protective role in the short term, and the fibrocartilaginous repair tissue gradually deteriorates. Therefore, microfracture augmentation, as a modified surgical technique, probably promotes the durability of newly generated tissue. Notwithstanding that bone marrow stimulation and augmentation are prevalent surgical procedures in cartilage repair, more studies are needed to investigate their long-term outcomes.

12.5.3 Osteochondral Autografts and Allografts

Osteochondral autografts and allografts have been regarded as effective and restorative approaches to repair cartilage injuries [58]. Osteochondral autograft plugs and mosaicplasty are harvested from low-weight-bearing sites and transferred to focal defects [59]. These autogenous grafts provide viable and minced hyaline cartilage-bone units that enable the formation of hyaline or hyaline-like repair tissue on defective cartilage surface through integrating with the surrounding tissue [59, 60]. In 1990s, osteochondral mosaicplasty under the use of arthroscope was first introduced to repair cartilage damage of femoral condyle by Hangody et al. and achieved good-to-excellent clinical results [61]. However, limited availability of osteochondral autografts constrains the clinical practice of this technique, which is more suitable for small chondral lesions ($<2–3\,cm^2$) [58, 59]. Osteochondral autograft transplantation is a clinically effective and cost-effective surgical approach to alleviate cartilage lesions at medium- to long-term postoperative period [62–65]. In 2003, Hangody et al. [63, 65] have reported that autologous osteochondral graft surgery resulted in significantly improved long-term clinical outcomes, and it is regarded as an alternative therapeutic option for small- and medium-sized focal lesions of weight-bearing surfaces of the knee and other synovial joints. In a recent systematic review, osteochondral autograft transfer showed favorable outcomes at a mean follow-up of 10.2 years [66]. For cartilage lesions of the tibial plateau, osteochondral autograft transfer exerts markedly elevated postoperative clinical outcomes as well [67, 68]. Additionally, Pareek et al. [69] found that the osteochondral autograft technique resulted in better clinical outcomes and activity levels compared to widely used microfracture. Besides, osteochondral autograft transplantation also significantly mitigated subjective symptoms and improved patient-reported outcomes in osteochondral lesions of the talus [70, 71].

One of the most prevalent adverse side effects of osteochondral autografts is coverage of fibrocartilage on donor sites [72], while the reduction of number or size of donor autografts and thus increased receptor gaps may enhance the occurrence of fibrous tissue on cartilage defects [73]. To counteract fibrous repair of chondral lesions, Hishimura et al. [73] found that osteochondral autografts combined with ultrapurified alginate (UPAL) gel significantly promote the restorative capacity of cartilage in rabbit models and potentially exert positive clinical outcomes. Similarly, another study found that osteochondral autografts with platelet-rich fibrin could maintain the hyaline cartilage repair in cartilage lesions and improve clinical outcomes [74]. Therefore, osteochondral autografts augmented with biomaterials may become an evolved technique for cartilage repair.

Osteochondral allograft transplantation is an effective and single-stage surgical approach to treat symptomatic, large, and complex cartilage defects. This technique utilizes the avascular characteristic-induced immune privilege of cartilage and nonlimitation of donor sites which are required in osteochondral autografts. In a recent study that involved 160 patients, Gilat et al. [75] have demonstrated that osteochondral allograft transplantation exhibits successful postoperative outcomes, and the revision and failure rate is about 10% with an average follow-up of 7.7 ± 2.7 years, indicating that osteochondral allograft is a feasible method for focal chondral or osteochondral defects in the long term. Moreover, osteochondral allografts also benefit athletes with cartilage defects [76]. In addition, variable preservation methods and storage time also affects the therapeutic effects of osteochondral allografts [77, 78].

Based on known studies, both osteochondral autografts and allografts obtained satisfactory clinical outcomes in medium and long terms. However, high-quality randomized controlled trails are insufficient in this promising research field.

12.5.4 Autologous Chondrocyte Implantation (ACI)

ACI, a cell-based biological technique for the treatment of large full-thickness cartilage repair, allows chondrocytes from a low-weight-bearing region to transplant into defective regions of cartilage surface [48]. Since the ACI technique was first used in the patch of cartilage lesions in 1994, it has evolved from ACI technique with periosteal flap (ACI-P) to matrix-induced autologous chondrocyte implantation (MACI) [79]. Generally, ACI requires two separate surgical procedures in which chondrocytes are isolated from relatively healthy articular cartilage of patients and subsequently implanted into debrided cartilage defects after expansion *in vitro* culture. Notably, intergenerational techniques adopt different covering materials. The recipient lesion of articular cartilage was originally covered by a tightly sutured periosteal patch [79]. However, this two-dimensional single-layer structure leads to cell dedifferentiation and hypertrophy of periosteal patch [80]. In addition, compared to microfracture, ACI-P seems to have no additional improvements in clinical outcome [81, 82]. MACI is a new iteration of ACI technique that isolated chondrocytes are seeded into scaffolds made of a 3D collagen (type I and type III) membrane [83]. Accumulating evidence suggests that MACI or the third-generation ACI technique is an effective and suitable treatment modality that improves knee functions, graft survivorship, and patient satisfaction in the long term. In a recent study, 87 patients received MACI operation for focal cartilage defects and assessed by clinical outcomes and MRI with an average age of 13.1 years postoperatively [84]. The results demonstrated that all patient-reported outcome measures (PROMs) were markedly ameliorated and only 9.1% of grafts were failed based on MRI outcome, indicating a favorable prognosis after MACI operation. Kreuz et al. [85] also reported that MACI technique improves the clinical and radiographic outcomes in 12 years after the transplantation [85]. Intriguingly, in a randomized controlled trial, 21 patients were incorporated and randomized into ACI-P group and MACI group, respectively [86]. During the long-term follow-up period, they were assessed at 1 year, 2 years, and final point (8–11 years) postoperatively based on subjective clinical outcome scores and MRI images. Finally, with an average follow-up of 9.6 years after surgery, 16 patients completed reassessment and results demonstrated that there is no significant difference in clinical outcomes and MRI scores between ACI-P and MACI groups versus preoperative status. Niethammer et al. [87] have reported that patients received third-generation ACI-gained significant improvement of clinical and radiographic outcomes in long-term assessment as well. Therefore, it is compelling that both the first-generation and third-generation techniques are effective treatments for full-thickness cartilage defects; however, the sample size was relatively small. Furthermore, as the third-generation technique, ACI with scaffolds made of type I/III collagen probably has better clinical outcomes for femoral lesions than patellar defects in a three-year follow-up [88]. Similarly, the third-generation ACI benefits athletes with full-thickness articular cartilage lesions as well in which patient-reported outcomes were significantly improved after two years of surgery [89].

Notably, ACI technique, as a second-line treatment for cartilage lesions, probably leads to unsatisfactory outcomes after failed bone marrow stimulation. In a systematic review, Schuette et al. [90] reported that after failed bone marrow stimulation, secondary ACI has a higher risk of treatment failure and worse patients subjective outcomes compared with primary ACI. Moreover, the third-generation ACI also resulted in inferior clinical results after the failed bone marrow stimulation procedure [91]. Thus, it is reasonable to choose a suitable surgical approach for cartilage repair according to more comprehensive consideration.

Accumulating studies have demonstrated that ACI is a clinically effective technique to restore cartilage defects; however, more large-sample comparative studies and randomized clinical trials are needed to further testify its long-term clinical outcomes and improvement of patients' quality of life.

12.6 Potential and Advanced Strategies of Cartilage Repair

Compared with the currently developed approaches for articular cartilage repair such as microfracture and ACI, several novel approaches in the midst of clinical trials or under laboratory studies are emerging, which may offer promise of better treatment methods in the near future (Table 12.1).

12.6.1 Tissue Engineering Approaches

The term "tissue engineering (TE)" was first coined in 1987. The first exploration of TE techniques was an attempt to culture skin tissue. TE is a promising technique using seed cells, scaffolds, and growth factors in order to create various human tissues with natural structures. The critical functions may be structural, barrier- and transport-related, or biochemical and secretory [92]. Whether there is an adequate source of living cells is always a primary problem. The source of donor tissue cells are often limited, whereas stem cells (e.g. adult stem cells) possess many attractive characteristics such as strong proliferative ability and feasible sources. In addition, it is necessary to constructing a biomimetic extracellular environment so as to make cells *in vitro* able to proliferate and differentiate as if they really grow up in the human body. The application of scaffolds is helping to facilitate this process as well to some extent. The final step is implanting the artificial tissue cultured *in vitro* into damaged sites to complete tissue repair [92, 93].

When it comes to the field of cartilage repair, TE aims at restoring both functional and mechanical properties of native articular cartilage and subchondral bone [94]. The graphical abstract of this process is as shown in Figure 12.5 [95].

The most investigated seed cells for cartilage TE are chondrocytes and stem cells [96]. It always occurs to us that chondrocytes are able to work as seed cells during TE and they can produce, maintain, and remodel the cartilage ECM *in vitro*

Table 12.1 Overview of pre-clinical methods for cartilage repair.

Pre-clinical methods for cartilage repair	Product name (up to three examples)	Advantages	Limitations
TE	BioCart II (ProChon Biotech) BioSeed-C (BioTissue SA) BST-CarGel (Piramal Healthcare)	Mimicking natural cartilage structure with biomechanical properties	Scaffold-associated toxicity, stress shielding, altered cell phenotypes, and hindrances to remodeling
Platelet-rich plasma (PRP)	Pure PRP Leukocyte-rich PRP (L-PRP) Pure platelet-rich fibrin (P-PRF) or platelet-rich fibrin matrix (PRFM)	Relieving symptoms caused by cartilage injuries; safety	Not effective for severe cartilage damage and unable to prevent long-term consequence
Gene editing	Viral vectors Nonviral vectors	Achieving cell-type specificity	Still needs to be developed

[97]. Besides, stem cells can also be applied so as to overcome the limited supply of primary chondrocytes and provide strong proliferation capability. There is no doubt that certain cytokines, such as growth factor (TGF-β), insulin-like growth factor 1 (IGF-1), bone morphogenetic proteins (BMPs), fibroblast growth factors (FGFs), and epidermal growth factor (EGF), would impact the anabolism and catabolism of chondrocytes [95]. Whether repairing articular cartilage *in vitro* or *in vivo*, the application of cytokines that promotes growth of cartilage will significantly enhance the therapeutic efficacy. Further, it is necessary to investigate which specific mechanical stimuli and their parameters could bring an optimal response of the cells in cultured constructs [97]. So far, many studies have been devoted to the application of scaffolds combined with bioactive factors for TE for cartilage repair. Jiang et al. [98] has shown the excellent effect of bioactive-stratified polymer ceramic-hydrogel scaffold for articular osteochondral repair. Qu et al. [99], through the successful ectopic osteochondral scaffold implantation, proved that PVA-n-HA/PA6 with induced bone marrow stem cells (BMSCs) is a possible cartilage repair strategies. Im et al. [100] got results suggesting that biphasic osteochondral scaffolds may help repair osteochondral defects in a large animal model. In addition, there is another biphasic scaffold based on collagen-GAG and nanocalcium phosphate to mimic the composition and structure of articular cartilage [94]. It is reported that a scaffold product (MaioRegen, Fin-Ceramica Faenza S.p.A., Italy) with hyaluronic acid nanoparticles has been developed for articular cartilage repair [94, 101]. This kind of scaffold can also function without any other fixation device [102]. Kon et al. [102] have studied the short-term efficacy of TE technique with scaffolds for cartilage defect from

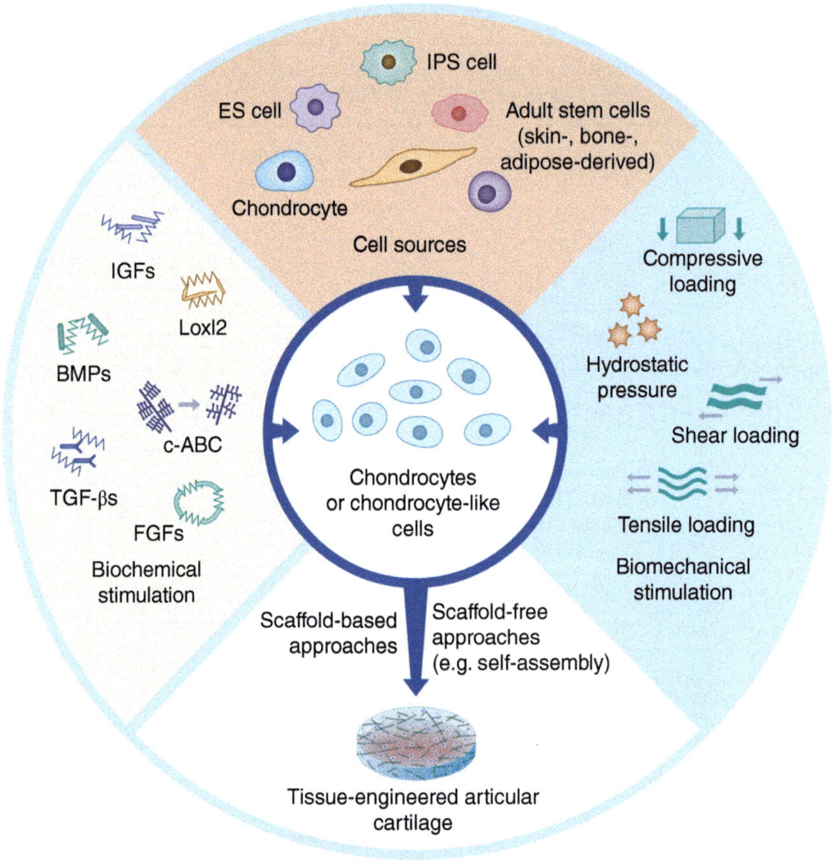

Figure 12.5 Tissue engineering approaches for articular cartilage repair. IPS, Induced pluripotent stem; ES, embryonic stem; Loxl2, lysyl oxidase-like 2; c-ABC, chondroitinase-ABC.

different sites, and the early postoperative adherence rate was observed at 5 and 25 weeks after implantation by high-resolution MRI. The results are encouraging in the short run but remain unclear in the long run to date [102]. There are a number of products under development for articular cartilage repair (e.g. BioCart II [ProChon Biotech], BioSeed-C [BioTissue SA], and BST-CarGel [Piramal Healthcare]) [48]. We believe that TE is bound to provide tremendous motivation for the development of articular cartilage repair in the near future. TE is a very complex subject that requires multidisciplinary collaboration, and mastering TE technology requires a good understanding of the various factors able to regulate cell, tissue, and organ development. Although this technology went through a wave and threefold, it would eventually benefit mankind.

12.6.2 Platelet-Rich Plasma (PRP) for Cartilage Repair

Platelet-rich plasma (PRP), which contributes to regenerative medicine, offers another new option for treatment planning. PRP treatment is effective because platelet growth factors (PGFs) in it would support wound healing and repair cascade [103]. Nowadays, there are many PRP formulations to choose from [103]. One of the earliest reviews on PRP came out in 2006 [104]. At that time, people began to get familiar with platelet function and underlying scientific rationale. Nowadays, PRP is becoming more and more widely used in different fields of medicine.

For damaged articular cartilage immersed in catabolic environment, PRP works by inhibiting the secretion of catabolic cytokines and meanwhile promoting cartilage matrix synthesis [105]. After intra-articular PRP injection, cartilage metabolism is modulated through affecting multiple cytokines (e.g. IL-1β, TNF-α, TGF-β, and FGFs) [106]. Huang et al. [107] found that PRP injection would bring significant reduction in pain and improvement in knee function in short-term follow-up because of its effect on cartilage repair. PRP is able to offer remarkable effects of anti-inflammation in human chondrocytes [108].

KOA is always characterized by cartilage degeneration. Over the past few decades, PRP injection has long been one of the conservative treatment options and its efficacy and safety seem to be great [109]. It is due to the potential of PRP for cartilage repair that PRP injection is effective for KOA. Besides, the application of PRP combined with stem cells may accelerate and augment the repair of injured menisci and cartilage in a clinical setting [110]. Moussa et al. [111] found that PRP could enhance protection of chondrocytes via increasing autophagy, anti-inflammatory effect, and decreasing apoptosis in cartilage. The remarkable advantage of PRP injection is symptom relief and high safety in the treatment of early osteoarthritis, for example, instead of end-stage osteoarthritis with severe cartilage lesions. Different products of PRP such as autologous conditioned plasma, autologous platelet gel, and leukocyte-rich PRP have been developed [103]. Xie et al. [112] reported that intra-articular PRP injections before strenuous exercises may be able to play a role in preventing articular cartilage damage. PRP can protect articular cartilage through multiple pathways. Nowadays, a growing number of studies indicate that PRP fulfills a much wider role in health and disease than we previously imagined [113]. Delaying knee joint degeneration through protecting cartilage is one of its functions. Under local catabolic environment, PRP works by inhibiting the catabolic cytokines such as IL-1β and TNF-α [114, 115] and by promoting factors such as FGFs and TGF-β [105, 116]. Activated PRP in chondrocytes could reduce the activity of NF-κB and the elevated levels of nitric oxide (NO), which is critical for inflammatory process [117, 118]. There are many remarkable advantages of PRP application. First of all, in the knee, PRP is versatile and has been used in patients with different disorders of knees such as articular cartilage pathology, ligamentous and meniscal injuries, and so on [119]. PRP can exert synergistic effects with other treatments. PRP can promote cartilage repair combined with the microfracture technique [120]. Intra-articular PRP injection could bring an overall improvement of the symptoms, especially moderate to severe pain even if pain at the site of injection may happen. Napolitano et al. [121] showed pain relief and improvements of knee function at six months after three leukocyte-rich platelet-rich plasma (L-PRP) injections in 27 patients with articular cartilage pathology. In a preliminary noncontrolled prospective study by Sánchez et al. [122], the outcomes of PRP injections in the treatment of hip OA were satisfactory just in a limited number of patients.

However, several limitations have gradually shown up, and there are still some controversies in efficacy. Sampson et al. [123] performed a pilot study which showed that three ultrasound-guided injections of L-PRP at four week intervals result in no significant differences in the cartilage thickness during the first six months. Filardo et al. [124] established that, in the long run, inferior outcomes tend to appear in patients with a higher body mass index (BMI) and female patients have a higher risk. Compared with hyaluronic acid (HA), autologous PRP showed stronger and longer efficacy than HA in relieving symptoms and improving articular function [125]. However, it is reported that we are prone to get worse results among patients with older age and severe cartilage degeneration [125]. PRP treatment turns out to be better outcomes in some cases such as patients with younger age, mild cartilage degeneration, male gender, low BMI, short-term follow-up, and multiple injection procedure [126].

12.6.3 Gene Editing for Cartilage Repair

The cartilage matrix and chondrocytes are important components of articular cartilage tissue, in which the cartilage matrix is a scaffolding part of articular chondrocytes, which plays an important role in the process of cartilage degeneration. The mutual influence between chondrocytes and the cartilage matrix always takes place. Previous studies have reported that cartilage tissue degeneration is mainly caused by the degradation of cartilage ECM and excessive apoptosis of chondrocytes, and miRNAs and coding genes are involved [127]. Chondrocytes undergo a series of alterations under pathological conditions that harbor complex, interconnected genomic events. The introduction of a gene fragment into the body aims to make the gene express stably, controllably, and chronically, thereby playing a role in repairing the damaged cartilage and protecting the cartilage from further injury [128]. Gene editing is one of the techniques aiming to deliver nucleic acids to target tissues by viral or nonviral vectors to prevent, treat, or cure disease [129]. The vector for gene editing is an essential part in the application of gene-editing technology to the clinic. Through viral or nonviral vectors, this method is able to achieve cell-type specificity repair, early diagnosis, and targeted therapy for cartilage. Gene therapy for cartilage repair mainly aims at relieving inflammation, preventing cartilage degradation, and increasing cartilage matrix synthesis to promote healing [130]. Bao et al. [131] showed that granzyme B (GZMB) gene silencing functions to inhibit the mitogen-activated protein kinase (MAPK) signaling pathway to protect cartilage tissue injury caused by RA. Ha et al. [132] injected TGF-β produced by genetically modified chondrocytes into the joint cavity, which resulted in an overall improvement in knee symptoms and function. Tao et al. [133] provided evidence for the benefits of co-transducing human bone marrow-derived mesenchymal stem cells via separate therapeutic rAAV vectors to significantly improve their chondro-reparative activities *in vitro*. Bougioukli et al. [134] utilized viral vectors to deliver TGF-β3 genes and collagen short hairpin RNA (shRNA) into chondrocytes and cartilage redifferentiation and decreased the collagen degradation happened. In this field, multigene therapy exhibited obvious superiority.

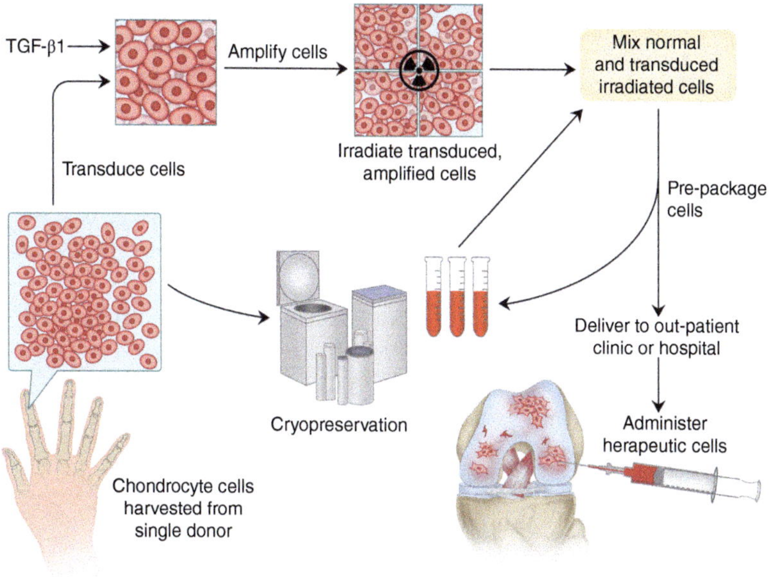

Figure 12.6 Overview of gene therapy protocol for cartilage repair.

To date, for both *in vivo* and *ex vivo*, gene delivery has been achieved in different animal models. Nowadays, it is workable that researchers make complementary DNAs (cDNAs) get inside the knee cavity by injection and express [135]. However, because of the short intra-articular dwell time of the injected cells and low probability to achieve long-term transgene expression, it is doubtful how much benefit can be obtained when this approach is applied [130]. Therefore, *in vivo* delivery has obvious advantages as gene therapies. Adeno-associated virus (AAV), with adequate levels and duration of gene expression when compared with nonviral vectors, serves as one of the outstanding representative viral vectors [136]. The great prospects of this kind of absolutely novel therapeutic ideas would result in rapid development of genetic strategies. An overview of gene therapy protocol for cartilage repair is represented in Figure 12.6.

12.7 Conclusion

This part focused on introducing some basic information of cartilage injury and the major repairing strategies from clinical prospective. Although there are still lack of methods that satisfy the fundamental requirements of successful cartilage healing, namely, one that embodies the appropriate structure–function relationship, ECM organization, bioactivity, and surgical logistics, efforts are still being made with the development of technology and bioactive materials, and we believed there will be more appropriate treatment options for curing cartilage injury and osteoarthritis in the coming future.

References

1 Krishnan, Y. and Grodzinsky, A.J. (2018). Cartilage diseases. *Matrix Biol.* 71, 72: 51–69.

2 Jiang, Y. and Tuan, R.S. (2015). Origin and function of cartilage stem/progenitor cells in osteoarthritis. *Nat. Rev. Rheumatol.* 11 (4): 206–212.

3 Usami, Y., Gunawardena, A.T., Iwamoto, M., and Enomoto-Iwamoto, M. (2016). Wnt signaling in cartilage development and diseases: lessons from animal studies. *Lab. Invest.* 96 (2): 186–196.

4 Chen, S., Fu, P., Wu, H., and Pei, M. (2017). Meniscus, articular cartilage and nucleus pulposus: a comparative review of cartilage-like tissues in anatomy, development and function. *Cell Tissue Res.* 370 (1): 53–70.

5 Browne, J.E. and Branch, T.P. (2000). Surgical alternatives for treatment of articular cartilage lesions. *J. Am. Acad. Orthop. Surg.* 8 (3): 180–189.

6 Punzi, L., Galozzi, P., Luisetto, R. et al. (2016). Post-traumatic arthritis: overview on pathogenic mechanisms and role of inflammation. *RMD Open* 2 (2): e000279.

7 Lieberthal, J., Sambamurthy, N., and Scanzello, C.R. (2015). Inflammation in joint injury and post-traumatic osteoarthritis. *Osteoarthritis Cartilage* 23 (11): 1825–1834.

8 Yamamoto, A., Levine, B.D., Padron, M., and Chung, C.B. (2020). Is there a role for cartilage imaging in athletes? *Semin. Musculoskelet. Radiol.* 24 (3): 246–255.

9 Murray, I.R., Benke, M.T., and Mandelbaum, B.R. (2016). Management of knee articular cartilage injuries in athletes: chondroprotection, chondrofacilitation, and resurfacing. *Knee Surg. Sports Traumatol. Arthrosc.* 24 (5): 1617–1626.

10 Mankin, H.J. (1971). Biochemical and metabolic aspects of osteoarthritis. *Orthop. Clin. North Am.* 2 (1): 19–31.

11 Chu, C.R., Williams, A., Tolliver, D. et al. (2010). Clinical optical coherence tomography of early articular cartilage degeneration in patients with degenerative meniscal tears. *Arthritis Rheum.* 62 (5): 1412–1420.

12 Kleemann, R.U., Krocker, D., Cedraro, A. et al. (2005). Altered cartilage mechanics and histology in knee osteoarthritis: relation to clinical assessment (ICRS grade). *Osteoarthritis Cartilage* 13 (11): 958–963.

13 McAlindon, T.E., Bannuru, R.R., Sullivan, M.C. et al. (2014). OARSI guidelines for the non-surgical management of knee osteoarthritis. *Osteoarthritis Cartilage* 22 (3): 363–388.

14 Tolk, J.J., Janssen, R.P.A., Prinsen, C. et al. (2019). Measurement properties of the OARSI core set of performance-based measures for hip osteoarthritis: a prospective cohort study on reliability, construct validity and responsiveness in 90 hip osteoarthritis patients. *Acta Orthop.* 90 (1): 15–20.

15 Slattery, C. and Kweon, C.Y. (2018). Classifications in brief: Outerbridge classification of chondral lesions. *Clin. Orthop. Relat. Res.* 476 (10): 2101–2104.

16 Dwyer, T., Martin, C.R., Kendra, R. et al. (2017). Reliability and validity of the arthroscopic international cartilage repair society classification system: correlation with histological assessment of depth. *Arthroscopy* 33 (6): 1219–1224.

17 Brittberg, M. and Winalski, C.S. (2003). Evaluation of cartilage injuries and repair. *J. Bone Joint Surg. Am.* 85-A (Suppl. 2): 58–69.

18 Mankin, H.J., Dorfman, H., Lippiello, L., and Zarins, A. (1971). Biochemical and metabolic abnormalities in articular cartilage from osteo-arthritic human hips. II. Correlation of morphology with biochemical and metabolic data. *J. Bone Joint Surg. Am.* 53 (3): 523–537.

19 Ostergaard, K., Andersen, C.B., Petersen, J. et al. (1999). Validity of histopathological grading of articular cartilage from osteoarthritic knee joints. *Ann. Rheum. Dis.* 58 (4): 208–213.

20 Waldstein, W., Perino, G., Gilbert, S.L. et al. (2016). OARSI osteoarthritis cartilage histopathology assessment system: a biomechanical evaluation in the human knee. *J. Orthop. Res.* 34 (1): 135–140.

21 Simon, W.H. (1970). Scale effects in animal joints. I. Articular cartilage thickness and compressive stress. *Arthritis Rheum.* 13 (3): 244–256.

22 Pritzker, K.P., Gay, S., Jimenez, S.A. et al. (2006). Osteoarthritis cartilage histopathology: grading and staging. *Osteoarthritis Cartilage* 14 (1): 13–29.

23 Glasson, S.S., Chambers, M.G., Van Den Berg, W.B., and Little, C.B. (2010). The OARSI histopathology initiative – recommendations for histological assessments of osteoarthritis in the mouse. *Osteoarthritis Cartilage* 18 (Suppl. 3): S17–S23.

24 Gerwin, N., Bendele, A.M., Glasson, S., and Carlson, C.S. (2010). The OARSI histopathology initiative – recommendations for histological assessments of osteoarthritis in the rat. *Osteoarthritis Cartilage* 18 (Suppl. 3): S24–S34.

25 Cook, J.L., Kuroki, K., Visco, D. et al. (2010). The OARSI histopathology initiative – recommendations for histological assessments of osteoarthritis in the dog. *Osteoarthritis Cartilage* 18 (Suppl. 3): S66–S79.

26 de Windt, T.S., Welsch, G.H., Brittberg, M. et al. (2013). Is magnetic resonance imaging reliable in predicting clinical outcome after articular cartilage repair of the knee? A systematic review and meta-analysis. *Am. J. Sports Med.* 41 (7): 1695–1702.

27 Kijowski, R., Demehri, S., Roemer, F., and Guermazi, A. (2020). Osteoarthritis year in review 2019: imaging. *Osteoarthritis Cartilage* 28 (3): 285–295.

28 Crema, M.D., Roemer, F.W., Marra, M.D. et al. (2011). Articular cartilage in the knee: current MR imaging techniques and applications in clinical practice and research. *Radiographics* 31 (1): 37–61.

29 Disler, D.G. and McCauley, T.R. (1998). Clinical magnetic resonance imaging of articular cartilage. *Top. Magn. Reson. Imaging* 9 (6): 360–376.

30 Potter, H.G., Linklater, J.M., Allen, A.A. et al. (1998). Magnetic resonance imaging of articular cartilage in the knee. An evaluation with use of fast-spin-echo imaging. *J. Bone Joint Surg. Am.* 80 (9): 1276–1284.

31 Van Breuseghem, I. (2004). Ultrastructural MR imaging techniques of the knee articular cartilage: problems for routine clinical application. *Eur. Radiol.* 14 (2): 184–192.

32 Marlovits, S., Mamisch, T.C., Vekszler, G. et al. (2008). Magnetic resonance imaging for diagnosis and assessment of cartilage defect repairs. *Injury* 39 (Suppl. 1): S13–S25.

33 Yulish, B.S., Montanez, J., Goodfellow, D.B. et al. (1987). Chondromalacia patellae: assessment with MR imaging. *Radiology* 164 (3): 763–766.

34 Forney, M., Subhas, N., Donley, B., and Winalski, C.S. (2011). MR imaging of the articular cartilage of the knee and ankle. *Magn. Reson. Imaging Clin. N. Am.* 19 (2): 379–405.

35 Pearle, A.D., Warren, R.F., and Rodeo, S.A. (2005). Basic science of articular cartilage and osteoarthritis. *Clin. Sports Med.* 24 (1): 1–12.

36 Hjelle, K., Solheim, E., Strand, T. et al. (2002). Articular cartilage defects in 1,000 knee arthroscopies. *Arthroscopy* 18 (7): 730–734.

37 Gomoll, A.H., Farr, J., Gillogly, S.D. et al. (2010). Surgical management of articular cartilage defects of the knee. *J. Bone Joint Surg. Am.* 92 (14): 2470–2490.

38 Reichenbach, S., Rutjes, A.W., Nüesch, E. et al. (2010). Joint lavage for osteoarthritis of the knee. *Cochrane Database Syst. Rev.* (5): Cd007320.

39 Kirkley, A., Birmingham, T.B., Litchfield, R.B. et al. (2008). A randomized trial of arthroscopic surgery for osteoarthritis of the knee. *N. Engl. J. Med.* 359 (11): 1097–1107.

40 Laupattarakasem, W., Laopaiboon, M., Laupattarakasem, P., and Sumananont, C. (2008). Arthroscopic debridement for knee osteoarthritis. *Cochrane Database Syst. Rev.* (1): Cd005118.

41 Totlis, T., Marín Fermín, T., Kalifis, G. et al. (2021). Arthroscopic debridement for focal articular cartilage lesions of the knee: a systematic review. *Surgeon* (1): CD005118.

42 Chaudhry, H., Madden, K., and Bhandari, M. (2014). Cochrane in CORR®: joint lavage for osteoarthritis of the knee. *Clin. Orthop. Relat. Res.* 472 (1): 22–27.

43 Popert, J. (1985). Rice-bodies, synovial debris, and joint lavage. *Br. J. Rheumatol.* 24 (1): 1–2.

44 Moseley, J.B., O'Malley, K., Petersen, N.J. et al. (2002). A controlled trial of arthroscopic surgery for osteoarthritis of the knee. *N. Engl. J. Med.* 347 (2): 81–88.

45 Lafita, F.R. and Castella, J.M. (2010). Review: Joint lavage does not reduce pain or improve function in knee osteoarthritis. *Ann. Intern. Med.* 153 (4): Jc2-2.

46 Gill, T.J., Asnis, P.D., and Berkson, E.M. (2006). The treatment of articular cartilage defects using the microfracture technique. *J. Orthop. Sports Phys. Ther.* 36 (10): 728–738.

47 Steadman, J.R., Rodkey, W.G., and Rodrigo, J.J. (2001). Microfracture: surgical technique and rehabilitation to treat chondral defects. *Clin. Orthop. Relat. Res.* 391: S362–S369.

48 Kwon, H., Brown, W.E., Lee, C.A. et al. (2019). Surgical and tissue engineering strategies for articular cartilage and meniscus repair. *Nat. Rev. Rheumatol.* 15 (9): 550–570.

49 Goyal, D., Keyhani, S., Lee, E.H., and Hui, J.H. (2013). Evidence-based status of microfracture technique: a systematic review of level I and II studies. *Arthroscopy* 29 (9): 1579–1588.

50 Knutsen, G., Engebretsen, L., Ludvigsen, T.C. et al. (2004). Autologous chondrocyte implantation compared with microfracture in the knee. A randomized trial. *J. Bone Joint Surg. Am.* 86 (3): 455–464.

51 Mithoefer, K., Williams, R.J. III, Warren, R.F. et al. (2005). The microfracture technique for the treatment of articular cartilage lesions in the knee. A prospective cohort study. *J. Bone Joint Surg. Am.* 87 (9): 1911–1920.

52 Mithoefer, K., McAdams, T., Williams, R.J. et al. (2009). Clinical efficacy of the microfracture technique for articular cartilage repair in the knee: an evidence-based systematic analysis. *Am. J. Sports Med.* 37 (10): 2053–2063.

53 Ibarra, C., Villalobos, E., Madrazo-Ibarra, A. et al. (2021). Arthroscopic matrix-assisted autologous chondrocyte transplantation versus microfracture: a 6-year follow-up of a prospective randomized trial. *Am. J. Sports Med.* 49 (8): 2165–2176.

54 Albright, J.C. and Daoud, A.K. (2017). Microfracture and microfracture plus. *Clin. Sports Med.* 36 (3): 501–507.

55 Strauss, E.J., Barker, J.U., Kercher, J.S. et al. (2010). Augmentation strategies following the microfracture technique for repair of focal chondral defects. *Cartilage* 1 (2): 145–152.

56 Cole, B.J., Haunschild, E.D., Carter, T. et al. (2021). Clinically significant outcomes following the treatment of focal cartilage defects of the knee with microfracture augmentation using cartilage allograft extracellular matrix: a multicenter prospective study. *Arthroscopy* 37 (5): 1512–1521.

57 Zanotto, G.M., Liesbeny, P., Barrett, M. et al. (2021). Microfracture augmentation with trypsin pretreatment and growth factor – functionalized self-assembling peptide hydrogel scaffold in an equine model. *Am. J. Sports Med.* 49 (9): 2498–2508.

58 Moran, C.J., Pascual-Garrido, C., Chubinskaya, S. et al. (2014). Restoration of articular cartilage. *J. Bone Joint Surg. Am.* 96 (4): 336–344.

59 Hangody, L., Ráthonyi, G.K., Duska, Z. et al. (2004). Autologous osteochondral mosaicplasty. Surgical technique. *J. Bone Joint Surg. Am.* 86-A (Suppl. 1): 65–72.

60 Krych, A.J., Harnly, H.W., Rodeo, S.A., and Williams, R.J. III (2012). Activity levels are higher after osteochondral autograft transfer mosaicplasty than after microfracture for articular cartilage defects of the knee: a retrospective comparative study. *J. Bone Joint Surg. Am.* 94 (11): 971–978.

61 Hangody, L., Kish, G., Kárpáti, Z. et al. (1997). Arthroscopic autogenous osteochondral mosaicplasty for the treatment of femoral condylar articular defects. A preliminary report. *Knee Surg. Sports Traumatol. Arthrosc.* 5 (4): 262–267.

62 Marcacci, M., Kon, E., Delcogliano, M. et al. (2007). Arthroscopic autologous osteochondral grafting for cartilage defects of the knee: prospective study results at a minimum 7-year follow-up. *Am. J. Sports Med.* 35 (12): 2014–2021.

63 Hangody, L., Vásárhelyi, G., Hangody, L.R. et al. (2008). Autologous osteochondral grafting – technique and long-term results. *Injury* 39 (Suppl. 1): S32–S39.

64 Miller, D.J., Smith, M.V., Matava, M.J. et al. (2015). Microfracture and osteochondral autograft transplantation are cost-effective treatments for articular cartilage lesions of the distal femur. *Am. J. Sports Med.* 43 (9): 2175–2181.

65 Hangody, L. and Füles, P. (2003). Autologous osteochondral mosaicplasty for the treatment of full-thickness defects of weight-bearing joints: ten years of experimental and clinical experience. *J. Bone Joint Surg. Am.* 85-A (Suppl. 2): 25–32.

66 Pareek, A., Reardon, P.J., Maak, T.G. et al. (2016). Long-term outcomes after osteochondral autograft transfer: a systematic review at mean follow-up of 10.2 years. *Arthroscopy* 32 (6): 1174–1184.

67 Yabumoto, H., Nakagawa, Y., Mukai, S. et al. (2017). Surgical technique and clinical outcomes of retrograde osteochondral autograft transfer for osteochondral lesions of the tibial plateau. *Arthroscopy* 33 (6): 1241–1247.

68 Wajsfisz, A., Makridis, K.G., and Djian, P. (2013). Arthroscopic retrograde osteochondral autograft transplantation for cartilage lesions of the tibial plateau: a prospective study. *Am. J. Sports Med.* 41 (2): 411–415.

69 Pareek, A., Reardon, P.J., Macalena, J.A. et al. (2016). Osteochondral autograft transfer versus microfracture in the knee: a meta-analysis of prospective comparative studies at midterm. *Arthroscopy* 32 (10): 2118–2130.

70 Zhang, L., Luo, Y., Zhou, X. et al. (2021). Outcomes from osteochondral autograft transplant or mosaicplasty in 26 patients with type V osteochondral lesions of the talus. *Med. Sci. Monit.* 27: e930527.

71 Park, K.H., Hwang, Y., Han, S.H. et al. (2018). Primary versus secondary osteochondral autograft transplantation for the treatment of large osteochondral lesions of the talus. *Am. J. Sports Med.* 46 (6): 1389–1396.

72 Iwasaki, N., Kato, H., Kamishima, T. et al. (2007). Donor site evaluation after autologous osteochondral mosaicplasty for cartilaginous lesions of the elbow joint. *Am. J. Sports Med.* 35 (12): 2096–2100.

73 Hishimura, R., Onodera, T., Hontani, K. et al. (2019). Osteochondral autograft transplantation technique augmented by an ultrapurified alginate gel enhances osteochondral repair in a rabbit model. *Am. J. Sports Med.* 47 (2): 468–478.

74 Maruyama, M., Satake, H., Suzuki, T. et al. (2017). Comparison of the effects of osteochondral autograft transplantation with platelet-rich plasma or platelet-rich fibrin on osteochondral defects in a rabbit model. *Am. J. Sports Med.* 45 (14): 3280–3288.

75 Gilat, R., Haunschild, E.D., Huddleston, H.P. et al. (2021). Osteochondral allograft transplant for focal cartilage defects of the femoral condyles: clinically significant outcomes, failures, and survival at a minimum 5-year follow-up. *Am. J. Sports Med.* 49 (2): 467–475.

76 Crawford, Z.T., Schumaier, A.P., Glogovac, G., and Grawe, B.M. (2019). Return to sport and sports-specific outcomes after osteochondral allograft transplantation in the knee: a systematic review of studies with at least 2 years' mean follow-up. *Arthroscopy* 35 (6): 1880–1889.

77 Merkely, G., Ackermann, J., Farina, E.M. et al. (2020). Shorter storage time is strongly associated with improved graft survivorship at 5 years after osteochondral allograft transplantation. *Am. J. Sports Med.* 48 (13): 3170–3176.

78 Stannard, J.P. and Cook, J.L. (2020). Prospective assessment of outcomes after primary unipolar, multisurface, and bipolar osteochondral allograft transplantations in the knee: a comparison of 2 preservation methods. *Am. J. Sports Med.* 48 (6): 1356–1364.

79 Brittberg, M., Lindahl, A., Nilsson, A. et al. (1994). Treatment of deep cartilage defects in the knee with autologous chondrocyte transplantation. *N. Engl. J. Med.* 331 (14): 889–895.

80 Peterson, L., Minas, T., Brittberg, M. et al. (2000). Two- to 9-year outcome after autologous chondrocyte transplantation of the knee. *Clin. Orthop. Relat. Res.* 374: 212–234.

81 Gou, G.H., Tseng, F.J., Wang, S.H. et al. (2020). Autologous chondrocyte implantation versus microfracture in the knee: a meta-analysis and systematic review. *Arthroscopy* 36 (1): 289–303.

82 Ossendorff, R., Franke, K., Erdle, B. et al. (2019). Clinical and radiographical ten years long-term outcome of microfracture vs. autologous chondrocyte implantation: a matched-pair analysis. *Int. Orthop.* 43 (3): 553–559.

83 Cherubino, P., Grassi, F.A., Bulgheroni, P., and Ronga, M. (2003). Autologous chondrocyte implantation using a bilayer collagen membrane: a preliminary report. *J. Orthop. Surg. (Hong Kong)* 11 (1): 10–15.

84 Ebert, J.R., Fallon, M., Wood, D.J., and Janes, G.C. (2021). Long-term prospective clinical and magnetic resonance imaging-based evaluation of matrix-induced autologous chondrocyte implantation. *Am. J. Sports Med.* 49 (3): 579–587.

85 Kreuz, P.C., Kalkreuth, R.H., Niemeyer, P. et al. (2019). Long-term clinical and MRI results of matrix-assisted autologous chondrocyte implantation for articular cartilage defects of the knee. *Cartilage* 10 (3): 305–313.

86 Barié, A., Kruck, P., Sorbi, R. et al. (2020). Prospective long-term follow-up of autologous chondrocyte implantation with periosteum versus matrix-associated autologous chondrocyte implantation: a randomized clinical trial. *Am. J. Sports Med.* 48 (9): 2230–2241.

87 Niethammer, T.R., Altmann, D., Holzgruber, M. et al. (2020). Patient-reported and magnetic resonance imaging outcomes of third-generation autologous chondrocyte implantation after 10 years. *Arthroscopy* 36 (7): 1928–1938.

88 Niethammer, T.R., Gallik, D., Chevalier, Y. et al. (2021). Effect of the defect localization and size on the success of third-generation autologous chondrocyte implantation in the knee joint. *Int. Orthop.* 45 (6): 1483–1491.

89 Niethammer, T.R., Altmann, D., Holzgruber, M. et al. (2021). Third generation autologous chondrocyte implantation is a good treatment option for athletic persons. *Knee Surg. Sports Traumatol. Arthrosc.* 29 (4): 1215–1223.

90 Schuette, H.B., Kraeutler, M.J., Schrock, J.B., and McCarty, E.C. (2020). Primary autologous chondrocyte implantation of the knee versus autologous chondrocyte implantation after failed marrow stimulation: a systematic review. *Am. J. Sports Med.* 49 (9): 2536–2541.

91 Müller, P.E., Gallik, D., Hammerschmid, F. et al. (2020). Third-generation autologous chondrocyte implantation after failed bone marrow stimulation leads to inferior clinical results. *Knee Surg. Sports Traumatol. Arthrosc.* 28 (2): 470–477.

92 Berthiaume, F., Maguire, T.J., and Yarmush, M.L. (2011). Tissue engineering and regenerative medicine: history, progress, and challenges. *Annu. Rev. Chem. Biomol. Eng.* 2: 403–430.

93 Langer, R. and Vacanti, J. (2016). Advances in tissue engineering. *J. Pediatr. Surg.* 51 (1): 8–12.

94 Panseri, S., Russo, A., Cunha, C. et al. (2012). Osteochondral tissue engineering approaches for articular cartilage and subchondral bone regeneration. *Knee Surg. Sports Traumatol. Arthrosc.* 20 (6): 1182–1191.

95 Kwon, H., Paschos, N.K., Hu, J.C., and Athanasiou, K. (2016). Articular cartilage tissue engineering: the role of signaling molecules. *Cell. Mol. Life Sci.* 73 (6): 1173–1194.

96 Chung, C. and Burdick, J.A. (2008). Engineering cartilage tissue. *Adv. Drug Delivery Rev.* 60 (2): 243–262.

97 Kock, L., van Donkelaar, C.C., and Ito, K. (2012). Tissue engineering of functional articular cartilage: the current status. *Cell Tissue Res.* 347 (3): 613–627.

98 Jiang, J., Tang, A., Ateshian, G.A. et al. (2010). Bioactive stratified polymer ceramic-hydrogel scaffold for integrative osteochondral repair. *Ann. Biomed. Eng.* 38 (6): 2183–2196.

99 Qu, D., Li, J., Li, Y. et al. (2011). Ectopic osteochondral formation of biomimetic porous PVA-n-HA/PA6 bilayered scaffold and BMSCs construct in rabbit. *J. Biomed. Mater. Res. Part B* 96 (1): 9–15.

100 Im, G.I., Ahn, J.H., Kim, S.Y. et al. (2010). A hyaluronate-atelocollagen/β-tricalcium phosphate-hydroxyapatite biphasic scaffold for the repair of osteochondral defects: a porcine study. *Tissue Eng. Part A* 16 (4): 1189–1200.

101 Kon, E., Muttini, A., Arcangeli, E. et al. (2010). Novel nanostructured scaffold for osteochondral regeneration: pilot study in horses. *J. Tissue Eng. Regener. Med.* 4 (4): 300–308.

102 Kon, E., Delcogliano, M., Filardo, G. et al. (2010). A novel nano-composite multi-layered biomaterial for treatment of osteochondral lesions: technique note and an early stability pilot clinical trial. *Injury* 41 (7): 693–701.

103 Everts, P., Onishi, K., Jayaram, P. et al. (2020). Platelet-rich plasma: new performance understandings and therapeutic considerations in 2020. *Int. J. Mol. Sci.* 21 (20).

104 Everts, P.A., Knape, J.T., Weibrich, G. et al. (2006). Platelet-rich plasma and platelet gel: a review. *J. Extra Corpor. Technol.* 38 (2): 174–187.

105 Kennedy, M.I., Whitney, K., Evans, T., and LaPrade, R.F. (2018). Platelet-rich plasma and cartilage repair. *Curr. Rev. Musculoskelet. Med.* 11 (4): 573–582.

106 Xie, X., Zhang, C., and Tuan, R.S. (2014). Biology of platelet-rich plasma and its clinical application in cartilage repair. *Arthritis Res. Ther.* 16 (1): 204.

107 Huang, P.H., Wang, C.J., Chou, W.Y. et al. (2017). Short-term clinical results of intra-articular PRP injections for early osteoarthritis of the knee. *Int. J. Surg.* 42: 117–122.

108 van Buul, G.M., Koevoet, W.L., Kops, N. et al. (2011). Platelet-rich plasma releasate inhibits inflammatory processes in osteoarthritic chondrocytes. *Am. J. Sports Med.* 39 (11): 2362–2370.

109 Zhao, J., Huang, H., Liang, G. et al. (2020). Effects and safety of the combination of platelet-rich plasma (PRP) and hyaluronic acid (HA) in the treatment of knee osteoarthritis: a systematic review and meta-analysis. *BMC Musculoskelet. Disord.* 21 (1): 224.

110 Liu, F., Xu, H., and Huang, H. (2019). A novel kartogenin-platelet-rich plasma gel enhances chondrogenesis of bone marrow mesenchymal stem cells in vitro and promotes wounded meniscus healing in vivo. *Stem Cell Res. Ther.* 10 (1): 201.

111 Moussa, M., Lajeunesse, D., Hilal, G. et al. (2017). Platelet rich plasma (PRP) induces chondroprotection via increasing autophagy, anti-inflammatory markers, and decreasing apoptosis in human osteoarthritic cartilage. *Exp. Cell Res.* 352 (1): 146–156.

112 Xie, X., Ulici, V., Alexander, P.G. et al. (2015). Platelet-rich plasma inhibits mechanically induced injury in chondrocytes. *Arthroscopy* 31 (6): 1142–1150.

113 Nurden, A.T. (2011). Platelets, inflammation and tissue regeneration. *Thromb. Haemost.* 105 (Suppl. 1): S13–S33.

114 Daheshia, M. and Yao, J.Q. (2008). The interleukin 1β pathway in the pathogenesis of osteoarthritis. *J. Rheumatol.* 35 (12): 2306–2312.

115 Chen, L.X., Lin, L., Wang, H.J. et al. (2008). Suppression of early experimental osteoarthritis by *in vivo* delivery of the adenoviral vector-mediated NF-κBp65-specific siRNA. *Osteoarthritis Cartilage* 16 (2): 174–184.

116 Boswell, S.G., Cole, B.J., Sundman, E.A. et al. (2012). Platelet-rich plasma: a milieu of bioactive factors. *Arthroscopy* 28 (3): 429–439.

117 Bendinelli, P., Matteucci, E., Dogliotti, G. et al. (2010). Molecular basis of anti-inflammatory action of platelet-rich plasma on human chondrocytes: mechanisms of NF-κB inhibition via HGF. *J. Cell. Physiol.* 225 (3): 757–766.

118 Farrell, A.J., Blake, D.R., Palmer, R.M., and Moncada, S. (1992). Increased concentrations of nitrite in synovial fluid and serum samples suggest increased nitric oxide synthesis in rheumatic diseases. *Ann. Rheum. Dis.* 51 (11): 1219–1222.

119 Shahid, M. and Kundra, R. (2017). Platelet-rich plasma (PRP) for knee disorders. *EFORT Open Rev.* 2 (1): 28–34.

120 Lee, G.W., Son, J.H., Kim, J.D., and Jung, G.H. (2013). Is platelet-rich plasma able to enhance the results of arthroscopic microfracture in early osteoarthritis and cartilage lesion over 40 years of age? *Eur. J. Orthop. Surg. Traumatol.* 23 (5): 581–587.

121 Napolitano, M., Matera, S., Bossio, M. et al. (2012). Autologous platelet gel for tissue regeneration in degenerative disorders of the knee. *Blood Transfus.* 10 (1): 72–77.

122 Sánchez, M., Guadilla, J., Fiz, N., and Andia, I. (2012). Ultrasound-guided platelet-rich plasma injections for the treatment of osteoarthritis of the hip. *Rheumatology (Oxford)* 51 (1): 144–150.

123 Sampson, S., Reed, M., Silvers, H. et al. (2010). Injection of platelet-rich plasma in patients with primary and secondary knee osteoarthritis: a pilot study. *Am. J. Phys. Med. Rehabil.* 89 (12): 961–969.

124 Filardo, G., Kon, E., Buda, R. et al. (2011). Platelet-rich plasma intra-articular knee injections for the treatment of degenerative cartilage lesions and osteoarthritis. *Knee Surg. Sports Traumatol. Arthrosc.* 19 (4): 528–535.

125 Kon, E., Mandelbaum, B., Buda, R. et al. (2011). Platelet-rich plasma intra-articular injection versus hyaluronic acid viscosupplementation as treatments for cartilage pathology: from early degeneration to osteoarthritis. *Arthroscopy* 27 (11): 1490–1501.

126 Marmotti, A., Rossi, R., Castoldi, F. et al. (2015). PRP and articular cartilage: a clinical update. *Biomed Res. Int.* 2015: 542502.

127 Zhang, W., Hsu, P., Zhong, B. et al. (2018). MiR-34a enhances chondrocyte apoptosis, senescence and facilitates development of osteoarthritis by targeting DLL1 and regulating PI3K/AKT pathway. *Cell. Physiol. Biochem.* 48 (3): 1304–1316.

128 Zhang, W., Zhong, B., Zhang, C. et al. (2018). miR-373 regulates inflammatory cytokine-mediated chondrocyte proliferation in osteoarthritis by targeting the P2X7 receptor. *FEBS Open Bio* 8 (3): 325–331.

129 Grol, M.W. and Lee, B.H. (2018). Gene therapy for repair and regeneration of bone and cartilage. *Curr. Opin. Pharmacol.* 40: 59–66.

130 Evans, C.H., Ghivizzani, S.C., and Robbins, P.D. (2018). Gene delivery to joints by intra-articular injection. *Hum. Gene Ther.* 29 (1): 2–14.

131 Bao, C.X., Chen, H.X., Mou, X.J. et al. (2018). GZMB gene silencing confers protection against synovial tissue hyperplasia and articular cartilage tissue injury in rheumatoid arthritis through the MAPK signaling pathway. *Biomed. Pharmacother.* 103: 346–354.

132 Ha, C.W., Cho, J.J., Elmallah, R.K. et al. (2015). A multicenter, single-blind, phase IIa clinical trial to evaluate the efficacy and safety of a cell-mediated gene therapy in degenerative knee arthritis patients. *Hum. Gene Ther. Clin. Dev.* 26 (2): 125–130.

133 Tao, K., Frisch, J., Rey-Rico, A. et al. (2016). Co-overexpression of TGF-β and SOX9 via rAAV gene transfer modulates the metabolic and chondrogenic activities of human bone marrow-derived mesenchymal stem cells. *Stem Cell Res. Ther.* 7: 20.

134 Bougioukli, S., Sugiyama, O., Alluri, R.K. et al. (2018). In vitro evaluation of a lentiviral two-step transcriptional amplification system using GAL4FF transactivator for gene therapy applications in bone repair. *Gene Ther.* 25 (4): 260–268.

135 Evans, C.H., Ghivizzani, S.C., and Robbins, P.D. (2013). Arthritis gene therapy and its tortuous path into the clinic. *Transl. Res.* 161 (4): 205–216.

136 Evans, C.H., Gouze, E., Gouze, J.N. et al. (2006). Gene therapeutic approaches – transfer in vivo. *Adv. Drug Delivery Rev.* 58 (2): 243–258.

13

Biofabrication for Cartilage Regeneration

Sien Lin, Boguang Yang, Zhixian Zong, and Gang Li

Department of Orthopaedics and Traumatology, Stem Cells and Regenerative Medicine Laboratory, Li Ka Shing Institute of Health Sciences, The Chinese University of Hong Kong, Shatin, Hong Kong, P.R. China

13.1 Introduction

Articular cartilage is one of the critical tissues in the musculoskeletal system, supporting smooth joint movement. Articular cartilage represents approximately 60–80 wt% of fluids, and the remainder is composed of type II collagen and glycosaminoglycans (GAGs). Figure 13.1 illustrates the hierarchical structure of articular cartilage. The structure of articular cartilage is usually divided into four distinct zones, i.e. superficial, middle, deep, and calcified cartilage [1]. The superficial zone occupies the upper 10–20% of the cartilage and has a small amount of proteoglycans and low permeability. Cartilage progenitor stem cells have been identified in this superficial zone. Those progenitor stem cells are not only responsible for cartilage repair but also responsible for osteoarthritis (OA) [2]. The next middle zone occupying the following 40–60% down is rich in proteoglycans but has a low number of cells. In addition, the arch-shaped middle zone has oriented collagen fibers and is highly compressive, allowing recovery from articular surface impacts. The deep zone has oriented collagen fibrils and cells perpendicular to the surface of the articular cartilage, and the fibrils are anchored onto the subchondral bone. The deep zone is also highly compressive as with the middle zone but has fewer proteoglycans and the least cells. The osteochondral interface is structurally connected between a hyaline cartilage layer and an underlying bone plate and plays a critical role in maintaining the cartilage integration. The osteochondral interface contains hypertrophic chondrocytes embedded in a mineralized cartilage matrix.

Cartilage is an avascular tissue with minimal supply of nutrients and progenitors from circulation and composed of a limited number of chondrocytes with low mitotic potential, making cartilage a poor self-regenerating tissue in response to injury. In cartilage, nutrients and waste exchange are achieved through synovial fluid perfusion, which also allow the delivery of various factors participating in healing. Scarce resident stem cells in cartilage are identified recently, which

Biofabrication for Orthopedics: Methods, Techniques, and Applications, First Edition.
Edited by Wenguo Cui, Xin Zhao, and Shen Liu.

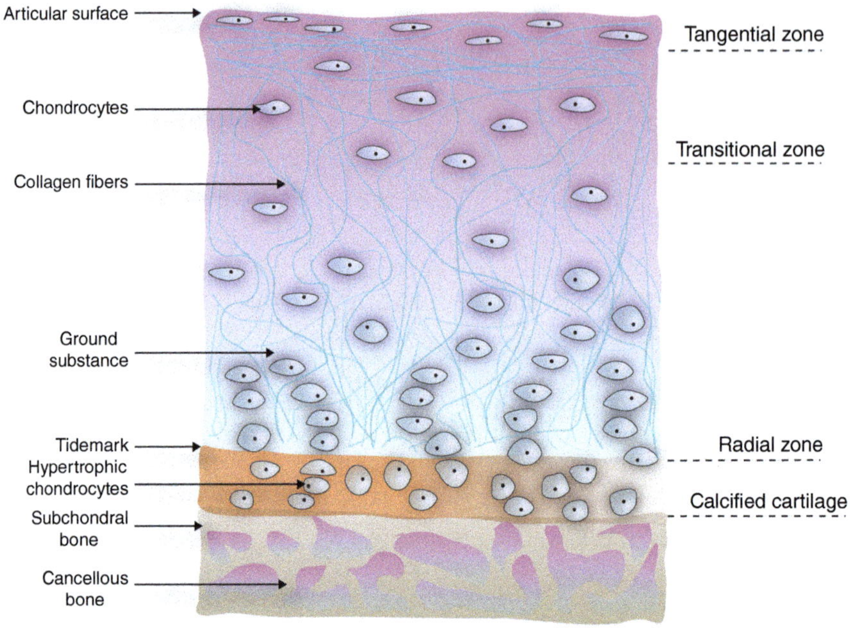

Figure 13.1 Osteochondral tissue structure. Structural/spatial organization of chondrocytes and collagen fibers in noncalcified articular cartilage. Tidemark and subchondral bone layers are shown.

require considerable manipulation efforts to generate cartilage *in vitro* [3]. Tiny defects are healed by migration of chondrocytes, while large defects are healed by formation of biomechanically incompetent fibrocartilage [4]. Hence, cartilage lesions seldom heal spontaneously and thus constitute one of the main causes of joint disease and disability [5, 6]. Given that persistent cartilage defects gradually lead to degeneration of the articular cartilage and osteoarthritis [7], the restoration of cartilage integrity through the promotion of cartilage regeneration has been a research question over the decades.

Cartilage injury (also called chondral injury) is known as the lesion within the cartilage layer, while osteochondral injury is the full-thickness lesion extending to the subchondral bone. Cartilage injury or osteochondral injury is common in sport injuries, road traffic accidents, and other trauma. An epidemiological study on 31 516 knee arthroscopies in the United States reported that 63% of patients had chondral lesions (averaging 2.7 lesions per knee) and 20% had full-thickness lesions, with 5% of these occurring in patients less than 40 years of age and 65% of them were accompanied with meniscal or ligament lesions, mostly anterior crucial ligament (ACL) tear [8, 9]. In subgroup analysis, 75% of young patients below 40 years old had solitary chondral lesions and the remaining 25% had multiple lesions. Another similar study conducted in Poland examining a total of 25 124 knee arthroscopies reported that chondral lesions were found in 60% of these patients. Medial meniscus tear (37%) and ACL injury (36%) were the most frequent associated factors [10].

Primary treatment options including protecting from further injury, ice cooling, and analgesic may help to settle the initial pain and swelling after acute cartilage injury. Further surgical treatments are subjected to the severity of cartilage lesions. Several surgical techniques are readily available to treat cartilage injuries of the knee upon different scenarios [11]. Overall, operations such as arthroscopic lavage, debridement, microfracture, osteochondral transplantation (OT), and autologous chondrocyte implantation (ACI) are most widely used as first-line surgical treatments nowadays encountering to cartilage lesions [12]. These reparative methods are tended to stimulate the formation of new fibrocartilage tissue by facilitating access to the vascular system and bringing new progenitor cells capable of chondrogenesis (e.g. microfracture procedure and drilling). Reconstructive methods fill up the defects with autologous, homologous, or other tissue (e.g. ACI and osteochondral autologous transplantation) [13, 14]. Such methods may associate with good outcomes after surgery, but according to a systematic review of level I and II studies on OT procedures and microfracture surgery, patients with small lesions who returned to higher demand activities had a higher progressive failure rate; only 52% of athletes returned to sports after receiving microfracture surgery, and 37% of them retained their same level of sports 10 years after operation [15, 16]. Besides, another systematic review reported by Filardo et al. revealed that 33.7% failure rate at a mean was a recorded follow-up of 8.5 years after ACI surgeries (5–12 years post-surgery) in 193 patients [17].

13.2 Biological Responses of Cartilage Injury

Articular cartilage homeostasis is a result of an intricate interplay between anabolic and catabolic, anti- and pro-inflammatory, and anti- and pro-apoptotic mediators. Chondrocytes, the only cell type in cartilage representing less than 5% of the tissue volume, are the versatile regulators of cartilage equilibrium [18]. They are highly sensitive and responsive to mechanical, biochemical, and other stressful stimuli. Traumatic cartilage injury leads to an irreversible cartilage loss since differentiated chondrocytes do not divide and hence do not compensate for these defects. Additionally, cartilage homeostasis is disturbed by the predominance of directly and indirectly catabolic acting factors such as pro-inflammatory cytokines, which can result in post-traumatic cartilage breakdown and secondary osteoarthritis (OA) [19]. The particular role of locally acting anti-inflammatory mediators and cytokines in cartilage and their influence on OA pathogenesis remain for the most part unclear. Meanwhile, it becomes more and more evident that, apart from typical immune cells, e.g. dendritic cells, nonimmune connective tissue cell types such as chondrocytes and fibroblasts are also able to produce anti-inflammatory cytokines. The anti-inflammatory cytokine interleukin (IL)-10 is involved in extracellular matrix remodeling by fibroblasts [20, 21]. In cartilage, IL-10 seems to possess a chondroprotective capacity, which makes it interesting as a supportive factor which may prove to be useful in combination with cartilage tissue engineering techniques for cartilage repair [22].

Cartilage trauma activates chondrocytes and synovial fibroblasts. Activated chondrocytes and synovial fibroblasts release pro-inflammatory cytokines such as tumor necrosis factor (TNF)-α, IL-1β, chemokines, as well as multiple catabolic mediators, e.g. matrix metalloproteinases (MMPs) or vascular endothelial growth factor (VEGF) or undergo apoptosis [19, 23–25]. It has been well known for many years that particularly TNF-α and IL-1β play a key role in OA pathogenesis [19, 26–28]: These cytokines are distributed in the joint by the synovial fluid. They directly inhibit cartilage-specific extracellular matrix (ECM) synthesis and induce multiple inflammatory and degradative secondary mediators including MMPs, cyclo-oxygenase (COX)-2, nitric oxide (NO) and aggrecanases such as A Disintegrin and Metalloproteinase with Thrombospondin motifs (ADAMTS) [29]. They also down-regulate tissue inhibitors of metalloproteinases (TIMPs) which protects cartilage from degeneration [29]. Additionally, cartilage trauma induces accelerated chondrocyte apoptosis [19, 25, 30–32].

13.3 Biofabrication: Traditional Cellular Products

As early as 1960s, Chesterman and Smith firstly described the successful transplantation of isolated chondrocytes in rabbits [33]. In 1994, Brittberg et al. reported the first ACI clinical trial in the world [5]. Autologous chondrocytes were harvested from cartilage biopsy samples and cultured, and then, the cells were implanted under a periosteal patch that is secured with resorbable sutures. Periosteal cambium layer cells may promote chondrocyte maturation by releasing appropriate growth factors and mesenchymal stem cells (MSCs) into the local milieu [34]. Modifications of the first-generation technique included the addition of fibrin glue to seal the periosteal patch. Later, second-generation and third-generation techniques have also been developed [35]. Second-generation ACI involves suspending chondrocytes covered with a membrane composed of type I or type III collagens. Third-generation techniques involve an extracellular collagen membrane or matrix to induce cell growth and maturation *in vitro*. These second-generation and third-generation modifications are also described as periosteal autologous chondrocyte implantation (P-ACI), collagen-covered autologous chondrocyte implantation (C-ACI), matrix-associated autologous chondrocyte implantation (MACI), and scaffold-based ACI. The efficiency of the second- and third-generation ACI has been proved by clinical case–control studies [36, 37]. Recently, fourth-generation ACI has been introduced as ACI by culturing autologous chondrocytes in collagen gel scaffolds before implantation. Hence, this cartilage product could be made and cultured in laboratories before implantation. The evolution of ACI is shown in Table 13.1. Among these, MACI has been approved by the US Food and Drug Administration (FDA) on the clinical use of autologous cultured chondrocytes on porcine collagen membranes for the repair of symptomatic, full-thickness cartilage defects of the knee in adult patients in 2016, which is indeed a positive movement to encourage further research on cell-based therapy for different cartilage problems [38].

Table 13.1 Evolution of autologous chondrocyte implantation (ACI).

Generations	Scaffolds	Cells
First generation	Periosteum flap	Suspending chondrocytes
Second generation	Collagen cap	Suspending chondrocytes
Third generation	Collagen membrane, matrix (type I/III collagen)	Chondrocytes seeded onto the membrane or matrix before implantation
Fourth generation	Collagen gel	Tissue-engineered cartilage grown from chondrocytes in collagen gel

Biomaterials for cartilage tissue engineering should meet the primary requirements including biocompatibility, bioactivity, biodegradation, and biomechanical properties [39, 40]. Biocompatibility means the materials do not elicit local or systemic immune responses or foreign body reactions to achieve a good implant–host integration. Biomaterials should be able to mimic the extracellular matrix (ECM) and the microenvironment. The biomaterial chosen should be able to enable cells to attach, grow, and proliferate at a similar natural rate, with an efficient gas and nutrient exchange. In addition, bioactive materials should be able to induce proliferation, migration, or chondrogenic differentiation of stem cells [41]. The biodegradation of materials during *in vivo* treatment should closely match tissue growth rates. Besides the chemical compositions and their functions, mechanical properties are very important for cartilage tissue engineering [39]. The mechanical properties of biomaterials must be able to bear the daily stress to which this tissue is subjected, as well as to support integration of the cells involved. Thus, the choice of the biomaterial for cartilage regeneration is extremely important. Different natural and synthetic materials have been employed to engineer cartilage repair, presenting advantages and limitations [41]. For instance, natural polymers are normally biocompatible and allow the interaction with cell receptors. However, safety concerns are usually an issue by using natural products. In contrast, synthetic materials with defined compositions can recapitulate key chemical and biophysical properties of living tissues, providing an important means to understanding the mechanisms by which cells sense and remodel their surrounding microenvironments [42]. Although biological cues lack in the synthetic materials, recent studies have focused on the biomimetic biomaterials by incorporating biological cues. In biomedical engineering, it is usually not possible to create the required physical and chemical properties in the scaffold using only natural materials or synthetic polymers. Instead, it is preferred that composites, or hybrid materials, such as polymer–polymer blends, polymer–ceramic blends, and co-polymers, be used [43]. In cartilage tissue engineering, scaffolds can be divided into four main groups accordingly to their compositions (Table 13.2). Among these scaffold materials, collagen has been intensively observed as a scaffold material in cartilage regeneration, as it is the most abundant protein in cartilage. Common resources of collagen for cartilage tissue engineering include

Table 13.2 Scaffolds used in autologous chondrocyte implantation (ACI).

Scaffold types	Components	Functions
Protein-based polymers	Collagen, fibrin, and gelatin	Support chondrocyte growth, retain cell phenotypes, and promotes cell adhesion
Carbohydrate-based polymers	Hyaluronan, chitosan, agarose, and polyethylene glycol (PEG)	Maintain chondrocytes shapes and morphology
Synthetic polymers	Polylactic acid (PLA), polyglycolic acid (PGA), polycaprolactone, polylactic-*co*-glycolic acid (PLGA), and polyurethane (PU)	Provide mechanical support
Composite polymers	Mixtures of the components above	Provide mechanical support, support chondrocyte growth, retain cell phenotypes, etc.

bovine tendon/skin, or porcine skin, which was decellularized from allogeneic or xenogeneic tissue.

Over the recent years of progress in tissue engineering, many commercial products for cartilage or osteochondral regeneration have been developed (Table 13.3) [35, 44]. For example, as a second generation of ACI, the MACI membrane is made up of a type I and type III collagen bilayer seeded with chondrocytes. One side of the bilayer has widely spaced collagen fibers, which creates a roughened surface and seeded with chondrocytes, whereas the other side of the bilayer is composed of tightly packed collagen fibers, which creates a smooth surface [45]. During surgery, the membrane was secured with fibrin glue, obviating the need for a sutured membrane. The advantages of MACI include (i) cells distribute evenly throughout the membrane, (ii) it only requires a minimal arthrotomy, and (iii) it acts as a barrier to invasion of the graft by fibroblasts, which would otherwise lead to a fibrous repair [35]. Another collagen-based implant ChondroMimetic™ (TiGenix NV) is an off-the-shelf bilayer (chondral and osseous layer) implant, which is used for the treatment of osteochondral defect. The chondral layer is composed of collagen and GAG, while the osseous layer is calcium phosphates. This implant supports the simultaneous repair of both articular cartilage and bone after implantation in patients underwent standard marrow stimulating procedures [46]. BST-CarGel (Smith & Nephew) has emerged as an advanced biodegradable and injectable chitosan hydrogel mixed with glycerophosphate and autologous blood to improve cartilage regeneration [47]. This off-the-shelf product can be used in conjunction with microfracture by directly mixing blood from the patient with the biomaterials. This BST-CarGel treatment has proved to be efficient in cartilage repair compared with microfracture alone, showing a high significance in clinical benefits over baseline levels in a multicenter-randomized controlled trial [48].

Table 13.3 Commercial products for cartilage or osteochondral regeneration.

Product	Scaffold composition	Applications	Cell sources
NeoCart (Histogenics)	Collagen	Cartilage regeneration (ACI)	Chondrocytes
Novocart 3D (TETEC AG)	Collagen	Cartilage regeneration (ACI)	Chondrocytes
Carticel (Genzyme/Vericel)	None	Cartilage regeneration (ACI)	Chondrocytes
MACI (Vericel)	Collagen	Cartilage regeneration (ACI)	Chondrocytes
ChondroCelect (TiGenix)	Collagen	Cartilage regeneration (ACI)	Chondrocytes
BioSeed-C (BioTissue)	Collagen	Cartilage regeneration (ACI)	Chondrocytes
ChondroCushion (Advanced Bio-Surfaces)	Polyurethane	Cartilage regeneration (ACI)	Chondrocytes
TruFit Plug (Smith & Nephew)	Biphasic scaffold containing polylactic-*co*-glycolic and calcium sulfate	Osteochondral regeneration	Stem cells by marrow stimulation
Chondrotissue (BioTissue)	Poly(glycolic acid) (PGA) and hyaluronan	Cartilage regeneration	Stem cells by marrow stimulation
BST-CarGel (Smith & Nephew)	Chitosan gel, glycerophosphate, and autologous blood	Cartilage regeneration	Stem cells by marrow stimulation
OsseoFit® plug (Smith & Nephew)	Type I collagen and 80% β-tricalcium phosphate (TCP) + 20% polylactic acid (PLA)	Osteochondral regeneration	Stem cells by marrow stimulation
TruFit (Smith & Nephew)	Calcium sulfate, poly(lactic-*co*-glycolic) acid/PGA (PLGA/PGA)	Osteochondral regeneration	Stem cells by marrow stimulation
Novocart Basic (B. Braun)	Collagen-based matrix	Osteochondral and cartilage regeneration	Stem cells by marrow stimulation
Agili-C (CartiHeal)	Calcium carbonate and aragonite with hyaluronic acid	Osteochondral regeneration	Stem cells by marrow stimulation
Collagraft (Nuecoll Inc.)	Collagen with granules of hydroxyapatite and β-TCP	Osteochondral regeneration	Stem cells by marrow stimulation
MaioRegen (Med&Care)	Type I collagen and magnesium-enriched HA	Osteochondral regeneration	Stem cells by marrow stimulation

(continued)

Table 13.3 (Continued)

Product	Scaffold composition	Applications	Cell sources
HYAFF® 11 (Anika Therapeutics)	Hyaluronic acid	Osteochondral and cartilage regeneration	Stem cells by marrow stimulation
SaluCartilage (SaluMedica)	Polyvinyl alcohol (PVA) hydrogel	Cartilage regeneration	Stem cells by marrow stimulation
ChondroMimetic (TiGenix)	Biphasic collagen/glycosaminoglycan and calcium phosphates	Osteochondral regeneration	Stem cells by marrow stimulation
Gel-One (Zimmer)	Hyaluronic acid	Osteoarthritis	None
BioCartilage (Arthrex)	Demineralized bone matrix (DBM)	Cartilage regeneration	Stem cells by marrow stimulation

Although new generations of ACI have shown their advantages in clinical outcomes, there are many limitations, including time consuming in the preparation of the cell-seeded implant, dedifferentiation phenotype of the chondrocytes after *in vitro* expansion, which leads to the formation of fibrocartilage [38]. An alternative cell source such as MSCs has been found. MSCs were first isolated from the stromal compartment of bone marrow, which could differentiate into bone tissue and bone marrow stroma *in vivo* by Friedenstein and coworkers [49]. MSCs were later isolated from many other tissues, including adipose tissue, skeletal muscle, tendon, meniscus, umbilical cord blood, Wharton's jelly, etc. These cells show multilineage differentiation potential, which can differentiate into bone, fat, cartilage, and muscle cells *in vitro*. The International Society for Cellular Therapy (ISCT) has proposed the criteria to classify MSCs, which includes fibroblast-like morphology; plastic-adherent property under standard culture conditions; differentiation potential into osteoblasts, adipocytes, and chondroblasts *in vitro*; expression of surface markers including cluster of differentiation 105 (CD105), CD73, and CD90; and lack of expression of CD45, CD34, CD14, or CD11b, CD79a, or CD19, and human leukocyte antigen – DR isotype (HLA-DR) [50]. MSCs are regarded as alternatives to chondrocytes, showing chondrogenic differentiation potential induced by specific growth factors and low immunogenicity, favoring allogeneic implantation [38]. Besides cartilage regeneration, MSCs are also applied in osteochondral regeneration, by promoting chondrogenic differentiation at the top layer, and osteogenic differentiation at the bottom layer when seeded in biphasic scaffolds.

13.4 Biofabrication: Bioprinting

Three-dimensional (3D) bioprinting is a tissue engineering fabrication method that uses spatial patterning of living cells and other biological materials assembling them in a layer-by-layer deposition approach for construction of living tissues and organ analogs [51]. The 3D bioprinting technology allows the fabrication of 3D scaffolds of any shape constructible in a layer-by-layer deposition way by a computer-aided design (CAD) digital model using the geometrical data obtained from the medical imaging techniques such as X-ray imaging, magnetic resonance imaging (MRI), and computerized tomography (CT) scan [52]. This technique integrates biomaterials, live cells, and controlled motor systems for creating complex structures with controlled permeability, mechanical properties, and biological cues. Precise biomimetic distribution of cells and bioactive factors in different zones achieved by 3D printing can provide the right cues to guide cell growth and differentiation, consequently affecting the deposit of extracellular matrix (ECM), and thus ultimately forming a functional tissue.

The principles of the 3D bioprinting process typically include the high resolution of the printed scaffold, small processing time, and compatibility with cells [53]. Parameters to be considered in 3D printing process for tissue engineering include material degradation rate, mechanical strength of the printed structures, morphology and topography of the printed products, and the capability of vascularization of the formed tissue [53]. One of the biggest challenges of 3D printing techniques is the limited types of bioink that can be used, which have led to thorough research on the development of materials for 3D printing.

Various 3D printing techniques have been developed, each associated with its own range of operating parameters and compatible selection of materials (Table 13.4). Current 3D printing techniques mainly include extrusion-based bioprinting (EBB),

Table 13.4 Working mechanisms, advantages, and disadvantages of different kinds of bioprinters.

	Extrusion-based bioprinting (EBB)	Droplet-based bioprinting (DBB)	Laser-based bioprinting (LBB)
Working mechanisms	Pressure-based extrusion through nozzle	Inkjet, acoustic-droplet-ejection, and microvalve to generate a droplet	Laser-based modalities to initiate a droplet release
Advantages	Greater deposition and printing speed	Generate highly complex heterocellular tissue constructs with different compositions of biologics, reasonable resolution and speed, and highly versatile and affordable	Avoid mechanical stress by eliminating the direct contact with the printer, high resolution, and increased precision
Disadvantages	Lower resolution and precision and higher solidification requirements	Clogging of orifice during printing and higher solidification requirements	Higher costs of the system and limited applications

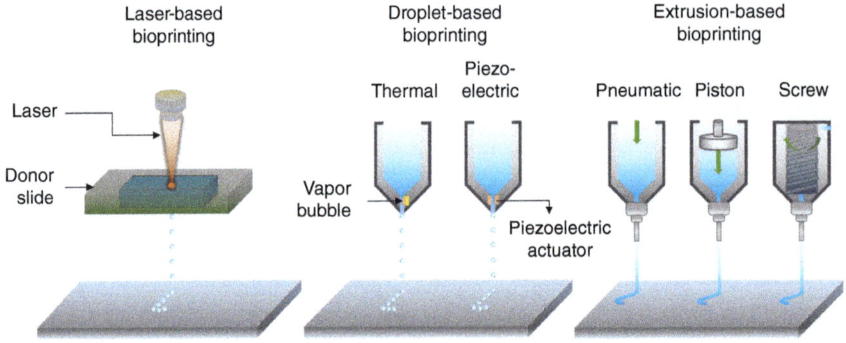

Figure 13.2 Current 3D printing techniques including laser-based, inkjet-based, or extrusion-based bioprinting (robotic dispensing).

droplet-based bioprinting (DBB), and laser-based bioprinting (LBB) (Figure 13.2) [51, 54]. EBB is a pressure-based bioprinting method that has grown in use over the past decade. Using a robotic system linked to a fluid-dispensing array, cells can be deposited in a 3D-shaped structure based on CAD modeling. This process is done by using the shear-thinning behavior of the bioink and is generally plotted in cylindrical lines through an attached nozzle. Bioinks in different forms, such as hydrogel, micro-carriers, and tissue spheroids, can be used in EBB systems, which could be extruded in a near solid state. Different solidification methods such as photocrosslinking, temperature change, and pH change can be adopted during material deposition to maintain the structure after extrusion [53]. EBB allows the deposition of cell-laden filaments and is regarded as the most suitable technique for the 3D bioprinting of variable constructs in different sizes. For the printing of cartilage constructs, EBB techniques are most often considered [55].

DBB comprises inkjet, acoustic-droplet-ejection (or simply acoustic), and micro-valve bioprinting [56]. Inkjet bioprinting physically manipulates a bioink solution to generate droplets. It leverages gravity, atmospheric pressure, and the fluid mechanics of the bioink solution to eject droplets onto a receiving substrate. Inkjet printing is further subclassified into continuous, drop-on-demand, and electrohydrodynamic printing methods [51]. Continuous inkjet printing forces the bioink under pressure through a nozzle, which then breaks into droplets as the potential energy is diminished. Drop-on-demand printing works to deliver single drops on demand by using a pressure pulse to push a drop through a nozzle that is held into place by surface tension. These pressure pulses are variable based on the system and can be piezoelectric, electrostatic, or thermal in nature. Electrohydro-dynamic jet bioprinting uses an electric field to pull the bioink droplets onto the substrate, which again creates the droplet by disrupting the surface tension at the nozzle tip. When the generation of the electric field overcomes the surface tension, the droplet is ejected. Acoustic bioprinting eject droplets from a pool by generation of an acoustic field. Microwave bioprinting uses an electromechanical valve to control droplet release. This process occurs by the generation of a magnetic field with the use of a solenoid coil that ejects bioink droplets from the gated microvalve.

To maintain desired phenotypes of chondrocytes or progenitor cells such as MSCs and to guide the cells to form a functional cartilage tissue structure, 3D bioprinting has many advantages in the field of cartilage tissue engineering [53]. Previous studies have shown the importance and success of using 3D scaffolds in supporting early formation of cartilage tissue, thus justifying the need for a strategy to fabricate precise and accurate 3D structures for cartilage regeneration applications [57, 58]. A development of biomaterials (bioinks) for 3D printing also enables the fabrication of structures that ultimately exhibit relevant properties of articular cartilage and can recapitulate the complexity of the ECM that is present *in vivo* because of the influence of biomechanical load and the microenvironment [52]. Combined with advanced imaging techniques, 3D printing could serve as a powerful tool to produce personalized patient-specific scaffolds for precise medicine applications.

Bioink plays a central role in 3D bioprinting. This bioink should be highly biocompatible to accommodate live cells, mechanically stable after printing, and it should provide high resolution during printing. The ideal bioink formulation should satisfy certain physical and biological requirements. The physical properties include printability, mechanical properties, biodegradation, modifiable functional groups on the surface, and post-printing maturation [52]. The printability of the bioink depends on the different parameters such as the viscosity of the solution, the surface tension of the bioink, the ability to cross-link on its own, and the surface properties of the printer nozzle. It should have the ability to self-retain the 3D printed structure after printing. The printing reliability and the live cell encapsulation highly depend on the hydrophilicity and the viscosity of the bioink solution. The hydrophilic surface of the bioink after printing provides good cell adhesion and a nontoxic environment for chondrocyte proliferation. The hydrophilic system could also allow incorporation of bioactive molecules such as growth factors or small molecules for cartilage tissue fabrication applications. The viscosity should be optimized to make it not only smoothly extrude from the nozzle during printing but also yield a stable structure after printing [52]. Different printers require different viscosities of bioink. Extrusion and droplet-based printing require high-viscous bioink formulations to generate a characteristic shear thinning property to compensate the high shear stress developed during the printing. The printed structure needs enough stiffness to retain the 3D structure as well as it should support the direct cellular behaviors. Biological requirements mainly include biocompatibility (not only nontoxic to the other tissues/cells but also live cells' viability inside bioink), cytocompatibility, and bioactivity of cells after printing [59]. The bioink materials should facilitate cell attachment, growth, and proliferation inside the 3D construct. In addition, it should be convenient to modify the functional groups of the biomaterials to include and deliver different biochemical signals or biomolecules. Apart from these properties, another important requirement is the stiffness of the print substrate, which directly affects the cell survival [60]. A summary of published bioinks is shown in Table 13.5.

Among the different biomaterials, hydrogels are most prominent materials, which are used as bioink in 3D bioprinting. This is mainly because of their ability to hold live cells, modifiable chemical structures, adjustable mechanical and biodegradation properties, and it can yield a good resolution during printing [52]. In the cartilage

Table 13.5 Bioinks and the related 3D printers in cartilage tissue engineering.

Polymer	Gelation/solidifying method	3D printer	Cell type	Comments	Reference
Collagen/alginate	Ionic at RT	3D Bioplotter system (Regenovo, Zhejiang, China)	Primary rat chondrocytes	Low mechanical stability, high cost, gelation slow, and clogging	Yang X. et al. [58]
Pluronic F127	Photopolymerization at 37 °C	BioFactory (REGENHU, Villaz-St-Pierre, Switzerland)	Bovine chondrocytes	Reversible polymer, synthetic	Müller M. et al. [61]
GelMA and Gellan gum	Photopolymerization at 15–37 °C	BioScaffolder dispensing system (SYS+ENG, Salzgitter-Bad, Germany)	Equine chondrocytes	Photopolymerization may affect cell and delayed gelation	Mouser V.H. et al. [62]
GelMA, PCL	Photopolymerization at RT	REGENHU 3D Discovery Bioprinter (REGENHU, Villaz-St-Pierre, Switzerland)	Sheep chondrocytes	Combination of mechanical and biological property in a single scaffold	Ruiz-Cantu L. et al. [63]
Alginate-gelatin	Combined enzymatic and ionic cross-linking	BioBot Basic 3D printer (Advanced Solutions, USA)	Human nasoseptal chondrocytes	Chondrocytes maintain their cartilage-specific phenotype	Schwarz S. et al. [64]
Hyaluronic acid -methacrylic acid (MA)	Ionic at RT	Custom-made robotic-assisted 3D bio-printing system	None	Highly appropriate for improving surgical procedure and promoting cartilage regeneration	Ma K. et al. [65]
Poly(vinyl alcohol) (PVA)–graphene oxide (GO)–hydroxyapatite	Solidifying freezing platform (−10 °C)	Bio-Architect®-Pro 3D Bioprinter (Regenovo, China)	None	The printed biomimetic gradient samples possessed suitable compressive and tribiological properties	Meng Y. et al. [66]

Water-based light-cured polyurethane (PU) and hyaluronic acid	Blue light digital stereolithography to cure individual layers and post-cured under UV light	MiiCraft DLP 3D printer (Young Optics, Taiwan)	Human Wharton's jelly mesenchymal stem cells	High cytocompatibility and mechanical properties favor chondrogenic differentiation	Shie M.Y. et al. [67]
PCL, PLA, and PLGA	Solidify at RT after extruding	3D bioprinter, (REGENHU, Switzerland)	Bone marrow-derived stem cells (BMSCs), chondrocytes, and fat pad-derived stem cells (FPSCs)	Support the development of vascularized endochondral bone overlaid by phenotypically stable cartilage	Critchley S. et al. [68]
Acrylonitrile butadiene styrene (ABS) and PLA	Solidify at RT after extruding	H-series H479 desktop 3D printer (Afinia 3D, USA)	Primary articular chondrocytes and nucleus pulposus cells	Support chondrocytes and NP cells proliferation	Rosenzweig D.H. et al. [69]
Collagen from rat tail tendon	Temperature controlled polymerization (37 °C)	Fab@Home 3D printer (Seraph Robotics, USA)	Fibrochondrocyte	Mechanically stable and were able to support and maintain cell growth	Rhee S. et al. [70]
Sodium alginate, $CaCl_2$, poly(ethylene glycol) diacrylate (PEGDA), I-2959; and hyaluronic acid -methacrylic acid (MA)	UV photopolymerization during printing	A modified 3D printer (Bio-Architect, Regenovo, China)	None	Bone and cartilage defects were restored perfectly *in situ* using 3D bioprinting	Li L. et al. [71]

tissue engineering field, hydrogels have been considered as a promising scaffold material for cell encapsulation because it offers a 3D environment for cell proliferation and differentiation and better diffusion of hydrophilic substrates such as growth factors and nutrients for living cells. In 3D bioprinting, a macroscale-printed porous structure is usually required in a hydrogel scaffold to facilitate the intrinsic diffusion. The polymers that are obtained as a biomaterial from natural resources are called as natural biomaterials in the biomedical fields. These natural materials have varying advantages over synthetic materials, primarily related to the biomimicking of ECM composition or structure, self-assembling ability, biocompatibility, and biodegradation properties. Various natural biomaterials such as alginate, collagen, agarose, and hyaluronic acid-based hydrogels are frequently used as bioinks. Natural materials such as fibrin, cellulose, silk, and ECM are also used as bioinks [52]. However, synthetic polymers provide their own advantageous properties that are not present in natural polymers such as controllability of mechanical stability, photocrosslinking ability, pH, temperature responses, etc. Chemical modification of the hydrogel is a common way to improve or expand its suitability for 3D printing cartilaginous tissue by enabling the materials to cross-link with other materials, thereby improving its mechanical strength or biological potential [53]. In addition, desirable enhancements in mechanical stability and printability may be achieved by the composite inks with two or more materials of different phases with significantly different properties [72].

Alginate is a natural biopolymer attained from brown algae. Alginates are negatively charged polysaccharides, which do not elucidate or provoke much inflammatory response when implanted *in vivo*. Alginate polymer has two monomers as a repeating unit, namely, (1–4)-β-D-mannuronic acid and α-L-guluronic acid. α-L-guluronic acid helps the gel formation, whereas (1–4)-β-D-mannuronic acid and a combination of (L→4)-β-d-mannuronic acid and α-L-guluronic acid aid in increasing the flexibility of the material [73, 74]. The alginate biopolymers can entrap water and other molecules by using capillary forces and still can allow it to diffuse from inside out. This characteristic property is ideal for 3D bioprinting bioinks [74, 75]. It was widely used in 3D bioprinting cartilage tissue engineering applications and showing encouraging results [76, 77]. Previous study reported that alginate best supported the development of hyaline-like cartilage by 3D bioprinting when compared to gelatin methacrylamide (GelMA) and BioINK™ [77]. Besides, it can be combined with other components for printing. Müller et al. used sodium alginate in combination with nanocellulose for 3D-printed cartilage [78]. It was also reported that the alginate-based 3D-bioprinted scaffold overlaid with agarose and the calcium salt of polyphosphate compounds significantly accelerated cell proliferation [79].

Collagen is a main component of ECM, which is obtained from natural biomaterials [80, 81]. Collagen has been used as a bioink material in 3D bioprinting either alone or in combination because of its excellent biocompatible properties. This biopolymer can be cross-linked using temperature or pH change or chemicals. Collagen cross-linking provides them with increased tensile strength and viscoelastic properties than the noncross-linked collagen [82]. However, the cross-linking or gelation of collagen requires a minimum of 30 minutes for gelation at 37 °C. Hence, usage of collagen directly in 3D printing is tough, and thus, combining with different other gelation materials may help to address this issue. Further, the mechanical properties of the collagen materials can be increased by adding different polymers in various proportions for using it in 3D bioprinting [58, 82]. Yang et al. [58] used collagen with sodium alginate as a bioink to develop 3D constructs with chondrocytes. They showed the effectiveness of the combination to suppress the dedifferentiation of the incorporated chondrocytes to any other cell phenotype and facilitated more cell attachment and proliferation. The results showed improved mechanical properties of the printed construct. Overall, they suggested that a combination of collagen and alginate can be preferred for cartilage tissue engineering applications.

Large molecules that are constituents of the native cartilage matrix, such as hyaluronic acid, can be used as promising biological cues for cartilage regeneration. HA is one of the prominent biomaterials that are used in 3D bioprinting for developing 3D structures. Many different blends of HA-based bioinks are reported till now. One of the works explains about the photo-cross-linked HA as a bioink to obtain increased rheological properties by using chemical modifications [83]. Like other natural polymers, HA has low mechanical properties and slow gelation behavior, considering the synthetic polymer hydrogels [84]. HA was combined with different synthetic polymers to obtain more stable structures with high cell viability. They demonstrated the ability of the hybrid 3D-printed structures to enhance chondrogenesis using a thiol-linked HA/polyglycidol gel with poly(ε-caprolactone) (PCL). Both chemical and photo-cross-linking were used in this study to enhance the functional properties of the bioinks [85]. The cell adhesion was enhanced by the addition of cell-adhesive oligopeptides in the hydrogels [86].

Synthetic materials that can be modified with photocrosslinkable side chains such as GelMA are good candidates for ultraviolet (UV)-based gelation process. GelMA and PCL composites have been shown to support cartilage tissue formation using chondrocytes, although this bioink appears to support the development of a more cartilaginous type of tissue with higher levels of collagen type II production [63]. The addition of hyaluronic acid and chondroitin sulfate to GelMA can also enhance chondrogenesis [87]. In addition, the incorporation of these components has the added benefit of increasing the bioink viscosity, which can enhance printability. In a similar approach, a synthetic thermosensitive hydrogel composed of a base-methacrylated poly *N*-(2-hydroxypropyl)methacrylamide (HPMA)-lac-polyethylene glycol (PEG) triblock copolymer was chemically combined with either methacrylated chondroitin sulfate (CSMA) or methacrylated hyaluronic acid (HAMA) for cartilage bioprinting [88].

Although there are many types of hydrogel materials for bioprinting, the current common bioprinting hydrogel materials still have some drawbacks. Most bioprinting hydrogel precursor gelation involves photopolymerization to form the elastic gel construct. In this case, it is particularly necessary to study and control the kinetics of the gelation reaction precisely; otherwise, the accumulated UV radiation may cause cytotoxicity to the cells. Bioinks can be cured in some alternative methods that do not involve exposure to light-activated reactions. For example, use temperature-sensitive bioinks, which are designed to perform temperature-induced phase changes (sol–gel) within a cell-compatible temperature range. Another option is to co-inject the polymer with a solution of cross-linker using a mixing injector or a coaxial injector [89]. This method involves the design of bioinks that undergo cross-linking from a two-part solution, with the additional requirement being a very rapid gelation reaction of the hydrogel precursor. Self-healing, shear-thinning biopolymers are ideal candidates for bioprinting, which can undergo phase transitions associated with higher shear as they are transported though the injector needle [86, 90–95]. HA functionalized with either hydrazide or aldehyde groups can form hydrogels by mixing the two constituents and shear thinning and self-healing was achieved because of the dynamic hydrazone bond [95]. In addition, the design of the interpenetrating double network can be used for photochemical cross-linking after injecting the bioink. This approach was effectively used to 3D print constructs with fibroblasts that retain >80% viability [95].

The new synthetic hydrogel uses more complex chemical modifications to achieve the goal of injectability. Although many studies provide insights into the mechanical properties of novel materials, cell response, biodegradability, and mechanical properties are all obtained through highly controlled laboratory settings, based on idealized gel conditions. Clinically destined cell-encapsulating hydrogel devices formulated using new cross-linking paradigms will need more elaborate testing for *in vivo* toxicity, genotoxicity, carcinogenicity, and biodegradation. Hydrogels cross-linked by light-activated reactions are subjected to scrutiny in terms of compliance with safety and reproducibility standards for medical products [96].

For bioink formulations, large quantities of materials are often needed to print tissues and organs on a physiologically relevant scale. Production and scale-up consideration will need to be addressed in order to make advances. In addition, as final hydrogel products make their way toward specific clinical applications, their "ease of use" will need to be taken into consideration. For example, photochemistry requires external light sources, an additional device in the operating theater, and calibration of this device before each use, in order to ensure proper light dosage. Undoubtedly, as the field moves from discovery phase toward the application of these materials as clinical products, much effort is needed to improve manufacturing, raw material supply, ease of use, and rigorous clinical testing reviews. As polymer chemistry and material design strategies continue to become more cell compatible, the modifications that are required for injectability will produce new biosynthetic hydrogel variants with higher clinical potential and a more pronounced clinical impact [96].

13.5 Biofabrication: Electrospinning

Electrospinning is a highly versatile technique that produces ultrafine fibers with a diameter in the nano- to micro-meter range by using electrostatic fields [97]. It has become popular in a wide range of biomedical and industrial applications, as it can produce fibrous mats with controlled orientations, sizes, porosity, mechanical properties, and high surface area-to-volume ratio. Until now, numerous natural, synthetic, and composite polymers have been electrospun and assessed for cartilage engineering; the critical issues for all of them are their compositional, structural, mechanical, degradation, and biocompatibility properties [98].

The degradation products of natural polymers, such as collagen, chitosan, alginate, gelatin, or hyaluronic acid, can be smoothly eliminated from the body, and that is why, they have been used extensively in cartilage repair and regeneration. However, their degradation by the harsh solvents used in the electrospinning process [99, 100] requires heavy cross-linking to stabilize them, which is frequently associated with cytotoxicity *in vitro* and foreign-body response *in vivo* [101], their fast degradation rate for a tissue that has slow recovery time, and their potential immune responses and microbial/viral contaminants have restricted the use of natural polymers in the fabrication of electrospun scaffolds for cartilage engineering [102, 103]. Synthetic polymers are in general stronger than natural polymers, can withheld the electrospinning process without any noticeable losses, and offer controllable biodegradability [104–106]; for these reasons, synthetic and composite polymers are extensively used in cartilage engineering (Table 13.6). Poly(α-hydroxy esters) (e.g. poly(glycolic acid) [PGA], poly(lactic acid) [PLA], PCL, poly(D,L-lactide-*co*-glycolide) [PLG], and their copolymers) are used extensively for tissue engineering applications, as they are well characterized and FDA approved for clinical use.

Although the use of PLA and PGA is limited for hard tissue regeneration, such as cartilage tissue, because of their relatively weak mechanical properties [104], one study showed that bidirectionally aligned and layered PLA electrospun mats loaded with human meniscus cells in an ECM hydrogel displayed five-fold higher tensile modulus to the randomly aligned scaffolds; they had comparable tensile modulus to the human meniscus in the circumferential direction, and they maintained physiological meniscus cell gene expression for COLA1A1, SOX9, and COMP [122]. In terms of cell proliferation, poly-L-lactic acid (PLLA) scaffolds promoted the highest rate of proliferation between all polymers when seeded with chondrocytes and human bone marrow derived stem cells (BMSCs) [123]. Because of its relatively cheap cost, high stability in ambient conditions, long degradation rate, and the long regeneration time of cartilage tissue, PCL is favored in cartilage repair and regeneration, with numerous studies having demonstrated that electrospun PCL scaffolds promote cartilage cell proliferation, cartilage ECM synthesis and deposition, and chondrogenic differentiation of various stem cell populations. Further, PCL nanofibrous scaffolds have shown higher chondrogenic differentiation, as judged by sulphated glycosaminoglycans (sGAG) synthesis, of BMSCs than cell pellet cultures in transforming growth factor (TGF)-β1 serum-free media [124, 125].

Over the years, numerous studies have assessed the influence of architectural features (e.g. fiber orientation, fiber diameter, and scaffold porosity) on cell fate. It has been shown that both aligned and random PLLA/PCL [126] and PCL/polylactic-*co*-glycolic acid (PLGA) [127] scaffolds support nasal septum-derived progenitors and human BMSCs, respectively, in their adhesion, proliferation, and chondrogenesis. However, their proliferation was higher on the

Table 13.6 Electrospun synthetic polymers used in cartilage engineering (with animal study).

Polymer	Solvent	Fiber diameter	Fiber orientation	Cell type	Animal model	Reference
PCL	Dimethylformamide/tetrahydrofuran	300–1 500 nm	Random	Swine chondrocytes Human BMSCs	Swine	Li W.J. et al. [107]
PLG	Dichloromethane	3 000–14 000 nm	Random	None	Rabbit	Toyokawa N. et al. [108]
Collagen type I/poly(L-lactide-co-ε-caprolactone) (PLCL)	Hexafluoroisopropanol	237 ± 65 nm	Random	Rabbit chondrocytes	Nude mice	He X. et al. [109]
PVA/PCL	Chloroform/dimethyl-formamide	300–800 nm	Random	Rabbit BMSCs	Rabbit	Shafiee A. et al. [110]
PCL/gelatin	Dichloromethane/dimethylbenzene/Span 20/formic acid/ethyl ester	305 ± 72 nm	Random	Mouse iPSCs	Rabbit	Liu J. et al. [111]
PVA methacrylate-chondroitin sulfate methacrylate	Water	410–500 nm	Random	Goat BMSCs	Rabbit	Coburn J.M. et al. [112]
Gelatin/PLLA	Methylene chloride/dimethylformamide	222 ± 14 nm	Random	Rabbit chondrocytes	Rabbit	Chen J.P. and Su C.H. [113]
PCL/collagen I/hyaluronic acid/tricalcium phosphate	Chloroform	6 480 ± 1640 nm	Aligned	Human BMSCs	Rabbit	Liu X. et al., 2014 [111]

PLGA/3,4,6-*O*-Bu_3GlcNAc	Dichloromethane	2–20 μm	Random	Human chondrocytes	Rat	Kim C. et al. [114]
PLA/gelatin/hyaluronic acid	Hexafluoroisopropanol	Non specified	Random	Rat chondrocytes	Rabbit	Chen W. et al. [115]
PCL/PLA	Chloroform/ethanol	1 430–3 160 nm	Random	None	Rabbit	Islas-Arteaga N.C. et al. [116]
PCL/PEO	Tetrahydrofuran/dimethylformamide/ethanol	471 ± 133 nm	Aligned	Rabbit synovial stem cells	Rabbit	Shimomura K. et al. [117]
PLA/gelatin/resveratrol	Hexafluoroisopropanol	200–2 200 nm	Random	None	Rat	Yu F. et al. [118]
PLLA/polydopamine/chondroitin sulfate	Dichloromethane	3–7 μm	Aligned	Rabbit chondrocytes/rabbit BMSCs	Rabbit	Ren X. et al. [119]
PCL/polytetrahydrofuran urethane/collagen I	Hexafluoroisopropanol	444 ± 67 nm	Random	Rat marrow stem cells	Rat	Jiang T. et al. [120]
Gelatin/PLGA	Hexafluoroisopropanol	~1 μm	Random	Rabbit chondrocytes	Nude mice	Chen W. et al. [121]

random scaffolds, while their differentiation was higher on the aligned scaffolds, rendering such conformation suitable for the superficial zone of the articular cartilage that exhibits an aligned orientation. Although both aligned nano- and micro-fibrous electrospun PCL scaffolds sustained the growth of human BMSCs, the nanofibrous scaffolds showed the highest chondrogenic activity, as judged by produced sGAG and collagen type II mRNA expression, suggesting that this combination may be a suitable form for the superficial zone, which normally shows the highest level of collagen type II than the any other zone [128]. Similar results were obtained with nanofibrous, as opposed to micro-fibrous or smooth (film) PLLA [123] or poly(L-lactide-co-D,L-lactide) (PLDLA) [129], scaffolds; the nanofibrous architecture maintained chondrocyte-like morphology and enhanced cartilage-specific mRNA expression and ECM synthesis. One should however note that not only the fiber size but also the pore size has an important role in chondrogenesis. For example, micro-size PLLA fibers of 5 and 9 μm in diameter and with pore sizes of 27 and 29 μm, respectively, were more chondrogenic (e.g. aggrecan, chondroadherin, SOX9, and collagen type II) than nanosize PLLA fibers of 300 and 600 to 1400 nm diameter and with pore sizes of 2 and 3 μm, respectively [130].

13.6 Biofabrication: Microfluidic

Microfluidic technology, which employs fluid flows at the micro-meter scale (that is, flow rates of microliters per minute or less and total volumes of microliters or less), has triggered extensive research in several scientific and engineering disciplines, including diagnostics, cell-based screening technologies, tissue engineering, biosensing, and micro- and nanofabrication [131]. To date, a wide range of microfabrication techniques that are enabled by microfluidics have been utilized in cartilage engineering. In this section, we will discuss the application of various microfluidic platforms to fabricate materials with specific shapes, structures, and/or functions in cartilage regeneration.

In the context of cell microencapsulation for medical and biotechnology applications, the microfabrication of micron and nanoparticles has been widely utilized in cartilage repair research. The microfluidic device provides an environment in which the flow characteristics can be fine-tuned to offer suitable conditions to ensure the uniformity of the formed micro–nanoparticle size. First of all, microfluidic synthesis of nanogels can be employed as a good drug carrier. The microfluidic-synthesized nanogels have a smaller size and higher monodispersity compared to conventional bulk synthesized nanogels. Microfluidic-fabricated uniform-sized nanogels showed a controlled release behavior for TGF-β3, resulting in better chondrogenic differentiation of MSCs [132]. In addition, microfluidic fabrication of macrogels can be applied as a scaffold for cartilage tissue engineering. Wang et al. have developed a new method to make a highly organized alginate scaffold using a microfluidic device. Chondrocytes encapsulated in the scaffold can maintain a normal phenotype, with high expression of proteoglycan and type II collagen. It is shown that

Table 13.7 Microfluidic fabrication used in cartilage tissue engineering.

Polymer	Material platform and size	Function	Cell type	Application	Reference
Alginate	Nanogels; 43–125 nm	Controlled release of cell growth factor	Chondrogenic differentiation of human mesenchymal stem cells (hMSCs)	Cartilage tissue engineering	Mahmoudi Z. et al. [132]
Gelatin norbornene (GelNB)/poly(ethylene glycol) (PEG)	Microgels; 300–600 μm	Cell microencapsulation	Chondrogenic differentiation of hMSCs	Articular cartilage tissue regeneration	Li F. et al. [137]
Alginate	Microgels; 250 μm	Cell microencapsulation	Porcine chondrocytes	Osteoarthritis treatment and cartilage tissue engineering	Wang C.C. et al. [133]
PEG-4-arm-poly(ethylene glycol)-maleimide/PLGA (PEG-4MAL/PLGA) nanoparticles (NPs)	Microgels; 50 μm	Controlled release of small molecules and peptides	Rabbits and human synoviocytes	Treatment of osteoarthritis	Mancipe Castro L.M. et al. [138]
PCL/decellularized extracellular matrix	Scaffold	Bioprinting of cell-laden constructs	Chondrogenic differentiation of MSCs	Cartilage tissue engineering	Pati F. et al. [134]
Polydimethylsiloxane (PDMS)/PEG functionalized with peptides	Organs on chips	Recapitulate organ functions and drug screen	Human chondrocytes	Treatment of osteoarthritis	Occhetta P. et al. [136]

(continued)

Table 13.7 (Continued)

Polymer	Material platform and size	Function	Cell type	Application	Reference
Alginate	Microfluidic tissue scaffolds	Control the chemical environment on a micrometer scale within a macroscopic scaffold could aid in engineering complex tissues	Bovine chondrocytes	Study the response of cells to spatial and temporal variations of soluble factors within well-defined 3D cultures	Choi N.W. et al. [139]
Gelatin methacrylamide (GelMA)/hyaluronic acid methacrylate (HAMA)	Coaxial bioprinting scaffold	Bioprinting of cell-laden constructs	Mesenchymal stem cells	Cartilage regeneration	Di Bella C. et al. [140]
PDMS	Microfluidic platform for 3D cultures	The development of *in vitro* models to screen the effect of different culture conditions on stem/progenitor cells	Human mesenchymal stromal cells	Microfluidic platform for 3D cultures of mesenchymal stem cells toward engineering developmental processes	Occhetta P. et al. [141]
PDMS/agarose	Steering simultaneous osteo/chondrogenesis of stem cells on a chip	Biomicrofluidic techniques to simultaneously steer distinct specialized differentiation of stem cells into chondrocytes and osteoblasts in one hydrogel	Human adipose-derived mesenchymal stem cells	Gradient-generating biomicrofluidic device	Shi X. et al. [142]

a highly organized alginate scaffold can be made with an economical microfluidic device, and this scaffold is effective in cartilage tissue regeneration [133].

Bioprinting of 3D constructs with a precisely controlled architecture can also be obtained by microfluidic fabrication, which have been widely used in cartilage tissue engineering. Pati et al. have developed a method for the bioprinting of cell-laden constructs with novel decellularized extracellular matrix (dECM) bioink capable of providing an optimized microenvironment conducive to the growth of 3D-structured cartilage tissue [134]. Organ-on-a-chip is a multichannel 3D microfluidic cell culture chip that simulates the activities, mechanics, and physiological response of entire organs and organ systems. Recent advances in microfabrication techniques allowed precise control of the 3D architecture of cellular compartments while integrating medium perfusion, the delivery of soluble factors, and accurate biophysical stimulation [135, 136]. These characteristics currently make organ-on-a-chip technology a promising candidate for preclinical drug screening [136]. A newly developed microphysiological cartilage-on-a-chip model was employed by applying strain-controlled compression to 3D articular cartilage microtissues [136]. It is shown that 30% compression recapitulates the mechanical factors of the pathogenesis of osteoarthritis and induces osteoarthritis traits. Using this chip, they also found that high physiological pressure triggers cartilage homeostasis to catabolism and inflammation, hypertrophy, and acquisition of gene expression profiles similar to those observed in clinical osteoarthritis tissues. The cartilage-on-a-chip model may enable the screening of disease-modifying osteoarthritis candidates (Table 13.7).

Acknowledgments

This work was partially supported by grants from the National Natural Science Foundation of China (82172430, 81772322, and 81874000); the University Grants Committee, the Research Grants Council of the Hong Kong Special Administrative Region, China (14121721, 14120118, 14108720, 14202920, C7030-18G, T13-402/17-N, and AoE/M-402/20); the Hong Kong Innovation Technology Commission Funds (PRP/050/19FX); and the Hong Kong Medical Research Funds (16170951, 17180831, and 08190416). This study also received support from the research funds from the Health@InnoHK program launched by Innovation Technology Commission of the Hong Kong SAR, P.R. China.

References

1 Simon, T.M. and Jackson, D.W. (2006). Articular cartilage: injury pathways and treatment options. *Sports Med. Arthrosc. Rev.* 14 (3): 146–154.

2 Pretzel, D., Linss, S., Rochler, S. et al. (2011). Relative percentage and zonal distribution of mesenchymal progenitor cells in human osteoarthritic and normal cartilage. *Arthritis Res. Ther.* 13 (2): R64.

3 Williams, R., Khan, I.M., Richardson, K. et al. (2010). Identification and clonal characterisation of a progenitor cell sub-population in normal human articular cartilage. *PLoS One* 5 (10): e13246.

4 Buckwalter, J.A. (2002). Articular cartilage injuries. *Clin. Orthop. Relat. Res.* 402: 21–37.

5 Brittberg, M., Lindahl, A., Nilsson, A. et al. (1994). Treatment of deep cartilage defects in the knee with autologous chondrocyte transplantation. *N. Engl. J. Med.* 331 (14): 889–895.

6 Caldwell, K.L. and Wang, J. (2015). Cell-based articular cartilage repair: the link between development and regeneration. *Osteoarthritis Cartilage* 23 (3): 351–362.

7 Mithoefer, K., McAdams, T.R., Scopp, J.M., and Mandelbaum, B.R. (2009). Emerging options for treatment of articular cartilage injury in the athlete. *Clin. Sports Med.* 28 (1): 25–40.

8 Curl, W.W., Krome, J., Gordon, E.S. et al. (1997). Cartilage injuries: a review of 31,516 knee arthroscopies. *Arthroscopy* 13 (4): 456–460.

9 Browne, J.E. and Branch, T.P. (2000). Surgical alternatives for treatment of articular cartilage lesions. *J. Am. Acad. Orthop. Surg.* 8 (3): 180–189.

10 Widuchowski, W., Widuchowski, J., and Trzaska, T. (2007). Articular cartilage defects: study of 25,124 knee arthroscopies. *Knee* 14 (3): 177–182.

11 Alford, J.W. and Cole, B.J. (2005). Cartilage restoration, part 1: basic science, historical perspective, patient evaluation, and treatment options. *Am. J. Sports Med.* 33 (2): 295–306.

12 Logerstedt, D.S., Scalzitti, D.A., Bennell, K.L. et al. (2018). Knee pain and mobility impairments: meniscal and articular cartilage lesions revision 2018. *J. Orthop. Sports Phys. Ther.* 48 (2): A1–A50.

13 Bekkers, J.E., Inklaar, M., and Saris, D.B. (2009). Treatment selection in articular cartilage lesions of the knee: a systematic review. *Am. J. Sports Med.* 37 (Suppl. 1): 148s–155s.

14 Mistry, H., Connock, M., Pink, J. et al. (2017). Autologous chondrocyte implantation in the knee: systematic review and economic evaluation. *Health Technol. Assess.* 21 (6): 1–294.

15 Goyal, D., Keyhani, S., Lee, E.H., and Hui, J.H. (2013). Evidence-based status of microfracture technique: a systematic review of level I and II studies. *Arthroscopy* 29 (9): 1579–1588.

16 Goyal, D., Keyhani, S., Goyal, A. et al. (2014). Evidence-based status of osteochondral cylinder transfer techniques: a systematic review of level I and II studies. *Arthroscopy* 30 (4): 497–505.

17 Filardo, G., Andriolo, L., Balboni, F. et al. (2015). Cartilage failures. Systematic literature review, critical survey analysis, and definition. *Knee Surg. Sports Traumatol. Arthrosc.* 23 (12): 3660–3669.

18 Pritzker, K.P. and Gahunia, H.K. (2020). *Articular cartilage: homeostasis, aging and degeneration*. In: *Articular Cartilage of the Knee* (ed. K.P. Pritzker, A.E. Gross, H.K. Gahunia, et al.), 99–122. Springer.

19 Burrage, P.S., Mix, K.S., and Brinckerhoff, C.E. (2006). Matrix metalloproteinases: role in arthritis. *Front. Biosci.* 11: 529–543.

20 Yamamoto, T., Eckes, B., and Krieg, T. (2001). Effect of interleukin-10 on the gene expression of type I collagen, fibronectin, and decorin in human skin fibroblasts: differential regulation by transforming growth factor-β and monocyte chemoattractant protein-1. *Biochem. Biophys. Res. Commun.* 281 (1): 200–205.

21 Moroguchi, A., Ishimura, K., Okano, K. et al. (2004). Interleukin-10 suppresses proliferation and remodeling of extracellular matrix of cultured human skin fibroblasts. *Eur. Surg. Res.* 36 (1): 39–44.

22 Moss, K.L., Jiang, Z., Dodson, M.E. et al. (2020). Sustained interleukin-10 transgene expression following intra-articular AAV5-IL-10 administration to horses. *Hum. Gene Ther.* 31 (1, 2): 110–118.

23 Pufe, T., Harde, V., Petersen, W. et al. (2004). Vascular endothelial growth factor (VEGF) induces matrix metalloproteinase expression in immortalized chondrocytes. *J. Pathol.* 202 (3): 367–374.

24 Kurz, B., Lemke, A.K., Fay, J. et al. (2005). Pathomechanisms of cartilage destruction by mechanical injury. *Ann. Anat.* 187 (5, 6): 473–485.

25 Stolberg-Stolberg, J., Sambale, M., Hansen, U. et al. (2020). Cartilage trauma induces necroptotic chondrocyte death and expulsion of cellular contents. *Int. J. Mol. Sci.* 21 (12): 4204.

26 Saklatvala, J. (1986). Tumour necrosis factor α stimulates resorption and inhibits synthesis of proteoglycan in cartilage. *Nature* 322 (6079): 547–549.

27 Fernandes, J.C., Martel-Pelletier, J., and Pelletier, J.P. (2002). The role of cytokines in osteoarthritis pathophysiology. *Biorheology* 39 (1, 2): 237–246.

28 Nesic, D., Whiteside, R., Brittberg, M. et al. (2006). Cartilage tissue engineering for degenerative joint disease. *Adv. Drug Delivery Rev.* 58 (2): 300–322.

29 Blanco Garcia, F.J. (1999). Catabolic events in osteoarthritic cartilage. *Osteoarthritis Cartilage* 7 (3): 308–309.

30 Borrelli, J. Jr. and Ricci, W.M. (2004). Acute effects of cartilage impact. *Clin. Orthop. Relat. Res.* 423: 33–39.

31 Vrahas, M.S., Mithoefer, K., and Joseph, D. (2004). The long-term effects of articular impaction. *Clin. Orthop. Relat. Res.* 423: 40–43.

32 Martin, J.A. and Buckwalter, J.A. (2006). Post-traumatic osteoarthritis: the role of stress induced chondrocyte damage. *Biorheology* 43 (3, 4): 517–521.

33 Chesterman, P.J. and Smith, A.U. (1968). Homotransplantation of articular cartilage and isolated chondrocytes. An experimental study in rabbits. *J. Bone Joint Surg. Br.* 50 (1): 184–197.

34 Brittberg, M., Sjögren-Jansson, E., Thornemo, M. et al. (2005). Clonal growth of human articular cartilage and the functional role of the periosteum in chondrogenesis. *Osteoarthritis Cartilage* 13 (2): 146–153.

35 Welch, T., Mandelbaum, B., and Tom, M. (2016). Autologous chondrocyte implantation: past, present, and future. *Sports Med. Arthrosc. Rev.* 24 (2): 85–91.

36 Ossendorf, C., Kaps, C., Kreuz, P.C. et al. (2007). Treatment of posttraumatic and focal osteoarthritic cartilage defects of the knee with autologous polymer-based three-dimensional chondrocyte grafts: 2-year clinical results. *Arthritis Res. Ther.* 9 (2): R41.

37 Zak, L., Albrecht, C., Wondrasch, B. et al. (2014). Results 2 years after matrix-associated autologous chondrocyte transplantation using the Novocart 3D scaffold: an analysis of clinical and radiological data. *Am. J. Sports Med.* 42 (7): 1618–1627.

38 Lee, W.Y. and Wang, B. (2017). Cartilage repair by mesenchymal stem cells: clinical trial update and perspectives. *J. Orthop. Translat.* 9: 76–88.

39 Seo, S.J., Mahapatra, C., Singh, R.K. et al. (2014). Strategies for osteochondral repair: focus on scaffolds. *J. Tissue Eng.* 5: 2041731414541850.

40 Zong, Z., Wu, X., Su, Z. et al. (2019). Cartilage tissue engineering: an update on multi-component approach. *J. Orthop. Rheumatol.* 6 (1): 9.

41 Wei, W., Ma, Y., Yao, X. et al. (2021). Advanced hydrogels for the repair of cartilage defects and regeneration. *Bioact. Mater.* 6 (4): 998–1011.

42 Zhang, K., Feng, Q., Fang, Z. et al. (2021). Structurally dynamic hydrogels for biomedical applications: pursuing a fine balance between macroscopic stability and microscopic dynamics. *Chem. Rev.* 121 (18): 11149–11193.

43 Del Bakhshayesh, A.R., Asadi, N., Alihemmati, A. et al. (2019). An overview of advanced biocompatible and biomimetic materials for creation of replacement structures in the musculoskeletal systems: focusing on cartilage tissue engineering. *J. Biol. Eng.* 13: 85.

44 Bicho, D., Pina, S., Reis, R.L., and Oliveira, J.M. (2018). Commercial products for osteochondral tissue repair and regeneration. *Adv. Exp. Med. Biol.* 1058: 415–428.

45 Kon, E., Gobbi, A., Filardo, G. et al. (2009). Arthroscopic second-generation autologous chondrocyte implantation compared with microfracture for chondral lesions of the knee: prospective nonrandomized study at 5 years. *Am. J. Sports Med.* 37 (1): 33–41.

46 Steinwachs, M. and Kreuz, P.C. (2007). Autologous chondrocyte implantation in chondral defects of the knee with a type I/III collagen membrane: a prospective study with a 3-year follow-up. *Arthroscopy* 23 (4): 381–387.

47 Stanish, W.D., McCormack, R., Forriol, F. et al. (2013). Novel scaffold-based BST-CarGel treatment results in superior cartilage repair compared with microfracture in a randomized controlled trial. *J. Bone Joint Surg. Am.* 95 (18): 1640–1650.

48 Shive, M.S., Stanish, W.D., McCormack, R. et al. (2015). BST-CarGel® treatment maintains cartilage repair superiority over microfracture at 5 years in a multicenter randomized controlled trial. *Cartilage* 6 (2): 62–72.

49 Friedenstein, A.J., Chailakhjan, R.K., and Lalykina, K.S. (1970). The development of fibroblast colonies in monolayer cultures of guinea-pig bone marrow and spleen cells. *Cell Tissue Kinet.* 3 (4): 393–403.

50 Dominici, M., Le Blanc, K., Mueller, I. et al. (2006). Minimal criteria for defining multipotent mesenchymal stromal cells. The International Society for Cellular Therapy position statement. *Cytotherapy* 8 (4): 315–317.

51 Dhawan, A., Kennedy, P.M., Rizk, E.B., and Ozbolat, I.T. (2019). Three-dimensional bioprinting for bone and cartilage restoration in orthopaedic surgery. *J. Am. Acad. Orthop. Surg.* 27 (5): e215–e226.

52 Gopinathan, J. and Noh, I. (2018). Recent trends in bioinks for 3D printing. *Biomater. Res.* 22: 11.

53 Guo, T., Lembong, J., Zhang, L.G., and Fisher, J.P. (2017). Three-dimensional printing articular cartilage: recapitulating the complexity of native tissue. *Tissue Eng. Part B* 23 (3): 225–236.

54 Moroni, L., Burdick, J.A., Highley, C. et al. (2018). Biofabrication strategies for 3D in vitro models and regenerative medicine. *Nat. Rev. Mater.* 3 (5): 21–37.

55 Mouser, V.H.M., Levato, R., Bonassar, L.J. et al. (2017). Three-dimensional bioprinting and its potential in the field of articular cartilage regeneration. *Cartilage* 8 (4): 327–340.

56 Gudapati, H., Dey, M., and Ozbolat, I. (2016). A comprehensive review on droplet-based bioprinting: past, present and future. *Biomaterials* 102: 20–42.

57 Hung, K.C., Tseng, C.S., and Hsu, S.H. (2014). Synthesis and 3D printing of biodegradable polyurethane elastomer by a water-based process for cartilage tissue engineering applications. *Adv. Healthcare Mater.* 3 (10): 1578–1587.

58 Yang, X., Lu, Z., Wu, H. et al. (2018). Collagen-alginate as bioink for three-dimensional (3D) cell printing based cartilage tissue engineering. *Mater. Sci. Eng., C* 83: 195–201.

59 Chen, C., Bang, S., Cho, Y. et al. (2016). Research trends in biomimetic medical materials for tissue engineering: 3D bioprinting, surface modification, nano/micro-technology and clinical aspects in tissue engineering of cartilage and bone. *Biomater. Res.* 20: 10.

60 Tirella, A., Vozzi, F., De Maria, C. et al. (2011). Substrate stiffness influences high resolution printing of living cells with an ink-jet system. *J. Biosci. Bioeng.* 112 (1): 79–85.

61 Müller, M., Becher, J., Schnabelrauch, M., and Zenobi-Wong, M. (2015). Nanostructured Pluronic hydrogels as bioinks for 3D bioprinting. *Biofabrication* 7 (3): 035006.

62 Mouser, V.H., Melchels, F.P., Visser, J. et al. (2016). Yield stress determines bioprintability of hydrogels based on gelatin-methacryloyl and gellan gum for cartilage bioprinting. *Biofabrication* 8 (3): 035003.

63 Ruiz-Cantu, L., Gleadall, A., Faris, C. et al. (2020). Multi-material 3D bioprinting of porous constructs for cartilage regeneration. *Mater. Sci. Eng., C* 109: 110578.

64 Schwarz, S., Kuth, S., Distler, T. et al. (2020). 3D printing and characterization of human nasoseptal chondrocytes laden dual crosslinked oxidized alginate-gelatin hydrogels for cartilage repair approaches. *Mater. Sci. Eng., C* 116: 111189.

65 Ma, K., Zhao, T., Yang, L. et al. (2020). Application of robotic-assisted *in situ* 3D printing in cartilage regeneration with HAMA hydrogel: an *in vivo* study. *J. Adv. Res.* 23: 123–132.

66 Meng, Y., Cao, J., Chen, Y. et al. (2020). 3D printing of a poly(vinyl alcohol)-based nano-composite hydrogel as an artificial cartilage replacement and the improvement mechanism of printing accuracy. *J. Mater. Chem. B* 8 (4): 677–690.

67 Shie, M.Y., Chang, W.C., Wei, L.J. et al. (2017). 3D printing of cytocompatible water-based light-cured polyurethane with hyaluronic acid for cartilage tissue engineering applications. *Materials (Basel)* 10 (2): 136.

68 Critchley, S., Sheehy, E.J., Cunniffe, G. et al. (2020). 3D printing of fibre-reinforced cartilaginous templates for the regeneration of osteochondral defects. *Acta Biomater.* 113: 130–143.

69 Rosenzweig, D.H., Carelli, E., Steffen, T. et al. (2015). 3D-printed ABS and PLA scaffolds for cartilage and nucleus pulposus tissue regeneration. *Int. J. Mol. Sci.* 16 (7): 15118–15135.

70 Rhee, S., Puetzer, J.L., Mason, B.N. et al. (2016). 3D bioprinting of spatially heterogeneous collagen constructs for cartilage tissue engineering. *ACS Biomater. Sci. Eng.* 2 (10): 1800–1805.

71 Li, L., Yu, F., Shi, J. et al. (2017). *In situ* repair of bone and cartilage defects using 3D scanning and 3D printing. *Sci. Rep.* 7 (1): 9416.

72 Ravanbakhsh, H., Bao, G., Luo, Z. et al. (2021). Composite inks for extrusion printing of biological and biomedical constructs. *ACS Biomater. Sci. Eng.* 7 (9): 4009–4026.

73 Nakamura, M., Kobayashi, A., Takagi, F. et al. (2005). Biocompatible inkjet printing technique for designed seeding of individual living cells. *Tissue Eng.* 11 (11–12): 1658–1666.

74 Moradali, M.F., Ghods, S., and Rehm, B.H. (2018). *Alginate biosynthesis and biotechnological production*. In: *Alginates and their Biomedical Applications* (ed. B.H. Rehm and M.F. Moradali), 1–25. Springer.

75 Axpe, E. and Oyen, M.L. (2016). Applications of alginate-based bioinks in 3D bioprinting. *Int. J. Mol. Sci.* 17 (12): 1976.

76 Costantini, M., Idaszek, J., Szöke, K. et al. (2016). 3D bioprinting of BM-MSCs-loaded ECM biomimetic hydrogels for *in vitro* neocartilage formation. *Biofabrication* 8 (3): 035002.

77 Daly, A.C., Critchley, S.E., Rencsok, E.M., and Kelly, D.J. (2016). A comparison of different bioinks for 3D bioprinting of fibrocartilage and hyaline cartilage. *Biofabrication* 8 (4): 045002.

78 Müller, M., Öztürk, E., Arlov, Ø. et al. (2017). Alginate sulfate–nanocellulose bioinks for cartilage bioprinting applications. *Ann. Biomed. Eng.* 45 (1): 210–223.

79 Neufurth, M., Wang, X., Schröder, H.C. et al. (2014). Engineering a morphogenetically active hydrogel for bioprinting of bioartificial tissue derived from human osteoblast-like SaOS-2 cells. *Biomaterials* 35 (31): 8810–8819.

80 Rodriguez-Pascual, F. and Slatter, D.A. (2016). Collagen cross-linking: insights on the evolution of metazoan extracellular matrix. *Sci. Rep.* 6: 37374.

81 Ferreira, A.M., Gentile, P., Chiono, V., and Ciardelli, G. (2012). Collagen for bone tissue regeneration. *Acta Biomater.* 8 (9): 3191–3200.

82 Mori, H., Shimizu, K., and Hara, M. (2013). Dynamic viscoelastic properties of collagen gels with high mechanical strength. *Mater. Sci. Eng., C* 33 (6): 3230–3236.

83 Highley, C.B., Prestwich, G.D., and Burdick, J.A. (2016). Recent advances in hyaluronic acid hydrogels for biomedical applications. *Curr. Opin. Biotechnol.* 40: 35–40.

84 Yoo, H.S., Lee, E.A., Yoon, J.J., and Park, T.G. (2005). Hyaluronic acid modified biodegradable scaffolds for cartilage tissue engineering. *Biomaterials* 26 (14): 1925–1933.

85 Stichler, S., Böck, T., Paxton, N. et al. (2017). Double printing of hyaluronic acid/poly (glycidol) hybrid hydrogels with poly (ε-caprolactone) for MSC chondrogenesis. *Biofabrication* 9 (4): 044108.

86 Ouyang, L., Highley, C.B., Rodell, C.B. et al. (2016). 3D printing of shear-thinning hyaluronic acid hydrogels with secondary cross-linking. *ACS Biomater. Sci. Eng.* 2 (10): 1743–1751.

87 Sun, M., Sun, X., Wang, Z. et al. (2018). Synthesis and properties of gelatin methacryloyl (GelMA) hydrogels and their recent applications in load-bearing tissue. *Polymers (Basel)* 10 (11): 1290.

88 Schuurmans, C.C.L., Mihajlovic, M., Hiemstra, C. et al. (2021). Hyaluronic acid and chondroitin sulfate (meth)acrylate-based hydrogels for tissue engineering: synthesis, characteristics and pre-clinical evaluation. *Biomaterials* 268: 120602.

89 Richard, C., Neild, A., and Cadarso, V.J. (2020). The emerging role of microfluidics in multi-material 3D bioprinting. *Lab Chip* 20 (12): 2044–2056.

90 Highley, C.B., Rodell, C.B., and Burdick, J.A. (2015). Direct 3D printing of shear-thinning hydrogels into self-healing hydrogels. *Adv. Mater.* 27 (34): 5075–5079.

91 Zhang, H., Cong, Y., Osi, A.R. et al. (2020). Direct 3D printed biomimetic scaffolds based on hydrogel microparticles for cell spheroid growth. *Adv. Funct. Mater.* 30 (13): 1910573.

92 Highley, C.B., Song, K.H., Daly, A.C., and Burdick, J.A. (2019). Jammed microgel inks for 3D printing applications. *Adv. Sci. (Weinh.)* 6 (1): 1801076.

93 Loebel, C., Rodell, C.B., Chen, M.H., and Burdick, J.A. (2017). Shear-thinning and self-healing hydrogels as injectable therapeutics and for 3D-printing. *Nat. Protoc.* 12 (8): 1521–1541.

94 Song, K.H., Highley, C.B., Rouff, A., and Burdick, J.A. (2018). Complex 3D-printed microchannels within cell-degradable hydrogels. *Adv. Funct. Mater.* 28 (31): 1801331.

95 Wang, L.L., Highley, C.B., Yeh, Y.C. et al. (2018). Three-dimensional extrusion bioprinting of single- and double-network hydrogels containing dynamic covalent crosslinks. *J. Biomed. Mater. Res. Part A* 106 (4): 865–875.

96 Birman, T. and Seliktar, D. (2021). Injectability of biosynthetic hydrogels: consideration for minimally invasive surgical procedures and 3D bioprinting. *Adv. Funct. Mater.* 31 (29): 2100628.
97 Xue, J., Wu, T., Dai, Y., and Xia, Y. (2019). Electrospinning and electrospun nanofibers: methods, materials, and applications. *Chem. Rev.* 119 (8): 5298–5415.
98 Yilmaz, E.N. and Zeugolis, D.I. (2020). Electrospun polymers in cartilage engineering – state of play. *Front. Bioeng. Biotechnol.* 8: 77.
99 Yang, L., Fitié, C.F., van der Werf, K.O. et al. (2008). Mechanical properties of single electrospun collagen type I fibers. *Biomaterials* 29 (8): 955–962.
100 Zeugolis, D.I., Khew, S.T., Yew, E.S. et al. (2008). Electro-spinning of pure collagen nano-fibres – just an expensive way to make gelatin? *Biomaterials* 29 (15): 2293–2305.
101 Delgado, L.M., Bayon, Y., Pandit, A., and Zeugolis, D.I. (2015). To cross-link or not to cross-link? Cross-linking associated foreign body response of collagen-based devices. *Tissue Eng. Part B* 21 (3): 298–313.
102 Schmidt, C.E. and Baier, J.M. (2000). Acellular vascular tissues: natural biomaterials for tissue repair and tissue engineering. *Biomaterials* 21 (22): 2215–2231.
103 Lavik, E. and Langer, R. (2004). Tissue engineering: current state and perspectives. *Appl. Microbiol. Biotechnol.* 65 (1): 1–8.
104 Cheung, H.-Y., Lau, K.-T., Lu, T.-P., and Hui, D. (2007). A critical review on polymer-based bio-engineered materials for scaffold development. *Composites, Part B* 38 (3): 291–300.
105 Zhang, L., Hu, J., and Athanasiou, K.A. (2009). The role of tissue engineering in articular cartilage repair and regeneration. *Crit. Rev. Biomed. Eng.* 37 (1, 2): 1–57.
106 Wen, Y.T., Dai, N.T., and Hsu, S.H. (2019). Biodegradable water-based polyurethane scaffolds with a sequential release function for cell-free cartilage tissue engineering. *Acta Biomater.* 88: 301–313.
107 Li, W.J., Chiang, H., Kuo, T.F. et al. (2009). Evaluation of articular cartilage repair using biodegradable nanofibrous scaffolds in a swine model: a pilot study. *J. Tissue Eng. Regener. Med.* 3 (1): 1–10.
108 Toyokawa, N., Fujioka, H., Kokubu, T. et al. (2010). Electrospun synthetic polymer scaffold for cartilage repair without cultured cells in an animal model. *Arthroscopy* 26 (3): 375–383.
109 He, X., Fu, W., Feng, B. et al. (2013). Electrospun collagen/poly(L-lactic acid-*co*-epsilon-caprolactone) hybrid nanofibrous membranes combining with sandwich construction model for cartilage tissue engineering. *J. Nanosci. Nanotechnol.* 13 (6): 3818–3825.
110 Shafiee, A., Soleimani, M., Chamheidari, G.A. et al. (2011). Electrospun nanofiber-based regeneration of cartilage enhanced by mesenchymal stem cells. *J. Biomed. Mater. Res. Part A* 99 (3): 467–478.
111 Liu, X., Liu, S., Liu, S., and Cui, W. (2014). Evaluation of oriented electrospun fibers for periosteal flap regeneration in biomimetic triphasic osteochondral implant. *J. Biomed. Mater. Res. Part B* 102 (7): 1407–1414.

112 Coburn, J.M., Gibson, M., Monagle, S. et al. (2012). Bioinspired nanofibers support chondrogenesis for articular cartilage repair. *Proc. Natl. Acad. Sci. U.S.A.* 109 (25): 10012–10017.

113 Chen, J.P. and Su, C.H. (2011). Surface modification of electrospun PLLA nanofibers by plasma treatment and cationized gelatin immobilization for cartilage tissue engineering. *Acta Biomater.* 7 (1): 234–243.

114 Kim, C., Shores, L., Guo, Q. et al. (2016). Electrospun microfiber scaffolds with anti-inflammatory tributanoylated *N*-acetyl-D-glucosamine promote cartilage regeneration. *Tissue Eng. Part A* 22 (7, 8): 689–697.

115 Chen, W., Chen, S., Morsi, Y. et al. (2016). Superabsorbent 3D scaffold based on electrospun nanofibers for cartilage tissue engineering. *ACS Appl. Mater. Interfaces* 8 (37): 24415–24425.

116 Islas-Arteaga, N.C., Rivera, A.R., Rendon, D.R.E. et al. (2019). Electrospun scaffolds with surfaces modified by plasma for regeneration of articular cartilage tissue: a pilot study in rabbit. *Int. J. Polym. Mater. Polym. Biomater.* 68 (18): 1089–1098.

117 Shimomura, K., Rothrauff, B.B., Hart, D.A. et al. (2019). Enhanced repair of meniscal hoop structure injuries using an aligned electrospun nanofibrous scaffold combined with a mesenchymal stem cell-derived tissue engineered construct. *Biomaterials* 192: 346–354.

118 Yu, F., Li, M., Yuan, Z. et al. (2018). Mechanism research on a bioactive resveratrol–PLA–gelatin porous nano-scaffold in promoting the repair of cartilage defect. *Int. J. Nanomed.* 13: 7845–7858.

119 Ren, X., Li, J., Li, J. et al. (2019). Aligned porous fibrous membrane with a biomimetic surface to accelerate cartilage regeneration. *Chem. Eng. J.* 370: 1027–1038.

120 Jiang, T., Kai, D., Liu, S. et al. (2018). Mechanically cartilage-mimicking poly(PCL-PTHF urethane)/collagen nanofibers induce chondrogenesis by blocking NF–kappa B signaling pathway. *Biomaterials* 178: 281–292.

121 Chen, W., Xu, Y., Liu, Y. et al. (2019). Three-dimensional printed electrospun fiber-based scaffold for cartilage regeneration. *Mater. Des.* 179: 107886.

122 Baek, J., Chen, X., Sovani, S. et al. (2015). Meniscus tissue engineering using a novel combination of electrospun scaffolds and human meniscus cells embedded within an extracellular matrix hydrogel. *J. Orthop. Res.* 33 (4): 572–583.

123 Li, W.J., Cooper, J.A. Jr., Mauck, R.L., and Tuan, R.S. (2006). Fabrication and characterization of six electrospun poly(α-hydroxy ester)-based fibrous scaffolds for tissue engineering applications. *Acta Biomater.* 2 (4): 377–385.

124 Li, W.J., Tuli, R., Okafor, C. et al. (2005). A three-dimensional nanofibrous scaffold for cartilage tissue engineering using human mesenchymal stem cells. *Biomaterials* 26 (6): 599–609.

125 Li, Y., Liu, Y., Xun, X. et al. (2019). Three-dimensional porous scaffolds with biomimetic microarchitecture and bioactivity for cartilage tissue engineering. *ACS Appl. Mater. Interfaces* 11 (40): 36359–36370.

126 Shafiee, A., Seyedjafari, E., Taherzadeh, E.S. et al. (2014). Enhanced chondrogenesis of human nasal septum derived progenitors on nanofibrous scaffolds. *Mater. Sci. Eng., C* 40: 445–454.

127 Zamanlui, S., Mahmoudifard, M., Soleimani, M. et al. (2018). Enhanced chondrogenic differentiation of human bone marrow mesenchymal stem cells on PCL/PLGA electrospun with different alignments and compositions. *Int. J. Polym. Mater. Polym. Biomater.* 67 (1): 50–60.

128 Wise, J.K., Yarin, A.L., Megaridis, C.M., and Cho, M. (2009). Chondrogenic differentiation of human mesenchymal stem cells on oriented nanofibrous scaffolds: engineering the superficial zone of articular cartilage. *Tissue Eng. Part A* 15 (4): 913–921.

129 Wimpenny, I., Ashammakhi, N., and Yang, Y. (2012). Chondrogenic potential of electrospun nanofibres for cartilage tissue engineering. *J. Tissue Eng. Regener. Med.* 6 (7): 536–549.

130 Shanmugasundaram, S., Chaudhry, H., and Arinzeh, T.L. (2011). Microscale versus nanoscale scaffold architecture for mesenchymal stem cell chondrogenesis. *Tissue Eng. Part A* 17 (5, 6): 831–840.

131 Sohrabi, S. and Moraveji, M.K. (2020). Droplet microfluidics: fundamentals and its advanced applications. *RSC Adv.* 10 (46): 27560–27574.

132 Mahmoudi, Z., Mohammadnejad, J., Bazaz, S.R. et al. (2020). Promoted chondrogenesis of hMCSs with controlled release of TGF-β3 via microfluidics synthesized alginate nanogels. *Carbohydr. Polym.* 229: 115551.

133 Wang, C.C., Yang, K.C., Lin, K.H. et al. (2011). A highly organized three-dimensional alginate scaffold for cartilage tissue engineering prepared by microfluidic technology. *Biomaterials* 32 (29): 7118–7126.

134 Pati, F., Jang, J., Ha, D.H. et al. (2014). Printing three-dimensional tissue analogues with decellularized extracellular matrix bioink. *Nat. Commun.* 5: 3935.

135 Ugolini, G.S., Visone, R., Redaelli, A. et al. (2017). Generating multicompartmental 3D biological constructs interfaced through sequential injections in microfluidic devices. *Adv. Healthcare Mater.* 6 (10): 1601170.

136 Occhetta, P., Mainardi, A., Votta, E. et al. (2019). Hyperphysiological compression of articular cartilage induces an osteoarthritic phenotype in a cartilage-on-a-chip model. *Nat. Biomed. Eng.* 3 (7): 545–557.

137 Li, F., Truong, V.X., Thissen, H. et al. (2017). Microfluidic encapsulation of human mesenchymal stem cells for articular cartilage tissue regeneration. *ACS Appl. Mater. Interfaces* 9 (10): 8589–8601.

138 Mancipe Castro, L.M., Sequeira, A., García, A.J., and Guldberg, R.E. (2020). Articular cartilage- and synoviocyte-binding poly(ethylene glycol) nanocomposite microgels as intra-articular drug delivery vehicles for the treatment of osteoarthritis. *ACS Biomater. Sci. Eng.* 6 (9): 5084–5095.

139 Choi, N.W., Cabodi, M., Held, B. et al. (2007). Microfluidic scaffolds for tissue engineering. *Nat. Mater.* 6 (11): 908–915.

140 Di Bella, C., Duchi, S., O'Connell, C.D. et al. (2018). *In situ* handheld three-dimensional bioprinting for cartilage regeneration. *J. Tissue Eng. Regener. Med.* 12 (3): 611–621.

141 Occhetta, P., Centola, M., Tonnarelli, B. et al. (2015). High-throughput microfluidic platform for 3D cultures of mesenchymal stem cells, towards engineering developmental processes. *Sci. Rep.* 5: 10288.

142 Shi, X., Zhou, J., Zhao, Y. et al. (2013). Gradient-regulated hydrogel for interface tissue engineering: steering simultaneous osteo/chondrogenesis of stem cells on a chip. *Adv. Healthcare Mater.* 2 (6): 846–853.

14

Intervertebral Disc Repair

Yong Cao[1,2], Tianding Wu[1,2], Chunyue Duan[1,2], JinSong Li[3], Yuxin Jing[1,2], Jiaqi Xu[1,2], Daudi R. Manini[1,2], and Jianzhong Hu[1,2]

[1] *Central South University, Xiangya Hospital, Department of Spine Surgery and Orthopaedics, Xiangya Road No. 87, Changsha 410008, P.R. China*
[2] *Key Laboratory of Organ Injury, Aging and Regenerative Medicine of Hunan Province, Xiangya Road No. 87, Changsha 410008, P.R. China*
[3] *The Third Xiangya Hospital, Department of Spine Surgery, Tongzipo Road No. 138, Changsha 410008, P.R. China*

14.1 Introduction

Intervertebral disc (IVD) disorder is accompanied with increased matrix degradation, abnormal matrix synthesis, inflammation as well as the in-growth of nociceptive nerves, blood vessel, and aneural and avascular tissue [1] (Figure 14.1). IVD disorder is highly associated with pain, neurological deficit, and incapacity in the affected individual. IVD disorder is one of the leading causes of physical disability worldwide, contributing to low back pain being ranked number one in disability-adjusted life years [2, 3]. The point prevalence of low back pain in the year 2017 was estimated to be 7.5% in globally, and around 577 million sick patients are attending the hospital to seek treatments [4]. Studies in developed countries indicate that the total costs associated with low back pain range between 0.1% and 2% of the gross domestic product, whereas the costs associated with low back pain in low- and middle-income countries (LMICs) are not fully studied. The costs associated with the loss of productivity in LMICs are estimated to be 52% among the workers [5, 6]. Pain due to IVD disease is also called discogenic pain. Discogenic pain contributes to back pain and neck pain in several ways with few available treatment strategies apart from surgery. Surgical treatment for discogenic pain, such as fusion or arthroplasty, is traumatic and expensive. Subsequently, a lot of work has been done to effectively and safely develop non-invasive treatment methods. Direct injection of active substances to lower or even reverse IVD degeneration is currently a hot research area. Several tactics to deliver these therapeutic biological materials including protein injection, viral, and non-viral gene transfer have been proposed and tested in preclinical trials [7–9]. These therapeutic techniques for IVD degeneration attempt to upregulate the production of important matrix

Biofabrication for Orthopedics: Methods, Techniques, and Applications, First Edition.
Edited by Wenguo Cui, Xin Zhao, and Shen Liu.

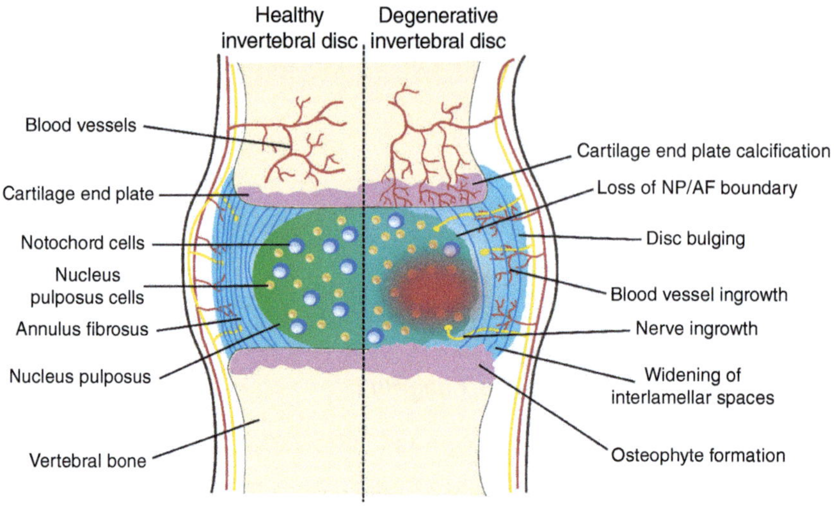

Figure 14.1 Schematic representation highlighting the hallmarks of human disc degeneration. Compared to the healthy intervertebral disc, the degenerative disc has increased inflammation, blood vessel and neuronal ingrowth, loss of the boundary between the NP and annulus fibrosus (AF), and widening of the interlamellar space between collagen bundles in the AF that often results in disc bulging. In advanced degeneration, the cartilage endplate calcifies and osteophytes form on the adjacent vertebral bones. Source: McCann and Séguin [1]/MDPI/CC BY 4.0.

protein-like proteoglycan and downregulate the catabolic phenomenon caused by proinflammatory cytokines, interleukin-1 (IL-1), and tumor necrosis factor-α (TNF-α). This ensures the balance between anabolism and catabolism of IVD cells; hence, maintenance homeostasis is achieved [10].

Many studies including clinical trials reporting the efficacy of injectable therapies such as cell-based, biological, or scaffold to treat IVD have been done, and yet no clinically validated method has been proved superior in preventing or reversing IVD degeneration. The lack of a successful treatment method highlights the great challenge in treating IVD degeneration. Cartilage repair has moved many steps ahead with a better outcome in orthopedic repair and regeneration relative to IVD; also, many types of research on cartilage repair have been done relative to IVD. This disparity in the research and consequently clinical trials can be used as a part of differences in finding a suitable cell source, mode of delivery, as well as a suitable microenvironment for IVD degeneration [11, 12].

In this review, we are focusing on the recent promising clinical methods used in IVD regeneration and repair in both experimental and clinical areas.

Currently, the treatment method of IVD degeneration falls into two major categories: IVD augmentation and IVD repair (Figure 14.2). IVD augmentation includes the addition of gel-like materials or resorbable polymer-based implants that are biologically inert or biologically active to a diseased disc which helps to restore its mechanical and biological functions. Currently, many different materials are available for void filling in IVD augmentation. The void filling largely includes

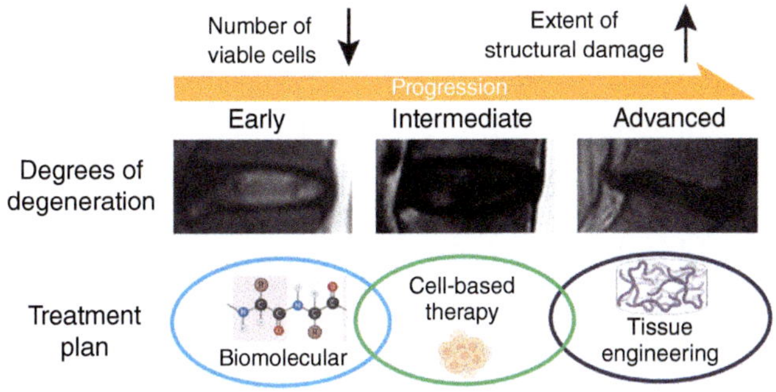

Figure 14.2 Treatment plan for biological intervertebral disc repair. As disc degeneration progresses, the disc loses a number of viable cells and integrity of structure. Biomolecular therapy will repopulate the disc with sufficient viable cells. If those cells are unviable, cell-based implantation can be a good option. If the integrity of structure is completely destroyed, bio tissue-engineered constructs that mimic the properties of native discs will help with biological reconstruction of the segment.

hydrogels, such as agarose, gelatin, and collagen. Sponges are manufactured from polyglycolic acid (PGA), polylactic acid (PLA), hyaluronic acid, silk ε-caprolactone, and glycosaminoglycan. These hydrogels are commonly injectable and able to conform to the various shapes and fill the defect sizes. Also, the hydrogel may be made for use in combination with biocompatibility cross-linkers, such as genipin or riboflavin for fibrin collagen cross-linkage, respectively.

The healthy IVD is characterized by three separate anatomical regions: the cartilaginous endplates, the annulus fibrosus (AF), and the nucleus pulposus (NP). Diffusion of materials occurs through capillary networks via the vertebral bodies. The degenerate disc is characterized by various structural and histological changes as well as an altered expression of extracellular matrix (ECM) markers and inflammatory markers. Degenerative changes are driven by inflammatory cytokines within the nucleus pulposus region, leading to altered matrix synthesis and reduced tissue integrity. Causes of low back pain in IVD degeneration are multifactorial; nevertheless, the occurrence of annular fissures provides an environment for the development of ectopic sensory nerve fibers and blood vessels, which are considered to be a major source of discogenic pain. A disintegrin and metalloproteinase with thrombospondin motifs (ADAMTS); collagen I, type I collagen; collagen II, type II collagen; and matrix metalloproteinase (MMP) can be detected in IVD degenerative changes (Table 14.1).

The IVD repair methods focus on boosting the repair and restoration of the growth capability of the diseased disc, incorporating bioengineering approaches combined with cell therapy, gene therapy, and protein-based therapy as a feasible option to regain disc homeostasis. Combination approaches use both methods to maximize the regeneration and repair of the IVD.

The number of viable cells present in the IVD and conducive microenvironment such as pH, nutrient supply, and oxygen delivery generally determine the regenerative potential of the disc. Therefore, the state of IVD degeneration is very important when planning the treatment strategy of IVD. Also, in advanced IVD degeneration,

Table 14.1 Commonly encountered changes on a degenerated IVD.

Other structural changes of degenerated IVD	Inflammatory markers produced by degenerated IVD	Altered expression of ECM markers		Histological changes of degenerated IVD
• Loss of disc height • Loss of water contents • Lowered ability to withstand the compressive force	Increased levels of cytokines and chemokines.	Increased matrix degeneration	Decreased matrix synthesis	1. Loss of demarcation between the annulus and nucleus pulposus 2. Presence of fissures 3. Increased ectopic nerve and blood vessel infiltration 4. Reduced cellular proliferation 5. High number of cellular senescence and apoptosis
	IL-1β, CCL3, CXCL8	↑Collagen I	↓Collagen II	
	TNF, CCL4, MCP3, IL-6, and CCL5	↑ADAMTSs	↓Aggrecan	
		↑MMPs	↓Elastin	

there is overexpression of proteases such as MMPs, ADAMTS, which can degrade various matrix components. These degrading activities are mediated by cytokines, such as in IL-1 and TNF-α [13, 14]. Generally, it argued that in "early IVD degeneration" with a high number of viable cells and less structure damage, molecular therapy or gene therapy can be applied as a treatment strategy, and an "intermediate stage of IVD degeneration" with a moderate number of viable cells and moderate structural damage cell-based therapy can be used as a treatment method, while in "late advanced stage of IVD degeneration" with the minimum number of viable cells and substantial structural damage, biomaterial therapy can be used as a treatment strategy [15].

14.2 Technique for Intervertebral Disc Therapy

14.2.1 Cell-Based Treatment Techniques

The number of health cells decreases with an advanced stage of IVD degenerations; cell-based therapies in the advanced stages is a good technique to increase the amount of viable cells within the IVD. The use of novel cell-based therapies aims to create the normal IVD microenvironmental homeostasis and eliminate the degenerative cascade by adding cells with anti-inflammatory effects. Several cell sources have been proposed and researched including mesenchymal stem cells (MSCs), nucleus pulposus cells, chondrocytes notochordal cells, induced pluripotent stem cells (iPSCs), and embryonic stem cells (ESCs) (refer to Table 14.2). There are several ongoing clinical trials or have been completed testing these cells on humans. The choice of cell types depends on understanding the microenvironmental biology of the IVD and the ultimate cellular changes associated with growth and degeneration. Factors such as oxygen and nutrient availability, biochemical changes, and mechanical stress have to be considered before cell transplantation as they may impair the survival of the cells. Autologous nucleus pulposus is the native cells of the disc, are phenotypically similar to chondrocytes, and are adapted to survive in the harsh avascular environment of the IVD. Furthermore, they are preprogrammed to produce tissue-specific ECM proteoglycans, making them a good option for transplantation. In an animal model, the transplanted disc chondrocytes produced an ECM that showed a composition similar to normal IVD tissue, including both type 1 and type 11 collagens [14]. Ju et al. [15] in their three-year prospective study to assess the safety and efficacy of activated NP cell transplantation in the degenerate lumbar IVD found improvement of IVD degeneration on magnetic resonance imaging (MRI), and no adverse effects during the three-year follow-up were observed. Binch et al. [21] in their prospective controlled randomized multicenter study to compare the safety and efficacy of autologous disc chondrocyte transplants in degenerated IVD found that disc chondrocytes have well integrated with the surrounding tissue, reduced low back pain, and improved disc hydration on MRI and excellent ECM production compared with controls. Nevertheless, harvesting the NP cells from the patient's disc is traumatic, technically difficult, and result

Table 14.2 Sources of cell-based therapy for IVD repair.

Type of cell	Tissue source	Molecular markers	Proposed mechanism for treating IVD
Nucleus pulposus cells [16, 17]	Nucleus pulposus	FOXF1, PAX1, laminin 332, GLUT1, keratin 9, keratin 18, HIF1α, and carbonic anhydrase 12. Aggrecan to type II collagen ratio 27 : 1 (measured as glycosaminoglycan [GAG]-to-hydroxyproline ratio)	Repopulate cellular compartment of nucleus pulposus tissue and produce native extracellular matrix
Chondrocytes [18]	Cartilage endplate	Type II collagen and aggrecan	Substitute cell type to reconstitute nucleus pulposus matrix
Mesenchymal stem cells [19]	Bone marrow, adipose tissue, or umbilical cord	CD73, THY1, and endoglin	Engraftment and differentiation to nucleus pulposus-like cells or chondrocytes; secretion of bioactive molecules, which exert immunomodulatory and reparative effects on host cells and extracellular matrix; modulation of host immune response
Notochordal cells [20]	Notochord	Brachyury protein, CD24, galectin 3, keratin 8, keratin 18, and keratin 19	Regenerate and restore the nucleus pulposus matrix and resident nucleus pulposus cell phenotype

in low cellular yields because of compromised tissue sources. However, a newly discovered nucleus pulposus progenitor might be the basis for a valuable additional approach [16].

All cell types that are used for therapeutic strategies in IVD disease including MSCs are related in terms of embryological origin and lineage. They have the potential to differentiate into all lines including chondrogenic and IVD cell lineage. Autologous MSCs are commonly used and are harvested from bone marrow or adipose tissues. They seem to offer a promising alternative for regeneration IVD therapies because of their ease of preparation, multilineage potential, self-renewal capacity, and immunosuppressive abilities [17, 18]. Theoretically, it is believed that injection of MSCs into a degenerated disc can differentiate into NP cells to prevent further degeneration and restore normal homeostasis (Figure 14.3). Several studies have proved the ability of MSCs to differentiate into NP cells, restore water contents, IVD

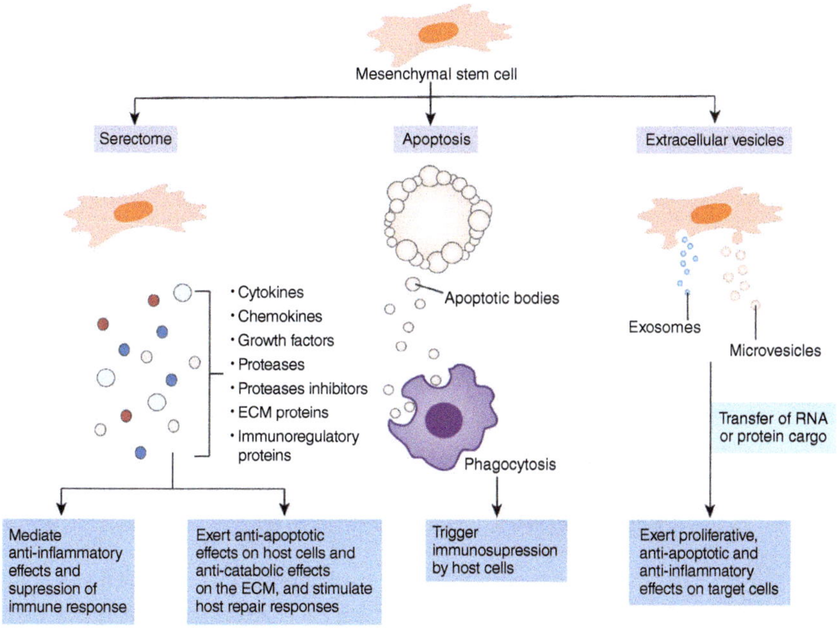

Figure 14.3 Potential mechanisms underlying mesenchymal stem cell-mediated repair. Several therapeutic mechanisms of action have been proposed for mesenchymal stem cell-mediated repair. These include paracrine effects associated with the transplanted cell secretome, in which the mesenchymal stem cells secrete bioactive factors (such as cytokines, chemokines, and immunoregulatory proteins) that mediate anti-inflammatory and immune-regulatory effects on host cells, as well as growth factors, extracellular matrix (ECM) proteins, proteases, and protease inhibitors that stimulate the host repair response by exerting the antiapoptotic effects on host cells and anticatabolic effects in the extracellular matrix. Rapid apoptosis of transplanted cells has been proposed as an important mechanistic element, whereby apoptotic bodies quickly elicit an immunomodulatory response by phagocytes, which in turn stimulate immune cells to provide a strong immunomodulatory response. Transplanted mesenchymal stem cells also shed extracellular vesicles, which exert proliferative, antiapoptotic, and anti-inflammatory effects on target cells. Source: Binch et al. [21]/with permission of Springer Nature.

height, and suppress the ongoing inflammatory cascade. Also, several authors have applied MSCs including bone marrow-derived mesenchymal stem cells (BM-MSCs), adipose-derived mesenchymal stem cells (ASCs), and umbilical cord-derived mesenchymal stem cells (UC-MSCs) to repair the degenerated IVD, either alone or joint combination with other biomaterial scaffolds with a promising outcome [19, 20, 22]. In humans, MSCs have been applied as a treatment method, and clinical trials have been done on MSCs as a treatment modality of IVD degeneration. Wallach et al. [23] in their pilot study to assess the feasibility and safety of MSCs in the treatment of chronic back pain patients due to IVD degeneration found 85% of pain improvement within the 3 month periods, and both disc height and water contents of the disc improved during the 12 months of follow-up. Also, Somia and Verma [24] in their study to assess the safety and feasibility of BM-MSCs in the treatment of low back pain secondary to IVD degeneration found good improvement of pain and IVD regeneration in BM-MSC-injected patients and 40% improvement to Pfirrmann grade I on MRI.

Autologous cells have less risk of disease transmission with no risk of immune reaction, but a separate traumatic harvesting procedure and an expensive culturing method are needed. Allogeneic cell sources do not need harvesting but carry a risk of disease transmission and a high risk of immune reaction. Studies have shown that both autologous and allogeneic cell sources can survive in the microenvironment of IVD in an animal model [25]. Also, most researchers on cell therapy for IVD degeneration have focused much on animal models. This has led difficult interpretation as it does not provide information easily translatable to humans. The small animal model brings difficulty in terms of translation to human disease because of their difference in the cell composition of the NP. Some species such as Canine and Ovine lose notochordal cells in early life similar to human beings, making them suitable candidates for study. Even some studies have proposed that the best animal model for IVD degeneration is Canine [26–28].

14.2.2 Patient Selection for Cell Therapy

Even the best treatment method will not work if applied to the wrong IVD; for maximum therapeutic effects, it is very necessary to categorize patients by using sophisticated MRI and quantification of circulating biomarkers to choose a good candidate for treatment. The suitable time for cell therapy is at an early stage of degeneration cascade before the onset of advanced degeneration coupled with endplate calcification and maximum annular degeneration. Therefore, a suitable candidate for cell therapy is that the disease should be localized to a single segment, moderate IVD degeneration, and Pfirrmann grade III or IV as seen on MRI. Patients with chronic back pain (>4 visual analogue scale [VAS]) and disability (>40 oswestry disability index [ODI]) for more than six months can be included for cell therapy. Patients with spinal abnormalities including spondylolisthesis, spondylosis, scoliosis, fracture, and severe kyphosis are excluded from cell therapy. Patients with a herniated disc may be included if cell therapy is applied in combination with AF repair. Recently, a team of experts agreed that there is no acceptable clinical standard for discogenic pain. New imaging techniques such as MRI (T1ρ and T2 relaxation times) and chemical exchange saturation transfer (CEST) have been developed to help localize pain and quantify therapeutic effects on the treated segments. This will help in the evaluation of IVD cellular therapy [29–31].

14.3 Gene Therapy for IVD

Gene therapy involves the transfer of exogenous genes to a target cell, resulting in gene expression and subsequent production of protein by the target cell (Figure 14.4). The major merit of gene therapy is its long-term efficacy; once the gene is injected into the target cells, they continue to produce the desired products (RNAs or proteins). The two commonly used methods for gene transfer are direct gene transfer method (*in vivo*) and indirect gene transfer (*ex vivo*). *In vivo*, gene

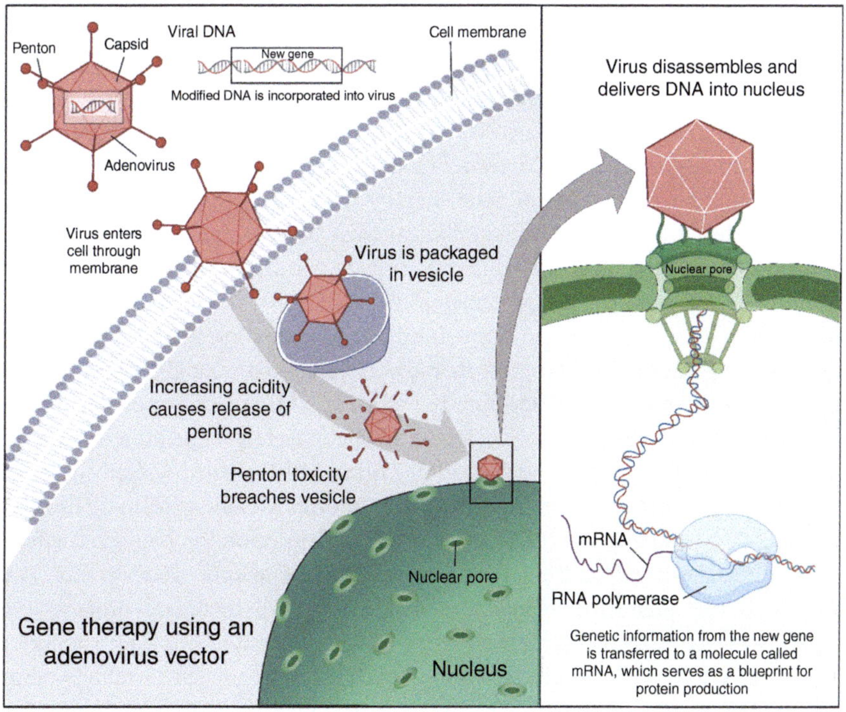

Figure 14.4 Schematic presentation of gene therapy flow. A new gene is inserted directly into a cell. A carrier called a vector is genetically engineered to deliver the gene. An adenovirus introduces the DNA into the nucleus of the cell, but the DNA is not integrated into a chromosome. Source: How does gene therapy work? National library of Medicine Retrieved from: https://medlineplus.gov/genetics/understanding/therapy/procedures/. Public Domain.

transfer vectors with the appropriate gene are transferred directly to the body of the target cell. While *ex vivo*, the gene transfer cells of the target gene are isolated, cultured, genetically altered *in vitro*, and then reimplanted to the target organ. The choice of the method for gene transfer depends on the anatomy, physiology, and pathophysiology of the disease; the vector of choice; and other variables. According to the biology of the IVD, most authors propose the use of the *in vivo* gene transfer method [22, 23]. The carrier for the altered genetic material can be viral or non-viral, but most researchers prefer to use viral for IVD degeneration; adenovirus is commonly used because of their higher transduction efficiency for slowly dividing cells. However, great care when using adenovirus as a vector shall be taken because of immunogenicity, a fatal anaphylactic reaction after the systematic injection has been reported [24]. To obviate the side effects of viral vector, some researchers have employed the use of non-viral vector as a carrier for the gene transfer. The non-viral vector is a plasmid-based system where no diseased agents are administered to the patient. Non-viral vectors have the following advantages: low immunogenicity, less toxicity, easy to prepare, and they are not restricted by the molecular size of the gene to be introduced. The most frequently used non-viral vectors are liposomes. They deliver the genetic material into the cell by fusing it with the cell's phospholipid membrane. The major drawback of the non-viral vector is transient transgene expression. Hence, they are considered less efficient. When they are injected into the target tissue, they promote gene expression in a limited period, usually a few days. This makes them insufficient for the treatment of chronic diseases such as IVD degeneration [22, 25, 26].

Up to date, the application of gene therapy for IVD regeneration is limited to animal studies. There are no human clinical trials based on gene therapy for IVD regeneration reported. Both anabolic regulators (e.g. transforming growth factor-β [TGF-β], LIM mineralization protein-1 [LMP-1], Sox-9), and anticatabolic regulators (e.g. anti-ADAMTS-5 and tissue inhibitor of metalloproteinase-1 [TIMP-1]) have given a great promise on gene therapy on the animal model [27]. Sangwook et al. [28] in their study to determine the effect of LMP-1 overexpression in IVD cells on the production of proteoglycans and bone morphogenetic proteins (BMPs) both *in vitro* and *in vivo* on a rabbit found that LMP-1 overexpression increases the disc cell production of proteoglycan, bone morphogenetic protein-2 (BMP-2), and BMP-7. Nishida et al. [29] in their study to determine the biological effects of the direct, adenovirus-mediated transfer of a therapeutic gene to the IVD found that the adenovirus-mediated transfer of TGF-β increased proteoglycan synthesis in the IVD of the rabbit. A study done by Ronjon et al. [30] reported that Sox-9 restored disc height and upregulated type II collagen in rabbits. Shoji et al. [31] observed a reduced IVD degeneration, improved MRI, and histological grading score on a degenerated IVD of a rabbit after injection of ADAMTS-5 siRNA. Also, Steven et al. [32] reported a decreased rate of IVD degeneration and improved histological tissues from the IVD of a rabbit which, was treated by adeno-associated virus serotype 2 (AAV2) vector carrying genes for either BMP-2 or TIMP-1 (Table 14.3).

Table 14.3 Commonly used carriers' vectors for gene therapy.

Viral vectors	Non-viral vectors
Adeno-associated virus	Liposomes
Adenovirus	Synthetics
Lentivirus	Naked DNA
Retrovirus	Gene gun
Herpes simplex virus	Electroporation

14.4 Molecular Therapy for IVD Degeneration

In molecular therapy also known as growth factor therapy, molecules or proteins are introduced into the IVD to reprogram the native architecture to stop the degeneration cascade. Many potential molecular therapies have been researched and fall into four major categories: anticatabolic regulators, mitogens, chondrogenic morphogens, and intracellular regulators. Each of these is explained as follows.

14.4.1 Anticatabolic Regulators

Anticatabolic regulators either increase the matrix synthesis or decrease the matrix degradation by the inhibition of the degrading enzymes in the IVD. Several degrading enzymes capable of breaking down the matrix molecules in the IVD have been reported. They include cathepsins, aggrecans, and MMPs. Among these catabolic enzymes, MMPs are key in the normal turnover of matrix molecules and are believed to be important in IVD degeneration. The important members of the MMP family are stromelysin (MMP 3), collagenase (MMPs 1, 8, and 13), and gelatinase (MMPs 2 and 9). MMP 3 is largely found in the NP, they destroy the core protein of proteoglycan. Furthermore, they gain access to the proteolytic cleavage sites, leaving isolated hyaluronate-binding regions, degraded proteoglycan aggregates, and glycosaminoglycan particles as the breakdown products [33, 34]. The second members of metalloproteinases called aggrecanases (ADAMTS4 and 5) play a major role in aggrecan and versican degradation [35]. Because of the powerful degrading ability of MMP treatment of degenerative IVD has focused much on tissue inhibitors of matrix metalloproteinases (TIMPs). Wallach et al. [36] in their study to assess the potential of TIMP-1 in modifying proteoglycan contents in degenerated IVD by delivering anticatabolic genes using an adenoviral vector found the increased amount of proteoglycan synthesis. Their research has proved the role of anticatabolic in the treatment of degenerative IVD. The net metabolic rate dictates the nutritional demand of the IVD cells. Cytokines such as IL-1 and TNF-α have important roles in disc digestion. Consequently, molecules such as interleukin-1 receptor antagonist (IL-1Ra) and infliximab, which block IL-1 and TNF-α, are very important in preventing disc degeneration.

14.4.2 Mitogens

Mitogens are molecules that are known to increase the rate of mitosis in the cell cycle. They contain the growth factors. In this study, we will differentiate mitogenic growth factors from chondrogenic molecules (morphogens). Mitogenic molecules include insulin-like growth factor-1 (IGF-1), epidermal growth factor (EGF), and fibroblast growth factor (FGF) [37]. Thompson et al. [38] in their *in vivo* study with canine disc cells have proved that mitogenic molecules can increase the rate of mitosis and proteoglycan synthesis to a different degree depending on the region of the disc where the cells were obtained. A similar finding of Thompson et al. has also been reported by other researchers [39]. Also, a study by Okuda et al. [40] proved an age-related decline in proteoglycan synthesis to IGF-1 in animal IVD. Based on their study, the restoration of IGF-1 in aging IVD may increase matrix synthesis. Some growth factors have been shown to protect the IVD cells from apoptosis. A significant reduction in the percentage of apoptotic disc cells after exposure to 50–500 ng/ml IGF-1 or exposure to 100 ng/ml platelet-derived growth factor has been reported [41]. These results broaden our knowledge on the biology of the disc cells and give a clue that cytokines can slow or prevent apoptosis of cell death *in vitro*. Also, IGF-1 has both anabolic and catabolic effects. IGF-1 has been shown to lower the levels of active TIMP-2 *in vitro* study, showing a complex effect on disc matrix metabolism by IGF-1. Further knowledge on IGF-1 is still needed because of its potential.

14.4.3 Chondrogenic Morphogens

Chondrogenic morphogens are cytokines with the mitogenic ability, and they are distinguished by their ability to increase the chondrocyte-specific patterns of the cell, which include secretion of type II collagen, Sox9, aggrecan, and sulfated glycosaminoglycans. Most studies have focused on chondrogenic morphogens such as TGF-β, BMPs, or growth and differentiation factors (GDFs). Chondrogenic morphogens can reverse the fibrotic phenotype of IVD cells to chondrocyte phenotype of disc nucleus cells in younger to more normal IVD discs. They are secretory molecules able to act in an autocrine, paracrine, and endocrine manner. The ability of the cytokine to diffuse makes it reach the target cells easily and function effectively. Their mechanism of action depends on the presence of receptor molecules on the surface of the responding cells and the competent intracellular messenger system. The IVD cells have been proved to express receptors for TGF and BMP molecules, and the expression of the disc receptor appears to decrease with age [42–44].

TGF-β1 was one of the first morphogenic molecules to be discovered. Studies have shown that TGF-β1 is both a mitogen and anabolic molecule with the maximum ability of proteoglycan synthesis. TGF-β1 has proved to be superior in proteoglycan synthesis compared to other growth factors EGF, IGF-1, platelet derived growth factor (PDGF), and FGF [38, 45]. TGF-β1 is very important, but its efficacy *in vivo* IVD degeneration has not been studied.

BMP-2 is also a chondrogenic morphogen. BMP-2 has been shown to increase the synthesis of proteoglycan aggrecan and type II collagen in the animal model. Also, BMP-2 has shown incomplete reverse of the inhibitory effects of nicotine in the synthesis of proteoglycan [46, 47]. BMP-2 accelerates the differentiation of osteoblast during bone formation, before it was thought that BMP-2 may also cause disc differentiation along an osteoblastic lineage. However, *in vitro* studies with human disc cells have shown that BMP-2 enhanced the expression of chondrocyte genes but not osteogenic genes [48]. At present, there are no published studies on the efficacy of BMP-2 in treating IVD degeneration in humans, so further research is needed.

BMP-7 is also another potential IVD disc morphogen known as osteogenic protein-1(OP-1); it can stimulate the growth of both annulus fibrosis and nucleus pulposus *in vitro* and in a dose-dependent manner. When challenged with BMP-7, the disc cells improved proteoglycan synthesis with maximum expression of chondrocyte genes aggrecan and type II collagen [49]. Zhang et al. [50] in their study to determine if cells isolated from three distinct zones in the bovine IVD differ in their response to growth factor OP-1 because of inherent biological differences. The authors found that OP-1 stimulates proteoglycan synthesis in the disc cells isolated from all the three zones of the bovine IVDs. Cells from all three zones proliferated significantly.

BMP-13 is sometimes called GDF-6 or cartilage-derived morphogenetic protein-2 (CDMP-2) and is a member of the BMP family with only half homology to the BMP-2 amino acid sequence. Studies [51, 52] have shown that BMP-13 increases the synthesis of proteoglycan in the IVD in the animal model but is inferior when compared to BMP-2. Both BMP-13 and BMP-2 have shown additive effects in proteoglycan synthesis or chondrocyte gene expression. BMP-13 has shown a good effect on tendon healing, but its use on IVD degeneration repair is not yet fully studied [53].

GDF-5 is sometimes called cartilage-derived morphogenetic protein-1 (CDMP-1); it mainly appears during precartilaginous mesenchymal condensation and throughout the cartilaginous cores of the developing long bones. In the animal study, GDF-5 was superior in increasing disc height and cellular synthesis compared to GDF-5, TGF-β1, and IGF-1 [39]. GDF-5 is currently in clinical use for spinal fusion surgery; we anticipate many studies few years to come.

Link protein is a link protein–peptide with a sequence DHLSDNYTLDHDRAIH, commonly known as "link N." Link N is formed by the breakage of human link protein by stromelysins 1 and 2, gelatinase A and B, and collagenase between histidine and isoleucine. Research has shown that link N acts as a growth factor and stimulates the synthesis of proteoglycans and collagen by chondrocytes of human articular cartilage [54]. However, the mechanism of action of link N is not clear; further studies are still needed in this area.

14.4.4 Intracellular Regulators

Intracellular regulators are a special group of molecules; they are neither cytokines nor growth factors. These classes of molecules play a different role in cell differentiation development and growth. Studies [55, 56] have found that Sma–Mad (Smad) proteins are intracellular molecules that mediate BMP-receptor signaling. Smad-1 and Smad-5 can promote the synthesis of both proteoglycan and type II collagen.

The Sox9 chondrocyte marker controls the transcription of type II collagen by mRNA. Sox9 has been found to increase the synthesis and production of type II collagen in the laboratory experiment. Also, in the animal model, Sox9 has been shown to prevent degeneration changes of the annulus fibrosis [46, 57].

LMP-1 is an intracellular regulator of bone formation. Upregulation of BMPs and stabilization of BMP/Smad signaling is the key mechanisms through which LMP-1 enhances osteogenesis [58]. A study done by Yoon et al. [28] found that LMP-1 upregulates the production of BMP-2, BMP-7, and proteoglycans *in vitro*; the authors further proved that the effect of LMP-1 involved a BMP-dependent mechanism.

14.5 Biomaterial Therapy

In the late stage of IVD disc degeneration, there is significant damage in the tissues accompanied with structure and chemical changes. Little cells are left with the potential to regenerate with cellular or molecular therapy. At this stage, tissue, engineering, and biomaterial strategies play a great role as a substitute and scaffold for the treatment of the degenerated IVD disease. Scientist commonly uses combined strategy by utilizing cell-loaded biomaterials. This offers both a structural environment for mechanical stability and cellular regenerative potential [59]. Several materials are currently available as substitutes or void fillers for IVD degeneration for both AF and NP. The ideal material to be used for tissue repair must meet the following criteria; manufacturing feasibility, mechanical properties (to offer short-term function without the disturbing long-term function of the tissue), low toxicity of degradation products (to local tissue and peripheral response), and good delivery drug compatibility. These materials include injectable hydrogel and synthetic polymers made from resources such as alginate, gelatin, collagen, PLA, hyaluronic acid, and PGA [59]. Studies [60, 61] have shown that cross-linked collagen hydrogels injected into AP in animal models restore the natural compressive loading mechanics of the IVD (Figure 14.5). Way et al. [63] in their multicenter study to assess the safety and efficacy of intradiscal fibrin sealant in adult chronic discogenic low back pain by volume- and pressure-controlled intradiscal delivery of BIOSTAT BIOLOGX® Fibrin Sealant found that fibrin sealant

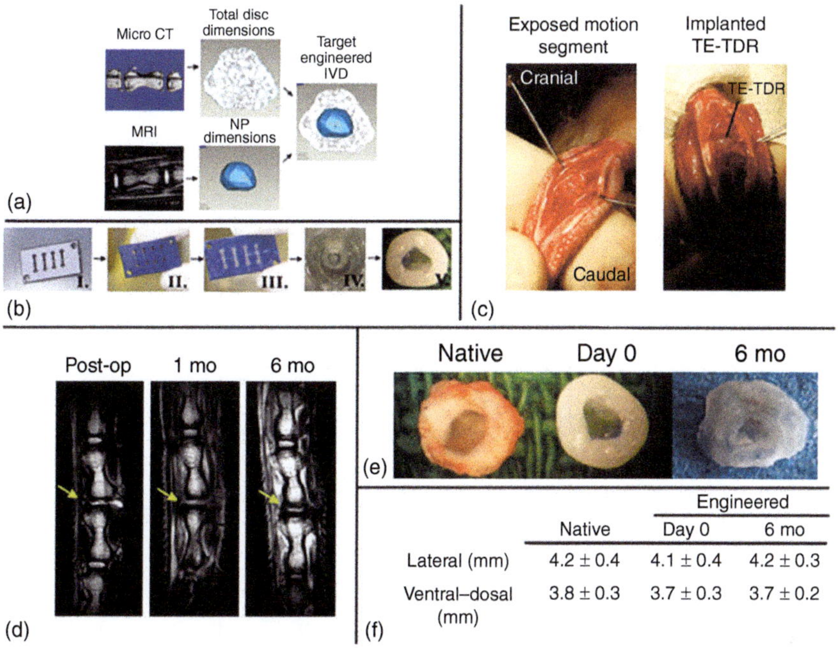

Figure 14.5 Anatomical tissue engineering total disc replacement (TE-TDR), designed from MRI and CT, survives in disc space for six months. (a) CT and MRI design procedure for obtaining TE-TDR dimensions. (b) Fabrication of TE-TDR. (i) NP dimensions used to design injection molds via computer-aided design. (ii) Injection mold 3D printed out of acrylonitrile butadiene styrene plastic. (iii) Cell-seeded alginate was injected into mold, removed, (iv) placed in the center of a 24-well plate, and cell-seeded collagen was poured around alginate NP. (v) After two weeks of culture, cell-seeded collagen contracts around the NP to form composite TE-TDR. (c) Intraoperative images showing exposed caudal 3/4 disc space and implanted TE-TDR. (d) T2-weighted MRI of implanted disc space (marked by yellow arrows) and adjacent native levels immediately postoperative, at one month, and six months after implantation. (e) History of TE-TDR in native disc space. Intraoperative photo showing explanted native IVD next to the TE-TDR (day 0) that was implanted in its place and TE-TDR after being implanted into native disc space for six months. (f) Size of engineered IVD compared to native IVD. Measurements were taken along the lateral and ventral–dorsal planes of the engineered and native IVD. Engineered IVD measurements were taken at day 0 before implantation ($n = 12$) and compared to explanted native discs ($n = 12$). Engineered IVD measurements were also taken after six months of implantation ($n = 12$). Source: Bowles et al. [62] (p. 02)/with permission of National Academy of Sciences.

downregulate the inflammatory cytokines, and there was a significant improvement of pain and disability after two years of follow-up. Also, Ulrich and Othmar [64] in their prospective study to investigate the use of NuCore Injectable Nucleus hydrogel (Spine Wave, Inc., Shelton, CT, USA) as a replacement for nuclear tissue lost to herniation and microdiscectomy found significant improvement for disc height, leg and back pain, as well as function scores. The author concluded that injectable cross-linked silk and elastin polymer has the potential to slow the degenerative cascade of the spinal segment over time.

Table 14.4 The ongoing clinical trials of biomaterial therapy for IVD degeneration.

ClinicalTrials.gov identifier	Location	Status	Biomaterials	Condition	Estimated sample size
NCT02070484	USA	Finished	Human amniotic tissue-derived allograft	Lumbar degenerative disc disease	60
NCT02379689	USA	Ongoing	Injectable placental tissue extract	Degenerative disc diseases	30
NCT02412735	USA	Ongoing	Rexlemestrocel-L combined with hyaluronic acid	Degenerative disc disease	330
NCT01640457	Australia and Germany	Ongoing	Biphasic and three-dimensional collagenous matrix	Intervertebral disc displacement and degeneration	120
NCT03197415	Taiwan	Ongoing	Autologous platelet-rich plasma gel	Lumbar degenerative disc disease and low back pain	30

Despite the promising, preclinical data, biomaterial therapy for degenerative IVD has not yet achieved clinical victory. Several clinical trials on the efficacy of biomaterial on degenerative IVD are still ongoing, and some are summarized in Table 14.4.

14.6 Summary

- Homeostasis of the IVD tissues is biologically controlled by the active maintenance of a balance between the anabolism and catabolism of disc cells. This is achieved by a complex and precise coordination of a variety of substances, including cytokines, growth factors, enzymes, and enzyme inhibitors, in a paracrine and/or autocrine manner. Whenever there is an imbalance in disc homeostasis, disc degeneration occurs.
- Currently, therapeutic approaches for IVD degeneration have included attempts to upregulate the production of key matrix proteins or to inhibit the catabolic events induced by the pro-inflammatory cytokines, IL-1, and TNF-α.
- Several treatment approaches have been proposed at different stages of IVD degeneration, as follows: molecular therapy is efficient during the early stage of disc degeneration, cell therapy performs better during the intermediate stage of IVD degeneration biomaterial, and tissue engineering is advocated during the late stage of IVD degeneration.
- All of the above treatment strategies are still in preclinical trial; none has been approved for clinical use.

Authorship Contribution Statement

Y. Cao, D.R. Manini, and T. Wu contributed equally in the preparation of this work. Y. Jing, J. Xu, and J. Li contributed to the collection of literature. The authors acknowledge professors J. Hu and C. Duan for designing this manuscript.

References

1 McCann, M.R. and Séguin, C.A. (2016). Notochord cells in intervertebral disc development and degeneration. *J. Dev. Biol.* 4 (1).

2 Murray, C.J.L. and Lopez, A.D. (2013). Measuring the global burden of disease. *N. Engl. J. Med.* 369 (5): 448–457.

3 Murray, C.J., Atkinson, C., Bhalla, K. et al. (2013). The state of US health, 1990–2010: burden of diseases, injuries, and risk factors. *JAMA* 310 (6): 591–608.

4 Wu, A., March, L., Zheng, X. et al. (2020). Global low back pain prevalence and years lived with disability from 1990 to 2017: estimates from the Global Burden of Disease Study 2017. *Ann. Transl. Med.* 8 (6): 299.

5 Olafsson, G., Jonsson, E., Fritzell, P. et al. (2018). Cost of low back pain: results from a national register study in Sweden. *Eur. Spine J.* 27 (11): 2875–2881.

6 Jackson, T., Thomas, S., Stabile, V. et al. (2016). A systematic review and meta-analysis of the global burden of chronic pain without clear etiology in low- and middle-income countries: trends in heterogeneous data and a proposal for new assessment methods. *Anesth. Analg.* 123 (3): 739–748.

7 Akeda, K., Ohishi, K., Masuda, K. et al. (2017). Intradiscal injection of autologous platelet-rich plasma releasate to treat discogenic low back pain: a preliminary clinical trial. *Asian Spine J.* 11 (3): 380–389.

8 An, H.S., Masuda, K., Cs-Szabo, G. et al. (2011). Biologic repair and regeneration of the intervertebral disk. *J. Am. Acad. Orthop. Surg.* 19 (7): 450–452.

9 Masuda, K. and An, H.S. (2004). Growth factors and the intervertebral disc. *Spine J.* 4 (6 Suppl.):: S330–S340.

10 Ahn, S.H., Cho, Y.W., Ahn, M.W. et al. (2002). mRNA expression of cytokines and chemokines in herniated lumbar intervertebral discs. *Spine (Phila Pa 1976)* 27 (9): 911–917.

11 Vedicherla, S. and Buckley, C.T. (2017). Cell-based therapies for intervertebral disc and cartilage regeneration – current concepts, parallels, and perspectives. *J. Orthop. Res.* 35 (1): 8–22.

12 Moriguchi, Y., Alimi, M., Khair, T. et al. (2016). Biological treatment approaches for degenerative disk disease: a literature review of in vivo animal and clinical data. *Global Spine J.* 6 (5): 497–518.

13 Wang, J., Markova, D., Anderson, D.G. et al. (2011). TNF-α and IL-1β promote a disintegrin-like and metalloprotease with thrombospondin type I motif-5-mediated aggrecan degradation through syndecan-4 in intervertebral disc. *J. Biol. Chem.* 286 (46): 39738–39749.

14 Tian, Y., Yuan, W., Fujita, N. et al. (2013). Inflammatory cytokines associated with degenerative disc disease control aggrecanase-1 (ADAMTS-4) expression in nucleus pulposus cells through MAPK and NF-κB. *Am. J. Pathol.* 182 (6): 2310–2321.

15 Ju, D.G., Kanim, L.E., and Bae, H.W. (2020). Intervertebral disc repair: current concepts. *Global Spine J.* 10 (2 Suppl.):: 130s–136s.

16 Thorpe, A.A., Binch, A.L., Creemers, L.B. et al. (2016). Nucleus pulposus phenotypic markers to determine stem cell differentiation: fact or fiction? *Oncotarget* 7 (3): 2189–2200.

17 Risbud, M.V., Schoepflin, Z.R., Mwale, F. et al. (2015). Defining the phenotype of young healthy nucleus pulposus cells: recommendations of the Spine Research Interest Group at the 2014 annual ORS meeting. *J. Orthop. Res.* 33 (3): 283–293.

18 Frazer, A., Bunning, R.A., Thavarajah, M. et al. (1994). Studies on type II collagen and aggrecan production in human articular chondrocytes *in vitro* and effects of transforming growth factor-β and interleukin-1β. *Osteoarthritis Cartilage* 2 (4): 235–245.

19 Dominici, M., Le Blanc, K., Mueller, I. et al. (2006). Minimal criteria for defining multipotent mesenchymal stromal cells. The International Society for Cellular Therapy position statement. *Cytotherapy* 8 (4): 315–317.

20 Watanabe, T., Sakai, D., Yamamoto, Y. et al. (2010). Human nucleus pulposus cells significantly enhanced biological properties in a coculture system with direct cell-to-cell contact with autologous mesenchymal stem cells. *J. Orthop. Res.* 28 (5): 623–630.

21 Binch, A.L.A., Fitzgerald, J.C., Growney, E.A., and Barry, F. (2021). Cell-based strategies for IVD repair: clinical progress and translational obstacles. *Nat. Rev. Rheumatol.* 17 (3): 158–175.

22 Sobajima, S., Kim, J.S., Gilbertson, L.G., and Kang, J.D. (2004). Gene therapy for degenerative disc disease. *Gene ther.* 11 (4): 390–401.

23 Wallach, C.J., Gilbertson, L.G., and Kang, J.D. (2003). Gene therapy applications for intervertebral disc degeneration. *Spine (Phila Pa 1976)* 28 (15 Suppl.):: S93–S98.

24 Somia, N. and Verma, I.M. (2000). Gene therapy: trials and tribulations. *Nat. Rev. Genet.* 1 (2): 91–99.

25 Patil, S., Gao, Y.G., Lin, X. et al. (2019). The development of functional non-viral vectors for gene delivery. *Int. J. Mol. Sci.* 20 (21).

26 van Haasteren, J., Li, J., Scheideler, O.J. et al. (2020). The delivery challenge: fulfilling the promise of therapeutic genome editing. *Nat. Biotechnol.* 38 (7): 845–855.

27 Dowdell, J., Erwin, M., Choma, T. et al. (2017). Intervertebral disk degeneration and repair. *Neurosurgery* 80 (3S): S46–S54.

28 Yoon, S.T., Park, J.S., Kim, K.S. et al. (2004). ISSLS prize winner: LMP-1 upregulates intervertebral disc cell production of proteoglycans and BMPs *in vitro* and *in vivo*. *Spine (Phila Pa 1976)* 29 (23): 2603–2611.

29 Nishida, K., Kang, J.D., Gilbertson, L.G. et al. (1999). Modulation of the biologic activity of the rabbit intervertebral disc by gene therapy: an in vivo study of adenovirus-mediated transfer of the human transforming growth factor beta 1 encoding gene. *Spine (Phila Pa 1976)* 24 (23): 2419–2425.

30 Paul, R., Haydon, R.C., Cheng, H. et al. (2003). Potential use of Sox9 gene therapy for intervertebral degenerative disc disease. *Spine (Phila Pa 1976)* 28 (8): 755–763.

31 Seki, S., Asanuma-Abe, Y., Masuda, K. et al. (2009). Effect of small interference RNA (siRNA) for ADAMTS5 on intervertebral disc degeneration in the rabbit anular needle-puncture model. *Arthritis Res. Ther.* 11 (6): R166.

32 Leckie, S.K., Bechara, B.P., Hartman, R.A. et al. (2012). Injection of AAV2-BMP2 and AAV2-TIMP1 into the nucleus pulposus slows the course of intervertebral disc degeneration in an in vivo rabbit model. *Spine J.* 12 (1): 7–20.

33 Liu, J., Roughley, P.J., and Mort, J.S. (1991). Identification of human intervertebral disc stromelysin and its involvement in matrix degradation. *J. Orthop. Res.* 9 (4): 568–575.

34 Vo, N.V., Hartman, R.A., Yurube, T. et al. (2013). Expression and regulation of metalloproteinases and their inhibitors in intervertebral disc aging and degeneration. *Spine J.* 13 (3): 331–341.

35 Guiot, B.H. and Fessler, R.G. (2000). Molecular biology of degenerative disc disease. *Neurosurgery* 47 (5): 1034–1040.
36 Wallach, C.J., Sobajima, S., Watanabe, Y. et al. (2003). Gene transfer of the catabolic inhibitor TIMP-1 increases measured proteoglycans in cells from degenerated human intervertebral discs. *Spine (Phila Pa 1976)* 28 (20): 2331–2337.
37 Risbud, M.V. and Shapiro, I.M. (2014). Role of cytokines in intervertebral disc degeneration: pain and disc content. *Nat. Rev. Rheumatol.* 10 (1): 44–56.
38 Thompson, J.P., Oegema, T.R. Jr.,, and Bradford, D.S. (1991). Stimulation of mature canine intervertebral disc by growth factors. *Spine (Phila Pa 1976)* 16 (3): 253–260.
39 Walsh, A.J., Bradford, D.S., and Lotz, J.C. (2004). *In vivo* growth factor treatment of degenerated intervertebral discs. *Spine (Phila Pa 1976)* 29 (2): 156–163.
40 Okuda, S., Myoui, A., Ariga, K. et al. (2001). Mechanisms of age-related decline in insulin-like growth factor-I dependent proteoglycan synthesis in rat intervertebral disc cells. *Spine (Phila Pa 1976)* 26 (22): 2421–2426.
41 Gruber, H.E., Norton, H.J., and Hanley, E.N. Jr., (2000). Anti-apoptotic effects of IGF-1 and PDGF on human intervertebral disc cells *in vitro*. *Spine (Phila Pa 1976)* 25 (17): 2153–2157.
42 Li, J., Yoon, S.T., and Hutton, W.C. (2004). Effect of bone morphogenetic protein-2 (BMP-2) on matrix production, other BMPs, and BMP receptors in rat intervertebral disc cells. *J. Spinal Disord. Tech.* 17 (5): 423–428.
43 Yonemori, K., Imamura, T., Ishidou, Y. et al. (1997). Bone morphogenetic protein receptors and activin receptors are highly expressed in ossified ligament tissues of patients with ossification of the posterior longitudinal ligament. *Am. J. Pathol.* 150 (4): 1335–1347.
44 Takae, R., Matsunaga, S., Origuchi, N. et al. (1999). Immunolocalization of bone morphogenetic protein and its receptors in degeneration of intervertebral disc. *Spine (Phila Pa 1976)* 24 (14): 1397–1401.
45 Lee, J.Y., Hall, R., Pelinkovic, D. et al. (2001). New use of a three-dimensional pellet culture system for human intervertebral disc cells: initial characterization and potential use for tissue engineering. *Spine (Phila Pa 1976)* 26 (21): 2316–2322.
46 Tim Yoon, S., Su Kim, K., Li, J. et al. (2003). The effect of bone morphogenetic protein-2 on rat intervertebral disc cells *in vitro*. *Spine (Phila Pa 1976)* 28 (16): 1773–1780.
47 Kim, K.S., Yoon, S.T., Park, J.S. et al. (2003). Inhibition of proteoglycan and type II collagen synthesis of disc nucleus cells by nicotine. *J. Neurosurg.* 99 (3 Suppl.):: 291–297.
48 Gilbertson, L., Ahn, S.H., Teng, P.N. et al. (2008). The effects of recombinant human bone morphogenetic protein-2, recombinant human bone morphogenetic protein-12, and adenoviral bone morphogenetic protein-12 on matrix synthesis in human annulus fibrosis and nucleus pulposus cells. *Spine J.* 8 (3): 449–456.

49 Masuda, K., Takegami, K., An, H. et al. (2003). Recombinant osteogenic protein-1 upregulates extracellular matrix metabolism by rabbit annulus fibrosus and nucleus pulposus cells cultured in alginate beads. *J. Orthop. Res.* 21 (5): 922–930.

50 Zhang, Y., An, H.S., Song, S. et al. (2004). Growth factor osteogenic protein-1: differing effects on cells from three distinct zones in the bovine intervertebral disc. *Am. J. Phys. Med. Rehabil.* 83 (7): 515–521.

51 Li, J., Kim, K.S., Park, J.S. et al. (2003). BMP-2 and CDMP-2: stimulation of chondrocyte production of proteoglycan. *J. Orthop. Sci.* 8 (6): 829–835.

52 Cheng, H., Jiang, W., Phillips, F.M. et al. (2003). Osteogenic activity of the fourteen types of human bone morphogenetic proteins (BMPs). *J. Bone Joint Surg. Am.* 85 (8): 1544–1552.

53 Helm, G.A., Li, J.Z., Alden, T.D. et al. (2001). A light and electron microscopic study of ectopic tendon and ligament formation induced by bone morphogenetic protein - 13 adenoviral gene therapy. *J. Neurosurg.* 95 (2): 298–307.

54 Mwale, F., Demers, C.N., Petit, A. et al. (2003). A synthetic peptide of link protein stimulates the biosynthesis of collagens II, IX and proteoglycan by cells of the intervertebral disc. *J. Cell. Biochem.* 88 (6): 1202–1213.

55 Nohe, A., Keating, E., Knaus, P., and Petersen, N.O. (2004). Signal transduction of bone morphogenetic protein receptors. *Cell. Signalling* 16 (3): 291–299.

56 Hatakeyama, Y., Nguyen, J., Wang, X. et al. (2003). Smad signaling in mesenchymal and chondroprogenitor cells. *J. Bone Joint Surg. Am.* 85-A (Suppl. 3): 13–18.

57 Aigner, T., Gebhard, P.M., Schmid, E. et al. (2003). SOX9 expression does not correlate with type II collagen expression in adult articular chondrocytes. *Matrix Biol.* 22 (4): 363–372.

58 Pan, H., Li, X., Wang, J. et al. (2015). LIM mineralization protein-1 enhances bone morphogenetic protein-2-mediated osteogenesis through activation of ERK1/2 MAPK pathway and upregulation of Runx2 transactivity. *J. Bone Miner. Res.* 30 (8): 1523–1535.

59 Fernandez-Moure, J., Moore, C.A., Kim, K. et al. (2018). Novel therapeutic strategies for degenerative disc disease: review of cell biology and intervertebral disc cell therapy. *SAGE Open Med.* 6: 2050312118761674.

60 Borde, B., Grunert, P., Härtl, R., and Bonassar, L.J. (2015). Injectable, high-density collagen gels for annulus fibrosus repair: an *in vitro* rat tail model. *J. Biomed. Mater. Res. Part A* 103 (8): 2571–2581.

61 Chik, T.K., Ma, X.Y., Choy, T.H. et al. (2013). Photochemically crosslinked collagen annulus plug: a potential solution solving the leakage problem of cell-based therapies for disc degeneration. *Acta Biomater.* 9 (9): 8128–8139.

62 Bowles, R.D., Gebhard, H.H., Härtl, R., and Bonassar, L.J. (2011). Tissue-engineered intervertebral discs produce new matrix, maintain disc height, and restore biomechanical function to the rodent spine. *Proc. Natl. Acad. Sci. U.S.A.* 108 (32): 13106–13111.

63 Yin, W., Pauza, K., Olan, W.J. et al. (2014). Intradiscal injection of fibrin sealant for the treatment of symptomatic lumbar internal disc disruption: results of a prospective multicenter pilot study with 24-month follow-up. *Pain Med.* 15 (1): 16–31.

64 Berlemann, U. and Schwarzenbach, O. (2009). An injectable nucleus replacement as an adjunct to microdiscectomy: 2 year follow-up in a pilot clinical study. *Eur. Spine J.* 18 (11): 1706–1712.

15

New Prospects in Skin Tissue Engineering and Fabrication

Lei Wang, Xiaonan Liu, and Ke Shen

Nanfang Hospital of Southern Medical University, Department of Orthopaedics, No. 1838, Guangzhou Avenue. North, Baiyun District, Guangzhou 510515, P.R. China

15.1 Skin Anatomy and Related Cell Types

The skin is the largest organ of the human body. It provides a role for external physical, chemical, or biological agents, and temperature regulation. The skin is a layered epithelial tissue derived from the embryo's ectoderm. It consists of three layers: epidermis, dermis, and subcutaneous tissue (subcutaneous layer) (Figure 15.1) [1].

The epidermis is the upper layer of the skin, which provides the human body with the first line of defense. Thanks to its waterproof properties, pathogens and other foreign bodies are isolated from the body. It is multi-layered, with cells extending from the basement membrane to the dermis. The basement membrane contains progenitor cells that differentiate into keratinocytes. Keratinocytes differentiate and mature as they move to the surface of the skin [2]. It is necessary that the cells are renewed quickly because the cells must be replaced immediately when they are lost. The keratinized layer of dead cells on the skin provides barrier properties for the skin [2]. There are skin pigments in the basal layer of the epidermis, that is, melanin produced by melanocytes. Melanin also helps filter out the ultraviolet rays of the sun. The epidermis also has Langerhans cells, part of the skin's immune system.

The dermis is the thickest of the three layers of skin and exists directly under the epidermis. It is a connective tissue composed of extracellular matrix (ECM), fibroblasts, vascular endothelial cells (ECs), hair follicles, sweat glands, sebaceous glands, blood vessels, and nerve endings [2]. Fibroblasts are the main population of the dermis, which is responsible for the production and maintenance of skin structural elements. The secreted collagen and elastin combine with non-fibrous substances (such as glycosaminoglycans [GAG]) to form the ECM. The ECM also supports the basement membrane to ensure the integrity of the dermal–epidermal junction (DEJ) [3]. Organized organizational renewal depends on ECM. Generally, the turnover rate of collagen is low, but it occurs more frequently during damage repair.

Lei Wang, Xiaonan Liu, and Ke Shen contributed equally.

Biofabrication for Orthopedics: Methods, Techniques, and Applications, First Edition.
Edited by Wenguo Cui, Xin Zhao, and Shen Liu.

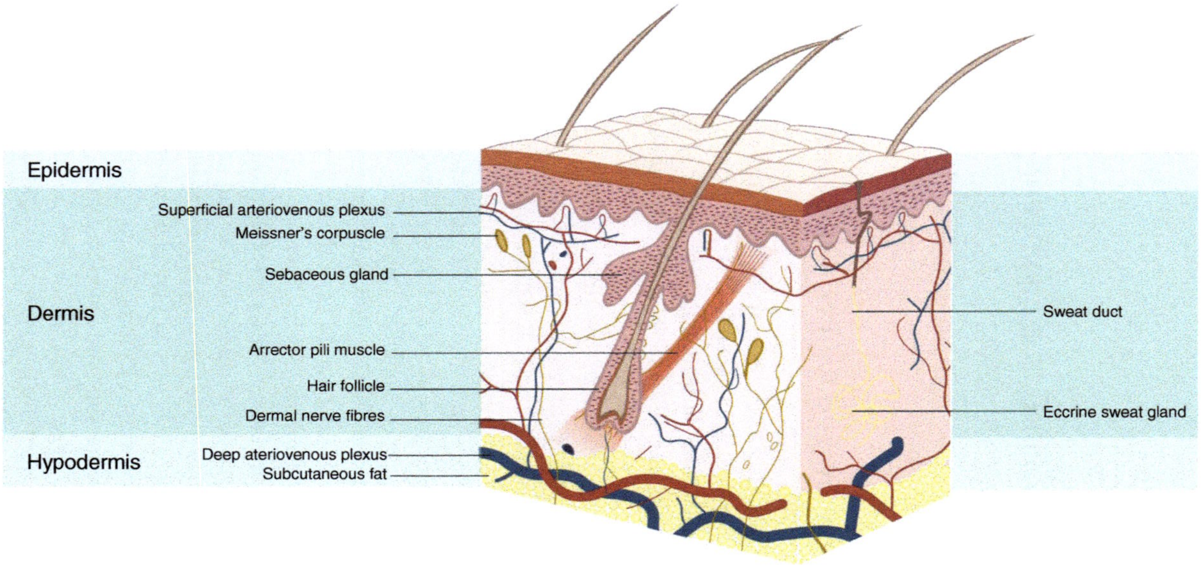

Figure 15.1 Schematic representation of the skin structure. Source: Weng et al. [1] / Springer Nature / CC BY-4.0.

Under the dermis is the subcutaneous layer of fatty tissue. It provides insulation and cushioning between the skin and skeletal structures (such as bones and muscles). It is also used as an energy storage area [2].

15.2 Types of Skin Wounds and Wound Healing Process

Skin damage is one of the most challenging problems in regenerative medicine. Wounds are generally considered injuries to the outer or inner surface of the body caused by burns, cuts, pressure, or trauma. Every trauma often results in acute skin wounds. After the repair process, these wounds often lead to a benign scar if not treated on time [3]. As a category, chronic wounds can be divided into four main types: diabetic ulcers, venous ulcers, pressure ulcers, and post-operative wounds [3–7].

In all organ systems, the normal mammalian response to injury occurs in three overlapping but distinct stages: inflammation, new tissue formation, and remodeling. In the first stage, inflammation occurs immediately after tissue damage (Figure 15.2a). Combining the coagulation cascade, inflammatory pathways, and the immune system components prevents the continuous loss of blood and body fluids, removes dead and inactivated tissues, and prevents infection. The process of hemostasis involves forming a platelet embolism, then forming a fibrin matrix, and becoming a scaffold for infiltrating cells. Neutrophils migrate to wounds under complement activation, degranulation of platelets, and recruitment of bacterial degradation products [2]. Monocytes appear in the wound and differentiate into macrophages in two to three days. Recent data indicate that redundancy in the inflammatory response can compensate for the lack of neutrophils and macrophage [8]. In the absence of both types of cells, the repair of minor wounds can still be carried out, and the scar formation reaction is even less [9].

As the second stage of wound repair, new tissue generation usually occurs 2–10 days after injury and is characterized by cell proliferation and migration of different cell types (Figure 15.2b). The first event is the migration of keratinocytes on the injured dermis (the inner layer of the skin). Then, new angiogenesis, the sprouting of capillaries associated with fibroblasts, and macrophages replace the fibrin matrix with granulation tissue, forming a new substrate for keratinocyte migration in the later repair process. The keratinocytes located behind the front edge proliferate and mature and eventually restore epithelium's barrier function. The most important positive regulators of angiogenesis in the process claimed by heart tissue are vascular endothelial growth factor A (VEGFA) and fibroblast growth factor 2 (FGF2; also known as bFGF) [10]. For example, the application of VEGFA alone on wounds in diabetic animal models (wound repair disorders) can normalize healing [11]. Studies have reported that angiogenesis can also result from the recruitment of bone marrow (BM)-derived endothelial progenitor cells. However, this contribution is small – at least in non-ischemic wounds (where the oxygen concentration is normal) [12]. In the latter part of the second stage, macrophages stimulate fibroblasts to differentiate into myofibroblasts [13]. Over time, the edges

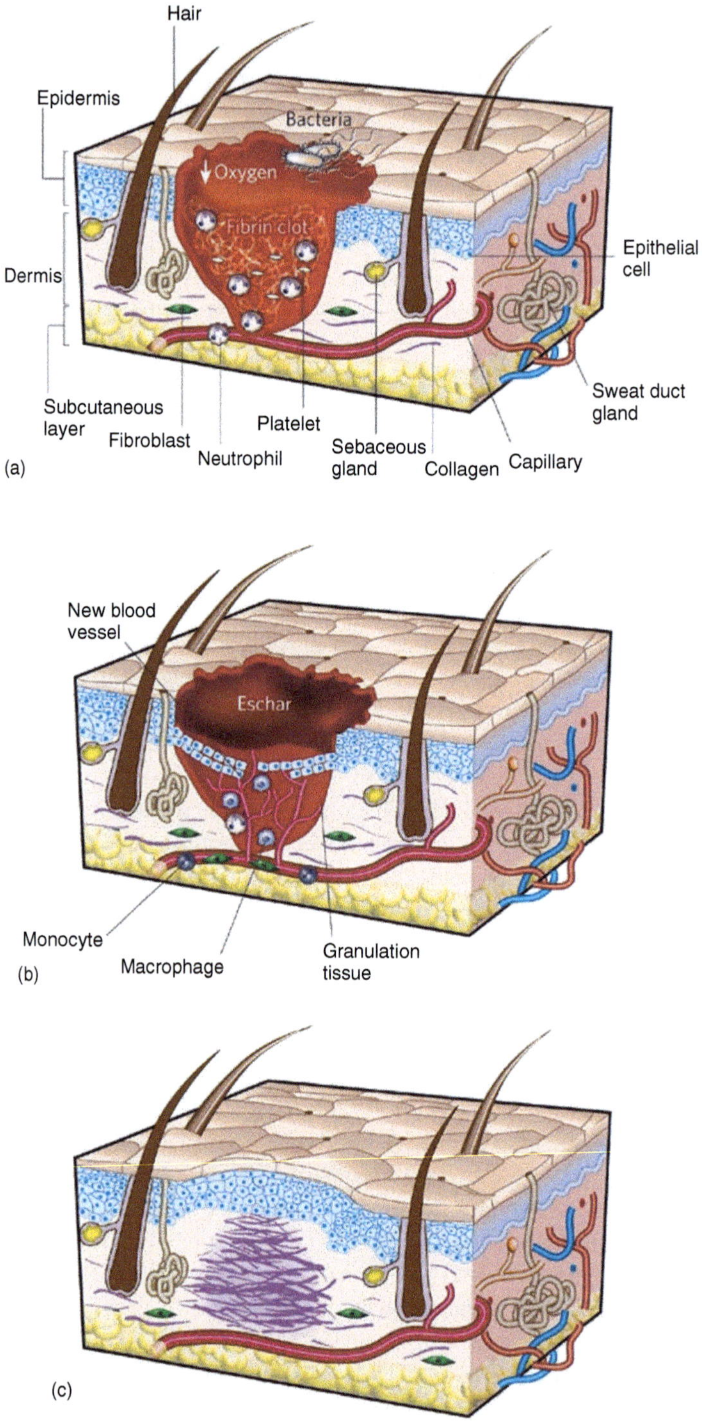

Figure 15.2 The various stages of skin wounds and wound healing. (a) Inflammation. (b) New tissue formation. (c) Remodeling. Source: Atiyeh et al. [5] / Reprinted with permission of Springer Nature.

of the wound will be gathered together by contractile myofibroblasts. The ECM produced by the interaction of fibroblasts and myofibroblasts mainly exists in the form of collagen, which eventually forms the majority of mature scars [14].

The third stage of wound repair, remodeling begins two to three weeks after the injury, and lasts for a year or more (Figure 15.2c). All processes activated after injury at this stage end and stop. Most endothelial cells, macrophages, and myofibroblasts will undergo apoptosis or flow out of the wound, and the remaining cells are mainly composed of collagen and other ECM proteins. Epithelial–mesenchymal interactions may continue to regulate skin integrity and homeostasis [15]. And there must be other feedback loops to maintain different cell types in the skin. In addition, within 6–12 months, the matrix metalloproteinases secreted by fibroblasts, macrophages, and endothelial cells promote the active remodeling of the acellular matrix from mainly type III collagen to mainly composed of type I collagen, enhancing the repaired organization [16]. However, the tissue will never recover the properties of the uninjured skin [17].

15.3 Treatments for Wound Healing

Regenerative medicine researchers have set the gold standard for wound treatment. It any therapy, three steps are essential: removing dead cells and necrotic tissue from the wound site, promoting angiogenesis in the wound bed, and inducing wound repair and epidermal re-epithelialization or complete replacement. Below we will enumerate the treatment methods in various directions to explain.

Inflammation regulation: The inflammation stage is one of the critical stages of wound repair [8]. Inflammation can prevent wound site infection and participate in the removal of apoptosis and necrotic cells [18]. Methods of applying inflammation modulators or anti-inflammatory strategies to promote wound healing have been proposed. Such as topical application of polyunsaturated fatty acids (i.e. docosahexaenoic acid and eicosapentaenoic acid), electrospun fiber scaffolds, or wound dressings containing anti-inflammatory agents [19–21], and macrophage/monocyte activation agent glucan [22]. In addition, Browne and Pandit reported the application of different biomaterials in delivery, including various anti-inflammatory drugs, such as cells (macrophages, mesenchymal stem cell [MSC]), various inflammatory mediators and cytokines (interleukin-10 [IL-10] and tumor necrosis factor alpha [TNF-α]), and polymer scaffolds, such as hyaluronic acid [18]. Many other researchers have also reported the use of biomaterials as delivery systems for anti-inflammatory agents in regenerative medicine [23, 24].

In the proliferation stage, another critical stage in the wound healing process, maintaining the balance between the proliferation and final maturation of these cells is essential for the complete healing of the wound [25]. The strategy to accelerate wound healing is to promote and/or maintain the proliferation of keratinocytes or inhibit the terminal differentiation of keratinocyte progenitor cells [26, 27]. Studies have found that Rho and ROCK signal pathways are involved in the regulation of epithelial cell proliferation and terminal differentiation [28, 29].

The results of Strudwick et al. [30] showed that the Y-27632 treatment of human keratinocytes can increase the proliferation of keratinocytes without changing the expression of differentiation markers. In another study, Tang et al. [31] showed that the topical application of Tiger-17 in a mouse full-thickness wound model can reduce the wound area from 67.3% on the second day after injury to 5.9% on the 9.67 day. The use of small molecules targeted to promote skin cell proliferation or inhibit skin damage can be a strategy for faster wound healing.

In the tissue engineering triangle, in addition to scaffolds and GFs, cells (differentiated and SCs) are one of the critical angles for engineering or regenerating damaged tissues [32–35]. Among the differentiated cells, keratinocytes and fibroblasts are the most common cells in wound-based cell therapy [36, 37]. Reverdin introduced in 1871 a method of using skin grafts to provide positive skin cells to treat skin damage [38]. However, limited by the donor site and immune rejection, in 1981, Burke et al. introduced a skin substitute for extensive treatment of burns [39]. From 1981 to the present, the development of skin substitutes for clinical use has been a topic of general concern among scientists in regenerative skin medicine and tissue engineering. The characteristics of an ideal skin substitute include anti-infection, prevention of water loss, tolerance to shear forces, cost–benefit ratio, easy availability, easy storage and long shelf life, lack of induced immune response, flexibility, and resistance to wounds. The surface is adaptable and convenient for clinicians to use. Currently, there are many types of commercial skin substitutes available for clinical applications: EpiDex [40], Apligraf [41], OrCel [42], and CellSpray [43]. However, the application of skin substitutes still has some limitations, such as high cost, low efficiency, and the inability to reconstruct the skin appendages altogether [44]. On the other hand, the application of SCs is a promising strategy for the development of cell-based therapies [45–47]. The most common types of SCs that can be used for chronic wound healing include MSC, KSC, hair follicle stem cells (HFSCs), and endothelial progenitor cells. However, the low survival rate of cells after transplantation and the possibility of immunogenicity and tumorigenicity are significant obstacles to the development of SC-based therapeutics [46]. The secretory group of SCs without cell therapy strategy is a complex collection of biologically active factors, such as hormones, cytokines, angiogenic factors, GF, ECM, and ECM proteases [48]. The secretion group released from SC contributes to cellular processes, such as cell signaling, cell growth, angiogenesis, and apoptosis [49, 50]. Despite this, more research is needed before clinical translation can be achieved.

EGF, nerve GF, fibroblast GF, platelet-derived GF, granulocyte-macrophage colony-stimulating factor, and vascular endothelial growth factor (VEGF) are some of the GFs that contribute to the wound healing process. Proliferation and migration of cell types (such as keratinocytes and fibroblasts) and induce angiogenesis; reduce inflammation [51–53]. Drug delivery systems have been used to solve short half-life, the possibility of bursting in the body, and the high price of recombinant GFs. Various kinds of scaffolds, constructed from natural, synthetic, or combinatorial biomaterials, have been developed by many researchers [36, 54]. Scaffolds are three-dimensional (3D) promising structures that improve the efficiency of tissue engineering approaches by mimicking the microenvironment condition around the cells [55]. The use of these biomaterial-based systems for cell delivery [56, 57], gene delivery [58, 59], and also bioactive drug delivery [60, 61] has been developed to restore or regenerate defective tissues. Many researchers have reported using scaffolds alone as skin equivalents and cells, GF, or a combination of the two can heal wounds [60, 62, 63]. Many alternative methods, include chemicals, plant extracts, and negative pressure wound therapy [64–67]. Each method has advantages and disadvantages, and the doctor will choose according to the actual situation in clinical practice.

15.4 Skin Regeneration Using Tissue Engineering

The skin is the largest organ of the human body. Because of the existence of stem cells, the injured epidermis can regenerate itself. However, many chronic wounds caused by severe injuries and burns cannot heal. Any loss of full-thickness skin more significant than 4 cm in diameter requires transplantation [68]. Conventional Treatments for Wound Healing usually lack healthy donor tissue, and the use of foreign bodies poses the risk of infection and immune rejection. Tissue-engineered skin (TES) substitutes are alternatives to traditional wound healing strategies and tissue regeneration, with many applications. TES substitutes for wound healing have evolved tremendously over the past couple of years. New advances have been made toward developing skin substitutes made up of artificial and natural materials [69–71]. Skin substitutes act as a temporary protective cover of the wound bed, thus protecting damaged regions from fluid loss and contamination and accelerating the wound-healing processes by promoting the release of cytokines and growth factors at the wound site [72]. TES is mainly of three types: cultured epidermal cells with no dermal components, with only dermal features, and a bilayer containing dermal and epidermal details [73]. The presence of many cells, especially stem cells, in TES enables the regeneration of native-like skin in burn patients. Although each has its own advantages and applications, none of them can completely replace native skin. Their limitations include reduced angiogenesis, poor mechanical integrity, failure to integrate, scarring, immune rejection, and high price [74]. Although there is an urgent need to improve tissue-engineered skin substitutes due to practical and therapeutic limitations, this field has come a long way in patient rehabilitation and is indeed expected to play a role in skin and wound healing soon. There are currently several skin substitutes, but skin substitutes composed of stem cells and biological materials are still a promising solution in the future.

15.4.1 Local Stem Cells

The skin provides the interface between the organism and the external environment, covering the entire surface area of the human body. Skin is a complex tissue that contains various epidermis [38], such as epidermis, dermal epidermis, and hair follicle-derived epidermis. The epidermis is characterized by a layered structure and acts as a protective layer, including controlling body temperature, sensory perception, maintaining hydration, and preventing environmental damage. Although the epidermis has the ability to repair and renew itself, serious skin injuries (such as burns) can impair the ability to regenerate and may be life-threatening in severe cases. As the skin is an organ with immune function, autologous skin transplantation is the only feasible method to cover severe skin wounds permanently. Despite the development of tissue-saving autograft techniques including Meek and mesh grafts, the number of skins that can be transplanted severely limits the treatment strategies for extensive skin injuries. In recent years, the application of EPSC in skin regeneration has attracted the attention of researchers. EPSC residing in the skin and its appendages, such as sweat glands and hair bulbs [75–77], are proliferative units that exhibit slow and limited divisions, can renew themselves, and are responsible for generating of various lineages that exist in mature skin [78]. At present, it is much easier to isolate and obtain specific EPSCs with clear characteristics by using fluorescence-activated cell sorting (FACS) and a set of biomarkers [79]. In addition to isolating EPSC from skin tissue, EPSC can also be produced from induced pluripotent stem cells (iPSCs) [80]. Recent investigations have reported that EPSC can be produced by human CD200+ and ITGA6+ iPSC to reconstruct hair follicle lineage and intravesicular epidermis [81]. In addition, improvements to the EPSC culture system, such as serum-free and feeder-free, medium with small molecules, and a 3D culture system are vital for expanding EPSC and maintaining EPSC homeostasis *in vitro*. The combined use of EPSC and biomaterial scaffolds provides a promising regeneration strategy for engineered skin, stem cell delivery, and regeneration of damaged skin tissue [82, 83].

Epidermal substitutes (some commercially available) appear in various forms, such as fusion or pre-fusion autologous or allogeneic keratinocytes on the delivery system or in combination with aerosol spray methods to facilitate delivery to the wound cells (see examples in Figure 15.3). Some examples include EpiCel and EpiDex [17]. In a retrospective study, EpiDex is composed of ORS keratinocytes derived from scalp hair follicles, used to treat chronic wounds, and can cure three-quarters of intractable chronic leg ulcers [8]. Cryderman is an example of an allogeneic keratinocyte sheet. In order to limit any delay *in vitro* expansion and maintain the proliferative phenotype, cells have also grown to a pre-confluent state on delivery systems that promote transplantation (such as Laserskin® and Myskin) [85, 86]. It has also been reported to use fibrin glue in combination with aerosol methods to deliver cell suspensions of keratinocytes to wounds in the form of BioSeed-S and CellSpray [87, 88]. The successful clinical application of these epidermal substitutes in the treatment of venous and diabetic ulcers and burn wounds have been described in various studies [81, 89, 90].

Because epidermal stem cells have the potential to regenerate the skin, they provide a convenient carrier for gene manipulation and offer an excellent new treatment option. A phase I clinical trial involving the application of retroviral infection delivery in patients with recessive dystrophic epidermolysis bullosa (a hereditary vesicular disease caused by mutations in the *COL7A1* gene) including corrected

A

Product name	Company	Description	Uses
Epidermal substitutes			
Epicel	Genzyme Corp., Cambridge, MA, USA	Confluent cultured autologous keratinocyte sheet delivered on petroleum gauze backing	Burn wounds, acute wounds; chronic wounds
Cryoskin	Altrika Ltd., Sheffield, UK	Confluent cultured allogenic keratinocyte sheet on silicone backing	Burn wounds, chronic wounds; donor-site wounds
CellSpray	Avita Medical, Northridge, CA, USA	Subconfluent suspension of proliferative keratinocytes applied to wounds via spraying	Partial-thickness wounds; donor-site wounds
EpiDex	Modex Therapeutics, Lausanne, Switzerland	Confluent cultured autologous keratinocyte sheet from ORS cells from hair follicles on silicone membrane	Full-thickness wounds; burn wounds; chronic wounds
MySkin	Altrika Ltd., Sheffield, UK	Subconfluent cultured autologous keratinocytes grown on silicone support treated with plasma polymer film	Partial-thickness wounds; burn wounds; chronic wounds; donor-site wounds
Celaderm	Celadon Science LLC, Brookline, MA, USA	Living foreskin-derived allogenic keratinocytes	Partial- and full-thickness wounds; burn wounds, chronic wounds
BioSeed-S	BioTissueTechnologies, Freiburg, Germany	Autologous keratinocytes in fibrin glue	Burn wounds; chronic wounds
Biobrane	Smith & Nephew, Hull, UK	Bilaminar membrane with silicone layer bonded to nylon coated with peptides derived from porcine collagen type I	Partial-thickness and full-thickness wounds; burn wounds; donor-site wounds
Suprathel	Stapleline GmbH, Bochum, Germany	Acellular synthetic co-polymer based on DL-lactide and contains triethylenecarbonate and ε-caprolactone	Burn wounds; donor-site wounds

Figure 15.3 *(Continued)*

Laserskin	Fidia Advanced Biopolymers, Abano Terme, Italy	100% esterified hyaluronic acid membrane with laser-drilled micropores seeded with autologous keratinocytes	Partial-thickness wounds; burn wounds, chronic wounds; vitiligo treatment

B

Product name	Company	Description	Uses
Dermal substitutes			
AlloDerm	LifeCell Corporation, The Woodlands, TX, USA	Human allogenic acellular dermis (cadaveric)	Full-thickness and burn wounds; chronic wounds; reconstruction
Hyalomatrix	Fidia Advanced Biopolymers, Abano Terme, Italy	Hyaluronic acid matrix with variable esterification attached to silicone membrane as temporary epidermis	Burns, acute and chronic wounds
Hyalograft-3D	Fidia Advanced Biopolymers, Abano Terme, Italy	Esterified hyaluronic acid matrix with autologous fibroblasts attached to silicone membrane as temporary epidermis	Partial- and full-thickness wounds; burns, acute; chronic wounds
Dermagraft	Organogenesis Inc., Canton, MA, USA	Bioabsorbable polyglactin mesh with living cultured allogenic neonatal foreskin-derived fibroblasts	Full-thickness wounds; burn wounds; chronic wounds; epidermolysis bullosa
Integra	Integra LifeSciences Corporation, Plainsboro, NJ, USA	Bovine collagen type I and shark chondroitin-6-sulfate attached to silicone membrane as temporary epidermis	Full-thickness wounds; burns wounds; acute wounds; chronic wounds
Matriderm	Medskin Solutions, Billerbeck, Germany	Acellular scaffold composed of bovine collagens types I, II, V, and elastin	Full-thickness wounds; burn wounds
Strattice	LifeCell Corporation, Bridgewater, NJ, USA	Allogenic porcine acellular dermis	Reconstruction

Figure 15.3 (*Continued*)

Trancyte	Advanced BioHealing, Inc., Westport, CT, USA	Nylon mesh coated with porcine collagen with non-viable cultured neonatal foreskin-derived fibroblasts attached to silicone membrane as temporary epidermis	Partial-thickness and full-thickness wounds; burn wounds
Ez-Derm	Brennen Medical, Inc., St. Paul, MN, USA	Perforated or non-perforated cross-linked porcine collagen	Partial-thickness wounds; burn wounds; chronic wounds
EpiFix	MiMedx Group Inc., Marietta, GA, USA	Human amniotic membrane	Full-thickness wounds; acute wounds; chronic wounds
Oasis	Smith & Nephew, Hull, UK	Porcine small intestinal submucosa	Partial-thickness, full-thickness; acute and chronic wounds

C

Product name	Company	Description	Uses
Composite skin equivalents			
Apligraf	Organogenesis Inc., Canton, MA, USA	Bilayered matrix composed of bovine collagen type I with living cultured allogenic neonatal foreskin-derived fibroblasts and keratinocytes	Full-thickness wounds, burn wounds; acute wounds; chronic wounds; donor-site wounds
OrCel	Forticell Bioscience, Inc., New York, NY, USA	Bilayered matrix composed of Bovine collagen type I with living cultured allogenic epidermal keratinocytes and dermal fibroblasts	Full-thickness wounds; burn wounds; chronic wounds; donor-site wounds; epidermolysis bullosa
TissueTech	Fidia Advanced Biopolymers, Abano Terme, Italy	Hyalograft 3D and Laserskin combination	Chronic wounds
Theraskin	Soluble Systems, Newport News, VA, USA	Human cadaveric allograft skin containing donor fibroblasts and keratinocytes	Chronic wounds

Figure 15.3 (*Continued*)

StrataGraft	Stratatech Corporation, Madison, WI, USA	Dermal equivalent containing human dermal fibroblasts and stratified epidermis derived from genetically stable, non-tumorigenic human keratinocyte progenitors, NIKS cells	Burn wounds; chronic wounds

Figure 15.3 Some examples of skin scaffolds. (A) Epidermal substitutes; (B) dermal substitutes; and (C) composite substitutes. Source: Ojeh et al. [84] / MDPI / CC BY-4.0.

treatment of chronic wounds of autologous epidermal flakes with wild-type VII collagen keratinocytes is still in progress [91, 92]. It is further proved that patients with bullous junctional epidermolysis can successfully produce phenotypically normal skin on severe combined immunodeficiency (SCID) mice by transfecting keratinocytes [93]. It was also reported that the transduction of primary keratinocytes from laminin B3 cDNA in patients with borderline epidermolytic bullae led to the successful completion of epidermal regeneration on both legs of the patient during the entire one-year follow-up [94]. Stem cell therapy is a developing therapy that has been proven to be beneficial for both acquired and hereditary diseases and therefore has broad prospects in the treating devastating and difficult-to-treat skin diseases.

In humans, other subpopulations of stem cells are located in hair follicles, such as melanocyte stem cells, mesenchymal-like stem cells derived from the dermal sheath (DS) and the dermal papilla (DP), and nestin-positive stem cells. The interstitial part of the hair follicle is composed of DP, an almond-shaped structure containing mesodermal-derived fibroblasts surrounded by the bulbous epithelium of the hair follicle and DS adjacent to the DP. It has been shown that DP and DS play a crucial role in the formation, growth, and support of hair follicles [95–98]. According to reports, pluripotent stem cells in the dermis originate from DP and DS [99] and can differentiate into smooth muscle cells, neurons, glial cells, and adipocytes [100]. In addition, it has been shown that DS and DP cells show adipogenic and osteogenic differentiation *in vitro* [101] and show hematopoietic activity *in vivo* and *in vitro* [102]. It is speculated that DP and DS cells may be fibroblast progenitor cells and participate in skin repair during wound healing [103]. According to reports, in human models, transplanting male DS onto female skin wounds can produce new follicles and fibers without any apparent signs of rejection [104].

Melanocyte stem cells are located in the swelling area and sprouting of hair follicles and express dopachrome tautomerase (Dct), Sox10, and matching box 3 (Pax3) [105–108]. Under a normal steady-state, melanocytes are present in the IFE of human skin and only in adult mouse skin hair follicles [109]. However, in response to skin damage or ultraviolet B (UVB) exposure, melanocyte stem cells have been shown to migrate up to IFE in a melanocortin 1 receptor (Mc1r)-dependent manner, where they differentiate into functional epidermal melanocytes to protect the skin from damage [109].

Nestin, a protein marker of neural progenitor cells, is also expressed in hair follicles [110]. Nestin-positive stem cells originate from the raised area and migrate to DP and surrounding tissues. These cells have broad differentiation potential and can differentiate into various cell lineages, such as neuronal, liver, pancreatic endocrine, cardiomyocytes, and mesenchymal/mesoderm cell lineages [111–115]. Using green fluorescent protein (GFP) tracing technology shows that the labeled nestin-expressing hFSCs in mice can differentiate into multiple lines, such as keratinocytes, melanocytes, neurons, glial cells, and smooth muscle cells [111, 116]. In addition, implanting these stem cells into the severed sciatic nerve gap area can stimulate nerve regeneration and restore nerve function, thereby indicating the complex interaction between the skin, its appendages, and neuronal stem cells [116].

15.4.2 Induced Pluripotent Stem Cells (iPSC)

The skin has a complex structure of appendages and many cell types. This complexity needs to be generalized to simulate human skin for physiological modeling and drug development. However, because of its limited availability and/or growth potential, it is impossible to grasp the complexity of primary cells fully [116]. The development of iPSC technology can overcome the limitations of cell availability because it can obtain cells from non-invasive sources (such as blood) [117] and use a limited number of somatic cells to generate a large number of cells with unlimited growth potential [117]. Then, these highly proliferative iPSCs can be differentiated into many cell lineages to reconstruct different organs including skin. Different groups have successfully constructed 3D human skin with fibroblasts, keratinocytes, and/or melanocytes wholly derived from iPSC, proving the reliability of iPSC-based technology in human skin models. In addition, we differentiated iPSCs into ECs, which were integrated into skin constructs and demonstrated their functions [118, 119]. Differentiation of other cell types has also been reported, such as sensory neurons and DP cells [120]. As the number of cell types from iPSCs increases, there are more opportunities to model the complexity of the skin.

iPSC technology opens up new avenues for possible medical applications, including cell therapy for hereditary skin diseases and full-thickness skin modeling for studying disease pathogenesis and drug screening. EB is an inherited skin fragility disease characterized by a mutation in a gene encoding a protein involved in epidermal–epidermal adhesion, which leads to mechanical blistering and scarring of the skin and mucous membranes [121]. One of the most severe forms is dystrophic EB (DEB). DEB is caused by a mutation in COL7A1 encoding type VII collagen (COL7), an ECM protein secreted by keratinocytes and dermal fibroblasts as homotrimers at the DEJ. Therefore, the sufficient and long-lasting recovery of COL7 that leads to the formation of AF at DEJ is the primary goal of effective treatment of this orphan skin disease. A novel treatment strategy for the treatment of DEB envisages the combination of iPSC derived from the patient's somatic cells with gene editing and tissue engineering techniques. The iPSC established by integration-free reprogramming can be easily differentiated into different skin cell types, including keratinocytes and fibroblasts, and can be used to produce autologous skin structures [122]. This method will avoid immune rejection due to allogeneic, thereby prolonging the life of skin constructs and allowing local production of missing proteins (such as DEB's COL7). In addition, since they have the highest self-renewal and proliferation capacity to withstand genetic manipulation, emerging gene-editing tools (such as CRISPR/Cas9) can quickly correct patient-derived iPSCs, which can be used to develop a personality for specific mutations chemical therapy. The previous results have demonstrated that iPSC fixed by CRISPR genes in DEB patient cells can differentiate into keratinocytes and fibroblasts [123]. In addition, these cells are also used to construct the skin structure, proving the use of iPSC to reconstruct the physiological function of the skin *in vitro/in vitro* [123]. In addition to skin replacement, it also provides other treatment strategies for EB patients, such as cell therapy and chemical drugs. The next generation of human skin structures can be a powerful test platform. To this end, the EB full-thickness skin model developed by iPSC from the patient's own cells will provide a physiologically relevant platform for preclinical screening of these potential drugs, and will further allow the identification of new patient-specific drug formulations to treat EB patients with higher precision [123, 124].

There are still notable limitations in iPSC technology, which may limit the scope of iPSC applications. These limitations include low reprogramming efficiency [125], retains the epigenetic memory of its parental cells [126], variability between iPSC lines [127], and genomic instability [128]. Attempts have been made to find alternative technologies to avoid the limitations associated with iPSC and the ethical issues inherent in embryonic stem cells (ESCs). For this reason, somatic cell nuclear transfer (SCNT) has been developed to convert somatic cells into stem cells by transferring human nuclei into non-nucleated human eggs [129]. This technique is technically challenging, time-consuming, and costly, so there is increasing attention to converting cells directly from one type to another. For example, fibroblasts have been transformed into sensory neurons [130, 131] and ECs [132]. However, there may not always be enough unit numbers for conversion.

Many restrictions related to iPSC may be circumvented. By targeting signal transduction pathways that regulate cell reprogramming, reprogramming efficiency can be improved [133]. Through extensive passaging, the epigenetic memory of parents can be eliminated, and the variation between iPSC lines can be avoided by using many iPSC colonies instead of selecting a single settlement. The instability and tumorigenicity of the genome have brought safety problems for clinical applications. However, many reagents can be used to obtain sufficient reprogramming efficiency without the overexpression of myc and the abolition of the p53 pathway that are commonly used [134]. A system was developed to define non-species, non-integrated, and non-feeding for iPSC reprogramming and differentiation to solve regulatory issues related to clinical applications [135]. Small molecules can also generate kinds of iPSC, although only a few types of iPSC have been successfully obtained from mouse cells [136]. Small molecules have great potential because they are easy to manipulate and avoid the variability associated with DNA, RNA, or protein-based reprogramming methods [136]. Advances in this field will enable the low-cost production of safe iPSCs for cell therapy, disease modeling, and drug development.

15.4.3 Bone Marrow Stem Cells (BMSC)

ESCs can cause cell types of all tissue lineages. However, their application in tissue engineering is limited because of the lack of basic understanding and control of their differentiation of the required specific tissue lineages, and ethical debates [137, 138]. BMSCs are easy to isolate and culture, with almost no ethical concerns. They are expanded *in vitro*, have a low level of immune rejection, and have a low risk of triggering teratoma formation [139]. Therefore, they seem to be an ideal cell source for skin and limb regeneration [38].

At least two different cell lineages in BM are hematopoietic and mesenchymal cells. Hematopoietic cells are produced by hematopoietic stem cells (HSCs)located in the BMSC market segment. Mesenchymal cells are thought to be derived from MSC in BM, although the certainty of MSC is still under investigation [140, 141]. MSC was initially defined as stem cells that can differentiate into a mesenchymal mass spectrum in culture, such as bone cells, chondrocytes, and adipocytes [142]. However, MSCs have also been shown to differentiate into other lineages, including neurons and epithelial cells [143, 144].

Many basic cell studies have emphasized the plastic relationship between skin and BM. In humans who have undergone bone marrow transplantation (BMT), donor cells that have differentiated into keratinocytes can be detected in the epidermis within at least three years [145]. BM cells also contribute to skin development: infusion of GFP BM cells in the uterus of mice will accumulate **a** subpopulation of GFP-positive cells in the uninjured skin dermis, especially related to developing hair follicles [146]. Regarding skin repair, it has been proven that BM from embryonic and postnatal sources can repair genetic defects in the host's skin. Mice lacking skin protein, type VII collagen (COL7), and adult mice lacking type XVII collagen and basement membrane components, can be normalized by stem cell infusion [146–148]. A clinical trial of allogeneic BMT in humans lacking COL7 resulted in a hereditary blistering skin disease, namely recessive dystrophic epidermolysis bullae [149], which has proven that BM cells can repair fragile skin and restores the expression of COL7 in the skin base membrane [150].

Regarding wound healing, many clinical trials have shown that BMSC can leave its niche market, migrate to remote tissues, and play a key role in wound repair and tissue regeneration. Recent studies have shown that BM cells administered systemically can migrate to injured skin and help its repair [151, 152]. When transplanted to burn wounds, engineered skin containing BMSCs showed better healing, less wound contraction, and better blood vessel development [153]. Tamai et al. identified specific BM cells that can be transformed into skin cells to repair damaged skin tissues [154] They discovered a naturally occurring BM cell mobilizer (high-mobility group 1 protein), which can attract BM cells to express platelet-derived growth factor receptor alpha (PDGFRα). Studies have shown that out of 450 BM cells, about one PDGFRα + cell could transform into skin cells and regenerate the skin. A recent investigation [155] showed that transplanted circulating PDGFRα+ cells derived from BM could restore COL7 in skin grafts of recessive dystrophic epidermolysis bullous mice.

The latest progress in the use of BMSC has been reported [156, 157]. GFP-labeled BMSCs were found in the bulging areas of hair follicles, epidermis, and sebaceous glands [157, 158]. Fik-1+ BMSC (H-2Kd, white) of fluorescently labeled CM-DiI BALB/c mice was transplanted into lethally irradiated C57BL/6 mice (H-2KB, black) [159]. Forty days later, the recipient mice grew gray hair primarily composed of donor-derived H-2Kd cells. Although compared with ESCs, BMSCs are easier to obtain and autologous, the properties of adult stem cells depend on culture conditions and donor factors. They may show loss of proliferation and pluripotency [160].

MSCs have great potential for the regeneration of sweat glands. BMSCs directly or indirectly co-cultured with heat shock fusion human sweat gland-like cells (SGCs) may result in the conversion of BMSCs to the SGCs phenotype, and ERK signaling is also involved in this process [156]. However, there are specific difficulties in using this method to regenerate and repair sweat glands. First, the patient's own BM needs to be extracted, which will increase the patient's mental and physical pain. Secondly, *in vitro* induction takes a long time, so it is difficult to obtain many SGCs derived from stem cells in time. Finally, because of the small sample size in clinical trials, it is still necessary to observe long-term clinical effects. Therefore, further clinical research requires whether SGCs can be directly and effectively applied to the esthetic bone resection wounds of patients with large-area deep burns to achieve sweat gland regeneration. The conversion of hUC-MSC to SGC, the effect of EGF on their conversion, and the role of the ERK signaling pathway in the conversion process have been studied [161, 162]. The SGCs obtained by this method can be used for sweat gland regeneration in the early stage of burn treatment. However, its detailed biological characteristics, safety, and whether it can successfully differentiate into sweat gland tissue after being applied to the human body need further research and confirmation. The heated SGC was incubated with BrdU/GFP-labeled MSC at 47 °C for 40 minutes [138]. After 72 hours of culture, GFP tracing and fluorescence staining analysis showed that the MSC had acquired the sweat gland cell phenotype. For further *in vivo* analysis, MSCs co-cultured with SGC were injected subcutaneously into the paws of nude mice. The results clearly show that transplanted MSCs that have undergone transdifferentiation are involved in the regeneration of sweat glands. In addition to these experiments in animal models, *in vivo* tests in humans were also performed. The co-cultured cell suspension is first covered with a piece of the decellularized allogeneic dermal matrix. Then, the granular autologous skin graft is evenly spread on the top of the sheet to heal the epidermis and leave enough space for the exit of the catheter. The regenerated SGCs showed the phenotype of SGC and the function of sweat [138]. Although the number of human trials conducted is relatively small, and the planting area is limited, these observations provide new insights and methods for further exploration of regenerative exocrine sweat gland repair. Another way to obtain the desired sweat glands is to form heterokaryons with MSCs and non-hematopoietic cells or to transfect MSCs with functional genes [163]. Ectodysplasin A (EDA), which can activate the NF-κB transcription factor through EDA receptors, plays a specific role in forming skin appendages [164, 165]. Sweat gland germ cells failed to form in Tabby mice with EDA mutations, but they were restored by EDA transgene or recombinant EDA [166]. Recently, Cai and colleagues used BM-MSC transfected with EDA, and then transplanted BM-MSC/EDA (+) cells into the injured area of an animal burn model [167]. Immunohistochemical analysis and sweat test showed that the combined BM-MSC improved the regeneration of sweat glands both in structure and function. The combination of gene therapy and stem cell therapy has brought new hope for the functional reconstruction of the skin [168]. In addition, adipose-derived MSCs, BMSCs, and iPSCs can be induced to differentiate into Schwann cells, thereby promoting nerve regeneration [169, 170].

Despite the knowledge of endogenous stem cell recruitment, most adult tissues still cannot regenerate perfectly after injury, indicating that these mechanisms can easily be overwhelmed. Attempts to enhance the participation of BMSCs by transplanting/injecting cultured cells have proven this principle [171], but it is generally not effective in various preclinical models [172, 173]. Poor retention of cells in a harsh injury environment and perhaps the use of inadequate numbers of autologous incompletely defined or heterogeneous cell populations may be possible factors that limit the clinical success of stem cell therapy [174, 175].

15.4.4 Summary

Stem cell-based treatment strategies provide a novel and powerful option for wound repair and management. Several cell types currently being explored deeply, such as local stem cells, iPSC and BMSC, have been proven to be quite effective in skin tissue regeneration and have great potential. They can not only regenerate tissues but also promote wound repair through paracrine methods. In addition, the risk of immune rejection is avoided. Nevertheless, stem cells need to be further studied and demonstrated in the reprogramming process, cells' differentiation potential, the difference between iPSCs and ESCs, induced pluripotency, and their future use in clinical therapy. We believe that these problems will be solved with cell biology development, tissue engineering, and regenerative medicine.

15.5 Skin Substitutes

Tissue engineering substitutes can support the regeneration of the epidermis, dermis, or both and are often used in patients with skin trauma. As replacements for animal models in cosmetic testing, skin substitutes are becoming more and more popular in the cosmetics and pharmaceutical industries. From cellular-level therapies (such as the delivery of MSCs or growth factors) to large-scale biomanufacturing technologies, including 3D printing, recent advances in biomedicine have enabled the implementation of unique strategies and new types of biomaterials to bring closer to a substitute for natural leather. We will describe here some of the latest methods of skin regeneration and bioprocessing using tissue engineering technology.

15.5.1 3D Printing

In 2014, Wood proposed the feasibility standard for any regenerative technology for skin regeneration into clinical application. The standard starts with the safety of the procedure, including the selection of cells and biological materials, at affordable health care system costs, to achieve superior quality of results [176].

3D bioprinting is the latest method of applying the wood standard. This technology customizes the combination of cells and extracellular proteins, which will serve as customized *in situ* scaffolds for patients. The printed constructs showed different characteristics from natural skin *in vitro* [177]. Another model framework uses keratinocytes and fibroblasts to print the skin, and the binding matrix is composed of collagen, fibrinogen, and thrombin. Based on Green's method, Ronfard et al. isolated and expanded the patient's keratinocytes in 1991 and used fibrin glue as a fixative for the cells, covering the burn wound completely closed and fused with the skin after one year of treatment. At the same time, research by other teams has also proved in experimental animals that 3D-bioprinted skin can be accepted by the animal body and promote wound closure [178]. On the other hand, 3D bioprinting plays an essential role in supporting skin appendages. One of the most critical challenges in skin regeneration is the lack of skin attachments. By incorporating melanocytes and stem cells into the printed cells, hair follicles with normal circulation can be generated *de novo*. Of course, intradermal hair transplantation may also be a subsequent step of printing [179]. Sweat glands are another vital accessory that keeps the skin moisturized and regulates body temperature. Liu et al. focused on printing 3D matrices with specific pore sizes and geometric shapes, which will promote the differentiation of epidermal progenitor cells into sweat gland-like structures. However, this technology still has apparent flaws; this structure is destroyed when the 3D architecture support is removed [180].

It is currently impossible to print large areas of skin. Lack of immediate blood supply and immune rejection are the two main reasons for the failure to integrate bioengineered skin from allogeneic sources [181, 182]. In addition, nozzle clogging, shear stress on cells, and multi-scale printing (such as vascular trees) are the main challenges faced by skin bioprinting [174, 183]. Although many scientists believe that bioprinting can be the future of reconstructive surgery, research is still needed to overcome the aforementioned obstacles [184]. It is expected that emerging technological advancements in the bioprinting process will soon provide solutions to most current problems [185].

15.5.2 Acellular Skin Substitute

Acellular skin substitutes first appeared in the late 1970s, when they were mainly used as skin substitutes for the epidermis or middle dermis of wounds and burns. Compared with the current skin substitutes, the acellular skin substitutes at that time were relatively simple, mainly composed of nylon mesh or collagen acting as the "dermis" and silicon membrane acting as the "epidermis." There are three common commercial acellular skins. Alternatives: Biobrane®, Integra®, and Alloderm

(Figure 15.5a). It can be used for diseases like epidermolysis bullosa [186] and purulent edema [187]. Detailed information about its composition can be found in Figure 15.4.

15.5.3 Cell Allogeneic Skin Substitutes

Cell allogeneic skin substitutes are mainly produced using neonatal foreskin fibroblasts with mesh or matrix. They have been successfully used in vestibular plasty after peeling off the mucosal-gingival junction and the upper periosteum [188, 189]. Cellular allogeneic skin substitutes also play an essential role in wound treatment in venous and diabetic ulcers, bullous epidermolysis, skin cancer, and burns [188, 189]. Commercial products include TRANSCYTE®, DERMAGRAFT®, Apligraf®, and Graftskin® (Figure 15.5d,f,g).

15.5.4 Cellular Autologous Skin Substitutes

Non-cellular or cellular allogeneic skin substitutes can only temporarily cover the surface of the original skin. If the wound is large, it must be replaced or re-transplanted with a split skin graft. It is necessary to epithelialize the wound surface for more minor injuries gradually. Autologous keratinocytes cultured according to the technique of Rheinwald and Green [190, 191] can provide long-lasting skin coverage for various types of wounds. There are two cellular autologous skin substitutes: cultured epidermal autograft (CEA) and sophisticated skin substitutes (CSS).

CEA requires the cultivation of autologous keratinocytes obtained from a patient's skin biopsy [188, 192] and requires specialized delivery systems or auxiliary dressings for treatment after a few weeks [189, 193]. Currently, commercially available CEAs have different delivery or transportation systems. A significant challenge is that keratinocytes alone may not be able to heal full-thickness wounds or burns. Another problem with CEA is that the junction between the dermis and the epidermis is not fully developed, and it is easy to cause blisters to grow under the action of friction. It may also lead to increased scars, contractures, and hyperkeratosis [188]. In addition, under the digestion of collagenase in the wound bed, the absorption of CEA is reduced to about 30%–80%. To overcome the problem of CEA uptake, another method has been developed [194]. It includes adapting the cadaver allogeneic skin to the wound bed for four days before cadaver transplantation and then placing autologous cells after removing the same epidermis.

CSS is an autograft with epidermal and dermal components. It has an excellent dermal-epidermal connection, is easy to handle, and can be used as a permanent covering. Still, it also has obvious disadvantages, that is, it is expensive and takes a long time to prepare [194]. Different dermal biosynthetic scaffolds are used to develop several types of CSS. For example, hyaluronic acid-derived alternatives are commonly used to prepare CSS. These types of CSS can stimulate the growth and movement of fibroblasts, control osmotic regulation and matrix hydration use, scavenge free radicals and regulate inflammation [195, 196].

Substitute type	Product	Company	Components
Acellular	Alloderm®	LifeCell Inc., Branchburg, NJ, USA	Human acellular lyophilized dermis
	SureDerm	HANS BIOMED Corporation, Daejeon, Korea	Human acellular lyophilized dermis
	OASIS Wound Matrix	Cook Biotech Inc., West Lafayette, IN, USA	Porcine acellular lyophilized small-intestine submucosa
	Biobrane®	Mylan Bertek Pharmaceuticals, USA	Ultrathin silicone as epidermal analog film and 3D nylon filament as dermal analog with type I collagen peptides
	Integra® DRT (dermal regeneration template)	Integra® LifeSciences Corp., USA	Dermal analog—bovine collagen and chondroitin-6-sulfate GAG; epidermal analog—silicone polymer polysiloxane
		Epidermal	
Autologous	EpiDex	Modex Therapeutiques, Lausanne, Switzerland	Cultured keratinocytes from outer root sheath of scalp hair follicles (confluent cell sheet)
	EPIBASE	Laboratoires Genevrier, Antibes, France	Cultured keratinocytes (confluent cell sheet)
	MySkin	CellTran Ltd., UK	Cultured keratinocytes (subconfluent cell sheet) silicone support layer with a specially formulated surface coating
	Bioseed-S	BioTissue Technologies GmbH, Germany	Cultured keratinocytes (subconfluent cell suspension) fibrin sealant
	CellSpray	Clinical Cell Culture (C3), Australia	Non-/cultured keratinocytes (subconfluent cell suspension)
	Epicel®	Genzyme Biosurgery, USA	Sheets of autologous keratinocytes attached to petrolatum gauze support
	Laserskin® or Vivoderm	Fidia Advanced Biopolymers Sri, Italy	Autologous keratinocytes and fibroblasts, grown on microperforated hyaluronic acid membranes

Figure 15.4 *(Continued)*

Substitute type	Product	Company	Components
	Autoderm (Autologous Inferior Dermal Sling)	XCELLentis NV, Belgium	Cultured keratinocytes
	TransDerm	XCELLentis NV, Belgium	Cultured keratinocytes
	Lyphoderm	XCELLentis NV, Belgium	Lyophilized neonatal keratinocytes
	Cryoceal	XCELLentis NV, Belgium	Cryopreserved keratinocytes
		Dermal	
Autologous	denovoDerm™	EUROSKINGRAFT, Switzerland	Autologous dermal substitute
	Pelnac Standard/Fortified	Gunze Ltd., Japan	Porcine-tendon-derived atelocollagen type I, sponge layer with silicone film
	HyalomatrixPA	Fidia Advanced Biopolymers, Italy	HYAFF (an ester of hyaluronic acid) layered on silicone membrane
	Hyalograft 3D	Fidia Advanced Biopolymers, Italy	Cultured fibroblasts hyaluronic acid membrane (HAM)
Allogenic	Dermagraft®	Advanced BioHealing, Inc., USA	Bioabsorbable polygalactin mesh matrix seeded with human neonatal fibroblasts and cryopreserved
	TransCyte®	Advanced BioHealing, Inc., USA	Collagen-coated nylon mesh seeded with allogenic neonatal human foreskin fibroblasts
	Terudermis	Olympus Terumo Biomaterial Corp., Japan	Silicone, bovine lyophilized crosslinked collagen sponge made of heat-denatured collagen
	Cyzact (ICX-PRO)	Intercytex, St John's Innovation Center, UK	Cultured allogeneic human dermal fibroblasts embedded in a human fibrin gel matrix
	ICX-SKN skin graft replacement	Intercytex, St John's Innovation Center, UK	Cultured dermal fibroblasts natural human collagen matrix
	Polycaprolactone collagen nanofibrous membrane	National University of Singapore, Singapore	Cultured dermal fibroblasts polycaprolactone-blended collagen electrospun nanofibrous membrane

Figure 15.4 (*Continued*)

Substitute type	Product	Company	Components
	Tegaderm-nanofiber construct	National University of Singapore, Singapore	Cultured dermal fibroblasts poly(e -caprolactone)/gelatin nanofibrous scaffold electrospun on polyurethane dressing
	Collagen-glycosaminoglycan-chitosan dermal matrix seeded with fibroblasts	INSERM, France	Cultured dermal fibroblasts bovine collagen I/chondroitin-4/6-sulfate/chitosan lyophilized dermal matrix
	Human hair keratincollagen sponge	Southern Medical University, China	Cryomilled porcine acellular diisocyanite cross-linked dermis
	Hyaluronan-FNfds hydrogel matrix	SUNY at Stony Brook, USA	Hyaluronan coupled with fibronectin functional domains
	Composite nano-titanium oxide-chitosan artificial skin (NTCAS)	Cardinal Tien College of Healthcare and Management, Taiwan	Composite nano-titanium oxide-chitosan with gelatin and hyaluronic acid
Xenogeneic	Permacol Surgical Implant	Tissue Science Laboratories pic, UK	Porcine acellular diisocyanite crosslinked dermis
	Matriderm	Dr Suwelack Skin and Healthcare AG, Germany	Bovine non-cross-linked lyophilized dermis, coated with a-elastin hydrolysate
	EZ DermTM	Brennen Medical Inc., USA	Porcine aldehyde cross-linked reconstituted dermal collagen
	Bovine collagen cross-linked with microbial transglutaminase	National University of Ireland, Ireland	Freeze-dried bovine collagen scaffold cross-linked with microbial transglutaminase
	Collatamp	SYNTACOLL AG, Switzerland	Multilayer bovine collagen matrix
Synthetic	Hybrid nanofibrous PLGA/chitosan membrane	Tianjin University, China	PLGA/chitosan hybrid electrospun nanofibrous membrane
	Biodegradable polyurethane microfibers	University of Delaware, USA	Biodegradable polyurethane microfibers

Figure 15.4 *(Continued)*

Substitute type	Product	Company	Components
		Epidermal/dermal (composite)	
Autologous	Permaderm™ (Cincinnati Shriners Skin Substitute)	Regeninic Inc .USA	Autologous fibroblasts and keratinocytes in culture with bovine collagen and GAG substrates
	Tiscover™ (A-Skin)	Advanced Tissue Medicinal Product, Netherlands	Autologous full-thickness cultured skin
	denovoSkin™	EUROSKIN-GRAFT, Univ, of Zurich, Switzerland	Autologous full-thickness substitute consisting of dermal and epidermal layers
	PolyActive	HC Implants BV, Netherlands	Cultured keratinocytes and fibroblasts polyethylene oxide terephthalate (PEO)/polybutylene terephthalate (PBT)
	TissueTech Autograft System (Laserskin and Hyalograft 3D)	Fidia Advanced Biopolymers, Italy	Cultured keratinocytes and fibroblasts microperforated hyaluronic acid membrane (HAM)
Allogenic	Apligraf®	Organogenesis Inc., USA	Bovine collagen matrix seeded with neonatal foreskin fibroblasts and keratinocytes
	OrCel®	Ortec International Inc., USA	Type I collagen matrix seeded with neonatal foreskin fibroblasts and keratinocytes
	Karoskin (Karocells)	Karocell Tissue Engineering AB, Sweden	Native human cadaver skin with dermal and epidermal cells
	CeladermTM	Celadon Science LLC, USA	Sheets of cells derived from neonatal (allogenic) foreskin
	StrataGraft™	Stratatech *Corporation,* USA	Full-thickness skin substitute with dermal and fully differentiated epidermal layers
	AcuDress	DFB Pharmaceuticals, Inc., USA	Cultured keratinocytes fibrin substrate
	Allox	DFB Pharmaceuticals, Inc., USA	Sprayed suspension of allogeneic keratinocytes and fibroblasts in fibrin substrate
Xenogeneic	Oasis®	Healthpoint Biotherapeutics, USA	Intact matrix from porcine small-intestine submucosa and intended for wound closure stimulation in acute, chronic, and burns wounds

Figure 15.4 Commercially available skin substitutes. Source: Based on Ojeh et al. [84].

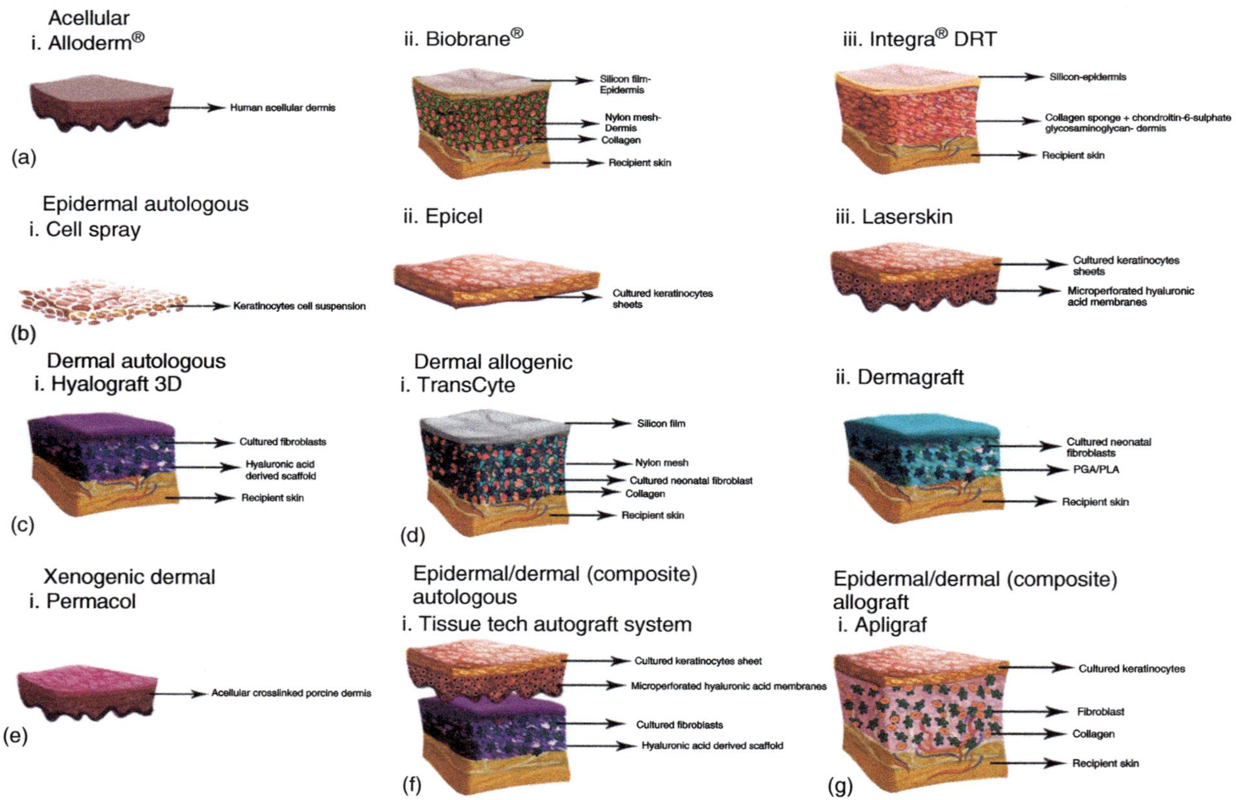

Figure 15.5 Tissue-engineered skin substitutes. (a) Acellular: i. Karoderm, ii. Biobrane, and iii. Integra. (b) Epidermal autologous: i. CellSpray, ii. Epicel, and iii. Laserskin. (c) Dermal autologous: i. Hyalograft 3D. (d) Dermal allogenic: i. TransCyte and ii. Dermagraft. (e) Dermal xenogenic: i. Permacol. (f) Epidermal/dermal (composite) autologous: i. Tissue tech Autograft system. (g) Epidermal/dermal (composite) allograft: i. Apligraf. Source: (a–g) Based on Ojeh et al. [84].

15.5.5 Summary

TES substitutes are important research objects for tissue regeneration and wound healing therapy and will be more widely used in clinical practice in the future. There are limitations in the currently marketed skin substitutes, such as reduced vascularization, poor mechanical integrity, failure to integrate, scarring, and immune rejection. Improving the vascularization of these alternatives to extend their lifespan and provide better integration with host tissues is one of the focuses of current research. One way to eliminate these problems is to simulate more skin characteristics in the body. Cell-based therapy, especially stem cells, has become a promising treatment. Because of their ability to regenerate tissues, MSCs are an excellent choice. Clinical data show that autologous MSC transplantation promotes healing at all stages of wound repair. However, there are difficulties in harvesting and isolating an optimized pool of MSCs with high purity. Therefore, the characterization of MSCs with niche-specific factors is still a challenge for researchers. In order to overcome these limitations, it is necessary to understand the cellular and molecular mechanisms of stem cell action, improve stem cell delivery methods and determine the ideal source. In addition, two to three weeks of cell culture is required before transplantation. Strict storage and preservation conditions, high costs, and the risk of malignant tumors limit its clinical promotion. Better products can be developed through further research on bioengineered skin substitutes.

Acknowledgments

L. Wang, X. Liu, and K. Shen contributed equally to this work. This work was supported in part by the National Natural Science Foundation of China (82002352), the Natural Science Foundation of Guangdong Province (no. 2020A1515010145).

References

1 Weng, T., Wu, P., Zhang, W. et al. (2020). Regeneration of skin appendages and nerves: current status and further challenges. *J. Transl. Med.* 18 (1): 53.
2 Wang, P.H., Huang, B.S., Horng, H.C. et al. (2018). Wound healing. *J. Chin. Med. Assoc.* 81 (2): 94–101.
3 Yazdanpanah, L., Nasiri, M., and Adarvishi, S. (2015). Literature review on the management of diabetic foot ulcer. *World J. Diabetes* 6 (1): 37–53.
4 Lal, B.K. (2015). Venous ulcers of the lower extremity: definition, epidemiology, and economic and social burdens. *Semin. Vasc. Surg.* 28 (1): 3–5.
5 Atiyeh, B.S., Ioannovich, J., Al-Amm, C.A., and El-Musa, K.A. (2002). Management of acute and chronic open wounds: the importance of moist environment in optimal wound healing. *Curr. Pharm. Biotechnol.* 3 (3): 179–195.

6 Tricco, A.C., Cogo, E., Isaranuwatchai, W. et al. (2015). A systematic review of cost-effectiveness analyses of complex wound interventions reveals optimal treatments for specific wound types. *BMC Med.* 13: 90.
7 Smith, M.E., Totten, A., Hickam, D.H. et al. (2013). Pressure ulcer treatment strategies: a systematic comparative effectiveness review. *Ann. Intern. Med.* 159 (1): 39–50.
8 Martin, P. and Leibovich, S.J. (2005). Inflammatory cells during wound repair: the good, the bad and the ugly. *Trends Cell Biol.* 15 (11): 599–607.
9 Martin, P., D'Souza, D., Martin, J. et al. (2003). Wound healing in the PU.1 null mouse – tissue repair is not dependent on inflammatory cells. *Curr. Biol.* 13 (13): 1122–1128.
10 Werner, S. and Grose, R. (2003). Regulation of wound healing by growth factors and cytokines. *Physiol. Rev.* 83 (3): 835–870.
11 Galiano, R.D., Tepper, O.M., Pelo, C.R. et al. (2004). Topical vascular endothelial growth factor accelerates diabetic wound healing through increased angiogenesis and by mobilizing and recruiting bone marrow-derived cells. *Am. J. Pathol.* 164 (6): 1935–1947.
12 Bluff, J.E., Ferguson, M.W., O'Kane, S., and Ireland, G. (2007). Bone marrow-derived endothelial progenitor cells do not contribute significantly to new vessels during incisional wound healing. *Exp. Hematol.* 35 (3): 500–506.
13 Opalenik, S.R. and Davidson, J.M. (2005). Fibroblast differentiation of bone marrow-derived cells during wound repair. *FASEB J.* 19 (11): 1561–1563.
14 Werner, S., Krieg, T., and Smola, H. (2007). Keratinocyte-fibroblast interactions in wound healing. *J. Invest. Dermatol.* 127 (5): 998–1008.
15 Szabowski, A., Maas-Szabowski, N., Andrecht, S. et al. (2000). c-Jun and JunB antagonistically control cytokine-regulated mesenchymal–epidermal interaction in skin. *Cell* 103 (5): 745–755.
16 Lovvorn, H.R., Cheung, D.T., Nimni, M.E. et al. (1999). Relative distribution and crosslinking of collagen distinguish fetal from adult sheep wound repair. *J. Pediatr. Surg.* 34 (1): 218–223.
17 Levenson, S.M., Geever, E.F., Crowley, L.V. et al. (1965). The healing of rat skin wounds. *Ann. Surg.* 161: 293–308.
18 Browne, S. and Pandit, A. (2015). Biomaterial-mediated modification of the local inflammatory environment. *Front. Bioeng. Biotechnol.* 3: 67.
19 Yuan, Z., Zhao, J., Zhu, W. et al. (2014). Ibuprofen-loaded electrospun fibrous scaffold doped with sodium bicarbonate for responsively inhibiting inflammation and promoting muscle wound healing *in vivo*. *Biomater. Sci.* 2 (4): 502–511.
20 Yuan, Z., Zhao, J., Chen, Y. et al. (2014). Regulating inflammation using acid-responsive electrospun fibrous scaffolds for skin scarless healing. *Mediators Inflamm.* 2014: 858045.
21 Ferreira, H., Silva, R., Matama, T. et al. (2016). A biologically active delivery material with dried-rehydrated vesicles containing the anti-inflammatory diclofenac for potential wound healing. *J. Liposome Res.* 26 (4): 269–275.

22 Zykova, S.N., Balandina, K.A., Vorokhobina, N.V. et al. (2014). Macrophage stimulating agent soluble yeast β-1,3/1,6-glucan as a topical treatment of diabetic foot and leg ulcers: a randomized, double blind, placebo-controlled phase II study. *J. Diabetes Invest.* 5 (4): 392–399.

23 Wang, Q.S., Cui, Y.L., Gao, L.N. et al. (2014). Reduction of the pro-inflammatory response by tetrandrine-loading poly(L-lactic acid) films *in vitro* and *in vivo*. *J. Biomed. Mater. Res. Part A* 102 (11): 4098–4107.

24 Friedrich, E.E., Sun, L.T., Natesan, S. et al. (2014). Effects of hyaluronic acid conjugation on anti-TNF-α inhibition of inflammation in burns. *J. Biomed. Mater. Res. A* 102 (5): 1527–1536.

25 Panjwani, N. (2014). Role of galectins in re-epithelialization of wounds. *Ann. Transl. Med.* 2 (9): 89.

26 Aslanova, A., Takagi, R., Yamato, M. et al. (2015). A chemically defined culture medium containing Rho kinase inhibitor Y-27632 for the fabrication of stratified squamous epithelial cell grafts. *Biochem. Biophys. Res. Commun.* 460 (2): 123–129.

27 Palechor-Ceron, N., Suprynowicz, F.A., Upadhyay, G. et al. (2013). Radiation induces diffusible feeder cell factor(s) that cooperate with ROCK inhibitor to conditionally reprogram and immortalize epithelial cells. *Am. J. Pathol.* 183 (6): 1862–1870.

28 Yugawa, T., Nishino, K., Ohno, S. et al. (2013). Noncanonical NOTCH signaling limits self-renewal of human epithelial and induced pluripotent stem cells through ROCK activation. *Mol. Cell. Biol.* 33 (22): 4434–4447.

29 Chapman, S., McDermott, D.H., Shen, K. et al. (2014). The effect of Rho kinase inhibition on long-term keratinocyte proliferation is rapid and conditional. *Stem Cell Res. Ther.* 5 (2): 60.

30 Strudwick, X.L., Lang, D.L., Smith, L.E., and Cowin, A.J. (2015). Combination of low calcium with Y-27632 rock inhibitor increases the proliferative capacity, expansion potential and lifespan of primary human keratinocytes while retaining their capacity to differentiate into stratified epidermis in a 3D skin model. *PLoS One* 10 (4): e0123651.

31 Tang, J., Liu, H., Gao, C. et al. (2014). A small peptide with potential ability to promote wound healing. *PLoS One* 9 (3): e92082.

32 Song, L., Baksh, D., and Tuan, R.S. (2004). Mesenchymal stem cell-based cartilage tissue engineering: cells, scaffold and biology. *Cytotherapy* 6 (6): 596–601.

33 Zhao, H.Y., Wu, J., Zhu, J.J. et al. (2015). Research advances in tissue engineering materials for sustained release of growth factors. *Biomed Res. Int.* 2015: 808202.

34 Petrof, G., Abdul-Wahab, A., and McGrath, J.A. (2014). Cell therapy in dermatology. *Cold Spring Harbor Perspect. Med.* 4 (6): a015156.

35 Naderi, H., Matin, M.M., and Bahrami, A.R. (2011). Review paper: critical issues in tissue engineering: biomaterials, cell sources, angiogenesis, and drug delivery systems. *J. Biomater. Appl.* 26 (4): 383–417.

36 Zhao, X., Lang, Q., Yildirimer, L. et al. (2016). Photocrosslinkable gelatin hydrogel for epidermal tissue engineering. *Adv. Healthcare Mater.* 5 (1): 108–118.

37 Blanpain, C. and Fuchs, E. (2006). Epidermal stem cells of the skin. *Annu. Rev. Cell Dev. Biol.* 22: 339–373.

38 Hagner, A. and Biernaskie, J. (2013). Isolation and differentiation of hair follicle-derived dermal precursors. *Methods Mol. Biol.* 989: 247–263.

39 Burke, J.F., Yannas, I.V., Quinby, W.J. et al. (1981). Successful use of a physiologically acceptable artificial skin in the treatment of extensive burn injury. *Ann. Surg.* 194 (4): 413–428.

40 Hafner, J., Kuhne, A., and Trueb, R.M. (2006). Successful grafting with EpiDex in pyoderma gangrenosum. *Dermatology* 212 (3): 258–259.

41 Waymack, P., Duff, R.G., and Sabolinski, M. (2000). The effect of a tissue engineered bilayered living skin analog, over meshed split-thickness autografts on the healing of excised burn wounds. The Apligraf Burn Study Group. *Burns* 26 (7): 609–619.

42 Still, J., Glat, P., Silverstein, P. et al. (2003). The use of a collagen sponge/living cell composite material to treat donor sites in burn patients. *Burns* 29 (8): 837–841.

43 Dunn, L., Prosser, H.C., Tan, J.T. et al. (2013). Murine model of wound healing. *J. Vis. Exp.* 75: e50265.

44 Hu, M.S., Leavitt, T., Malhotra, S. et al. (2015). Stem cell-based therapeutics to improve wound healing. *Plast. Surg. Int.* 2015: 383581.

45 Brignier, A.C. and Gewirtz, A.M. (2010). Embryonic and adult stem cell therapy. *J. Allergy Clin. Immunol.* 125 (2 Suppl. 2): S336–S344.

46 Herberts, C.A., Kwa, M.S., and Hermsen, H.P. (2011). Risk factors in the development of stem cell therapy. *J. Transl. Med.* 9: 29.

47 Kim, J.Y. and Suh, W. (2010). Stem cell therapy for dermal wound healing. *Int. J. Stem Cells* 3 (1): 29–31.

48 Chen, Y.X., Zeng, Z.C., Sun, J. et al. (2015). Mesenchymal stem cell-conditioned medium prevents radiation-induced liver injury by inhibiting inflammation and protecting sinusoidal endothelial cells. *J. Radiat. Res.* 56 (4): 700–708.

49 Tran, C. and Damaser, M.S. (2015). Stem cells as drug delivery methods: application of stem cell secretome for regeneration. *Adv. Drug Delivery Rev.* 82, 83: 1–11.

50 Bollini, S., Gentili, C., Tasso, R., and Cancedda, R. (2013). The regenerative role of the fetal and adult stem cell secretome. *J. Clin. Med.* 2 (4): 302–327.

51 Dinh, T., Braunagel, S., and Rosenblum, B.I. (2015). Growth factors in wound healing: the present and the future? *Clin. Podiatr. Med. Surg.* 32 (1): 109–119.

52 Mann, A., Niekisch, K., Schirmacher, P., and Blessing, M. (2006). Granulocyte-macrophage colony-stimulating factor is essential for normal wound healing. *J. Invest. Dermatol. Symp. Proc.* 11 (1): 87–92.

53 Park, J.W., Hwang, S.R., and Yoon, I.S. (2017). Advanced growth factor delivery systems in wound management and skin regeneration. *Molecules* 22 (8): 1259.

54 Cheng, L., Sun, X., Zhao, X. et al. (2016). Surface biofunctional drug-loaded electrospun fibrous scaffolds for comprehensive repairing hypertrophic scars. *Biomaterials* 83: 169–181.

55 Hasse, S., Duong, T.T., Hahn, O. et al. (2016). Induction of proliferation of basal epidermal keratinocytes by cold atmospheric-pressure plasma. *Clin. Exp. Dermatol.* 41 (2): 202–209.

56 Naderi-Meshkin, H., Andreas, K., Matin, M.M. et al. (2014). Chitosan-based injectable hydrogel as a promising in situ forming scaffold for cartilage tissue engineering. *Cell Biol. Int.* 38 (1): 72–84.

57 Garg, T., Singh, S., and Goyal, A.K. (2013). Stimuli-sensitive hydrogels: an excellent carrier for drug and cell delivery. *Crit. Rev. Ther. Drug Carrier Syst.* 30 (5): 369–409.

58 Makadia, H.K. and Siegel, S.J. (2011). Poly lactic-*co*-glycolic acid (PLGA) as biodegradable controlled drug delivery carrier. *Polymers (Basel)* 3 (3): 1377–1397.

59 Merino, S., Martin, C., Kostarelos, K. et al. (2015). Nanocomposite hydrogels: 3D polymer-nanoparticle synergies for on-demand drug delivery. *ACS Nano* 9 (5): 4686–4697.

60 Shooter, G.K., Van Lonkhuyzen, D.R., Croll, T.I. et al. (2015). A pre-clinical functional assessment of an acellular scaffold intended for the treatment of hard-to-heal wounds. *Int. Wound J.* 12 (2): 160–168.

61 Ramji, K. and Shah, R.N. (2014). Electrospun soy protein nanofiber scaffolds for tissue regeneration. *J. Biomater. Appl.* 29 (3): 411–422.

62 Arasteh, S., Kazemnejad, S., Khanjani, S. et al. (2016). Fabrication and characterization of nano-fibrous bilayer composite for skin regeneration application. *Methods* 99: 3–12.

63 Geesala, R., Bar, N., Dhoke, N.R. et al. (2016). Porous polymer scaffold for on-site delivery of stem cells – protects from oxidative stress and potentiates wound tissue repair. *Biomaterials* 77: 1–13.

64 Sung, S.H., Park, S.H., Song, S.Y. et al. (2011). Epidermal regeneration by *ent*-16α, 17-dihydroxy-kauran-19-oic acid isolated from *Siegesbeckia pubescens*. *Cell Prolif.* 44 (6): 527–536.

65 Schluter, H., Paquet-Fifield, S., Gangatirkar, P. et al. (2011). Functional characterization of quiescent keratinocyte stem cells and their progeny reveals a hierarchical organization in human skin epidermis. *Stem Cells* 29 (8): 1256–1268.

66 Vollmers, A., Wallace, L., Fullard, N. et al. (2012). Two- and three-dimensional culture of keratinocyte stem and precursor cells derived from primary murine epidermal cultures. *Stem Cell Rev. Rep.* 8 (2): 402–413.

67 Xu, G.M., Shi, X.M., Cai, J.F. et al. (2015). Dual effects of atmospheric pressure plasma jet on skin wound healing of mice. *Wound Repair Regen.* 23 (6): 878–884.

68 Herndon, D.N., Barrow, R.E., Rutan, R.L. et al. (1989). A comparison of conservative versus early excision. Therapies in severely burned patients. *Ann. Surg.* 209 (5): 547–552; discussion 552–3.

69 Kroner, E., Kaiser, J.S., Fischer, S.C., and Arzt, E. (2012). Bioinspired polymeric surface patterns for medical applications. *J. Appl. Biomater. Funct. Mater.* 10 (3): 287–292.

70 Alonso, L. and Fuchs, E. (2003). Stem cells of the skin epithelium. *Proc. Natl. Acad. Sci. U.S.A.* 100 (Suppl. 1): 11830–11835.

71 Clark, R.A., Ghosh, K., and Tonnesen, M.G. (2007). Tissue engineering for cutaneous wounds. *J. Invest. Dermatol.* 127 (5): 1018–1029.

72 de Mel, A., Seifalian, A.M., and Birchall, M.A. (2012). Orchestrating cell/material interactions for tissue engineering of surgical implants. *Macromol. Biosci.* 12 (8): 1010–1021.

73 Shevchenko, R.V., James, S.L., and James, S.E. (2010). A review of tissue-engineered skin bioconstructs available for skin reconstruction. *J. R. Soc. Interface* 7 (43): 229–258.

74 MacNeil, S. (2007). Progress and opportunities for tissue-engineered skin. *Nature* 445 (7130): 874–880.

75 Ohyama, M., Terunuma, A., Tock, C.L. et al. (2006). Characterization and isolation of stem cell-enriched human hair follicle bulge cells. *J. Clin. Invest.* 116 (1): 249–260.

76 Tiede, S., Kloepper, J.E., Bodo, E. et al. (2007). Hair follicle stem cells: walking the maze. *Eur. J. Cell Biol.* 86 (7): 355–376.

77 Nagel, S., Rohr, F., Weber, C. et al. (2013). Multipotent nestin-positive stem cells reside in the stroma of human eccrine and apocrine sweat glands and can be propagated robustly *in vitro*. *PLoS One* 8 (10): e78365.

78 Yang, R., Liu, F., Wang, J. et al. (2019). Epidermal stem cells in wound healing and their clinical applications. *Stem Cell Res. Ther.* 10 (1): 229.

79 Kumar, S., Poojan, S., Verma, V. et al. (2014). Rapid isolation of integrin rich multipotent stem cell pool and reconstruction of mouse epidermis equivalent. *Am. J. Stem Cells* 3 (1): 27–36.

80 Sugiyama-Nakagiri, Y., Fujimura, T., and Moriwaki, S. (2016). Induction of skin-derived precursor cells from human induced pluripotent stem cells. *PLoS One* 11 (12): e0168451.

81 Lazic, T. and Falanga, V. (2011). Bioengineered skin constructs and their use in wound healing. *Plast. Reconstr. Surg.* 127 (Suppl. 1): 75S–90S.

82 Asti, A. and Gioglio, L. (2014). Natural and synthetic biodegradable polymers: different scaffolds for cell expansion and tissue formation. *Int. J. Artif. Organs* 37 (3): 187–205.

83 Hinderer, S., Layland, S.L., and Schenke-Layland, K. (2016). ECM and ECM-like materials – biomaterials for applications in regenerative medicine and cancer therapy. *Adv. Drug Deliv. Rev.* 97: 260–269.

84 Ojeh, N., Pastar, I., Tomic-Canic, M., and Stojadinovic, O. (2015). Stem cells in skin regeneration, wound healing, and their clinical applications. *Int. J. Mol. Sci.* 16 (10): 25476–25501.

85 Moustafa, M., Bullock, A.J., Creagh, F.M. et al. (2007). Randomized, controlled, single-blind study on use of autologous keratinocytes on a transfer dressing to treat nonhealing diabetic ulcers. *Regen. Med.* 2 (6): 887–902.

86 Andreassi, L., Pianigiani, E., Andreassi, A. et al. (1998). A new model of epidermal culture for the surgical treatment of vitiligo. *Int. J. Dermatol.* 37 (8): 595–598.

87 Atiyeh, B.S. and Costagliola, M. (2007). Cultured epithelial autograft (CEA) in burn treatment: three decades later. *Burns* 33 (4): 405–413.
88 Vanscheidt, W., Ukat, A., Horak, V. et al. (2007). Treatment of recalcitrant venous leg ulcers with autologous keratinocytes in fibrin sealant: a multinational randomized controlled clinical trial. *Wound Repair Regen.* 15 (3): 308–315.
89 Kamel, R.A., Ong, J.F., Eriksson, E. et al. (2013). Tissue engineering of skin. *J. Am. Coll. Surg.* 217 (3): 533–555.
90 Wong, V.W. and Gurtner, G.C. (2012). Tissue engineering for the management of chronic wounds: current concepts and future perspectives. *Exp. Dermatol.* 21 (10): 729–734.
91 Siprashvili, Z., Nguyen, N.T., Bezchinsky, M.Y. et al. (2010). Long-term type VII collagen restoration to human epidermolysis bullosa skin tissue. *Hum. Gene Ther.* 21 (10): 1299–1310.
92 Uitto, J., McGrath, J.A., Rodeck, U. et al. (2010). Progress in epidermolysis bullosa research: toward treatment and cure. *J. Invest. Dermatol.* 130 (7): 1778–1784.
93 Robbins, P.B., Lin, Q., Goodnough, J.B. et al. (2001). *In vivo* restoration of laminin 5 β 3 expression and function in junctional epidermolysis bullosa. *Proc. Natl. Acad. Sci. U.S.A.* 98 (9): 5193–5198.
94 Mavilio, F., Pellegrini, G., Ferrari, S. et al. (2006). Correction of junctional epidermolysis bullosa by transplantation of genetically modified epidermal stem cells. *Nat. Med.* 12 (12): 1397–1402.
95 Jahoda, C.A., Horne, K.A., and Oliver, R.F. (1984). Induction of hair growth by implantation of cultured dermal papilla cells. *Nature* 311 (5986): 560–562.
96 Oliver, R.F. (1967). The experimental induction of whisker growth in the hooded rat by implantation of dermal papillae. *J. Embryol. Exp. Morphol.* 18 (1): 43–51.
97 Xing, L. and Kobayashi, K. (2001). Ability of transplanted cultured epithelium to respond to dermal papillae. *Tissue Eng.* 7 (5): 535–544.
98 Reynolds, A.J. and Jahoda, C.A. (1992). Cultured dermal papilla cells induce follicle formation and hair growth by transdifferentiation of an adult epidermis. *Development* 115 (2): 587–593.
99 Biernaskie, J., Paris, M., Morozova, O. et al. (2009). SKPs derive from hair follicle precursors and exhibit properties of adult dermal stem cells. *Cell Stem Cell* 5 (6): 610–623.
100 Toma, J.G., Akhavan, M., Fernandes, K.J. et al. (2001). Isolation of multipotent adult stem cells from the dermis of mammalian skin. *Nat. Cell Biol.* 3 (9): 778–784.
101 Jahoda, C.A., Whitehouse, J., Reynolds, A.J., and Hole, N. (2003). Hair follicle dermal cells differentiate into adipogenic and osteogenic lineages. *Exp. Dermatol.* 12 (6): 849–859.
102 Lako, M., Armstrong, L., Cairns, P.M. et al. (2002). Hair follicle dermal cells repopulate the mouse haematopoietic system. *J. Cell Sci.* 115 (Pt. 20): 3967–3974.

103 Jahoda, C.A. and Reynolds, A.J. (2001). Hair follicle dermal sheath cells: unsung participants in wound healing. *Lancet* 358 (9291): 1445–1448.

104 Reynolds, A.J., Lawrence, C., Cserhalmi-Friedman, P.B. et al. (1999). Trans-gender induction of hair follicles. *Nature* 402 (6757): 33–34.

105 Harris, M.L., Buac, K., Shakhova, O. et al. (2013). A dual role for SOX10 in the maintenance of the postnatal melanocyte lineage and the differentiation of melanocyte stem cell progenitors. *PLoS Genet.* 9 (7): e1003644.

106 Nishimura, E.K., Jordan, S.A., Oshima, H. et al. (2002). Dominant role of the niche in melanocyte stem-cell fate determination. *Nature* 416 (6883): 854–860.

107 Osawa, M., Egawa, G., Mak, S.S. et al. (2005). Molecular characterization of melanocyte stem cells in their niche. *Development* 132 (24): 5589–5599.

108 Lang, D., Lu, M.M., Huang, L. et al. (2005). Pax3 functions at a nodal point in melanocyte stem cell differentiation. *Nature* 433 (7028): 884–887.

109 Chou, W.C., Takeo, M., Rabbani, P. et al. (2013). Direct migration of follicular melanocyte stem cells to the epidermis after wounding or UVB irradiation is dependent on Mc1r signaling. *Nat. Med.* 19 (7): 924–929.

110 Li, L., Mignone, J., Yang, M. et al. (2003). Nestin expression in hair follicle sheath progenitor cells. *Proc. Natl. Acad. Sci. U.S.A.* 100 (17): 9958–9961.

111 Amoh, Y., Mii, S., Aki, R. et al. (2012). Multipotent nestin-expressing stem cells capable of forming neurons are located in the upper, middle and lower part of the vibrissa hair follicle. *Cell Cycle* 11 (18): 3513–3517.

112 Kadam, S.S. and Bhonde, R.R. (2010). Islet neogenesis from the constitutively nestin expressing human umbilical cord matrix derived mesenchymal stem cells. *Islets* 2 (2): 112–120.

113 Zulewski, H., Abraham, E.J., Gerlach, M.J. et al. (2001). Multipotential nestin-positive stem cells isolated from adult pancreatic islets differentiate ex vivo into pancreatic endocrine, exocrine, and hepatic phenotypes. *Diabetes* 50 (3): 521–533.

114 Yashiro, M., Mii, S., Aki, R. et al. (2015). From hair to heart: nestin-expressing hair-follicle-associated pluripotent (HAP) stem cells differentiate to beating cardiac muscle cells. *Cell Cycle* 14 (14): 2362–2366.

115 Mignone, J.L., Roig-Lopez, J.L., Fedtsova, N. et al. (2007). Neural potential of a stem cell population in the hair follicle. *Cell Cycle* 6 (17): 2161–2170.

116 Amoh, Y., Li, L., Katsuoka, K. et al. (2005). Multipotent nestin-positive, keratin-negative hair-follicle bulge stem cells can form neurons. *Proc. Natl. Acad. Sci. U.S.A.* 102 (15): 5530–5534.

117 Loh, Y.H., Agarwal, S., Park, I.H. et al. (2009). Generation of induced pluripotent stem cells from human blood. *Blood* 113 (22): 5476–5479.

118 Gledhill, K., Guo, Z., Umegaki-Arao, N. et al. (2015). Melanin transfer in human 3D skin equivalents generated exclusively from induced pluripotent stem cells. *PLoS One* 10 (8): e0136713.

119 Abaci, H.E., Guo, Z., Coffman, A. et al. (2016). Human skin constructs with spatially controlled vasculature using primary and iPSC-derived endothelial cells. *Adv. Healthc. Mater.* 5 (14): 1800–1807.

120 Gnedeva, K., Vorotelyak, E., Cimadamore, F. et al. (2015). Derivation of hair-inducing cell from human pluripotent stem cells. *PLoS One* 10 (1): e0116892.

121 Fine, J.D., Bruckner-Tuderman, L., Eady, R.A. et al. (2014). Inherited epidermolysis bullosa: updated recommendations on diagnosis and classification. *J. Am. Acad. Dermatol.* 70 (6): 1103–1126.

122 Okita, K., Matsumura, Y., Sato, Y. et al. (2011). A more efficient method to generate integration-free human iPS cells. *Nat. Methods* 8 (5): 409–412.

123 Itoh, M., Kiuru, M., Cairo, M.S., and Christiano, A.M. (2011). Generation of keratinocytes from normal and recessive dystrophic epidermolysis bullosa-induced pluripotent stem cells. *Proc. Natl. Acad. Sci. U.S.A.* 108 (21): 8797–8802.

124 Tolar, J., Xia, L., Riddle, M.J. et al. (2011). Induced pluripotent stem cells from individuals with recessive dystrophic epidermolysis bullosa. *J. Invest. Dermatol.* 131 (4): 848–856.

125 Schlaeger, T.M., Daheron, L., Brickler, T.R. et al. (2015). A comparison of non-integrating reprogramming methods. *Nat. Biotechnol.* 33 (1): 58–63.

126 Kim, K., Zhao, R., Doi, A. et al. (2011). Donor cell type can influence the epigenome and differentiation potential of human induced pluripotent stem cells. *Nat. Biotechnol.* 29 (12): 1117–1119.

127 Lister, R., Pelizzola, M., Kida, Y.S. et al. (2011). Hotspots of aberrant epigenomic reprogramming in human induced pluripotent stem cells. *Nature* 471 (7336): 68–73.

128 Laurent, L.C., Ulitsky, I., Slavin, I. et al. (2011). Dynamic changes in the copy number of pluripotency and cell proliferation genes in human ESCs and iPSCs during reprogramming and time in culture. *Cell Stem Cell* 8 (1): 106–118.

129 Tachibana, M., Amato, P., Sparman, M. et al. (2013). Human embryonic stem cells derived by somatic cell nuclear transfer. *Cell* 153 (6): 1228–1238.

130 Wainger, B.J., Buttermore, E.D., Oliveira, J.T. et al. (2015). Modeling pain in vitro using nociceptor neurons reprogrammed from fibroblasts. *Nat. Neurosci.* 18 (1): 17–24.

131 Blanchard, J.W., Eade, K.T., Szucs, A. et al. (2015). Selective conversion of fibroblasts into peripheral sensory neurons. *Nat. Neurosci.* 18 (1): 25–35.

132 Kim, K.L., Song, S.H., Choi, K.S., and Suh, W. (2013). Cooperation of endothelial and smooth muscle cells derived from human induced pluripotent stem cells enhances neovascularization in dermal wounds. *Tissue Eng. Part A* 19 (21–22): 2478–2485.

133 Ebrahimi, B. (2015). Reprogramming barriers and enhancers: strategies to enhance the efficiency and kinetics of induced pluripotency. *Cell Regen.* 4: 10.

134 Huangfu, D., Osafune, K., Maehr, R. et al. (2008). Induction of pluripotent stem cells from primary human fibroblasts with only Oct4 and Sox2. *Nat. Biotechnol.* 26 (11): 1269–1275.

135 Lu, H.F., Chai, C., Lim, T.C. et al. (2014). A defined xeno-free and feeder-free culture system for the derivation, expansion and direct differentiation of transgene-free patient-specific induced pluripotent stem cells. *Biomaterials* 35 (9): 2816–2826.

136 Hou, P., Li, Y., Zhang, X. et al. (2013). Pluripotent stem cells induced from mouse somatic cells by small-molecule compounds. *Science* 341 (6146): 651–654.

137 Wang, Y., Kim, H.J., Vunjak-Novakovic, G., and Kaplan, D.L. (2006). Stem cell-based tissue engineering with silk biomaterials. *Biomaterials* 27 (36): 6064–6082.

138 Sheng, Z., Fu, X., Cai, S. et al. (2009). Regeneration of functional sweat gland-like structures by transplanted differentiated bone marrow mesenchymal stem cells. *Wound Repair Regen.* 17 (3): 427–435.

139 Li, H. and Fu, X. (2012). Mechanisms of action of mesenchymal stem cells in cutaneous wound repair and regeneration. *Cell Tissue Res.* 348 (3): 371–377.

140 Prockop, D.J. (2009). Repair of tissues by adult stem/progenitor cells (MSCs): controversies, myths, and changing paradigms. *Mol. Ther.* 17 (6): 939–946.

141 Orkin, S.H. and Zon, L.I. (2008). Hematopoiesis: an evolving paradigm for stem cell biology. *Cell* 132 (4): 631–644.

142 Pittenger, M.F., Mackay, A.M., Beck, S.C. et al. (1999). Multilineage potential of adult human mesenchymal stem cells. *Science* 284 (5411): 143–147.

143 Tropel, P., Platet, N., Platel, J.C. et al. (2006). Functional neuronal differentiation of bone marrow-derived mesenchymal stem cells. *Stem Cells* 24 (12): 2868–2876.

144 Kang, S.K., Putnam, L.A., Ylostalo, J. et al. (2004). Neurogenesis of Rhesus adipose stromal cells. *J. Cell Sci.* 117 (Pt. 18): 4289–4299.

145 Yamanaka, S. (2007). Strategies and new developments in the generation of patient-specific pluripotent stem cells. *Cell Stem Cell* 1 (1): 39–49.

146 Takahashi, K., Tanabe, K., Ohnuki, M. et al. (2007). Induction of pluripotent stem cells from adult human fibroblasts by defined factors. *Cell* 131 (5): 861–872.

147 Shumakov, V.I., Onishchenko, N.A., Rasulov, M.F. et al. (2003). Mesenchymal bone marrow stem cells more effectively stimulate regeneration of deep burn wounds than embryonic fibroblasts. *Bull. Exp. Biol. Med.* 136 (2): 192–195.

148 He, L., Nan, X., Wang, Y. et al. (2007). Full-thickness tissue engineered skin constructed with autogenic bone marrow mesenchymal stem cells. *Sci. China C Life Sci.* 50 (4): 429–437.

149 Kamolz, L.P., Kolbus, A., Wick, N. et al. (2006). Cultured human epithelium: human umbilical cord blood stem cells differentiate into keratinocytes under in vitro conditions. *Burns* 32 (1): 16–19.

150 Roh, C. and Lyle, S. (2006). Cutaneous stem cells and wound healing. *Pediatr. Res.* 59 (4 Pt. 2): 100R–103R.

151 Woodley, D.T., Remington, J., Huang, Y. et al. (2007). Intravenously injected human fibroblasts home to skin wounds, deliver type VII collagen, and promote wound healing. *Mol. Ther.* 15 (3): 628–635.

152 Chino, T., Tamai, K., Yamazaki, T. et al. (2008). Bone marrow cell transfer into fetal circulation can ameliorate genetic skin diseases by providing fibroblasts to the skin and inducing immune tolerance. *Am. J. Pathol.* 173 (3): 803–814.

153 Liu, P., Deng, Z., Han, S. et al. (2008). Tissue-engineered skin containing mesenchymal stem cells improves burn wounds. *Artif. Organs* 32 (12): 925–931.

154 Tamai, K., Yamazaki, T., Chino, T. et al. (2011). PDGFRα-positive cells in bone marrow are mobilized by high mobility group box 1 (HMGB1) to regenerate injured epithelia. *Proc. Natl. Acad. Sci. U.S.A.* 108 (16): 6609–6614.

155 Iinuma, S., Aikawa, E., Tamai, K. et al. (2015). Transplanted bone marrow-derived circulating PDGFRα+ cells restore type VII collagen in recessive dystrophic epidermolysis bullosa mouse skin graft. *J. Immunol.* 194 (4): 1996–2003.

156 Li, H., Fu, X., Ouyang, Y. et al. (2006). Adult bone-marrow-derived mesenchymal stem cells contribute to wound healing of skin appendages. *Cell Tissue Res.* 326 (3): 725–736.

157 Badiavas, E.V., Abedi, M., Butmarc, J. et al. (2003). Participation of bone marrow derived cells in cutaneous wound healing. *J. Cell. Physiol.* 196 (2): 245–250.

158 Fu, X., Qu, Z., and Sheng, Z. (2006). Potentiality of mesenchymal stem cells in regeneration of sweat glands. *J. Surg. Res.* 136 (2): 204–208.

159 Deng, W., Han, Q., Liao, L. et al. (2005). Engrafted bone marrow-derived Flk-1^+ mesenchymal stem cells regenerate skin tissue. *Tissue Eng.* 11 (1, 2): 110–119.

160 Mistriotis, P. and Andreadis, S.T. (2013). Hair follicle: a novel source of multipotent stem cells for tissue engineering and regenerative medicine. *Tissue Eng. Part B Rev.* 19 (4): 265–278.

161 Bai, T., Liu, F., Zou, F. et al. (2017). Epidermal growth factor induces proliferation of hair follicle-derived mesenchymal stem cells through epidermal growth factor receptor-mediated activation of ERK and AKT signaling pathways associated with upregulation of cyclin D1 and downregulation of p16. *Stem Cells Dev.* 26 (2): 113–122.

162 Xu, Y., Hong, Y., Xu, M. et al. (2016). Role of keratinocyte growth factor in the differentiation of sweat gland-like cells from human umbilical cord-derived mesenchymal stem cells. *Stem Cells Transl. Med.* 5 (1): 106–116.

163 Fu, X., Sun, X., Li, X., and Sheng, Z. (2001). Dedifferentiation of epidermal cells to stem cells in vivo. *Lancet* 358 (9287): 1067–1068.

164 Mustonen, T., Pispa, J., Mikkola, M.L. et al. (2003). Stimulation of ectodermal organ development by ectodysplasin-A1. *Dev. Biol.* 259 (1): 123–136.

165 Cui, C.Y. and Schlessinger, D. (2006). EDA signaling and skin appendage development. *Cell Cycle* 5 (21): 2477–2483.

166 Srivastava, A.K., Durmowicz, M.C., Hartung, A.J. et al. (2001). Ectodysplasin-A1 is sufficient to rescue both hair growth and sweat glands in Tabby mice. *Hum. Mol. Genet.* 10 (26): 2973–2981.

167 Cai, S., Pan, Y., Han, B. et al. (2011). Transplantation of human bone marrow-derived mesenchymal stem cells transfected with ectodysplasin for regeneration of sweat glands. *Chin. Med. J. (Engl.)* 124 (15): 2260–2268.

168 Gurtner, G.C., Werner, S., Barrandon, Y., and Longaker, M.T. (2008). Wound repair and regeneration. *Nature* 453 (7193): 314–321.

169 Kaewkhaw, R., Scutt, A.M., and Haycock, J.W. (2011). Anatomical site influences the differentiation of adipose-derived stem cells for Schwann-cell phenotype and function. *Glia* 59 (5): 734–749.

170 Wu, W., Zhang, S., Chen, Y., and Liu, H. (2017). Biological function and mechanism of bone marrow mesenchymal stem cells-packed poly (3,4-ethylenedioxythiophene) (PEDOT) scaffolds for peripheral nerve injury: the involvement of miR-21-Notch signaling pathway. *Curr. Neurovasc. Res.* 14 (1): 19–25.

171 Zhao, W., Li, J.J., Cao, D.Y. et al. (2012). Intravenous injection of mesenchymal stem cells is effective in treating liver fibrosis. *World J. Gastroenterol.* 18 (10): 1048–1058.

172 Malliaras, K., Kreke, M., and Marban, E. (2011). The stuttering progress of cell therapy for heart disease. *Clin. Pharmacol. Ther.* 90 (4): 532–541.

173 Lee, J.W., Fang, X., Krasnodembskaya, A. et al. (2011). Concise review: mesenchymal stem cells for acute lung injury: role of paracrine soluble factors. *Stem Cells* 29 (6): 913–919.

174 Teng, M., Huang, Y., and Zhang, H. (2014). Application of stems cells in wound healing – an update. *Wound Repair Regener.* 22 (2): 151–160.

175 Phinney, D.G. (2012). Functional heterogeneity of mesenchymal stem cells: implications for cell therapy. *J. Cell. Biochem.* 113 (9): 2806–2812.

176 Wood, F.M. (2014). Skin regeneration: the complexities of translation into clinical practise. *Int. J. Biochem. Cell Biol.* 56: 133–140.

177 Lee, V., Singh, G., Trasatti, J.P. et al. (2014). Design and fabrication of human skin by three-dimensional bioprinting. *Tissue Eng. Part C Methods* 20 (6): 473–484.

178 Ronfard, V., Broly, H., Mitchell, V. et al. (1991). Use of human keratinocytes cultured on fibrin glue in the treatment of burn wounds. *Burns* 17 (3): 181–184.

179 Mandrycky, C., Wang, Z., Kim, K., and Kim, D.H. (2016). 3D bioprinting for engineering complex tissues. *Biotechnol. Adv.* 34 (4): 422–434.

180 Liu, N., Huang, S., Yao, B. et al. (2016). 3D bioprinting matrices with controlled pore structure and release function guide in vitro self-organization of sweat gland. *Sci. Rep.* 6: 34410.

181 Bi, H. and Jin, Y. (2013). Current progress of skin tissue engineering: seed cells, bioscaffolds, and construction strategies. *Burns Trauma* 1 (2): 63–72.

182 Yoo, S.S. (2015). 3D-printed biological organs: medical potential and patenting opportunity. *Expert Opin. Ther. Pat.* 25 (5): 507–511.

183 Jablonska, A. and Lukomska, B. (2011). Stroke induced brain changes: implications for stem cell transplantation. *Acta Neurobiol. Exp. (Wars)* 71 (1): 74–85.

184 Sigaux, N., Pourchet, L., Albouy, M. et al. (2017). Is 3D bioprinting the future of reconstructive surgery? *Plast. Reconstr. Surg. Glob. Open* 5 (3): e1246.

185 Gudapati, H., Dey, M., and Ozbolat, I. (2016). A comprehensive review on droplet-based bioprinting: past, present and future. *Biomaterials* 102: 20–42.

186 Shakespeare, P. and Shakespeare, V. (2002). Survey: use of skin substitute materials in UK burn treatment centres. *Burns* 28 (4): 295–297.

187 Melkun, E.T. and Few, J.W. (2005). The use of biosynthetic skin substitute (Biobrane) for axillary reconstruction after surgical excision for hidradenitis suppurativa. *Plast. Reconstr. Surg.* 115 (5): 1385–1388.

188 Supp, D.M. and Boyce, S.T. (2005). Engineered skin substitutes: practices and potentials. *Clin. Dermatol.* 23 (4): 403–412.

189 Horch, R.E., Kopp, J., Kneser, U. et al. (2005). Tissue engineering of cultured skin substitutes. *J. Cell. Mol. Med.* 9 (3): 592–608.

190 Rheinwald, J.G. and Green, H. (1975). Serial cultivation of strains of human epidermal keratinocytes: the formation of keratinizing colonies from single cells. *Cell* 6 (3): 331–343.

191 Rheinwald, J.G. and Green, H. (1977). Epidermal growth factor and the multiplication of cultured human epidermal keratinocytes. *Nature* 265 (5593): 421–424.

192 Watt, F.M. (1994). Studies with cultured human epidermal keratinocytes: potential relevance to corneal wound healing. *Eye (Lond.)* 8 (Pt. 2): 161–162.

193 Daugherty, S. and Spear, M. (2015). Skin and skin substitutes – an overview. *Plast. Surg. Nurs.* 35 (2): 92–97.

194 Bello, Y.M., Falabella, A.F., and Eaglstein, W.H. (2001). Tissue-engineered skin. Current status in wound healing. *Am. J. Clin. Dermatol.* 2 (5): 305–313.

195 Zaulyanov, L. and Kirsner, R.S. (2007). A review of a bi-layered living cell treatment (Apligraf) in the treatment of venous leg ulcers and diabetic foot ulcers. *Clin. Interv. Aging* 2 (1): 93–98.

196 Price, R.D., Myers, S., Leigh, I.M., and Navsaria, H.A. (2005). The role of hyaluronic acid in wound healing: assessment of clinical evidence. *Am. J. Clin. Dermatol.* 6 (6): 393–402.

16

Techniques for Biofabrication of Vascular and Vascularized Tissue

Zhe Zhang[1], *Qiang Yang*[2], *and Chunming Wang*[1]

[1] *University of Macau, Institute of Chinese Medical Sciences, Department of Pharmaceutical Sciences, Faculty of Health Science, Avenida da Universidade, Taipa, Macau SAR, China*
[2] *Tianjin University, Tianjin Hospital, Department of Spine Surgery, No. 406 Jiefangnan Road, Hexi District, Tianjin 300211, China*

16.1 Introduction

For tissue regeneration, reconstruction of the circulatory system is of great importance. Regardless of the regeneration of bones, muscles, or cardiovascular tissues, vascular networks are needed to exchange nutrients, oxygen, carbon dioxide, metabolites, and cytokines. The generally designed blood vessel network is about 100–200 μm, and the structure of a single layer of endothelial cells (ECs) is formed inside [1]. On the one hand, when blood vessel diseases occur, such as thrombosis and atherosclerosis, we need biologically manufactured blood vessel stents to replace the original blood vessel network. On the other hand, vascularization is also crucial in the regeneration of different tissues. Therefore, biofabrication of blood vessels and vascularized tissues becomes vital.

The diameter of blood vessels in the human body ranges from 5 μm to 30 mm [2]. According to the structure and function of blood vessels, they can be divided into arteries, capillaries, and veins. Arteries and veins are composed of three tissue layers. The outermost layer (tunica externa or tunica adventitia) of blood vessels is made up of thick connective tissue. The middle layer (media) is more viscous, comprising circularly aligned elastic fibers, connective tissue, as well as smooth muscle cells (SMCs). There is more contractile tissue in arteries than in veins. The inner layer (inner membrane) is the thinnest and consists of a single layer of endothelium, which is supported by the subendothelial layer. The capillaries are only composed of a single layer of endothelium and related connective tissue [3]. The development of blood vessels is complex through two major but different mechanisms – vasculogenesis and angiogenesis. Vasculogenesis refers to the differentiation of precursor cells into ECs and the formation of new blood vessels, which occurs during embryonic development to form the heart and primitive vascular plexus. Angiogenesis refers to the process that ECs sprout from an existing vascular network to shape new capillaries. In different stages of tissue development

Biofabrication for Orthopedics: Methods, Techniques, and Applications, First Edition.
Edited by Wenguo Cui, Xin Zhao, and Shen Liu.

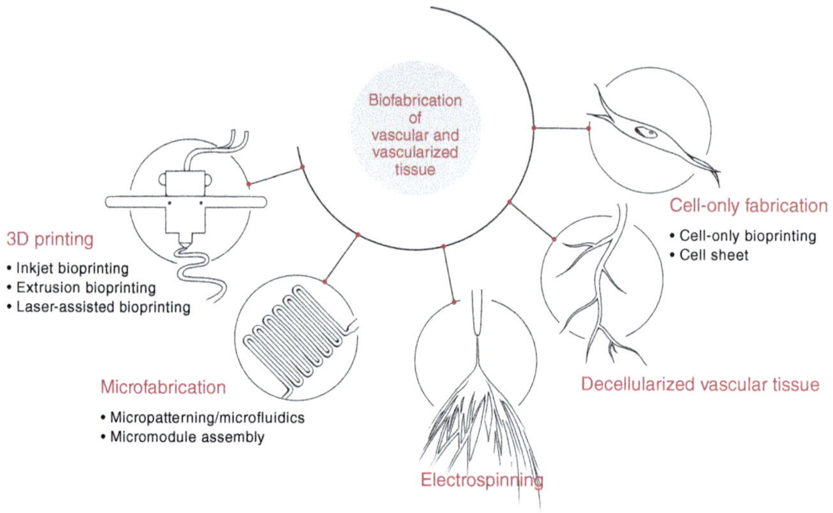

Figure 16.1 Illustration of techniques for biofabrication of vascular and vascularized tissue.

and repair, a variety of growth factors coordinately regulate vasculogenesis and angiogenesis [4].

In recent years, there have been notable breakthroughs in the biofabrication of blood vascular and vascularized tissues. For example, three-dimensional (3D) printing technology can quickly fabricate a tissue engineering vascularized scaffold with or without cells. In particular, laser-assisted biological 3D printing can improve printing accuracy to produce more complex and diverse vascular structures. Microfluidic/micropatterning technology can create a complex network of vascular structures to meet tissue vascularization needs. Moreover, as a more established technology, electrospinning can produce nano/micro-sized fibrous substrates for the development of vascular stents. Various methods of seeding ECs in decellularized vascular tissue – or printing cells alone – also show excellent potential (Figure 16.1).

16.2 3D Printing

3D bioprinting is a technology that combines computational science, biology, and materials engineering. In 1981, the Japanese inventor Hideo Kodama used additive technology to perform 3D printing for the first time. He created a product that uses ultraviolet light to stiffen polymeric materials and build up solid objects. Hull invented a technology similar to 3D printing, called stereolithography, to create smaller objects through layer-by-layer printing, solvent washing, and eventually hardening with ultraviolet light. Computer-aided design (CAD) is also employed for creating a 3D model. In 2003, Boland at the University of El Paso designed the ancestor of the bioprinter currently in use. He modified the office inkjet printer that has enabled much more extensive research in biomaterial printing. The manufacturing of 3D vascularized structures for tissue engineering has been extensively explored. This scalable, repeatable, and high-precision fabrication technology provides great potential for achieving vascularization in printed tissues, which is an essential milestone for organ printing in the foreseeable future [5].

16.2.1 Inkjet Bioprinting

Compared with cell seeding after fabrication, the bioprinting of cells can ensure a higher cell density in the scaffold. Many studies have concluded that the higher cell density inside the scaffold promotes tissue production by secreting large amounts of biomolecules. Therefore, the researchers emphasized that 3D bioprinting integrates many cell populations into the scaffold in a controlled and well-distributed manner. For the bioprintable mixture of cells and hydrogels, many studies have explored methods based on extrusion [6].

Inkjet printers and 3D bioplotters are now widely used for stent and vascular network printing. In inkjet printing, the bio-ink droplets are distributed layer by layer, while the bioplotting system extrudes a continuous filament, following the computer-generated digital (CAD) instructions, to create a predefined structure. Generally, hydrogels that can undergo fast gelation are used in inkjet systems to achieve high printing speeds. In addition, inkjet printers have the advantages of depositing multiple cell types under control to simulate the developmental process of multi-cell distribution in native tissue.

Many studies have used different inkjet printing to study alginate or its composite bio-ink and calcium chloride crosslinking as agents to create a vascular network that grows *in vitro* and *in vivo* [7–11]. In one study, researchers used sodium alginate to load ATDC5 cells (murine teratocarcinoma cell line), used inkjet 3D printing technology, printed on polyetherimide (PEI), and used calcium chloride for crosslinking (Figure 16.2) [10]. Another similar study used sodium alginate and gelatin as bio-ink, inkjet canine SMCs, human amniotic fluid-derived stem cells, and bovine aortic ECs layer by layer at the same time [12]. Similarly, the use of bio-ink composed of human microvascular endothelial cells (HMVECs) and fibrinogen, which is distributed using an inkjet printer, can arrange and proliferate HMVEC and form a capillary-like tubular structure [13]. Although inkjet 3D printers are economical and have many attractive features, they still have very significant drawbacks, such as cavitation bubbles, nozzle clogging, and cell death during printing [14].

16.2.2 Extrusion Bioprinting

Several studies have explored bioprinting technology based on indirect and direct extrusion in the manufacturing of complex vascular networks. In indirect printing, tissue-specific hydrogels are employed to encapsulate the pre-fabricated sacrificial vascular network, and then the sacrificial vascular network is removed in an appropriate solvent. Next, the injection/microfluidic method is used to inoculate

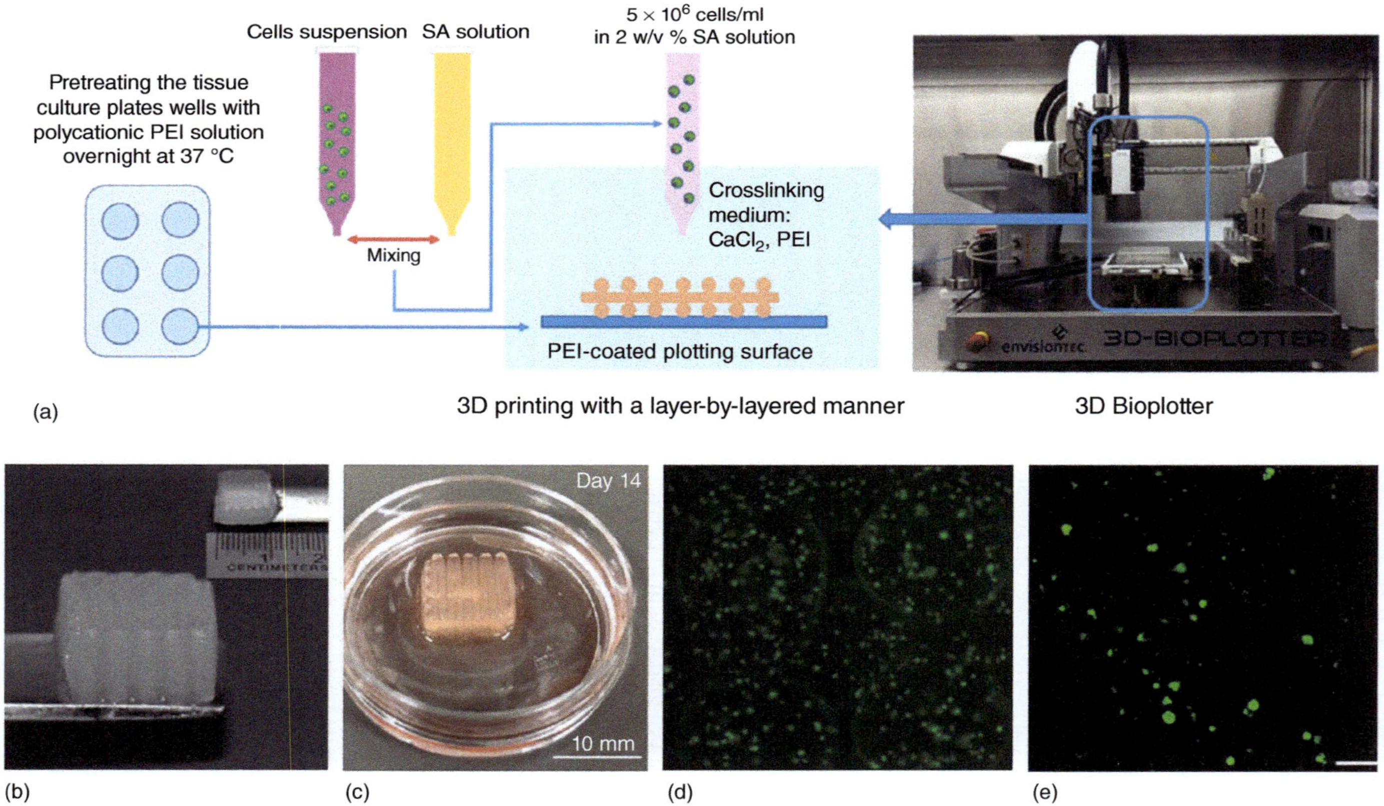

Figure 16.2 Inkjet 3D printing. (a) Schematic of the printing process using an inkjet 3D printing technique; (b) fabricated porous hydrogel scaffold; and (c) digital photos of ATDC5 cell-laden hydrogel after 14 days. (d) Fluorescent surface images of the ATDC5 cell-laden hydrogel constructs showing live (green) after 14s days and (e) accumulation of type II collagen in ATDC5 cell-laden hydrogel constructs after 28 days. SA = sodium alginate. Source: You et al. [10]. Reprinted with permission of ACS Publications.

ECs around the capillary channels. The capillary network and the perfusion of medium/blood form a single layer of ECs around the capillaries within a few days, promoting diffusion and mass transfer, and finally remodeling into the blood vessels.

Researchers have explored various cell-compatible materials as sacrificial templates for preventing the use of toxic solvents or harsh processing conditions to eliminate sacrificial filaments in engineering structures [15]. For instance, carbohydrate glass filaments were printed at 110 °C to draw a 3D network, which was then encapsulated in an agarose polymer containing primary rat hepatocytes and fibroblasts. The cell culture medium washes away the soluble sacrificial sugar compound filaments and inoculates human umbilical vein endothelial cells (HUVECs) in the vessel lumen to perfuse the blood in the vascular network. Compared with the structure without vascularization, this vascularized structure supports the primary metabolic function and maintains high albumin secretion [16]. In another study, capillaries and various cells were printed side by side to simulate natural tissues or organs. The vascular network prepared by volatile ink (Pluronics F127) was encapsulated in gelatin-methacryloyl (GelMA) and then liquefied and taken out at 4 °C. HUVEC was injected into the lumen left after the volatile ink was removed and then gently shaken. A fused vascular EC layer was visible 48 hours after inoculation, and the cell survival rate exceeded 95% [17].

Combining the coaxial or shell/chip nozzle system with the extrusion 3D printing system can directly print hollow fibers with thin tubes to form a blood vessel network [18, 19]. In particular, the flow of ionic crosslinking agents (such as $CaCl_2$) is maintained on the core side of the coaxial nozzle; meanwhile, the hydrogel flow is kept on the shell side [20]. The flow's rheology and concentration significantly affect the core diameter, wall thickness, and microfluidic channel geometry. Changes in the core/shell size significantly affect the final compressive strength and Young's modulus of the produced hollow fibers [21]. Generally, microfluidic channels made by coaxial nozzle extrusion systems are used to inoculate ECs and then perfused with culture medium or blood to promote capillary formation. However, this 3D microfluidic network can encapsulate multiple cell types in one hydrogel. Besides seeding, ECs can also be encapsulated in hollow channel shells during the biofabrication process to generate high cell densities [22]. There are also studies using multiple print heads to simultaneously distribute different cell spheres in the space between two continuous hollow filaments [23]. Studies have used the coaxial 3D printing method to print out the microtubule structure with nutrient transmission function and verified its good mass transferability (Figure 16.3) [24]. Another study used a layer of agarose sticks as a scaffold and SMC spheres or fibroblast cell spheres as discrete units. After printing, the discrete units are fused to form a blood vessel network, realizing a material-free tissue engineering scaffold [25].

16.2.3 Laser-Assisted Bioprinting

Laser-assisted bioprinting, including laser-induced forward transfer (LIFT), laser-guided direct writing (LGDW), and matrix-assisted pulsed laser evaporation

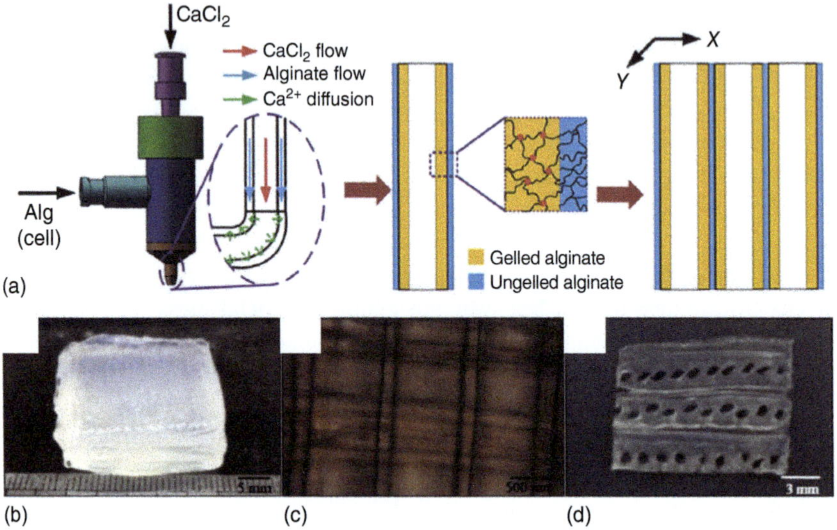

Figure 16.3 Extrusion bioprinting. (a) Fabrication of a 3D alginate structure with built-in microchannels; (b) example of a printed cuboid structure comprising six layers of hollow filaments; (c) microscopic view of the longitudinal section; and (d) macroscopic observation of the cross section. Source: Gao et al. [24]. Reprinted with permission of Elsevier.

direct writing (MAPLEDW), provides useful tools to construct two-dimensional (2D) and 3D cell patterns [26]. This printing technology has outstanding features, including non-clogging of nozzles and a printing unit with high resolution and accuracy. Among them, LIFT employs a pulsed laser to force the material from a thin donor layer onto a receiving substrate. LGDW uses a laser to stack cell aggregates layer by layer and place a hydrogel layer on each cell aggregate to print a 3D blood vessel network. MAPLEDW technology relies on low-power pulsed lasers to deposit multiple cell types. The bubbles caused by laser pulses generate shock waves, forcing cells to flow to the collector substrate.

Many studies have used laser-based bioprinting technology to print structures with blood vessel cells. For instance, researchers use LIFT-based cell printing technology to print HUVEC and human mesenchymal stem cells (hMSC) on a heart patch and reported an increased capillary density and improved infarcted heart function (Figure 16.4a, b) [27]. In addition, a study used LGDW technology to model HUVEC in 2D and 3D with micron accuracy on Matrigel. Afterward, the co-cultivation of vascular structures and hepatocytes resulted in aggregated tubular structures similar to hepatic sinusoids in tissues [29]. However, several shortcomings – including long preparation time, laser-induced cell damage, as well as low scalability – hamper the extension of this technology to a broader field of applications including tissue vascularization.

Stereolithography has been widely used to generate complex 3D blood vessel patterns using photosensitive materials [30]. Digital light projection (DLP) and laser-based stereolithography (LS) can print complex structures based on scanning

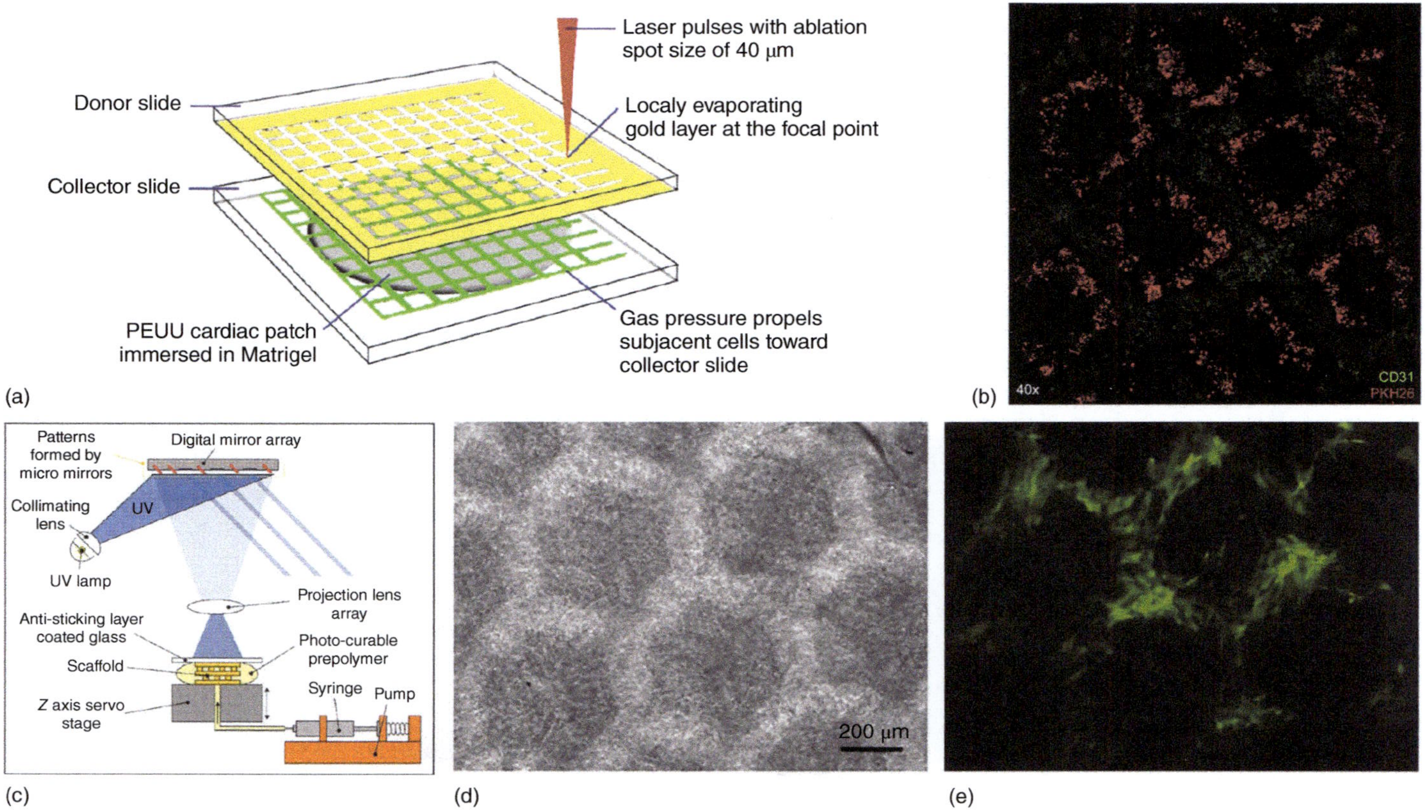

Figure 16.4 Examples of LIFT and DLP bioprinting. (a) Illustration of a LIFT setup; (b) view of the transferred human MSC cells through LIFT after 24 hours (red: PKH26; green: Pecam1); (c) schematic representation of the DLP apparatus; and (d, e) hexagonal scaffolds were seeded with HUVEC-GFP. Source: (a) Gao at al. [24]/with permission of Elsevier. (b) Gaebel et al. [27]. Reprinted with permission of Elsevier. (c) Gauvin et al. [28]/with permission of Elsevier. (d) Gauvin et al. [28]. Reprinted with permission of Elsevier.

information from CAD software, computer tomography, and magnetic resonance imaging (MRI) [31]. The digital signal generates a 2D pixel pattern in the DLP system by adjusting a device's rotation containing millions of tiny mirrors projected onto a light-cured biological material to obtain a complex 3D structure. In one study, the DLP chip generated active and reflective dynamic photomasks based on CAD drawings, copied the cross-sectional image of the 3D microstructure from the photomask, and used an ultraviolet (UV) light source to project the image onto the methacrylate (GelMA) solution. When 3D complex patterns are seeded with HUVEC, HUVEC forms a fused monolayer and remains for four days after seeding (Figure 16.4c–e) [28]. Similarly, another study reported that HUVEC formed a cord-like structure after four days of culture in a scaffold made with the DLP system [32]. However, high cost, less detailed printing of large structures, and cytotoxicity are shortcomings for the DLP technology. LS can eliminate the need to assemble a photomask and multiple 2D planes to form a 3D vascular network. Although LS is suitable for printing extensive and detailed blood vessel structures, the printing speed of LS is lower than that of DLP technology. In this technology, a computer-controlled ultraviolet laser beam generates patterns on photosensitive materials according to CAD designs [33]. Because short-wavelength lasers can cause apparent cytotoxicity, two or more laser systems are often used in practical applications to print 3D delicate structures [34]. However, even if multiple laser systems can reduce cytotoxicity compared to short-wavelength lasers, they still significantly impact cell viability [35]. Researchers used relatively low energy (<2 nJ) ultrafast laser pulses to create 2D/3D multi-scale patterns in soft hydrogels made of silk proteins, without chemical crosslinking agents [36]. Many studies use laser stereolithography to construct vascular structures, resulting in tubes and bifurcated tube systems less than 2 mm in diameter [37]. Similarly, another study used organic–inorganic hybrid polymer and epoxy-based SU8 materials to prepare 3D vascular structures [38].

16.3 Microfabrication

Natural tissues contain large blood vessels and capillaries that provide nutrients, oxygen, and other biomolecules to large cell populations. Similarly, the engineered tissue scaffolds should arrange the encapsulated cells into 3D structures and promote their growth and maturation. Nevertheless, many challenges remain, especially in producing functional vasculature to provide an affluent mass exchange. Microfabrication technology has been widely used to pattern cell-loaded hydrogels to study basic cell biology. Current micromachining technologies, including microfluidics, micropatterning, and microassembly, also serve to fabricate vascular and vascularized tissue.

16.3.1 Micropatterning/Microfluidics

Many studies use micropatterned substrates to build up capillary-like structures that simulate natural tissues. The formation of capillaries in the micropatterned substrate is affected by physical factors such as stiffness, gaps, and shear flow. Studies have shown that interstitial flow regulates capillary morphology [39] and shear stress, enhances angiogenesis [40], and material stiffness controls the development of the capillary network structure. For example, in one study, ECs seeded in rigid collagen gel formed a broad and deep vascular network with a larger vascular cavity, while ECs seeded in a flexible hydrogel created a thin and dense vascular network with a small vascular niche [41]. Generally, plasma etching [42], laser ablation [43], soft lithography [44], and copy molding [45] are used to create microfluidic patterns on certain biocompatible substrates. Laser direct writing technology is also helpful to prepare microtubules to mimic the structure of capillaries in the body accurately [46].

In one study, ECs, SMCs, and pericytes were seeded in a monolayer matrix according to a microfluidic method and then cultured for a while to promote stable capillary formation [47]. Several researchers claimed that microfluidics could regulate cell distribution, culture media, enzymes, and biomolecules in micropatterns [48]. In addition, microfluidics help to form a patterned cell distribution in a co-culture system and induce endothelialization through shear forces [49]. Especially, the shear press and biomolecules applied to the microchannel can affect the morphology, arrangement, differentiation, and remodeling of ECs, which substantially affect the formation of capillaries [50]. To adjust the shear stress, scientists can apply computational microfluidics and different fluid flow modes (such as laminar flow, pulsation, and turbulence) to induce the formation of an endothelial monolayer around the microchannel [51, 52]. In addition, researchers use technologies such as direct writing assembly (robot deposition, fused deposition, and two-photon polymerization) [53], extrusion printing based on sacrificial materials, modular assembly [54], omnidirectional printing [55], and electrostatic discharge printing [56] to form complex structures rapidly and embed tissue-specific cells. This method can create microfluidic channels to generate intricate 3D blood vessel patterns. Because ECs seeded on synthetic materials might have sub-optimal biological functions, such as poor barrier capacity and low oxygen and nutrient transportability, many studies have used silk fibroin, Matrigel, type I collagen, and fibrin hydrogel to form endothelial tube 3D scaffolds [57]. For example, the cell-seeded fibrin gel microfluidic system 3D microvessels showed desirable barrier functions and long-term stability *in vitro* [58].

16.3.2 Micromodule Assembly

The construction of a macro-scale tissue-engineered scaffold needs to meet several requirements, including a mass diffusion network within the range of 100–200 μm from cells, a reasonably uniform distribution of multiple cell types, and the ECs and vasculature [59]. To solve this problem, researchers use self-assembled microstructures or modules to engineer large structures. The size of these modules enables nutrients, oxygen, and essential biomolecules to diffuse and exchange [60]. So far, there have been several methods to generate microstructures or modules, including

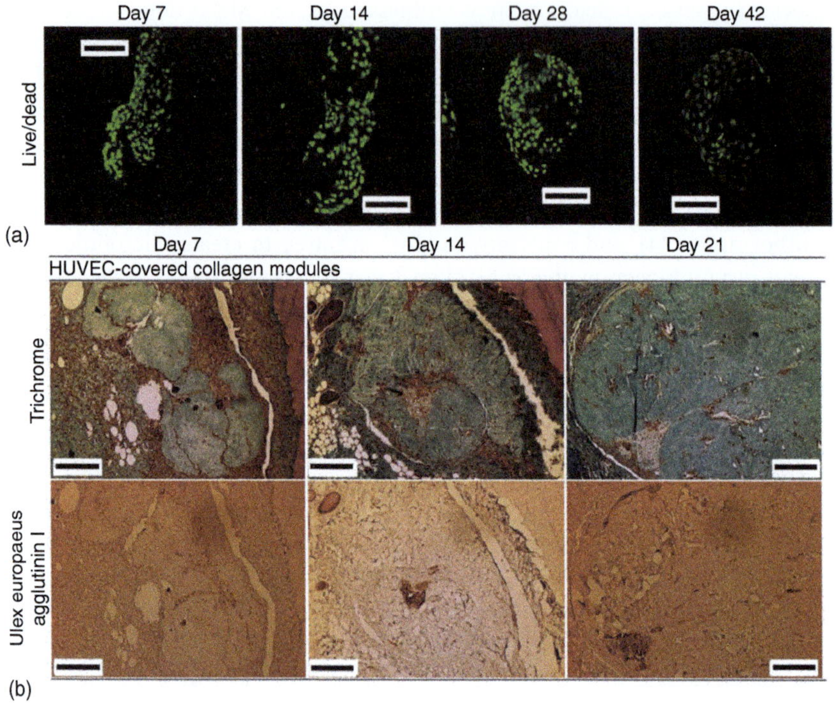

Figure 16.5 Micromodule assembly. (a) Assessment of the viability of HUVEC-covered collagen modules *in vitro* by live/dead assay. Scale bar = 200 μm. (b) Histological analysis of HUVEC-covered collagen modules, at 7, 14, and 21 days after subcutaneous implantation into SCID/Bg mice. Source: Cooper et al. [65]. Reprinted with permission of Elsevier.

micromolding, UV crosslinking, directional self-assembly, and gravity self-assembly methods [61–63].

During the preparation process, tissue-specific cells are usually encapsulated in micro modules, and the outer surface is covered by the HUVEC layer at the same time [64]. In some studies, before EC inoculation, the outer shell is further coated with protein molecules to improve function. For example, compared with collagen modules without fibronectin coating, microcollagen modules coated with fibronectin formed mature and perfused capillaries two weeks after implantation in mice [65] (Figure 16.5). In addition, microgels with stem cells or progenitor cells embedded in the EC coating can form capillaries more stable than general modules [66].

In terms of packaging methods, many studies have used techniques such as random packaging, directional, or sequential assembly to prepare cell loading micromodules of various shapes and sizes to establish macro-scale vascularized tissue structures [67, 68]. In random packaging, the EC-coated modules are perfused with a blood or culture medium to connect and develop channels in the interstitial spaces of the modules. Depending on the surfactant concentration, aspect ratio, stirring rate, and time, unidirectionally assembled, shape-controlled microgels automatically form aggregates in a variety of patterns – including offset, linear, branched, or random ones [68]. To better control microgel organization within internal microchannels, researchers also developed a sequential assembly method. The microgel is prepared by photolithography and then assembled into a tubular structure to form an interconnected and bifurcated structure similar to the natural vasculature [64]. This method has practical advantages – such as being economical and applicable, but the disadvantages of poor tissue integration and capillary network formation may limit its application.

16.4 Electrospinning

Because nanofibers have a similar morphology to ECM, researchers focused on using different electrospinning techniques to prepare random, aligned, and core/shell nanofibers to build blood vessel structures for tissue vascularization [69]. A range of natural and synthetic materials – including chitosan, gelatin, collagen, poly(ethylene oxide) (PEO), polyvinyl alcohol (PVA), poly(L-lactic acid) (PLLA), poly(lactic-co-glycolic acid) (PLGA), and polycaprolactone (PCL) – have been used in nanobiofabrication applications [70]. Compared with natural fibers, synthetic nanofibers show higher mechanical stability, but their hydrophobicity, lack of cell-binding sites, and poor biodegradability restrict their broader applications. On the contrary, nature polymer materials are hydrophilic and biocompatible and have cell-binding sites, but high degradability and mechanical properties are the main shortcomings of nanobiofabrication. Therefore, researchers have explored composite materials, copolymers, and hybrid biopolymers to make nanofibers and have remarkable tissue vascularization [71]. In some studies, random or aligned fibers affect the formation of the vasculature. For example, electrospun random polycaprolactone/collagen (rPCL/Col) and aligned PCL/collagen-PEO (aPCL/Col-PEO) nanofibers are implanted into the rat arteriovenous ring, the latter has more blood vessels and the capillary density is higher. In addition, in the rPCL/Col fiber group, more small capillaries were formed, while in the aPCL/Col-PEO fiber group, blood vessels were evenly distributed throughout the stent [72].

Pre-vascularized and aligned nanofibers promote vascularization. Random or arranged poly(ε-caprolactone)/cellulose electrospun nanofibers can form network structures layer by layer. HUVEC seeded on arranged nanofibers promotes revascularization *in vitro* by forming a capillary-like structure compared with randomly arranged nanofibers. When pre-vascularized and neatly arranged nanofibers are implanted subcutaneously in rats, the nanofibers can be connected with the host's blood vessel network [73]. In a study, chitosan/poly ε-caprolactone (CS/PCL) nanofibers fabricated by sequential electrospinning technology were heparinized to loading vascular endothelial growth factor (VEGF) for preventing blood coagulation in small-diameter vessels (Figure 16.6) [74].

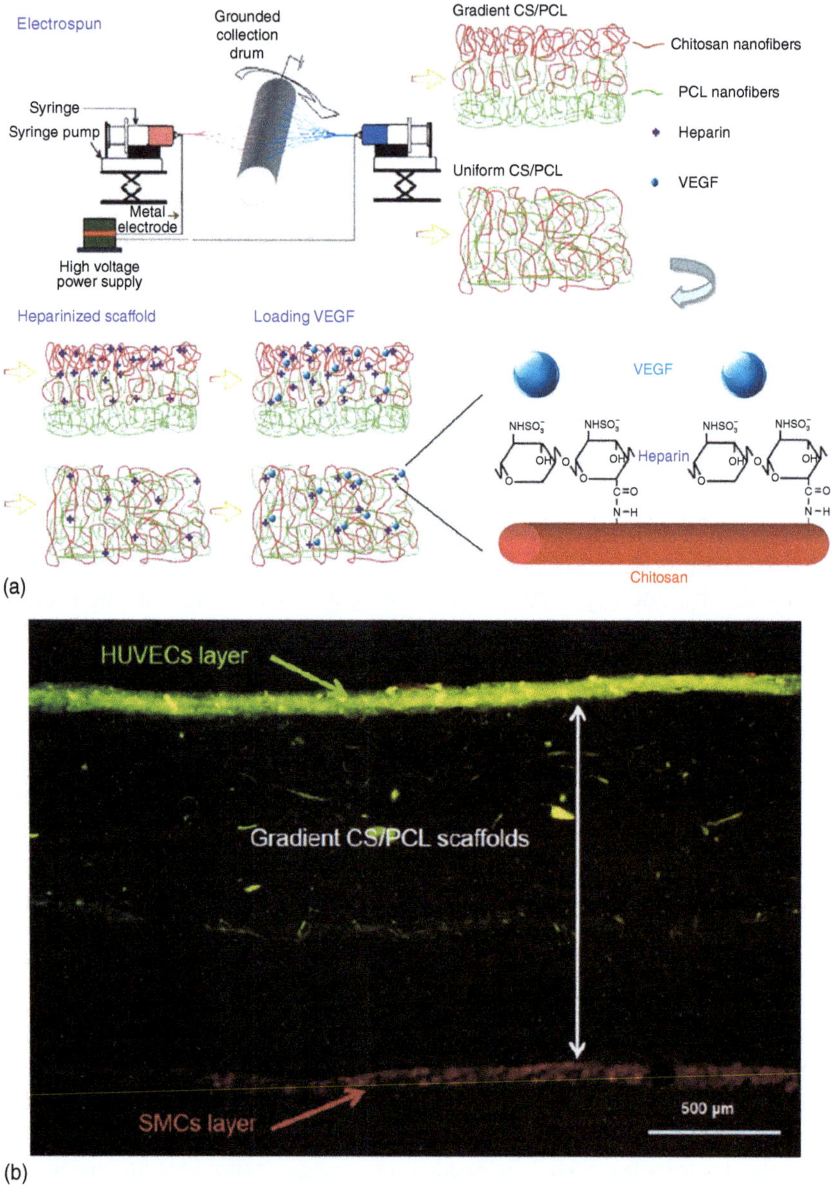

Figure 16.6 Electrospinning nanofibrous vascular scaffolds. (a) Schematic diagram of uniformed and gradient nanofibrous scaffolds for vascular tissue engineering; (b) observation of the cross section of the scaffold–cell "sandwich" structure (green: Von Willebrand factor antibody [vWF]-FITC for HUVEC; red: calponin-PE for vascular smooth muscle cell [vSMC]). Source: Du et al. [74]. Reprinted with permission of Elsevier.

16.5 Decellularized Vascular Tissue

Tissue-specific acellular matrix has significant advantages, including the original tissue's biological, structural, functional, and topographical characteristics. Previous studies have used chemical (acid, alkali, detergent, hypotonic, and hypertonic solutions), biological (such as trypsin, dispase, etc.), or physical (such as temperature, pressure, electroporation, stirring, pressure gradient, etc.) methods for the preparation of decellularized vascular tissue. However, these processes usually destroy the ultrastructure of ECM, remove growth factors, and trigger immune responses [75–77]. In addition, it can affect the differentiation and proliferation of refilled vascular cells by changing the signal pathway, phenotype, and mechanical transduction [78, 79].

Many studies have successfully cultured functional large blood vessels with the acellular matrix. For example, in one study, decellularized tissue-engineered blood vessels with autologous endothelial progenitor cells (EPC) can prevent blood clotting and intimal hyperplasia for 30 days in a pig model [80]. When the iliac arteries of decellularized rats inoculated with ECs were implanted into the abdominal aorta of rats, natural vascular-like structures were observed at three months [81]. Similarly, the inner layer of decellularized porcine arteries was repopulated with human ECs and implanted in isolated porcine arteries. It can be observed that EC covers the inner layer of the cellular matrix, and no thrombosis was reported after 70 days [82].

In addition to EC implant, the efficacy of acellular matrix vascularization relies mainly upon implementing physical stimulation. For example, the decellularized rat heart is repopulated with cardiac cells or EC and then cultured in a bioreactor for 28 days to maintain simulated cardiac physiological functions. In larger and smaller coronary vessels, the EC layer was identified seven days after recellularization. Interestingly, the cell-seeded heart tissue exhibited contraction and expansion under electrical stimulation eight days after the cell seeding [83]. In another study, ECs were implanted into decellularized lungs with intact internal structures and cultured in a bioreactor. When this lung is transplanted and perfused through the blood circulation in the recipient, regular gas exchange can be observed for six hours [84].

16.6 Cell-Only Fabrication

Classical approaches have focused on developing scaffolds with biological, chemical, and mechanical properties to instruct cell differentiation and 3D orchestration. Furthermore, most tissue engineering scaffolds rely on exogenous biocompatible materials, in which cells are seeded, grow, and mature *in vitro* or *in vivo*. As a result, scaffold-based tissue engineering has achieved remarkable results in reconstructing various tissues and organs. However, solutions based on biomaterials still face common challenges. Immunogenicity, degradation rate, the toxicity of degradation products, inflammatory response from the host, formation of fibrous capsules, and mechanical mismatch with surrounding tissues are still significant issues. Therefore, scientists have developed various techniques to transform tissues without any scaffolds, which consist only of cells and the matrix they secrete. These technologies are being applied even in the areas where scaffolds have achieved notable successes, such as in skin, bone, or cartilage engineering.

16.6.1 Cell-Only Bioprinting

In the body, tissues or organs are formed as a result of precisely regulated self-assembly and organization of multiple cell types. Many studies have reported using bioprinted cell aggregates to create pre-designed tissue structures through self-organization and tissue fusion [85]. Cell aggregates composed of single cells or multiple cell types can be bioprinted as cell particles, tissue balls, or tissue filaments. Researchers prepare tissue balls in a controlled manner to avoid clogging and cell damage during the bioprinting process [86]. In contrast, the bioprinting of cell particles can ensure high cell density without causing nozzle clogging or cell damage. However, the biodistribution of cell particles requires supportive hydrogels to accumulate, organize, and fuse to form tissues [87]. The tissue scaffold can also be formed by injecting cell pellets into a tubular mold prepared with a coaxial nozzle system. In one study, cell particles injected in permeable alginate capsules were organized and fused to form tissue chains *in vitro*. Load the tissue trend into a custom biographer and then extrude it layer by layer to form a predefined 3D structure [88]. These cell aggregation methods can be used to create tissues with vascular networks.

16.6.2 Cell Sheet

Many studies use cell aggregates – rather than 3D tissue scaffolds – to generate tissue that is vascularized and functional. However, scaffold materials can inhibit cell–cell interactions and cause various complications, including inflammation and cytotoxicity. Particularly, in the cell sheet technology, tissue-specific cells are cultivated on temperature-responsive materials to form a cell–ECM matrix. To form vascularized tissues, single-layer ECs can be combined with sandwich cell sheets, or vascular cell sheets can be co-cultured with tissue-specific cells to generate vascularized multi-cell sheets. Generally, cell sheets can be connected to and separated from temperature-sensitive Petri dishes at higher (37 °C) or lower temperatures (25 °C). The nanoscale poly(*N*-isopropyl acrylamide) coating on the surface of polystyrene can control the detachment of ECs and liver cells by changing from a hydrophobic state to a hydrophilic state at a temperature below 25 °C[89]. The vascularized cell sheet prepared in this way can be compiled into thick tissue. Although cell sheets can form vascularized tissues, the mechanical strength and poor tissue integrity of stacked cell sheets are the main challenges and require further research.

16.7 Conclusion and Outlook

In this chapter, we have introduced the representative approaches for biofabricating blood vessels and vascularized tissues. We have analyzed the advantages and challenges of each method. In summary, 3D inkjet bioprinting is a quick and cheap method for biofabrication but may cause cell damage and needle blockage. 3D extrusion printing is suitable for using sacrificial materials to make large pieces of vascularized tissue, but it will significantly increase the printing process. The recently emerged 3D coaxial printing technology has attracted the attention of researchers because it can form vascular structures at one time. With the development of laser technology in the past few decades, laser-based 3D bioprinting technology can improve printing accuracy and have the printing of complex vascular structures, but its cytotoxicity problem has not been effectively solved.

With the development of microfluidic technology, more and more research on seed ECs in microfluidic lumen face insufficient strength of the formed blood vessel network and a lack of biological functions. To satisfy self-assembly to form a vascular network instead of designing a vascular network in advance, EC-coated micromodules are also used as a tool to manufacture the vascular network in the research. Unfortunately, there is a big difference between such a blood vessel network and the blood vessel network in the body.

Because of the importance of ECM to tissues, many researchers use electrospinning technology to prepare nanofibers as vascular scaffolds. The structure similar to blood vessels in the body ensures the adhesion and development of ECs. There are also good results in animal experiments. There are also many studies using vascular decellularization matrix to prepare vascular stents. The decellularized matrix has an excellent physical structure, mechanical strength, and biological structure to support cell growth. However, because decellularization and seed cells are involved, its operation is more complicated, and it is easy to produce an immune response. Biomaterials face many challenges, including immune response, toxicity, and biocompatibility. Therefore, some researchers use uninfected cells to construct vascular networks. However, the mechanical strength of the vascular network constructed in this way is unsatisfactory, limiting its application.

The reconstruction and regeneration of large-scale tissues and organs are very dependent on the establishment of mature, stable, and biologically functional vascular networks. With the continuous development and maturity of new biological biofabrication technologies, new research has also shifted from the pursuit of constructing high-fidelity structures to the pursuit of rich biological functions that can regulate cells and tissues. Thus, in the future, the biological biofabrication of blood vessels and vascularized tissues will inevitably have higher precision and more biological functions to promote the regeneration and functional reconstruction of tissues.

Acknowledgments

The writing of this chapter was financially supported by The National Natural Science Foundation of China (no. 32022088) and the Qi-Huang Scholarship by the State Administration of Traditional Chinese medicine of China, both to Prof. Chunming Wang. The authors thank Dr. Yiming Niu and Mr. Zhencheng Liao for their constructive discussion.

References

1 Sarker, M.D., Naghieh, S., Sharma, N.K. et al. (2018). *J. Pharm. Anal.* 8 (5): 277–296.
2 Burton, A.C. (1954). *Physiol. Rev.* 34 (4): 619–642.
3 Tennant, M. and McGeachie, J.K. (1990). *Aust. N. Z. J. Surg.* 60 (10): 747–753.
4 Patan, S. (2004). *Angiogenesis in Brain Tumors*, 3–32. Springer New York, NY.
5 Thayer, P., Martinez, H., and Gatenholm, E. (2020). *3D Bioprinting*, 3–18. New York, NY: Humana.
6 Li, X., Liu, B., Pei, B. et al. (2020). *Chem. Rev.* 120 (19): 10793–10833.
7 Naghieh, S., Karamooz-Ravari, M.R., Sarker, M.D. et al. (2018). *J. Mech. Behav. Biomed. Mater.* 80: 111–118.
8 Sarker, M. and Chen, X.B. (2017). Modeling the flow behavior and flow rate of medium viscosity alginate for scaffold fabrication with a three-dimensional bioplotter. *J. Biofabrication Sci. Eng.* 139: 8.
9 Izadifar, Z., Chang, T., Kulyk, W. et al. (2016). *Tissue Eng. Part C* 22 (3): 173–188.
10 You, F., Wu, X., Zhu, N. et al. (2016). *ACS Biomater. Sci. Eng.* 2 (7): 1200–1210.
11 Sarker, M., Izadifar, M., Schreyer, D. et al. (2018). *J. Biomater. Sci., Polym. Ed.* 29 (10): 1126–1154.
12 Xu, T., Zhao, W., Zhu, J.-M. et al. (2013). *Biomaterials* 34 (1): 130–139.
13 Cui, X. and Boland, T. (2009). *Biomaterials* 30 (31): 6221–6227.
14 Yanagawa, F., Sugiura, S., and Kanamori, T. (2016). *Regener. Ther.* 3: 45–57.
15 Naghieh, S., Ravari, M.R.K., Badrossamay, M. et al. (2016). *J. Mech. Behav. Biomed. Mater.* 59: 241–250.
16 Miller, J.S., Stevens, K.R., Yang, M.T. et al. (2012). *Nat. Mater.* 11 (9): 768–774.
17 Kolesky, D.B., Truby, R.L., Gladman, A.S. et al. (2014). *Adv. Mater.* 26 (19): 3124–3130.
18 Zhang, Y., Yu, Y., Chen, H. et al. (2013). *Biofabrication* 5 (2): 025004.
19 Naghieh, S., Sarker, M.D., Izadifar, M. et al. (2018). *J. Mech. Behav. Biomed. Mater.* 78: 298–314.
20 Zhang, Y., Yu, Y., and Ozbolat, I.T. (2013). Direct bioprinting of vessel-like tubular microfluidic channels. *J. Nanotechnol. Eng. Med.* 4 (2): 020902. (7 pages).
21 Luo, Y., Lode, A., and Gelinsky, M. (2013). *Adv. Healthcare Mater.* 2 (6): 777–783.

22 Yu, Y., Zhang, Y., Martin, J.A. et al. (2013). *J. Biomech. Eng.* 135 (9): 091011. (9 pages).

23 Ozbolat, I.T. and Yu, Y. (2013). *IEEE Trans. Biomed. Eng.* 60 (3): 691–699.

24 Gao, Q., He, Y., Fu, J.-z. et al. (2015). *Biomaterials* 61: 203–215.

25 Norotte, C., Marga, F.S., Niklason, L.E. et al. (2009). *Biomaterials* 30 (30): 5910–5917.

26 Sadr, N., Zhu, M., Osaki, T. et al. (2011). *Biomaterials* 32 (30): 7479–7490.

27 Gaebel, R., Ma, N., Liu, J. et al. (2011). *Biomaterials* 32 (35): 9218–9230.

28 Gauvin, R., Chen, Y.-C., Lee, J.W. et al. (2012). *Biomaterials* 33 (15): 3824–3834.

29 Nahmias, Y. et al. (2005). *Biotechnol. Bioeng.* 92 (2): 129–136.

30 Melchels, F.P.W., Feijen, J., Grijpma, D.W. et al. (2010). *Biomaterials* 31 (24): 6121–6130.

31 Mankovich, N.J., Samson, D., Pratt, W. et al. (1994). *Otolaryngol. Clin. North Am.* 27 (5): 875–889.

32 Zhang, A.P., Qu, X., Soman, P. et al. (2012). *Adv. Mater.* 24 (31): 4266–4270.

33 Lee, J.W. et al. (2009). *Microelectron. Eng.* 86 (4–6): 1465–1467.

34 Xing, J.-F., Zheng, M.-L., and Duan, X.-M. (2015). *Chem. Soc. Rev.* 44 (15): 5031–5039.

35 Torgersen, J. et al. (2013). *Adv. Funct. Mater.* 23 (36): 4542–4554.

36 Applegate, M.B., Coburn, J., Partlow, B.P. et al. (2015). *Proc. Natl. Acad. Sci. U.S.A.* 112 (39): 12052–12057.

37 Meyer, W., Engelhardt, S., Novosel, E. et al. (2012). *J. Funct. Biomater.* 3 (2): 257–268.

38 Ovsianikov, A. et al. (2007). *J. Tissue Eng. Regener. Med.* 1 (6): 443–449.

39 Ng, C.P., Helm, C.-L.E., and Swartz, M.A. (2004). *Microvasc. Res.* 68 (3): 258–264.

40 Ueda, A., Koga, M., Ikeda, M. et al. (2004). *Am. J. Physiol. Heart Circ. Physiol.* 287 (3): H994–H1002.

41 Yamamura, N., Sudo, R., Ikeda, M. et al. (2007). *Tissue Eng.* 13 (7): 1443–1453.

42 Weston, D.F., Smekal, T., Rhine, D.B. et al. (2001). *J. Vac. Sci. Technol., B* 19 (6): 2846–2851.

43 Hsieh, Y.-K., Chen, S.-C., Huang, W.-L. et al. (2017). *Polymers* 9 (7): 242.

44 Hansen, A.S., Hao, N., and O'shea, E.K. (2015). *Nat. Protoc.* 10 (8): 1181–1197.

45 He, J., Chen, R., Lu, Y. et al. (2013). *Adv. Healthcare Mater.* 2 (8): 1108–1113.

46 Lim, D., Kamotani, Y., Cho, B. et al. (2003). *Lab Chip* 3 (4): 318–323.

47 Myers, D.R., Sakurai, Y., Tran, R. et al. (2012). *J. Visualized Exp.* 64: e3958.

48 Takayama, S., Ostuni, E., LeDuc, P. et al. (2001). *Nature* 411 (6841): 1016–1016.

49 Kaji, H., Camci-Unal, G., Langer, R. et al. (2011). *Biochim. Biophys. Acta, Gen. Subj.* 1810 (3): 239–250.

50 Tardy, Y., Resnick, N., Nagel, T. et al. (1997). *Arterioscler. Thromb. Vasc. Biol.* 17 (11): 3102–3106.

51 Shao, J., Wu, L., Wu, J. et al. (2009). *Lab Chip* 9 (21): 3118–3125.

52 Song, J.W., Gu, W., Futai, N. et al. (2005). *Anal. Chem.* 77 (13): 3993–3999.

53 Therriault, D., White, S.R., and Lewis, J.A. (2003). *Nat. Mater.* 2 (4): 265–271.

54 Chen, A. and Pan, T. (2011). *Biomicrofluidics* 5 (4): 046505.

55 Wu, W., DeConinck, A., and Lewis, J.A. (2011). *Adv. Mater.* 23 (24): H178–H183.
56 Huang, J.-H., Kim, J., Agrawal, N. et al. (2009). *Adv. Mater.* 21 (35): 3567–3571.
57 Baker, B.M., Trappmann, B., Stapleton, S.C. et al. (2013). *Lab Chip* 13 (16): 3246–3252.
58 Kim, S., Lee, H., Chung, M. et al. (2013). *Lab Chip* 13 (8): 1489–1500.
59 Kim, B.-S., Putnam, A.J., Kulik, T.J. et al. (1998). *Biotechnol. Bioeng.* 57 (1): 46–54.
60 Corstorphine, L. and Sefton, M.V. (2011). *J. Tissue Eng. Regener. Med.* 5 (2): 119–129.
61 Dean, D.M., Napolitano, A.P., Youssef, J. et al. (2007). *FASEB J.* 21 (14): 4005–4012.
62 Kelm, J.M., Djonov, V., Ittner, L.M. et al. (2006). *Tissue Eng.* 12 (8): 2151–2160.
63 Yeh, J., Ling, Y., Karp, J.M. et al. (2006). *Biomaterials* 27 (31): 5391–5398.
64 McGuigan, A.P. and Sefton, M.V. (2006). *Proc. Natl. Acad. Sci. U.S.A.* 103 (31): 11461–11466.
65 Cooper, T.P. and Sefton, M.V. (2011). *Acta Biomater.* 7 (3): 1072–1083.
66 Chamberlain, M.D., Gupta, R., and Sefton, M.V. (2012). *Tissue Eng. Part A* 18 (3, 4): 285–294.
67 McGuigan, A.P., Leung, B., and Sefton, M.V. (2006). *Nat. Protoc.* 1 (6): 2963–2969.
68 Du, Y., Lo, E., Ali, S. et al. (2008). *Proc. Natl. Acad. Sci. U.S.A.* 105 (28): 9522–9527.
69 He, W. et al. (2009). *J. Biomed. Mater. Res. Part A* 90 (1): 205–216.
70 Smith, I.O., Liu, X.H., Smith, L.A. et al. (2009). *Wiley Interdiscip. Rev. Nanomed. Nanobiotechnol.* 1 (2): 226–236.
71 Ghasemi-Mobarakeh, L., Prabhakaran, M.P., Morshed, M. et al. (2008). *Biomaterials* 29 (34): 4532–4539.
72 Klumpp, D. et al. (2012). *J. Biomed. Mater. Res. Part A* 100 (9): 2302–2311.
73 Cui, L., Li, J., Long, Y. et al. (2017). *RSC Adv.* 7 (19): 11462–11477.
74 Du, F., Wang, H., Zhao, W. et al. (2012). *Biomaterials* 33 (3): 762–770.
75 Yang, M., Chen, C.-.Z., Wang, X.-.N. et al. (2009). *J. Biomed. Mater. Res. Part B* 91 (1): 354–361.
76 Crapo, P.M., Gilbert, T.W., and Badylak, S.F. (2011). *Biomaterials* 32 (12): 3233–3243.
77 Fu, R.-H., Wang, Y.-C., Liu, S.-P. et al. (2014). *Cell Transplant.* 23 (4, 5): 621–630.
78 Badylak, S.F., Taylor, D., and Uygun, K. (2011). *Annu. Rev. Biomed. Eng.* 13: 27–53.
79 Chiquet, M. (1999). *Matrix Biol.* 18 (5): 417–426.
80 Quint, C. et al. (2011). *Proc. Natl. Acad. Sci. U.S.A.* 108 (22): 9214–9219.
81 Dall'Olmo, L., Zanusso, I., Di Liddo, R. et al. (2014). *Biomed Res. Int.* 2014.
82 Pellegata, A.F., Dominioni, T., Ballo, F. et al. (2014). *Cells Tissues Organs* 200 (6): 363–373.
83 Ott, H.C., Matthiesen, T.S., Goh, S.-K. et al. (2008). *Nat. Med.* 14 (2): 213–221.
84 Ott, H.C., Clippinger, B., Conrad, C. et al. (2010). *Nat. Med.* 16 (8): 927–933.

85 Jakab, K., Norotte, C., Damon, B. et al. (2008). *Tissue Eng. Part A* 14 (3): 413–421.

86 Marga, F., Neagu, A., Kosztin, I. et al. (2007). *Birth Defects Res. C Embryo Today* 81 (4): 320–328.

87 Owens, C.M., Marga, F., Forgacs, G. et al. (2013). *Biofabrication* 5 (4): 045007.

88 Yu, Y., Moncal, K.K., Li, J. et al. (2016). *Sci. Rep.* 6 (1): 1–11.

89 Okano, T., Yamada, N., Okuhara, M. et al. (1995). *Biomaterials* 16 (4): 297–303.

17

Vascularization

Xiao Lu[1], Jianxi Lu[1], Fangchun Jin[2], and Yajie Lu[3]

[1]*Shanghai Bio-lu Biomaterials Co., Ltd., Xinjunhuan Road, 188, Building 9, 4F-402, Shanghai 201114, P.R. China*
[2]*Xin Hua Hospital Affiliated to Shanghai Jiao Tong University, School of Medicine, Department of Pediatric Orthopedics, 1665 Kongjiang Road, Yangpu District, Shanghai 200092, P.R. China*
[3]*Air Force Medical University of PLA, The First Affiliated Hospital (Xijing Hospital), Department of Orthopaedics, 127 Changle West Road, Xi'an 710032, P.R. China*

17.1 Introduction

Grief never mended no broken bones... said renowned writer Charles Dickens, and if he was still among us today, he might probably add *...but biomaterials can.*

Advances in regenerative medicine during the past decades provided researchers and clinicians with a powerful tool to repair or replace damaged organs: tissue engineering. As the building blocks of this new discipline, biomaterials have been extensively studied for over 50 years. In this chapter, we will focus on one of the prerequisites for tissue repair, the vascularization of said biomaterials. In tissue engineering, the scaffold acts as the favorable ground where seeded cells and growth factors (GFs) will form living tissues *ex vivo* or *in vivo* to replace the defected ones. The implant thus provides an appropriate three-dimensional (3D) space where cells attach, proliferate, and differentiate. At the same time, blood vessels form inside the scaffold to restore microcirculation, providing vital compounds to sustain cellular activity. Although not studied in this chapter, biocompatibility, biodegradability, and mechanical support are obviously important characteristics for a successful clinical application. A high biocompatibility and adequate absorption rate would optimize the biological effects of the implant, whereas its mechanical properties directly impact the biomechanics of repaired zone, hence its clinical outcomes.

Vascularization, new bone formation, implant biodegradation, and bone remodeling are the four basic steps of bone repair after the implantation of biomaterials. Vascularization is the crucial process, providing essential nutrients for the growth of both cells and tissues. It also influences new bone formation, especially in the case of large bone defects, where it impacts both cell viability and motility. While repairing bone defects, the implant acts as a tissue engineering scaffold, where cellular proliferation provides the building blocks for new bone formation. This is why

Biofabrication for Orthopedics: Methods, Techniques, and Applications, First Edition.
Edited by Wenguo Cui, Xin Zhao, and Shen Liu.

vascularization plays such an important role in guaranteeing appropriate cellular nutrition. *In vitro*, bone tissue engineering is obtained by seeding cells on the scaffold in the presence of growth factors, both cellular nutrition and metabolism being achieved by renewing culture media. *In vivo*, vascularization of the scaffold is not immediate, even when implanted in rich vascularized tissues; thus, nutrition and metabolism can only be achieved through exchange with the tissue fluid at the early stages. Naturally, cellular activity is hindered without the appropriate nutrient supply, potentially leading to cell death. In this case, the scaffold does not integrate into the surrounding tissues, and without new bone growth, tissue engineering is meaningless. When it comes to vascularization and new bone formation, seed cells and growth factors play the roles of prerequisites and promoters, but the scaffold is the essential base on which tissue growth takes place. That is why, the repair of defected tissue can only be achieved if a functional vascularization can be rapidly restored. So far, no consensus was reached on how to effectively promote formation of new blood vessels, which is by far one of the biggest challenges of tissue engineering.

Nowadays, the research on vascularization of orthopedic implants is mainly focused on three areas: growth factors, coculture of vascular endothelial cells (ECs) with osteoblasts, and blood vessel grafting. Nevertheless, studies have shown the promotion of vascularization, and bone growth is dependent on the 3D structure of the scaffold. Whether used directly as a bone graft or as a tissue engineering scaffold, the implanted biomaterial must possess a porous structure to allow swift vascularization, leading to an integration into surrounding tissues. Lu et al. suggested that the biological interactions of the "cells–material–blood supply trinity" are directly impacted by the porous macrostructure of the scaffold. In their theory, cells, implants, and blood supply could be compared to their respective agricultural counterparts: seeds, soil, and fertilizers. Only seeded in the right soil, with the nutrients from the fertilizer, will a plant grow. Similarly, the 3D porous structure of the scaffold is the base on which blood vessels will grow, guaranteeing proper cell development. That is why, the complementarity of each parameter should be taken into account to optimize the biological effects of tissue engineering (Figure 17.1). Unfortunately, many manufacturing techniques lead to irregular porous structures, uneven pore or interconnection sizes, making quantitative study impossible. To solve this problem, Lu et al. described a method allowing accurate control on the porous structure (pore and interconnection sizes, porosity) of bioceramics, and the systemic study of their influence on vascularization, including biological behavior of cells, cytokine, receptor expression, and histological features. In this chapter, we will describe the link between porous structures and vascularization, its mechanism of action, formation pattern, and quantity effect relation.

17.2 Is Porous Structure Needed for Vascularization?

Vascularization refers to the formation of new blood vessels through angiogenesis and vasculogenesis, which are two distinct mechanisms [1, 2]. Angiogenesis is

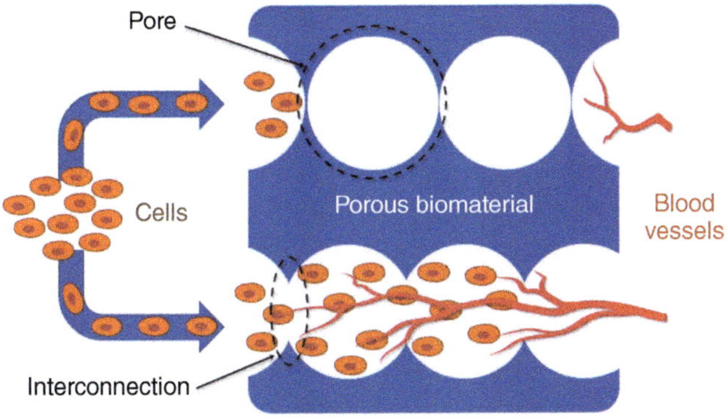

Figure 17.1 Biological interactions of the "cells–material–blood supply trinity."

defined as the growth of new capillaries from pre-existing blood vessels, whereas vasculogenesis is defined as the differentiation of precursor cells (angioblasts) into ECs and the *de novo* formation of a primitive vascular network. The basic steps of angiogenesis include enzymatic degradation of capillary basement membrane, EC proliferation, directed migration of ECs, tubulogenesis, and vessel pruning, with a growth rate of tens of micrometers per day. Vascularization in bone tissue engineering refers to the process where surrounding blood vessels grow into the scaffold; therefore, it is the angiogenesis process which is involved.

The repair of bone defect is triggered by hypoxic conditions, but many *in vitro* bone tissue engineering models are done under normoxic conditions, which do not replicate properly the environment of *in vivo* tissue engineering. The defect, devoid of vasculature, prompts the activation of hypoxia inducible factor-1 (HIF-1) genes [3], which in turn induces the transcription of vascular endothelial growth factor (VEGF), basic fibroblast growth factor (bFGF), platelet-derived growth factor (PDGF), and angiopoietin (Ang), starting the vascularization process. Therefore, HIF-1 is a key element for controlling defect-induced vascularization but is often missing in research models. As can be seen, vascularization is under the simultaneous influence of scaffolds, cells, and growth factors; thus, any model should consider these parameters as a whole and not independently.

Studies on the influence of porous structure upon vascularization started with autologous cancellous bone grafting. Cancellous bone is a highly interconnected porous structure, whereas the dense cortical bone consists of closely packed Haversian canals, surrounded by lamellae, interconnected by Volkmann's canals. Early stages of bone grafting are similar for both types of bones, with the formation of a hematoma and then a fibrocartilaginous matrix. As the open cell porous network of cancellous bone grafts facilitates the ingrowth of capillaries, resulting in an abundantly ramified vessels network, their vascularization starts after a couple of hours and can be completed in two weeks. In contrast, full vascularization of cortical bone grafts takes as long as two months since the Haversian/Volkmann's systems offer far less room for the formation of blood vessels. Yang et al. showed superior vascularization of iliac bone graft (IBG) compared to cranial bone graft (CBG) in rats [4]. After seven days, blood vessel formation was clearly observed in IBG, whereas almost no blood vessel penetrated through the cortical bone in CBG. At 14 days, a vascular network was formed inside the porous IBG, while only a few blood vessels penetrated the diploë. Albrektsson reported the growth rate of blood vessels in cancellous and cortical bones to be, respectively, 0.2–0.4 and 0.15–0.30 mm/day [5]. Our studies showed that the average vascularization speed in cancellous bone defects was 59.88 μm/day, separated into three distinct phases: the first growth peak was reached within 4 weeks (753 μm/week), followed by a plateau from 4 to 8 weeks (110 μm/week), then a second growth peak from 8 to 12 weeks (395 μm/day). The open cell porous structure of cancellous bones does not only provide a path for blood vessels ingrowth but is also the support on which new bone formation takes place. Even if autologous cancellous bone contains a large amount of stem cells and osteogenic factors, the key point of defect healing after grafting remains the ability of the porous structure to promote rapid vascularization. Therefore, scaffolds for bone tissue engineering should be designed with a similar porous framework to enhance vascularization, promoting the implant incorporation into the surrounding bone. The 3D porous properties (pore and interconnection sizes, porosity) of the scaffold are crucial for the ingrowth of ECs and the formation of a functional vascular network [6, 7]. Large pores and higher porosity provide more space for cell migration and proliferation, increasing nutrient supply and metabolite excretion, thus enhancing vascularization.

So far, there is no consensus regarding the influence of scaffolds on vascularization, the debate being mainly focused on the impact of both pore and interconnection sizes. It is generally acknowledged that only scaffolds of high porosity will enable the ingrowth of blood vessels, supply of nutrients, and removal of wastes. On the other hand, excessive porosity will decrease the mechanical strength of the scaffold and accelerate its degradation. Clinically, the implant would entirely degrade before the bone defect is fully repaired. Chiu et al. observed full vascularization in polyethylene glycol hydrogel scaffolds of larger pore sizes (50–100 and 100–150 μm), whereas only surface vascularization was achieved for scaffolds of smaller pore sizes (25–50 μm) [6]. Therefore, the growth rate and depth of new blood vessels are mainly impacted by the pore diameter. Druecke et al. compared the vascularization process of three porous polyethylene scaffolds with different pore sizes (20–75, 75–212, and 250–300 μm) [8]. At 8 and 12 days, larger pore sizes showed higher vascular density and erythrocyte sedimentation rate than smaller pore sizes. At 20 days, the vascularization action remains highly active. According to Kuboki et al., porous hydroxyapatite (HA) bioceramics could promote not only vascularization but also adhesion, proliferation, and differentiation of osteoblasts [9]. Their studies showed that the critical pore diameter is 300–400 μm. On HA scaffolds of smaller pore sizes (90–120 μm), cartilage was formed first and then ossification took place, whereas with pore sizes >350 μm, direct ossification was observed. Another study showed that a porosity of 90% and a pore diameter of 150 μm enhance greatly vascularization of the implant [10].

The aforementioned studies overlooked the direct effect of interconnections on cell migration and nutrient supply. If pores are not connected with each other, newly formed bone would also not connect, lacking the continuity and integration necessary to repair the bone defect. Blood vessels would also not be able to grow into these closed pores (pores without any connection to the surface or other pores). That is why, the influence of interconnections might be even more significant than pore sizes. Maddalena et al. prepared two types of HA porous scaffolds with different porosities, pores, and interconnection sizes, seeded with bone marrow stem cells (BMSCs) and implanted subcutaneously into rats [11]. Histological results showed that both scaffolds displayed good vascularization, and abundant blood vessels filled all the pores. Interestingly, a positive correlation was observed between the diameter of blood vessels and the interconnection size, indicating that the latest is a limiting parameter for vascular ingrowth. Similarly, Lu et al. observed cell penetration and proliferation in porous bioceramic scaffolds when the interconnection diameter was over 20 μm, formation of chondroid tissue inside macropores for interconnections between 20 and 50 μm, and formation of mineralized bone above 50 μm [12].

It is clear that both pore and interconnection play an important role in the vascularization of porous biomaterials. Nevertheless, the aforementioned studies showed some limitations. First, because of the inability to accurately control the porous architecture of the scaffolds, materials used showed random pore geometry, a wide range of porosities, and pores of various sizes with smaller interconnections or even without interconnection. Second, the preparation methods did not allow independent control on pore and interconnection sizes. Obviously, to evaluate the influence of a single structural parameter on the biological properties of the implant, other parameters must be kept identical. Few studies specifically and systematically investigated the role of these two critical structural features in neovascularization in porous biomaterials; that is why, there has been no consensus regarding the optimal porous structure for vascularization so far, and quantitative results from these studies are seldom obtained.

In order to overcome the limitations of imprecise structural characteristics, we developed a porous β-tricalcium phosphate (β-TCP) scaffold with accurately controlled pore and interconnection sizes by using slip casting shaping and polymeric beads as a porogen agent [13]. An organic skeleton constituted of polymethylmethacrylate (PMMA) beads was carried by a chemical forming treatment to establish a connection between PMMA beads. This reaction was accompanied with significant shrinkage of the organic skeleton, which was correlated with the interconnection size between beads. The skeleton was then impregnated by an aqueous slurry of β-TCP; after drying, the porogen skeleton was eliminated by debinding and sintering to obtain the porous bioceramic. Scaffolds obtained with this technique were highly homogeneous, with a consistent porous structure and showed no significant variation in mean pore and interconnection sizes. This precise control of the structural parameters allowed us to assess independently the effect of pore and interconnection sizes on vascularization. Results showed that both pore and interconnection affect the diameter of blood vessels. The pores provide the space and support for newly formed blood vessels, whereas interconnections act as the door for the ingrowth of blood vessels. That is why, their sizes determine the diameter of the blood vessels infiltrating through the macropores in the scaffold. We postulated that the upper limit of pore size for vascularization is 400 μm, as no significant difference in vascularization of scaffolds with pore size above 400 μm was observed. On the opposite, when the size of interconnection increased, both the size and number of blood vessels formed in the scaffold increased, demonstrating a superior influence of the interconnection size on vascularization in porous scaffolds compared to the pore size. Thus, we established a quantitative correlation between structural parameters (pore and interconnection sizes) and vascularization of porous bioceramics, the optimal specifications for tissue ingrowth and vascularization being 400–500 μm pore diameter and 120 μm interconnection size.

17.3 Promotion of Vascularization in Biomaterials

17.3.1 Vascular Graft

After implantation, vascular bundles proliferate inside the porous biomaterial to form capillaries. This simple method does not need specific procedures such as separation of the vascular pedicle or vascular anastomosis. Grafted blood vessels will promote bone formation at the bundle/defect interface through capillary growth inside the hematoma. A new vascular network is then formed, which will in turn boost the proliferation of osteoblasts as the hematoma is resorbed, starting the bone formation process. Vascularized bone grafts (with attached artery and veins) have been used in the treatment of avascular necrosis of the femoral head, showing promising results as new blood vessels growing inside the surrounding bones have been observed [14]. Vacanti and Bonassar successfully generated a vascularized pedicle of tissue-engineered bone by molding a biomaterial scaffold, seeded with periosteal cells, around a vascular bundle [15].

The process triggered by vascular bundle grafting is similar to the regeneration of the vascular bed on site and is well adapted to the physiological specificity of blood supply in the bone. It does not only shorten the time needed for vascularization of the implant but may also serve as a vascularized pedicle for tissue-engineered bone [16]. Enhanced neovascularization and osteogenesis were clearly observed after ectopic intramuscular implantation of osteogenic composites. The said materials were prepared by coating coralline HA bioceramics with BMSCs and then combined with a vascular pedicle [17]. *In vivo* vascularization of three different tube-shaped bioceramics was compared, respectively, the bioceramics alone, those coated with VEGF, and those combined with a vascular pedicle in the center [18]. At eight weeks, all the groups showed vascular ingrowth on the surface layer, and the implant was fully vascularized for the pedicle group. At 12 weeks, the pedicle group showed further improvement in vascularization, whereas implants of the other two groups were still not fully vascularized. Zeng et al. prepared BMSC-seeded β-TCP bioceramic tubes

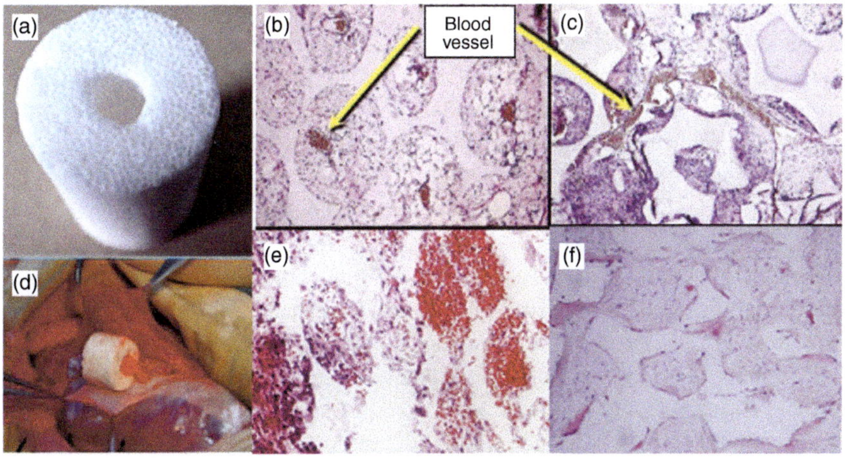

Figure 17.2 Histological analysis (HE, ×100) of a tube-shaped bioceramic (a) showed blood vessel formation in the center at four weeks (b) and full vascularization at eight weeks (c) for the vascular graft group (d). In the control group, no blood vessel was observed at four weeks (e) and only a few at eight weeks (f).

with a side groove, where a vein bundle was inserted, covered with a fascial pedicle [19]. After implantation in a macaque tibial defect model, the experimental group showed significative improvement in vascularization, bone formation, and material resorption compared to the bioceramics alone. We have also prepared porous β-TCP bioceramic tubes with an artery inside (Figure 17.2). Two pockets in bilateral perivertebral lumbodorsal fascia of New Zealand white rabbits were created, and artery inserted material or bioceramic alone were implanted in each pocket. After four weeks, a large amount of blood vessels was observed in the bioceramics with the artery, whereas the control group only showed blood vessel penetration in the peripheral areas of the implants. At eight weeks, the experimental group was fully vascularized, with the newly formed vascular network running from the outer layer to the center of the material. Blood vessel formation was also observed in the center of the scaffolds from the control group, however, in fewer number and with a smaller vessel diameter.

17.3.2 Growth Factors

Growth factors, especially vascular ECs, have an important influence on vascularization as they are capable of stimulating a variety of cellular processes during tissue development and healing. This includes cell proliferation, migration, differentiation, and multicellular morphogenesis. Growth factors are widely used in tissue engineering to enhance vascularization, and the most common molecules include VEGF, bFGF, PDGF, hepatocyte growth factor (HGF), transforming growth factor-β1 (TGF-β1), angiopoietin-1 (Ang-1), and angiopoietin-2 (Ang-2), each of them playing a different role in the vascularization process. VEGF is undoubtedly the most commonly used but also one of the most efficient growth factors to enhance vascularization. This growth factor promotes cell migration and proliferation by binding to specific receptors on the surface of ECs. This process causes them to dimerize and become activated through transphosphorylation, promoting the formation of functional new blood vessels [20]. Murphy et al. prepared a localized VEGF releasing scaffold to repair rat cranium critical defects. Regeneration of organs involves a complex interplay between multiple signal types, leading to coordinated growth of multiple tissue types. Vascular regeneration is closely linked to the presence of growth factors such as BMP, FGF, TGF-α, TGF-β, PDGF, tumor necrosis factor (TNF), VEGF, and Ang, which will promote the migration, proliferation, and differentiation of perivascular cells and smooth muscle cells. Results on cranium defect model showed a high density of blood vessels interspersed throughout the implant in VEGF releasing scaffolds, whereas few blood vessels were observed in the control group [21]. PDGF promotes the angiogenic capacity of ECs, including proliferation, migration, adhesion, and tube formation, and thereby also contributes to the vascularization [22]. bFGF is a potent mitogenic factor for ECs, by binding to its receptors FGFR-1, FGFR-2, FGFR-3, and FGFR-4 to activate signaling cascades that mediate EC activities. TGF-β can induce macrophages to produce VEGF, PDGF, and bFGF, further increasing the vascularization properties. Ang-1 and Ang-2 binds to the extracellular domain of Tie2 in ECs. It results in the activation of several intracellular signaling pathways leading to EC migration, tube formation, and sprouting, all crucial for angiogenesis and vascular maintenance.

17.3.2.1 Vascular Endothelial Growth Factor

VEGF is a potent angiogenic factor and was first described as an essential growth factor for vascular ECs. VEGF stimulates ECs to degrade extracellular matrix, proliferate, migrate, and form tubes, also acting as an EC survival factor. Additionally, it increases vascular permeability, leading to its alternative name, vascular permeability factor. As a major angiogenic factor, VEGF has been widely used to promote vascularization in bone tissue engineering, the studies being mainly focused on three aspects:

1) ECs, endothelial progenitor cells (EPCs), and/or mesenchymal stem cells (MSCs) are seeded into a scaffold. After implantation, VEGF secreted by ECs will promote vascularization of the tissue-engineered bone.
2) Stem cells and VEGF releasing carriers are combined with a porous scaffold. Promotion of vascularization in the tissue-engineered bone is achieved through progressive release of VEGF for two to three weeks.
3) Through gene transfer, stem cells are genetically modified to express VEGF, before being seeded inside the tissue-engineered bone to enhance vascularization.

Hu et al. seeded a suspension of $VEGF_{165}$ and vascular ECs on a pre-wetted β-TCP scaffold [23]. Results showed that the growth factor clearly increased cell adhesion and proliferation, with the formation of vascular endothelium after one week of culture. Street et al. developed a VEGF-loaded polylactic acid (PLA) microsphere system by dry freezing [24]. This system allowed a continuous local delivery of the protein after implantation in a murine femoral fracture model. Vascularity, callus formation, and mineralization in VEGF-treated femur fractures were higher than those of the control group, as VEGF-stimulated vascular ECs and osteoblast migration and differentiation. Nevertheless, the amount of VEGF released could not be controlled in these local delivery systems. Research into genetic modification of stem cells to express VEGF is ongoing and might be a promising technique to increase vascularization in bone tissue engineering; nevertheless, most are still at an early experimental stage. Cell co-transplantation requires coculture and then transplantation of two types of cells, and decreased adhesion rate of stem cell has been observed *in vitro*. *In vivo*, a nutrient competition between both cells may appear, inevitably decreasing cell viability, differentiation, and bone formation capacity. Present day genetic engineering also raised concerns about carcinogenicity. In contrast, VEGF releasing systems seem to be a more promising application prospect.

Vascularization in bone tissue engineering is a complex process, involving multiple growth factors. That is why, systems where only VEGF is loaded might show therapeutic limitations and hinder their clinical application.

17.3.2.2 Bone Morphogenetic Protein

Bone morphogenetic protein, originally identified as osteoinductive components in extracts derived from bone, is now known to play an important role in promoting vascular growth through stimulating the production of VEGFs by osteoblasts as VEGFs do not only enhance angiogenesis but also act simultaneously as paracrine factors, modulating endothelial and osteoclast functions, and as autocrine factors, modulating osteoblast differentiation [25]. Peng et al. suggested that although VEGF alone did not improve bone regeneration, it acted synergistically with BMP-4 to increase the recruitment of MSCs, to enhance cell survival, and to increase cartilage formation in the early stages of endochondral bone formation [26]. Bouletreau et al. hypothesized that a reciprocal regulatory and functional relationship occurs between VEGF and BMP during the bone repair process [27]. At the fracture site, hypoxia increases VEGF expression in osteoblasts, increasing the differentiation and proliferation of ECs, thus formation of new blood vessels. Simultaneously, ECs respond to stimulation by VEGF and, in turn, produce BMP-2. As osteoblasts are affected by BMP-2, this growth factor does not only increase the VEGF expression but can also promote osteogenesis, creating a positive regulation mechanism. Jia and coworkers prepared BMP-2- and VEGF-doped fibrin glue coated on porous titanium and evaluated their angiogenetic properties *in vivo* [28]. Four weeks post-implantation, they noticed significantly enhanced angiogenesis for single- or dual-factor-doped implants, and a synergistic effect was observed for dual factor combination. That is why, BMP and VEGF have a synergic effect on vascularization in bone tissue engineering. Sun et al. observed superior neovascularization in 3D bioprinted scaffolds loaded with BMSCs, RAW264.7 macrophages, and BMP-4-loaded mesoporous silica nanoparticles compared to those without the addition of BMP-4 [29]. Similarly, Bhattarai et al. demonstrated that sequential release of tauroursodeoxycholic acid and BMP-2 from coaxially electrospun PLA fibers enhanced new blood vessel formation when implanted in rabbits [30].

17.3.2.3 Basic Fibroblast Growth Factor

bFGF can induce angiogenesis by binding to Sp1 sites in the core VEGF promoter, whose activation mediates regulation of VEGFs. Perets et al. prepared a porous alginate scaffold that incorporated poly(lactic-*co*-glycolic acid) (PLGA) microspheres capable of controlling the release of encapsulated bFGF [31]. After implantation on the mesenteric membrane in rat peritoneum, bFGF-loaded scaffolds induced the formation of larger and more mature blood vessels compared to those without bFGF.

17.3.2.4 Multifactor Combination

In vivo vascularization is a non-linear process under the combined influence of many growth factors, where cascading and looping activations take place simultaneously. Using high dose of a single growth factor is risky and might cause serious side effects, such as angioma or vascular hyperpermeability for VEGF. On the opposite, combining multiple growth factors in low dose would not only guarantee a safe and steady release but would also take advantage of their potential synergy. Ramoshebi and Ripamonti noticed that the single application of relatively high doses of TGF-β, bFGF, or BMP in the chick chorioallantoic membrane assay elicited a prominent angiogenic response as expected [32]. Interestingly, the binary application of low doses of BMP with bFGF or TGF-β resulted in a distinctly more pronounced angiogenic effect than the single application of high doses of the respective factors. These results confirm a potential synergistic enhancement of vascularization when combining growth factors. Elçin et al. described the use of biocompatible chitosan–albumin microspheres and fibers as a potential delivery system for the controlled and localized release of endothelial cell growth factor (ECGF) [33]. This system was able to sustain the release of ECGF for three weeks *in vitro*. Implantation in rat groin fascia showed a high degree of neovascularization for both shapes starting after seven days. Control animals that received ECGF injection did not show any significant neovascularization after the same period of time. Huang et al. developed a scaffold system capable of sustained and localized presentation of BMP-4, VEGF, and human BMSCs to promote bone formation at an ectopic site [34]. After implantation into the subcutaneous tissue of SCID mice, concerted delivery of the three molecules promoted greater bone formation relative to any single factor or a combination of two factors. Sun et al. studied the synergistic effect of multiple angiogenic growth factors (VEGF+ Ang-1 and VEGF+ insulin-like growth factor [IGF]+ stromal cell-derived factor-1) loaded on a hydrogel [35]. The combination of all GFs dramatically increased the size and number of newly formed functional vessels compared to any individual growth factor.

Multifactor combination is clearly a promising technique to enhance vascularization in tissue engineering, but further studies are necessary in order to determine the optimal combination and dose of said growth factors.

17.3.3 Stem Cells

In tissue engineering, the prerequisite to build any functional vascular network is the presence of stem cells with angiogenic potential. ECs, MSCs, embryonic stem cells (ESCs), and induced pluripotent stem cells (iPSCs) are among the most commonly used angiogenic stem cells.

17.3.3.1 Endothelial Cells and Endothelial Progenitor Cells

The bone vasculature is crucial in the development, remodeling, and repair of bone, as both intramembranous and endochondral ossification occur in association with blood capillaries. Osteoblast–ECs (from the vascular bed) crosstalk is central to bone formation. ECs line blood vessels and form the barrier between vessels and other tissues. That is why, any method to induce vascularization will directly or indirectly influence their migration, aggregation, and proliferation. When cultured *in vitro*, ECs seeded into scaffolds could form vascular-like networks. These implants showed enhanced cell infiltration and vascularization upon *in vivo* implantation.

Peters et al. investigated whether sustained and localized delivery of VEGFs combined with transplantation of human microvascular endothelial cells (HMVECs), incorporated in porous PLGA matrices, could be used to engineer new vascular networks [36]. After implantation into SCID mice, immature human-derived vessels were present within three days and functional vessels were observed within seven days. In tissue engineering, the easiest way to obtain ECs is from the differentiation of EPCs. Compared to ECs, not only are the conditions for EPCs culture less demanding but EPCs also have better cell adhesion and proliferation rate and are less prone to cell aging. EPCs and ECs coculture enhanced their angiogenic abilities as shown by Koga et al. [37]. They seeded EPCs isolated from rat bone marrow on top of a confluent EC monolayer on a collagen gel scaffold. Quantitative analyses revealed that the total length, number, and depth of vascular networks were significantly enhanced with the addition of EPCs. Furthermore, ECs can also be cocultured with other types of cells to promote vascularization, such as vascular smooth muscle cells (VSMCs). Vascular smooth muscles (VSMs) are composing the majority of the wall of blood vessels. They contract or relax to change both the volume of blood vessels and the local blood pressure, allowing the redistribution of blood. Hegen et al. seeded primary HMVECs and VSMCs at different ratios in poly-L-lactic acid (PLLA) scaffolds enriched with basement membrane proteins (Matrigel), before implanting them subcutaneously into immunocompromised mice [38]. Vascularization in the tissue engineering context was strongly enhanced after one week, with the formation of a microvasculature within the scaffolds that anastomosed with the host circulation. Similarly, coculture of ECs and human osteoblasts also enhances the angiogenic properties of ECs. Unger et al. grew human dermal microvascular endothelial cells (HDMECs) on 3D bone biomaterials but observed no migration of HDMECs to form microcapillary-like structures [39]. When HDMECs and primary human osteoblast cells or human osteoblast-like cells MG-63 were cocultured on these biomaterials, a tissue-like self-assembly of cells occurred with time, with ECs forming microcapillary-like structures containing a lumen and giving a strong

PECAM-1 expression at cell interfaces. The life span of cocultured HDMECs was also significantly enhanced (42 days) compared to that of the monoculture of HDMECs (one week).

Wang et al. cocultured human osteoblasts with high alkaline phosphatase (ALP) activity and human umbilical vein endothelial cells (HUVEC) with little ALP activity [40]. They observed increased levels of expression of VEGF receptor genes in ECs, which was in turn followed by increased production of osteotropic growth factors, such as IGF-1 and endothelin-1. Kaigler et al. hypothesized the potential of BMSCs to modulate the growth and differentiation activities of ECs by their secretion of VEGFs, enhancing survival and differentiation of ECs *in vitro*, and promoting angiogenesis *in vivo* [41]. By using muscle-derived stem cells genetically engineered to express BMP-4 and VEGF to repair rat cranium critical defect, Peng et al. demonstrated that the optimal VEGF/BMP-4 ratio was 1 : 5, with better vascularization and bone formation compared to higher concentration of VEGF [26]. That is why, the ratio of growth factors in cocultures should be carefully controlled to generate beneficial effects and prevent detrimental reactions in bone tissue engineering. Wenger et al. grew HUVECs as 3D multicellular spheroids and then seeded them in a collagen matrix, before incorporating human osteoblasts into the ECs spheroids to form heterogeneous cospheroids [42]. The cospheroids displayed alignment of HUVECs in a circular structure after four days. Sprouting from ECs was promoted by angiogenic stimulation of VEGF and bFGF; however, direct contact or close proximity between human osteoblasts seemed to override said stimulation, potentially influenced by an altered compilation of ECM proteins from human osteoblasts (osteocalcin). Nevertheless, Yang noticed that coculture of ECs, and osteoblasts showed no contact inhibition, increased proliferation of osteoblasts, or production type I collagen, ALP, and osteocalcin [43].

Ideally, ECs would act as angiogenic stem cells; once seeded in the biologically active microenvironment of a living organism, they would promote capillarization. A microvasculature would then be formed inside the tissue-engineered scaffold, progressively integrated to the host vascular system. Many studies showed the capacity of ECs to secrete IGF, prostaglandin, and colony-stimulating factor, promoting the proliferation and differentiation of osteoblasts. Osteoblasts could also secrete bFGF and VEGF, as shown by Street et al. in a murine femoral fracture model [24]. Through injection of a VEGF inhibitor, it was shown that VEGF did not only enhance blood vessel formation but also promote the proliferation of osteoblasts, ossification, and new bone maturation.

It is clear that the cells involved in vascularization affect and regulate each other through the secretion of growth factors. Thus, by directly or indirectly controlling the presence and dose of the said factors, it would be possible to enhance tissue regeneration, especially in bone tissue engineering.

17.3.3.2 Mesenchymal Stem Cells

MSCs are multipotent stem cells found in the bone marrow. They are crucial for making and repairing skeletal tissues, such as cartilage, bone, and the fat found in bone marrow. Sometimes referred as the "other" BMSC because of their versatility,

MSCs can also differentiate into ECs to form blood vessels, thus having the dual function of osteogenesis and angiogenesis [1]. MSCs are among the most used stem cells in bone tissue engineering, as they are widely available and can be easily isolated. They can be extracted from locations as varied as bone marrow, umbilical cord blood, adipose tissue, and so on. Results from Zhou et al. showed that by co-seeding MSCs and MSC-derived ECs into a porous β-TCP scaffold in rabbit ulnar defects model, vascularization, osteogenesis, and mechanical properties were all improved [44]. Zhou et al. seeded MSC-derived osteoblasts into β-TCP bioceramic granules before implanting the resulting tissue-engineered bone under the periosteum of rabbit femur [45]. Ectopic tissue growth was observed as a large amount of blood vessels, and new bones were present in 75% of the implants.

17.3.3.3 Embryonic Stem Cells

ESCs are pluripotent stem cells derived from the inner cell mass of a blastocyst, distinguished by their ability to differentiate into any embryonic cell type and to self-renew *in vitro* and *in vivo*. ESCs can be efficiently and effectively differentiated into ECs [46], quickly engrafted into the host vasculature, enhancing neovascularization [47]. Ferreira et al. proposed to enhance vascular differentiation of human ESCs by encapsulation in a dextran-based hydrogel [48]. The fraction of cells expressing VEGF receptor KDR/fetal liver kinase-1 (Flk-1) increased up to 20-fold compared to spontaneously differentiated embryoid bodies (EBs). Furthermore, once removed from the hydrogels and cultured, the number of vascular cells was higher than the number obtained through EBs. Even if researchers are currently focusing heavily on the therapeutic potential of ESCs, with clinical use being the goal for many laboratories, the major concerns remain such as their potential to form tumors after transplantation, including teratoma. Because of their embryonic nature, harvesting them necessitates the destruction of the embryo, raising ethical debates and limiting their clinical use.

17.3.3.4 Induced Pluripotent Stem Cells

iPSCs are pluripotent stem cells generated from adult cells by reprogramming. iPSCs have the same properties as ESCs and can therefore self-renew and differentiate into almost any cell types [49]. Rufaihah et al. initiated iPSC differentiation into ECs by adding BMP-4 and VEGF to the culture media [50]. The ratio of blood perfusion increased when iPSC–ECs were delivered by intramuscular injection into the ischemic hind limb of SCID mice, as well as the total number of capillaries. Suzuki et al. obtained fetal liver kinase-1 positive (Flk-1+) cells from iPSCs and injected the into ischemic hind limbs of athymic nude mice [51]. Revascularization of the ischemic hind limb was accelerated, and the expression of VEGF mRNA in ischemic tissue increased in mice transplanted with Flk-1+ cells compared with control mice. Similar to ESCs, iPSCs can greatly enhance vascularization; nevertheless, their propensity to form tumors limits their clinical use. All the genes that have been shown to promote iPSC formation have also been linked to cancer in one way or another [52].

Currently, many stem cells are studied for their angiogenic and osteogenic properties in tissue engineering, each with its advantages and drawbacks. With no obvious "ideal" candidate so far, the present-day research is focused on increasing the availability and proliferation capacity of stem cells, while decreasing their immunogenic risks and solving potential ethical issues.

17.3.4 Microfabricated Vasculature

Fabrication of the intricate anatomy of vasculature within artificial tissues remains one of the key challenges facing the field of tissue engineering today. Without proper vascularization, tissue-engineered organs rely mainly on host vasculature for oxygen, nutrients, and waste removal [53, 54]. As a result, the complexity and the thickness of these tissues are limited by the mass transfer properties of the scaffold material. Without an intrinsic capillary network, the maximal thickness of the engineered tissue is approximately 150–200 μm because of oxygen diffusion limitations [55].

Soft lithography is one of the most promising techniques for fabricating microvasculature in bone tissue engineering. It is achieved by constructing a predesigned matrix on an elastomeric material, before replicating it onto the chosen biomaterial. Once this template is seeded with stem cells, controlled tubulogenesis can be achieved by varying its geometry, guiding the formation of branched tubes [56]. A biomimetic hydrodynamic process can be obtained by replicating the 3D geometry of the vascular network and connecting it to a perfusion system. This technique can be extensively used to create artificial microcirculatory networks. Rivron et al. used soft lithography and micromolding to form non-adherent agarose templates, before seeding them with a coculture of HUVECs and MSCs to construct self-remodeling microfabricated tissues [57]. The coculture produced VEGFs, mimicking the early events of angiogenesis. They linked the contraction and deformation of microfabricated tissues to the formation of gradients of angiogenic microenvironments and angiogenesis. Fidkowski et al. etched a capillary pattern onto silicon wafers by microelectromechanical system techniques, before transferring it by micromolding onto a poly(glycerol sebacate) (PGS) film [58]. The obtained microfabricated capillary networks were then endothelialized under flow conditions, and parts of the capillary networks were covered by nearly confluent ECs within 14 days.

Even if we are still at the early research stages of artificial microvasculatures, as their current use is mainly limited to the *in vitro* studies of vascularization and cellular properties, the future development of this technology may be a breakthrough in tissue engineering. By finding the right degradable biomaterial with appropriate mechanical properties, and perfecting pattern replication techniques, it is not impossible to imagine manufacturing one day fully functional biomimetic vascular networks. They would be similar to natural vasculatures, with physiological flow and mass transport patterns, providing the necessary oxygen, nutrients, and waste removal for the mesenchymal cells. The resultant constructs could potentially lead to implantable vascularized tissue-engineered organs.

17.4 Vascularization in Porous Biomaterials

The porous macrostructure of a bone repair material is a determining factor of its clinical outcomes, as it impacts directly or indirectly the mechanical and biological properties of the implant. Some of these properties are independent of the material chosen and can be generalized regardless of its chemical composition. Two of the most influential parameters are the pore size and the interconnection diameter, regulating cell adhesion and proliferation, as well as vascularization and tissue regeneration. Albeit the porosity of a material is also important, being correlated with its mechanical strength, high porosity can be easily obtained for modern biomaterials. That is why, we no longer consider porosity as the limiting factor for biological properties.

17.4.1 Influence of the Pore Diameter

Dense bioceramics were among the earliest grafting materials used to repair bone defects. One of the first reports dates back to the end of the nineteenth century when a German physician Dreesman successfully used plaster of paris to treat tuberculous osteomyelitis. Nevertheless, dense bone grafts have shown mixed clinical outcomes because of their weak tissue penetration. To overcome this limitation, the concept of using porous materials has been investigated in the early 1970s. Highly porous bioceramics are not only desirable for diffusion of nutrients and waste removal but also allow vascularization and oxygenation. As such, pores are vital constructs where cell penetration and adhesion take place and provide 3D substrates to guide tissue formation. The pore size directly affects both physical and biological properties of the material, such as mechanical strength, cell behavior, and osteogenesis. Variation of the pore diameter correlates with vascularization of the implant, especially regarding the number, diameter, or length of the newly formed blood vessels. Generally speaking, larger pores lead to better vascularization. Druecke et al. compared vascularization of three poly(ether ester) block copolymer scaffolds of different pore sizes *in vivo* [8]. At 8 and 12 days, scaffolds with wider pores showed a larger vessel diameter and a higher red blood cell (RBC) velocity, resulting in increased functional vessel density in both the edge and the center of the implants. After 20 days, neovascularization was still active in the center, whereas microcirculatory parameters on the edge returned to values similar to those found in the surrounding tissue. We compared *in vivo* vascularization of four β-TCP porous scaffolds with similar interconnection size (120 μm) and different pore sizes (300–400, 400–500, 500–600, and 600–700 μm) (Figure 17.3) [13]. Large blood vessels full of red blood cells were observed for larger pore sizes (Figure 17.4). Even if there was no increase in the number of blood vessels for the 600–700 μm group, their diameter was significantly higher. Vessels with diameter above 100 and 200 μm, respectively, accounted for 25.9% and 4.98% of the total amount of blood vessels. Comparatively, the percentage of blood vessels above 100 μm for the 300–400 μm group was only 14.7%, and no vessels above 200 μm were observed. Nevertheless, we noticed that pore diameter over 400 μm led to higher vascularization, but the difference between the three

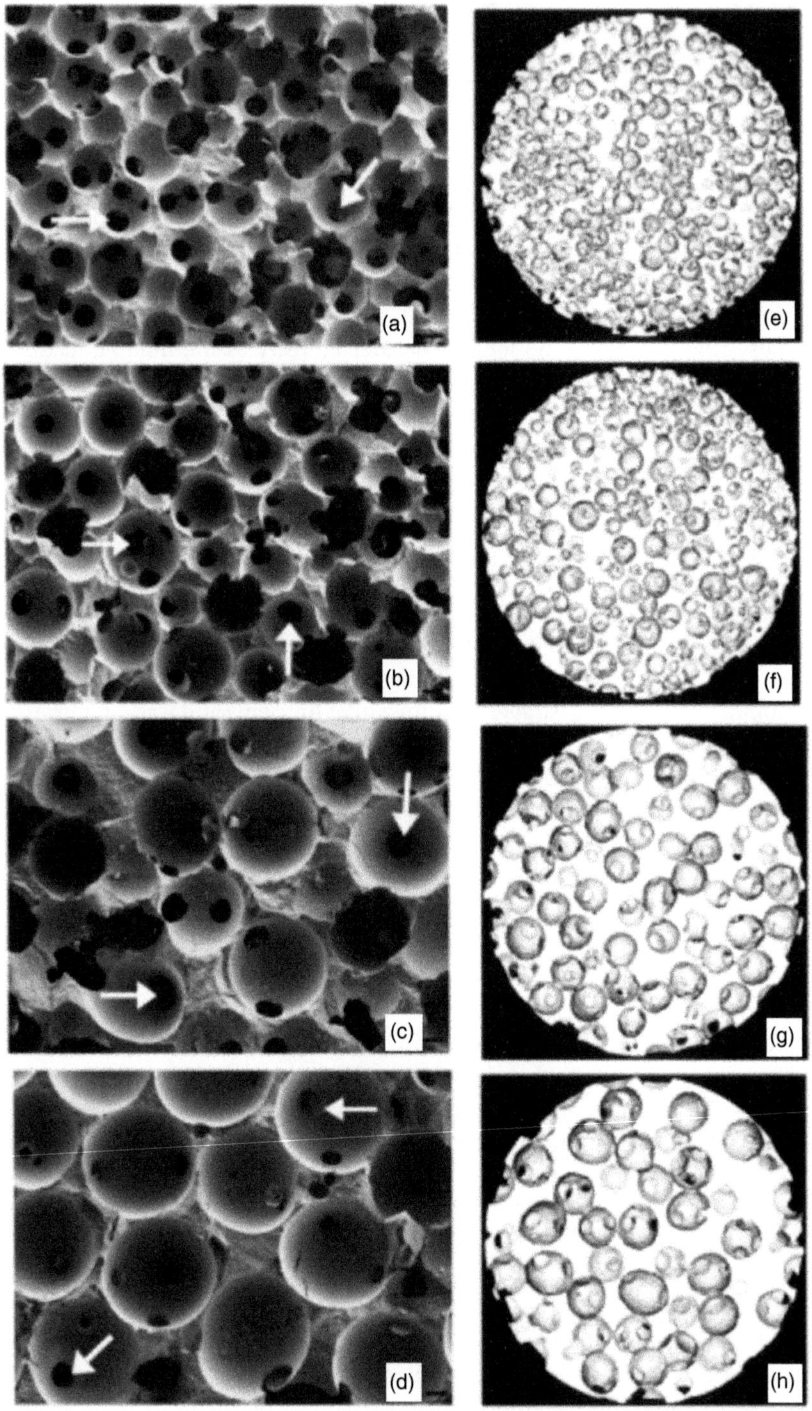

Figure 17.3 Scaffolds of identical interconnection size (120 μm) and different pore sizes (300–400, 400–500, 500–600, and 600–700 μm) in SEM (a–d) and micro-CT (e–h). Source: Reprinted with permission from Bai et al. [13] Mary Ann Liebert, Inc.

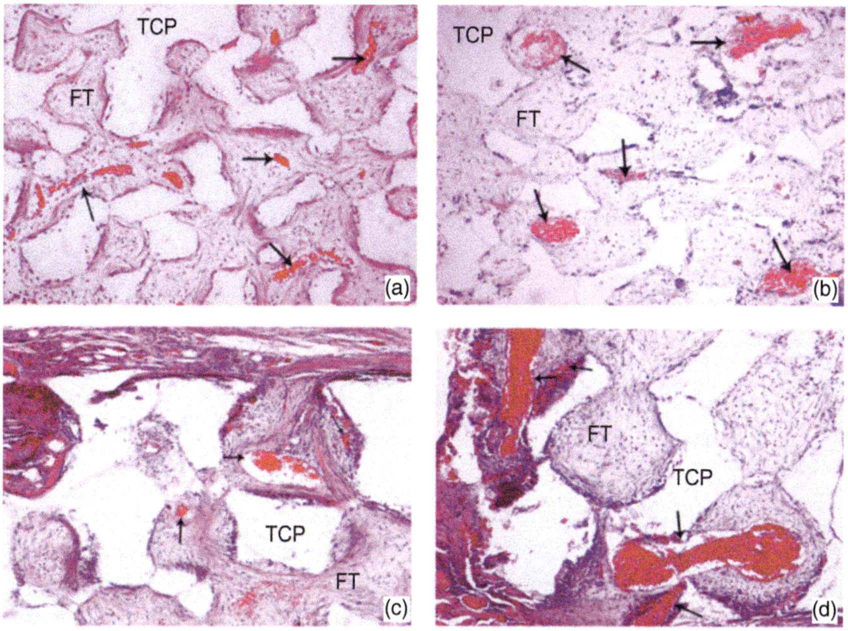

Figure 17.4 Histological results at four weeks: 300–400 (a), 400–500 (b), 500–600 (c), and 600–700 μm (d) pore size. Scale: 100 μm. FT, fibrous tissue; TCP, tricalcium phosphate scaffold, black arrows. Source: Reprinted with permission from Bai et al. [13] Mary Ann Liebert, Inc.

groups with larger pore size was not statistically significant (Figure 17.5). Choi et al. reached a similar conclusion when comparing the *in vivo* neovascularization of four poly(D,L-lactide-*co*-glycolide) (PLG) inverse opal scaffolds with uniform pore sizes of 79, 147, 224, and 312 μm in diameter [59]. Histology analyses revealed that scaffolds with large pores (>200 μm) favored the formation of vascular networks with large blood vessels and deeper penetration.

Somo et al. noticed that discrepancies remained between different studies regarding the optimal pore size. They suggested that these results may be confounded by differences in the interconnectivity of the pores in the scaffolds investigated [60]. Interconnections should not be viewed as a binary factor for describing whether pores are connected or not, and their precise description allows a better understanding of oxygen diffusion, nutrient distribution, cell migration, and other biological processes taking place during tissue formation. It is easy to imagine that there would be no cell penetration inside a highly porous scaffold without interconnected pores, even if said pores were large enough. Cellular activities would only take place on the surface of the material; no blood vessels would grow into the implant, never mind vascularization. Many of the early studies on the vascularization of tissue engineering scaffolds were done on materials with various porous structures. Without a precise structural control, studied scaffolds were highly heterogeneous, with random pore geometry, uneven porosity, large range of pore diameter, or lack of interconnections. Research often focused solely

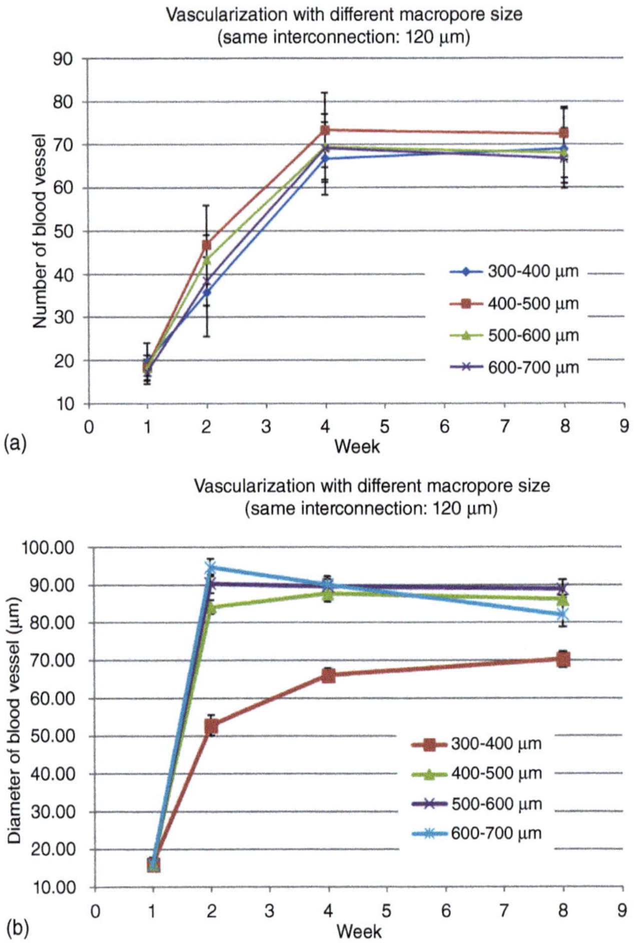

Figure 17.5 Positive correlation was observed between the pore size and vascularization on both number (a) and diameter (b) of blood vessels.

on one parameter, omitting to describe the others, leading to discrepancies in the final results. When studying the influence of one single structural variable, it seems logical to keep all other parameters unchanged. That is why, a precise and independent control of the porous architecture is necessary to assess its biological properties.

17.4.2 Influence of the Interconnection Diameter

The interconnectivity of the porous structure in tissue engineering is one of the prerequisites for vascularization. If we draw an analogy between a porous scaffold and a house, pores could be assimilated to rooms, in which cells adhere and proliferate, whereas interconnections would be the doors connecting each room. Their presence and size are determinant to cell migration, regulating the

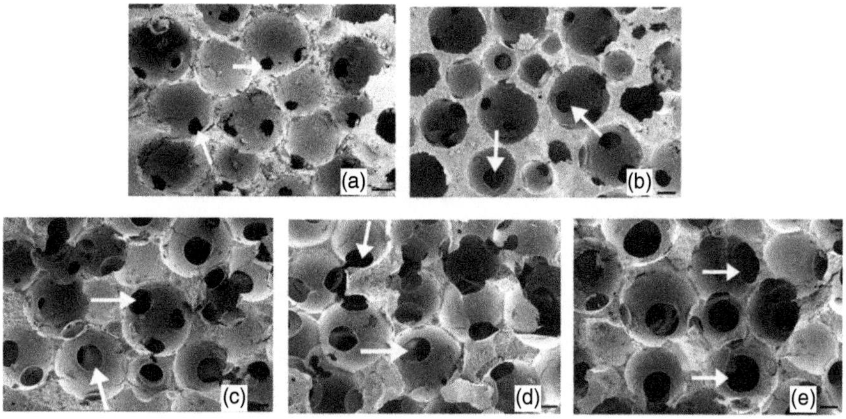

Figure 17.6 SEM pictures of porous bioceramics with identical pore size (300–400 μm) and different interconnections ((a) 70, (b) 100, (c) 120, (d) 150, and (e) 200 μm). Scale: 100 μm, white arrows indicate the interconnections between pores. Source: Reprinted with permission from Bai et al. [13] Mary Ann Liebert, Inc.

growth of blood vessels and other tissues. As seen in previously cited studies, no consensus has been reached regarding the optimal pore size for vascularization. The discrepancies in these finding might be explained by the differences in the interconnectivity of the studied scaffolds, as well as variation in the interconnection size.

We investigated the effect of interconnections on vascularization by preparing five groups of porous β-TCP bioceramics, implanted in the perivertebral lumbodorsal fascia of New Zealand white rabbits [13]. All the scaffolds had identical pore size (300–400 μm) and different interconnection diameter (70, 100, 120, 150, and 200 μm) (Figure 17.6). Vascularization was slower for smaller interconnection diameters (70 and 100 μm) as no blood vessel was observed in the center of the scaffolds at four weeks. Implants with interconnection diameter above 100 μm showed full vascularization at the same time point. A "rosary" pattern was observed on the newly formed blood vessels as their lumen was smaller when passing through interconnections, whereas the vessels enlarged when extending into the pores afterward (Figure 17.7). Larger connection sizes (150 and 200 μm) demonstrated improved vascularization compared to the three other groups, with larger diameter and number of blood vessels (Figure 17.8). Similar to cell migration, interconnections are the doors through which blood vessels run, and therefore, they act as bottlenecks, limiting the number and diameter of newly formed vessels. Additionally, the neovasculature was denser and more homogeneous in scaffolds with larger interconnections, whereas for the smaller ones, a negative gradient in the density, number, and size of blood vessels was observed from the edge toward the center of the material. Somo et al. generated porous poly(ethylene glycol) diacrylate (PEGD) hydrogel scaffolds with controlled pore interconnectivity by using PMMA microspheres as a sacrificial agent [60]. *In vitro* results showed interconnectivity-influenced PDGF-BB transport

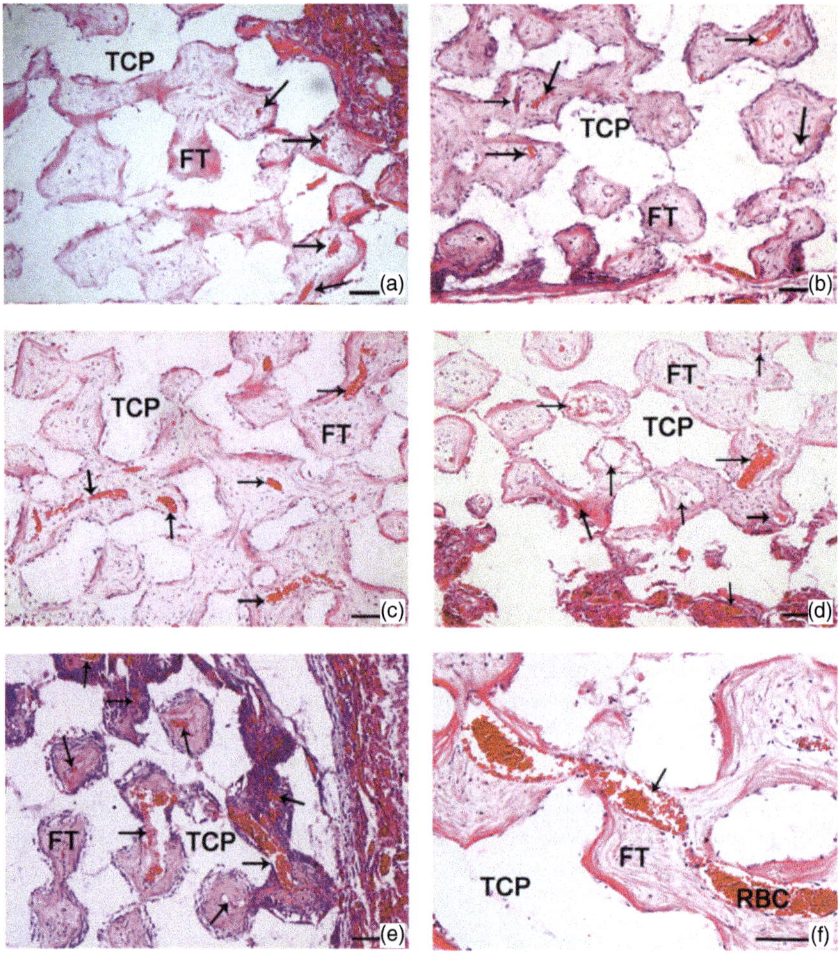

Figure 17.7 Histological results at four weeks: 70 (a), 100 (b), 120 (c), 150 (d), and 200 μm (e) interconnection size. Large blood vessels run through several pores (f). Scale: 100 μm. FT, fibrous tissue; TCP, tricalcium phosphate scaffold; RBC, red blood cell; black arrows indicate newly formed blood vessels. Source: Reprinted with permission from Bai et al. [13] Mary Ann Liebert, Inc.

within the scaffolds, and vascularized tissue invasion varied with interconnectivity in the rodent subcutaneous model. Authors deducted that pore interconnectivity could independently influence vascularization of the implants.

17.5 Vascularization of Porous Biomaterials in Bones

The influence of the porous structure on vascularization in soft tissues has been extensively studied. Nevertheless, one of the main applications of the said scaffolds, especially bioceramics, is bone tissue engineering for defect repair. That is why,

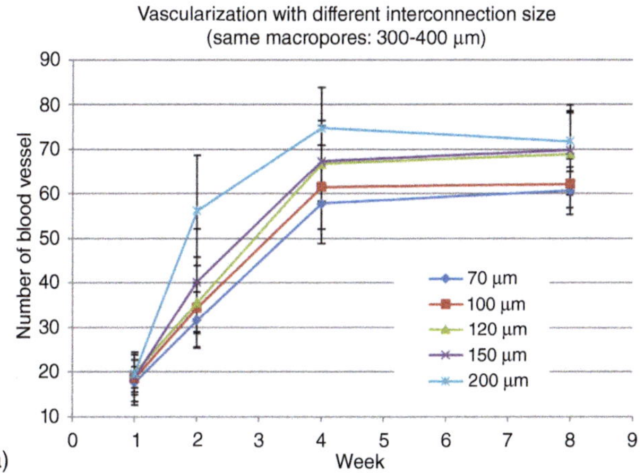

(a)

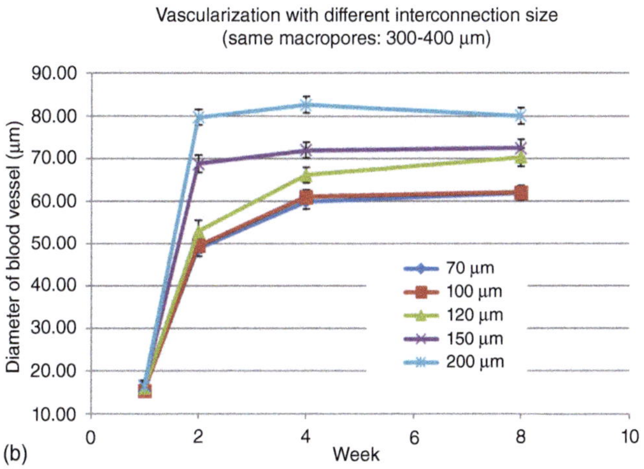

(b)

Figure 17.8 Positive correlation was observed between the interconnection size and vascularization on both number (a) and diameter (b) of blood vessels.

vascularization of porous biomaterials in the bone microenvironment has to be assessed and compared to soft tissue implantation.

17.5.1 Influence of the Porous Structure

Klenke et al. implanted dense and porous (pore sizes 40–70, 70–140, 140–210, and 210–280 μm) biphasic calcium phosphate particles into critical-sized cranial defects in BALB/c mice [61]. At four weeks, the functional capillary density was significantly higher with particles whose pore sizes exceeded 140 μm compared to the three other groups. Interestingly, the difference in vascularization between 140–210 μm group and 210–280 μm group was not statistically significant. A positive correlation between the volume of newly formed bone and pore size was also noticed. The authors deducted that the pore size is a critical parameter governing the dynamic processes of vascularization and osseointegration of bone substitutes. We prepared

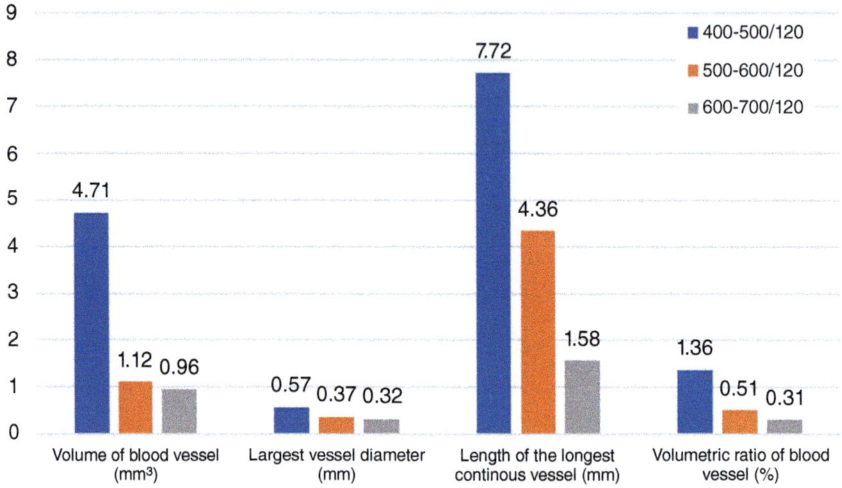

Figure 17.9 Influence of the pore size on vascularization.

three porous β-TCP bioceramics of identical interconnection size (120 μm) and different pore sizes (400–500, 500–600, and 600–700 μm) before implantation into rabbit femoral condyle defects. The amount, diameter, and density of newly formed blood vessels as well as the length of longest continuous vessel were compared to determine vascularization properties of the scaffolds. Surprisingly, scaffolds with a pore diameter of 400–500 μm showed the best vascularization with regard to all the four values. The two other pore sizes displayed no obvious difference apart from the length of the longest continuous blood vessel where the 500–600 μm group performed better (Figure 17.9).

From the aforementioned results, we can deduct that the effect of pore size on vascularization follow similar trends whether the scaffold is implanted in soft tissues or in the bone. Nevertheless, the research of Klenke et al. has some limitations as interconnections are omitted when describing the scaffolds. Lu et al. suggested that the interconnectivity of porous scaffolds was a determining factor for vascular ingrowth after implantation in the bone [62]. If the pores are not connected, the formation of blood vessels is limited to the surface of the material, and no functional vascular network can be formed. Although pore size has a strong impact on vascularization, interconnection size is more critical for biomaterial scaffolds as inadequate values will constrain the overall biological system, limiting blood vessel invasion. We prepared a series of porous β-TCP bioceramics with the same pore sizes (300–400 μm) and variable interconnection sizes (100, 120, and 150 μm) to evaluate how the interconnection size influenced vascularization [63]. After implantation in a femur defect rabbit model, the formation of new blood vessels was analyzed at 4 and 12 weeks by 3D reconstruction from micro-CT scans (Figure 17.10). The blood vessel areas of scaffolds with 150 μm interconnection diameter were significantly higher than those of smaller sizes at four weeks. Nevertheless, no statically significant difference was noticed for the diameter of the newly formed blood vessels within all the

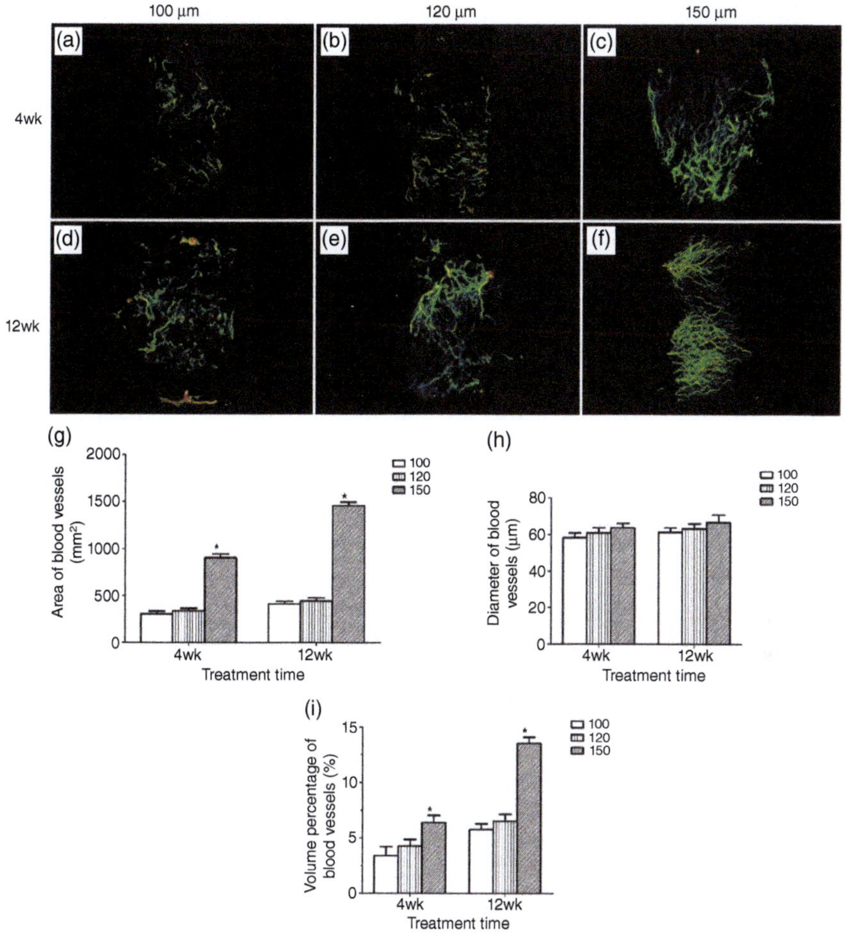

Figure 17.10 3D reconstructions of neovascularization (a–f) and statistical analysis (g–i) for different interconnection sizes, # difference with 100 μm group was statistically significant ($P < 0.05$), ★ difference with the 120 μm group was statistically significant ($P < 0.05$).

three groups. After 12 weeks, all the scaffolds showed vascularization in their center. Comparatively, the number, diameter, and density of new blood vessels were higher for 150 μm interconnections than for 100 or 120 μm. Maddalena et al. also concluded that pore interconnectivity acts as a bottleneck regulating vascular ingrowth inside porous implants [11]. Those results suggest that the interconnectivity between pores affect both the diameter and the number of newly formed blood vessels, a similar trend to what was found in soft tissue models.

17.5.2 Guided Vascular Growth in Porous Biomaterials

We have established that biomaterials with an appropriate porous structure can promote vascularization *in vivo*. Blood vessels grow inside the pores through the interconnections, supplying newly formed tissues with stem cells, nutrients, and oxygen. Some human bones have unique features in regard of its internal blood flow, such as the vascular network of the femoral head. The majority of its blood supply comes from the extracapsular arterial ring around the neck of the femur, which itself formed by the anastomose of the medial and lateral circumflex arteries. Many small arteries branch off the ring to perfuse the femoral head. Because of the limited collateral circulation, disruption of the vascular supply to the femoral head may lead to ischemia and subsequent necrosis, a disease called osteonecrosis of the femoral head (ONFH). Rapid restoration of blood supply is necessary to avoid death of osteocytes and collapse of the joint, which might be achieved using the angiogenetic properties of porous biomaterials through guided vascularization.

A new model was developed to simulate the vascular heterogenicity found in ONFH [64]. We wrapped a 5 mm wide titanium membrane around the extremity of a porous β-TCP cylinder (diameter 6 mm, length 12 mm, pore size 500–600 μm, and interconnection size 120 μm), affixed by photopolymerization of an adhesive resin. The wrapping area simulates the ischemic environment of ONFH and prevents the ingrowth of peripheral blood vessels. Vascularization of this blind end may only be achieved through guided angiogenesis from exposed areas of the bioceramic. Membrane-wrapped scaffolds were then implanted into the rabbit femoral condyle defect model, and vascularization was assessed at 4, 8, and 12 weeks. After Microfil casting, the distal femur was harvested, cleaned, and decalcified with EDTA for one month, before micro-CT scanning, 3D reconstruction, and quantitative assessment. At four weeks, stereoscopic visualization of the vascular network showed that reparative angiogenesis induced formation of numerous new blood

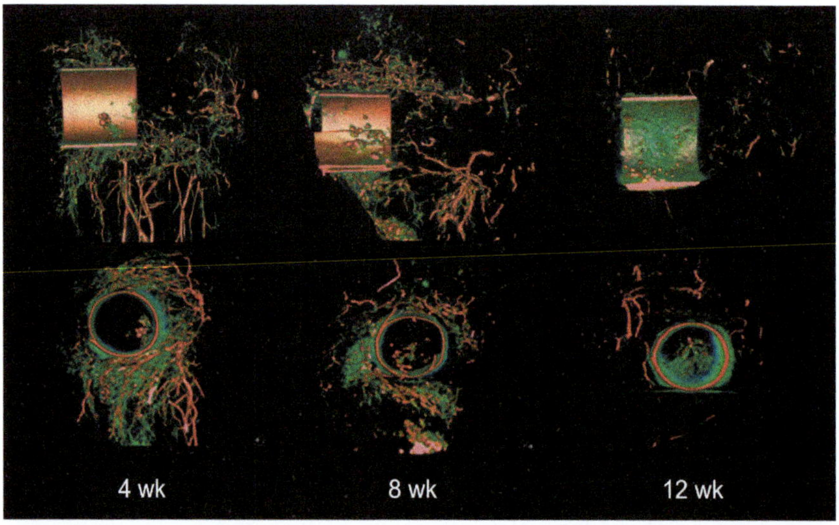

Figure 17.11 Micro-CT scans of vascularization in the scaffold. Reparative angiogenesis induces rapid vascular growth the first four weeks at the interface of the defect.

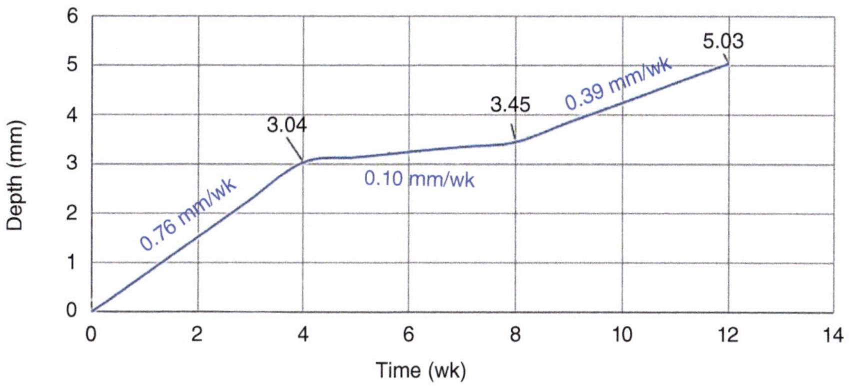

Figure 17.12 Depth of directed vascularization (black) and growth rate (blue).

vessels around the drilled defect. Interestingly, some blood vessels even started to grow into the wrapped area. At eight weeks, neovascularization around the defect increased, and abundant blood vessel clusters were present in the wrapped extremity. After 12 weeks, neovascularization decreased around the defect, but the wrapped area was almost fully vascularized. Results of statistical assessment showed significant differences between the depth, number, and diameter of blood vessels (Figure 17.11). These results illustrate the ability of porous biomaterials to guide vascularization from a non-ischemic area to the ischemic zone. The growth rate of blood vessels can be separated into three phases. Probably because of the acute inflammatory response-induced reparative angiogenesis, the growth speed during the first four weeks is extremely high (0.76 mm/week). A plateau appears during the next four weeks (0.10 mm/week), potentially caused by the maturation of the newly formed capillary network, before a new increase in growth rate from the 8th to the 12th week (0.39 mm/week) (Figure 17.12). With this new model, we demonstrated the ability of porous bioceramics for guided regeneration, laying a strong foundation for the development of new treatments for ONFH.

17.5.3 Clinical Application of Guided Vascularization with Porous Biomaterials

ONFH, also called aseptic necrosis of the femoral head, is a common bone and joint disease and a worldwide medical challenge. Caused by the disturbance of the vascularization of the femoral head, this disease ultimately leads to the collapse of the joint, painful arthritis, and loss of mobility. More than 30 million patients are afflicted with ONFH worldwide, including 10 million patients in China.

Treatments for ONFH can be divided into arthroplasty (hip replacement) and joint preservation techniques. Modern hip arthroplasty is a well-documented procedure and has been extensively used since the 1950s [65] but is generally not indicated for early stages of the disease or younger patients. Complications such as infection [66], dislocation [67], loosening, or implant failure [68] are still reported even with a

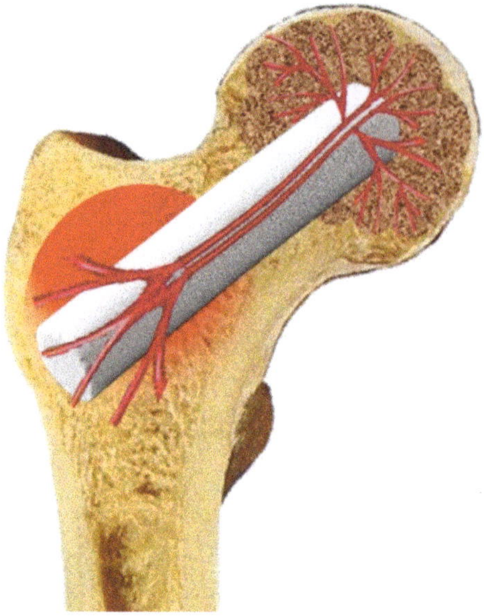

Figure 17.13 The principle of the procedure is to restore the blood supply of the ischemic femoral head from the healthy trochanteric region by guided vascularization.

newer implant material such as alumina or zirconia ceramics. The goal of hip preservation is to delay or avoid the onset of arthritis and the need for a hip replacement in the future by repairing or revising the afflicted joint. Preservative surgeries have been increasingly studied to overcome the limitations of joint replacement [69]. Commonly employed procedures include non-operative treatments, core decompression, stem cell therapy, osteotomy, and bone grafting [70].

As demonstrated previously, porous biomaterials have the ability to guide vascularization and could potentially restore the functional blood supply of necrosed femoral heads. Therefore, we developed a minimally invasive technique to implant porous β-TCP bioceramics for the joint-preserving treatment of ONFH [71]. The goal of this procedure is to guide the vascularization of the ischemic femoral head from the healthy trochanteric region (Figure 17.13). A set of ancillary instruments was designed to simplify the removal of necrotic tissue. Bone sludge, including bone marrow, were retrieved during drilling and then reimplanted to increase the bioactivity.

The implanted β-TCP bioceramics are composed of dense particles (mean size 1.0–3.0 mm), porous particles (mean size 1.0–3.5 mm), and a porous cylindrical rod (diameter 10 mm and length 80 mm). All porous materials have a pore size of 500–600 μm and an interconnection size of 120 μm.

To allow better understanding of the procedure, we will briefly summarize the key steps. Readers interested in more detailed description can refer to our published work [71]. With the patient in supine position, a 3 mm-diameter K-wire was inserted from 2 cm below the trochanter major through the center of the femur neck (Figure 17.14a). A 12 mm bone channel was drilled to achieve core decompression (Figure 17.14b), during which bone sludge containing MSCs, stromal cells, and blood cells was collected and then mixed with the implants to increase their bioactivity. Necrosis debridement was carried out with an expandable

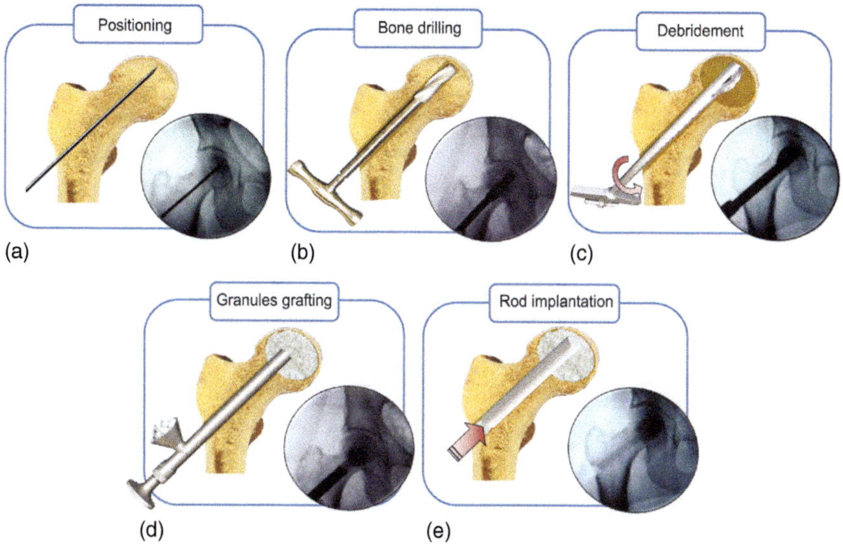

Figure 17.14 Description of the surgical procedure. (a) Positioning, (b) bone drilling, (c) debridement, (d) granules grafting, and (e) rod implantation.

reamer (Figure 17.14c). The mix of bone sludge, porous, and dense granules was impaction grafted afterward (Figure 17.14d) before inserting the bioceramic rod (Figure 17.14e) and closing the wound.

A multicenter retrospective study on 232 hips, with an average follow-up time of 22.7 months, showed that the hip survival rate (without further hip replacement) was above 93% [64]. According to the imaging evaluation, over 79% of operated hips showed no progression of the disease or recovered. The joint function assessed with Harris hip scores also significantly increased (38%) and nearly two-third of the patients had excellent hip function. The case of a 42 years-old male patient with ARCO stage IIIA osteonecrosis of the left hip is shown below (Figure 17.15a–c). Six months after the surgery, X-rays showed bone ingrowth inside the bioceramic rod and progressive biodegradation of the granules inside the femoral head (Figure 17.15d). One year post-operation, the granules were fully integrated into the surrounding bone (Figure 17.15e). After two years, most of the bioceramics were replaced by new bone through creeping substitution (Figure 17.15f). Similarly, we observed on magnetic resonance imaging (MRI) vascularization of the bioceramic rod and the femoral head, respectively, at two and nine months. These results are consistent with previous results in animal models, as blood vessels are guided by the porous bioceramic from the abundantly vascularized trochanteric region toward the ischemic femoral head.

17.6 Conclusion

During the past decades, tissue engineering has become the focal point of regenerative medicine. Instead of the traditional harvest of samples from biological

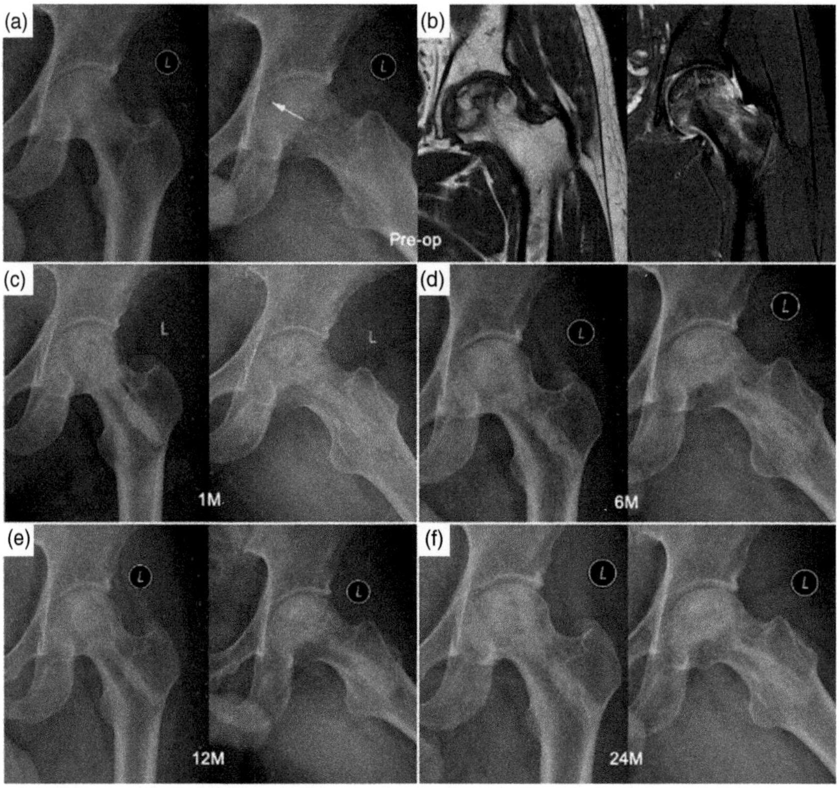

Figure 17.15 42-year-old male patient with ARCO stage IIIA osteonecrosis of the left hip. Crescent sign is visible pre-operatively on both radiography (a) and MRI (b). Implant degradation and progressive osteointegration can be observed 1 (c), 6 (d), 12 (e), and 24 months (f) after the operation.

sources, this new technology allowed *ex vivo* and *in vivo* regeneration of tissues from cells with the support of biomaterials and growth factors. Nevertheless, functional vascularization of the implant remained one of the biggest challenges, as the lack of a sufficient blood supply hinders cellular activity. By a precise control on structural parameters such as pore and interconnection sizes, highly interconnected porous biomaterials can promote neovascularization both *in vitro* and *in vivo*. Seeded with angiogenic cells, formation of new blood vessels can be further enhanced using growth factors and vascular grafting. Microfabricated vasculature is also a promising technology and might lead one day to the construction of fully functional organs *ex vivo*. Angiogenetic properties of bioceramics are particularly suited to treat bone defects and diseases, including ONFH. Clinical results showed a high success rate, with restoration of the blood supply in the ischemic zone by guided vascularization. More research is needed to understand better the interactions between cells, growth factors, biomaterials, and vascularization. We can imagine to create one day tissues identical to their biological counterparts in a lab, by printing a predesigned matrix on porous biomaterials, where stem cells would rapidly form a functional neovasculature under the simultaneous stimulation of multiple growth factors.

References

1 Harris, G.M., Rutledge, K., Cheng, Q. et al. (2013). *Curr. Pharm. Des.* 19: 3456–3465.

2 Hendrickx, B., Vranckx, J.J., and Luttun, A. (2011). *Tissue Eng. Part B* 17: 13–24.

3 Adams, J., Difazio, L., Rolandelli, R. et al. (2009). *Acta Physiol. Hung.* 96: 19–28.

4 Yang, B., Zhao, M., and Liu, Z. (1999). *Chin. J. Plast. Surg. Burns* 15: 196–198.

5 Albrektsson, T. (1980). *Scand. J. Plast. Reconstr. Surg.* 14: 1–12.

6 Chiu, Y.-C., Cheng, M.-H., Engel, H. et al. (2011). *Biomaterials* 32: 6045–6051.

7 Keskar, V., Marion, N.W., Mao, J.J., and Gemeinhart, R.A. (2009). *Tissue Eng. Part A* 15: 1695–1707.

8 Druecke, D., Langer, S., Lamme, E. et al. (2004). *J. Biomed. Mater. Res. Part A* 68A: 10–18.

9 Kuboki, Y., Jin, Q., Kikuchi, M. et al. (2002). *Connect. Tissue Res.* 43: 529–534.

10 Kasoju, N., Bhonde, R.R., and Bora, U. (2009). *J. Tissue Eng. Regener. Med.* 3: 539–552.

11 Mastrogiacomo, M., Scaglione, S., Martinetti, R. et al. (2006). *Biomaterials* 27: 3230–3237.

12 Lu, J.X., Flautre, B., Anselme, K. et al. (1999). *J. Mater. Sci. - Mater. Med.* 10: 111–120.

13 Bai, F., Wang, Z., Lu, J. et al. (2010). *Tissue Eng. Part A* 16: 3791–3803.

14 Hori, Y., Tamai, S., Okuda, H. et al. (1979). *J. Hand Surg.* 4: 23–33.

15 Vacanti, C.A. and Bonassar, L.J. (1999). *Clin. Orthop.* 367: S375–S381.

16 Fan, Y.E., Qiao, J.M., Wang, R.G., and Liang, D.Y. (1988). *Chin. J. Microsurg.* 11: 206–207.

17 Pelissier, P., Villars, F., Mathoulin-Pelissier, S. et al. (2003). *Plast. Reconstr. Surg.* 111: 1932–1941.

18 Yang, Z.M. and Jie, H.Q. (2000). *Chin. J. Microsurg.* 23: 21–22.

19 Zeng, X.L., Pei, J., and Tang, G.H. (2005). *Chin. J. Orthop. Trauma* 7: 353–357.

20 Singh, S., Wu, B.M., and Dunn, J.C.Y. (2011). *Tissue Eng. Part A* 17: 1819–1830.

21 Murphy, W.L., Simmons, C.A., Kaigler, D., and Mooney, D.J. (2004). *J. Dent. Res.* 83: 204–210.

22 Chim, S.M., Tickner, J., Chow, S.T. et al. (2013). *Cytokine Growth Factor Rev.* 24: 297–310.

23 Hu, X.F., Wang, Z., Zhang, R.Q., and Wang, H.Y. (2005). *Orthop. J. China* 13: 921–923.

24 Street, J., Bao, M., deGuzman, L. et al. (2002). *Proc. Natl. Acad. Sci. U.S.A.* 99: 9656–9661.

25 Deckers, M.M., Karperien, M., van der Bent, C. et al. (2000). *Endocrinology* 141: 1667–1674.

26 Peng, H., Wright, V., Usas, A. et al. (2002). *J. Clin. Invest.* 110: 751–759.
27 Bouletreau, P.J., Warren, S.M., Spector, J.A. et al. (2002). *Plast. Reconstr. Surg.* 109: 2384–2397.
28 Lv, J., Xiu, P., Tan, J. et al. (2015). *Biomed. Mater. Bristol Engl.* 10: 035013.
29 Sun, X., Ma, Z., Zhao, X. et al. (2021). *Bioact. Mater.* 6: 757–769.
30 Bhattarai, D.P., Kim, M.H., Park, H. et al. (2020). *Chem. Eng. J.* 389: 123470.
31 Perets, A., Baruch, Y., Weisbuch, F. et al. (2003). *J. Biomed. Mater. Res. Part A* 65A: 489–497.
32 Ramoshebi, L.N. and Ripamonti, U. (2000). *Anat. Rec.* 259: 97–107.
33 Elçin, Y.M., Dixit, V., and Gitnick, G. (1996). *Artif. Cells, Blood Substitues, Immobilization Biotechnol.* 24: 257–271.
34 Huang, Y.-C., Kaigler, D., Rice, K.G. et al. (2004). *J. Bone Miner. Res.* 20: 848–857.
35 Sun, G., Shen, Y.-I., Kusuma, S. et al. (2011). *Biomaterials* 32: 95–106.
36 Peters, M.C., Polverini, P.J., and Mooney, D.J. (2002). *J. Biomed. Mater. Res.* 60: 668–678.
37 Koga, M., Sudo, R., Abe, Y. et al. (2009). *Tissue Eng. Part A* 15: 2727–2739.
38 Hegen, A., Blois, A., Tiron, C.E. et al. (2011). *J. Tissue Eng. Regener. Med.* 5: e52–e62.
39 Unger, R.E., Sartoris, A., Peters, K. et al. (2007). *Biomaterials* 28: 3965–3976.
40 Wang, D.S., Yamazaki, K., Nohtomi, K. et al. (1996). *J. Am. Soc. Bone Miner. Res.* 11: 472–479.
41 Kaigler, D., Krebsbach, P.H., Polverini, P.J., and Mooney, D.J. (2003). *Tissue Eng.* 9: 95–103.
42 Wenger, A., Stahl, A., Weber, H. et al. (2004). *Tissue Eng.* 10: 1536–1547.
43 Yang, Z.M. (2004). *Fundamentals and Clinical Application of Tissue Engineering*, 1ste. Sichuan Science and Technology Press.
44 Zhou, J., Lin, H., Fang, T. et al. (2010). *Biomaterials* 31: 1171–1179.
45 Zhou, Q., Wang, Y., Yu, J., and Yu, Y.H. (2011). *China J. Orthop. Traumatol.* 24: 838–840.
46 Kane, N.M., Meloni, M., Spencer, H.L. et al. (2010). *Arterioscler. Thromb. Vasc. Biol.* 30: 1389–1397.
47 Huang, N.F., Niiyama, H., Peter, C. et al. (2010). *Arterioscler. Thromb. Vasc. Biol.* 30: 984–991.
48 Ferreira, L.S., Gerecht, S., Fuller, J. et al. (2007). *Biomaterials* 28: 2706–2717.
49 Takahashi, K. and Yamanaka, S. (2006). *Cell* 126: 663–676.
50 Rufaihah, A.J., Huang, N.F., Jamé, S. et al. (2011). *Arterioscler. Thromb. Vasc. Biol.* 31: e72–e79.
51 Suzuki, H., Shibata, R., Kito, T. et al. (2010). *BMC Cell Biol.* 11: 72.
52 Zhang, G., Shang, B., Yang, P. et al. (2012). *Stem Cells Dev.* 21: 955–964.
53 Vacanti, J.P. and Langer, R. (1999). *Lancet Lond. Engl.* 354 (Suppl. 1): S32–S34.
54 Langer, R.S. and Vacanti, J.P. (1999). *Sci. Am.* 280: 86–89.
55 Colton, C.K. (1995). *Cell Transplant.* 4: 415–436.
56 Raghavan, S., Nelson, C.M., Baranski, J.D. et al. (2010). *Tissue Eng. Part A* 16: 2255–2263.

57 Rivron, N.C., Vrij, E.J., Rouwkema, J. et al. (2012). *Proc. Natl. Acad. Sci. U.S.A.* 109: 6886–6891.

58 Fidkowski, C., Kaazempur-Mofrad, M.R., Borenstein, J. et al. (2005). *Tissue Eng.* 11: 302–309.

59 Choi, S.-W., Zhang, Y., Macewan, M.R., and Xia, Y. (2013). *Adv. Healthcare Mater.* 2: 145–154.

60 Somo, S.I., Akar, B., Bayrak, E.S. et al. (2015). *Tissue Eng. Part C* 21: 773–785.

61 Klenke, F.M., Liu, Y., Yuan, H. et al. (2008). *J. Biomed. Mater. Res. Part A* 85: 777–786.

62 Lu, J., Flautre, B., and Anselme, K. (1997). *Biomaterials* 10: 583–586.

63 Xiao, X., Wang, W., Liu, D. et al. (2015). *Sci. Rep.* 5: 9409.

64 Lu, Y., Wang, Z., Lu, X. et al. (2019). *Chin. J. Reparative Reconstr. Surg.* 33: 1291–1298.

65 Knight, S.R., Aujla, R., and Biswas, S.P. (2011). *Orthop. Rev.* 3: e16.

66 Senthi, S., Munro, J.T., and Pitto, R.P. (2011). *Int. Orthop.* 35: 253–260.

67 D'Angelo, F., Murena, L., Zatti, G., and Cherubino, P. (2008). *Indian J. Orthop.* 42: 252–259.

68 Johnsen, S.P., Sørensen, H.T., Lucht, U. et al. (2006). *J. Bone Joint Surg. Br.* 88: 1303–1308.

69 Iyer, M. (2020). *Hip Preservation Techniques*. CRC Press.

70 Cao, H., Guan, H., Lai, Y. et al. (2016). *J. Orthop. Transl.* 4: 57–70.

71 Lu, Y., Lu, X., Li, M. et al. (2018). *Int. Orthop.* 42: 1567–1573.

18

Muscle Repair

Rong Bao, Wei Wang, Cuiyi Fan, and Shen Liu

Shanghai Jiao Tong University Affiliated Sixth People's Hospital, Department of Orthopedics, 600 Yishan Road, Shanghai 200233, China

18.1 Introduction

Skeletal muscles account for over 40% of body mass in humans, and its importance is beyond doubt. Owing to the presence of a pool of resident stem cells that orchestrate a series of repair programs, the skeletal muscles possess the intrinsic ability to regenerate after minor injury, to be specific, less than 20% of the whole muscle portions. However, for muscle loss, over 20%, the outcomes are usually destructive [1]. The abrupt removal of a certain portion of muscle, which is generally named as volumetric muscle loss (VML), leads to the loss of regenerative components, compromising the self-repairing ability of skeletal muscles [2]. In addition, the excessive deposition of nonfunctional fibrotic tissue and the degeneration of fatty muscle are also responsible for the destructive outcomes of VML [3]. Of note, apart from traumatic injuries, VML could also result from debridement surgeries and tumor-associated muscle reconstruction [4].

Although there is no accurate data of the incidence of VML, it is generally accepted that VML is pervasive in open soft tissue injury and musculoskeletal trauma [4]. About 250 000 civilians encounter open fracture in the United States alone, indicating a high incidence of VML [5]. As for military populations, the prevalence of unsolved VML is much higher. A survey conducted in the United States suggested that about 14 500 servicemen encountered musculoskeletal trauma during 2001 and 2013, most of them accompanied by open tissue loss injury [6, 7]. VML leads to disability and significantly compromised the life quality of the involved individuals. As shown in a retrospective study, VML accounted for 65% of the permanent disability rate in a cohort of medically retired soldiers who suffered from type III open tibia fracture [8]. The VML-related injuries costed the United States over 400 billion dollars per year (2010 data), and the estimated cost for a VML-induced lifetime disabled patient was about 400 000 dollars [4, 9].

The management for VML remains a challenge, and its therapeutic efficiency is severely limited. Currently, the clinical treatments mainly include autologous tissue

Biofabrication for Orthopedics: Methods, Techniques, and Applications, First Edition.
Edited by Wenguo Cui, Xin Zhao, and Shen Liu.

transplantation, also known as functional free muscle transfer surgery, and the use of bracing [2]. Free muscle transfer surgery is to date the "gold standard" treatment for VML. Lin et al. reported the successful application of such surgery in treating extensive tissue defects in lower extremities [10]. However, such complex and invasive operation calls for a highly skilled surgical team, and meanwhile, it brings about the side effects such as donor site morbidity and potential postoperative infections [11, 12]. VML injuries below the knee could also be treated with advanced bracing strategies, which offer a nonsurgical option to VML. However, such bracing could not address VML above the knee and could not essentially solve the VML challenge. Moreover, the bracing is custom-made and very expensive [2].

With the fast advances in biomaterials, there are various studies on repairing VML with bio-fabricated scaffolds. An acellular scaffold that is composed of an extracellular matrix (ECM) was applied to bridge the muscle defect in humans [13]. The researchers observed actively regenerating muscle cells, newly formed blood vessels, and nerve fibers in the regenerated muscle tissue. The patient who received scaffold implantation also exhibited significant improvement in terms of functional recovery. This suggested the huge potential of bio-fabricated products as an alternative therapy for VML.

18.2 Anatomy of Skeletal Muscle

18.2.1 Morphological Characteristics of Skeletal Muscle

Skeletal muscle is one of the striped muscles attached to the bone. A skeletal muscle refers to multiple bundles of cells called muscle fibers (fascicles). The fibers and muscles are surrounded by connective tissue layers called fascia. Skeletal muscle consists of bundles of muscle cells. The length of muscle cell is different, and the cells are arranged closely and complement each other in length. Each cell is covered with a slender reticular membrane called the endomembrane. Muscle bundles are wrapped in the connective tissue membrane and mixed with glial fibers and elastic fibers, called muscle fascicles. There is a thick connective tissue outside the muscle, called the epimysium. The connective tissue of the membrane is continuous, and blood vessels and nerves are distributed along the connective tissue membrane to the muscle.

18.2.2 Composition of Skeletal Muscle

Humans have a large number of muscles, but the basic structure is similar. A typical muscle consists of venter in the middle and tendons at both ends. The muscle venter is the main part of the muscle. It is composed of clusters of striated muscle fibers. It is red in color and soft and has the ability to contract. The tendon attaches to the bone and is firmly woven with the periosteum. The abdomens and tendons of the platysma are membranous, and the tendons of this muscle are called fascia. The surface of the muscle belly is covered by perimysium, which fuses with tendon tissue at both ends.

In addition to the muscle abdomen, tendons, and fascia, there are many blood vessels and nerves in the skeletal muscle. Skeletal muscle is a voluntary muscle distributed in the trunk and limbs. The muscle fibers are slender and cylindrical, with multiple to hundreds of nuclei located at the periphery of the fibers. The outside of the muscle is the perimysium composed of connective tissue. The perimysium contains blood vessels and nerves and extends into the muscle, dividing the muscle into several muscle bundles. Muscle bundles form the fascia, enclose the muscle bundles, and then extend to the periphery of each muscle fiber to form an endometrium rich in capillaries and nerve fibers.

18.3 Pathophysiology After Skeletal Muscle Injury

There are two types of skeletal muscle injuries: small scale one and VML. However, skeletal muscle regeneration contains three common phases: inflammatory, repair, and remodeling. Their differences focus on the size of the injury.

18.3.1 Inflammatory Phase

After injury, myofibers, blood vessels, and neurons that exist in the damaged skeletal muscle experience necrosis because of the influx of extracellular calcium, which triggers autodigestion via calcium-dependent proteases. Thanks to a specific structure called the contraction band, namely, a kind of condensed cytoskeletal material, the necrosis will be isolated merely around the wound. During the first two weeks, growth factor from ECM such as hepatocyte growth factor (HGF), fibroblast growth factor 2 (FGF2), and transforming growth factor-β (TGF-β can also be released by platelets) is released from the defect of basement membrane. The growth factor and activation of the complement cascade stimulate the mononuclear cells in the muscle tissue, which recruit circulating leucocytes to the injury site. Especially, neutrophils reach in 24 hours. They act both as proinflammatory cytokines and as the generator of oxygen free radicals, which finally induce tissue damage. TGF-β attracts and activates monocytes and macrophages to the wound. Macrophages have an important part in the inflammatory phase, which is categorized into two types: classically activated proinflammatory (M1) and alternatively activated anti-inflammatory (M2) macrophages. M1 macrophages phagocytose debris and release reactive oxygen species and proinflammatory cytokines, which stimulate satellite cell (SC) proliferation and mobilization. M2 macrophages replaced M1 macrophages soon due to promoting the SC exiting from the cell cycle, a commitment to differentiation.

18.3.2 Repair Phase

The repair phase occurs one to four weeks post-injury, which include the activation and proliferation of SCs and their differentiation into mature muscle tissue.

Firstly, damaged basement membrane initiates the release of growth factor such as HGF, FGF2, and TGF-β. HGF is released timely upon injury, and its presence in muscle wounds peaks two to four days post-injury, which is responsible for promoting SC proliferation and migration by HGF's receptor c-MET on SCs. FGF2 is released two to eight days post-injury and peaks about six to eight days, which is bound to the proliferation and migration of activated SCs. These growth factors also play an important role in promoting revascularization and innervation of skeletal muscle. It is noticed that these growth factors inhibit subsequent differentiation into mature muscle in the late repair phase, contrasting to their function described above in the early phase, which demonstrates the essentiality of the highly regulated and temporal expression of growth factors. After the activation from growth factors, SCs initiate to differentiate into myoblasts. Then, undifferentiated stem SCs begin to proliferate by 24 hours, providing new SCs by asymmetric cell division for future needs of regeneration. Then, the myoblasts fuse to form myotubes within few days. Regenerating myofibers take the place of ruptured myofibers and penetrate into the connective tissue scar between the gap of the ruptured myofibers. Revascularization and reinnervation of the tissue are also essential, occurring during the repair phase. Nerve provides electrical stimulation to the tissue. Revascularization of muscle tissue provides the oxygen required for aerobic metabolism. This promotes the regeneration of ruptured muscle.

18.3.3 Remodeling Phase

The remodeling phase occurs two to six weeks post-injury, which consists of regenerating myofiber maturation, ECM remodeling, and formation and remodeling of scar tissue. The basement membrane promotes myofibers to mature muscle tissue, while the formation of small fiber clusters or forked fibers is found because the fusion of fibers is not complete all the time. Myofiber maturation is also highly dependent on revascularization and the generation of neuromuscular junctions. ECM stimulated by many growth factors provides the stability for the regenerating muscle tissue and enhance the muscle tensile strength as well as myofiber–tendon junctions. However, the ECM remodels into scar tissues when it is overproduced, which influence the regeneration of muscle. Especially, the injury is large scale. This case is more severe when it comes to be VML. The basement membrane is destroyed to such an extent that it cannot promote SCs to facilitate generation. It means the inflammatory, revascularization, and reinnervation phase derived from SCs are limited. The myoblasts differentiated from SCs cannot fuse to form myotubes. Instead, fibroblast-mediated collagen I deposition dominates the phase and produces a nonfunctional scar tissue in the wound, which finally restricts torque production and range of motion. As described, revascularization and reinnervation are essential to muscle regeneration. VML causes their decrease. While SCs mediated proliferation and differentiation take place in denervated muscle tissue, injured muscle tissue lacks action potential-mediated muscle contractions without reinnervation. Less revascularization can impede the ingrowth of neurons and vasculature and limit oxygen diffusion, resulting in denervated and ischemic muscle.

18.4 Biofabrication for Skeletal Muscle Regeneration

18.4.1 Materials for Fabrication (Natural, Synthetic, and Hybrid)

Muscle tissue is a soft tissue that is important in power generation, body movement, posture support, and internal organ function. At present, muscle tissue engineering has been considered as a promising therapeutic strategy for the treatment of this problem, which has attracted great attention [14–16]. Biomaterials have the function of imitating natural ECM and are [17] important factors in tissue engineering. Therefore, the development of biomaterials for muscle tissue engineering is of great significance. Here is a summary of recent research advances in biomaterials used in muscle regeneration, including natural materials, synthetic materials, mixed materials. These biomaterials have their own advantages and disadvantages. How to use them and reduce unexpected disadvantages in material synthesis is a great challenge.

Naturally derived polymers, such as collagen, chitosan [18, 19], and gelatin [19, 20], have excellent ability to support cell attachment. However, they also have drawbacks, such as immunogenicity and complex structural composition, which limit their application [17] *in vivo*. On the other hand, synthetic polymers have more controllable structures and fewer immunological problems, but their biocompatibility may be a challenge [17, 21]. Fortunately, researchers continue to study conductive polymer-based biomaterials and try to find a balance between the use of naturally derived and synthetic polymers.

Conductive biomaterials have become an ideal candidate for muscle tissue engineering scaffolds because of their good electrical conductivity and promoting muscle tissue formation. Different kinds of conductive materials show different properties. Conductive polymers can easily bind to other polymers to form the conductive scaffold [21].

Conductive polymer is a new conductive material proposed in 1970s with similar electrical and optical properties to metal and inorganic semiconductors. Compared with conventional conductive materials, they are easier to synthesize and show better processing performance. Conductive polymers themselves can be assembled into the unique two morphology. Liu et al. [22] developed a simple and general method for axial explosion prevention of oxidation template assemblies to generate a large number of conductive polymers with a nanoclamped morphology. Other forms of conducting polymers have also been developed, such as spindle shaped, sand glass shaped, and flower shaped. In 1997, conductive polymer polypyrrole (PPy) was used as a matrix [23] for cell culture in tissue engineering. However, most conductive polymers are insoluble in water and cannot degrade *in vivo*. Oligomers of conductive polymers such as oligoaniline overcome the nondegradability, but their conductivity is lower than that of conductive polymers [24]. Carbon nanomaterials including carbon nanotubes, graphene, and other carbon-based nanomaterials have high electrical conductivity, but they are insoluble in water, and dispersion in solution is also a problem [21]. Metallic nanomaterials, including gold and silver nanoparticles, also exhibit good electrical conductivity, but they are not degradable *in vivo* and may have

Table 18.1 Advantages and shortcomings of different types of polymers.

Classification	Advantages	Shortcomings
Naturally derived polymers	Good ability of supporting cell attachment	Immunogenicity and complex structural composition, which limit their application *in vivo*
Synthetic polymers	With a more controllable structure and fewer immunological problems	Biocompatibility may be a challenge
Mixed conductive biological materials	Conductivity similar to natural muscle; easy to bind to other polymers to form conductive scaffolds	Insoluble in water, unable to degrade *in vivo*

higher cytotoxicity [25] than other conductive materials. Therefore, the blending of conductive polymers with natural or synthetic polymers has been widely studied and applied in the field of biomaterials (Table 18.1).

18.4.2 Biochemical Substances for Fabrication (Cytokines, Genetic Drugs, Micromolecules, Etc.)

Sustained presentation of VEGF and IGF-1 to ischemic skeletal muscle tissue was shown to enhance myogenesis in injured muscle tissue while simultaneously promoting tissue survival and function with new capillary and nerve growth [26]. Sustained delivery of VEGF, in particular, has been shown to promote regrowth of damaged axons by upregulating the expression of nerve growth factor and glial-derived neurotrophic factor in ischemic tissue [27]. Controlled delivery of IGF-1 has also been demonstrated to recruit Pax7+ muscle cells into a scaffold, resulting in higher muscle formation [28].

Dex significantly promotes muscle regeneration via promoting the kinesin-1 motor activity and directly increasing the kinesin-1 motor activity, which is required for the expression of a myogenic marker (muscle myosin heavy chain 1/2) and also for the process of myoblast fusion and the formation of polarized myotubes. Upon differentiation, kinesin-1 mediates the recruitment of integrin β1 onto microtubules, allowing delivery of the protein into focal adhesions. Integrin β1-mediated focal adhesion signaling then guides myoblast fusion toward a polarized morphology. By imposing geometric constrains via micropatterns, research has proved that cell adhesion is able to rescue the defects caused by kinesin-1 inhibition during the process of myogenesis [29].

Muscle extracellular matrix (MEM) and induced skeletal myogenic progenitor cells (iMPCs) are integrated within thermally drawn fiber-based microchannel scaffolds. The microchannel fibers decorated with MEM enhance the differentiation and maturation of iMPCs. Furthermore, engraftment of these bioengineered hybrid muscle constructs induce *de novo* muscle regeneration accompanied with microvessel and neuromuscular junction formation in a VML mouse model, ultimately leading to functional recovery of muscle activity [30].

Cellular repressor of E1A stimulated genes (CREG) protein is used in medicine to promote skeletal muscle regeneration. E1A activating *gene* repressor (cellular repressor of E1A stimulated *gene*, CREG) protein is a transcriptional regulator expressed in many tissues and cells and plays an important role in maintaining cell maturation and differentiation. CREG plays an important role in regulating the regeneration of skeletal muscle injury. CREG protein or its active fragments were prepared to promote the regeneration and repair of various skeletal muscle injuries and to prevent and treat the weakening of skeletal muscle regeneration caused by various diseases or physiological phenomena [31].

SVVYGLR (SV), osteopontin-derived multifunctional short peptide, plays a potent role in facilitating skeletal muscle regeneration by promoting the migration and differentiation of myogenic precursor and progenitor cells. In a rat model with volumetric loss of masticatory muscle, immunohistochemical staining of regenerating muscle tissue immediately after injury demonstrated an increase of the number of both MyoD- and myogenin-positive cells in the SV peptide-treated group [32].

Macroporous, alginate scaffolds modified with the RGD motif, a peptide that can bind to integrin β1, have been shown to promote the survival and outward migration of primary myoblasts when the scaffolds delivered growth factors for activating the cells [33, 34] (Table 18.2).

18.4.3 Acellular Scaffolds

Acellular scaffolds for VML mainly include ECM-based scaffolds, synthetic polymeric scaffolds, and hybrid scaffolds. The ECM-based scaffolds take advantage of the ECM harvested from different tissue sources (e.g. small intestinal submucosa and urinary bladder matrix). The polymeric scaffolds are constructed with various synthetic materials such as polypropylene, poly(lactic acid) (PLA), poly(glycolic acid)

Table 18.2 Examples of substances and functions for biofabrication.

Substances	Examples	Effect
Cytokines	VEGF and IGF-1	Myogenesis, nerve growth, and angiogenesis
		Maintenance of nerve innervation and promotion of reperfusion
Drugs	Dexamethasone	Myogenesis
Macromolecules	Muscle extracellular matrix CREG protein	Myogenesis
	SVVYGLR (SV)	
	Alginate	

(PGA), poly(ε-caprolactone) (PCL), and polyurethanes (PU). The hybrid scaffolds are a combination of the above-mentioned elements. This subsection mainly discusses the applications of ECM-based scaffolds, the ECM-mediated muscle repair process, and the current challenges for ECM-based scaffolds.

Among the acellular scaffolds mentioned above, the ECM-based scaffolds are widely studied for the treatment of VML. ECM contains collagen molecules, laminin, fibronectin, glycosaminoglycans, and a variety of growth factors. It was generally considered that ECM was responsible for the structural support of native tissues. To date, the deep understanding of the ECM-mediated tissue remodeling reveals that ECM not only provides mechanical support for muscle regeneration but also conveys a series of signals and triggers a cascade of biology responses. The therapeutic efficiency of the ECM-mediated scaffolds was tested in different animal models (e.g. rat, rabbit, and canine), different VML loss size (e.g. 30–70%), and of course different defect sites (e.g. quadriceps and gastrocnemius). The preclinical assessment of the ECM-based scaffolds proves efficiency and safety of such scaffolds. Of note, the ECM-based scaffold has been implanted into human body. Mase et al. reported a case that a serviceman suffering from traumatic muscle defect received ECM-based scaffold implantation after scar tissue debridement [35]. Sixteen weeks postoperatively, the patient significantly restored motor functionals and improved the life quality. In addition, Sicari et al. studied the biopsies from the ECM-based scaffold-treated VML patent [13]. The authors observed stem cell activation, robust angiogenesis, and neurogenesis within the implantation site. Subsequently, the patient who received ECM-based scaffold implantation exhibited comparatively satisfactory functional restoration. Taken together, the ECM-based scaffolds hold great promise for the treatment of VML.

The ECM scaffold-mediated muscle repair involves multiple biology processes. Firstly, the ECM-based scaffold recruits stem or progenitor-like cells. By definition, an acellular scaffold refers to a biomaterial without cells; thus, the cells accounting for the formation of a new muscle tissue must originate from the host. Dellavalle et al. identified a critical role of multipotent perivascular stem cells in the muscle repair process after injury [36]. Perivascular stem cells are CD146-positive and NG2-positive cells found in capillaries. They serve as progenitors of skeletal muscle myoblasts and actively participate in the ECM-mediated muscle regeneration process. The macrophage behavior after muscle injury is also strongly associated with the regeneration process. M1 macrophages, generally considered as a proinflammation subtype, first respond to ECM scaffold implantation after VML [37]. Deng et al. reported that the macrophage phenotypic switch from M1 to M2 was indispensable for efficient muscle regeneration following injury [38]. Ruffell et al. took advantage of transgenic mice to inhibit M2 macrophage activity, and the authors found that the muscle fiber regeneration was severely compromised [39]. The therapeutic efficiency after the ECM scaffold implantation partially depend on the degradation products from ECM, which could induce M2 polarization of the recruited macrophages and therefore promote the myogenesis process [40]. Notably, the immune responses and the degradation products also contribute to muscle regeneration via recruiting endothelial cells (ECs) and modulating angiogenesis [41, 42]. Moreover, the innervation degree is also strongly correlated with the functional recovery after ECM-based scaffold implantation. Finally, the regenerated muscle should be able to contract; thus, the researchers apply the maximal contractile force to assess the therapeutic efficiency of the ECM-based scaffold. Valentin et al. implanted the ECM-based scaffold into a rat abdominal wall muscle defect and found that the regenerated muscle could generate a similar maximal contractile force to native tissue [43].

Despite the merits discussed above, the therapeutic efficiency of the ECM-based scaffold remains controversial. In a patient suffering from quadricep VML, ECM-based scaffold implantation led to 68% strength defect four months postoperatively, while the pre-surgery strength defect was 72% [35, 44]. The data suggested that significant functional improvements are yet to be made while the ECM-based scaffold could repair VML. Current data showed that the limited efficiency of ECM scaffold-mediated muscle regeneration was due to the deficiency of SCs, in other words, Pax7+ cells [4]. Of note, SCs have been proven to be indispensable for muscle regeneration after injury. Researchers took advantage of pax7 knockout mice to demonstrate the vital role of SCs in muscle regeneration [45, 46]. The low presence of Pax7+ cells coincided with unsatisfactory muscle fiber regeneration mediated by the ECM scaffold, while the pax7 expression was much higher in the autologous tissue transplantation group [47].

To conclude, the acellular scaffold showed multiple advantages in muscle tissue regeneration; however, they cannot guarantee a satisfactory functional recovery after implantation (Table 18.3).

18.4.4 Cellular Scaffolds for Skeletal Muscle Regeneration

Seeded cells and biomimetic scaffolds could be used in skeletal muscle regeneration, either separately or in combination. Although acellular scaffolds are relatively safe to cell-based therapies, they are inadequate to stimulate efficient host cell migration and activity. Mesenchymal stem cells (MSCs) have the capacities of self-renewing and secreting proregenerative growth factors. More importantly, on the appropriate induction or stimulation, MSCs develop and differentiate along specific lineages and promote constructive therapeutic effect for muscle regeneration [49, 50]. Growth factors are essential to induce the differentiation of MSCs *in vitro* studies [51]. Bioactive supplements (hepatocyte growth factor and insulin-like growth factor-1) play vital roles in the myogenic regeneration [52]. Growth factor-incorporated scaffolds can propagate the myogenic differentiation of bone marrow-derived mesenchymal stem cells (BMSCs) [53]. Besides, Witt et al. discovered that co-cultivation of MSCs and myoblasts in the nanofiber scaffolds created synergistic effects [49]. In their research, the ECM scaffolds not only myogenically differentiated MSCs but also converted the proinflammatory M1 macrophages into the anti-inflammatory M2 macrophages [49]. Similarly, Merritt et al. also reported the promyogenic and immune-regulatory effects of the MSC-seeded ECM scaffolds [54]. Adipose-derived stem cells have greater tendency for myogenic differentiation compared to BMSCs [55]. Besides, adipose-derived stem cells (ADSCs) improved

Table 18.3 Acellular ECM scaffold could improve the functional outcomes of VML injury with physical therapy. The table presents the strength measured in different time points after injury and corresponding physical therapy.

Subject	Injury site (side)	Activity	Baseline force measurement (lb)	6–8 wk post-surgery (%)	10–12 wk post-surgery (%)	24–28 wk post-surgery (%)
1	Anterior tibial compartment (left)	Dorsiflexion		0	0	0
2	Anterior tibial compartment (left)	Dorsiflexion	0	0	0	0
3	Quadriceps (left)	Knee extension	6	−10.0	18.3	20
4	Quadriceps (right)	Knee extension	6.1	127.9	149.2	136.1
5	Anterior/lateral tibial compartment (left)	Dorsiflexion	3.6	−33.3	16.7	33.3
6	Brachialis (left)	Biceps flexion	35.8	NT	−19.5	−17.9
7	Biceps (right)	Wrist supination strength	42	66.7	102.4	126.2
		Biceps flexion	38.1	12.3	7.6	16.8
8	Quadriceps (left)	Knee extension	10.3	15	12	64.1
9	Quadriceps (right)	Knee extension	33.3	19	27	61.9
10	Rectus femoris (right)	Knee extension	6.6	11	30	86.4
11	Biceps/deltoid (left)	Shoulder abduction	69.2	−4.6	−4.1	20.1
		Shoulder flexion	46.6	41.9	42.5	104.1
		Shoulder extension	51.3	13.3	22.6	46.8
		Elbow flexion	66.9	0	−0.3	−4.0
		Elbow extension	49	−8.2	31	1.6
12	Sartorius (left)	Hip flexion	68.1	NT	−15.6	−3.5
		Knee extension	92		−28.0	−1.1
13	Hamstring (left)	Knee flexion	53.5	11.8	11	−3.4
		Knee extension	99.2	−33.0	−33.0	0.5
			Average	15.2	21.1	37.3
			s.e.m.	12.6	12.2	12.4

Source: Data source from Ref. [48]/Springer Nature/CC BY-4.0.

Table 18.4 Four types of cells seeded in scaffolds and its functions.

Cell types	Condition	Effect	References
BMSC	Monoculture Co-cultured with myoblast	Myogenesis and immune regulation	[6, 10]
ADSC	Monoculture	Myogenesis, angiogenesis, and immune regulation	[11, 12]
Satellite cells	Monoculture	Myogenesis	[13, 67]
Myoblast	Monoculture Co-cultured on aligned scaffolds with endothelial cells	Engineered skeletal muscle and microvascularization formation	[68–70]

the angiogenic capacity of ECM scaffolds in the treatment of VML [56]. Huang et al. fabricated the ADSC-loaded collagen hydrogel and confirmed its angiogenic, anti-inflammatory, and myogenic potential [57]. Shen Liu and coworkers discovered that MMP-2-responsive unidirectional hydrogel-electrospun patch loading TGF-β siRNA polyplex inhibits fibroblast proliferation through the accurate space–time release of TGF-β siRNA triggered by MMP-2 microenvironment for peritendinous anti-adhesion [58]. An undifferentiated myogenic stem cell population known as SCs also resides in the plasma membrane of skeletal muscle [59]. During muscle repair, the resident SCs proliferate and differentiate into myoblasts to orchestrate muscle regeneration [60]. Synthetic scaffolds can provide sustained SC delivery as a biomimetic cell niche. Rossi et al. invented SC-loaded hyaluronan hydrogel and repaired the partially ablated tibialis anterior muscles both structurally and functionally [61]. Pollot et al. demonstrated that different natural polymeric materials might lead to different myogenic effects of SCs [62]. In the context of muscle damage, SCs proliferate and then exit the cell cycle to differentiate into myoblasts, a more mature type of muscle stem cells. Myoblasts displayed excellent interaction with fibrin scaffolds and exhibited better performance in the rat model of traumatized muscle [63, 64]. Dziki et al. clarified that mechanical loading was important to the performance of myoblast-loaded ECM scaffolds [65]. Mechanically loaded myoblasts showed greater ability to promote the M2 phenotype of macrophages [65]. Myoblasts co-cultured with endothelial cells on aligned scaffolds differentiated into myotube and finally formed highly organized myofibers and microvasculature [66] (Table 18.4).

18.4.5 Tissue Engineering for Muscle Repair

In vivo tissue engineering strategies make the cells susceptible to inactivity, retention, and immune rejection [71]. *In vitro* tissue engineering enables cell delivery through the development and implantation of functional tissue engineered muscle grafts (TEMG) combining scaffolds, biological factors, and cells, creating cultured constructs *in vitro* that permit cell differentiation into myofibers. It can be used for disease modeling and drug screening [72] *in vitro*. TEMG functions are better before implantation than constructs developed through *in situ* and *in vivo* tissue engineering strategies. Aligned constructs, mechanical and electrical stimulation, angiogenesis, and innervation are frequent tissue engineering technologies.

Usually, functional TEMG constructs use scaffolds with anisotropic surface characteristics to guide the orientation and differentiation of myoblasts. Fibrous scaffolds and arranged pores are most common in TEMG constructs [66, 73]. The Grayson lab [74] developed hydrogel microfibers by electrospinning co-extruded solutions of fibrinogen and sodium alginate, and defects treated with myoblasts TEMG and acellular fibrin microfibrous scaffolds showed complete functional recovery [63]. *In vitro*, co-cultured aligned nanofibrillar scaffolds generated constructs with longer myotubes, more synchronized contractility, and higher secretion of angiogenic cytokines compared to randomly oriented scaffolds.

The use of external stimuli such as electrical or mechanical stimulation in generating mature and contractile TEMGs is commonly employed. Mechanical and electrical stimulation promote myoblast fusion, alignment, hypertrophy, and contractile TEMGs *in vitro*. They create culture conditions that mimic *in vivo* stimuli and initiate the activation of intracellular signaling pathways involved in skeletal muscle development [75]. Electrical stimulation has been used to generate TEMGs that promote myofiber formation when implanted *in vivo*. While electrical stimulation is a commonly exploited technique for generating mature muscle tissue *in vitro*, more studies evaluating the ability of these constructs to facilitate functional muscle regeneration upon *in vivo* implantation must be conducted [76]. Cyclic mechanical conditioning is also commonly utilized to generate mature TEMGs before implantation into VML injuries [77]. It has been reported that the recovery of contractile force in muscle defects implanted with pretreated TEMG was significantly higher than that of non-celled scaffolds or untreated controls [78]. Static mechanical stimulation is also used to provide uniaxial tension to TEMGs *in vitro* [79].

Vascularization is important for TEMG to promote their survival [64]. Multiple cell types are used to generate TEMGs instead of myoblasts alone [80]. Nerve activity has been shown to be essential for skeletal muscle maturation because it provides electrical stimulation and promotes the switch from fast to slow MyHC in regenerating muscle [81].

Electrospinning is a versatile technique to produce polymer nanofibers (diameter from 40 to 2000 nm) forming 3D scaffolds and replicate aligned muscle architecture. Electrospinning is performed when the electric force of the mother liquid surface exceeds the surface tension and initiates an electric spark provoking the solution to be ejected from a syringe, and as jet flows, the nanofiber is produced. Electrospinning fibers have also been used for muscle cell growth in a recent paper. GO-PCL composites have been produced by Chaudhuri et al. [82] to create a mesh with ~85% porosity for C2C12 cell growth.

Skin, fat, adipose, and cardiac muscle can now be produced by 3D printing technology [83]. PLA 3D-printed disks can induce differentiation because of the

Table 18.5 Characterization of tissue-engineered muscle grafts and fabricate technique.

Type	Technique
Tissue engineered muscle grafts	Aligned constructs
	Mechanical and electrical stimulation
	Angiogenesis and innervation
Versatile technique	Electrospinning
	3D cell technique

proximity of cells in the print channels [84]. PU bioink-differentiated stiffness can be obtained by co-printing with fibroblast-containing PCL with muscle cells [85]. So far, both PU and PCL have been successfully combined to graphene derivatives [86]. Kang et al. [87] created a 3D integrated tissue–organ-containing mouse myoblasts. The retrieved muscle constructs showed well-organized muscle and as nerve contacts and vascularization two weeks later. PEG/fibrinogen/alginate constructs also generate a totally maturated muscle-like tissue [88]. 3D printing has been employed to create artificial muscles that can be moved and embedded in prosthesis and allow performing complex motions driven by laser light stimulation on account of the photothermal properties of graphene [89, 90] (Table 18.5).

18.5 Application and Prospect

18.5.1 Promising Strategies for Skeletal Muscle Regeneration

The ultimate goal of skeletal muscle tissue engineering is to treat injured or diseased muscles by transplantation *in vivo*. With the development of *in vitro* engineered skeletal muscle constructs, many constructs have been proven to be safe and effective in preclinical practice. One area of preclinical translation is the area of VML. Acellular biological scaffolds have been tested in small and large VML animal models. For example, an acellular small intestinal submucosa scaffold has been tested in a canine tendon-muscle VML model. At six months after transplantation, the stent placement site was characterized by vascularized and innervated muscles, and the contractile response was significantly improved, approaching 50% of the uninjured tendon connection. In a similar study in a rat model of VML, delivery of the SIS scaffold resulted in complete recovery of maximum contractile force after 26 weeks compared to natural tissue. The benefits of these treatments may be due to the ability of scaffolds to provide structural support and promote cell growth. However, contradictory findings that acellular stents have no significant therapeutic effect have also been reported. The differences in these results are due in part to potential differences in scaffold material preparation, VML surgical models, animal species, and age of the animals (Table 18.6).

Table 18.6 Examples of tissue engineering approaches for the treatment of volumetric muscle loss.

Treatment category	Target tissue for ablation	Treatment	Major findings
Acellular scaffolds	Canine gastrocnemius muscle and Achilles	Porcine small intestinal submucosa (SIS) decellularized scaffold for 6 mo	Vascularized and innervated skeletal muscle had formed at the implantation site
	Tendon bundle		Scaffold-treated group had a contractile force that was 48% of the contralateral musculotendinous junction
	Rat abdominal muscle	Porcine SIS decellularized scaffold for 26 wk	Complete recovery in maximal contractile force to native tissue
	Rat tibialis anterior muscle	Porcine urinary bladder matrix	Limited muscle formation and scaffold-mediated fibrosis
	Mouse tibialis anterior muscle	Porcine urinary bladder matrix or nanofibrillar rat collagen scaffold	No benefit of decellularized matrix or collagen scaffolds on muscle or vascular regeneration
Cell-seeded scaffolds	Mouse latissimus dorsi muscle	Muscle-derived cells (MDCs) seeded on porcine urinary bladder matrix	Higher max tetanic force compared to the control ablated muscle group at 1 mo. At 2 mo, cell-seeded scaffolds with prolonged bioreactor conditioning + second application of MDCs showed sustained functional benefit
	Mouse tibialis anterior muscle	Muscle stem cells, endothelial cells (ECs), and muscle resident cells in decellularized muscle scaffold	Addition of ECs and other support cells promoted higher force production in the ablated muscle compared to scaffolds without ECs
	Mouse latissimus dorsi muscle	Rat muscle progenitors in bladder acellular matrix with stretch pretreatment	Significant recovery of force generation after 2 mo when compared to the no treatment control group

Source: Data source from Ref. [91]/John Wiley & Sons.

18.5.2 Challenges and Future Perspectives

Severe muscle trauma that results in VML often leads to the formation of nonfunctional fibrous tissue and degeneration of fatty muscle [1]. Although free functional muscle transfer (FFMT) serves as the current standard of care, this approach does not carry a guarantee of full restoration of muscle function to preinjury conditions. Toward the clinical goal of long-term functional integration and recovery, tissue engineering and regenerative medicine offer a promising therapeutic approach to engineer musculoskeletal replacements that can augment muscle function, stimulate regeneration, and improve quality of life [4].

Current bioengineering approaches for skeletal muscle tissue engineering are composed of small molecules and growth factors, mechanical conditioning, scaffold topography, genetic modification, and electrical stimulation. Despite the excellent performance of natural polymer-based scaffolds during the *in vitro* analysis, the *in vivo* approach for VML remains a major challenge especially on the larger animals and human patients. The approach with the greatest near-term potential for clinical use is decellularized scaffolds, largely because they can convey a series of signals and trigger a cascade of biological responses without seeding cells or other bioactive molecules. Emerging clinical studies using decellularized scaffolds have shown promising results. For example, Sicari et al. implanted acellular porcine bladder ECM into sites of muscle injury in patients and demonstrated the regeneration of skeletal muscle [13]. In addition, scaffolds made of piezoelectric biomaterials are an attractive choice as they exhibit electromechanical behavior that transforms mechanical energy into electric polarization without external voltage. Piezopolymers and piezoceramics have been widely used for different tissue repair applications, particularly in bone repair, where mechanical stress charges have enhanced bone formation. It is speculated that the electrical and mechanical stimuli transduced by piezoelectric biomaterials may induce myogenic differentiation and functional maturation of muscle like tissue [92].

Although tissue engineering approaches to treat muscle injury or diseases remain promising, a number of challenges will need to be resolved to enable clinical practice. Concerns about biocompatibility of the exogenous scaffolds always hinder their application in clinical practice. Physical cross-linking is worth exploring. Nevertheless, extensive and thorough studies are needed to optimize the efficacy of the scaffolds for *in vivo* application. Batch-to-batch variation is another major drawback of employing natural polymers. Variability from batch to batch of the scaffolds, fabricated from natural polymers, is inevitable. Moreover, batch-to-batch variation may also provoke an immune response when implanted *in vivo* because of the xenogeneic polymer source. The synthesis of directed short-chain polysaccharides and polypeptide via polycondensation or classical solution-phase peptide synthesis could be the solution to control the batch variation in natural scaffolds. Moreover, mutagenesis could also be the solution for batch variation. The directed synthesis of the natural polymer would also eliminate the concern of complication from zoonotic pathogen infection. Since skeletal muscle injuries and diseases are complex, with large variability in the spatial geometry of ablated muscle, the development of smaller modular-sized tissue constructs may enable the flexibility for clinicians to tailor to the unique spatial requirements of each patient's muscle injury. Scalability is a major challenge, as most studies are performed in rodents, with limited studies that have been investigated using porcine models of muscle injury. As cell-based therapies may require autologous cells to eliminate immune rejection, the potential manufacturing cost may be a hindrance. With the advent of CRISPR gene-editing technology, it is predictable that allogeneic cells can develop without consideration of immunogenicity. Although many challenges remain to be overcome, tissue engineering-based skeletal muscle engineering and regeneration approaches are highly promise.

References

1 Scharner, J. and Zammit, P.S. (2011). The muscle satellite cell at 50: the formative years. *Skeletal Muscle* 1 (1): 28.

2 Grogan, B.F., Hsu, J.R., and Skeletal Trauma Research Consortium (2011). Volumetric muscle loss. *J. Am. Acad. Orthop. Surg.* 19 (Suppl. 1): S35–S37.

3 Han, W.M., Anderson, S.E., Mohiuddin, M. et al. (2018). Synthetic matrix enhances transplanted satellite cell engraftment in dystrophic and aged skeletal muscle with comorbid trauma. *Sci. Adv.* 4 (8): eaar4008.

4 Corona, B.T. and Greising, S.M. (2016). Challenges to acellular biological scaffold mediated skeletal muscle tissue regeneration. *Biomaterials* 104: 238–246.

5 Court-Brown, C.M., Bugler, K.E., Clement, N.D. et al. (2012). The epidemiology of open fractures in adults. A 15-year review. *Injury* 43 (6): 891–897.

6 Armed Forces Health Surveillance Center (AFHSC) (2013). Medical evacuations from Afghanistan during Operation Enduring Freedom, active and reserve components, U.S. Armed Forces, 7 October 2001–31 December 2012. *MSMR* 20 (6): 2–8.

7 Belmont, P.J. Jr., McCriskin, B.J., Hsiao, M.S. et al. (2013). The nature and incidence of musculoskeletal combat wounds in Iraq and Afghanistan (2005–2009). *J. Orthop. Trauma* 27 (5): e107–e113.

8 Corona, B.T., Rivera, J.C., Owens, J.G. et al. (2015). Volumetric muscle loss leads to permanent disability following extremity trauma. *J. Rehabil. Res. Dev.* 52 (7): 785–792.

9 Corona, B.T., Wu, X., Ward, C.L. et al. (2013). The promotion of a functional fibrosis in skeletal muscle with volumetric muscle loss injury following the transplantation of muscle-ECM. *Biomaterials* 34 (13): 3324–3335.

10 Lin, C.H., Lin, Y.T., Yeh, J.T., and Chen, C.T. (2007). Free functioning muscle transfer for lower extremity posttraumatic composite structure and functional defect. *Plast. Reconstr. Surg.* 119 (7): 2118–2126.

11 Doi, K., Hattori, Y., Tan, S.H., and Dhawan, V. (2002). Basic science behind functioning free muscle transplantation. *Clin. Plast. Surg.* 29 (4): 483–495.

12 Sass, F.A., Fuchs, M., Pumberger, M. et al. (2018). Immunology guides skeletal muscle regeneration. *Int. J. Mol. Sci.* 19 (3): 835.

13 Sicari, B.M., Rubin, J.P., Dearth, C.L. et al. (2014). An acellular biologic scaffold promotes skeletal muscle formation in mice and humans with volumetric muscle loss. *Sci. Transl. Med.* 6 (234): 234ra58.

14 Weinberger, F., Mannhardt, I., and Eschenhagen, T. (2017). Engineering cardiac muscle tissue: a maturating field of research. *Circ. Res.* 120 (9): 1487–1500.

15 Chen, S., Nakamoto, T., Kawazoe, N., and Chen, G. (2015). Engineering multi-layered skeletal muscle tissue by using 3D microgrooved collagen scaffolds. *Biomaterials* 73: 23–31.

16 Manchineella, S., Thrivikraman, G., Khanum, K.K. et al. (2016). Pigmented silk nanofibrous composite for skeletal muscle tissue engineering. *Adv. Healthcare Mater.* 5 (10): 1222–1232.

17 Guo, B.L. and Ma, P.X. (2014). Synthetic biodegradable functional polymers for tissue engineering: a brief review. *Sci. China Chem.* 57 (4): 490–500.

18 Guo, B., Qu, J., Zhao, X., and Zhang, M. (2019). Degradable conductive self-healing hydrogels based on dextran-graft-tetraaniline and *N*-carboxyethyl chitosan as injectable carriers for myoblast cell therapy and muscle regeneration. *Acta Biomater.* 84: 180–193.

19 Zarrintaj, P., Urbanska, A.M., Gholizadeh, S.S. et al. (2018). A facile route to the synthesis of anilinic electroactive colloidal hydrogels for neural tissue engineering applications. *J. Colloid Interface Sci.* 516: 57–66.

20 Bagher, Z., Atoufi, Z., Alizadeh, R. et al. (2019). Conductive hydrogel based on chitosan-aniline pentamer/gelatin/agarose significantly promoted motor neuron-like cells differentiation of human olfactory ecto-mesenchymal stem cells. *Mater. Sci. Eng., C* 101: 243–253.

21 Saberi, A., Jabbari, F., Zarrintaj, P. et al. (2019). Electrically conductive materials: opportunities and challenges in tissue engineering. *Biomolecules* 9 (9): 448.

22 Liu, Z., Zhang, X.Y., Poyraz, S. et al. (2010). Oxidative template for conducting polymer nanoclips. *J. Am. Chem. Soc.* 132 (38): 13158–13159.

23 Schimidt, C.E., Shastri, V.R., Vacanti, J.P., and Langer, R. (1997). Stimulation of neurite outgrowth using an electrically conducting polymer. *Proc. Natl. Acad. Sci. U.S.A.* 94 (17): 8948–8953.

24 Zarrintaj, P., Bakhshandeh, B., Rezaeian, I. et al. (2017). A novel electroactive agarose-aniline pentamer platform as a potential candidate for neural tissue engineering. *Sci. Rep.* 7 (1): 17187.

25 Daraee, H., Eatemadi, A., Abbasi, E. et al. (2016). Application of gold nanoparticles in biomedical and drug delivery. *Artif. Cells Nanomed. Biotechnol.* 44 (1): 410–422.

26 Borselli, C., Storrie, H., Benesch-Lee, F. et al. (2017). Functional muscle regeneration with combined delivery of angiogenesis and myogenesis factors. *Sci. Rep.* 7 (1): 17187.

27 Shvartsman, D., Storrie-White, H., Lee, K. et al. (2014). Sustained delivery of VEGF maintains innervation and promotes reperfusion in ischemic skeletal muscles via NGF/GDNF signaling. *Mol. Ther.* 22 (7): 1243–1253.

28 Ju, Y.M., Atala, A., Yoo, J.J., and Lee, S.J. (2014). In situ regeneration of skeletal muscle tissue through host cell recruitment. *Acta Biomater.* 10 (10): 4332–4339.

29 Lin, J.W., Huang, Y.M., Chen, Y.Q. et al. (2021). Dexamethasone accelerates muscle regeneration by modulating kinesin-1-mediated focal adhesion signals. *Cell Death Discovery* 7 (1): 35.

30 Jin, Y., Shahriari, D., Jeon, E. et al. (2021). Functional skeletal muscle regeneration with thermally drawn porous fibers and reprogrammed muscle progenitors for volumetric muscle injury. *Adv. Mater.* 33: e2007946.

31 Ghobrial, G., Araujo, L., Jinwala, F. et al. (2018). The structure and biological function of CREG. *Front. Cell Dev. Biol.* 6: 136.

32 Tanaka, S., Fujishita, Y., Kawaguchi, N. et al. (2021). The synthetic peptide SVVYGLR promotes cell motility of myogenic cells and facilitates differentiation in skeletal muscle regeneration. *Dental Mater. J.* 40 (3): 766–771.

33 Hill, E., Boontheekul, T., and Mooney, D. (2006). Regulating activation of transplanted cells controls tissue regeneration. *Proc. Natl. Acad. Sci. U.S.A.* 103 (8): 2494–2499.

34 Hill, E., Boontheekul, T., and Mooney, D. (2006). Designing scaffolds to enhance transplanted myoblast survival and migration. *Tissue Eng.* 12 (5): 1295–1304.

35 Mase, V.J. Jr., Hsu, J.R., Wolf, S.E. et al. (2010). Clinical application of an acellular biologic scaffold for surgical repair of a large, traumatic quadriceps femoris muscle defect. *Orthopedics* 33 (7): 511.

36 Dellavalle, A., Sampaolesi, M., Tonlorenzi, R. et al. (2007). Pericytes of human skeletal muscle are myogenic precursors distinct from satellite cells. *Nat. Cell Biol.* 9 (3): 255–267.

37 Tidball, J.G. and Villalta, S.A. (2010). Regulatory interactions between muscle and the immune system during muscle regeneration. *Am. J. Physiol. Regul. Integr. Comp. Physiol.* 298 (5): R1173–R1187.

38 Deng, B., Wehling-Henricks, M., Villalta, S.A. et al. (2012). IL-10 triggers changes in macrophage phenotype that promote muscle growth and regeneration. *J. Immunol.* 189 (7): 3669–3680.

39 Ruffell, D., Mourkioti, F., Gambardella, A. et al. (2009). A CREB-C/EBPβ cascade induces M2 macrophage-specific *gene* expression and promotes muscle injury repair. *Proc. Natl. Acad. Sci. U.S.A.* 106 (41): 17475–17480.

40 Sicari, B.M., Dziki, J.L., Siu, B.F. et al. (2014). The promotion of a constructive macrophage phenotype by solubilized extracellular matrix. *Biomaterials* 35 (30): 8605–8612.

41 Tidball, J.G., Dorshkind, K., and Wehling-Henricks, M. (2014). Shared signaling systems in myeloid cell-mediated muscle regeneration. *Development* 141 (6): 1184–1196.

42 Li, F., Li, W., Johnson, S. et al. (2004). Low-molecular-weight peptides derived from extracellular matrix as chemoattractants for primary endothelial cells. *Endothelium* 11 (3, 4): 199–206.

43 Valentin, J.E., Turner, N.J., Gilbert, T.W., and Badylak, S.F. (2010). Functional skeletal muscle formation with a biologic scaffold. *Biomaterials* 31 (29): 7475–7484.

44 Gentile, N.E., Stearns, K.M., Brown, E.H. et al. (2014). Targeted rehabilitation after extracellular matrix scaffold transplantation for the treatment of volumetric muscle loss. *Am. J. Phys. Med. Rehabil.* 93 (11 Suppl. 3): S79–S87.
45 Lepper, C., Partridge, T.A., and Fan, C.M. (2011). An absolute requirement for Pax7-positive satellite cells in acute injury-induced skeletal muscle regeneration. *Development* 138 (17): 3639–3646.
46 von Maltzahn, J., Jones, A.E., Parks, R.J., and Rudnicki, M.A. (2013). Pax7 is critical for the normal function of satellite cells in adult skeletal muscle. *Proc. Natl. Acad. Sci. U.S.A.* 110 (41): 16474–16479.
47 Aurora, A., Roe, J.L., Corona, B.T., and Walters, T.J. (2015). An acellular biologic scaffold does not regenerate appreciable de novo muscle tissue in rat models of volumetric muscle loss injury. *Biomaterials* 67: 393–407.
48 Dziki, J., Badylak, S., Yabroudi, M. et al. (2016). An acellular biologic scaffold treatment for volumetric muscle loss: results of a 13-patient cohort study. *npj Regener. Med.* 1: 16008.
49 Witt, R., Weigand, A., Boos, A.M. et al. (2017). Mesenchymal stem cells and myoblast differentiation under HGF and IGF-1 stimulation for 3D skeletal muscle tissue engineering. *BMC Cell Biol.* 18 (1): 15.
50 Rybalko, V., Hsieh, P.L., Ricles, L.M. et al. (2017). Therapeutic potential of adipose-derived stem cells and macrophages for ischemic skeletal muscle repair. *Regener. Med.* 12 (2): 153–167.
51 Kraemer, W.J., Ratamess, N.A., Hymer, W.C. et al. (2020). Growth hormone(s), testosterone, insulin-like growth factors, and cortisol: roles and integration for cellular development and growth with exercise. *Front. Endocrinol. (Lausanne)* 11: 33.
52 Miller, K.J., Thaloor, D., Matteson, S., and Pavlath, G.K. (2000). Hepatocyte growth factor affects satellite cell activation and differentiation in regenerating skeletal muscle. *Am. J. Physiol. Cell Physiol.* 278 (1): C174–C181.
53 Beier, J.P., Bitto, F.F., Lange, C. et al. (2011). Myogenic differentiation of mesenchymal stem cells co-cultured with primary myoblasts. *Cell Biol. Int.* 35 (4): 397–406.
54 Merritt, E.K., Cannon, M.V., Hammers, D.W. et al. (2010). Repair of traumatic skeletal muscle injury with bone-marrow-derived mesenchymal stem cells seeded on extracellular matrix. *Tissue Eng. Part A* 16 (9): 2871–2881.
55 De La Garza-Rodea, A.S., Van Der Velde-van Dijke, I., Boersma, H. et al. (2012). Myogenic properties of human mesenchymal stem cells derived from three different sources. *Cell Transplant.* 21 (1): 153–173.
56 Aurora, A., Wrice, N., Walters, T.J. et al. (2018). A PEGylated platelet free plasma hydrogel based composite scaffold enables stable vascularization and targeted cell delivery for volumetric muscle loss. *Acta Biomater.* 65: 150–162.
57 Huang, H., Liu, J., Hao, H. et al. (2017). Preferred M2 polarization by ASC-based hydrogel accelerated angiogenesis and myogenesis in volumetric muscle loss rats. *Stem Cells Int.* 2017: 2896874.

58 Cai, C., Wang, W., Liang, J. et al. (2021). MMP-2 responsive unidirectional hydrogel-electrospun patch loading TGF-β1 siRNA polyplexes for peritendinous anti-adhesion. *Adv. Funct. Mater.* 31: 2008364.

59 Mauro, A. (1961). Satellite cell of skeletal muscle fibers. *J. Biophys. Biochem. Cytol.* 9 (2): 493–495.

60 Yin, H., Price, F., and Rudnicki, M.A. (2013). Satellite cells and the muscle stem cell niche. *Physiol. Rev.* 93 (1): 23–67.

61 Rossi, C.A., Flaibani, M., Blaauw, B. et al. (2011). *In vivo* tissue engineering of functional skeletal muscle by freshly isolated satellite cells embedded in a photopolymerizable hydrogel. *FASEB J.* 25 (7): 2296–2304.

62 Pollot, B.E., Rathbone, C.R., Wenke, J.C., and Guda, T. (2018). Natural polymeric hydrogel evaluation for skeletal muscle tissue engineering. *J. Biomed. Mater. Res. Part B* 106 (2): 672–679.

63 Gilbert-Honick, J., Iyer, S.R., Somers, S.M. et al. (2018). Engineering functional and histological regeneration of vascularized skeletal muscle. *Biomaterials* 164: 70–79.

64 Page, R.L., Malcuit, C., Vilner, L. et al. (2011). Restoration of skeletal muscle defects with adult human cells delivered on fibrin microthreads. *Tissue Eng. Part A* 17 (21, 22): 2629–2640.

65 Dziki, J.L., Giglio, R.M., Sicari, B.M. et al. (2018). The effect of mechanical loading upon extracellular matrix bioscaffold-mediated skeletal muscle remodeling. *Tissue Eng. Part A* 24 (1, 2): 34–46.

66 Nakayama, K.H., Quarta, M., Paine, P. et al. (2019). Treatment of volumetric muscle loss in mice using nanofibrillar scaffolds enhances vascular organization and integration. *Commun. Biol.* 2: 170.

67 Dumont, N.A., Bentzinger, C.F., Sincennes, M.C., and Rudnicki, M.A. (2015). Satellite cells and skeletal muscle regeneration. *Compr. Physiol.* 5 (3): 1027–1059.

68 Mahdy, M.A.A. (2019). Skeletal muscle fibrosis: an overview. *Cell Tissue Res.* 375 (3): 575–588.

69 Carnes, M.E. and Pins, G.D. (2020). Skeletal muscle tissue engineering: biomaterials-based strategies for the treatment of volumetric muscle loss. *Bioengineering (Basel)* 7 (3): 85.

70 Fernandes, T.L., Pedrinelli, A., and Hernandez, A.J. (2011). Muscle injury – physiopathology, diagnosis, treatment and clinical presentation. *Rev. Bras. Ortop.* 46 (3): 247–255.

71 Sicari, B.M., Londono, R., and Badylak, S.F. (2015). Strategies for skeletal muscle tissue engineering: seed vs. soil. *J. Mater. Chem. B* 3 (40): 7881–7895.

72 Vandenburgh, H., Shansky, J., Benesch-Lee, F. et al. (2008). Drug-screening platform based on the contractility of tissue-engineered muscle. *Muscle Nerve* 37 (4): 438–447.

73 Kroehne, V., Heschel, I., Schügner, F. et al. (2008). Use of a novel collagen matrix with oriented pore structure for muscle cell differentiation in cell culture and in grafts. *J. Cell. Mol. Med.* 12 (5a): 1640–1648.

74 Gilbert-Honick, J. and Grayson, W. (2020). Vascularized and innervated skeletal muscle tissue engineering. *Adv. Healthcare Mater.* 9 (1): e1900626.

75 Rangarajan, S., Madden, L., and Bursac, N. (2014). Use of flow, electrical, and mechanical stimulation to promote engineering of striated muscles. *Ann. Biomed. Eng.* 42 (7): 1391–1405.

76 Serena, E., Flaibani, M., Carnio, S. et al. (2008). Electrophysiologic stimulation improves myogenic potential of muscle precursor cells grown in a 3D collagen scaffold. *Neurol. Res.* 30 (2): 207–214.

77 Mintz, E.L., Passipieri, J.A., Franklin, I.R. et al. (2020). Long-term evaluation of functional outcomes following rat volumetric muscle loss injury and repair. *Tissue Eng. Part A* 26 (3, 4): 140–156.

78 Passipieri, J.A., Hu, X., Mintz, E. et al. (2019). *In silico* and *in vivo* studies detect functional repair mechanisms in a volumetric muscle loss injury. *Tissue Eng. Part A* 25 (17, 18): 1272–1288.

79 Gholobova, D., Decroix, L., Van Muylder, V. et al. (2015). Endothelial network formation within human tissue-engineered skeletal muscle. *Tissue Eng. Part A* 21 (19, 20): 2548–2558.

80 Shandalov, Y., Egozi, D., Koffler, J. et al. (2014). An engineered muscle flap for reconstruction of large soft tissue defects. *Proc. Natl. Acad. Sci. U.S.A.* 111 (16): 6010–6015.

81 Kalhovde, J.M., Jerkovic, R., Sefland, I. et al. (2005). 'Fast' and 'slow' muscle fibres in hindlimb muscles of adult rats regenerate from intrinsically different satellite cells. *J. Physiol.* 562 (Pt. 3): 847–857.

82 Chaudhuri, B., Mondal, B., Kumar, S., and Sarkar, S.C. (2016). Myoblast differentiation and protein expression in electrospun graphene oxide (GO)-poly (ε-caprolactone, PCL) composite meshes. *Mater. Lett.* 182: 194–197.

83 Li, J., Liu, X., Crook, J.M., and Wallace, G.G. (2019). 3D graphene-containing structures for tissue engineering. *Mater. Today Chem.* 14: 100199.

84 Rimington, R.P., Capel, A.J., Christie, S.D.R., and Lewis, M.P. (2017). Biocompatible 3D printed polymers via fused deposition modelling direct C_2C_{12} cellular phenotype *in vitro*. *Lab Chip* 17 (17): 2982–2993.

85 Choi, Y.J., Kim, T.G., Jeong, J. et al. (2016). 3D cell printing of functional skeletal muscle constructs using skeletal muscle-derived bioink. *Adv. Healthcare Mater.* 5 (20): 2636–2645.

86 Palmieri, V., Lattanzi, W., Perini, G. et al. (2020). 3D-printed graphene for bone reconstruction. *2D Mater.* 7 (2): 022004.

87 Kang, H.W., Lee, S.J., Ko, I.K. et al. (2016). A 3D bioprinting system to produce human-scale tissue constructs with structural integrity. *Nat. Biotechnol.* 34 (3): 312–319.

88 Costantini, M., Testa, S., Mozetic, P. et al. (2017). Microfluidic-enhanced 3D bioprinting of aligned myoblast-laden hydrogels leads to functionally organized myofibers *in vitro* and *in vivo*. *Biomaterials* 131: 98–110.

89 Zhou, L.Y., Gao, Q., Fu, J.Z. et al. (2019). Multimaterial 3D printing of highly stretchable silicone elastomers. *ACS Appl. Mater. Interfaces* 11 (26): 23573–23583.

90 Han, B., Zhang, Y.L., Zhu, L. et al. (2019). Plasmonic-assisted graphene oxide artificial muscles. *Adv. Mater.* 31 (5): e1806386.

91 Nakayama, K.H., Shayan, M., and Huang, N.F. (2019). Engineering biomimetic materials for skeletal muscle repair and regeneration. *Adv. Healthcare Mater.* 8 (5): e1801168.

92 Yoon, J.-K., Misra, M., Yu, S.J. et al. (2017). Thermosensitive, stretchable, and piezoelectric substrate for generation of myogenic cell sheet fragments from human mesenchymal stem cells for skeletal muscle regeneration. *Adv. Funct. Mater.* 27 (48): 1703853.

19

Rotator Cuff Repair

Chongyang Wang[1] *and Yaohua He*[1,2]

[1]*Shanghai Jiao Tong University Affiliated Sixth People's Hospital, Department of Orthopedics, 600 Yishan Road, Shanghai 200233, P.R. China*
[2]*Jinshan Branch of Shanghai Sixth People's Hospital Affiliated to Shanghai University of Medicine & Health Sciences, Department of Orthopedics, 147 Jiankang Road, Shanghai 201503, P.R. China*

19.1 Introduction

The rotator cuff is a complex that includes four muscles: supraspinatus, infraspinatus, subscapularis, and teres minor (Figure 19.1) [1]. They arise from the scapula and end with tendons that attach to the tuberosities of the humerus. The rotator cuff tendons play a significant role in lifting the arm, maintaining the stability, and coordinating the movement of the shoulder joint.

The rotator cuff tear is an injury of one or more of the rotator cuff tendons at the tendon-to-bone insertion site (enthesis) (Figure 19.2) [2] and frequently causes shoulder pain, disability, and dysfunction (Figure 19.3). Although the pathogenesis is still unclear, it is mainly induced by intrinsic factors (age-related degeneration, inflammation, oxidative stress, and hypovascularity) and extrinsic factors (subacromial impingement, internal impingement, acute traumatic injury, shoulder instability, and repetitive microtrauma) [3–5]. There is no accurate prevalence of rotator cuff tear because of the lack of symptoms in some patients; however, it is generally recognized that the incidence increases with age and more than 40% in the adults over the age of 50 years [6, 7]. Because of the high incidence, more than 2 million people per year in the United States make consultation to doctors because of rotator cuff tear, and over 460 000 repair surgeries per year are performed [8].

Clinicians often first prescribe conservative therapies such as rest, physical therapy, oral nonsteroidal anti-inflammatory drugs, and local steroid injections to treat patients with symptomatic rotator cuff tears. When all the conservative therapies fail to alleviate the patients' symptom, clinicians would suggest rotator cuff repair surgery. Despite the different surgical techniques (single-row, double-row, transosseous sutures, or transosseous-equivalent techniques), the final target is to reattach the torn tendons to the humeral tuberosity. Although the technique of rotator cuff repair surgery is mature, the retear rate remains high [9, 10]. Previous

Biofabrication for Orthopedics: Methods, Techniques, and Applications, First Edition.
Edited by Wenguo Cui, Xin Zhao, and Shen Liu.

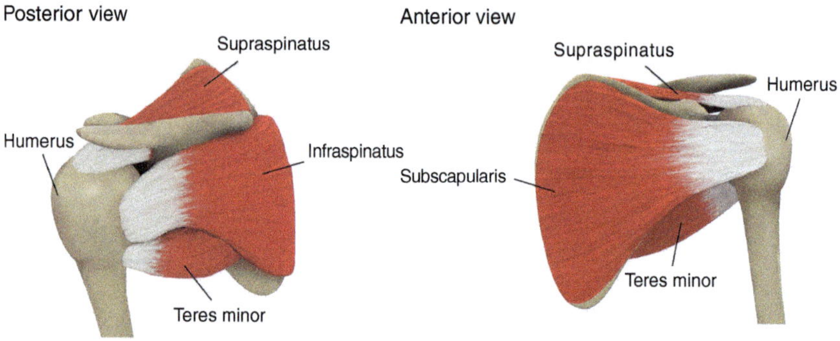

Figure 19.1 Rotator cuff anatomy and the involved tendons including supraspinatus, infraspinatus, teres minor, and subscapularis. Source: Saveh-Shemshaki et al. [1]/with permission of Elsevier.

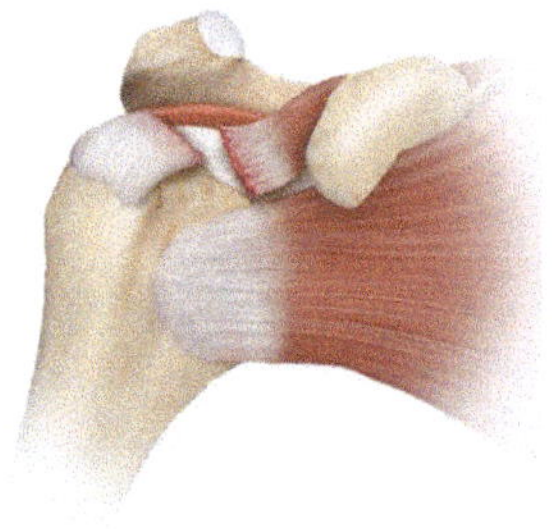

Figure 19.2 Illustration of rotator cuff tendon tears. Source: Millett et al. [2]/with permission of Elsevier.

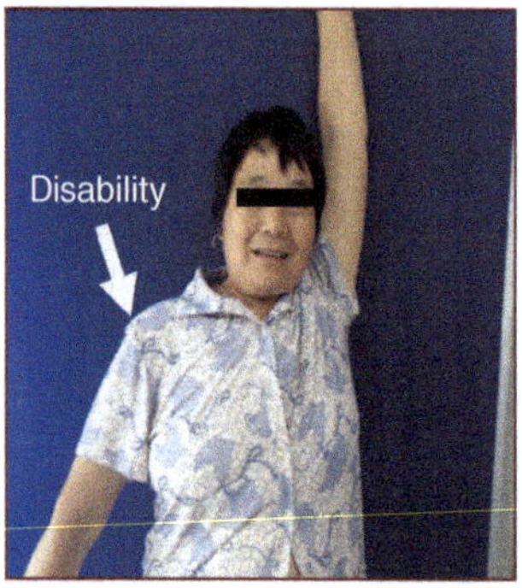

Figure 19.3 Shoulder disability resulted by rotator cuff tears.

studies have reported that the retear rate of large and massive repaired rotator cuff ranges from 20% to 94% [11]. Hence, researchers more recently have begun to explore various biofabrication techniques for rotator cuff repair to enhance tendon-to-bone healing [12]. Before giving an overview of the biofabrication techniques used in manufacturing applications for rotator cuff repair, it is necessary to understand the histology of tendon-to-bone enthesis and risk factors associated with high failure rates. Without the basic knowledge, it is impossible to design desirable biofabrication techniques for rotator cuff repair.

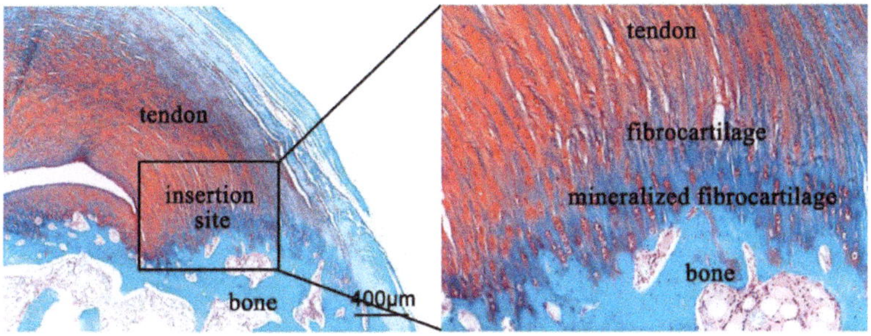

Figure 19.4 Histological picture of the rabbit supraspinatus tendon-to-bone fibrocartilaginous enthesis.

19.2 Histology of the Rotator Cuff Tendon-to-Bone Enthesis

Histologically, the tendon-to-bone enthesis is characterized as a fibrocartilaginous structure [13]. It has been separated into four zones, including tendon, uncalcified/non-mineralized fibrocartilage, calcified/mineralized fibrocartilage, and bone (Figure 19.4). The tendon zone is characterized by the presence of parallel collagen fibers, which contains interspaced and elongated fibroblasts organized in arrays [14]. The extracellular matrix (ECM) of tendon zone mainly composed of collagen type I and few amounts of proteoglycans [15]. The histological structure of uncalcified fibrocartilage zone is characterized by round fibrochondrocytes distributed in rows. The ECM of uncalcified fibrocartilage primarily consists of collagen types II and III, few collagen types X and I, and small amounts of proteoglycans (mainly aggrecan) and chondroitin 4- and 6-sulfate glycosaminoglycans (GAGs). The calcified fibrocartilage zone is characterized by round and big hypertrophic fibrochondrocytes. Its ECM is primarily composed of collagen type II and great amounts of collagen type X and aggrecan. The bone zone mainly contains osteoblasts, osteocytes, and osteoclasts. Its ECM mainly consists of collagen type I and hydroxyapatite (HA). The cellular and ECM composition of the rotator cuff tendon-to-bone enthesis is summarized in Table 19.1 and Figure 19.5. In the enthesis site, Rossetti et al. [16] found that there is a transition zone (spanning over 500 μm) where the wider tendon fibers (105 ± 21 μm) unravel into thinner interface fibers (13 ± 4 μm) (Figures 19.6 and 19.7). Mechanically, the splayed interface fibers and interdigitated attachments can make themselves more resilient by increasing the force transduction area and reducing stress concentrations [17, 18].

19.3 Risk Factors Associated with High Failure Rates of Rotator Cuff Repair

Previous studies have reported that the risk factors associated with high failure rates of rotator cuff repair include initial tear size, fatty infiltration of rotator cuff muscles,

Table 19.1 Cellular and ECM composition of the rotator cuff tendon-to-bone enthesis.

Zone	Cell type	ECM component
Tendon	Fibroblasts	Collagen type I, few proteoglycans
Uncalcified fibrocartilage	Fibrochondrocytes	Collagen type II and III, few collagen types X and I, small amounts of aggrecan and GAGs
Calcified fibrocartilage	Hypertrophic fibrochondrocytes	Collagen type II and great amounts of collagen type X and aggrecan
Bone	Osteoblasts, osteocytes, and osteoclasts	Collagen type I and hydroxyapatite

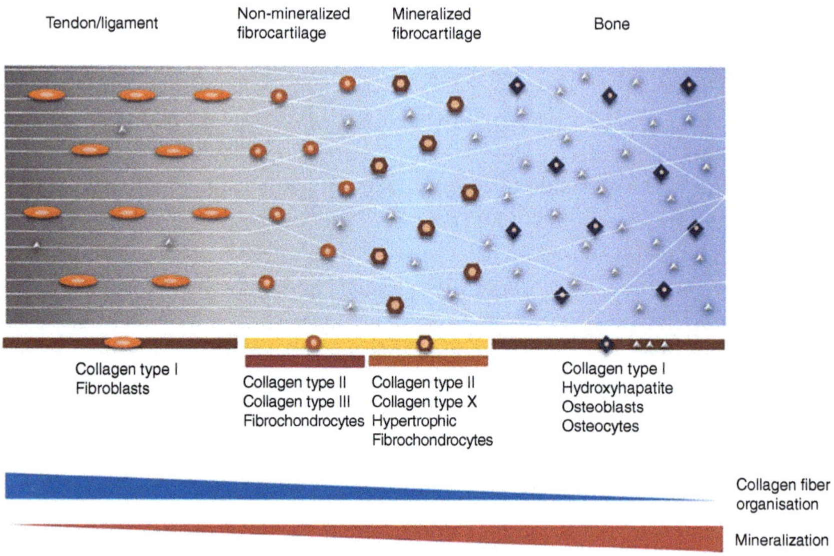

Figure 19.5 Structure and composition of fibrocartilaginous entheses. Fibrocartilaginous entheses are composed of four different zones: tendon/ligament, non-mineralized fibrocartilage, mineralized fibrocartilage, and bone. The main extracellular matrix components and cellular types of each zone are indicated. Of note is the reversal gradient on collagen fiber organization and mineral content across the enthesis. This structural gradient turns into a gradation in mechanical properties that ensures smooth stress distribution across the insertion and decreases the risk of fracture. Source: Tellado et al. [13]/with permission of Elsevier.

failure to regenerate the microstructure of normal tendon-to-bone enthesis, and low bone mineral density (BMD) of the humeral head. In this section, we will provide an overview of these risk factors to help readers to better understand the principles in designing biofabrication for rotator cuff tear.

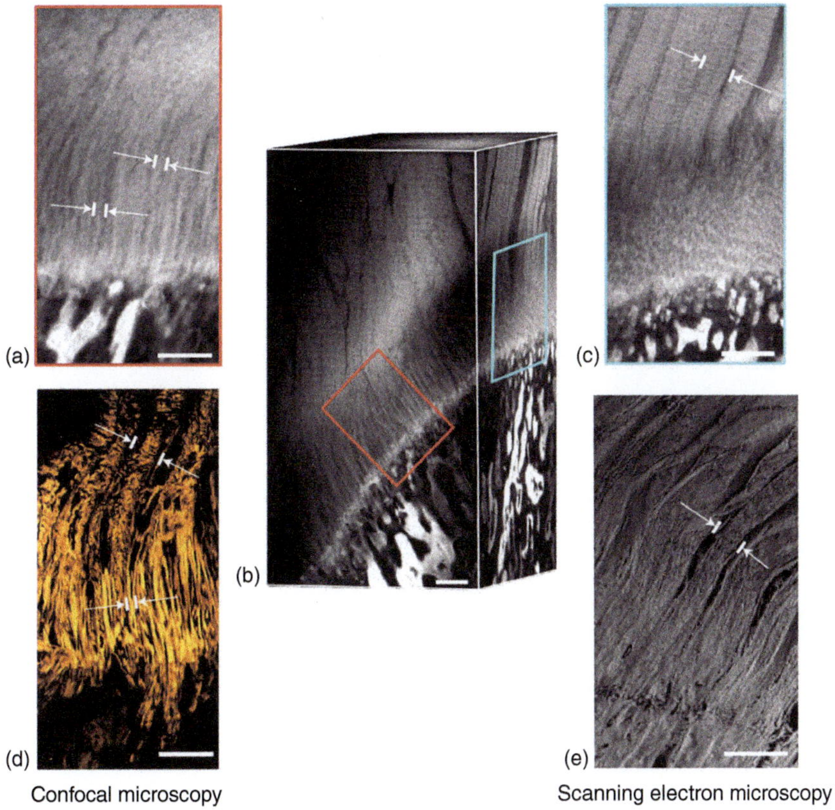

Figure 19.6 The structure of tendon fibers changes before attaching to bone. Enthesis high-resolution microcomputed tomography (μCT) and confocal reflection imaging as well as scanning electron microscopy show that fibers continue from tendon toward bone. Fibers unravel into thinner interface fibers upon entering the interface region, which spans ~500 μm between tendon and bone. (a) Interface fibers observed with high-resolution μCT. (b) Rendered volume of μCT acquisition showing both interface fibers and tendon fibers. (c) Tendon fibers in μCT. (d) Tendon fibers and interface fibers imaged with confocal reflection microscopy of enthesis cryocut sections. (e) Tendon fibers observed with scanning electron microscopy. White arrows mark fiber width. Scale bars, 250 μm. Source: Rossetti et al. [16] Reprinted with permission Springer Nature.

19.3.1 Rotator Cuff Tear Size

Generally, the rotator cuff tear size is classified into four types, including small (<1 cm), medium (1–3 cm), large (3–5 cm), or massive (>5 cm) [19]. It is well acknowledged that the greater the tear size is, the greater the likelihood of causing postsurgical retear. For clinicians, the tear size impacts their choice of surgical approach. The first option to repair a torn rotator cuff most intends to reattach the tendon to the humerus. In this situation, most biofabrication techniques

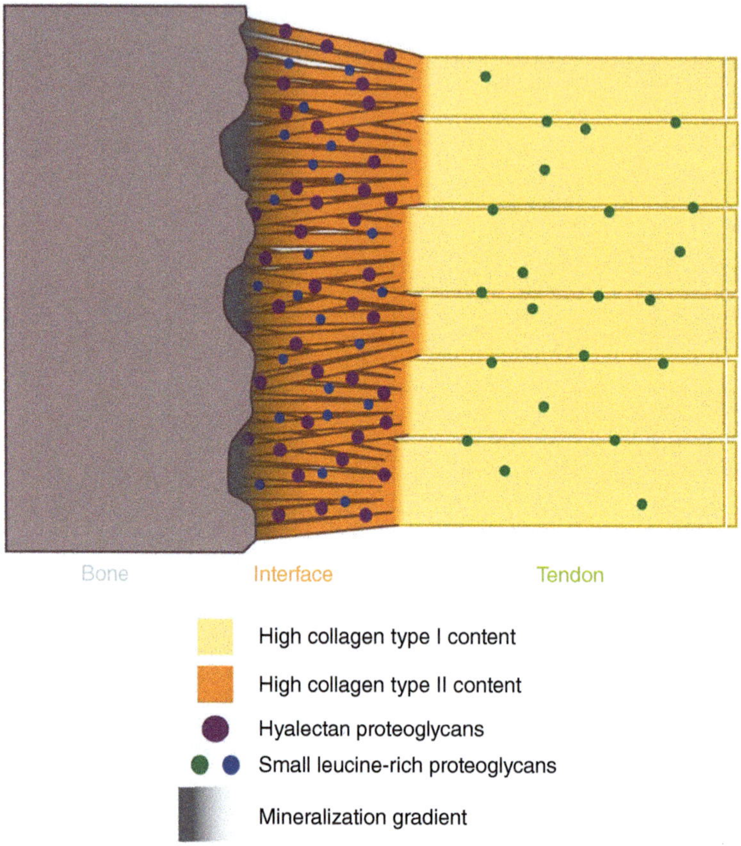

Figure 19.7 Fundamental structural and molecular components of the enthesis. Collagen type I-rich tendon fibers run toward the bone and undergo a transition in both geometry and composition ~500 μm before attaching to the bone. Collagen-type II-rich fibers at the interface are supported by a distinct extracellular matrix rich in hyalectan proteoglycans such as aggrecan and versican. Different small leucine-rich proteoglycans are enriched in the interface and tendon, respectively. The tissue is gradually mineralized toward the bone. Source: Rossetti et al. [16]/with permission of Springer Nature.

are used to manufacture applications to improve the reconstruction of natural tendon-to-bone enthesis. However, some large or massive rotator cuff tears cannot be directly reattached to the humerus. In this case, biofabrication techniques are usually designed to connect the torn tendon and humerus, or to replace the torn cuffs.

19.3.2 Fatty Infiltration of Rotator Cuff Muscles

Fatty infiltration of chronically disrupted muscle is commonly recognized as a consequence of massive rotator cuff tears [20]. Although fatty infiltration is closely associated with poor postsurgical outcomes, however, current repair surgeries fail to prevent or reverse its progression. No effective treatment for fatty infiltration ascribes to the very little known about the pathophysiology behind it. Therefore, deeper understanding of the underlying molecular and cellular mechanism helps to develop a novel method to treat muscle fatty infiltration of rotator cuff tears.

The source of adipocytes in rotator cuff muscle fatty infiltration is still unclear. Current studies speculated that the major cellular origins include adipogenic differentiation of resident stem cells or recruitment from extramuscular sources [20–22]. Liu et al. [21] demonstrated that PDGFRα^+ fibro/adipogenic progenitor cells (FAPs) are the major source of adipocytes in rotator cuff muscle fatty infiltration. Itoigawa et al. [23] ascribed the emergence of adipocytes to the adipogenic differentiation of muscle progenitors. Furthermore, they found that the inhibition of Wnt10b can promote muscle fatty infiltration in torn cuffs. Meanwhile, the lithium chloride (LiCl)-activated Wnt signaling can prevent the fatty infiltration. Besides stem cells, Gibbons et al. [22] hypothesized that the recruited lipid-laden macrophages could differentiate into adipocytes.

19.3.3 Failure to Regenerate the Microstructure of Normal Tendon-to-Bone Enthesis

Because of the complex microstructure of the tendon-to-bone enthesis, traditional rotator cuff repair surgeries are unable to regenerate the native enthesis and usually induce reactive fibrous scar tissue formation. Mechanically, the fibrous scar tissue is inferior to the native enthesis and is easy to be ruptured. For the scaffolds without mineral (e.g. collagen, polyglycolic acid [PGA], PLA, polylactid-*co*-glycolid acid [PLGA], etc.) [24], they are capable of regenerating cartilaginous transition but generally fail to regenerate the graded morphology of the mature tendon-to-bone insertion site. For scaffolds with gradient mineral [25], they were reported effective in promoting gradient-mineralized cartilage formation. However, the regenerated mineralized cartilage is still different with the native mineral gradation.

19.3.4 Low BMD of the Humeral Head

In rotator cuff tear patients who have diminished BMD of the humeral head, the repair surgery might fail because of the suture anchor pullout of bone or transosseous sutures cutting through bone before tendon healing [26, 27]. Chung et al. [28] reported that patients with low BMD had 7.25 times more chance of failed rotator cuff repair surgery than patients who had normal BMD.

19.4 Biofabrication for Rotator Cuff Repair

Generally, an optimal biofabrication may include three general components: (i) a biomimetic scaffold for transplantation and support; (ii) cells to form a functional matrix; and (iii) bioreactive molecules, such as growth factors and small molecules [29]. Previous studies usually classify the commonly used biofabrication for rotator cuff repair according to their composition, such as natural polymers (collagen, silk, etc.), synthetic degradable polymers (PGA, poly-L-lactic acid [PLLA], PLGA, etc.), synthetic non-degradable polymers (polytetrafluoroethylene [PTFE], polycarbonate and polyurethane, polyester, etc.), decellularized native tendon/ligament matrices, and hybrid of these biomaterials [13, 30]. Next, we will first separately discuss the current rotator cuff biofabrication techniques according to the four design concepts:

(1) promote the regeneration of enthesis structure:
  ① interplate between the tendon and bone;
  ② suture to the superficial surface of rotator cuff tendon;
(2) replace/regenerate the torn tendons;
(3) inhibit fat infiltration of rotator cuff muscles;
(4) increase BMD.

19.4.1 Biofabrication Promotes the Regeneration of Enthesis Structure

19.4.1.1 Biofabrication Interplate Between the Tendon and Bone

Biofabrication with monolayer structure is easier to be fabricated and has been well developed to improve the healing of repaired tendon-to-bone enthesis. For instance, Yoon et al. [31] inserted a transforming growth factor beta 1 (TGF-β1)-loaded alginate scaffold between the tendon and bone. At 12 weeks after surgery, the scaffold significantly improved the biomechanical and histological outcomes. However, they only compared the TGF-β1 group with the TGF-β1-sustained release group, which failed to determine the effect of alginate scaffold itself on rotator cuff tendon-to-bone healing.

Because of the complex structure of tendon-to-bone enthesis, monolayer biofabrication usually fail to reconstruct the native enthesis site. Hence, researchers have begun to focus on multilayer biofabrication. To mimic non-mineralized fibrocartilage and mineralized fibrocartilage, Li et al. [32] fabricated a dual-layer organic/inorganic flexible bipolar fibrous membrane (BFM). They utilized electrospinning to fabricate PLLA fibrous membrane as the upper layer and induced apatite nanoparticle layer on the nanohydroxyapatite-poly-L-lactic acid (nHA-PLLA) fibrous membrane as the lower layer. By *in vivo* implanting between the tendon and bone, the BFM significantly improved cartilage regeneration, collagen organization, ultimate load-to-failure, and stiffness than the simplex fibrous membrane of PLLA and merely repair.

To mimic the whole gradient structure of tendon-to-bone region, Soo Kim et al. [33] generated a human collagen-based scaffold composed of four layers: (i) a tendon layer composed of collagen, (ii) a uncalcified fibrocartilage layer composed of collagen and chondroitin sulfate, (iii) a calcified fibrocartilage layer composed of collagen and less apatite, and (iv) a bone layer composed of collagen and apatite. The mechanical testing showed a gradual increase in Young's modulus and a decrease in elongation capacity, which revealed that the chemical composition of each layer accounts for its mechanical properties. By *in vitro* incubating with the resuspended human synovial membrane fibroblasts, chondrocytes, and osteoblasts

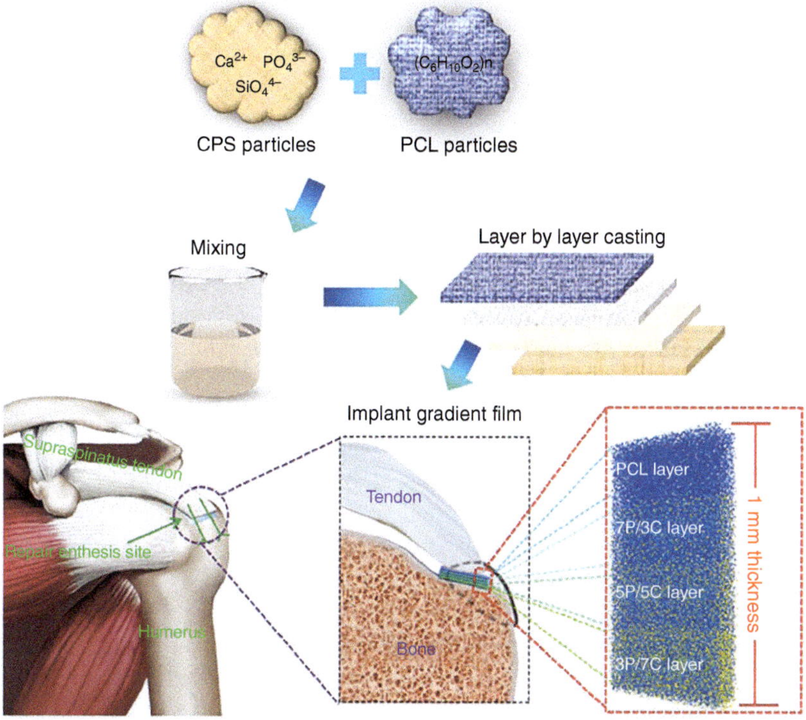

Figure 19.8 Schematic illustration of fabrication of G-P/C composite film by tape casting technique and its application in shoulder tendon enthesis site. The PCL phase of G-P/C composite film was adjacent to the tendon edge, and the CPS phase was adjacent to the bone edge. The yellow area indicates CPS, and the blue area indicates PCL. Source: Su et al. [25] Reprinted with permission of Elsevier.

for 21 days, the proliferation of fibroblasts on the tendon layer, chondrocytes on the uncalcified fibrocartilage layer, and osteoblasts on the bone layer were the highest. These results highlighted the significance of matrix composition for cell proliferation and survival. In a similar approach, Su et al. [25] fabricated a four-layer gradient-poly(ε-caprolactone)/calcium phosphate silicate (G-P/C) composite films with tape casting technique. They combined biodegradable poly(ε-caprolactone) (PCL) polymers with gradient calcium phosphate silicate (CPS) content (1P/0C, 7P/3C, 5P/5C, and 3P/7C) to fabricate the G-P/C film with four consecutive layers (Figure 19.8). *In vitro* co-cultured with osteoblasts showed that the CPS ingredient could improve the adhesion, proliferation, and mineralization of osteoblasts. By *in vivo* implanting between the supraspinatus tendon and the humerus, the histological and biomechanical properties of the G-P/C group showed significant improvement than the control (merely repair), PCL, or 5P-5C groups.

19.4.1.2 Biofabrication Overlays the Repaired Rotator Cuff Tendon

In primary rotator cuff repair, the retears have been reported in the initial three months postoperatively. Hence, augmenting the repair with a patch material can theoretically provide acute stability. Santoni et al. [34] fixed the polyurethane scaffold mesh (PSM) on the top of the repaired infraspinatus tendon. Three months after surgery, the PSM was not grossly visible and completely encompassed by the fibrous reparative tissue. Biomechanically, the PSM significantly increased the failure force with 74.2% than the solely repair group.

In addition to provide acute stability, the overlayed biofabrication may also play a significant role in promoting the regeneration of tendon insertion, inhibiting fatty infiltration or elevating humerus BMD. For instance, Bokor et al. [35, 36] attached a highly oriented and highly porous (85–90% porosity) collagen scaffolds, made from bovine tendons by propriety method, to the bursal surface of the supraspinatus tendon for treating patients with full-thickness or partial-thickness tear. The magnetic resonance imaging (MRI) performed at three months postoperatively showed significant new tissue formation. Twenty-four months later, MRI revealed normal footprint anatomy. For histological evaluation at six months after surgery, the newly generated tissue showed the histological appearance of a tendon and no evidence of any remnants of the implant [37]. In a similar study, Barber et al. [38] used acellular human dermal matrix to augment arthroscopic repair of two-tendon rotator cuff tears (greater than 3 cm). After a mean follow-up of 24 months (range, 12–38 months), the gadolinium-enhanced MRI showed the augmented group with 85% intact repair and 40% of the nonaugmented group.

Cell sheet, a cell delivery method, has also been used to cover the repaired cuff tendons. For instance, Shin et al. [39] isolated adipose-derived mesenchymal stem cells (ASCs) from the inguinal region of Sprague–Dawley rats and fabricated cell sheet to cover the repaired supraspinatus tendons. They found that repair and cell sheet transplantation showed improvement on histomorphological staining, bone volume/total volume values, and biomechanical properties. In a similar study, Harada et al. [40] isolated human rotator cuff-derived cells and utilized them to fabricate cell sheet. They repaired the rat torn infraspinatus tendons by the transosseous method and covered the tendons with cell sheet. Four weeks after surgery, the cell sheet promoted cartilage regeneration and angiogenesis at the enthesis site. Eight weeks after surgery, the cell sheet-covered samples showed superior mechanical strength compared with the merely repaired ones.

19.4.2 Biofabrication Bridges the Torn Tendon and Bone

In 1909, Codman first performed the rotator cuff repair surgery and described a gap between the supraspinatus and greater tuberosity, which made it difficult to reattach the torn supraspinatus tendon to the greater tuberosity [41]. Even if tendon-to-bone attachment can be achieved, high tension of the torn tendon is prone to retear [42, 43]. Since then, researchers have explored various biofabrication processes to bridge the large gaps between the torn tendons and humerus.

In 1984, Ozaki et al. [41] implanted Teflon felt consisting of PTFE to repair torn segments of the musculotendinous cuff in 25 patients. During surgery, the proximal edge of the torn cuff was cut back to expose healthy tendinous area, and the cuff edge was firmly sutured with the felt. The lateral site of the felt was anchored around the tuberosities. After an average follow-up of 2.1 years, 23 patients gained satisfactory function. In a similar approach, Yokoya et al. [44] compared the effects of PGA material and poly-L-lactate-epsilon-caprolactone (PLC) material in repairing massive rotator cuff tears in a rabbit model. The scanning electron microscopic (SEM) photograph of the PGA sheet showed a density of 200 g/m^2, and the SEM photograph of PLC showed a lactate-to-caprolactone ratio of 7 : 3. To determine the optimal scaffold material for the regeneration of structures within an articular joint, they firstly performed a pilot study by suturing the PTFE, PLC, or PGA sheets to the surface of the medial joint capsules. Twenty-four weeks later, the implants were harvested and stained with hematoxylin and eosin (H&E) for histological observation. The PTFE sheet, a nonabsorbable biomaterial, still existed and caused substantial foreign body reaction. There were no fibrous cells to invade the inside of the PTFE fibers. The PLC, a slowly absorbable material, persisted at 24 weeks postoperatively and resulted in scant fibrous tissue production. Besides, the numerous multinuclear cells around the remaining PLC fibers revealed the causation of a significant foreign body reaction. In contrast, the PGA has been completely absorbed and replaced with an abundant fibrous tissue layer. The fibroblasts were matured and arranged in the shape of a spindle with wave formation. Meanwhile, only a mild foreign body reaction was observed. After the pilot study, the PLC and PGA sheets were cut into $10 \times 5 \times 1$ mm as a scaffold for rotator cuff regeneration. To create a full-thickness defect of the rotator cuff, a segment of 5 mm in width and 10 mm in length was completely resected at the insertion of the infraspinatus into bone. Then, the PGA or PLC sheet graft was used to connect the large gap between infraspinatus tendon and humerus. The distal end of the scaffold was fixed into the bony. The proximal end of the scaffold was sutured to the infraspinatus tendon to bridge the large gap. Sixteen weeks later, some chondrocytes could be seen in the bony insertion site of PLC implants but not arranged along the long axis. In the PGA group, a well-arranged fibrocartilage was regenerated at the bony insertion site. For biomechanical testing, the PGA showed significantly higher values in the maximum failure load, tensile strength, and Young's modulus. In addition, biodegradable bovine two-layer type I collagen-based scaffold [45], chitosan-based hyaluronan hybrid fiber scaffold [46], and electrospun poly(D,L-lactide-*co*-glycolide) scaffolds [47] have also been demonstrated to be effective to bridge the large gap between tendon and bone.

In spite of aforementioned biofabrication, limited healing between bone and graft or between tendon and graft is still a problem [48]. Cong et al. [49] fabricated a triple-layer enthesis-mimicking (EM) scaffold consisting of an aligned PCL-Col I layer (aPCL-Col I, upper layer), a nonaligned PCL-Col I layer (nPCL-Col I, middle layer), and a nonaligned PCL-nanohydroxyapatite layer (nPCL-nHA, bottom layer). After the fabrication of the aligned PCL-Col I layer, it was covered with a plastic plate with only 3 mm length exposed to electrospinning the middle and the bottom layers. The same technique was applied to fabricate control scaffolds with the upper layer composed of aligned PCL-Col I and the middle and bottom layers composed of nonaligned PCL. They created 2 mm-long tendon defects close to the enthesis site and allowed it to be retracted medially to form a 5 mm gap. The triple-layer side was fixed on the footprint of the rotator cuff, and the monolayer side was sutured with the tendon stump. At eight weeks after surgery, the EM scaffolds promoted the regeneration of enthesis-like structure with transitional layers, which showed higher graft-bone healing score compared to the control group. For graft-tendon healing, the EM group showed more regularly arranged cells, denser tissue, and a significantly higher tendon maturing score than the control group.

19.4.3 Biofabrication Inhibits Fat Infiltration of Rotator Cuff Muscles

Besides the difficulty in reconstructing tendon-to-bone enthesis, fatty infiltration in torn rotator cuff muscles is one of the critical challenges for restoring the physiological and biomechanical functions of repaired rotator cuffs [22]. To address this problem, Huang et al. [50] incorporated lithium-containing mesoporous silica into electrospun poly(ester urethane) urea (PEUU) nanofibers (Li^+@MSNs/PEUU). The chronic rat rotator cuff tears model was established at six weeks after the detachment of supraspinatus tendon and then the sutured supraspinatus tendon was covered with Li^+@MSNs/PEUU (Figure 19.9). The released lithium$^+$ ions inhibited fatty infiltration by activating the canonical Wnt/β-catenin signaling pathway.

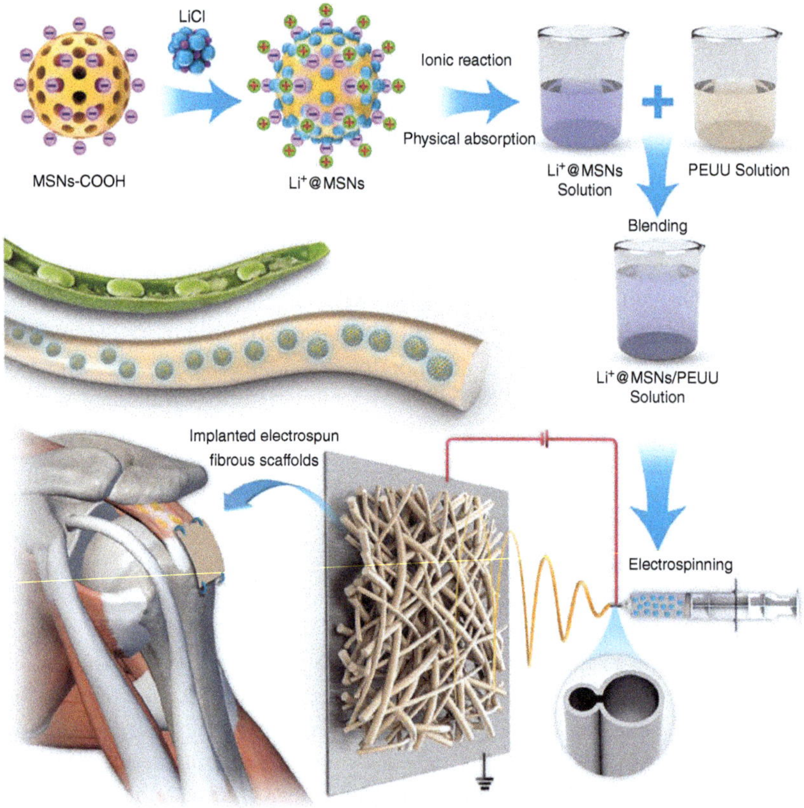

Figure 19.9 A schematic drawing of the lithium-containing mesoporous silica incorporated electrospun poly(ester urethane) urea (PEUU) nanofibers (Li^+@MSNs/PEUU). Source: Huang et al. [50] Reprinted with permission of Elsevier.

19.4.4 Biofabrication Increases Bone Mineral Density

The BMD of the proximal part of the humerus is highly associated with the pullout strength of suture anchors in rotator cuff repair [51]. Hence, an increase in BMD may reduce the possibility of anchor pullout. In Zhao's study, they used and compared the tricalcium phosphate CPS ceramic and hydroxyapatite (HA) ceramic in the repair of rat rotator cuff tears. The microcomputed tomography (μCT) results showed remarkable osteogenic activity and osteoconductivity of CPS ceramic and HA ceramic on promoting new bone formation and ingrowth, with CPS ceramic showing superior results than the HA ceramic. The superior effects of CPS ceramic ascribe to its better biocompatibility and biodegradability *in vivo*. Improved new bone formation in the CPS and HA group showed a significant increase in biomechanical properties with higher maximum ultimate load to failure, stiffness, and stress than the control group. In this study, they used 3-0 Ethibond suture (Ethicon) without the application of suture anchor. Hence, it failed to directly demonstrate the impact of increased new bone density on anchor pullout strength. However, it is well acknowledged that increased bone density can prevent anchor loosening and reduce failure of the rotator cuff repair [28, 52]. In a similar approach, Thangarajah et al. [53] reported that a combination of demineralized bone matrix (DBM) and rat bone marrow-derived mesenchymal stem cells (BMSCs) for rotator cuff repair can restore BMD of the enthesis to its preinjury levels. In addition, the engineered cell sheet that delivers ASCs also improved the bone volume/total volume values in rotator cuff repair surgery [39].

19.5 Biological Augmentation for Rotator Cuff Repair

In the past decades, biofabrication techniques have made tremendous advances for rotator cuff repair and regeneration. Generally, the scaffold is an engineered replacement of the native ECM that supports cell adhesion, provides mechanical strength, and guides tissue formation [54]. However, the scaffold alone is generally insufficient to reconstruct the native rotator cuff tendon-to-bone structure. Hence, a combination of scaffolds with biological augmentation factors (e.g. cell and growth factor) might better improve the healing process after surgical repair of rotator cuff tears [55].

19.5.1 Cell-Based Strategies

Cells are indispensable in tissue engineering repair of the rotator cuff. Various cell sources, cell types, and culture strategies have been explored to regenerate the complex cellular composition of native enthesis [13]. Four different cell culture approaches are most frequently used: mesenchymal stem cells (MSCs) alone, culture of different differentiated cells, and culture of MSCs together with differentiated cells (e.g. fibroblasts, chondrocytes, and osteoblasts) (Figure 19.10).

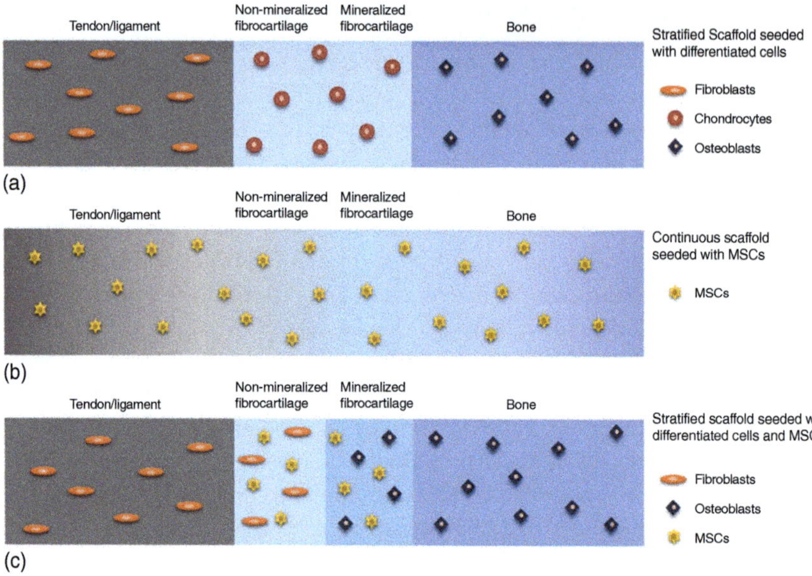

Figure 19.10 Cell and co-culture strategies for the generation of a tendon/ligament–bone construct. The three main cell culture strategies for enthesis tissue engineering are shown. (a) Co-culture strategy: co-culture of fibroblasts, chondrocytes, and osteoblasts in a stratified scaffold. This strategy follows the hypothesis that heterotypic interactions between chondrocytes and fibroblasts/osteoblasts lead to the differentiation of chondrocytes to the fibrochondrocytes and hypertrophic fibrochondrocytes that form the enthesis. (b) Multipotent cell strategy: culture of MSCs in a scaffold with a continuous gradation of structural properties. This strategy follows the hypothesis that a gradation in structural, biochemical, and mechanical cues along the scaffold can stimulate MSCs to differentiate into the different cell types that constitute the insertion site. (c) Combinatory strategy: co-culture of MSCs with fibroblasts and osteoblasts. This strategy follows the hypothesis that heterotypic interactions between MSCs and fibroblasts/osteoblasts can promote the differentiation of MSCs into fibrochondrocytes and hypertrophic fibrochondrocytes. Source: (a–c) Tellado et al. [13]/with permission of Elsevier.

Seeding MSCs into scaffolds is the easiest strategy because of their ease of expansion and multidifferentiation potential. Numerous studies have demonstrated the great capacity of MSCs to differentiate into different musculoskeletal tissue types, such as osteocytes, chondrocytes, tenocytes, and fibroblasts. Besides, bioactive molecules secreted from MSCs can provide a regenerative microenvironment [56]. In recent years, BMSCs [53, 57–59], ASCs [39, 60, 61], muscle-derived mesenchymal stem cells (M-MSCs) [62], and perivascular stem cells [63] have been demonstrated to be effective on promoting rotator cuff healing. Compared with BMSCs, the ASCs isolated from adipose tissue through lipoaspiration have higher cell yield and lower associated pain [64, 65]. Besides, *in vitro* culture of ASCs shows greater proliferation capacity and lower levels of cellular senescence compared to BMSCs [66, 67]. To stimulate MSCs differentiate into the various cell types of tendon-to-bone enthesis, researchers focused on fabricating scaffolds with spatiotemporal control of both biochemical and mechanical stimuli as well as biomaterial–cell interactions for generating graded yet continuous changes in cell phenotypes and ECM composition

[13, 68]. For instance, Min et al. [69] fabricated a PCL porous membrane loaded with reverse gradient of platelet-derived growth factor B (PDGF-BB) and BMP-2 to generate a gradient of molecular stimuli, which was used to regulate the seeded ASCs for specific differentiation into multiple cell phenotypes.

In addition to MSCs, differentiated cells are another cell source. For instance, Kwon et al. [70] injected allogenic dermal fibroblasts (ADFs) to repair torn supraspinatus tendons in a rabbit model of chronic rotator cuff tears. Twelve weeks after surgery, the biomechanical and histological properties were significantly elevated compared to the control group. One type differentiated cell approach was generally used for tendon regeneration. For the reconstruction of the complex structure of tendon-to-bone enthesis, researchers usually applied co-culture of different differentiated cells, which was based on the hypothesis that heterotypic interactions between different cell types can result in the regeneration of a graded tissue that is similar to native enthesis [68, 71]. As far as we know, no studies have reported the combination of different differentiated cells with scaffolds for rotator cuff repair. In spite of this, this approach was demonstrated feasible in the regeneration of anterior cruciate ligament-to-bone interface. For example, Spalazzi et al. [72] fabricated a triphasic scaffold system co-cultured with relevant cell types: phase A (soft phase) composed of a PLGA mesh and fibroblasts, phase B (interface phase) composed of PLGA microspheres and chondroblasts, and phase C (bone phase) composed of PLGA/bioglass composite microspheres and osteoblasts. Being eight weeks in a subcutaneous athymic rat model, the controlled phase-specific matrix heterogeneity was induced on the scaffold with distinct fibrocartilage-like and mineral tissue formed. Hence, further study needs to be performed to explore the seeding of different differentiated cells in scaffolds for rotator cuff repair. However, the isolation of mature chondrocytes, fibroblasts, and osteoblasts poses practical challenges for clinical use.

Co-culture MSCs together with differentiated cells is another strategy, which is based on the hypothesis that cellular cross-talk can promote the multidifferentiation of MSCs via intercellular signals and paracrine factors [71]. For instance, He et al. [73] established a trilineage co-culture system (osteoblasts–BMSCs–fibroblasts) on a hybrid silk scaffold. After *in vitro* culture, a gradual transition from the uncalcified to the calcified region was formed in the co-cultured BMSCs from the region that directly interacted with fibroblasts to the region that directly interacted with osteoblasts. However, the difficulty in acquiring enough differentiated cells has always restricted the application of this strategy.

19.5.2 Cell-Derived Hybrid Biological Factors

In recent decades, researchers found that the cell-derived bioactive factors also play a significant role in tissue regeneration. Platelet-rich plasma (PRP) [74, 75], platelet-rich fibrin [76, 77], mesenchymal stem cell-conditioned medium (MSC-CM) [78, 79], and mesenchymal stem cell-derived extracellular vesicles (MSC-EVs) [80–82] were reported to be effective in promoting rotator cuff healing by directly injecting or controlled releasing. The requirement for autologous platelets restricts the application of PRP and platelet-rich fibrin [83]. The MSC-EVs are generally classified into microparticles/microvesicles (100–1000 nm), exosomes (30–150 nm), and apoptotic bodies based on their biogenesis, release pathways, size, content, and function [84]. Fafián-Labora et al. [85] dissected the conditioned medium (CM) into soluble factors (SFs), exosomes, and microparticles. They found that the exosomes exert superior effects on ameliorating senescence-related tissue damage compared to the SFs, while the microparticles displayed no function. One possible reason ascribes to the SFs (e.g. growth factors, cytokines. and nucleic acid) in MSC-CM is prone to be rapidly degraded, therapeutic molecules (proteins and nucleic acid) encapsulated in exosomes are protected from degradation [86]. With similar functional properties to parental cells, MSC-Exos have attracted researchers' huge attention in recent years. As far as we know, Wang et al. [80, 81] first reported the effects of MSC-Exos in preventing muscle degeneration and enhancing tendon-to-bone healing in chronic rotator cuff tears. As a novel cell-free approach for rotator cuff healing, further study needs to be performed to characterize the underlying mechanism, which helps to promote its clinical transformation [87].

19.5.3 Growth Factors

Different from the uncertain types and doses of growth factors in PRP, platelet-rich fibrin, MSC secretome, and MSC-Exos, one or multiple growth factors produced by genetic engineering have been investigated in augmenting rotator cuff repair. According to the regenerated structure in rotator cuff repair, growth factors can be classified into three categories: bone regeneration (BMP-2 [88–90], BMP-7 [91, 92], and VEGF [93]), tendon regeneration (growth differentiation factor-7 (GDF-7) [94], transforming growth factor beta 3 (TGF-β3) [94] and basic fibroblast growth factor (bFGF) [95–98]), and enthesis regeneration (TGF-β1 [31, 99], TGF-β3 [100–102], and FGF2 [97, 103–105]). Previously, researchers usually applied one growth factor by directly injecting or incorporating into scaffolds. Lately, researchers have begun to explore the application of multiple growth factors with grading concentrations, which aimed to achieve gradual and sequential release in a spatiotemporally regulated manner so as to provide the seeded cell populations with differential stimuli. For instance, Min et al. [69] fabricated a PCL/Pluronic F127 membrane with reverse gradients of dual PDGF-BB and BMP-2 concentrations using a diffusion method (Figure 19.11). After *in vitro* cultured with ASCs for one and two weeks, the membrane sections with higher PDGF-BB and lower BMP-2 concentrations were better for promoting ASC tenogenesis, while the membrane sections with lower PDGF-BB and higher BMP-2 concentrations provided a better environment for osteogenesis. This study suggested that biofabrication with reverse gradients of different growth factors may be a promising approach for tendon-to-bone repair.

19.5.4 Drugs or Small Molecules

In addition to the aforementioned biological augmentation, hyaluronic acid [106], kartogenin [107, 108], fatty acid-binding protein 4 (FABP4) inhibitor [109], and zoledronic acid [110] have also shown effective in promoting rotator cuff repair.

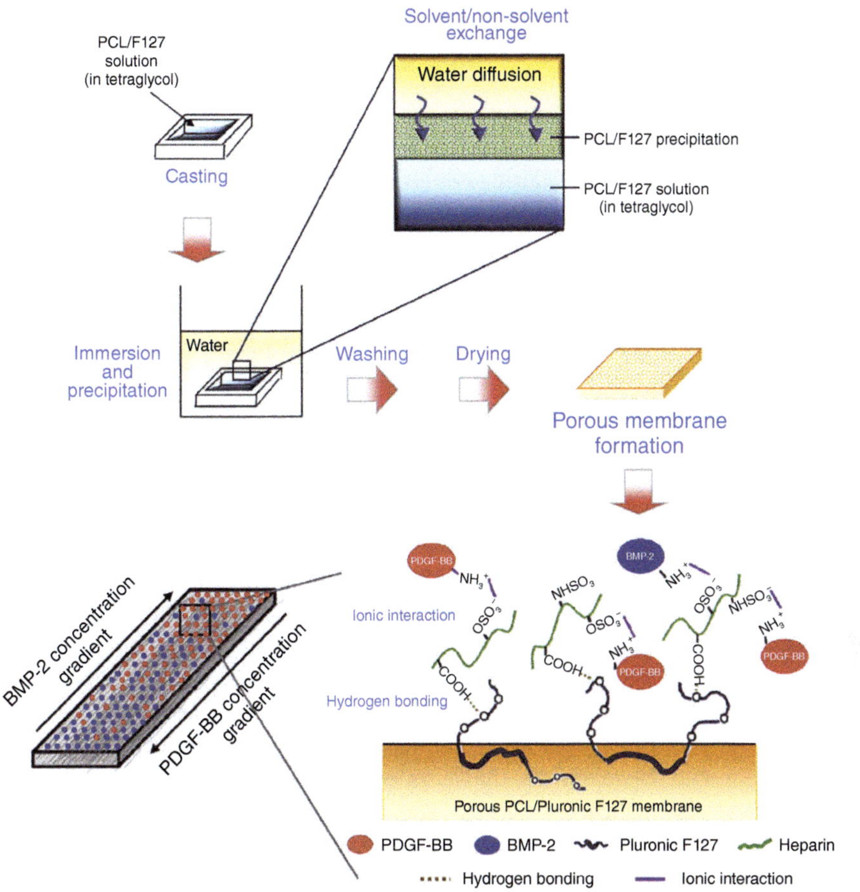

Figure 19.11 Schematic diagram illustrating the formation of PCL/Pluronic F127 membrane with reverse gradients of PDGF-BB and BMP-2 and the successive binding of heparin and growth factors onto the membrane surface. Source: Min et al. [69]/with permission of Elsevier.

Compared with cell-derived biological factors and traditional growth factors, these strategies are less prone to be degraded both *in vitro* and *in vivo*. Hence, their effects may last longer in the body.

To conclude, exploring the different biological factor combination approaches, extending biological factor release period, and achieving sequential delivery to precisely control the fate of cells are the major challenges for rotator cuff tissue engineering in the next years [13].

19.6 Future Perspectives of Biofabrication for Rotator Cuff Repair

Despite the utilization of various surgical techniques for rotator cuff repair, the repaired structure is still prone to fail to heal or retear. The high failure rate ascribes to the difficulty in the reconstruction of the transitional enthesis structure as well as repair of massive tears. Hence, there is an urgent need for biofabrication to provide acute stability in the initial period after repair surgery, serve as an interpositional device for irreparable tears, and promote the reconstruction of native tendon-to-bone enthesis structure [24]. Despite the fast progress of previous reported biofabrication in repairing rotator cuff tears, there is still no biofabrication which can be used for clinical transformation to regenerate the functional enthesis with native structure and mechanical properties.

As a significant component of biofabrication, scaffolds can not only mimic the rotator cuff structure but also provide support for cells and biological factors. To fabricate desirable scaffolds, the selection of polymer is of the first and most significance. As FDA-approved polymers, PLA, PCL, PGA, and PLGA have been frequently used in rotator cuff engineering because of their excellent biocompatibility, degradation properties, and tunable mechanical properties. The major challenge regarding the application of these polymers in rotator cuff repair is the acidic by-products (e.g. lactic acid and glycolic acid) that reduce the pH at the implantation site and induce inflammatory responses, which usually induces osteolysis and repair failure at the tendon-to-bone interface [111]. To prevent the reduction of pH, researchers have explored various additives (e.g. bioactive glass and hydroxyapatite) to alter the degradation kinetics. In addition, other materials (e.g. collagen, silk, hyaluronic acid, and tendon-derived matrix) have also been utilized alone or as additives to fabricate scaffolds. In the following years, the decision on selecting the polymers for rotator cuff scaffolds should be taken on the basis of their biocompatibility, degradation profile, and mechanical properties. For mechanical properties, the scaffolds generally work as mechanical augmentation to maintain acute stability in the initial period (12 weeks) after rotator cuff repair [112, 113]. To provide adequate mechanical support, previous studies reported that scaffolds should achieve the minimum mechanical properties with approximately the stiffness (~200 N/mm), modulus (~150 MPa), and ultimate load (~800 N) [114, 115]. However, the mechanical properties vary with the degradation of the scaffolds. Hence, it is a great challenge to balance the mechanical properties and degradation. Otherwise, it would be almost impossible to fabricate desirable scaffolds for clinical transplantation.

Besides scaffolds, the biological factors also play a significant role in rotator cuff repair. When biological factors are administrated to the tear site, keeping them in the tear region is the major challenge [116]. By incorporating the biological factors into the scaffolds, researchers need to achieve an accurate control of spatial and temporal growth factor release, so as to precisely regulate the fate of host cells or seeded cells to form native cell types of tendon-to-bone enthesis [13]. Meanwhile, the best cell source is also still unclear. The combination of different cell populations or a single type of multipotent cell have been explored. In our personal perspective, the MSCs have greater clinical transformation potential. In the following next years, further research should be done to determine which MSC source (e.g. BMSCs, ASCs, M-MSCs, etc.) is superior. Besides, researchers should devote their energies on exploring the combined molecular and mechanical stimulations to precisely drive MSC differentiate into the complex cellular composition at the enthesis site.

In addition to the effects to design a desirable fabrication technique for rotator cuff repair, basic research explores the deeper understanding of the molecular and mechanical mechanism in governing the development, homeostasis, and regeneration of native enthesis, which is indispensable to point out future directions for fabricating the rotator cuff biofabrication.

All in all, the reconstruction of native rotator cuff tendon-to-bone enthesis is a great challenge; however, the continuous progress of biofabrication will eventually overcome this challenge.

Acknowledgments

This work was supported by the National Natural Science Foundation of China (82072401 and 81871755), the Shanghai Outstanding Academic Leaders Plan of Shanghai Municipal Science and Technology Committee (19XD1402800), and the Major Project of Shanghai Municipal Education Commission's Scientific Research and Innovation Program (2021-01-07-00-02-E00082).

References

1 Saveh-Shemshaki, N., Nair, L.S., and Laurencin, C.T. (2019). *Acta Biomater.* 94: 64.
2 Millett, P.J., Hussain, Z.B., Fritz, E.M. et al. (2017). *Arthrosc. Tech.* 6 (4): e1075.
3 Nho, S.J., Yadav, H., Shindle, M.K., and Macgillivray, J.D. (2008). *Am. J. Sports Med.* 36 (5): 987.
4 Matava, M.J., Purcell, D.B., and Rudzki, J.R. (2005). *Am. J. Sports Med.* 33 (9): 1405.
5 Tsiridis, E., Velonis, S., Limb, D., and Giannoudis, P.V. (2008). *Connect. Tissue Res.* 49 (6): 455.
6 Sher, J.S., Uribe, J.W., Posada, A. et al. (1995). *J. Bone Joint Surg. Am.* 77 (1): 10.
7 Oliva, F., Osti, L., Padulo, J., and Maffulli, N. (2014). *Muscles Ligaments Tendons J.* 4 (3): 309.
8 Abraham, A.C., Shah, S.A., Golman, M. et al. (2019). *Sci. Transl. Med.* 11 (481): eaav4319.
9 Kovacevic, D. and Rodeo, S.A. (2008). *Clin. Orthop. Relat. Res.* 466 (3): 622.
10 Uhthoff, H.K., Trudel, G., and Himori, K. (2003). *J. Orthop. Sci.* 8 (3): 449.
11 Chung, S.W., Kim, J.Y., Kim, M.H. et al. (2013). *Am. J. Sports Med.* 41 (7): 1674.
12 Goldenberg, B.T., Lacheta, L., Dekker, T.J. et al. (2020). *Orthop. Res. Rev.* 12: 151.

13 Tellado, S.F., Balmayor, E.R., and Van Griensven, M. (2015). *Adv. Drug Delivery Rev.* 94: 126.

14 Shaw, H.M. and Benjamin, M. (2007). *Scand. J. Med. Sci. Sports* 17 (4): 303.

15 Yang, P.J. and Temenoff, J.S. (2009). *Tissue Eng. Part B* 15 (2): 127.

16 Rossetti, L., Kuntz, L.A., Kunold, E. et al. (2017). *Nat. Mater.* 16 (6): 664.

17 Hu, Y., Birman, V., Deymier-Black, A. et al. (2015). *Biophys. J.* 108 (2): 431.

18 Balijepalli, R.G., Begley, M.R., Fleck, N.A. et al. (2016). *Int. J. Solids Struct.* 85, 86: 160.

19 Cofield, R.H. (1982). *Surg. Gynecol. Obstet.* 154 (5): 667.

20 Kang, J.R. and Gupta, R. (2012). *J. Shoulder Elbow Surg.* 21 (2): 175.

21 Liu, X., Ning, A.Y., Chang, N.C. et al. (2016). *Muscles Ligaments Tendons J.* 6 (1): 6.

22 Gibbons, M.C., Singh, A., Anakwenze, O. et al. (2017). *J. Bone Joint Surg. Am.* 99 (3): 190.

23 Itoigawa, Y., Kishimoto, K.N., Sano, H. et al. (2011). *J. Orthop. Res.* 29 (6): 861.

24 Chainani, A. and Little, D. (2016). *Tech. Orthop.* 31 (2): 91.

25 Su, W., Guo, J., Xu, J. et al. (2021). *Chem. Eng. J.* 404: 126473.

26 Entezari, V. and Lazarus, M. (2019). Surgical considerations in managing osteoporosis, osteopenia, and vitamin D deficiency during arthroscopic rotator cuff repair. *Orthop. Clin. North Am.* 50 (2): 233.

27 Tingart, M.J., Apreleva, M., Zurakowski, D., and Warner, J.J. (2003). *J. Bone Joint Surg. Am.* 85 (11): 2190.

28 Chung, S.W., Oh, J.H., Gong, H.S. et al. (2011). *Am. J. Sports Med.* 39 (10): 2099.

29 de Isla, N., Huselstein, C., Jessel, N. et al. (2010). *Bio-Med Mater. Eng.* 20 (3): 127.

30 Zhao, S., Su, W., Shah, V. et al. (2017). *Colloids Surf., B* 157: 407.

31 Yoon, J.P., Lee, C.-H., Jung, J.W. et al. (2018). *Am. J. Sports Med.* 46 (6): 1441.

32 Li, X., Cheng, R., Sun, Z. et al. (2017). *Acta Biomater.* 61: 204.

33 Soo Kim, B., Ji Kim, E., Suk Choi, J. et al. (2014). *J. Biomed. Mater. Res. Part A* 102 (11): 4044.

34 Santoni, B.G., McGilvray, K.C., Lyons, A.S. et al. (2010). *Am. J. Sports Med.* 38 (4): 679.

35 Bokor, D.J., Sonnabend, D., Deady, L. et al. (2015). *Muscles Ligaments Tendons J.* 5 (3): 144.

36 Bokor, D.J., Sonnabend, D., Deady, L. et al. (2016). *Muscles Ligaments Tendons J.* 6 (1): 16.

37 Arnoczky, S.P., Bishai, S.K., Schofield, B. et al. (2017). *Arthroscopy* 33 (2): 278.

38 Barber, F.A., Burns, J.P., Deutsch, A. et al. (2012). *Arthroscopy* 28 (1): 8.

39 Shin, M.J., Shim, I.K., Kim, D.M. et al. (2020). *Am. J. Sports Med.* 48 (13): 3347.

40 Harada, Y., Mifune, Y., Inui, A. et al. (2017). *J. Orthop. Res.* 35 (2): 289.

41 Ozaki, J., Fujimoto, S., Masuhara, K. et al. (1986). *Clin. Orthop. Relat. Res.* 202: 173.

42 Flury, M. (2016). *Orthopade* 45 (2): 136.

43 Petri, M., Greenspoon, J.A., Bhatia, S., and Millett, P.J. (2015). *Arthrosc. Tech.* 4 (5): e487.

44 Yokoya, S., Mochizuki, Y., Nagata, Y. et al. (2008). *Am. J. Sports Med.* 36 (7): 1298.

45 Roßbach, B.P., Gülecyüz, M.F., Kempfert, L. et al. (2020). *Am. J. Sports Med.* 48 (2): 450.

46 Funakoshi, T., Majima, T., Iwasaki, N. et al. (2015). *Am. J. Sports Med.* 33 (8): 1193.

47 Inui, A., Kokubu, T., Mifune, Y. et al. (2012). *Arthroscopy* 28 (12): 1790.

48 D'Ambrosi, R., Ragone, V., Comaschi, G. et al. (2019). *Cell Tissue Bank.* 20 (1): 1.

49 Cong, S., Sun, Y., Lin, J. et al. (2020). *Am. J. Sports Med.* 48 (8): 1826.

50 Huang, K., Su, W., Zhang, X. et al. (2020). *Chem. Eng. J.* 392: 123671.

51 Kimura, A., Aoki, M., Fukushima, S. et al. (2003). *J. Bone Joint Surg. Br.* 85 (2): 282.

52 Kirchhoff, C., Braunstein, V., Milz, S. et al. (2010). *Am. J. Sports Med.* 38 (3): 564.

53 Thangarajah, T., Sanghani-Kerai, A., Henshaw, F. et al. (2018). *Am. J. Sports Med.* 46 (1): 98.

54 Sahoo, S., Teh, T., He, P. et al. (2011). *Ann. Acad. Med. Singapore* 40 (5): 245.

55 Patel, S., Gualtieri, A.P., Lu, H.H., and Levine, W.N. (2016). *Ann. N. Y. Acad. Sci.* 1383 (1): 97.

56 Rothrauff, B.B. and Tuan, R.S. (2014). *Organogenesis* 10 (1): 13.

57 Gulotta, L.V., Kovacevic, D., Ehteshami, J.R. et al. (2009). *Am. J. Sports Med.* 37 (11): 2126.

58 Yokoya, S., Mochizuki, Y., Natsu, K. et al. (2012). *Am. J. Sports Med.* 40 (6): 1259.

59 Gulotta, L.V., Kovacevic, D., Packer, J.D. et al. (2011). *Am. J. Sports Med.* 39 (6): 1282.

60 Kim, Y.S., Sung, C.H., Chung, S.H. et al. (2017). *Am. J. Sports Med.* 45 (9): 2010.

61 Kaizawa, Y., Franklin, A., Leyden, J. et al. (2019). *J. Orthop. Res.* 37 (4): 877.

62 Pelinkovic, D., Lee, J.Y., Engelhardt, M. et al. (2003). *Tissue Eng.* 9 (1): 143.

63 Eliasberg, C.D., Dar, A., Jensen, A.R. et al. (2017). *J. Bone Joint Surg. Am.* 99 (4): 331.

64 Zhu, Y., Liu, T., Song, K. et al. (2008). *Cell Biochem. Funct.* 26 (6): 664.

65 Liao, H.-T. and Chen, C.-T. (2014). *World J. Stem Cells* 6 (3): 288.

66 Izadpanah, R., Trygg, C., Patel, B. et al. (2006). *J. Cell. Biochem.* 99 (5): 1285.

67 Kern, S., Eichler, H., Stoeve, J. et al. (2006). *Stem Cells* 24 (5): 1294.

68 Smith, L., Xia, Y., Galatz, L.M. et al. (2012). *Connect. Tissue Res.* 53 (2): 95.

69 Min, H.K., Oh, S.H., Lee, J.M. et al. (2014). *Acta Biomater.* 10 (3): 1272.

70 Kwon, J., Kim, Y.H., Rhee, S.-M. et al. (2018). *Am. J. Sports Med.* 46 (8): 1901.

71 Sahoo, S., Toh, S.L., and Goh, J.C. (2010). *Biomaterials* 31 (11): 2990.

72 Spalazzi, J.P., Dagher, E., Doty, S.B. et al. (2008). *J. Biomed. Mater. Res. Part A* 86 (1): 1.

73 He, P., Ng, K.S., Toh, S.L., and Goh, J.C. (2012). *Biomacromolecules* 13 (9): 2692.

74 Jo, C.H., Lee, S.Y., Yoon, K.S. et al. (2018). *Am. J. Sports Med.* 46 (13): 3142.

75 Han, L., Fang, W.L., Jin, B. et al. (2019). *Eur. Rev. Med. Pharmacol. Sci.* 23 (20): 9075.

76 Hurley, E.T., Lim Fat, D., Moran, C.J., and Mullett, H. (2019). *Am. J. Sports Med.* 47 (3): 753.

77 Hasan, S., Weinberg, M., Khatib, O. et al. (2016). *Int. J. Sports Med.* 37 (1): 36–42.

78 Sevivas, N., Teixeira, F.G., Portugal, R. et al. (2017). *Am. J. Sports Med.* 46 (2): 449.

79 Sevivas, N., Teixeira, F.G., Portugal, R. et al. (2016). *Am. J. Sports Med.* 45 (1): 179.

80 Wang, C., Song, W., Chen, B. et al. (2019). *Am. J. Sports Med.* 47 (13): 3247.

81 Wang, C., Hu, Q., Song, W. et al. (2020). *Am. J. Sports Med.* 48 (6): 1456.

82 Huang, Y., He, B., Wang, L. et al. (2020). *Stem Cell Res. Ther.* 11 (1): 496.

83 Tao, S.C., Yuan, T., Rui, B.Y. et al. (2017). *Theranostics* 7 (3): 733.

84 Doyle, L.M. and Wang, M.Z. (2019). *Cells* 8 (7): 727.

85 Fafián-Labora, J.A., Rodríguez-Navarro, J.A., and O'Loghlen, A. (2020). *Cell Metab.* 32 (1): 71.

86 Basu, J. and Ludlow, J.W. (2016). *Expert Opin. Biol. Ther.* 16 (4): 489.

87 Ludwig, N., Whiteside, T.L., and Reichert, T.E. (2019). *Int. J. Mol. Sci.* 20 (19): 4684.

88 Lee, K.W., Lee, J.S., Kim, Y.S. et al. (2017). *J. Biomed. Mater. Res. Part B* 105 (7): 1840.

89 Chen, C.H., Chang, C.H., Wang, K.C. et al. (2011). *Knee Surg. Sports Traumatol. Arthrosc.* 19 (9): 1597.

90 Hirakawa, Y., Manaka, T., Orita, K. et al. (2018). *J. Shoulder Elbow Surg.* 27 (5): 894.

91 Kabuto, Y., Morihara, T., Sukenari, T. et al. (2015). *Tissue Eng. Part A* 21 (13, 14): 2025.

92 Klatte-Schulz, F., Pauly, S., Scheibel, M. et al. (2013). *PLoS One* 8 (6): e67209.

93 Huang, Y., Pan, M., Shu, H. et al. (2020). *J. Cell. Biochem.* 121 (3): 2343.

94 Han, B., Jones, I.A., Yang, Z. et al. (2020). *Arthroscopy* 36 (3): 629.

95 Zhao, S., Zhao, J., Dong, S. et al. (2014). *Int. J. Nanomed.* 9: 2373.

96 Peterson, D.R., Ohashi, K.L., Aberman, H.M. et al. (2015). *J. Shoulder Elbow Surg.* 24 (11): 1764.

97 Yonemitsu, R., Tokunaga, T., Shukunami, C. et al. (2019). *Am. J. Sports Med.* 47 (7): 1701.

98 Tokunaga, T., Shukunami, C., Okamoto, N. et al. (2015). *Am. J. Sports Med.* 43 (10): 2411.

99 Arimura, H., Shukunami, C., Tokunaga, T. et al. (2017). *Am. J. Sports Med.* 45 (10): 2394.

100 Reifenrath, J., Wellmann, M., Kempfert, M. et al. (2020). *Int. J. Mol. Sci.* 21 (3): 1046.

101 Kovacevic, D., Fox, A.J., Bedi, A. et al. (2011). *Am. J. Sports Med.* 39 (4): 811.
102 Manning, C.N., Kim, H.M., Sakiyama-Elbert, S. et al. (2011). *J. Orthop. Res.* 29 (7): 1099.
103 Tokunaga, T., Karasugi, T., Arimura, H. et al. (2017). *J. Shoulder Elbow Surg.* 26 (10): 1708.
104 Ide, J., Kikukawa, K., Hirose, J. et al. (2009). *J. Shoulder Elbow Surg.* 18 (3): 391.
105 Kataoka, T., Mifune, Y., Inui, A. et al. (2019). *J. Shoulder Elbow Surg.* 28 (11): e389.
106 Honda, H., Gotoh, M., Kanazawa, T. et al. (2017). *Am. J. Sports Med.* 45 (14): 3322.
107 Wang, D., Tan, H., Lebaschi, A.H. et al. (2018). *Arthroscopy* 34 (9): 2579.
108 Zhu, Q., Ma, Z., Li, H. et al. (2019). *RSC Adv.* 9 (27): 15582.
109 Lee, Y.-S., Kim, J.-Y., Kim, K.I. et al. (2019). *Am. J. Sports Med.* 47 (13): 3089.
110 Schanda, J.E., Keibl, C., Heimel, P. et al. (2020). *Am. J. Sports Med.* 48 (9): 2151.
111 Lipner, J., Shen, H., Cavinatto, L. et al. (2015). *Tissue Eng. Part A* 21 (21, 22): 2766.
112 Kim, J.H., Hong, I.T., Ryu, K.J. et al. (2014). *Am. J. Sports Med.* 42 (11): 2606.
113 McCarron, J.A., Derwin, K.A., Bey, M.J. et al. (2013). *Am. J. Sports Med.* 41 (1): 134.
114 Ricchetti, E.T., Aurora, A., Iannotti, J.P., and Derwin, K.A. (2012). *J. Shoulder Elbow Surg.* 21 (2): 251.
115 Lake, S.P., Miller, K.S., Elliott, D.M., and Soslowsky, L.J. (2009). *J. Orthop. Res.* 27 (12): 1596.
116 Rak Kwon, D., Jung, S., Jang, J. et al. (2020). *Am. J. Sports Med.* 48 (4): 947.

20

Tendon Repair

Xiao Wang

Johns Hopkins University, Department of Orthopaedic Surgery, Ross 229, 720 Rutland Ave, Baltimore, MD 21205, USA

20.1 Introduction

One of the important features in the disease area for clinical doctors and biomedical scientists is to understand and appreciate the unbelievable explosion in work that is likely to be coming our way. In 2016, the Global Burden of Disease, Injury, and Risk Factor study looked not at that disease as a cause of death or premature death but disease as a cause of disability and describing not just in terms of the magnitude and severity of the disability but the duration of the disability through one's life and creating ranking of what causes people live with disease for years [1]. The musculoskeletal causes of years lived with disease are prominent, occupying 4 of the leading 15 causes of disability in both 2005 and 2015 [1]. This is expected to get worse as the society ages and more people participate in sports. There is a huge need to address this problem if we can push out healthy years for longer to minimize the impacts.

Tendon is a vital component of the musculoskeletal system. It connects muscles and bones for force transition and motion production. It also acts as a buffering cushion to prevent stress concentrations caused by direct muscle-to-bone connections. Moreover, tendon could store energy and buffer energy dissipation by muscle, resulting in enhanced muscle performance for many activities [2, 3]. Tendon injuries can affect people of all ages and impair people's activity in their work environment or sports activities. Injury and degeneration of tendon can be extremely debilitating and can result in considerable pain and disability. It is the most common reason for consultation with a musculoskeletal specialist [4–7], accounting for approximately 30% of general practice consultations for musculoskeletal pain [5]. The incidence of tendon injuries and the costs to treat them have increased in the past three decades, contributing to the rapidly increasing economic impact of musculoskeletal conditions [5, 8]. The pathologies of tendon range from acute to chronic injuries with partial or complete tendon rupture [9, 10]. Tendinopathy, the chronic tendon injury, is the most common overuse/overload tendon injury, causing pain

Biofabrication for Orthopedics: Methods, Techniques, and Applications, First Edition.
Edited by Wenguo Cui, Xin Zhao, and Shen Liu.

and impairment of function [4–6, 11]. However, the knowledge on the pathogenesis of tendinopathy is limited. Moreover, the role of inflammation in tendinopathy has never been confirmed [12]. Although it has been reported that tendinopathy could result in tendon rupture, the causative relationship between tendinopathy and tendon rupture is unknown. Tendon undergoes a healing process following injuries, comprising inflammation and formation and remodeling of the extracellular matrix (ECM). Again, the cellular and molecular mechanisms of tendon healing largely remain unknown.

20.2 Tendon Biology

Tendons connect muscles to bones, transmitting muscle-generated forces to the skeleton to maintain posture and make locomotion possible (Figure 20.1). Ligaments and fasciae have a similar structure to tendons and link bones with bones and connect muscles to muscles, respectively. A structurally and functionally mature tendon is formed through multiple independent steps during development [9]. First, fibril intermediates are assembled in a series of extracellular compartments. Second, the immature fibril intermediates are incorporated into linear fibers. Third, the longitudinal intermediates fuse laterally to form mature fibrils. This process is regulated by tenocytes, which play a major role in the interactions of fibrils with extracellular macromolecules including collagens I, III, V, VII, and IX; the proteoglycans biglycan, lumican, fibromodulin, and decorin; and the transcription and growth factors such as scleraxis [13, 14] and transforming growth factor-β (TGF-β) [12]. Collagens are the primary content of the ECM of a mature tendon tissue. The most abundant form is collagen I, accounting for approximately 90–95% of the tendon collagen [9, 15]. Collagen III accounts for 0–10% of the tendon collagen [16]. Small amounts of proteoglycans and glycoproteins are also present, while collagen II is rarely seen [16].

The collagen fiber bundles in mature tendons are orientated in a hierarchically structured manner (Figure 20.2). The molecular structure of a tendon is composed of repeating amino acid sequences Gly-X-Y, where the glycine residue enables the α-chains to coil around each other. Proline and 4-hydroxyproline residues appear

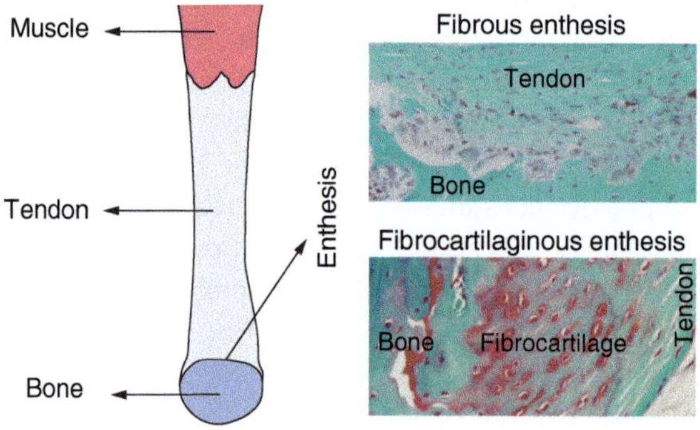

Figure 20.1 Tendons link muscles to bones and are fixed to bones by the enthesis, the fibrous enthesis, and the fibrocartilaginous enthesis.

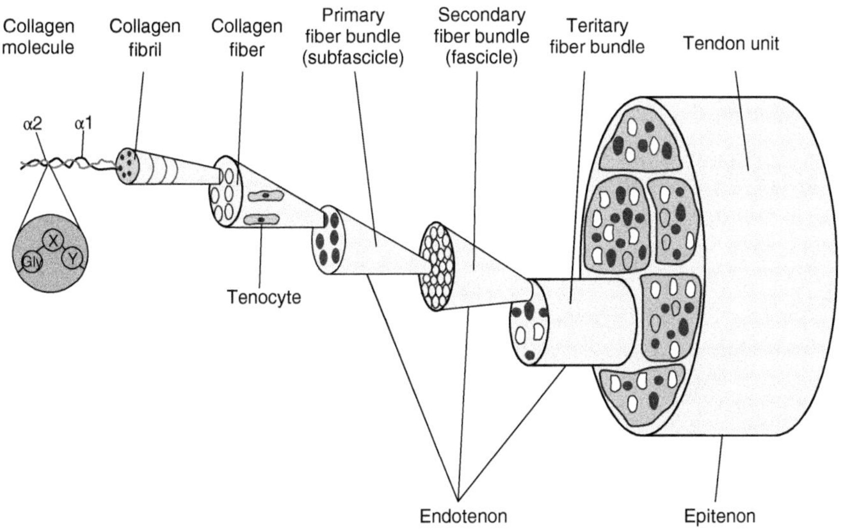

Figure 20.2 The tendon has a multi-unit hierarchical structure composed of collagen molecules, fibrils, fiber bundles, fascicles, and tendon units that run parallel to the tendon's long axis.

frequently at the X- and Y-positions, respectively, and promote the formation of intermolecular cross-links. Triple-helical type I collagen molecules assemble into fibrils, varying from 10 to 500 nm, depending on species, age, and sample location. The fibril is the smallest tendon structural unit. Young animals have uniformly small fibrils, whereas mature animals typically have both small and large fibrils. The characteristic specific D-periodic cross-striated fibril is typically 64–68 nm. The collagen fibrils are parallel to the bone–muscle axis. They are the fundamental unit of the tendon and combine to form fibers (50–100 μm). Tendon fascicles (20–200 μm), tube-like structures oriented along the long axis of the tendon, are bundles of collagen fibers with tenocytes residing in. Each fascicle is covered by a connective tissue called the endotenon [17, 18]. Fascicles are bound together forming a tendon unit (100–500 μm), which is encapsulated by epitenon [17, 18]. Both endotenon and epitenon contain blood vessels, lymphatics, and nerves [17, 19].

Only 5% of tendon content is cells, where tenocytes account for 95% for the synthesis of type I collagen [16]. Although there is no unique marker that selectively distinguishes tenocytes at all stages of development, a number of molecules have been considered as markers. Scleraxis (Scx) has been used to identify tendon or ligament cells at all stages of their development (even though Scx is also necessary for the development of other mesodermal tissues). A second marker candidate is tenomodulin (Tnmd), a molecule whose expression is induced by Scx. It regulates tenocyte proliferation and plays a role in the maturation of collagen fibrils. There is also tenascin-C, which is expressed by tenocytes in response to mechanical stress, but again is not specific for tendons alone, for it is also present in bone, smooth muscle, and healing fibroblasts.

Tendon stem/progenitor cells (TSPCs) have been isolated from the tendon proper and tendon sheaths of human and rodent tendons [20–22]. TSPCs are originally from mesoderm and therefore share many same markers with bone marrow-derived mesenchymal stem cells (MSCs). They exhibit classical characteristics, having typical MSC surface antigens, self-renewability, clonogenicity, and three-lineage differentiation capacity. In contrast to MSCs, TSPCs also express tendon-related genes such as Scx, Tnmd, cartilage oligomeric protein (Comp), and tenascin-C and are able to form tendon-like tissues when implanted *in vivo*.

One unique anatomic structure of tendon is the enthesis, where tendon is fixed to bone (Figure 20.1). There are two types of enthesis: the fibrous enthesis and the fibrocartilaginous enthesis. At the fibrous enthesis, the tendon attaches to the bone directly without any transitional zone. At the fibrocartilaginous enthesis, there is a transitional zone of calcified fibrocartilage and uncalcified fibrocartilage, which distributes mechanical loads. The enthesis can bear tensile, compressive, and shear forces. The tendon enthesis initially forms as an attachment unit with cartilage in utero and then mineralizes postnatally. The enthesis progenitors are distinct from chondrocytes and tenocyte progenitor cells and positive for both Sox9, a transcription factor necessary for the differentiation of MSCs into chondrocytes, and Scx, a transcription factor specific for tenocytes and tendon stem/progenitor cells and essential for tendon development. Several factors, including hedgehog (Hh), Runx1, Klf2/Klf4, etc., drive developmental patterning that is critical for enthesis development.

A unique parameter in tendon biology is the mechanical properties, which are nonlinear, viscoelastic, anisotropic, and heterogeneous, depending on the strain, rate of strain, loading direction, and location. The mechanical stimulations can convert to biomedical signals, including activation of the integrins and TGF-β receptors and release of active TGF-β, which induces downstream signaling and leads to tenogenic differentiation [23–25]. The mechanical stimuli were reported to open the calcium channel and increase the inflow of Ca^{2+}, which regulates tenogenic differentiation [26]. Mechanical stimulation during development is critical to successful limb and musculoskeletal tissue formation during embryogenesis. The absence of mechanical force due to the loss of muscle impairs tendon formation during tendon development [14, 27–29]. The mechanical stimulation is also vital for adult tendon homeostasis [29, 30]. The loss of mechanical loading results in decreased tendon size and reduction in the expression of tendon markers, including Scx, and changes the ECM composition in mice [30, 31]. By contrast, an increase in mechanical loading upregulates the synthesis of collagen and other ECM components and the expression levels of type I collagen, Scx and Tnmd [32–34].

20.3 Tendon Injury

Tendon pathologies include acute and chronic injuries with partial or complete tendon rupture [35]. Acute injuries typically occur after an isolated, overloading event. Chronic tendon injury, or tendinopathy, is the most common overuse tendon injury, resulting from over time through repetitive, excessive loading events [36, 37].

Tendinopathy is a condition characterized by pain and impaired performance, commonly involving patellar tendon (jumper's knee), Achilles tendon, and the origin of the extensor carpi radialis brevis (tennis elbow), accounting for 7% the musculoskeletal clinic visits in the United States [38]. It is characterized by the loss of normal tendon microstructure, morphological changes and apoptosis of tenocytes, disrupted fibril distribution, neovascularization, and innervation [39, 40]. These cellular and molecular modifications further changed the mechanical properties of tendon. However, the pathogenesis of tendinopathy is not fully understood, resulting in no disease-modified treatment. Acute tendon injury includes partial or complete rupture, leading to tendon discontinuity and loss of motion. Additionally, mechanical imperfection may lead to the histopathological changes of tendon, including the production of metalloproteinases, growth factors, prostaglandins, and aberrant differentiation of TSPCs into non-tendon cells [33, 41, 42].

20.4 Tendon Healing

Tendons are able to heal naturally but at a much slower rate than other connective tissues. The tensile strength of a ruptured tendon remains inferior to that of an uninjured tissue because its pre-injury conditions are not fully restored. Tendon healing can take months, and the repaired tendon can regain only 60% at best of its original function [43]. The poor mechanical properties after natural tendon healing may result from predominantly small fibrils relative to the normal distribution of both large and small fibrils observed in normal adult human tendons [44, 45].

The natural healing process following tendon injury can occur intrinsically by the proliferation of epitenon and endotenon tenocytes, or extrinsically, via invasion of cells from the surrounding sheath and synovium [46]. Scar tissues are formed through three overlapping phases: inflammation, modeling (matrix production), and remodeling [46, 47]. The earliest inflammatory phase is characterized by the development of a fibrin clot to stabilize the site, hemostasis, migration of erythrocytes and inflammatory cells, particularly neutrophils, and subsequent angiogenesis. The modeling stage commences when tenocytes migrate to the wound and type III collagen synthesis begins. The final remodeling stage is initiated approximately four to six weeks after injury. The healing tissue is resized and reshaped. The tenocyte metabolism, density, and vascularity decline at this stage, and scar tissues are formed. A limitation of tendon healing is peritendinous adhesion formation (PAF), posing a major clinical problem [48]. Following the injury of the synovial sheath at the time of the injury or surgery, predominately exogenous cells from the surrounding tissue invade the repair site over endogenous tenocytes, allowing the granulation tissue to attach to the repair site and resulting in PAF.

20.5 Tendon Repair

After tendon injury, operative intervention is often necessary to restore function. Previous work has shown inferior healing when operative repair is delayed following injury [49, 50]. While current therapies produce functional outcomes in the short term, long-term repair outcome varies with respect to the type of injury, injury location, and severity [47, 51].

20.5.1 Physical Modalities

The application of physical modalities is a routine in the daily practice in the management of tendon disorders. Almost all of the physical modalities are used to induce either vasoconstriction or vasodilation. However, there is a lack of strong supporting evidence for the current treatment. Most of the evidence is still pre-clinical and, at times, anecdotally successful.

Heat modalities transmit heat in three manners: conduction (i.e. hot packs), convection (i.e. air, water, or whirlpool), and radiation (i.e. infrared heat). It creates several physiologic changes via vasodilation, resulting in an increased level of nutrients and oxygen at the cellular level and removal of waste products. It also gives an analgesic effect, theoretically acting on the free nerve endings and peripheral nerve fibers to increase the pain threshold. It may also decrease the muscle spasm and decrease the metabolites that stimulate the nerve fibers.

Cold modalities, also known as cryotherapy, are used in the treatment of tendinopathy. It is widely used in the immediate care and rehabilitation of soft tissue injuries. The localized use of cold controls edema and facilitates therapeutic exercise. More than 6000 military patients who suffered soft tissue injuries received cold modalities, and more than 80% of the patients were able to return to active duty within three treatments [52].

Therapeutic ultrasound, known as sonotherapy, penetrates human tissues by ultrasound energy, which gives the thermal effect. One mHz ultrasound will easily heat 5 cm tissues, while a 90 kHz ultrasound will heat tissues at a depth of 10 cm. The temperature of muscle rises 4 °C with a 1 MHz continuous ultrasound at 1.5 W/cm^2 [53]. Fabrizio et al. reported that ultrasound (1.0 MHz, 1.0 and 1.5 W/cm^2) increased the blood flow velocity of triceps surae [54]. However, Robinson and Buono [55] applied continuous ultrasound to the volar forearm (1.5 W/cm^2, five minutes) and accessed the skeletal muscle blood flow. They reported no differences for up to 30 minutes post-treatment [55].

Extracorporeal shock wave therapy (ESWT) was firstly applied in patients to disintegrate kidney stones [56], offering the possibility to disintegrate pathological neocalcifications in musculoskeletal disorders. Although no study exists reporting on calcification disappearance [57], the success rate of the analgesic effect of ESWT is high, ranging from 60% to 80% in epicondylitis, plantar fasciitis, cuff tendinitis, trocanteritis, Achilles tendinitis, or jumper's knee in humans [58, 59]. The experimental findings confirm that ESWT resulted in neovascularization and an increase in the angiogenesis-related markers such as nitric oxide synthase and vascular endothelial growth factor (VEGF) [60], elevated levels of TGF-β1 in the early stage and persistently elevated levels of insulin-like growth factor-1 (IGF-1) [61], and increased levels of glycosaminoglycan (GAG) and protein syntheses [62]. However, caution should be exercised as ESWT has been found to depend on the flux density of energy: low energy level with low impulses showed positive stimulatory effects, while high energy level with high impulses showed inhibitory effects. Additionally, dose-dependent tendon damage, including fibrinoid necrosis, fibrosis, and inflammation, has been reported [63].

Pulsed magnetic fields (PMF) have been shown to enhance the healing of tendinopathy in a double-blind controlled clinical trial [64]. PMF improved collagen fiber alignment and increased force to breakage after Achilles tendinopathy in rats [65, 66]. Controversially, Huegel et al. found that pulsed electromagnetic field (PEMF) showed no clear beneficial effects in completely or partially tenotomized Achilles healing [67].

Direct current electrical stimulation is assumed to augment regeneration of various tissues. In an *in vitro* rabbit flexor tendon culture, direct electrical current increased proline and hydroxyproline activities [68], type I collagen production, and decreased adhesion formation [69].

Low-energy laser phototherapy has been shown to be effective as analgesics [70] and to accelerate the healing of injured tissues in several clinical studies [71–75]. It increased stimulation of collagen production in healing rabbit Achilles tendons [76] and reduced postoperative edema but provided no improvement in pain relief, grip strength, or functional results in a placebo-controlled, double-blind, prospective study of patients with digital flexor tendon repairs [77].

Radiofrequency coblation is an application of bipolar radiofrequency energy to break up the scar tissue of tendon pathologies, especially in a painfully inflamed Achilles tendon. This procedure is simple and minimally invasive and would be an alternative to an open treatment of tendinopathy. It has been reported that radiofrequency coblation stimulated angiogenesis in rabbit Achilles tendons [78] and rapidly releases pain in patients with tendinopathy of the Achilles tendon, patellar tendon, and common extensor origin [78], resulting in complete or near-complete resolution of tendinopathy in 50% of patients [78].

20.5.2 Mechanical Loading and Mobilization

Stretching and strengthening, especially eccentric strengthening, are a common component for tendinopathy therapy programs [79–81]. It has been shown that eccentric strengthening program was effective for treating patellar and Achilles tendinopathy in recreational athletes [80, 82] and more efficiently releases pain than concentric training program with thinning and normalization of the tendon structure both on ultrasound and on magnetic resonance imaging (MRI) [83, 84]. In animal experiments, tendons with training showed improved tensile strength, elastic stiffness, weight, and cross-sectional area relative to immobilized tendons [85]. The eccentric strengthening produced better outcome in the mid portion of the tendinopathic Achilles compared with insertional tendinopathy [86]. Eccentric strengthening programs have also been successful in the treatment of lateral epicondylitis with regard to improvement in pain relief, strength, and function [87], while another study showed no beneficial effects [88].

In addition to the results on eccentric strengthening, a formal mobilization has been reported to be beneficial for rotator cuff tendinopathy with decreased 24 hour pain and pain with subacromial compression test in patients with primary shoulder impingement syndrome [89]. It is benefit of early mobilization following tendon repair, and several postoperative mobilization protocols have been advocated [90–94]. However, prolonged immobilization following musculoskeletal injury may have detrimental effects. Immobilized rabbit patellar tendons displayed lower tensile strength and strain at failure of collagen fascicles [95]. Prolonged immobilization has been reported to decrease the water and proteoglycan contents of tendons and increase collagen cross-links [96, 97], resulting in tendon atrophy [98].

20.5.3 Cytokines and Growth Factors

Cytokine and growth factor therapies are promising for tendon repair; thus, they have been extensively studied *in vitro*, using stem cell cultures, and *in vivo*, in tendon injury models. The application of VEGF and platelet-derived growth factor (PDGF) has been shown to increase cell proliferation and promote angiogenesis in injured tendons [99, 100]. However, addition of VEGF did not increase Tnmd expression in tendons [101] and showed harmful effects on tendon repair in animal models [99]. By contrast, recombinant platelet derived growth factor BB (PDGF-BB) promoted tendon repair in animal models [100] presumably via its angiogenic and mitogenic ability [100]. IGF-1 has shown an anabolic effect on tendons from humans and mice [102, 103] and promotes collagen synthesis in engineered human tendon and ligament tissues [104, 105].

Bone morphogenetic protein-12 (BMP-12), found at sites of tendon remodeling, has been reported to increase the expression of type-I and III procollagen genes in human patellar tenocytes [106]. High expression of BMP-12 in tenocytes showed increased type-I collagen synthesis, tensile strength, and load to failure in chicken flexor tendons [107].

TGF-β and fibroblast growth factor (FGF) have been reported as the main growth factors involved in vertebrate tendon development [108, 109]. Transcriptomic analysis of mouse limb tendon cells during development also showed that TGF-β–Smad2/3 (suppressor of mothers against decapentaplegic) and FGF–extracellular signal-regulated kinase (ERK)/mitogen-activated protein kinase (MAPK) signaling pathways are essential for tendon development [110]. TGF-β ligand systematically promotes the tenogenic differentiation of stem cells [12, 111–115] and enhance the tendon repair [99, 116]. However, other studies showed that neutralizing antibody to TGF-β1 increases postoperative tendon and enthesopathy healing [117, 118]. Ectopic application of basic FGF is reported to increase tenocyte proliferation in the early stages of tendon repair [99] but decrease the mechanical strength of injured tendons [119]. By contrast, another study reported that exogenous FGF improved Scx expression and tendon strength [119, 120]. It is worth to mention that the use of cytokines and growth factors remains largely experimental and has not yet been approved for tendon healing in clinical practice.

In addition to application of a single growth factor, delivery of several molecules is available using platelet-rich plasma (PRP) to enhance the tendon repair process [121, 122]. The use of PRP could therefore be considered as a cost-effective alternative means of delivering high concentrations of autologous growth factors for cell differentiation *in vitro* [123, 124]. Again, results from animal studies showed contradictory outcomes of PRP application for tendon repair, which may be due to the variability in the PRP preparations between studies [125, 126].

20.5.4 Tissue Engineering

The past few decades have seen significant progress in tendon tissue engineering [127–131]. This technique usually involves suitable cells [132–140], biocompatible materials [141–148], and administration of active factors [149–157].

MSCs have the unique feature of "self-renewal," which is the ability to proliferate while avoiding apoptosis and differentiation [158–160], making them the most studied cells for tendon tissue engineering [161–163]. Although MSCs offer an intriguing and potentially viable approach for future reconstructive investigation, there are still certain caveats for fully understanding the means by which MSC populations can be driven to differentiate into tendon-like tissue. It is therefore plausible to suggest that tenocytes themselves could be used as the main cellular components for tendon tissue engineering. However, a major problem for use of tendon cells is their limited supply.

A suitable scaffold with appropriate properties should also be carefully selected. Surgical suturing, biological grafts [164], and permanent artificial prostheses [165–167] are commonly applied [168]. One of the main disadvantages of these approaches is that the healing process of surgically repaired tendons can take up to six weeks, which could lead to prolonged immobilization, and may ultimately result in joint contracture and stiffness. In cases of biological grafts, further complications may prolong and limit the tendon healing process, e.g. donor site-related infection, disease transmission, and tissue rejection. In cases where artificial prostheses have been used, further complications may include poor long-term clinical outcomes owing to mechanical mismatch, material durability, limited integration with the host tissues, and a high incidence of fatigue failures [169, 170]. Thus, the scaffold should not only be designed to promote optimal local cellular activity [171] but it should also be able to withstand the local mechanical forces. The rate of biodegradation of the scaffold is also important. If the scaffold biodegrades too rapidly, it may cause the loss of mechanical integrity of the tendon, whereas if the biodegradation process is too long, it may have detrimental effects on tendon remodeling. Furthermore, to allow perfusion of nutrients to the cells, the three-dimensional structure of the scaffold could play a pivotal role. Natural polymers such as collagen, alginate, chitin, hyaluronic acid, silk, fibrin, and cellulose are attractive materials for biomedical applications because of their biocompatibility and capacity to structurally mimic the native tendon ECM [163, 172–179]. The synthetic polymers used in the patch are already in common use in surgery as sutures and have an excellent safety record. Several biodegradable polymer scaffolds that mimic a normal tendon with proper mechanical strength have been reported [180].

20.5.5 Prevention of Adhesions

Investigations from patients and experimental animals showed that peritendinous adhesion is commonly formed in the area where two tendons glide within a tendon synovial sheath. It causes stiffness, loss of function, and pain and greatly limits the range of motion of patients. The standard treatment of PAF is tenolysis. In addition, many attempts have been made to reduce PAF using materials as mechanical barriers such as polyethylene or silicone or by pharmacological agents such as indomethacin and ibuprofen [181–184]. Hyaluronate is reported to decrease PAF in repaired rabbit flexor tendons [185, 186] but not in a rat Achilles tendon model [187]. This may be due to the absence of a synovial membrane around the Achilles tendon. 5-Fluorouracil, an antimetabolite with anti-inflammatory properties, could effectively preserve tendon gliding function at a dose of 25 mg/ml while produced inferior results at a dose of 50 mg/ml [188].

20.6 Conclusion

Tendon injuries substantially affect people's daily life, and at present, there are only a limited number of scientifically proven management modalities. How to better treat tendon injuries with surgical management as well as nonsurgical management is still a challenge. It is crucial for clinical doctors and biomedical scientists to understand the etiology and pathogenesis of tendon injuries and the treatment targets. A better understanding of tendon development, biology, function, and healing is prerequisite to the design of novel and relevant therapies and will allow developing optimized treatment strategies.

References

1 Feigin, V. and GBD 2015 Disease and Injury Incidence and Prevalence Collaborators (2016). Global, regional, and national incidence, prevalence, and years lived with disability for 310 diseases and injuries, 1990–2015: a systematic analysis for the Global Burden of Disease Study 2015. *Lancet* 388: 1545–1602.

2 Roberts, T.J. and Konow, N. (2013). How tendons buffer energy dissipation by muscle. *Exercise Sport Sci. Rev.* 41: 186–193.

3 Biewener, A.A. (2009). Muscle and tendon energy storage. In: *Encyclopedia of Neuroscience* (ed. M.D. Binder, N. Hirokawa and U. Windhorst), 2492–2496. Berlin/Heidelberg: Springer.

4 Forde, M.S., Punnett, L., and Wegman, D.H. (2005). Prevalence of musculoskeletal disorders in union ironworkers. *J. Occup. Environ. Hyg.* 2: 203–212.

5 Kaux, J.F., Forthomme, B., Goff, C.L. et al. (2011). Current opinions on tendinopathy. *J. Sports Sci. Med.* 10: 238–253.

6 Riley, G. (2008). Tendinopathy – from basic science to treatment. *Nat. Clin. Pract. Rheumatol.* 4: 82–89.

7 Scott, A. and Ashe, M.C. (2006). Common tendinopathies in the upper and lower extremities. *Curr. Sports Med. Rep.* 5: 233–241.

8 Yelin, E., Weinstein, S., and King, T. (2016). The burden of musculoskeletal diseases in the United States. *Semin. Arthritis Rheum.* 46: 259–260.

9 Zhang, G., Young, B.B., Ezura, Y. et al. (2005). Development of tendon structure and function: regulation of collagen fibrillogenesis. *J. Musculoskelet. Neuronal Interact.* 5: 5–21.

10 Birk, D.E. and Mayne, R. (1997). Localization of collagen types I, III and V during tendon development. Changes in collagen types I and III are correlated with changes in fibril diameter. *Eur. J. Cell Biol.* 72: 352–361.

11 James, R., Kesturu, G., Balian, G., and Chhabra, A.B. (2008). Tendon: biology, biomechanics, repair, growth factors, and evolving treatment options. *J. Hand Surg. Am.* 33: 102–112.

12 Pryce, B.A., Watson, S.S., Murchison, N.D. et al. (2009). Recruitment and maintenance of tendon progenitors by TGFβ signaling are essential for tendon formation. *Development* 136: 1351–1361.

13 Cserjesi, P., Brown, D., Ligon, K.L. et al. (1995). Scleraxis: a basic helix-loop-helix protein that prefigures skeletal formation during mouse embryogenesis. *Development* 121: 1099–1110.

14 Schweitzer, R., Chyung, J.H., Murtaugh, L.C. et al. (2001). Analysis of the tendon cell fate using Scleraxis, a specific marker for tendons and ligaments. *Development* 128: 3855–3866.

15 ChandraRajan, J. (1978). Separation of type III collagen from type I collagen and pepsin by differential denaturation and renaturation. *Biochem. Biophys. Res. Commun.* 83: 180–186.

16 Kannus, P. (2000). Structure of the tendon connective tissue. *Scand. J. Med. Sci. Sports* 10: 312–320.

17 Kastelic, J., Galeski, A., and Baer, E. (1978). The multicomposite structure of tendon. *Connect. Tissue Res.* 6: 11–23.

18 Benjamin, M., Kaiser, E., and Milz, S. (2008). Structure-function relationships in tendons: a review. *J. Anat.* 212: 211–228.

19 Ochiai, N., Matsui, T., Miyaji, N. et al. (1979). Vascular anatomy of flexor tendons. I. Vincular system and blood supply of the profundus tendon in the digital sheath. *J. Hand Surg. Am.* 4: 321–330.

20 Bi, Y., Ehirchiou, D., Kilts, T.M. et al. (2007). Identification of tendon stem/progenitor cells and the role of the extracellular matrix in their niche. *Nat. Med.* 13: 1219–1227.

21 Zhang, J. and Wang, J.H. (2010). Characterization of differential properties of rabbit tendon stem cells and tenocytes. *BMC Musculoskelet. Disord.* 11: 10.
22 Mienaltowski, M.J., Adams, S.M., and Birk, D.E. (2013). Regional differences in stem cell/progenitor cell populations from the mouse Achilles tendon. *Tissue Eng. Part A* 19: 199–210.
23 Dunn, S.L. and Olmedo, M.L. (2016). Mechanotransduction: relevance to physical therapist practice – understanding our ability to affect genetic expression through mechanical forces. *Phys. Ther.* 96: 712–721.
24 Wang, T., Thien, C., Wang, C. et al. (2018). 3D uniaxial mechanical stimulation induces tenogenic differentiation of tendon-derived stem cells through a PI3K/AKT signaling pathway. *FASEB J.* 32: 4804–4814.
25 Goncalves, A.I., Rotherham, M., Markides, H. et al. (2018). Triggering the activation of Activin A type II receptor in human adipose stem cells towards tenogenic commitment using mechanomagnetic stimulation. *Nanomedicine* 14: 1149–1159.
26 Riehl, B.D., Park, J.H., Kwon, I.K., and Lim, J.Y. (2012). Mechanical stretching for tissue engineering: two-dimensional and three-dimensional constructs. *Tissue Eng. Part B* 18: 288–300.
27 Eloy-Trinquet, S., Wang, H., Edom-Vovard, F., and Duprez, D. (2009). Fgf signaling components are associated with muscles and tendons during limb development. *Dev. Dyn.* 238: 1195–1206.
28 Kardon, G. (1998). Muscle and tendon morphogenesis in the avian hind limb. *Development* 125: 4019–4032.
29 Bonnin, M.A., Laclef, C., Blaise, R. et al. (2005). *Six1* is not involved in limb tendon development, but is expressed in limb connective tissue under Shh regulation. *Mech. Dev.* 122: 573–585.
30 Wang, J.H., Guo, Q., and Li, B. (2012). Tendon biomechanics and mechanobiology – a minireview of basic concepts and recent advancements. *J. Hand Ther.* 25: 133–140; quiz 141.
31 Maeda, T., Sakabe, T., Sunaga, A. et al. (2011). Conversion of mechanical force into TGF-β-mediated biochemical signals. *Curr. Biol.* 21: 933–941.
32 Heinemeier, K.M., Olesen, J.L., Haddad, F. et al. (2009). Effect of unloading followed by reloading on expression of collagen and related growth factors in rat tendon and muscle. *J. Appl. Physiol.* 106: 178–186.
33 Zhang, J. and Wang, J.H. (2013). The effects of mechanical loading on tendons – an *in vivo* and *in vitro* model study. *PLoS One* 8: e71740.
34 Mendias, C.L., Gumucio, J.P., Bakhurin, K.I. et al. (2012). Physiological loading of tendons induces scleraxis expression in epitenon fibroblasts. *J. Orthop. Res.* 30: 606–612.
35 Docheva, D., Müller, S.A., Majewski, M., and Evans, C.H. (2015). Biologics for tendon repair. *Adv. Drug Delivery Rev.* 84: 222–239.
36 Sandrey, M.A. (2003). Acute and chronic tendon injuries: factors affecting the healing response and treatment. *J. Sport Rehabil.* 12: 70–91.
37 Maganaris, C.N., Narici, M.V., Almekinders, L.C., and Maffulli, N. (2004). Biomechanics and pathophysiology of overuse tendon injuries: ideas on insertional tendinopathy. *Sports Med.* 34: 1005–1017.

38 Rutland, M., O'Connell, D., Brismee, J.M. et al. (2010). Evidence-supported rehabilitation of patellar tendinopathy. *North Am. J. Sports Phys. Ther.* 5: 166–178.

39 Maffulli, N., Khan, K.M., and Puddu, G. (1998). Overuse tendon conditions: time to change a confusing terminology. *Arthroscopy* 14: 840–843.

40 Arnoczky, S.P., Lavagnino, M., and Egerbacher, M. (2007). The mechanobiological aetiopathogenesis of tendinopathy: is it the over-stimulation or the under-stimulation of tendon cells? *Int. J. Exp. Path.* 88: 217–226.

41 Thornton, G.M. and Hart, D.A. (2011). The interface of mechanical loading and biological variables as they pertain to the development of tendinosis. *J. Musculoskelet. Neuronal Interact.* 11: 94–105.

42 Davis, M.E., Gumucio, J.P., Sugg, K.B. et al. (2013). MMP inhibition as a potential method to augment the healing of skeletal muscle and tendon extracellular matrix. *J. Appl. Physiol.* 115: 884–891.

43 Pennisi, E. (2002). Tending tender tendons. *Science* 295: 1011.

44 Matthew, C.A. and Moore, M.J. (1991). Collagen fibril morphometry in transected rat extensor tendons. *J. Anat.* 175: 263–268.

45 Williams, I.F., Craig, A.S., DAD, P. et al. (1985). Development of collagen fibril organization and collagen crimp patterns during tendon healing. *Int. J. Biol. Macromol.* 7: 275–282.

46 Sharma, P. and Maffulli, N. (2006). Biology of tendon injury: healing, modeling and remodeling. *J. Musculoskelet. Neuronal Interact.* 6: 181–190.

47 Sharma, P. and Maffulli, N. (2005). Tendon injury and tendinopathy: healing and repair. *J. Bone Joint Surg. Am.* 87: 187–202.

48 Manske, P.R. (1988). Flexor tendon healing. *J. Hand Surg. Br.* 13: 237–245.

49 Thomopoulos, S., Genin, G.M., and Galatz, L.M. (2010). The development and morphogenesis of the tendon-to-bone insertion – what development can teach us about healing. *J. Musculoskelet. Neuronal Interact.* 10: 35–45.

50 Galatz, L.M., Rothermich, S.Y., Zaegel, M. et al. (2005). Delayed repair of tendon to bone injuries leads to decreased biomechanical properties and bone loss. *J. Orthop. Res.* 23: 1441–1447.

51 Lin, T.W., Cardenas, L., and Soslowsky, L.J. (2004). Biomechanics of tendon injury and repair. *J. Biomech.* 37: 865–877.

52 Grant, A.E. (1964). Massage with ice (cryokinetics) in the treatment of painful conditions of the musculoskeletal system. *Arch. Phys. Med. Rehabil.* 45: 233–238.

53 Rose, S., Draper, D.O., Schulthies, S.S., and Durrant, E. (1996). The stretching window part two: rate of thermal decay in deep muscle following 1-MHz ultrasound. *J. Athl. Train.* 31: 139–143.

54 Fabrizio, P.A., Schmidt, J.A., Clemente, F.R. et al. (1996). Acute effects of therapeutic ultrasound delivered at varying parameters on the blood flow velocity in a muscular distribution artery. *J. Orthop. Sports Phys. Ther.* 24: 294–302.

55 Robinson, S.E. and Buono, M.J. (1995). Effect of continuous-wave ultrasound on blood flow in skeletal muscle. *Phys. Ther.* 75: 145–149; discussion 149–150.

56 Chaussy, C., Brendel, W., and Schmiedt, E. (1980). Extracorporeally induced destruction of kidney stones by shock waves. *Lancet* 2: 1265–1268.

57 Yalcin, E., Keskin Akca, A., Selcuk, B. et al. (2012). Effects of extracorporal shock wave therapy on symptomatic heel spurs: a correlation between clinical outcome and radiologic changes. *Rheumatol. Int.* 32: 343–347.
58 Notarnicola, A. and Moretti, B. (2012). The biological effects of extracorporeal shock wave therapy (eswt) on tendon tissue. *Muscles Ligaments Tendons J.* 2: 33–37.
59 Gerdesmeyer, L., Wagenpfeil, S., Haake, M. et al. (2003). Extracorporeal shock wave therapy for the treatment of chronic calcifying tendonitis of the rotator cuff: a randomized controlled trial. *JAMA* 290: 2573–2580.
60 Wang, C.J., Wang, F.S., Yang, K.D. et al. (2003). Shock wave therapy induces neovascularization at the tendon-bone junction. A study in rabbits. *J. Orthop. Res.* 21: 984–989.
61 Chen, Y.J., Wang, C.J., Yang, K.D. et al. (2004). Extracorporeal shock waves promote healing of collagenase-induced Achilles tendinitis and increase TGF-β1 and IGF-I expression. *J. Orthop. Res.* 22: 854–861.
62 Bosch, G., Lin, Y.L., van Schie, H.T. et al. (2007). Effect of extracorporeal shock wave therapy on the biochemical composition and metabolic activity of tenocytes in normal tendinous structures in ponies. *Equine Vet. J.* 39: 226–231.
63 Rompe, J.D., Kirkpatrick, C.J., Kullmer, K. et al. (1998). Dose-related effects of shock waves on rabbit tendo Achillis: a sonographic and histological study. *J. Bone Joint Surg. Br.* 80: 546–552.
64 Binder, A., Parr, G., Hazleman, B., and Fitton-Jackson, S. (1984). Pulsed electromagnetic field therapy of persistent rotator cuff tendinitis: a double-blind controlled assessment. *Lancet* 1: 695–698.
65 Lee, E.W., Maffulli, N., Li, C.K., and Chan, K.M. (1997). Pulsed magnetic and electromagnetic fields in experimental Achilles tendonitis in the rat: a prospective randomized study. *Arch. Phys. Med. Rehabil.* 78: 399–404.
66 Owoeye, I., Spielholz, N.I., Fetto, J., and Nelson, A.J. (1987). Low-intensity pulsed galvanic current and the healing of tenotomized rat Achilles tendons: preliminary report using load-to-breaking measurements. *Arch. Phys. Med. Rehabil.* 68: 415–418.
67 Huegel, J. and Boorman-Padgett, J.F. (2020). Effects of pulsed electromagnetic field therapy on rat Achilles tendon healing. *J. Orthop. Res.* 38: 70–81.
68 Nessler, J.P. and Mass, D.P. (1987). Direct-current electrical stimulation of tendon healing in vitro. *Clin. Orthop. Relat. Res.* 217: 303–312.
69 Fujita, M., Hukuda, S., and Doida, Y. (1992). The effect of constant direct electrical current on intrinsic healing in the flexor tendon in vitro: an ultrastructural study of differing attitudes in epitenon cells and tenocytes. *J. Hand Surg. Br.* 17: 94–98.
70 Wedlock, P., Shephard, R.A., Little, C., and McBurney, F. (1996). Analgesic effects of cranial laser treatment in two rat nociception models. *Physiol. Behav.* 59: 445–448.
71 Abergel, R.P., Meeker, C.A., Lam, T.S. et al. (1984). Control of connective tissue metabolism by lasers: recent developments and future prospects. *J. Am. Acad. Dermatol.* 11: 1142–1150.

72 Abergel, R.P., Dwyer, R.M., Meeker, C.A. et al. (1984). Laser treatment of keloids: a clinical trial and an in vitro study with Nd:YAG laser. *Lasers Surg. Med.* 4: 291–295.

73 Lyons, R.F., Abergel, R.P., White, R.A. et al. (1987). Biostimulation of wound healing in vivo by a helium-neon laser. *Ann. Plast. Surg.* 18: 47–50.

74 Enwemeka, C.S. (1991). Connective tissue plasticity: ultrastructural, biomechanical, and morphometric effects of physical factors on intact and regenerating tendons. *J. Orthop. Sports Phys. Ther.* 14: 198–212.

75 Romanos, G.E., Pelekanos, S., and Strub, J.R. (1995). Effects of Nd:YAG laser on wound healing processes: clinical and immunohistochemical findings in rat skin. *Lasers Surg. Med.* 16: 368–379.

76 Reddy, G.K., Stehno-Bittel, L., and Enwemeka, C.S. (1998). Laser photostimulation of collagen production in healing rabbit Achilles tendons. *Lasers Surg. Med.* 22: 281–287.

77 Ozkan, N., Altan, L., Bingol, U. et al. (2004). Investigation of the supplementary effect of GaAs laser therapy on the rehabilitation of human digital flexor tendons. *J. Clin. Laser Med. Surg.* 22: 105–110.

78 Tasto, J.P., Cummings, J., Medlock, V. et al. (2003). The tendon treatment center: new horizons in the treatment of tendinosis. *Arthroscopy* 19 (Suppl. 1): 213–223.

79 Alfredson, H., Pietila, T., Jonsson, P., and Lorentzon, R. (1998). Heavy-load eccentric calf muscle training for the treatment of chronic Achilles tendinosis. *Am. J. Sports Med.* 26: 360–366.

80 Mafi, N., Lorentzon, R., and Alfredson, H. (2001). Superior short-term results with eccentric calf muscle training compared to concentric training in a randomized prospective multicenter study on patients with chronic Achilles tendinosis. *Knee Surg. Sports Traumatol. Arthrosc.* 9: 42–47.

81 Silbernagel, K.G., Thomee, R., Thomee, P., and Karlsson, J. (2001). Eccentric overload training for patients with chronic Achilles tendon pain – a randomised controlled study with reliability testing of the evaluation methods. *Scand. J. Med. Sci. Sports* 11: 197–206.

82 Jonsson, P. and Alfredson, H. (2005). Superior results with eccentric compared to concentric quadriceps training in patients with jumper's knee: a prospective randomised study. *Br. J. Sports Med.* 39: 847–850.

83 Ohberg, L., Lorentzon, R., and Alfredson, H. (2004). Eccentric training in patients with chronic Achilles tendinosis: normalised tendon structure and decreased thickness at follow up. *Br. J. Sports Med.* 38: 8–11; discussion 11.

84 Shalabi, A., Kristoffersen-Wilberg, M., Svensson, L. et al. (2004). Eccentric training of the gastrocnemius-soleus complex in chronic Achilles tendinopathy results in decreased tendon volume and intratendinous signal as evaluated by MRI. *Am. J. Sports Med.* 32: 1286–1296.

85 Kannus, P., Jozsa, L., Natri, A., and Jarvinen, M. (1997). Effects of training, immobilization and remobilization on tendons. *Scand. J. Med. Sci. Sports* 7: 67–71.

86 Fahlstrom, M., Jonsson, P., Lorentzon, R., and Alfredson, H. (2003). Chronic Achilles tendon pain treated with eccentric calf-muscle training. *Knee Surg. Sports Traumatol. Arthrosc.* 11: 327–333.

87 Croisier, J.L., Foidart-Dessalle, M., Tinant, F. et al. (2007). An isokinetic eccentric programme for the management of chronic lateral epicondylar tendinopathy. *Br. J. Sports Med.* 41: 269–275.

88 Martinez-Silvestrini, J.A., Newcomer, K.L., Gay, R.E. et al. (2005). Chronic lateral epicondylitis: comparative effectiveness of a home exercise program including stretching alone versus stretching supplemented with eccentric or concentric strengthening. *J. Hand Ther.* 18: 411–419; quiz 420.

89 Conroy, D.E. and Hayes, K.W. (1998). The effect of joint mobilization as a component of comprehensive treatment for primary shoulder impingement syndrome. *J. Orthop. Sports Phys. Ther.* 28: 3–14.

90 Buckwalter, J.A. (1995). Activity vs. rest in the treatment of bone, soft tissue and joint injuries. *Iowa Orthop. J.* 15: 29–42.

91 Buckwalter, J.A. (1996). Effects of early motion on healing of musculoskeletal tissues. *Hand Clin.* 12: 13–24.

92 Chow, J.A., Thomes, L.J., Dovelle, S. et al. (1988). Controlled motion rehabilitation after flexor tendon repair and grafting. A multi-centre study. *J. Bone Joint Surg. Br.* 70: 591–595.

93 Cullen, K.W., Tolhurst, P., Lang, D., and Page, R.E. (1989). Flexor tendon repair in zone 2 followed by controlled active mobilisation. *J. Hand Surg. Br.* 14: 392–395.

94 Elliot, D., Moiemen, N.S., Flemming, A.F. et al. (1994). The rupture rate of acute flexor tendon repairs mobilized by the controlled active motion regimen. *J. Hand Surg. Br.* 19: 607–612.

95 Yamamoto, E., Hayashi, K., and Yamamoto, N. (1999). Mechanical properties of collagen fascicles from stress-shielded patellar tendons in the rabbit. *Clin. Biomech. (Bristol, Avon)* 14: 418–425.

96 Akeson, W.H., Woo, S.L., Amiel, D. et al. (1973). The connective tissue response to immobility: biochemical changes in periarticular connective tissue of the immobilized rabbit knee. *Clin. Orthop. Relat. Res.* 93: 356–362.

97 Akeson, W.H., Amiel, D., Mechanic, G.L. et al. (1977). Collagen cross-linking alterations in joint contractures: changes in the reducible cross-links in periarticular connective tissue collagen after nine weeks of immobilization. *Connect. Tissue Res.* 5: 15–19.

98 Maffulli, N. and King, J.B. (1992). Effects of physical activity on some components of the skeletal system. *Sports Med.* 13: 393–407.

99 Halper, J. (2014). Advances in the use of growth factors for treatment of disorders of soft tissues. *Adv. Exp. Med. Biol.* 802: 59–76.

100 Hee, C.K., Dines, J.S., Solchaga, L.A. et al. (2012). Regenerative tendon and ligament healing: opportunities with recombinant human platelet-derived growth factor BB-homodimer. *Tissue Eng. Part B* 18: 225–234.

101 Kaux, J.F., Janssen, L., Drion, P. et al. (2014). Vascular Endothelial Growth Factor-111 (VEGF-111) and tendon healing: preliminary results in a rat model of tendon injury. *Muscles Ligaments Tendons J.* 4: 24–28.

102 Hansen, M., Boesen, A., Holm, L. et al. (2013). Local administration of insulin-like growth factor-I (IGF-I) stimulates tendon collagen synthesis in humans. *Scand. J. Med. Sci. Sports* 23: 614–619.

103 Nielsen, R.H., Clausen, N.M., Schjerling, P. et al. (2014). Chronic alterations in growth hormone/insulin-like growth factor-I signaling lead to changes in mouse tendon structure. *Matrix Biol.* 34: 96–104.

104 Herchenhan, A., Bayer, M.L., Eliasson, P. et al. (2015). Insulin-like growth factor I enhances collagen synthesis in engineered human tendon tissue. *Growth Horm. IGF Res.* 25: 13–19.

105 Hagerty, P., Lee, A., Calve, S. et al. (2012). The effect of growth factors on both collagen synthesis and tensile strength of engineered human ligaments. *Biomaterials* 33: 6355–6361.

106 Fu, S.C., Wong, Y.P., Chan, B.P. et al. (2003). The roles of bone morphogenetic protein (BMP) 12 in stimulating the proliferation and matrix production of human patellar tendon fibroblasts. *Life Sci.* 72: 2965–2974.

107 Lou, J., Tu, Y., Burns, M. et al. (2001). BMP-12 gene transfer augmentation of lacerated tendon repair. *J. Orthop. Res.* 19: 1199–1202.

108 Schweitzer, R., Zelzer, E., and Volk, T. (2010). Connecting muscles to tendons: tendons and musculoskeletal development in flies and vertebrates. *Development* 137: 2807–2817.

109 Tozer, S. and Duprez, D. (2005). Tendon and ligament: development, repair and disease. *Birth Defects Res. C Embryo Today* 75: 226–236.

110 Havis, E., Bonnin, M.A., Olivera-Martinez, I. et al. (2014). Transcriptomic analysis of mouse limb tendon cells during development. *Development* 141: 3683–3696.

111 Goncalves, A.I., Rodrigues, M.T., Lee, S.J. et al. (2013). Understanding the role of growth factors in modulating stem cell tenogenesis. *PLoS One* 8: e83734.

112 Barsby, T. and Guest, D. (2013). Transforming growth factor beta3 promotes tendon differentiation of equine embryo-derived stem cells. *Tissue Eng. Part A* 19: 2156–2165.

113 Kapacee, Z., Yeung, C.Y., Lu, Y. et al. (2010). Synthesis of embryonic tendon-like tissue by human marrow stromal/mesenchymal stem cells requires a three-dimensional environment and transforming growth factor β3. *Matrix Biol.* 29: 668–677.

114 Farhat, Y.M., Al-Maliki, A.A., Chen, T. et al. (2012). Gene expression analysis of the pleiotropic effects of TGF-β1 in an *in vitro* model of flexor tendon healing. *PLoS One* 7: e51411.

115 Barsby, T., Bavin, E.P., and Guest, D.J. (2014). Three-dimensional culture and transforming growth factor beta3 synergistically promote tenogenic differentiation of equine embryo-derived stem cells. *Tissue Eng. Part A* 20: 2604–2613.

116 Majewski, M., Porter, R.M., Betz, O.B. et al. (2012). Improvement of tendon repair using muscle grafts transduced with TGF-β1 cDNA. *Eur. Cell Mater.* 23: 94–101; discussion 101–102.

117 Chang, J., Thunder, R., Most, D. et al. (2000). Studies in flexor tendon wound healing: neutralizing antibody to TGF-β1 increases postoperative range of motion. *Plast. Reconstr. Surg.* 105: 148–155.

118 Wang, X., Xie, L., Crane, J. et al. (2018). Aberrant TGF-β activation in bone tendon insertion induces enthesopathy-like disease. *J. Clin. Invest.* 128 (2): 846–860.

119 Tang, J.B., Chen, C.H., Zhou, Y.L. et al. (2014). Regulatory effects of introduction of an exogenous FGF2 gene on other growth factor genes in a healing tendon. *Wound Repair Regener.* 22: 111–118.

120 Tang, J.B., Cao, Y., Zhu, B. et al. (2008). Adeno-associated virus-2-mediated bFGF gene transfer to digital flexor tendons significantly increases healing strength. an in vivo study. *J. Bone Joint Surg. Am.* 90: 1078–1089.

121 Baksh, N., Hannon, C.P., Murawski, C.D. et al. (2013). Platelet-rich plasma in tendon models: a systematic review of basic science literature. *Arthroscopy* 29: 596–607.

122 Wang, X., Qiu, Y., Triffitt, J. et al. (2012). Proliferation and differentiation of human tenocytes in response to platelet rich plasma: an in vitro and in vivo study. *J. Orthop. Res.* 30: 982–990.

123 Marx, R.E. (2001). Platelet-rich plasma (PRP): what is PRP and what is not PRP? *Implant Dent.* 10: 225–228.

124 Alsousou, J., Thompson, M., Hulley, P. et al. (2009). The biology of platelet-rich plasma and its application in trauma and orthopaedic surgery: a review of the literature. *J. Bone Joint Surg. Br.* 91: 987–996.

125 Wang, J.H. (2014). Can PRP effectively treat injured tendons? *Muscles Ligaments Tendons J.* 4: 35–37.

126 Guevara-Alvarez, A., Schmitt, A., Russell, R.P. et al. (2014). Growth factor delivery vehicles for tendon injuries: mesenchymal stem cells and platelet rich plasma. *Muscles Ligaments Tendons J.* 4: 378–385.

127 Arosarena, O. (2005). Tissue engineering. *Curr. Opin. Otolaryngol. Head Neck Surg.* 13: 233–241.

128 Muschler, G.F., Nakamoto, C., and Griffith, L.G. (2004). Engineering principles of clinical cell-based tissue engineering. *J. Bone Joint Surg. Am.* 86: 1541–1558.

129 Butler, D.L., Grood, E.S., Noyes, F.R., and Zernicke, R.F. (1978). Biomechanics of ligaments and tendons. *Exercise Sport Sci. Rev.* 6: 125–181.

130 Guilak, F. (2002). Functional tissue engineering: the role of biomechanics in reparative medicine. *Ann. N. Y. Acad. Sci.* 961: 193–195.

131 Woo, S.L., Hildebrand, K., Watanabe, N. et al. (1999). Tissue engineering of ligament and tendon healing. *Clin. Orthop. Relat. Res.* (367 Suppl): S312–S323.

132 Yin, Z., Chen, X., Chen, J.L., and Ouyang, H.W. (2010). Stem cells for tendon tissue engineering and regeneration. *Expert Opin. Biol. Ther.* 10: 689–700.

133 Sahoo, S., Toh, S.L., and Goh, J.C. (2010). A bFGF-releasing silk/PLGA-based biohybrid scaffold for ligament/tendon tissue engineering using mesenchymal progenitor cells. *Biomaterials* 31: 2990–2998.

134 Liu, W., Chen, B., Deng, D. et al. (2006). Repair of tendon defect with dermal fibroblast engineered tendon in a porcine model. *Tissue Eng.* 12: 775–788.

135 Ouyang, H.W., Goh, J.C., Thambyah, A. et al. (2003). Knitted poly-lactide-*co*-glycolide scaffold loaded with bone marrow stromal cells in repair and regeneration of rabbit Achilles tendon. *Tissue Eng.* 9: 431–439.

136 Juncosa-Melvin, N., Boivin, G.P., Gooch, C. et al. (2006). The effect of autologous mesenchymal stem cells on the biomechanics and histology of gel-collagen sponge constructs used for rabbit patellar tendon repair. *Tissue Eng.* 12: 369–379.

137 Pacini, S., Spinabella, S., Trombi, L. et al. (2007). Suspension of bone marrow-derived undifferentiated mesenchymal stromal cells for repair of superficial digital flexor tendon in race horses. *Tissue Eng.* 13: 2949–2955.

138 Tischer, T., Vogt, S., Aryee, S. et al. (2007). Tissue engineering of the anterior cruciate ligament: a new method using acellularized tendon allografts and autologous fibroblasts. *Arch. Orthop. Trauma Surg.* 127: 735–741.

139 Chen, J.M., Willers, C., Xu, J. et al. (2007). Autologous tenocyte therapy using porcine-derived bioscaffolds for massive rotator cuff defect in rabbits. *Tissue Eng.* 13: 1479–1491.

140 Kryger, G.S., Chong, A.K., Costa, M. et al. (2007). A comparison of tenocytes and mesenchymal stem cells for use in flexor tendon tissue engineering. *J. Hand Surg. Am.* 32: 597–605.

141 Altman, G.H., Diaz, F., Jakuba, C. et al. (2003). Silk-based biomaterials. *Biomaterials* 24: 401–416.

142 Wang, Y., Kim, H.J., Vunjak-Novakovic, G., and Kaplan, D.L. (2006). Stem cell-based tissue engineering with silk biomaterials. *Biomaterials* 27: 6064–6082.

143 Ahmed, T.A., Dare, E.V., and Hincke, M. (2008). Fibrin: a versatile scaffold for tissue engineering applications. *Tissue Eng. Part B* 14: 199–215.

144 Ingram, J.H., Korossis, S., Howling, G. et al. (2007). The use of ultrasonication to aid recellularization of acellular natural tissue scaffolds for use in anterior cruciate ligament reconstruction. *Tissue Eng.* 13: 1561–1572.

145 Bagnaninchi, P.O., Yang, Y., Zghoul, N. et al. (2007). Chitosan microchannel scaffolds for tendon tissue engineering characterized using optical coherence tomography. *Tissue Eng.* 13: 323–331.

146 Androjna, C., Spragg, R.K., and Derwin, K.A. (2007). Mechanical conditioning of cell-seeded small intestine submucosa: a potential tissue-engineering strategy for tendon repair. *Tissue Eng.* 13: 233–243.

147 Sahoo, S., Ouyang, H., Goh, J.C. et al. (2006). Characterization of a novel polymeric scaffold for potential application in tendon/ligament tissue engineering. *Tissue Eng.* 12: 91–99.

148 Gentleman, E., Lay, A.N., Dickerson, D.A. et al. (2003). Mechanical characterization of collagen fibers and scaffolds for tissue engineering. *Biomaterials* 24: 3805–3813.

149 Murray, M.M., Spindler, K.P., Abreu, E. et al. (2007). Collagen-platelet rich plasma hydrogel enhances primary repair of the porcine anterior cruciate ligament. *J. Orthop. Res.* 25: 81–91.

150 Kondo, E., Yasuda, K., Yamanaka, M. et al. (2005). Effects of administration of exogenous growth factors on biomechanical properties of the elongation-type anterior cruciate ligament injury with partial laceration. *Am. J. Sports Med.* 33: 188–196.

151 Wei, X.L., Lin, L., Hou, Y. et al. (2008). Construction of recombinant adenovirus co-expression vector carrying the human transforming growth factor-β1 and vascular endothelial growth factor genes and its effect on anterior cruciate ligament fibroblasts. *Chin. Med. J. (Engl.)* 121: 1426–1432.

152 Spindler, K.P., Murray, M.M., Detwiler, K.B. et al. (2003). The biomechanical response to doses of TGF-β 2 in the healing rabbit medial collateral ligament. *J. Orthop. Res.* 21: 245–249.

153 Spindler, K.P., Dawson, J.M., Stahlman, G.C. et al. (2002). Collagen expression and biomechanical response to human recombinant transforming growth factor beta (rhTGF-β2) in the healing rabbit MCL. *J. Orthop. Res.* 20: 318–324.

154 Hildebrand, K.A., Woo, S.L., Smith, D.W. et al. (1998). The effects of platelet-derived growth factor-BB on healing of the rabbit medial collateral ligament. An in vivo study. *Am. J. Sports Med.* 26: 549–554.

155 Joshi, S.M., Mastrangelo, A.N., Magarian, E.M. et al. (2009). Collagen-platelet composite enhances biomechanical and histologic healing of the porcine anterior cruciate ligament. *Am J. Sports Med.* 37: 2401–2410.

156 Murray, M.M., Spindler, K.P., Devin, C. et al. (2006). Use of a collagen-platelet rich plasma scaffold to stimulate healing of a central defect in the canine ACL. *J. Orthop. Res.* 24: 820–830.

157 Fulzele, S., Arounleut, P., Cain, M. et al. (2010). Role of myostatin (GDF-8) signaling in the human anterior cruciate ligament. *J. Orthop. Res.* 28: 1113–1118.

158 Satija, N.K., Gurudutta, G.U., Sharma, S. et al. (2007). Mesenchymal stem cells: molecular targets for tissue engineering. *Stem Cells Dev.* 16: 7–23.

159 Caplan, A.I. (2007). Adult mesenchymal stem cells for tissue engineering versus regenerative medicine. *J. Cell. Physiol.* 213: 341–347.

160 Caplan, A.I. (1991). Mesenchymal stem cells. *J. Orthop. Res.* 9: 641–650.

161 Kanaya, A., Deie, M., Adachi, N. et al. (2007). Intra-articular injection of mesenchymal stromal cells in partially torn anterior cruciate ligaments in a rat model. *Arthroscopy* 23: 610–617.

162 Watanabe, N., Woo, S.L., Papageorgiou, C. et al. (2002). Fate of donor bone marrow cells in medial collateral ligament after simulated autologous transplantation. *Microsc. Res. Tech.* 58: 39–44.

163 Young, R.G., Butler, D.L., Weber, W. et al. (1998). Use of mesenchymal stem cells in a collagen matrix for Achilles tendon repair. *J. Orthop. Res.* 16: 406–413.

164 Milthorpe, B.K. (1994). Xenografts for tendon and ligament repair. *Biomaterials* 15: 745–752.

165 Dong, D.Y. and Sheng, Z.Y. (1988). Experimental permanent artificial tendon for the hand. *J. Biomed. Eng.* 10: 406–410.

166 Dong, D.Y. and Sheng, C.Y. (1988). Study of permanent artificial tendon. *Biomater. Artif. Cells Artif. Organs* 16: 927–944.

167 Murray, G.A. and Semple, J.C. (1979). A review of work on artificial tendons. *J. Biomed. Eng.* 1: 177–184.

168 Liu, Y., Ramanath, H.S., and Wang, D.A. (2008). Tendon tissue engineering using scaffold enhancing strategies. *Trends Biotechnol.* 26: 201–209.

169 Guidoin, M.F., Marois, Y., Bejui, J. et al. (2000). Analysis of retrieved polymer fiber based replacements for the ACL. *Biomaterials* 21: 2461–2474.

170 Bercovy, M., Goutallier, D., Voisin, M.C. et al. (1985). Carbon-PGLA prostheses for ligament reconstruction. Experimental basis and short-term results in man. *Clin. Orthop. Relat. Res.* 159–168.

171 Butler, D.L. and Awad, H.A. (1999). Perspectives on cell and collagen composites for tendon repair. *Clin. Orthop. Relat. Res.* (367 Suppl): S324–S332.

172 Leong, N.L., Petrigliano, F.A., and McAllister, D.R. (2014). Current tissue engineering strategies in anterior cruciate ligament reconstruction. *J. Biomed. Mater. Res. Part A* 102: 1614–1624.

173 Lamberti, P.M. and Wezeman, F.H. (2002). Biologic behavior of an in vitro hydrated collagen gel–human tenocyte tendon model. *Clin. Orthop. Relat. Res.* 397: 414–423.

174 Noth, U., Schupp, K., Heymer, A. et al. (2005). Anterior cruciate ligament constructs fabricated from human mesenchymal stem cells in a collagen type I hydrogel. *Cytotherapy* 7: 447–455.

175 Ide, A., Sakane, M., Chen, G.P. et al. (2001). Collagen hybridization with poly(L-lactic acid) braid promotes ligament cell migration. *Mater. Sci. Eng., C* 17: 95–99.

176 Koob, T.J. and Hernandez, D.J. (2002). Material properties of polymerized NDGA-collagen composite fibers: development of biologically based tendon constructs. *Biomaterials* 23: 203–212.

177 Funakoshi, T., Majima, T., Suenaga, N. et al. (2006). Rotator cuff regeneration using chitin fabric as an acellular matrix. *J. Shoulder Elbow Surg.* 15: 112–118.

178 Funakoshi, T., Majima, T., Iwasaki, N. et al. (2005). Application of tissue engineering techniques for rotator cuff regeneration using a chitosan-based hyaluronan hybrid fiber scaffold. *Am. J. Sports Med.* 33: 1193–1201.

179 Wang, X., Qiu, Y., Carr, A.J. et al. (2011). Improved human tenocyte proliferation and differentiation *in vitro* by optimized silk degumming. *Biomed. Mater.* 6: 035010.

180 Silva, M., Ferreira, F.N., Alves, N.M., and Paiva, M.C. (2020). Biodegradable polymer nanocomposites for ligament/tendon tissue engineering. *J. Nanobiotechnol.* 18: 23.

181 Hunter, J.M. and Salisbury, R.E. (1971). Flexor-tendon reconstruction in severely damaged hands: a two-stage procedure using a silicone-dacron reinforced gliding prosthesis prior to tendon grafting. *J. Bone Joint Surg. Am.* 53: 829–858.

182 Nishimura, K., Nakamura, R.M., and diZerega, G.S. (1984). Ibuprofen inhibition of postsurgical adhesion formation: a time and dose response biochemical evaluation in rabbits. *J. Surg. Res.* 36: 115–124.

183 Nishimura, K., Shimanuki, T., and diZerega, G.S. (1984). Ibuprofen in the prevention of experimentally induced postoperative adhesions. *Am. J. Med.* 77: 102–106.

184 Szabo, R.M. and Younger, E. (1990). Effects of indomethacin on adhesion formation after repair of zone II tendon lacerations in the rabbit. *J. Hand Surg. Am.* 15: 480–483.

185 Hagberg, L. and Gerdin, B. (1992). Sodium hyaluronate as an adjunct in adhesion prevention after flexor tendon surgery in rabbits. *J. Hand Surg. Am.* 17: 935–941.

186 Wiig, M., Abrahamsson, S.O., and Lundborg, G. (1997). Tendon repair – cellular activities in rabbit deep flexor tendons and surrounding synovial sheaths and the effects of hyaluronan: an experimental study *in vivo* and *in vitro*. *J. Hand Surg. Am.* 22: 818–825.

187 Tuncay, I., Ozbek, H., Atik, B. et al. (2002). Effects of hyaluronic acid on postoperative adhesion of tendo calcaneus surgery: an experimental study in rats. *J. Foot Ankle Surg.* 41: 104–108.

188 Moran, S.L., Ryan, C.K., Orlando, G.S. et al. (2000). Effects of 5-fluorouracil on flexor tendon repair. *J. Hand Surg. Am.* 25: 242–251.

21

Nerve Regeneration

Zhiwen Yan[1], Wei Chen[2], Yun Qian[1†], and Cunyi Fan[1]*

[1] *Shanghai Jiao Tong University Affiliated Sixth People's Hospital, Department of Orthopedics, 600 Yishan Road, Shanghai 200233, P.R. China*
[2] *Tufts Medical Center, Department of Anesthesiology and Perioperative Medicine, 800 Washington Street, Boston, MA 02111, USA*

21.1 Peripheral Nerve Regeneration

21.1.1 Introduction

Although the peripheral nervous system (PNS) possesses remarkable regenerative capacity, in clinical practices, only a few patients who encountered peripheral nerve injury (PNI) can successfully regain satisfactory motor and sensory functions [1]. The prognosis of PNI is closely associated with the patients' age, injury site, and especially injury severity [2]. Severe PNI usually leads to nerve gaps, and nerve gaps smaller than 8 mm can be treated with epineural coaptation. However, for nerve gaps larger than 3 cm, clinicians cannot suture the nerves directly because this will give rise to unbearable nerve tension. Thus, an autograft, which is a nerve trunk of less important function, is harvested from another part of the body and grafted to the injury site to bridge the extended nerve gap. Although autograft transplantation is now considered the gold standard therapy for severe PNI, it inevitably brings negative consequences such as numbness in donor sites, lack of sources, size mismatch, and neuroma formation [3]. Nerve guide conduit (NGC) implantation, an alternative therapy, holds great promise for the treatment of PNI. Owing to the advances in bio-fabricating techniques, the cutting-edge NGC achieves gratifying therapeutic effects in laboratory work (Table 21.1).

However, the NGC implantation is still less effective than the autograft transplantation [11]. For the lack of extracellular matrix (ECM), basal laminin, and cellular support, the nerve gap poses a serious challenge to NGC. The two nerve stumps spontaneously form new tissues to bridge the nerve gap, but this process is extremely vulnerable because of the complex and disturbed microenvironment [12]. Although NGCs provide physical support for regenerating tissues, the

*cyfan@sjtu.edu.cn
†lollipopcloudland@foxmail.com

Biofabrication for Orthopedics: Methods, Techniques, and Applications, First Edition.
Edited by Wenguo Cui, Xin Zhao, and Shen Liu.

Table 21.1 Summarize table of representative bio-fabricated products for peripheral nerve repair.

Material	Bio-fabricating methods	Functionalized agents	Injury model	Microenvironment rebalance approach	Reference
PCL	3D printing and layer-by-layer casting	Graphene, PDA, and RGD	Rat 15 mm sciatic nerve defect	Restore the bioelectrical activity	[4]
PCL	Concentrically integrative layer-by-layer bio-assembling	Black phosphorus	Rat 20 mm sciatic nerve defect	Restore the bioelectrical activity Reconstruct the microcirculation Regulate the immune responses	[5]
PCL	Multilayer molding	Gold nanoparticles and PDA	Rat 15 mm sciatic nerve defect	Restore the bioelectrical activity Reconstruct the microcirculation	[6]
PCL	Integration molding	Graphene oxide	Rat 15 mm sciatic nerve defect	Restore the bioelectrical activity Reconstruct the microcirculation	[7]
PCL	Multilayer molding	Melatonin	Rat 15 mm sciatic nerve defect	Regulate the immune responses Rectify the metabolic disorder	[8]
PCL	Asymmetrical 3D manufacture	Nanoceria and collagen	Rat 15 mm sciatic nerve defect	Reconstruct the microcirculation Regulate the immune responses	[9]
PCL	Multilayered spraying and gradient dotting	Nanodiamond	Rat 20 mm sciatic nerve defect	Regulate the immune responses Reconstruct the microcirculation	[10]

PCL, polycaprolactone; PDA, polydopamine; RGD, arginylglycylaspartic acid.

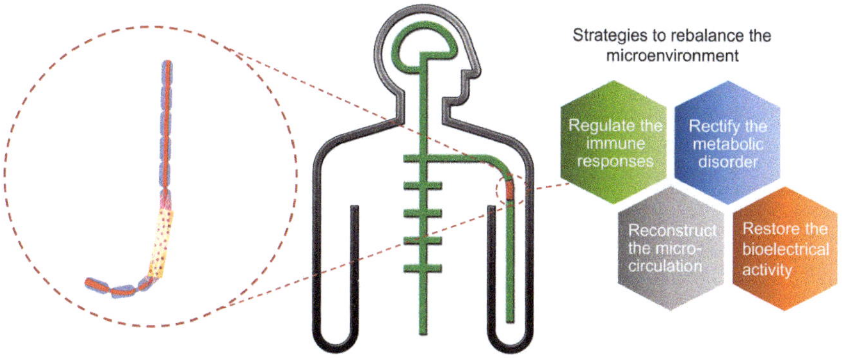

Figure 21.1 Illustration of strategies to rebalance the postinjury microenvironment and bio-fabrication for peripheral nerve repair.

imbalanced microenvironment critically compromises the therapeutic effects of NGC implantation. In order to bio-fabricate a rational and reliable NGC, it is essential for researchers to understand what the microenvironmental imbalance is after PNI and how the functionalized NGCs help rebalance the microenvironment (Figure 21.1).

This chapter briefly describes the microenvironment imbalance after PNI and assesses the reported NGCs which rectify such imbalance and benefit neural regeneration.

21.1.2 Microenvironmental Imbalance

Qian and coworker first proposed the novel concept that the microenvironment imbalance after PNI can be generally categorized into four aspects: (i) the overactive immune responses, (ii) the inefficient metabolic processes, (iii) the insufficient microcirculatory support, and (iv) the loss of bioelectrical signal conduction [13].

Right after PNI, Schwann cells, the main glia of PNS, sense the damaged axons via the neuregulin 1 (NRG1)–ErbB2 signaling pathway [14, 15]. What follows is the dedifferentiation of Schwann cells into a proregenerative phenotype [16]. This process involves the activation of extracellular regulated protein kinases (ERK)/mitogen-activated protein kinase (MAPK), c-Jun N-terminal kinase (JNK)/c-Jun, Notch, and Janus tyrosine kinase (JAK)-signal transducer and activator of transcription (STAT) signaling pathway [17]. Among them, the Ras-associated factor (Raf)/MEK/ERK signaling pathway plays a central role in aiding peripheral nerve regeneration [18]. Apart from promoting neurite outgrowth, angiogenesis, cell migration, cell proliferation, and tissue remodeling, Raf activation in Schwann cells also leads to transcription of inflammation-related agents: amphiregulin, chemokine ligand 2, chemokine ligand 10, interleukin (IL)-11, kit ligand, macrophage migration inhibitory factor, matrix metalloproteinase 14, multiple epidermal growth factor-like domain protein 10, plasminogen activator urokinase, and small inducible cytokine subfamily E. Furthermore, Raf signal in Schwann cells also triggers the breakdown of blood–nerve barrier, thereby leading to robust inflammatory responses. Macrophages participate in the immune microenvironment imbalance as well. After axotomy, the nerve injury site and Schwann cells express Ccl2, recruiting blood-borne macrophages to the injury microenvironment [19]. During the early regeneration stage, macrophages selectively sense the hypoxic microenvironment in the injury site and excrete vascular endothelial growth factor A (VEGF-A), inducing blood vessel formation. This process is essential for a successful peripheral nerve regeneration [20]. In addition, these proinflammatory macrophages are thought to help clear the damaged axons and myelin debris, providing a favored environment for regeneration [21]. Despite the positive effects of macrophages in the early regeneration stage, the prolonged and overactive inflammatory responses mediated by macrophages are detrimental to the regeneration process [22]. Thus, modulation of this "double-edged sword" helps ameliorate the prognosis of PNI.

Metabolic reprogramming happens during PNI, and the Schwann cells undergo a metabolic shift from oxidative phosphorylation to glycolysis [23]. This metabolic shift helps feed the energy-demanding dedifferentiated Schwann cells, and the underlying mechanism involves the mTORC1–Hif1α–c-Myc signaling pathway [24]. However, to our knowledge, the glycolysis-based energy synthesis is extremely inefficient. Such glycolysis shift may possibly result from the hypoxic microenvironment, which is inhibitory to oxidative phosphorylation and may be a compensatory response to the energy-demanding regenerating process. Yang et al. demonstrated that the local energy deficit led to pathological axonal death under the control of MAPK cascade after traumatic nerve injury [25]. Han et al. showed that increased mitochondrial density and ATP production in the injured site critically support axonal regeneration [26]. To conclude, efficient and sufficient energy supply to the injured nerve is the cornerstone of successful peripheral nerve regeneration.

Traumatic PNI destroys the vasculature structures in the injured site, blocking the microcirculatory support. New vasculature formation benefits peripheral nerve regeneration not only for its nutritive function but also for its role in guiding Schwann cell-mediated axonal regeneration [20]. Cattin et al. observed significantly reduced blood vessel formation and compromised axonal regeneration in transgenic mice lacking vascular endothelial growth factor (VEGF) expression [20]. Zhao et al. augmented angiogenesis via VEGF gene therapy and observed enhanced neurogenesis after sciatic nerve injury [27]. In short, peripheral nerves cannot regenerate without the support of microvasculature.

Healthy peripheral nerves conduct upstream electrical pulses to targeted effectors. PNI cuts off such bioelectrical signaling and strategies that help restore such neuron–effector communication are reported to promote peripheral nerve regeneration [4–7]. In addition, artificial electrical stimulation enhances functional recovery after PNI [28, 29]. English et al. revealed that electrical stimulation activated tyrosine kinase B (TrkB) and its ligand, thereby upregulating axonal growth-related proteins such as α1-tubulin, growth-associated protein 43 (GAP-43), and brain-derived neurotrophic factor (BDNF) [30]. Eberhardt et al. showed that electrical stimulation increased the level of motor nerve-specific HNK-1 carbohydrate, a mediator of BDNF/TrkB signaling axis [31]. Accelerated muscle reinnervation was observed, and subsequently, functional recovery after PNI was improved. In brief, the reconstruction of bioelectrical active microenvironment plays a positive role in repairing damaged peripheral nerves.

21.1.3 Microenvironmental Rebalance

As previously described, the PNI-induced microenvironment imbalance impedes neural regeneration. Functionalized NGCs help rectify such imbalance and therefore prevail over commercially available basic NGCs. The advances in modern bio-fabricating techniques enable the production of properly functionalized NGCs. Researchers took advantage of bio-fabricating techniques and engineered a variety of high-performance NGCs [32, 33]. Biological assessments indicated that the gratifying therapeutic effects can be attributed to the rebalanced microenvironment.

The sustained and excessive stimulation of proinflammatory agents hinders the formation of a regeneration favorable microenvironment. Multiple studies have demonstrated that the suppression of overactive immune responses facilitated neural regeneration. With the aid of modern bio-fabricating techniques, researchers successfully produced immune modulating neural regenerative scaffolds. Qian et al. invented a nanodiamond/polycaprolactone (PCL) NGC via multilayered spraying and gradient dotting method [10]. The authors bridged a long-range 20 mm sciatic nerve defect with the NGC in a rat model. The macrophage phenotype shift may explain the NGC's superior therapeutic effect. The authors dissected the regenerated nerve tissue within the NGC and found that macrophage recruitment measured by CD68 was attenuated 3 and 14 days after injury. The expression of typical biomarkers of M1 macrophages including IL-6 and tumor necrosis factor (TNF)-α was significantly reduced at the same time point. Compared with normal PCL NGCs, the nanodiamond/PCL NGCs mediated *in vivo* macrophage polarization by upregulating CD206 and IL-10 expression levels. Interestingly, in the normal PCL NGC-treated group, expression of M1 macrophage biomarkers remained at a high level at four months postoperatively, whereas CD68 and M2 macrophage biomarkers showed no differences compared to autograft transplantation, indicating a potential role of overactive and prolonged inflammatory responses in the compromised therapeutic effects of non-functionalized NGCs. Another study also verified the necessity of rebalancing immune microenvironment after PNI. Chen et al. utilized electrospinning technique to fabricate a multilayered NGC, allowing the controlled release of melatonin and Fe_3O_4 magnetic nanoparticles [34]. The authors identified IL-6, IL-10, TNF-α, interferon (IFN)-γ, and neuronal nitric oxide synthase (nNOS) as proinflammatory biomarkers and IL-10 as anti-inflammatory biomarkers. Similar results were observed and further confirmed the effectiveness of immune modulating therapy for treating PNI.

PNI-induced metabolic disorder can be characterized by mitochondrial dysfunction and the subsequent reactive oxygen species (ROS) production. According to previous publications, the restoration of mitochondrial function and the mitigation of ROS stress are closely associated with superior therapeutic effects [8, 9, 26, 35]. Qian et al. bridged a 15 mm rat sciatic nerve defect with a melatonin-loaded PCL NGC [8]. The authors assessed the mitochondrial function of the defect site by quantifying complex I and ATP synthase expression at 12 weeks postinjury. The melatonin-loaded NGC upregulated the expression of mitochondrial function-related proteins and thereby effectively boosting neural regeneration. In addition, melatonin-mediated oxidative stress alleviation was strongly corrected to the improved morphological and functional recovery after PNI. In the melatonin-loaded NGC-treated group, the anti-oxidant biomarkers manganese superoxide dismutase (MnSOD), glutamate cysteine ligase (GCLC), and heme oxygenase-1 (HO-1) showed a significant increase at 12 weeks postinjury. In another study, researchers fabricated an asymmetrical NGC composing of collagen, nanoceria, and PCL [9]. This NGC activated the nuclear factor-like 2 (Nrf2) anti-oxidative signaling axis and unsurprisingly repaired a 15 mm rat sciatic nerve defect. Given the above, an ideal bio-fabricated product for peripheral nerve repair should be able to remedy the impaired mitochondrial function and to restore redox homeostasis.

Peripheral nerves cannot regenerate without the support and guidance of microvessels [20]. Although the microcirculation spontaneously restores after PNI, the process is usually slow and vulnerable. Thus, bio-fabricated products that boost angiogenesis favor neural regeneration. Qian et al. created a novel three-dimensional (3D) integration molding technique to invent porous and biocompatible graphene oxide/PCL NGCs [7]. The authors demonstrated the angiogenic capability of this NGC by measuring CD34, CD31, microvessel density, and vessel-like structure density. Furthermore, the potential mechanism underlying this phenomenon is the activation of AKT/eNOS/VEGF signaling axis. In another study, the authors constructed a polydopamine (PDA)-coated gold/PCL NGC with a multilayer molding method [6]. The researchers performed CD31 immunohistochemistry to analyze the microvessel density of the injured site and found that the incorporation of gold nanoparticle and PDA is beneficial to the angiogenic process. Moreover, the researchers tested the angiogenic potential of cell-loaded NGCs. Schwann cells and bone marrow mesenchymal stem cells (BMSCs) excrete VEGF under certain conditions and loading them on NGCs is a promoting strategy to enhance angiogenesis during peripheral nerve regeneration. The microvessel density was higher in the cell-loaded NGC group than the control group. To sum up, the restoration of microcirculation characterized by microvessel formation is a vital factor for peripheral nerve regeneration and bio-fabricated products, which facilitate that such processes are highly possible to bridge a nerve gap.

Electrically conductive bio-fabricated products hold great promise for treating PNI. A variety of studies supported the positive role of conductive biomaterials in promoting peripheral nerve regeneration [36, 37]. Graphene, known for its extraordinary conductivity, was loaded on a PCL NGC with the aid of an integrated layer-by-layer casting method and 3D printing [4]. The researchers assessed the performance of such NGC thoroughly and proved its effectiveness in bridging extended nerve gaps. In order to test the clinical translation potential of the graphene-embedded NGC, Qian et al. performed a preclinical assessment on the long-term biosafety *in vivo* and confirmed that the graphene-embedded NGC could repair lengthy peripheral nerve defect while exerting negligible toxicity [37]. In another study, the same team utilized integrative layer-by-layer bio-assembling technique to fabricate a black phosphorus/PCL NGC [5]. With the load of 0.5% black phosphorus, the NGC demonstrated an electrical conductivity of 9.81×10^{-3} S/cm and satisfactory biocompatibility. As is expected, the conductive black phosphorus/PCL NGC enabled both morphological and functional recovery in a rat long-range sciatic defect model. The authors further illustrated that the conductive black phosphorus nanoparticles elevated intracellular free Ca^{2+} concentrations, thereby modulating the transcription of c-fos and BDNF. Apart from restoration of bioelectrical signaling with conductive biomaterials, application of artificial electrical stimulation is another strategy to rebalance bioelectrical microenvironment. Cheng et al. developed a self-powered polyvinylidene fluoride/PCL NGC with a cast/annealing-solvent displacement method [28]. This NGC spontaneously and wirelessly generates electrical stimulation by utilizing ambient mechanical energy. The authors observed that this electrical stimulating therapy attenuated amyotrophy by slowing down the fast to slow muscle phenotype switch. To be specific, the electrical stimulation elevated the expression level of fast-type myosin in the targeted muscle and thereby leading to the increased muscle mass and fiber size. In another study, the piezoelectric zinc oxide was incorporated into NGCs [29]. This "nanogenerator" NGC not only provided physical guidance but also electrically stimulated the regenerating nerve tissues. The accelerated nerve conducting velocity and improved motor function restoration proved the effectiveness of such electrical stimulation therapy. In conclusion, the reconstruction of the bioelectrical microenvironment is critical to the regeneration of electroactive tissues such as peripheral nerves.

The microenvironment imbalance and rebalance theory highlighted above provides guidance for the rational design of high-performance neural regenerative scaffolds. With the help of advanced bio-fabricating techniques, researchers may possibly investigate a multifaceted microenvironment rebalance therapy for PNI and push the boundaries of the currently limited therapeutic effects.

21.2 Spinal Cord Regeneration

21.2.1 Introduction

Different from PNS, the central nervous system (CNS) does not have intrinsic regenerative capacity after injury. Therefore, individuals after spinal cord injury (SCI) always lose substantial essential physiological functions.

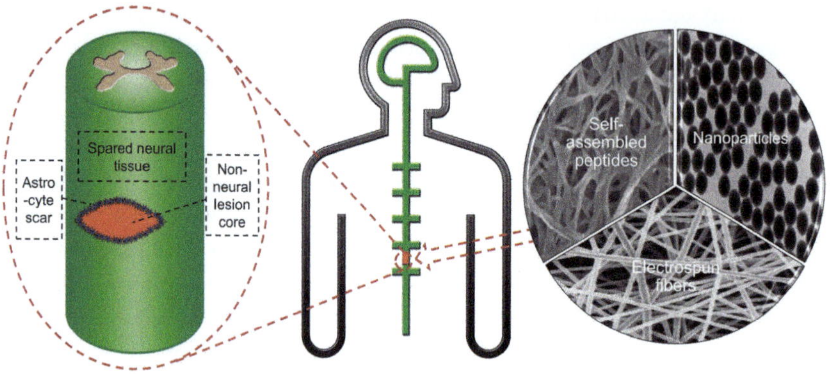

Figure 21.2 Illustration of pathophysiological compartments of an injured spine and bio-fabrication for spinal cord repair.

In clinical practices, patients who are diagnosed with SCI usually receive corticosteroid pulse therapy, namely, 5.4 mg/(kg h) of methylprednisolone (MP) for over 24 hours [38]. There are still a number of limitations, although the current therapy helps alleviate early phase inflammatory responses and benefit the functional recovery. The therapeutic effect of MP administration is limited to treating mild SCIs and is overall unsatisfactory. Furthermore, the side effects such as corticosteroid myopathy and infections are usually associated with uncertain prognoses [39]. Therefore, it is of urgent need to seek for alternative therapies. The rapid advances of bio-fabrication methods brought hope to the influenced individuals, and recent studies have proved that biomaterials help augment spinal cord repair.

This chapter briefly introduces the pathophysiological process after SCI and assesses bio-fabricated scaffolds for spinal cord regeneration (Figure 21.2).

21.2.2 Pathophysiological Process

Expanding the mechanistic knowledge about the question why the spinal cord fails to regenerate across the lesion site helps researchers design more reliable neural regenerative scaffolds.

The postinjury pathophysiological process can be generally categorized as early and late phase reaction [40].

Early phase reaction: Right after traumatic SCI, the neuronal axons are transected, and the cellular membrane is disrupted. The myelin sheath breaks down and the injury site swells. What follows is the breakdown of blood–spinal cord barrier and the subsequent infiltration of proinflammatory agents such as monocytes and neutrophils.

Late phase reaction: Astrocytes, the glia of CNS, activate and secrete cytokines. The peripheral fibroblasts and Schwann cells are recruited to the injury site. The cell-derived glial scar along with the acellular fibrotic scar contributes to the formation of a non-neural (stromal) lesion core. In addition, astrocytes migrate and take up residence alongside the lesion core, forming the astrocyte scar border.

In terms of the functional outcomes and mechanisms of regeneration, there is a huge difference between incomplete and complete SCIs, two different modes of injury [41, 42].

Anatomically incomplete SCI: There are some supraspinal axons directly and indirectly connected with spared neural tissues. New neural detour forms and neural circuit remodels spontaneously after injury, leading to a degree of functional recovery. This mode of injury may benefit from augmenting the reorganization of spared neural connections.

Anatomically complete SCI: The connections between supraspinal and downstream neural tissues are completely cut off in the lesion site. The functional loss of this injury mode is usually complete, and its recovery requires interventions that help build neural connections across the lesion.

Despite the differences between two injury modes, the spinal cord lesion can be generally divided into three compartments [41, 43].

Non-neural lesion core: Known as fibrotic scar, the non-neural lesion core mainly comprises residual inflammatory cells and ECM such as fibronectins, collagens, laminins, etc. SCI generates cellular debris and causes potential toxicity. The local inflammatory responses are then activated, recruiting residual and blood-borne inflammatory cells to the lesion site. After that, the extravasated cells recede and a non-active lesion core matures.

Astrocyte scar: During the debris clearance process mentioned above, astrocytes, the glia of CNS, respond to the inflammatory microenvironment and form a limitans border to protect neural tissues from inflammatory damage.

Spared neural tissue: This compartment contains functional neurons and glia, including astroglia, oligodendroglia, and microglia. These nerve tissues are of vital importance to spinal cord repair for their ability to reorganize neural circuits. Such reorganization is driven by spontaneous synapse loss and replacement, and intervention promoting this process could greatly benefit functional recovery after SCI.

It will be greatly advantageous to design rational bio-fabricated neural regenerative scaffolds by understanding the reasons behind the failure of spinal cord in regeneration. Neuroscientists have explored this question for over a decade and the notions remain controversial. The historical and contemporary prevailing perspectives are listed as follows [42]:

The injured spinal cord cannot produce adequate neurotrophic chemoattractants. Furthermore, the intrinsic growth potential of adult spinal cord is limited, and the lack of matrix support hinders axonal regeneration.

The presence of inhibitors such as perineuronal nets, chondroitin sulfate proteoglycans (CSPGs), tenascins, semaphorins, and netrins critically contribute to the failure of spinal cord regeneration.

Astroglial scars form a physical barrier and obstruct the axons from regrowing across the non-neural lesion core.

21.2.3 Bio-fabrication for Spinal Cord Regeneration

Systematic administration of MP, the current clinical therapy for SCI, has dose-dependent side effects [39]. Chondroitinase ABC (ChABC), a glial scar remover, effectively promotes spinal cord regeneration at room temperature in laboratory work but is unstable at 37 °C [44, 45]. A properly bio-fabricated drug carrier can address these issues, and a suitable subcellular topography also benefits spinal cord regeneration [46]. Although researchers have made gratifying achievements in labs, there are still no clinical trials to assess the performance of these bio-fabricated products for spinal cord repair.

In this chapter, three main bio-fabricated products for spinal cord regeneration are reviewed, namely, nanoparticles, electrospun fibers, and self-assembled peptides (Table 21.2). The therapeutic strategies for these products to treat SCI can be generally divided into three aspects [53]: (i) to alleviate the inflammatory stress, (ii) to overcome the inhibitory effects, and (iii) to promote axonal regeneration.

Nanoparticles for spinal cord repair include polymers, silica, metals, and their derivatives. Among them, polymers are widely used as carriers for local drug delivery in a controlled manner.

SCI breaks down the blood–spinal cord barrier. The circulating neutrophils and monocytes respond to the injury signal and recruit to the injury site. Proinflammatory cytokines excreted by these cells such as IL-1, IL-6, and TNF-α harm the reactive neurons and glia [54]. Thus, nanoparticles encapsulated with anti-inflammatory drugs benefit spinal cord regeneration. Chvatal et al. encapsulated MP in poly(lactic-*co*-glycolic acid) (PLGA)-based nanoparticles and embedded them in agarose hydrogels [47]. The MP infiltrated the injured spinal cord effectively and significantly decreased the number of activated macrophages. Moreover, the down-regulated expression of Calpain and inducible nitric oxide synthase (iNOS) led to the decreased lesion volume after spinal cord contusion injury. Papa et al. loaded minocycline, a well-known anti-inflammatory drug, in poly-ε-caprolactone and polyethylene glycol (PEG) polymeric nanoparticles [55]. Interestingly, these nanoparticles selectively targeted activated macrophages and diminished the expression of IL-6 and CD68. The authors also took advantage of the same technique to deliver minocycline to microglia [56]. These drug-loading nanoparticles attenuated the detrimental proinflammatory responses, generated a proregenerative milieu, and therefore ameliorated the functional outcomes after SCIs. Conventional nanoparticles require surgical intervention to distribute them to the injury site. Saxena et al. investigated liposome-like PEGylated nanoparticles with a certain size to load Chicago Sky Blue, an alternative to MP [57]. These nanoparticles were administered intravenously and specifically accumulated in the injured spinal cord. The therapeutic effect was evidenced by the improved vascular and white matter integrity.

After SCI, the myelin sheath breakdowns and forms Nogo-A, myelin-associated glycoprotein, and oligodendrocyte myelin glycoprotein. These three myelin-derived inhibitory factors lead to actin stabilization and growth cone collapse via the Rho-dependent pathway [58]. Among them, Nogo-A is usually considered the

Table 21.2 Summarize table of representative bio-fabricated products for spinal cord repair.

Material	Bio-fabricated product	Drug load	Injury model	Performance	Reference
PLGA	Nanoparticles	Methylprednisolone	Rat T10 contusion	Reduce the lesion volume 7 d after injury	[47]
PLGA/lipid	Nanoparticles/ microspheres	NEP1–40/ChABC	Rat T8 hemisection	Promote axonal elongation	[48]
				Limit glial scar formation 2 wk after injury	
Polylactide	Electrospun fibers	IL-4 plasmid/nerve growth factor	Rat T9 hemisection	Improve the BBB and inclined plane test scores 8 wk after injury	[49]
Poly-L-lactic acid	Electrospun fibers	N/A	Rat T8 laminectomy	Improve the BBB scores	[50]
				Enhance axonal regeneration 4 wk after injury	
Dextran sulfate	Metal ion-assisted self-assembled peptides	Methylprednisolone	Rat C5 laminectomy	Reduce lesion volume	[51]
				Protect axons and myelin	
				Modulate inflammatory responses 6 wk after injury	
IKVAV	Self-assembled peptides	N/A	Mice T10 compression	Attenuate astrogliosis	[52]
				Reduce apoptotic cell death	
				Improve the BBB scores 9 wk after injury	

PLGA, poly(lactic-*co*-glycolic acid); NEP, Nogo-A extracellular peptide residues; ChABC, chondroitinase ABC; IL, interleukin; BBB, Basso, Beattie, Bresnahan; N/A, Not applicable; IKVAV, isoleucine–lysine–valine–alanine–valine.

primary inhibitory agent. Moreover, the glial scar composing of CSPGs initially limits the spread of inflammatory agents but blocks the extension of regenerating axons physically for their dense meshwork [59, 60]. Lee et al. used lipid microtubes loaded with thermostabilized ChABC to treat T10 hemisection injury rats [61]. Bioactive ChABC was released continuously and sustainedly for over two weeks, and the functional recovery of SCI rats was improved. Wilems and Sakiyama-Elbert engineered PLGA microspheres to deliver the Nogo-A extracellular peptide residues (NEP)1–40 peptide, a competitive inhibitor of Nogo-A [48]. A rat T8 compression model was applied to assess the efficiency of this treatment, and the authors found that their controlled delivery system benefited axonal regrowth after SCI. Baumann et al. intrathecally injected a PLGA and hyaluronic acid (HA)-based hydrogel to treat SCI, bypassing the blood–spinal cord barrier [62]. Incorporating with drug-loaded nanoparticles, their intrathecally injection system potentially helps localize the therapeutic agents to the injured spinal cord.

Apart from alleviating the inflammatory responses and overcoming the inhibitory factors, facilitating the axonal growth is another key to successful spinal cord regeneration [63, 64]. Santhosh et al. used recombinant human neuregulin-1β1 peptide to treat rat T7 compression injury by encapsulating them in PLGA microparticles [65]. This therapy not only mitigated neuroinflammation but also boosted neurite outgrowth and myelination *in vivo*. Khaing et al. embedded BDNF-loaded PLGA microparticles in hydrogel composing of HA and methylcellulose to treat complete translation injury of the rat spine [66]. *In vivo* study showed that the controlled release of BDNF helped train new neural circuits and foster spinal learning, thereby enhancing the functional recovery. Ansorena et al. treated rat spinal cord hemisection injury with glial cell line-derived neurotrophic factor (GDNF)-controlled delivery therapy in a similar approach [67]. More neurofilaments were observed in the injury site, leading to the superior behavioral recovery.

Electrospinning is a versatile and viable technique for producing nanoscale and biomimetic fibers. Extensive research studies have been carried out to explore its application in nerve tissue engineering.

In order to modify the complex immune microenvironment after SCI, Xi et al. constructed pH-responsive electrospun fibers for triggered release of IL-4 plasmid-loaded liposomes during the acid-producing early phase reaction [49]. In a rat T9 3 mm spinal cord hemisection defect model, significant immune cell subtype shift was observed, thereby elevating the Basso, Beattie, Bresnahan (BBB) score, which is the measurement of motor function.

In terms of reducing detrimental glial scars, Xia et al. employed electrospun fibers to deliver dibutyryl cyclic adenosine monophosphate (dbcAMP) to the injury site [68]. The authors measured the thickness of glial scar four weeks after rat spinal hemisection injury and found that the glial scar formation was significantly attenuated.

Electrospun fibers could mimic the ECM and provide nanoscale topological cues, which benefit neurite outgrowth. Hurtado et al. constructed an electrospun poly-L-lactic acid scaffold with aligned microfibers [50]. After four weeks of implantation into a 3 mm T8 rat spinal cord defect, the scaffold was successfully infiltrated by rostrocaudal axons, and the aligned microfibers played a critical role in guiding axonal regeneration. Milbreta et al. produced 3D aligned poly(ε-caprolactone-*co*-ethyl ethylene phosphate) nanofibers with electrospinning technique and combined them with collagen hydrogels [69]. In a rat C5 spinal cord incision injury model, the hybrid scaffold with nanofibers parallel to the spinal cord length demonstrated a significantly improved ability in supporting axonal regeneration.

The self-assembling technique for spinal cord repair usually refers to the construction of nanofibrous hydrogels via ionic complementarity and amphiphilicity of natural amino acid sequences. Wang et al. took advantage of the metal ion-assisted self-assembling technique to mix minocycline hydrochloride with dextran sulfate, enabling a stable and low-dose release of this anti-inflammatory drug for 37 days [51]. The authors treated rat unilateral C5 spinal contusion injury models with their self-assembled complexes and found a decrease in CD68-expressing macrophage number six weeks after injury, thereby reducing the lesion volume and improving the behavioral outcomes. Wu et al. designed self-assembled nanoparticles for spinal cord regeneration with a hydrophobic ferulic acid core and a hydrophilic glycol chitosan shell [70]. In a rat T10 spinal contusion injury model, histological analysis revealed that these self-assembled nanoparticles suppressed the inflammatory responses, and the significantly elevated BBB locomotor score indicated the potential of this technique for spinal cord.

Self-assembled peptides also demonstrate the ability of reducing glial scars after SCI. Silva et al. reported that the self-assembled laminin epitope, isoleucine–lysine–valine–alanine–valine (IKVAV), selectively induced the neuronal differentiation of neural progenitor cells, while limiting the activation of astrocytes *in vitro* [71]. The same group then assessed the *in vivo* performance of this self-assembled pentapeptide epitope in treating mice T10 spinal cord compression injury [52]. Results showed that this therapy could reduce astrogliosis measured by glial fibrillary acidic protein (GFAP) immunofluorescence and facilitate both motor and sensory axonal regrowth.

Self-assembled nanofibrous structures can provide both physical and biological guidance for axonal regeneration. Guo et al. seeded Schwann cells on RADA16-I peptides and transplanted them into the rat C6/7 hemisection injury site [72]. Successful axonal regeneration into the scaffold was observed, indicating the potential of self-assembled peptides in treating SCI. Sun et al. functionalized RADA16-I peptides with cell adhesion-related peptide arginylglycylaspartic acid (RGD) and neuroprotective peptide IKVAV [73]. In a rat T8 2 mm spinal cord defect model, the grafted peptides enabled robust axonal regeneration across the injury site.

To conclude, the advances in modern bio-fabricated products revolutionize conventional therapies and bring new hope for the inefficient management of SCI.

21.3 Clinical Application of Nerve Conduit

21.3.1 Clinical Classification of Nerve Injuries

The goal for surgical nerve repair is to achieve best functional regains. The likeliness of the peripheral nerve repair relies on the extent of injury. The most widely accepted grading systems are those created by Seddon [74] and Sunderland [75]. Seddon categorized nerve injuries into neurapraxia, axonotmesis, and neurotmesis by severity from mildest to worst. Seddon classification is based on the nature of nerve injury and extent of loss of nerve continuity. Neurapraxia means the injured axon remains its continuity but loses function. This often results from compressive or crush injury to the nerve where partial conduction is blocked without Wallerian degeneration. On the molecular biology level, this type of injury is likely because of the local ion-induced conduction block at the injury site, although subtle alterations in the myelin structure have also been found [76]. It is transient, and recovery of function is evident in three to six weeks. Axonotmesis occurs after complete transection of the axon and part of surrounding supporting structures while most layers remain intact. As for neurotmesis, all layers of the nerve trunk are completely disconnected.

Sunderland further stratified the Seddon classification into five categories. This results from further recognition of the histological structure of a nerve trunk, and Sunderland classification allows a better prediction of outcomes (Table 21.3 and Figure 21.3). The first-degree injury is equivalent to neurapraxia. The second-degree injury to fourth-degree injury is a subgroup of axonotmesis, which respectively involves partial injuries of myelin, endoneurium, and perineurium. The fifth-degree injury is complete severance of the nerve as neurotmesis. The prospect of repair of axonotmesis is promising because the continuity of nerve trunk remains, and the preserved epineurium creates a network as a path for subsequent extension of nerve fibers. On the contrary, neurotmesis has the worst prognosis because of complete loss of continuity. Recovery is nearly impossible without surgical intervention, which requires the development in microsurgical procedures for effective repair strategies.

Frankel scale was developed to classify SCIs [77]. However, it has many limitations and is not used anymore. In 1982, the American Spinal Injury Association (ASIA) published their first version of the International Standards for Neurological Classification of Spinal Injury (ISNCSI), which has been modified and updated for 22 years, and the most recent updated one was released in April 2019. Compared to Frankel scale, the ISNCSI identifies key muscle groups and sensory functions that precisely point out the levels of SCI. In addition, it describes in detail of each sensory and motor grade based on annual updates, which provides accurate characterization of SCIs (Figure 21.4a,b). ISNCSI aims to (i) standardize SCIs in detail, (ii) provide further guidelines on imaging and treatment, and (iii) determine the completeness of SCIs, which is subtle in neurologic distinction with prognostic implications [78]. It has been widely accepted internationally as a classification and scaling system for the evaluation of SCIs.

Table 21.3 Simplified Seddon and Sunderland classification.

Seddon classification	Sunderland classification	Injury	Intact layers
Neurapraxia	Grade I	Regional demyelination	All
Axonotmesis	Grade II	Axon and myeline damaged	Endoneurium, perineurium, and epineurium
Axonotmesis	Grade III	Axon, myeline, and endoneurium damaged	Perineurium, epineurium
Axonotmesis	Grade IV	Axon, myeline, endoneurium, and perineurium damaged	Epineurium
Neurotmesis	Grade V	Complete nerve damaged	None

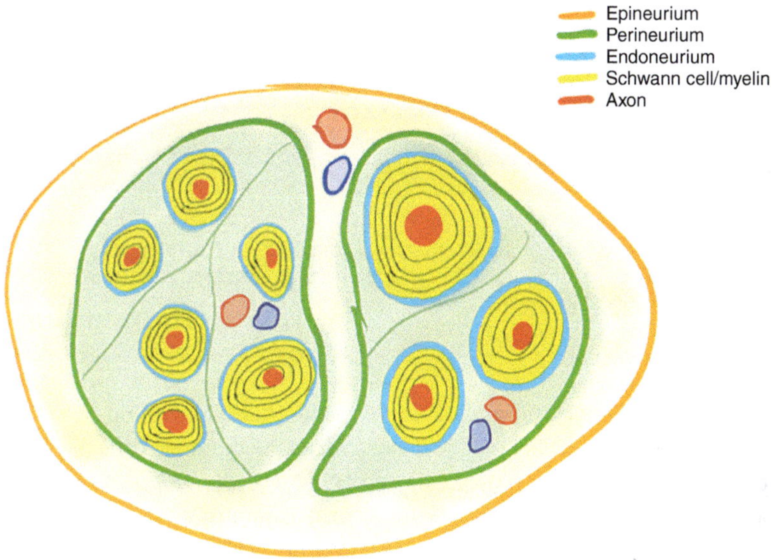

Figure 21.3 The illustration demonstrates the structure of the peripheral nerve. A transverse section is enlarged and colored to show the relation of the nerve fibers to the surrounding connective tissue (endoneurium, perineurium, and epineurium).

21.3.2 Clinical Evaluation

Evaluation of the patient with nerve injuries starts with "how did you get injured?" This is to determine the mechanism of trauma. Crush, stretch, twisting, or compression injuries, such as Saturday night palsy, often lead to neurapraxia. Laceration or penetration can result in partial or full nerve transection. Gunshot trauma may lead to complicated nerve deficits. Blunt trauma, if the energy is high enough to cause nerve deficits, may lead to a variety of injury patterns, and further examination is required to determine the prognosis with careful observation.

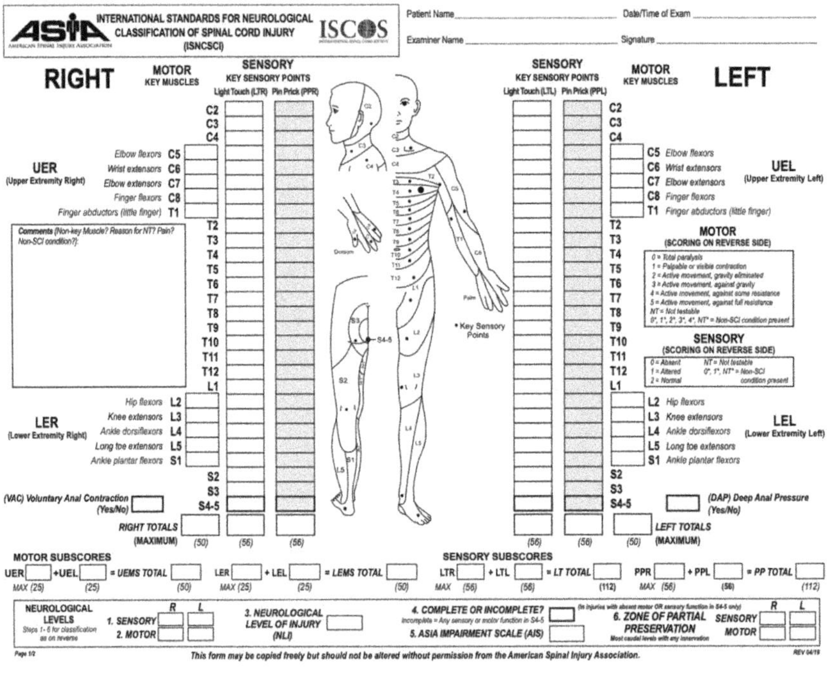

(a)

Figure 21.4 The 2019 revision of the International Standards for Neurological Classification of Spinal Cord Injury (ISNCSCI) was released at ASIA's Honolulu meeting in Spring, 2019. (a) Page 1 and (b) page 2.

Diagnostic studies are helpful to differentiate acute injury to chronic injury, and they may assist to determine the type of nerve deficits. Electrodiagnostic tools such as nerve conduction studies are used as a screening test. It can detect the reduction of action potentials and determine the severity of injuries [79]. For example, gradual spontaneous recovery of sensorimotor function indicates an injury with a neurapraxic or axonotmetic component, which relaxes the timing of interventional treatment and further observation may be the choice. Spontaneous recovery of nerve injury may take a few months [80]. Ultrasound and magnetic resonance imaging (MRI) are the imaging exam of choice for nerve injury. Ultrasound is handy to be used and gets increasing popularity in the diagnosis. On the other hand, MRI can show high resolution of fascicular patterns and nerve edema. It provides useful information for surgical planning as well.

Physical examination is used to evaluate the sensorimotor function, level of injury, and recovery in SCI. Imaging tools such as MRI are also commonly used for most SCIs. It provides information about the cord and vertebral components, helps generate a plan for surgery if needed, and provides prognostic implications [81]. One of the disadvantages of MRI is that the resolution to complex arrays of directionally oriented nerve fibers in the spinal cord is not adequate enough for

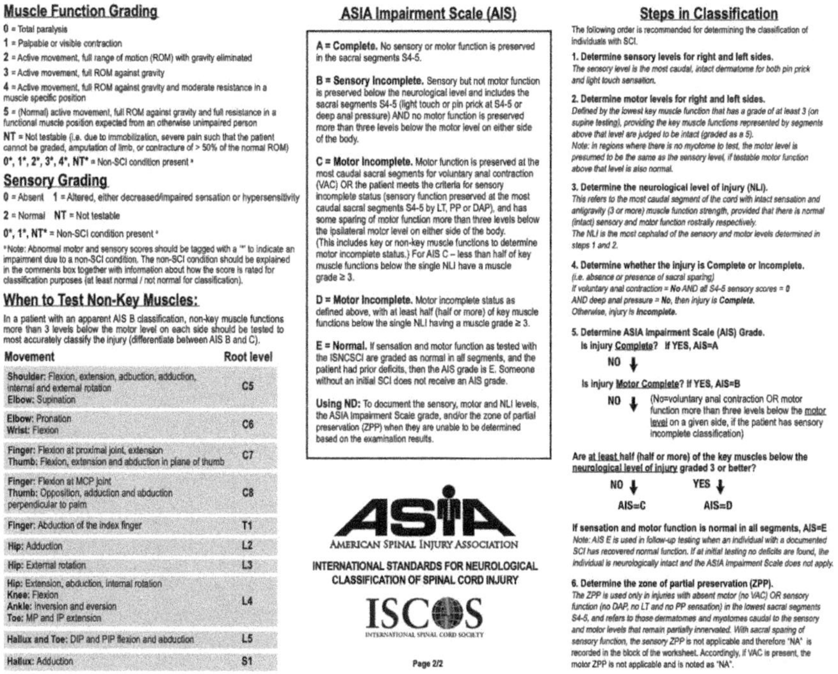

Muscle Function Grading

0 = Total paralysis

1 = Palpable or visible contraction

2 = Active movement, full range of motion (ROM) with gravity eliminated

3 = Active movement, full ROM against gravity

4 = Active movement, full ROM against gravity and moderate resistance in a muscle specific position

5 = (Normal) active movement, full ROM against gravity and full resistance in a functional muscle position expected from an otherwise unimpaired person

NT = Not testable (i.e. due to immobilization, severe pain such that the patient cannot be graded, amputation of limb, or contracture of > 50% of the normal ROM)

0*, 1*, 2*, 3*, 4*, NT* = Non-SCI condition present *

Sensory Grading

0 = Absent **1** = Altered, either decreased/impaired sensation or hypersensitivity

2 = Normal **NT** = Not testable

0*, 1*, NT* = Non-SCI condition present *

Note: Abnormal motor and sensory scores should be tagged with a "" to indicate an impairment due to a non-SCI condition. The non-SCI condition should be explained in the comments box together with information about how the score is rated for classification purposes (at least normal / not normal for classification).

When to Test Non-Key Muscles:

In a patient with an apparent AIS B classification, non-key muscle functions more than 3 levels below the motor level on each side should be tested to most accurately classify the injury (differentiate between AIS B and C).

Movement	Root level
Shoulder: Flexion, extension, adbuction, adduction, internal and external rotation **Elbow:** Supination	C5
Elbow: Pronation **Wrist:** Flexion	C6
Finger: Flexion at proximal joint, extension **Thumb:** Flexion, extension and abduction in plane of thumb	C7
Finger: Flexion at MCP joint **Thumb:** Opposition, adduction and abduction perpendicular to palm	C8
Finger: Abduction of the index finger	T1
Hip: Adduction	L2
Hip: External rotation	L3
Hip: Extension, abduction, internal rotation **Knee:** Flexion **Ankle:** Inversion and eversion **Toe:** MP and IP extension	L4
Hallux and Toe: DIP and PIP flexion and abduction	L5
Hallux: Adduction	S1

ASIA Impairment Scale (AIS)

A = Complete. No sensory or motor function is preserved in the sacral segments S4-5.

B = Sensory Incomplete. Sensory but not motor function is preserved below the neurological level and includes the sacral segments S4-5 (light touch or pin prick at S4-5 or deep anal pressure) AND no motor function is preserved more than three levels below the motor level on either side of the body.

C = Motor Incomplete. Motor function is preserved at the most caudal sacral segments for voluntary anal contraction (VAC) OR the patient meets the criteria for sensory incomplete status (sensory function preserved at the most caudal sacral segments S4-5 by LT, PP or DAP), and has some sparing of motor function more than three levels below the ipsilateral motor level on either side of the body. (This includes key or non-key muscle functions to determine motor incomplete status.) For AIS C – less than half of key muscle functions below the single NLI have a muscle grade ≥ 3.

D = Motor Incomplete. Motor incomplete status as defined above, with at least half (half or more) of key muscle functions below the single NLI having a muscle grade ≥ 3.

E = Normal. If sensation and motor function as tested with the ISNCSCI are graded as normal in all segments, and the patient had prior deficits, then the AIS grade is E. Someone without an initial SCI does not receive an AIS grade.

Using ND: To document the sensory, motor and NLI levels, the ASIA Impairment Scale grade, and/or the zone of partial preservation (ZPP) when they are unable to be determined based on the examination results.

ASIA

AMERICAN SPINAL INJURY ASSOCIATION

INTERNATIONAL STANDARDS FOR NEUROLOGICAL CLASSIFICATION OF SPINAL CORD INJURY

ISCOS

INTERNATIONAL SPINAL CORD SOCIETY

Page 2/2

Steps in Classification

The following order is recommended for determining the classification of individuals with SCI.

1. Determine sensory levels for right and left sides.
The sensory level is the most caudal, intact dermatome for both pin prick and light touch sensation.

2. Determine motor levels for right and left sides.
Defined by the lowest key muscle function that has a grade of at least 3 (on supine testing), providing the key muscle functions represented by segments above that level are judged to be intact (graded as a 5).
Note: in regions where there is no myotome to test, the motor level is presumed to be the same as the sensory level, if testable motor function above that level is also normal.

3. Determine the neurological level of injury (NLI).
This refers to the most caudal segment of the cord with intact sensation and antigravity (3 or more) muscle function strength, provided that there is normal (intact) sensory and motor function rostrally respectively.
The NLI is the most cephalad of the sensory and motor levels determined in steps 1 and 2.

4. Determine whether the injury is Complete or Incomplete.
(i.e. absence or presence of sacral sparing)
*If voluntary anal contraction = **No** AND all S4-5 sensory scores = **0** AND deep anal pressure = **No**, then injury is **Complete**.*
*Otherwise, injury is **Incomplete**.*

5. Determine ASIA Impairment Scale (AIS) Grade.
Is injury Complete? If YES, AIS=A
NO ↓
Is injury Motor Complete? If YES, AIS=B
NO ↓ (No=voluntary anal contraction OR motor function more than three levels below the motor level on a given side, if the patient has sensory incomplete classification)

Are at least half (half or more) of the key muscles below the neurological level of injury graded 3 or better?
NO ↓ **YES ↓**
AIS=C **AIS=D**

If sensation and motor function is normal in all segments, AIS=E
Note: AIS E is used in follow-up testing when an individual with a documented SCI has recovered normal function. If at initial testing no deficits are found, the individual is neurologically intact and the ASIA Impairment Scale does not apply.

6. Determine the zone of partial preservation (ZPP).
The ZPP is used only in injuries with absent motor (no VAC) OR sensory function (no DAP, no LT and no PP sensation) in the lowest sacral segments S4-5, and refers to those dermatomes and myotomes caudal to the sensory and motor levels that remain partially innervated. With sacral sparing of sensory function, the sensory ZPP is not applicable and therefore "NA" is recorded in the block of the worksheet. Accordingly, if VAC is present, the motor ZPP is not applicable and is noted as "NA".

(b)

Figure 21.4 (*Continued*)

regular MRI machines, which depicts the white matter as uniform tissue [82]. A new MRI technology – diffusion tensor imaging (DTI) – that takes advantage of the anisotropic nature of water diffusion in biological tissue is able to show specific fibers such as the corticospinal tract (CST) [83]. To evaluate the connectivity of spinal cord, the gold standard is to use neural tracers, either chemical neural tracers such as biotinylated dextran amine (BDA) or neurotropic viruses such as herpes simplex virus, and lentivirus [84–86]. These technologies can establish the connectivity at the histological level. Electrodiagnostic tools such as electromyograms (EMGs), motor evoked potentials (MEPs), or somato-sensory evoked potentials (SEPs) are non-invasive and focus on the functional connectivity of spinal cord and nerve fibers, which are widely applied in clinical practice [87].

21.3.3 Surgical Repair of Peripheral Nerve Injuries

End-to-end neurorrhaphy, which approximate nerve fascicles by epineural suturing, is a conventional direct suturing technique. There are two difficulties of this technique. (i) It is not applicable to stump gap >5 mm because high tension at the anastomosis site impairs anatomical recovery and functional regain [88]. (ii) It may cause malrotation of the transected axonal stumps and malalignment of nerve fibers. Some studies suggest that motor functions are hardly achieved because regenerating axons must reach the target tissues within a time window, beyond which recovery is compromised [89]. Neurorrhaphy is thus highly dependent on the injury location on the axon and its distance to the enervated tissue [90]. Fibrin glue is used as an adjunct to end-to-end neurorrhaphy, and many surgeons think it is quick and easy to use [91]. Studies showed that fibrin glue may improve bridging of gaps [92].

The "gold standard" of surgical repair is autologous nerve grafting in order to bridge the defect. The graft harvest usually comes from cutaneous nerves such as lateral cutaneous nerve of the thigh, sural nerve, and lateral cutaneous nerve of the forearm (radial forearm flap). It has been performed since 1876 but hardly achieved ideal results at the beginning. Since the 1960s, with the development of vascularized nerve grafts and fascicular suturing, autologous graft provided a reliable method to bridge long defects with predictable results. There are several reasons why this method has become so universally accepted: (i) There is no concern for immunogenicity. (ii) The autologous graft has a high success rate and usually restores the majority of functionality to the damaged location. (iii) This graft fills the nerve gap and achieves tension-free repair [93]. With the advancement of acellular graft, immunosuppression is eliminated. The decellularization technology maintains the internal structure and basal structures of the nerve and forms a path for axon sprouting, myeline regrowth, and proliferation [94]. Nerve transfer is another surgical technique that redirects fascicle of a healthy nerve to the distal portion of an injured nerve. It prolongs the time window for surgical nerve repair for cases where patients need to wait for a certain medical management. This technique is gaining more attention for brachial plexus injuries and other injuries to upper extremities.

Application of NGC is an option for cases where nerve injuries with gap are more than 5 mm [95]. There are multiple reasons that nerve conduits may potentially improve nerve repair and function regain. First, a tensionless repair has been reported as one major factor to successful recovery of sensorimotor function [2]. Using NGCs do not require end-to-end suturing. In some animal studies, it was reported that the gapping can be as long as 3–5 cm dependent on the type of the conduit [96]. Secondly, NGCs theoretically direct the advancing of proximal nerve end and create an environment and surface for axonal growth. NGCs can be categorized into autologous vessel grafts and artificial conduits. NGC manufacturing technologies have seen tremendous development. First-generation conduits were made of synthetic materials such as silicone or polytetrafluoroethylene and are non-absorbable at all. Second-generation NGCs were made of polyglycolic acid (PGA), PLGA, and collagen with biodegradability of four to eight months. Third-generation conduits are tailored in such a way that NGCs are made of biodegradable materials, and Schwann cells or other ECM are incorporated to increase trophic support for axonal growth. Three main types of bioabsorbable conduits have been approved by the US Food and Drug Administration (FDA). These NGCs, however, have various features and yet perfect to achieve excellent clinical outcomes. Therefore, more emerging materials are being studied such as biomaterials with eletro-conducting elements, polypyrrole, poly(3,4-ethylenedioxythiophene), graphene, and its derivatives.

A systemic review published in 2019 [97] retrieved 6767 articles, with only 27 studies retained accounting for 1022 patients. Among those, 10 studies were randomized controlled trials, 15 case series, and 2 cohort studies. In total, 10 types of NGCs were described with a variety of outcome evaluation for nerve function recovery and postoperative complications. The Semmes–Weinstein monofilament test and the static and moving two-point discrimination (2PD) test were the most common assessment tools. Outcomes demonstrated no significant difference between groups and synthetic conduits had more complications. Therefore, the author concluded that use of NGCs is preferable over suture repair and nerve grafting, as the functional recovery rates are above 80% but use of synthetic conduits is discouraged. It is noted that the systemic review has lots of limitations on methodology of search and the variabilities of conduits that were used. In terms of assessment tools, the widely used 2PD for sensory function check is subjective and remains non-standardized [95].

Collagen, especially type-I and type-IV collagen, is the most commonly used biodegradable material for nerve conduit design. Some animal studies with collagen conduits showed equivalent effect as autograft [98, 99]. Only case reports or case series are available with no level-I studies published. Bushnell et al. reported installment of collagen NGCs on digital nerve repair over a two-year period. It involved 12 patients in total. Nine patients had a follow-up of at least one year, with an average follow-up time of 15 months (12–22 months). No intraoperative or postoperative complications were seen in these patients. Outcomes were evaluated with modified American Society for Surgery of the Hand guidelines, and 2PD was good or excellent in eight out of nine patients. Semmes–Weinstein testing was performed in five patients. The Average Quick Disabilities of the Arm, Shoulder, and Hand (DASH) score was 10.86 overall. The study showed comparable results with various types of nerve repairs and reconstructions, suggesting the effectiveness of collagen NGCs for motor and sensory function restoration [98].

The PGA NGCs have favorable physical and chemical properties. They are flexible and porous in nature, which can easily allow diffusion of oxygen and nutrients. In the Mackinnon and Dellon study, 15 patients had gaps of 0.5–3.0 cm (1.7 cm on average) in digital nerve reconstruction. 33% achieved excellent functional sensation recovery, 53% achieved good sensory recovery, and 14% had poor or no recovery with the use of a PGA NGC [100]. In this study, sensory function was evaluated with the British Medical Research Council grading scale with moving and static 2PD. The author concluded that PGA NGCs could produce equivalent outcomes compared with the classic nerve graft technique while avoiding donor-site morbidity.

In a randomized prospective study conducted by Weber et al. [101], they compared PGA NGCs with primary repair or autograft for digital nerve injuries. The overall results showed no significant differences between the two groups. The study concluded that improved sensation when a conduit repair is used for nerve gaps of

4 mm or less, compared with standard repair. Also, PGA conduit repair also produces better results for larger nerve gaps, and it eliminates the donor-site morbidity associated with nerve graft harvesting.

Poly(DL-lactide-ε-caprolactone) NGCs are another type of second-generation products. In a multicenter, blinded, randomized controlled trial by Bertleff et al. [102], the authors studied 34 nerve lesions in 30 patients. Patients were randomly distributed to the standard reconstruction group versus poly(DL-lactide-ε-caprolactone) NGCs group. The nerve defect gaps ranged from 6 to 8 mm, and outcomes were assessed by moving and static 2PD. Results showed that the sensory function recovery was comparable in both groups, although more complications in the experimental group were observed. In the Chiriac et al. study, it did report disadvantage of poly(DL-lactide-ε-caprolactone) such as formation of neuroma in 1 out of 23 patients.

Relative indications for NGCs based on the reported studies include lengthened nerve gaps which were hard to achieve tension-free anastomosis in the standard neurorrhaphy and patient's and surgeon's preference to avoid harvesting.

21.3.4 Surgical Repair of Spinal Cord Injuries

Although the exact reason why adult CNS is not able to regenerate has not been established, there are some hypotheses under discussion. Nogo proteins were discovered and have been extensively studied. These proteins are expressed by myelin and prevent repair of fiber tracts in the CNS [103]. Secondly, formation of glial scar is viewed as a barrier to CNS regeneration [104]. Astrocytes are the most abundant cells in the CNS compared to Schwann cells in the PNS. It plays a crucial role in spinal cord repair by reactive gliosis. In the subacute phase (from one to two weeks after injury), reactive astrocytes migrate to the lesion epicenter and seclude inflammatory cells leading to tissue repair [105–107]. This process indeed restores the construction of injured spinal cord at the cost of function impairment at the same time. Thirdly, CSPGs are considered a major chemophysical barrier to axonal regrowth after SCI [108]. CSPGs are growth inhibitory ECM glycoproteins that are widely expressed in the CNS. The inflammatory response following SCI stimulates the secretion of CSPGs from astrocytes and other supporting tissues for the purpose of preventing further damage. However, accumulated CSPGs stay at injured sites and further obstruct the regrowth of nerve fibers [109–111].

In addition, there is obvious difference in size between peripheral nerve and spinal cord, which makes revascularization, axonal sprouting, and accurate reinnervation harder. Syringomyelia is unique to SCI. With overwhelming cell death and tissue degeneration, the loss of parenchymal tissue without timely repair process can form cystic cavities referred to as syringomyelia [112]. One of the applications of nerve conduit is to serve as implant constructs inside the syringomyelia and to restore continuous tissue connection, as vehicle to deliver neural stem cells (NSCs), or drugs to help regrowth [113].

There are many different types of NGCs for SCI: (i) natural polymers, (ii) synthetic biodegradable polymers, and (iii) synthetic non-degradable polymers. Many biomaterial substrates that are being studied as for peripheral nerve repair are also applicable to SCI, such as collagen, poly(2-hydroxyethyl methacrylate) (pHEMA), poly(*N*-(2-hydroxypropyl) methacrylamide) (pHPMA), and PLGA [114]. With these NGCs as scaffolds, guidance for aligned axon growth through the injured site for axonal sprouting and structuring is hoped to occur. Multiple studies are focused on combinational therapy with NGCs, stem cells, and growth factors. There are also emerging manufacture technologies to bioengineer them into a unity. However, these are at the stage of molecular and animal studies and have not yet been used on human beings.

21.3.5 Future Development of NGCs

3D printing is a new technology for medical instruments. NGCs are its potential area of products with lots of ongoing research. Biomaterial inks such as hydrogels, thermoplastics, and thermoset polyesters, and composites are selected for NGCs. The 3D-printed NGCs have (i) various mechanical, physical, chemical, and biological properties; (ii) conductive properties with polymers and carbon-based nanomaterials; (iii) high precision and speed of production; and (iv) embedded stem cells or drug delivery system from many different sources to help nerve regeneration [114].

Conductive conduit with electric stimulation reveals promising data in both *in vitro* and *in vivo* studies. Nerve cells possess electrical properties such as resistance and conductance. Pulsatile electric stimulus may facilitate nerve growth and reinnervation into muscular tissues. Common biomaterials include graphene and its derivative, carbon-based chemicals, and conductive polymers such as polyaniline, polypyrrole, polythiophene, and gold. A report that used graphene and PCL hybrid conduits demonstrated successful axonal regrowth and remyelination both *in vitro* and *in vivo* [4]. However, most of these materials are non-biodegradable and need to increase their biocompatibility.

21.4 Conclusion

Clinical application of nerve guidance conduits for PNI or SCI is mostly at the stage of bench work with a limited number of clinical studies. FDA has approved three NGCs for clinical trials; however, large multicenter, randomized controlled trials are either undergoing or yet to be published. From level-II to level-IV studies as described above, NGCs have demonstrated equivalent sensory or motor function recovery compared to end-to-end neurorrhaphy or grafting. However, NGCs were rarely proved to be superior to the standard surgical techniques. Bio-fabrication of new materials is a topic of interest in this field, and more manufacture technologies, conduit design, and hybrid unities such as scaffolds fabricated with stem cells, growth factors, or ECM will be studied.

Acknowledgments

The study was sponsored by National key R&D Program of China (No. 2021YFC2400801) Projects of National Natural Science Foundation of China (grant nos. 82002290 and 81830076), Young Elite Scientist Sponsorship Program by Cast (No. YESS20200153), the Shanghai Sailing Program (no. 20YF1436000), the Municipal Hospital Newly-developing Cutting-edge Technologies Joint Research Program of Shanghai Shenkang Hospital Development Center (no. SHDC12018130), The Second Round of Three-year Action Plan to Promote Clinical Skills and Clinical Innovation in Municipal Hospitals (no. SHDC2020CR6019-002), the Shanghai Engineering Technology Research Center and Professional Technology Service Platform Project of 2020 "Science and Technology Innovation Action Plan" of Shanghai (no. 20DZ2254100), the Special Fund for Research on People's Livelihood (Medical Treatment and Public Health) of Shanghai Pudong Science, Technology and Economic Commission Scientific and Technological Development Fund (no. PKJ2018-Y52), and the Shanghai Pudong Health Commission Special Program for Clinical Research in the Health Industry (no. PW2018E-01). We appreciate the support from Base for Interdisciplinary Innovative Talent Training, Shanghai Jiao Tong University and Youth Science and Technology Innovation Studio of Shanghai Jiao Tong University School of Medicine.

References

1 Ruijs, A.C.J., Jaquet, J.B., Kalmijn, S. et al. (2005). *Plast. Reconstr. Surg.* 116: 484.
2 Sunderland, S. (1952). *Brain* 75: 19.
3 Grinsell, D. and Keating, C.P. (2014). *Biomed Res. Int.* 2014: 698256.
4 Qian, Y., Zhao, X., Han, Q. et al. (2018). *Nat. Commun.* 9: 323.
5 Qian, Y., Yuan, W.E., Cheng, Y. et al. (2019). *Nano Lett.* 19: 8990.
6 Qian, Y., Song, J., Zheng, W. et al. (2018). *Adv. Funct. Mater.* 28: 1707077.
7 Qian, Y., Song, J., Zhao, X. et al. (2018). *Adv. Sci. (Weinh.)* 5: 1700499.
8 Qian, Y., Han, Q., Zhao, X. et al. (2018). *J. Pineal Res.* 65: e12516.
9 Qian, Y., Han, Q., Zhao, X. et al. (2019). *iScience* 12: 216.
10 Qian, Y., Cheng, Y., Ouyang, Y. et al. (2019). *NPG Asia Mater.* 11: 36.
11 Fadia, N.B., Bliley, J.M., DiBernardo, G.A. et al. (2020). *Sci. Transl. Med.* 12: eaav7753.
12 Cattin, A.L. and Lloyd, A.C. (2016). *Curr. Opin. Neurobiol.* 39: 38.
13 Yan, Z.W. and Qian, Y. (2020). *Adv. Mater. Lett.* 11: 20101562.
14 Corfas, G., Velardez, M.O., Ko, C.P. et al. (2004). *J. Neurosci.* 24: 9250.
15 Guertin, A.D., Zhang, D.P., Mak, K.S. et al. (2005). *J. Neurosci.* 25: 3478.
16 Clements, M.P., Byrne, E., Guerrero, L.F.C. et al. (2017). *Neuron* 96: 98.
17 Newbern, J.M. and Snider, W.D. (2012). *Neuron* 73: 623.
18 Napoli, I., Noon, L.A., Ribeiro, S. et al. (2012). *Neuron* 73: 729.
19 Zigmond, R.E. and Echevarria, F.D. (2019). *Prog. Neurobiol.* 173: 102.

20 Cattin, A.L., Burden, J.J., Emmenis, L.V. et al. (2015). *Cell* 162: 1127.
21 Barrette, B., Hébert, M.A., Filali, M. et al. (2008). *J. Neurosci.* 28: 9363.
22 Stoll, G., Jander, S., and Schroeter, M. (2002). *Adv. Exp. Med. Biol.* 513: 87.
23 Trimarco, A. and Taveggia, C. (2020). *Nat. Neurosci.* 23: 1179.
24 Babetto, E., Wong, K.M., and Beirowski, B. (2020). *Nat. Neurosci.* 23: 1215.
25 Yang, J., Wu, Z., Renier, N. et al. (2015). *Cell* 160: 161.
26 Han, S.M., Baig, H.S., and Hammarlund, M. (2016). *Neuron* 92: 1308.
27 Zhao, X., Qian, Y., Cheng, Y. et al. (2018). *Biomater. Sci.* 6: 2059.
28 Cheng, Y., Xu, Y., Qian, Y. et al. (2020). *Nano Energy* 69: 104411.
29 Qian, Y., Cheng, Y., Song, J. et al. (2020). *Small* 16: e2000796.
30 English, A.W., Schwartz, G., Meador, W. et al. (2007). *Dev. Neurobiol.* 67: 158.
31 Eberhardt, K.A., Irintchev, A., Al-Majed, A.A. et al. (2006). *Exp. Neurol.* 198: 500.
32 Yan, J., Wu, R., Liao, S. et al. (2020). *Front. Bioeng. Biotechnol.* 8: 590998.
33 Jiang, H., Qian, Y., Fan, C., and Ouyang, Y. (2020). *Front. Bioeng. Biotechnol.* 8: 582646.
34 Chen, X., Ge, X., Qian, Y. et al. (2020). *Adv. Funct. Mater.* 30: 2004537.
35 Qian, Y., Yao, Z., Wang, X. et al. (2020). *Cell Prolif.* 53: e12730.
36 Qian, Y., Xu, Y., Yan, Z. et al. (2021). *Nano Energy* 83: 105779.
37 Qian, Y., Wang, X., Song, J. et al. (2021). *npj Regener. Med.* 6: 31.
38 Bracken, M.B., Shepard, M.J., Collins, W.F. et al. (1990). *N. Engl. J. Med.* 322: 1405.
39 Gerndt, S.J., Rodriguez, J.L., Pawlik, J.W. et al. (1997). *J. Trauma* 42: 279.
40 McDonald, J.W. and Sadowsky, C. (2002). *Lancet* 359: 417.
41 O'Shea, T.M., Burda, J.E., and Sofroniew, M.V. (2017). *J. Clin. Invest.* 127: 3259.
42 Sofroniew, M.V. (2018). *Nature* 557: 343.
43 Burda, J.E. and Sofroniew, M.V. (2014). *Neuron* 81: 229.
44 Bradbury, E.J., Moon, L.D.F., Popat, R.J. et al. (2002). *Nature* 416: 636.
45 Tester, N.J., Plaas, A.H., and Howland, D.R. (2007). *J. Neurosci. Res.* 85: 1110.
46 Song, Y.H., Agrawal, N.K., Griffin, J.M., and Schmidt, C.E. (2019). *Adv. Drug Delivery Rev.* 148: 38.
47 Chvatal, S.A., Kim, Y.T., Bratt-Leal, A.M. et al. (2008). *Biomaterials* 29: 1967.
48 Wilems, T.S. and Sakiyama-Elbert, S.E. (2015). *J. Controlled Release* 213: 103.
49 Xi, K., Gu, Y., Tang, J. et al. (2020). *Nat. Commun.* 11: 4504.
50 Hurtado, A., Cregg, J.M., Wang, H.B. et al. (2011). *Biomaterials* 32: 6068.
51 Wang, Z., Nong, J., Shultz, R.B. et al. (2017). *Biomaterials* 112: 62.
52 Tysseling-Mattiace, V.M., Sahni, V., Niece, K.L. et al. (2008). *J. Neurosci.* 28: 3814.
53 Anderson, M.A., O'Shea, T.M., Burda, J.E. et al. (2018). *Nature* 561: 396.
54 Nakamura, M., Houghtling, R.A., MacArthur, L. et al. (2003). *Exp. Neurol.* 184: 313.
55 Papa, S., Rossi, F., Ferrari, R. et al. (2013). *ACS Nano* 7: 9881.
56 Papa, S., Caron, I., Erba, E. et al. (2016). *Biomaterials* 75: 13.
57 Saxena, T., Loomis, K.H., Pai, S.B. et al. (2015). *ACS Nano* 9: 1492.

58 Fournier, A.E., Takizawa, B.T., and Strittmatter, S.M. (2003). *J. Neurosci.* 23: 1416.

59 Fawcett, J.W. and Asher, R.A. (1999). *Brain Res. Bull.* 49: 377.

60 Crespo, D., Asher, R.A., Lin, R. et al. (2007). *Exp. Neurol.* 206: 159.

61 Lee, H., McKeon, R.J., and Bellamkonda, R.V. (2010). *Proc. Natl. Acad. Sci. U.S.A.* 107: 3340.

62 Baumann, M.D., Kang, C.E., Tator, C.H., and Shoichet, M.S. (2010). *Biomaterials* 31: 7631.

63 Bradke, F., Fawcett, J.W., and Spira, M.E. (2012). *Nat. Rev. Neurosci.* 13: 183.

64 He, Z. and Jin, Y. (2016). *Neuron* 90: 437.

65 Santhosh, K.T., Alizadeh, A., and Abdolrezaee, S.K. (2017). *J. Controlled Release* 261: 147.

66 Khaing, Z.Z., Agrawal, N.K., Park, J.H. et al. (2016). *J. Mater. Chem. B* 4: 7560.

67 Ansorena, E., Berdt, P.D., Ucakar, B. et al. (2013). *Int. J. Pharm.* 455: 148.

68 Xia, T., Ni, S., Li, X. et al. (2013). *Brain Res.* 1538: 41.

69 Milbreta, U., Nguyen, L.H., Diao, H. et al. (2016). *ACS Biomater. Sci. Eng.* 2: 1319.

70 Wu, W., Lee, S.Y., Wu, X. et al. (2014). *Biomaterials* 35: 2355.

71 Silva, G.A., Czeisler, C., Niece, K.L. et al. (2004). *Science* 303: 1352.

72 Guo, J., Su, H., Zeng, Y. et al. (2007). *Nanomedicine* 3: 311.

73 Sun, Y., Li, W., Wu, X. et al. (2016). *ACS Appl. Mater. Interfaces* 8: 2348.

74 Seddon, H.J. (1943). *Brain* 66: 237.

75 Sunderland, S. (1990). *Muscle Nerve* 13: 771.

76 Burnett, M.G. and Zager, E.L. (2004). *Neurosurg. Focus* 16: E1.

77 Frankel, H.L., Hancock, D.O., Hyslop, G. et al. (1969). *Paraplegia* 7: 179.

78 Kirshblum, S.C., Waring, W., Biering-Sorensen, F. et al. (2011). *J. Spinal Cord Med.* 34: 547.

79 Strandberg, E.J., Mozaffar, T., and Gupta, R. (2007). *J. Hand Surg. Am.* 32: 1280.

80 Taylor, C.A., Braza, D., Rice, J.B., and Dillingham, T. (2008). *Am. J. Phys. Med. Rehabil.* 87: 381.

81 Miyanji, F., Furlan, J.C., Aarabi, B. et al. (2007). *Radiology* 243: 820.

82 Stroman, P.W., Bosma, R.L., Kornelsen, J. et al. (2012). *Clin. Neurol. Neurosurg.* 114: 460.

83 Cheran, S., Shanmuganathan, K., Zhuo, J. et al. (2011). *J. Neurotrauma* 28: 1881.

84 Kuang, R.Z. and Kalil, K. (1990). *J. Comp. Neurol.* 302: 461.

85 Wickersham, I.R., Finke, S., Conzelmann, K., and Callaway, E.M. (2007). *Nat. Methods* 4: 47.

86 Lo, L. and Anderson, D.J. (2011). *Neuron* 72: 938.

87 Feng, B., Qiu, G., Shen, J. et al. (2012). *J. Spinal Disord. Tech.* 25: E108.

88 Isaacs, J. and Cochran, A.R. (2019). *Bone Joint J.* 101: 124.

89 Ma, C.H.E., Omura, T., Cobos, E.J. et al. (2011). *J. Clin. Invest.* 121: 4332.

90 Raza, C., Riaz, H.A., Anjum, R., and Shakeel, N.U.A. (2020). *Life Sci.* 243: 117308.

91 Whitlock, E.L., Kasukurthi, R., Yan, Y. et al. (2010). *Microsurgery* 30: 218.
92 Ornelas, L., Padilla, L., Silvio, M.D. et al. (2006). *J. Reconstr. Microsurg.* 22: 123.
93 Schmidt, C.E. and Leach, J.B. (2003). *Annu. Rev. Biomed. Eng.* 5: 293.
94 Hudson, T.W., Liu, S.Y., and Schmidt, C.E. (2004). *Tissue Eng.* 10: 1346.
95 Meek, M.F. and Coert, J.H. (2008). *Ann. Plast. Surg.* 60: 466.
96 Song, J., Sun, B., Liu, S. et al. (2016). *Front. Mol. Neurosci.* 9: 117.
97 Silva, J.B., Marchese, G.M., Cauduro, C.G., and Debiasi, M. (2017). *Hand Surg. Rehabil.* 36: 71.
98 Bushnell, B.D., McWilliams, A.D., Whitener, G.B., and Messer, T.M. (2008). *J. Hand Surg. Am.* 33: 1081.
99 Waitayawinyu, T., Parisi, D.M., Miller, B. et al. (2007). *J. Hand Surg. Am.* 32: 1521.
100 Mackinnon, S.E. and Dellon, A.L. (1990). *Plast. Reconstr. Surg.* 85: 419.
101 Weber, R.A., Breidenbach, W.C., Brown, R.E. et al. (2000). *Plast. Reconstr. Surg.* 106: 1036.
102 Bertleff, M.J.O.E., Meek, M.F., and Nicolai, J.P.A. (2005). *J. Hand Surg. Am.* 30: 513.
103 Schwab, M.E. (2010). *Nat. Rev. Neurosci.* 11: 799.
104 Katoh, H., Yokota, K., and Fehlings, M.G. (2019). *Front. Cell. Neurosci.* 13: 248.
105 Okada, S., Nakamura, M., Katoh, H. et al. (2006). *Nat. Med.* 12: 829.
106 Herrmann, J.E., Imura, T., Song, B. et al. (2008). *J. Neurosci.* 28: 7231.
107 Wanner, I.B., Anderson, M.A., Song, B. et al. (2013). *J. Neurosci.* 33: 12870.
108 Tran, A.P., Warren, P.M., and Silver, J. (2018). *Physiol. Rev.* 98: 881.
109 Jones, L.L., Margolis, R.U., and Tuszynski, M.H. (2003). *Exp. Neurol.* 182: 399.
110 Andrews, E.M., Richards, R.J., Yin, F.Q. et al. (2012). *Exp. Neurol.* 235: 174.
111 Anderson, M.A., Burda, J.E., Ren, Y. et al. (2016). *Nature* 532: 195.
112 Seki, T. and Fehlings, M.G. (2008). *J. Neurosurg. Spine* 8: 365.
113 Donaghue, I.E., Tam, R., Sefton, M.V., and Shoichet, M.S. (2014). *J. Controlled Release* 190: 219.
114 Yu, X., Zhang, T., and Li, Y. (2020). *Polymers (Basel)* 12: 1637.

Index

a

abiotic signals 279
acellular intestinal submucosa 525
ACI periosteal flap technique (ACI-P) 351–352
activating gene-associated therapy 274–275
active micromixing 190
acute tendon injury 563
ADA-GEL hydrogel 62
additive manufacturing (AM) 55
  technology 272
  VAM 81
adipose-derived mesenchymal stem cells (ADMSCs) 31
adipose-derived stem cells (ADSCs) 95–97, 99, 106–108, 112, 115, 273, 275, 521, 523
adult MSCs 95, 102
agarose 12, 37, 255, 273, 380, 403, 465, 494, 590
  BCP composite scaffolds 37
aggrecan 165, 167, 370, 386, 411–413, 537, 540
aggrecanases 370, 411
akermanite 33–35, 41, 43, 44
  implants 35
akermanite ($Ca_2MgSi_2O_7$) 33, 57, 83
alginate 6, 62, 217, 380
  based hydrogel beads 6
  mesoporous silica nanoparticles (MSN) 277
alginate-dialdehyde (ADA) hydrogels 62
alkali-H^+ ion exchange 211
alkaline phosphatase (ALP) 135, 200, 252, 253, 277, 492
  activity 29
AlloDerm 146, 440
allografts 55, 115, 265, 266, 283, 291, 348–351
American Orthopaedic Foot and Ankle Society (AOFAS) scores 325
American Spinal Injury Association (ASIA) 118, 121, 594
2-aminoethyl methacrylated chondroitin sulphate (CS-AEMA) 77
3-aminopropyltriethoxysilane (APTES) 228
angiogenesis 213–214, 461–462, 482–483
anterior crucial ligament (ACL) 368
  reconstruction 117
anti-BMP-2 antibodies 147
antibody-mediated and immunoregulatory response 14
anticatabolic 410–412
antimicrobial activity, bioactive glass 214–215
articular cartilage 367
  anatomy and histology of 342
  chondroprogenitor cells 62
  homeostasis 369
  regional classification of 161–162
  role of 160–161
artificial biolubricants 160, 168–174
artificial joint prosthesis 174–177
artificial joint replacement 159, 174

Biofabrication for Orthopedics: Methods, Techniques, and Applications, First Edition.
Edited by Wenguo Cui, Xin Zhao, and Shen Liu.

ascorbic acid 102, 106
aseptic necrosis 505
Asp@Lipo/BFP-1 65
astrocyte scar 589
autogenous bone graft 145, 294
autograft 55, 109, 115, 265, 266, 283, 291, 349–351, 430, 441, 581, 599
autologous BMSC-enriched β-tricalcium phosphate scaffold 109
autologous chondrocyte implantation (ACI) 347, 348, 351–352, 369, 371, 372
autologous chondrocytes 94, 110, 370
automated scaffold design (ASD) 9

b

balanced chromosomal EWS/ETS translocations 282
basic fibroblast grow factor (bFGF) 31, 33, 36, 43, 97, 106, 107, 149, 214, 425, 483, 487, 488, 490, 492, 550
bioabsorbability 335
bioactive ceramics 25, 26, 68
45S5 bioactive glass 27, 211
bioactive glass (BG) 25–30, 292
  alternative 215–216
  angiogenesis 213–214
  antimicrobial activity 214–215
  in composite and hybrid materials 216–230
  first 211–212
  osteogenesis 212–213
bioactive glass ceramics (BGC) 37, 39
bioactive glass (SiO_2-CaO) nanofibers 142
bioceramic scaffolds 25, 38, 42, 249, 251, 252, 255, 294, 485
biodegradable ceramics
  bioactive glass 26–30
  calcium phosphate cements (CPCs) 32
  immunomodulation 40–42
  mechanical properties 36–38
  protein adsorption 39–40
  silicate bioceramics 32–36
  β tricalcium phosphate 30–32
  vascularization 42–44
biodegradable crosslinkable poloxamer (C-PCaP) 62, 67, 77
biodegradable synthetic polymers 219
bioengineered tissue structures 3
biofabrication 291
  bioprinting 374–382
  electrospinning 382–386
  microfluidic technology 386–389
  for rotator cuff repair
    bone mineral density 547
    enthesis structure regeneration 542–543
    fat infiltration of rotator cuff muscles 546
    future perspectives 552
    tendon-to-bone attachment 544–546
  traditional cellular products 370–374
Bio-Gide 146
bioglass (BG) 221, 229, 230, 336
bioinert ceramics 25, 44
BioINK™ 380
bioinks 5, 377, 381
  bone-biomimetic ceramic 62
  C_2C_{12} myoblasts-embedded alginate 65
  cell-laden alg/mc 62
  collagen and hyaluronic acid hydrogel 63
  collagen-based 18
  gelatin-based 9
  HA-based 381
  hyaluronic acid-based osteoblast-laden hydrogel 63
  osteocyte-laden hybrid 63
  temperature-sensitive 9
biological augmentation, for rotator cuff repair 547
  cell-based strategies 547–549
  cell-derived hybrid biological factors 549–550
  drugs/small molecules 550–551
  growth factors 550

biomaterial inks
calcium phosphates (CaP) ceramic inks 67–68
challenges and future directions 85–86
extrusion-based (bio)printing 71–77
gelatin 60–62
hydrogel biomaterial inks 59–64
inkjet-based 3D (bio)printing technique 77–81
light-based (bio)printing 81–85
supramolecular and stimuli-responsive inks 68–71
thermoplastic polymer inks 64–67
biomaterials associated infections 214
biomedical engineering 182, 283, 371
biomedical μ-TAS technology 193
biomimetic nanospheres 172
biomimicry 9, 323, 324
biomineralization 25, 267–268, 336
bionic cartilage acellular matrix microsphere (BCAMM) 337
bionic fixation
advantages 338
bionic materials
application and designation of 335–337
bionic design 337
calcaneal fracture 331–332, 335
definition of 323
DTS 324–326
intertrochanteric fracture 328, 330–331
MIAP 327–329
bionics
definition 323
MC 336
plate design 332
bioprinted hBMSC
alg/mc bioink 68
laden PEG-peptide scaffolds 80
bioprinted PEG-peptide scaffolds 80
bioprinted tissues 18
bioprinters 15, 18, 77, 258, 375, 462
bioprinting techniques 55, 56, 58, 67, 71–85
bioresorbable ceramics 25
bio-scaffolds 105, 106
BIOSTAT BIOLOGX®Fibrin Sealant 414
biphasic calcium phosphates (BCPs) 31, 32, 37, 40, 68, 294, 501
biphasic osteochondral scaffolds 72, 353
bisphosphonates 287
black phosphorus (BP) nanosheets 69, 279, 288, 294
blood vascular and vascularized tissues
biofabrication 462
cell-only fabrication 473–474
decellularized vascular tissue 473
electrospinning 471–472
microfabrication
micro module assembly 469–471
micro-patterning/microfluidics 468–469
3D printing
extrusion bioprinting 463, 465, 466
inkjet bioprinting 463, 464
laser-assisted bioprinting 465–468
blood vessels 461, 485
ECs line 491
formation of 502–503
grafted 486
peripheral 504
BMSC-laden CEDPs 106
body-on-a-chip 184
B_2O_3-incorporated β-$CaSiO_3$ bioceramics 36
BonAlive 212, 221, 229
bone
biomimetic ceramic bioink 62
cancer treatments
biofabrication 291–295
chemotherapy 283–284
cytotoxic agents 286
diagnosis 282
drug delivery 286–287
gene therapy 289
immunotherapy 285
metastatic tumors 282

bone (*contd.*)
molecular targeted therapy 284–285
nanotechnology 285–286
phototherapy 287–289
primary bone cancers 281
radiotherapy (RT) 284
surgery 283
traditional approaches 282
cells 267
cements 25, 272
fillers 25
grafting 265
guided vascular growth in porous biomaterials 503–505
marrow 203
ONFH 505–506
porous structure 501–503
bone demineralized matrix (BDM) 223
bone-forming peptide-1 (BFP-1) 65
bone marrow-derived mesenchymal stem cells (BM-MSCs) 408, 439
bone marrow stem cells ((BMSCs) 29, 30, 83, 94, 95, 106, 107, 109, 115, 117, 118, 274, 279, 383, 437–439, 486, 492, 586
bone marrow stimulation technique 347, 348
bone marrow stromal cells 29, 30, 153, 203, 214
bone marrow transplants 93, 94, 110, 437
bone mesenchymal stem cells (BMSCs) 59, 68, 72, 83, 94, 106, 107, 109, 110, 114, 115, 117, 118, 279, 437–439, 486, 492
bone morphogenetic proteins (BMPs) 96, 268, 412, 489
BMP-2 32, 34, 79, 105, 137, 276, 413
BMP-7 413
BMP-12 566
BMP-13 413
bone-on-a-chip technology
bone marrow 203–204
cancer 204–206
mineralization and osteogenesis 198–201
osteochondral (OC) tissue 201–203
bone regeneration
bone biology 266–267
field of 266
first generation of biomaterials 266
implants role
activating gene-associated therapy 274–275
architectural and microstructural properties 270
biological coatings and cell strategies 271
converging multiple synergistic strategies 279, 280
electrical stimuli 277–278
ESCs 274
iPSCs 274
macro level 270
mechanical stimuli 276–277
MSCs 273–274
orthobiologics 275–276
osteoprogenitor cells 273
photothermal 278
structural support 271–272
tissue engineering 270
ultrasound 278
osteogenesis and bone healing 267–268
second generation of biomaterials 266
third generation of biomaterials 266
bone scaffolds 25, 66, 83, 249, 251, 294
bone sialoprotein (BSP) 35, 36, 200, 267
bone structural units (osteons) 266, 267
bone-targeted alendronate 289
bone tissue engineering 229
orthopedics 3D printing 246–247
scaffolds 25
in vitro 482, 483
bone tissue repair 105, 144–147
bone trabeculae 250, 328, 330
boron containing 45S5 modified glass 214

boron-incorporated calcium silicate ($Ca_{11}Si_4B_2O_{22}$) 42
borosilicate glasses 215, 221–223
boundary lubrication 159, 162–165
branched polymer 6
bredigite ($Ca_7MgSi_4O_{16}$) 33, 34, 36, 43
BST-Cargel 354, 372

c

calcaneal fracture 331–335
calcified cartilage 77, 82, 201, 202, 367
calcium magnesium silicates 32, 33
calcium phosphate bioceramics 40
calcium phosphate bone cements 272
calcium phosphate cement (CPC) 25, 32, 33, 58, 62, 63, 68, 72, 74, 272
calcium phosphate ceramics 42, 293
calcium phosphates (CaP) ceramic inks 67–68
calcium phosphate silicate (CPS) 37, 543, 547
calcium-polyphosphate microparticles (Ca-polyP-MP) 65, 66
calcium sensing receptor (CaSR) 30–31
calcium silicates 32, 33, 41–44, 68, 252, 294
calcium sulfate 247
calcium zinc silicates 32
calcium zirconium silicates 32
cancellous bone 36, 37, 69, 197, 252, 336, 483, 484
cancer 204–206
  biofabrication for 291–295
  primary bone 204
carbon dots/poly(ethylene glycol)/chitosan (CDs/PEG/CS) composite gel 169
cartilage categories 341
cartilage/chondrocyte derived progenitor cells (CDPCs) 101, 106, 122
cartilage-derived morphogenetic protein-1 (CDMP-1) 413
cartilage-derived morphogenetic protein-2 (CDMP-2) 413
cartilage-derived MSCs 101
cartilage-derived stem/progenitor cells 101
cartilage ECM-derived particles (CEDPs) 106
cartilage forming 106
cartilage injury 247, 368, 369
  arthroscopic evaluation of 343–344
  clinical characteristics of 343
  histopathology evaluation of 344–345
  magnetic resonance imaging assessment 345–346
  pathology of 342
cartilage/osteochondral regeneration 372, 373
cartilage progenitor stem cells 367
cartilage repair 148–149, 369, 371, 402
  arthroscopic debridement 347
  autologous chondrocyte implantation 351–352
  bone marrow stimulation 348–349
  cartilage shaving 347
  gene editing for 356–357
  joint lavage 347
  microfracture augmentation 349
  osteochondral allograft transplantation 350
  osteochondral autografts 349–350
  platelet-rich plasma 354–356
  tissue engineering approaches 352–354
cartilage stem/progenitor cells (CSPCs) 101, 106
cartilage tissue engineering 247–248, 255, 369, 371, 377–379–381, 386, 387, 389
cartilage trauma activates chondrocytes 370
cartilaginous joints 160
β-catenin 108
$Ca_2ZnSi_2O_7$ nano-grains 37, 38
C_2C_{12} myoblasts-embedded alginate bioink 65
cell-adhesive peptide sequences 192
cell-based strategies 547–549

cell-derived hybrid biological factors 549–550
cell differentiation 12–13, 26, 96, 105, 108, 151, 228, 277, 414, 473, 524, 567
cell-laden alg/mc bioink 62, 72
cell microencapsulation 386
cell-only fabrication 473–474
  cell only bioprinting 474
  cell sheet 474
cell sheet 474, 544, 547
  technique 105
cells-material-blood supply trinity 482, 483
cellular repressor of E1A stimulated genes (CREG) protein 519
cellular scaffolds 521, 523
cell viability 5, 8, 11, 12, 15, 18, 63, 67, 68, 72, 77, 80, 82, 222, 223, 228, 248, 251, 286, 381, 468, 481, 489
cementum attachment protein (CAP) 36
cementum protein 1 (CEMP1) 36
ceramic calcium silicate ($CaSiO_3$) 252
ceramic-integrated MEW meshes 62, 75
ceramic material 176–177, 252, 253
check-valve micropump 189, 190
chemical and mechanical etching 181
chemical crosslinking 9, 10, 468
chemical vapor deposition (CVD) 192
chemodynamic therapy (CDT) 290
chemokines 40, 108, 370, 407
chemotherapy 283–285, 287, 290, 294, 295
chitin/chitosan derivatives 335
chitosan 105
  PCL nanofibers 471
chondral injury 368
chondrocyte marker (COL2) 101, 414
chondrogenesis 42, 99, 102, 103, 106, 248, 273, 369, 381, 383, 386
chondrogenic morphogens 411–414
chondroitinase ABC (ChABC) 354, 590, 592
chondroitin sulfate proteoglycans (CSPGs) 589, 592, 600
ChondroMimetic™ 372
Chondrosarcoma 281, 282, 284
chronic tendon injury 559, 563
coatings on bone implants 25
co-axial adapter 82
coaxial electrospinning 134, 136–138, 140, 146
cobalt-chromium-molybdenum alloy 175
co-culture MSCs 549
Co^{2+}-doped MBG 29
cold modalities 564
collagen 59, 192, 380, 560
  based-bioink 18
  bioceramic composites 37
  and hyaluronic acid hydrogel bioink 63
  organic matrix 267
  protein polysaccharides 163
  sponge-based scaffold 105
  β-TCP composite 38
collagenase 96, 411, 413, 441
collagen I 6, 200, 403, 516, 560
collagen membrane (C-ACI) 370
colony-forming unit fibroblasts (CFU-f) 94, 99
colony stimulating factor 1 (CSF-1) receptor/CSF1R 103
compact bone 197, 271
composite materials 6, 34, 220–225, 335–337, 471
composites fabrication methods 224–225
composites in clinics 229–230
computational topology design (CTD) 271
computer-aided design (CAD) 5, 58, 245, 375, 462
  digital model 375
concentration gradient 190–192
conducting polymers (CPs) 277, 517
conductive biomaterials 517, 586, 587
conductive polymer 16, 277, 517, 518, 601

conjugated electrospinning 134, 138–139
connective tissue growth factor (CTGF) 103, 107, 146
continuous inkjet printing 376
contraction band 515
control bionics 323
Conventional Treatments for Wound Healing 429
copper-containing mesoporous bioactive glass (Cu-MBG) 29
coral scaffolds 273
cortical bone 26, 31, 34–37, 65, 66, 97, 197, 250, 266, 270, 294, 483, 484
cranial bone graft (CBG) 484
cross-linked ADA-gelatin (ADA-GEL) multi-component hydrogel 62
cryderman 430
cryo-scanning electron microscopy 169
cryotherapy 564
Cu-doped bioactive glasses 214
C-X-C chemokine receptor type 4 (CXCR-4 or CD184) 107
cyclic monomer ε-caprolactone 220
cytokines 108
  and growth factors 566–567
  pro-inflammatory 40
cytotoxic agents 286

d
decellularized extracellular matrix (dECM) 389
decellularized vascular tissue 462, 473
deep zone 201, 202, 367
delayed type hypersensitivity 13
denosumab 285
deposition
  of curable material 181
  CVD 192
  of electrosprayed microparticles on the fibers 141, 142
dermis 423–425, 434, 437, 440, 441
desmin 107
destabilization of the medial meniscus (DMM) model 109
dexamethasone (DEX) 34, 102
D-glucosamine sulfate (GAS) 171
dicalcium phosphate dihydrate (DCPD) 32, 215
dicalcium silicate (Ca_2SiO_4) 32, 33
different solidification methods 376
diffuser micropumps 189, 190
digital light processing (DLP) 58, 81–83, 245, 466
1,25-dihydroxyvitamin D_3 102
diopside ($CaMgSi_2O_6$) 33, 34
dipalmitoylphosphatidylcholine (DPPC) liposomes 168
direct current electrical stimulation 565
direct genes transfer method 409
direct ink writing (DIW) 71, 75
Disabilities of the Arm, Shoulder, and Hand (DASH) 599
discogenic pain 401, 403, 409
distal tibiofibular syndesmosis (DTS) injury 324–326
1,2-distearoyl-sn-glycero-3-phospho-choline (DSPC) liposomes 171
double-edged sword 584
droplet-based bioprinting (DBB) 7, 8, 376
drop-on-demand printing 376
drug delivery systems 428
3ds Max 58
dual-channel bioink system 64
dual-channel 3D bone-bioprinting 63
dual-layer organic/inorganic flexible biopolar fibrous membrane 542
Dulbecco's Modified Eagles Medium (DMEM) 97, 98, 102
dynamic fluid electrospinning 134
dynamic hip screw (DHS) 330
dynamic liquid electrospinning 139–140

e
early phase reaction 588, 592
eccentric strengthening programs 565
ECM/GelMA/exosome scaffolds 84
elastofluid theory 163
elastohydrodynamic theory 163

electrically conductive bio-fabricated products 586
electrical stimuli 277
electrohydrodynamic jet bioprinting 376
electromagnetic microvalves 188
electron beam melting (EBM) 245, 246, 248
electrospinning techniques 225, 382, 471, 524, 592
  bone tissue repair 144–147
  cartilage repair 148–150
  coaxial electrospinning 137
  conjugated electrospinning 138–139
  dynamic liquid electrospinning 139–140
  electrospun fibers integrated with post treatments 141–144
  emulsion electrospinning 138
  four stages 133
  multi-fluid electrospinning 140–141
  tendon and ligament repair 149, 151–153
  tendon-bone junction interface repair 153–155
  traditional electrospinning 134–136
  vascularization 147–148
electrospray 141
electrospun
  fibers 590
  nanofibers 147
  polyacrylonitrile nanofibers 141
  synthetic polymers 384
electrostatic microvalves 188, 189
embryoid bodies (EBs) 493
embryonic stem cell-induced mesenchymal stem cells (ESC-MSCs) 102, 106, 109
embryonic stem cells (ESC) 93, 274, 405, 436, 491, 493
emulsion electrospinning 134, 138
emulsion-electrospun nanofibers 138
endochondral bone formation 267, 268, 273, 489
endomembrane 514
endothelial cells (ECs) 15, 31, 85, 102, 195, 273, 423, 427, 461, 482, 483, 491–492
endothelial progenitor cells (EPCs) 273, 425, 428, 473, 488, 491–492
enzymes 96, 103, 108, 347, 411, 417, 469
epidermal growth factor (EGF) 102, 276, 285, 353, 412
epidermal growth factors receptor (EGFR) signaling 102
epidermal stem cells 430
epidermis 423, 429, 430, 437–441
epimysium 514
EPSC 430
etching 181
Ewing sarcoma 281, 282, 284
exosomes 108–109
extracellular matrix (ECM) 5, 267, 371, 375
  markers 403
  proteins 192
  of tendon zone 537
extracellular matrix stiffness 105
extracellular signal-related kinases (ERK) pathway 30, 35
extracorporeal shock wave therapy (ESWT) 564
extrusion and droplet-based printing 377
extrusion-based bioprinter 77
extrusion-based bioprinting (EBB) 7, 8, 63, 66, 71, 255, 375, 376
extrusion-based 3D bioprinting approach 62
extrusion-based 3D printer 255, 257
extrusion-based 3D printing 249, 252–254
extrusion-based printing 8, 59, 67, 73, 255
extrusion bioprinting 463, 465, 466

f

fascia 136, 487, 490, 499, 514, 515
fascial pedicle 487

fatty infiltration, of rotator cuff muscles 540–541
$Fe\text{-}CaSiO_3$ 294
$FeOOH/g\text{-}C_3N_4$ submicron particles 141
fetal bovine serum (FBS) 97, 337
fetal calf serum (FCS) 97
fetal MSCs 95
FGF-2 102
fiber deposition 71, 72
fibroblast growth factors (FGFs) 15, 97, 102, 147, 149, 412, 425, 515, 550, 566
fibrocartilaginous entheses 562
  structure and composition of 538
fibronectin 141, 142, 192, 196, 203, 204, 267, 279, 470, 520, 589
fibrous enthesis 560, 562
fibrous joints 160
finite element analysis 327, 330, 331, 333, 334
first bioactive glass 211–212
fluid controlling components 187–192
fluorescein isothiocyanate-dextran 138
fluorine (F) substitution 34
focal adhesion kinase (FAK) 105
4D (bio)printing techniques 69
four-layer gradient-poly(ε-caprolactone)/ calcium phosphate silicate (G-P/C) composite film fabrication 543
fourth generation ACI 370
fracture healing 266–268, 273, 330
Frankel scale 594
free functional muscle transfer (FFMT) 527
freeze-drying 143, 223, 224, 337
friction 159
  coefficient 162
functional bionics 323
fused deposition modeling (FDM) 71, 225, 245
fused deposition printing 245

g

Gamma nail 330, 331
gas foaming 42, 224
gelatin 12, 60, 217
  based bionics 9
  cell-adherent arginine-glycine-aspartate (RGD) peptides 62
  PLA nanofibers 143, 149
  PLGA nanofibers 142–144
gelatin methacrylamide/methacryloyl (GelMA) 9, 11, 18, 61–64, 71, 73, 77, 80, 380, 381, 465, 468
  GelMA-based ink 60
  microspheres 279
Gene Activated Matrices (GAMs) 274
gene editing, for cartilage repair 356–357
gene therapy protocol, for cartilage repair 357
geometric code (GCode) 58
glial fibrillary acidic protein (GFAP) 593
glutamate cysteine ligase (GCLC) 586
β-glycerophosphate 102
3-glycidoxypropyltrimethoxysilane (GPTMS) 227, 228
glycosaminoglycans (GAGs) 63, 367, 537
gold nanoparticles 286, 288, 586
grafted blood vessels 486
granular hybrid PCL/Ca-polyP-MP 65
green fluorescent protein (GFP) 435, 437, 438
growth and differentiation factors (GDFs) 412
growth/differentiation factor 5 (GDF-5) 269, 413
growth factors, basic fibroblast growth factor 490
α-L-guluronic acid 380
gunshot trauma 595

h

Haversian bone-mimicking scaffolds 83
Haversian systems 266
Haversian/Volkmann's systems 483
hBMSC-encapsulated acrylated PEG hydrogels 80
healthy bones 268, 293
healthy peripheral nerves 584

heat modalities 564
hematopoietic cells 96, 437, 438
hematopoietic stem cells (HSC) 93, 103, 197, 203, 204, 267, 437
heterogeneity 14, 107, 549
hierarchical Haversian bone structured biomimetic scaffolds 82–83
high-dose therapy followed by stem cell transplant (HDT/SCT) 285
highly cross-linked polyethylene 175, 176
human adipose-derived stem cells (hADSCs/hASCs) 35, 66, 71, 223, 253
human aortic endothelial cells (HAECs) 34
human bone marrow stromal cells (hBMSCs) 30, 35, 44, 59, 62, 63, 65, 66, 68, 75, 80, 96, 99, 105
human collagen-based scaffold composition 542
human dental pulp cells (hDPCs) 36
human dermal microvascular endothelial cells (HDMECs) 491, 492
human induced pluripotent stem-MSCs (hiPS-MSCs) 102
human mesenchymal stem cells (hMSCs) 12, 32, 105, 277, 337, 466
human periodontal ligament cells (HPDLCs) 30
human platelet lysate (hPL) 97, 101
human umbilical vein endothelial cells (HUVECs) 32, 33, 43, 44, 198, 465, 466, 468, 470, 471, 492, 494
hyaluronic acid 63, 164
  aggrecan complex 167
  based osteoblast-laden hydrogel bioink 63
hyaluronic acid methacrylamide (HAMA) 9, 63, 64, 77, 381
hybrid inorganic-organic materials 226
hybrid scaffolds 226, 227, 276, 519, 520, 593
hydration lubrication 159, 163–164, 168, 173, 174
hydrodynamic lubrication theory 162, 163
hydrogels 6, 105, 272, 276
  ADA-GEL 62
  biomaterial inks 59
  hBMSC-encapsulated acrylated PEG 80
  nanocomposite 60
  nSi-incorporated GelMA 61
  NuCore Injectable Nucleus 415
  PNAGA 69
  poly(ethylene glycol) (PEG)-clay nanocomposite 63
  pristine GelMA 60
  SELP 275
  stimuli-responsive 272
  supramolecular polymer 68
  3D-bioprint cell-encapsulated 59
  3D-bioprinted osteocyte-laden 64
hydrogenated soy phosphatidylcholine (HSPC) 168
hydroxyapatite (HA) 25, 105, 198, 247, 286
  based bioinks 381
  based ceramic bone-on-a-chip 200
  based multi-layered films 26, 27
  bioceramics 484
  nanocrystals 59
hydroxyapatite-coated PLGA/poly(L-lactic acid) (HA-PLGA/PLLA) scaffolds 106
hypoxia-inducible factor (HIF-1α) 29
hypoxia inducible factor-1 (HIF-1) genes 483

i

icariin 147
ICRS scoring system 343, 344
iliac bone graft (IBG) 484
iliosacral screw (ISS) 327
Image ware 13.1 58
immunomodulation 26, 39–42, 95
immunomodulatory and growth factors 108
implanted β-TCP bioceramics 506

Indian hedgehog (IHH)/Patched1 (PTCH1) signaling 102
indirect gene transfer 409
indoleamine 2,3-dioxygenase (IDO) 97
induced pluripotent stem cell (iPSC) 93, 274, 435
  iMSCs 105
induced skeletal myogenic progenitor cells (iMPCs) 202, 518
inflammation regulation 427
inflammatory cytokines (TNF-α) 31
infrapatellar fat pad-derived MSCs (IF-MSCs) 99, 100
inkjet-based 3D (bio)printing technique 71, 77, 80
inkjet bioprinting 7, 376, 463, 475
inkjet 3D printing 251, 252, 254–255, 463, 464
inkjet printing 245, 255, 256, 376, 463
innate immunity 13
innervation 16–18, 521, 524, 563
insulin-like growth factor-1 (IGF-1) 30, 102, 108, 268, 353, 412, 413, 492, 518, 521, 565, 566
insulin-transferrin-selenium premix 102
integrated circuits 181
integrated organ printing (IOP) 71, 75, 80–82
integrated tissue and organ printer (ITOP) 18
intercellular adhesion molecule 1 (ICAM-1) 95, 204
interleukin-8 (IL-8) 31, 96
interleukin-10 (IL-10) 41, 42, 97, 369, 427, 585
International Society for Cell Therapy (ISCT) 96, 99, 374
International Standards for Neurological Classification of Spinal Cord Injury (ISNCSCI) 596
interstitial hydraulics 161
intertrochanteric fracture 328, 330–331
intervertebral disc (IVD) degeneration
  biomaterial therapy 414–417
  cell-based therapy for 406
  cell-based treatment techniques 405, 407–408
  clinical trials 402
  definition 401
  degenerative changes 403, 404
  gene therapy for 409–411
  healthy 403
  intracellular regulators 414
  mitogens 412
  molecular therapy for 411–414
  therapeutic techniques for 401
  treatment method of 402
  treatment plan for 403
intramembranous bone formation 268
in-vitro bone tissue engineering models 482, 483
ion-doped mesoporous bioactive glass 28
iron-containing MBG (Fe-MBG) 29
3-isocyanatopropyltriethoxysilane (ICPTES) 228
isoelectric points (IEP) 39, 276
IVD augmentation 402

j

Janus guided bone regeneration membrane 146
joint lubrication behaviour
  component 164
  lubrication performance 164–168
joint lubrication mechanism
  hydration lubrication 163–164
  tribological theory 162–163

k

kappa-carrageenan (κ-CA) 60, 61
keratinocytes 423, 425, 427, 428, 430, 434–437, 440, 441
k-means clustering algorithm 9
knee osteoarthritis (KOA), PRP injection of 355

l

laminin B3 cDNA 434
large bone defects 29, 36, 110, 246, 291, 294, 481

large molecules 161, 381
laser-assisted bioprinting (LAB) 256, 465–468
laser-based bioprinting (LBB) 14, 376, 466
laser-guided direct writing (LGDW) 465, 466
laser-induced forward transfer (LIFT) 82, 256, 465–467
late phase reaction 588
Li-Fraumeni syndrome 282
light-based (bio)printing 81–85
LIM mineralization protein-1 (LMP-1) 410, 414
Li+@MSNs/PEUU 546
lip nanoparticles 172
lipopolysaccharide (LPS) 40
liquid-crystal display (LCD) 81, 82
lithium calcium silicate 41
lithography 181, 185–187, 469, 494
LMP-1 overexpression 410
LockActiv 230
lotus root-like biomimetic scaffolds 43
low-and middle-income countries (LMICs) 401
low-dose Cobalt (Co) 29
low energy laser phototherapy 565
low temperature deposition manufacturing (LDM) 225
low-viscous alginate 10
L-proline 102
lubrication performance 164–168, 171, 173
lubricin 165–168
lysozyme (LSZ) 39

m

Mach3 58
macrophage inflammatory protein-1alpha (MIP-1a, CCL3) 96
Major Steele 323
manganese (Mn)-doped β-TCP 31
(1–4)-β-D-mannuronic acid 380
material bionics 323, 324
matrix-assisted pulsed laser evaporation direct writing (MAPLEDW) 465–466
matrix induced autologous chondrocyte implantation (MACI) 348, 351, 370, 372
matrix metalloproteinases (MMPs) 13, 370, 403, 405, 411, 427
mechanical bionics 323
mechanically reinforced scaffolds 143
medical bionics 323–324
melanocyte stem cells 434
melt electrowriting (MEW) PCL microfiber 62
melt-plotted PCL struts 65, 73
membrane-associated autologous chondrocyte implantation (MACI) 351, 370, 372
merwinite ($Ca_3MgSi_2O_8$) 33, 35, 36
mesenchymal precursor cells (MPCs) 117
mesenchymal stem cell-derived extracellular vesicles (MSC-EVs) 108–109, 549–550
mesenchymal stem cells (MSCs) 93, 197, 267, 273, 274, 288, 374, 488, 492
  bone non-union and defects 109–110
  considerations of stem cells and derivations for clinical usage 118–122
  derived exosomes 108–109, 122
  isolation, identification and expansion of 96–98
  MSC-EVs and exosomes 108–109
  osteochondral defect and osteoarthritis (OA) 110–115
  osteogenesis imperfecta (OI) 110
  sourced bioactive molecules 108
  spine 117–118
  tendon and ligament injury 115, 117–118
  tenogenesis 106–107
  tissue sources 94–96
  in vitro and *in vivo* 102–108

mesenchyme-derived stem/progenitor cells 95
mesoporous bioactive glasses (MBGs) 27, 28
mesoporous silica nanoparticles (MSNs@lip) 171
messenger RNAs 108
metabolic reprogramming 584
metal implants 291
methacrylated chondroitin sulphate (CSMA) 381
methacrylated gelatin 6, 9
2-methacryloxyethyl phosphorylcholine (MPC) 173
methylcellulose (mc) 62, 592
Mg-based biodegradable 252
microelectromechanical systems (MEMS) 181, 187, 494
microenvironmental imbalance 583–585
microfabricated vasculature 494, 508
microfabrication
  micromodule assembly 469–471
  micropatterning/microfluidics 468–469
  techniques 181
  technology 468
microfluidic 3D bioprinting system 82
microfluidic/micropatterning technology 462
microfluidic organ-on-a-chip' technology
  biological features of 194–196
  cell/tissue culture 193–194
microfluidic printhead 82
microfluidic pump 189
microfluidics
  cell/tissue culture 192–193
  fluid controlling components 187–192
  soft lithography 185–187
  technology 182, 386
microfracture augmentation 349
micromixers 187, 190, 192
micro-patterned HA nanoparticles 26
micropatterning/microfluidics 468–469
microRNAs 108
microspheres (MSs) 34, 72, 223, 275, 276, 279, 287, 337, 488, 490, 499, 549, 592
microvalves 186–188–190, 195
mild heat stimulation 278
mineralization and osteogenesis 198–201
mineralization module formation 36
mineralized Col-I (MC) 59
mineralized collagen (MC) 60, 336
minimally invasive adjustable plate (MIAP) 327–329
α-minimum essential medium (α-MEM) 97, 98
mitogen-activated protein kinase (MAPK) 41, 356, 566, 584
mitomycin-C 151, 152
mixing index heat map 82
modern bionics 323
Moneyprint 58
monocalcium silicate ($CaSiO_3$) 32
monocyte chemoattractant protein-1 (MCP-1) 96
monticellite ($CaMgSiO_4$) 33, 34, 36
moxifloxacin hydrochloride 147
mTORC1–Hif1α–c-Myc signaling pathway 584
multi-armed oligomers/polymers 6
multifactor combination 490
multi-fluid electrospinning 140–141
multilayered 3D osteochondral construct 63
multilayered PCL struts 65
multiple synergistic strategies 279–280
multi-potency 93, 94, 273
multiwalled carbon nanotubes (MWCNTs) 34
muramyl tripeptide phosphatidylethanolamine (MTP-PE) 285
muscle extracellular matrix (MEM) 518
muscle fascicles 514
muscle fibers (fascicles) 514, 515
myoblast determination protein (MyoD) 107, 519

myogenesis 62, 102, 107–108, 518, 520
myogenic factor 5 (Myf-5) 107

n

nagelschmidtite (NAGEL, $Ca_7Si_2P_2O_{16}$) scaffolds 38
nanobioceramics 25
nanocomposite hydrogels 60, 63
nanofibrous scaffolds 135, 137, 138, 141–144, 149, 151, 386, 472
nanogenerator 587
nano-hydroxyapatite (nHA) 37, 59, 62, 67, 73, 77, 137
nanoparticles 590
  gold 286, 288, 586
  lip 172
  micro-patterned HA 26
  MSN 277
  PSPMA-g-HSNPs 173, 174
  PSPMK 171
  silver 286
  UCNPs 288, 290
nanotechnology 169, 266, 270, 281, 285–291
nanoyarns 138–140
natural biomaterials 380
natural bone tissue 336–337
natural joint structure
  regional classification of articular cartilage 161–162
  role of articular cartilage 160–161
natural materials 134–136, 371, 380, 429, 517
natural polymer materials 253, 523
natural polymers 59, 60, 217–219, 224, 229, 253, 371, 380, 381, 383, 523, 527, 542, 567, 600
nature bone 37
N-(3,4-dihydroxyphenethyl) methacrylamide (DMA) 173
negative-charged nanofibers 138
NEL-like type I molecular-1 DNA 72
nerve conduit
  clinical classification 594–595
  clinical evaluation 595–597
  surgical repair of peripheral nerve injuries 597–600
  surgical repair of spinal cord injuries 600–601
nerve guide conduit (NGC) 581–583, 585–587, 598, 599
nestin 101, 435
neuregulin 1 (NRG1)-ErbB2 signaling pathway 583
next-generation drug screening platform 185
non-adherent hematopoietic cells 96, 438
non-doped and Mg-doped beta-TCP surfaces 31
nonpolymer materials 137
Noraker 229, 230
notch intracellular domain (NICD) 108
Novabone 212, 229–230, 247
Novabone Macroform® 230
nSi-incorporated GelMA hydrogels 61
nuclear factor-like 2 (Nrf2) 586
nucleic acid-based therapeutics 289
NuCore Injectable Nucleus hydrogel 415

o

OctoPrint 58
oncostatin M (OSM) 41
organ-on-a-chip 183–185, 192–206, 389
orthopedics 3D printing
  biomaterials for
    bioceramics 251–253
    metal 251–252
    polymer materials 253
  and bioprinting techniques 56
  bone tissue engineering 246–247
  cartilage tissue engineering 247–248
  characteristics 254
  extrusion-based 3D printer 255–256
  future prospects 256–258
  inkjet 3D printing 254–255
  LAB 256
  micro extrusion print 257
  process of 245, 246
  structural requirements of

pore size 248–249
pore structure 249–250
porosity 249
tissue engineering scaffolds 245
osseointegration 86, 271, 272, 291, 501
osteoarthritis (OA) 159, 369
knee 94, 355
osteochondral defect 110–115
temporomandibular joint 109
Osteoarthritis Research Society International (OARSI) scoring system 344, 346
osteoblastogenesis 104
osteocalcin (OCN) 35, 103, 105, 135, 200, 267, 268, 277, 279, 492
osteochondral allograft transplantation 348, 350
osteochondral autograft transplantation 350
osteochondral defect 110–115
osteochondral interface 77, 82, 367
osteochondral (OC) tissue 201, 202
osteoclastogenesis 103, 104
osteocyte-laden hybrid bioinks 63
osteogenesis
bioactive glass 212–213
mineralization 198–201
osteogenesis imperfecta (OI) 94, 109, 110
osteogenesis-on-a-chip 200
osteogenic peptide 69, 294
osteonecrosis of the femoral head (ONFH) 504–506, 508
osteons 197, 199, 205, 247, 266
osteopontin (OPN) 35, 36, 103, 105, 135, 200, 204, 267, 279
osteosarcoma 281, 282, 284–289
Osterix 103
Outerbridge score 343

p

particle replication in non-wetting templates (PRINT) 289
particulate HA/polyethylene composites 220
particulate-leaching 224
passive micromixing 190, 191
patient-specific, implantable 3D scaffolds 62
Pax3 108, 434
PCaP pastes 77
PCL/Pluronic F127 membrane fabrication 550, 551
percutaneous bone marrow grafting 109
perfusion-based maturation assessment 13
periodontal ligament cells 29, 34
Perioglass 212
periosteal autologous chondrocyte implantation (P-ACI) 370
periosteal cambium layer cells 370
periosteum-mimic membranes 26
peripheral nerve injury (PNI) 581, 583–587
peripheral nerve regeneration
microenvironmental imbalance 583–585
microenvironmental rebalance 585–587
peripheral nerve repair 582, 583, 586, 594, 601
peritendinous adhesion formation (PAF) 563, 568
phosphate buffered saline (PBS) 40, 173
phosphate glasses 26, 215, 221, 222
phosphatidylinositol 3 kinase/protein kinase B (PI3K/AKT) 40, 65
photodynamic agents 288
photodynamic therapy (PDT) 287, 288, 290
photolithography 181, 185, 186, 471
photopolymerize (bio)inks 81
photosensitizer (PS) 288–290
phototherapy 287–290, 565
photothermal therapy (PTT) 287, 288, 290, 294, 295
physical modalities, tendon repair 564–565
Piezo1/2 105
piezoelectric actuation 188

PI3K/protein kinase B (Akt) pathways 30
platelet-derived growth factors (PDGF) 15, 102, 106, 154, 268, 275, 285, 413, 483, 488, 566
platelet endothelial cell adhesion molecule (PECAM) 196, 492
platelet-rich plasma (PRP) 273, 567
  for cartilage repair 354–356
  treatments 107
pleiotrophin 204
pluri-potency 36, 93, 274, 438
pluripotent stem cells 12, 102, 434, 493
PNAGA-clay composite hydrogel scaffolds 69
pneumatic 3D extrusion-based bioprinting 66
pneumatic microvalves 188
point-of-care diagnostics 182
point-of-care technology (POCT) 182
poly(D,L-lactide-co-glycolide) (PLG) 383, 497, 545
poly(ethylene glycol) (PEG) 6, 8
  clay nanocomposite hydrogel 63
  GelMA hybrid scaffold 80
poly(glycolic acid) (PGA) 219, 220, 383, 385, 403, 414, 519–520, 541, 545, 552, 598–600
poly(L-lactic acid) (PLLA) 37, 337, 383, 386, 471, 491, 542
poly(lactic acid) (PLA) 37, 58, 63, 66, 71, 141, 147, 151, 153, 154, 219–221, 273, 336, 383, 403, 414, 488, 490, 519, 541, 552
poly(lactide-co-glycolide) 12, 34
poly(methacryloxyethyltrimethyl ammonium chloride-2-Aminoethyl2-methylacrylate hydrochloride) 146
poly (vinyl alcohol) (PVA) 37
poly(glycolic) acid (PGA) 219
poly(lactic) acid (PLA) 219
polycaprolactone (PCL) 58, 65, 135, 220, 383
poly(diallydimethylammonium) chloride 203
poly(ethylene glycol)-diacrylate (PEGDA)/laponite XLG nanoclay 63
polydimethylsiloxane (PDMS) 185
poly-DL-lactic acid (PDLLA) 37
polyetheretherketone (PEEK) 66
polyetherimide (PEI) 463
polyethylene oxide (PEO) 65, 137
polyglycolic acid (PGA) 135, 599
poly (2-methacryloyloxyethyl phosphorylcholine) grafted mesoporous silica nanospheres 173
poly(L-lactic acid) (PDLA)-HA composite scaffolds 37
poly(N-acryloyl glycinamide) (PNAGA) hydrogel 69
Polyjet 3D printing technique 250
polylactic acid (PLA) 37, 58, 135, 273
polylactic-co-glycolic acid (PLGA) 75, 105, 106, 135, 138, 143, 151, 219, 220, 223, 253, 275, 279, 471, 490, 491, 541, 549, 552, 590, 592, 598, 601
poly-L-lactate-epsilon-caprolactone (PLC) material 545
polymerized high internal phase emulsion technique (polyHIPE) 200, 201
polymethylmethacrylate (PMMA) 253, 272, 289, 485, 499
poly(N-isopropylacrylamide) (PNIPAAm) microgels 173, 174
poly (DL-lactide-ε-caprolactone) NGCs 600
poly(3-sulfopropyl methacrylate potassium salt) [PSPMA]-g-HSNPs 173, 174
polytetrafluoroethylene (PTFE) sheet 175, 542, 545, 598
polyvinyl alcohol (PVA) 37, 72, 471
porous akermanite implants 35
porous bioceramic scaffolds 25, 485
porous diopside microspheres 34

porous scaffolds 72, 139, 140, 143, 201, 223, 225, 245, 250, 252, 255, 294, 485, 486, 488, 497, 498, 502
positive-charged nanofibers 138
posterior pelvic ring 324, 327, 328
powder bed fusion (PBF) process 84
powder melt extrusion 71
primary bone cancer 281, 282
primary isolated MSCs 97
pristine GelMA hydrogel 60
pro-inflammatory cytokines 40, 369, 370
proliferating cell nuclear antigen (PCNA) 107
proliferation stage 427
prostaglandin E2 (PGE2) 97
protein adsorption 25, 26, 39–40, 219, 230
protein-incorporated nanofibrous scaffolds 137
proteoglycan production 99
proteoglycans 161, 202, 367, 405, 410, 413, 414, 537, 560
proximal femoral nail anti-rotation (PFNA) 330
pseudowollastonite 32
PSPMK nanoparticles 171
pulsed magnetic fields (PMF) 565

q

quantum dots (QDs) 288

r

radiofrequency coblation 565
radiotherapy (RT) 284, 285, 287
reactive oxygen species (ROS) 286, 288–290, 294, 295, 515, 586
receptor activator of nuclear factor-κB ligand (RANKL) 33, 103, 267, 285
reconstructive methods 369
regenerative medicine 3, 18, 93, 229, 245, 246, 354, 425, 427, 439, 481, 507, 527
regorafenib 285
reparative methods 369
Repetier Host 58
resorbable PLA nanofibers 151
Reynolds' lubrication theory 162
RGD cyclic peptide-conjugated gold nanoparticles (RGNPs) 71–72
Rhinoceros 5.0 58
rigid screw fixation 324
RNA-sequencing (RNA-Seq) analysis 65, 101
rotator cuff anatomy 535, 536
rotator cuff repair
  biofabrication for 542–543
  biological augmentation for 547–551
  failure rates of 537–538
    fatty infiltration 540–541
    low BMD of humeral head 541
    microstructure of tendon-to-bone enthesis 541
    tear size 539–540
rotator cuff tear 115, 117
rotator cuff tendons
  role 535
rotator cuff tendons tears 535, 536
  conservative therapies 535
rotator cuff tendon-to-bone enthesis
  cellular and ECM composition 538
  histology of 537
runt-related transcription factor 2 (RUNX2) 34, 103, 337

s

scaffold-based ACI 370
scaffold fabrication methods 5, 42
scaffold properties 3, 219
scaffolds topography 3
scarce resident stem cells 367
Schwann cell-mediated axonal regeneration 584
scleraxis (Scx) 101, 560, 561
sclerostin 267
screw-based extrusion system 66
second generation ACI 370
Seddon classification 594
selective laser melting (SLM) technique 84

selective laser sintering (SLS) 84, 225, 245, 246, 250, 294
selenium-doped hydroxyapatite nano-particles (Se-HANs) 286, 287
self-assembling peptide (SAP) 273, 590, 593
severe combined immunodeficiency (SCID) 273, 434, 490, 491, 493
short carbon fibers (SCFs) 67
shoulder disability 536
signal transduction pathways 436
silica shell 6
silicate bioactive glass 27
silicate bioceramics 32–36, 41, 44, 252
silicates 36
  calcium 32, 33, 41–44, 68, 252, 294
  calcium magnesium 32, 33
  calcium zinc 32
  calcium zirconium 32
silk-elastin like protein polymer (SELP) hydrogels 275
silk fibroin 137, 146, 469
silver nanoparticles 85, 252, 286, 517
Simplify3D 58
sintering additives 36
$45SiO_2$–$24.5Na_2O$–$24.5CaO$–$6P_2O_5$ 211
skeletal muscle
  composition of 514–515
  morphological characteristics of 514
skeletal muscle injury, pathophysiology, pathophysiology
  inflammatory phase 515
  remodeling phase 516
  repair phase 515–516
skeletal muscle regeneration
  application and prospect 525–528
  biofabrication
    acellular scaffold 519–521
    cellular scaffolds 521–523
    cytokine, genetic drugs, micromolecule 518–519
    natural, synthetic and hybrid 517–518
  tissue engineering for muscle repair 523–525
skeletal muscles, definition 513
SketchUp 58
skin
  anatomy and related cell types 423–425
  wound healing treatments 427–429
  wounds and wound healing process 425–427
skin regeneration, tissue engineering
  BMSC 437–439
  iPSC technology 435–437
  local stem cells 429–435
Slic3r 58
small unilamellar vesicles (SUVs) 168
Sma-Mad (Smad) proteins 414
sodium hyaluronate (NaHA) 169
soft lithography 185–187, 469, 494
solid free-form fabrication (SFF) 271
solid or solvent-based extrusion free-forming (SEF) 71
Solidworks 58
solvent casting 224
somatic cell nuclear transfer (SCNT) 436
sonotherapy 564
Sox9 chondrocyte marker 414
spared neural tissue 589
S53P4 bioactive glass 212
spheroids 193, 376, 492
spinal cord injury (SCI) 117, 119, 587, 596
spinal cord regeneration
  bio-fabrication 590–593
  pathophysiological process 588–589
spinal fusion 117, 413
spine 117–118, 327, 588
spongy bone 197
Sr-doped calcium polyphosphate (SCPP) 31
Sr-doped $Ca_2ZnSi_2O_7$ scaffold 38
SRY-Box Transcription Factor 7 (SOX7) 107
Standard Triangle Language (STL) file format 58

stereolithography (SLA) 58, 63, 200, 225, 245, 462, 466
stimuli-responsive hydrogel 272
stimuli-responsive materials 59, 85
stress shielding 270, 291, 327, 330, 335
stromal derived factor-1 (SDF-1) 105, 107, 108, 273
strontium-containing mesoporous bioactive glass (Sr-MBG) 29
structural bionics 323–335
subchondral bone 67, 115, 116, 162, 201, 202, 247, 248, 341, 343, 344, 346, 349, 352, 367, 368
Sunderland Classification 594, 595
superlubricated drug-loaded nanospheres 172, 173
supramolecular and stimuli-responsive inks 68–71
supramolecular polymer hydrogel 68
surface charge 39, 60, 63, 277
surface topography 39–41
SVVYGLR (SV) 519
SYLGARD® 185
synovial fibroblasts 370
synovial fluid-derived mesenchymal stemcells (SF-MSCs) 101
synovial joints 160, 162–164, 350
synovium-derived mesenchymal stem cells (SM-MSCs) 100, 101
synthetic polymers 37, 66, 134–136, 194, 219–220, 224, 225, 253, 266, 272, 371, 380, 381, 383, 384, 414, 517–519, 567

t

β-TCP/poly(lactic acid-co-trimethylene carbonate) photothermal-responsive shape-memory scaffold 69
temperature-sensitive bioink 9, 382
tenascin-C 561, 562
tendinopathy 559, 560, 563–566
tendon and ligament injury 115, 117
tendon and ligament repair 149, 151–153
tendon biology 560–562
tendon-bone junction interface repair 153–155
tendon enthesis 543, 562
tendon healing 107, 413, 541, 560, 563, 567
tendon injuries 562–563, 568
  incidence 559
tendon repair
  cytokine and growth factor therapies 566–567
  mechanical loading and mobilization 565–566
  physical modalities 564–565
  tissue engineering 567–568
tendon stem/progenitor cells (TSPCs) 151, 562
tendon to-bone enthesis, histology of 537
tenolysis 568
tenomodulin (Tnmd) 101, 561
tension band plate (TBP) 327
tetraethyl orthosilicate (TEOS) 226
texture-stored CDs/PEG/CS composite gel 171
therapeutic ultrasound 564
thermoplastic polymer-based biomaterials 58
thermoplastic polymer inks 64
third-generation techniques 370
3D bioplotting 71
3D-bioprint cell-encapsulated hydrogels 59
3D-bioprinted graded scaffold 82
3D-bioprinted osteocyte-laden hydrogel 64
3D bioprinting 374
  cell differentiation 12
  cell selection and optimization 7
  cell viability 11
  choice of bioink 5
  design and post processing 9
  immune response 13
  innervation 16

3D bioprinting (*contd.*)
  printing resolution and mechanical stability 8
  process 375
  process of 5
  remodeling and maturation 13
  steps in 4
  technology 375
  vascularization 14
3D collagen-glycosaminoglycan scaffold 105
3D extrusion-based bioprinting 63
3D functional tissue models 3
3D HC-PMNFs 83
3D hierarchical scaffolds 143
3D hybrid (bio)printing technique 62
3D-melt-plotted PCL struts 65
3D (bio)plotting 72
3D porous properties 484
3D-print bioactive glass/chitosan nanoparticle (BG/CSn) composite porous scaffolds 72
3D-printed ECM/GelMA/exosome construct 83
3D-printed hybrid PCL grafts 79
3D-printed hyperelastic bone 79
3D-printed (electron beam melting-enabled sintering) implants 85
3D-printed microscale PCL struts 72
3D-printed PCL scaffolds 65
3D-printed PEEK grafts 67
3D-printed PEEK prosthesis 67
3D-printed PLA scaffold 66
3D-printed PLA scaffolds 66
3D-printed porous β-TCP/CS hybrid scaffold 68
3D-printed scaffolds 59
3D printed transplantable tissues 18
3D printing 55, 225
3D (bio)printing techniques 69
3D scaffolds 144
3D tissue model 193
3D tissue regeneration 139
Ti-6Al-4V alloy 35
TightRope flexible fixation 326
tiny defects 368
tissue barrier functions 194
tissue-derived collagen 146
tissue engineered muscle grafts (TEMG) 523
tissue engineered skin (TES) 429
tissue engineering 567
  for articular cartilage repair 352
  scaffolds 245
  triangle 428
tissue inhibitor of metalloproteinase 1 (TIMP1) 410
tissue specific stem/progenitor cells (TSPCs) 94, 95
tissue specific stem/progenitors 101
traditional electrospinning 134
transforming growth factor (TGF) 15
  BMPs signalling 103
  TGF-β1 34, 106, 268
  TGF-β3 102, 106
transforming growth factor-β (TGF-β) 96, 97, 412, 566
  BMP signaling 102
transglutaminase cross-linked gelatin (TG-Gel) 279
traumatic PNI destroys 584
triaxial electrospinning 140
triboelectric nanogenerator (TENG) 176
tribological theory 162–163
tricalcium phosphate (TCP) 25, 30, 293
  bioceramics 40, 41
  β-TCP 30–32, 68, 109, 117, 247, 485
tricalcium silicate (Ca_3SiO_5) 32, 33, 252
Trichostatin A molecules 151
triphasic scaffolds 75, 80, 549
triple-helical type I collagen molecules 561
triple layer enthesis-mimicking (EM) scaffolds 545
true micro total analysis system (μ-TAS) 183
trypsin 96, 162, 349, 473
2D electrospun nanofibers 143
2D Nb2C MXene 292–293

2D Ti_3C_2 MXenes 292
two-phase lubrication mechanism 163
type I collagen (Col I) 34

u

ultra-high molecular weight polyethylene (UHMWPE) 159, 169, 170, 175
uniaxially aligned PCL nanofibers 141, 153
upconversion nanoparticles (UCNPs) 288, 290
U.S. Food and Drug Administration (FDA) 65, 370, 598

v

vascular bundles grafting 486
vascular cell adhesion protein 1 (VCAM-1) 95, 97
vascular endothelial (VE)-cadherin 196
vascular endothelial growth factor (VEGF) 15, 29, 96, 268, 488–489
  bone morphogenetic protein 489–490
  expression 584
vascular graft 486–487, 508
vascularization 14, 42, 147, 524
  basic fibroblast growth factor 490
  biological effects 482
  definition 481
  growth factors
    bone morphogenetic protein 489–490
    multifactor combination 490
    vascular endothelial growth factor 488–489
  in-vitro, bone tissue engineering 482
  microfabricated vasculature 494
  orthopedic implants 482
  in porous biomaterials
    bone 501
    interconnection diameter 498–500
    pore diameter 495–498
  porous structure 482–486
  stem cells
    endothelial cells and endothelial progenitor cells 491–492
    ESCs 493
    induced pluripotent stem cells 493–494
    mesenchymal stem cells 492–493
  vascular graft 486–487
vascular smooth muscle cell (vSMC) 14, 491
vasculogenesis 461, 462, 482, 483
vinyl-capped PDMS 185
volumetric additive manufacturing (VAM) 81, 82
volumetric muscle loss (VML) 513–516, 519–527

w

Wnt antagonist 103
Wnt/β-catenin signaling pathway 36, 103
WNT signaling 30, 40, 102, 103, 541
wollastonite (β-$CaSiO_3$) 32, 33, 35–37

x

xenografts 265, 266

y

Yes-associated protein (YAP1) 105
Yes-associated protein/transcriptional coactivator with PDZ-binding domain (YAP/TAZ) 277

z

zinc 30, 288, 335
zinc phthalocyanine (ZnPc) 288
Zn-doped β-TCP 30